Pearson Advanced Chemistry Series

The need for innovation, adaptability, and discovery is more glaring in our world today than ever. Globally, we all look to "thought leaders" for progress, many of whom were, are, or will be students of science. Whether these students were inspired by a book, a teacher, or technology, we at Pearson Education want to do our part to support their studies. The new **Advanced Chemistry Series** supports upper-level course work with cutting-edge content delivered by experienced authors and innovative multimedia. We realize chemistry can be a difficult area of study and we want to do all we can to encourage not just completion of course work, but also the building of the foundations of remarkable scholarly and professional success. Pearson Education is honored to be partnering with chemistry instructors and future STEM majors. To learn more about Pearson's **Advanced Chemistry Series**, explore other titles, or access materials to accompany this text and others in the series, please visit *www.pearsonhighered.com/advchemistry*.

Books available in this series include:

Analytical Chemistry and Quantitative Analysis
by David S. Hage *University of Nebraska Lincoln* and
James R. Carr *University of Nebraska Lincoln*

Forensic Chemistry
by Suzanne Bell *West Virginia University*

Inorganic Chemistry
by Gary Miessler *St. Olaf College,* Paul Fischer *Macalester College,*
Donald Tarr *St. Olaf College*

Medicinal Chemistry: The Modern Drug Discovery Process
by Erland Stevens *Davidson College*

Physical Chemistry: Quantum Chemistry and Molecular Interactions
by Andrew Cooksy *University of California San Diego*

Physical Chemistry: Thermodynamics, Statistical Mechanics, and Kinetics
by Andrew Cooksy *University of California San Diego*

Physical Chemistry
by Thomas Engel *University of Washington* and Philip Reid *University of Washington*

Physical Chemistry: Principles and Applications in Biological Sciences
by Ignacio Tinoco Jr. *University of California Berkeley*, Kenneth Sauer *University of California Berkeley*, James C. Wang *Harvard University*, Joseph D. Puglisi *Stanford University*, Gerard Harbison *University of Nebraska Lincoln*, David Rovnyak *Bucknell University*

Quantum Chemistry
by Ira N. Levine *Brooklyn College, City College of New York*

Periodic Table of the Elements

Key (legend):

6	— Atomic number
C	— Element symbol
12.01	— Atomic weight*
Carbon	— Element name

Noble gases

Period	1A	2A	3B	4B	5B	6B	7B	8B	8B	8B	1B	2B	3A	4A	5A	6A	7A	8A
1	1 **H** 1.008 Hydrogen																	2 **He** 4.003 Helium
2	3 **Li** 6.941 Lithium	4 **Be** 9.012 Beryllium											5 **B** 10.81 Boron	6 **C** 12.01 Carbon	7 **N** 14.01 Nitrogen	8 **O** 16.00 Oxygen	9 **F** 19.00 Fluorine	10 **Ne** 20.18 Neon
3	11 **Na** 22.99 Sodium	12 **Mg** 24.31 Magnesium											13 **Al** 26.98 Aluminum	14 **Si** 28.09 Silicon	15 **P** 30.97 Phosphorus	16 **S** 32.07 Sulfur	17 **Cl** 35.45 Chlorine	18 **Ar** 39.95 Argon
4	19 **K** 39.10 Potassium	20 **Ca** 40.08 Calcium	21 **Sc** 44.96 Scandium	22 **Ti** 47.867 Titanium	23 **V** 50.94 Vanadium	24 **Cr** 52.00 Chromium	25 **Mn** 54.94 Manganese	26 **Fe** 55.85 Iron	27 **Co** 58.93 Cobalt	28 **Ni** 58.70 Nickel	29 **Cu** 63.546 Copper	30 **Zn** 65.39 Zinc	31 **Ga** 69.72 Gallium	32 **Ge** 72.64 Germanium	33 **As** 74.92 Arsenic	34 **Se** 78.96 Selenium	35 **Br** 79.90 Bromine	36 **Kr** 83.80 Krypton
5	37 **Rb** 85.47 Rubidium	38 **Sr** 8762 Strontium	39 **Y** 88.91 Yttrium	40 **Zr** 91.22 Zirconium	41 **Nb** 92.90 Niobium	42 **Mo** 95.94 Molybdenum	43 **Tc** (98) Technetium	44 **Ru** 101.07 Ruthenium	45 **Rh** 102.9 Rhodium	46 **Pd** 106.4 Palladium	47 **Ag** 107.9 Silver	48 **Cd** 112.4 Cadmium	49 **In** 114.8 Indium	50 **Sn** 118.7 Tin	51 **Sb** 121.8 Antimony	52 **Te** 127.6 Tellurium	53 **I** 126.9 Iodine	54 **Xe** 131.3 Xenon
6	55 **Cs** 132.9 Cesium	56 **Ba** 137327 Barium	71 **Lu** 175.0 Lutetium	72 **Hf** 178.5 Hafnium	73 **Ta** 180.9479 Tantalum	74 **W** 183.84 Tungsten	75 **Re** 186.2 Rhenium	76 **Os** 190.23 Osmium	77 **Ir** 192.2 Iridium	78 **Pt** 195.1 Platinum	79 **Au** 197.0 Gold	80 **Hg** 200.6 Mercury	81 **Tl** 204.4 Thallium	82 **Pb** 207.2 Lead	83 **Bi** 209.0 Bismuth	84 **Po** (209) Polonium	85 **At** (210) Astatine	86 **Rn** (222) Radon
7	87 **Fr** (223) Francium	88 **Ra** (226) Radium	103 **Lr** (262) Lawrencium	104 **Rf** (261) Rutherfordium	105 **Db** (262) Dubnium	106 **Sg** (266) Seaborgium	107 **Bh** (264) Bohrium	108 **Hs** (269) Hassium	109 **Mt** (268) Meitnerium	110 **Ds** (281) Darmstadtium	111 **Rg** (272) Roentgenium	112 **Cn** (285) Copernicium	113 (284)	114 **Fl** (289) Flerovium	115 (288)	116 **Lv** (293) Livermorium	117 (293)	118 (294)

Lanthanide series:

57 **La** 138.9 Lanthanum	58 **Ce** 140.1 Cerium	59 **Pr** 140.9 Praseodymium	60 **Nd** 144.2 Neodymium	61 **Pm** (145) Promethium	62 **Sm** 150.4 Samarium	63 **Eu** 152.0 Europium	64 **Gd** 157.3 Gadolinium	65 **Tb** 158.9 Terbium	66 **Dy** 162.5 Dysprosium	67 **Ho** 164.9 Holmium	68 **Er** 167.3 Erbium	69 **Tm** 168.9 Thulium	70 **Yb** 173.0 Ytterbium

Actinide series:

89 **Ac** (227) Actinium	90 **Th** 232.0 Thorium	91 **Pa** (231) Protactinium	92 **U** 238.0 Uranium	93 **Np** (237) Neptunium	94 **Pu** (244) Plutonium	95 **Am** (243) Americium	96 **Cm** (247) Curium	97 **Bk** (247) Berkelium	98 **Cf** (251) Californium	99 **Es** (252) Einsteinium	100 **Fm** (257) Fermium	101 **Md** (258.10) Mendelevium	102 **No** (259) Nobelium

*Numbers in parentheses are mass numbers of the most stable or best-known isotope of radioactive elements.

pK_a values for functional groups and molecules commonly encountered in pharmaceuticals and biology (*http://research.chem.psu.edu/brpgroup/ pKa_compilation.pdf*)

Entry	Acid (name)	Conjugate Base (name)	pK_a
1	H_3O^+ (hydronium ion)	H_2O (water)	−1.7
2	H_3PO_4 (phosphoric acid)	$H_2PO_4^-$ (dihydrogen phosphate)	2.0
3	$PhCO_2H$ (benzoic acid)	$PhCO_2^-$ (benzoate)	4.2
4	$PhNH_3^+$ (anilinium)	$PhNH_2$ (aniline)	4.6
5	CH_3CO_2H (acetic acid)	$CH_3CO_2^-$ (acetate)	4.8
6	 (pyridinium)	 (pyridine)	5.1
7	H_2CO_3 (carbonic acid)	HCO_3^- (bicarbonate)	6.4
8	$H_2PO_4^-$ (dihydrogen phosphate)	HPO_4^{-2} (hydrogen phosphate)	6.8
9	 (imidazolium)	 (imidazole)	7.0
10	NH_4^+ (ammonium)	NH_3 (ammonia)	9.2
11	$PhOH$ (phenol)	PhO^- (phenolate)	10.0
12	HCO_3^- (bicarbonate)	CO_3^{-2} (carbonate)	10.3
13	CH_3CH_2SH (ethanethiol)	$CH_3CH_2S^-$ (ethyl thiolate)	10.5
14	HPO_4^{-2} (hydrogen phosphate)	PO_4^{-3} (phosphate)	12.5
15	CH_3OH (methanol)	CH_3O^- (methoxide)	15.5
16	H_2O (water)	HO^- (hydroxide)	15.7

Medicinal Chemistry

The Modern Drug Discovery Process

Erland Stevens

Davidson College, North Carolina

PEARSON

Boston Columbus Indianapolis New York San Francisco Upper Saddle River
Amsterdam Cape Town Dubai London Madrid Milan Munich Paris Montréal Toronto
Delhi Mexico City São Paulo Sydney Hong Kong Seoul Singapore Taipei Tokyo

Editor in Chief: Adam Jaworski
Executive Editor: Jeanne Zalesky
Senior Marketing Manager: Jonathan Cottrell
Project Editor: Jessica Moro
Editorial Assistant: Lisa Tarabokjia
Marketing Assistant: Nicola Houston
Media Producer: Erin Fleming
Managing Editor, Chemistry and Geosciences:
  Gina M. Cheselka
Production Project Manager: Edward Thomas
Full Service/Composition: PreMediaGlobal

Full Service Project Manager: Tracy Duff, PreMediaGlobal
Illustrations: PreMediaGlobal
Copyeditor: Sarah Wales-McGrath
Proofreader: Rebecca Roby
Manager of Permissions: Beth Wollar
Permissions Researcher: Liz Kincaid, PreMediaGlobal
Design Manager: Mark Ong
Interior Design: Gary Hespenheide
Cover Design: Jodi Notowitz
Operations Specialist: Jeffrey Sargent
Cover Photo Credit: Shutterstock

Credits and acknowledgments borrowed from other sources and reproduced, with permission, in this textbook appear on the same page as the borrowed material.

1 2 3 4 5 6 7 8 9 10—EBM—16 15 14 13 12

ISBN-10: 0-321-89270-4; ISBN-13: 978-0-321-89270-6

Brief Contents

Contents

Chapter 3	**A Trip through the Body** 34

Chapter 4	**Enzymes as Drug Targets** 62

Chapter 11 **Lead Optimization: Traditional Methods**   273

Chapter 12 **Lead Optimization: Hansch Analysis**   298

Chapter 13 **Aspects in Pharmaceutical Synthesis**   321

To the Reader

In fall 2001, I taught medicinal chemistry for the first time. My training is in synthetic organic chemistry, and I learned many of the class topics about a week before my students. In summer 2002, I attended the Residential School on Medicinal Chemistry at Drew University. The Drew course was a fantastic experience. For me, the value of the course was less about the material and more about seeing practitioners of medicinal chemistry talk about the drug discovery process in their own language. In this textbook, I have tried to be true to the ideas and attitudes that seem inherent to those who are actively involved in the pharmaceutical industry.

THE ROLE OF MEDICINAL CHEMISTRY IN THE UNDERGRADUATE CURRICULUM

This book is primarily an undergraduate text, and it reinforces essential chemistry skills through the study of medicinal chemistry. Medicinal chemistry overlaps with organic, physical, and biological chemistry. Under the American Chemical Society educational guidelines, an upper-level medicinal chemistry course serves as an in-depth course that bridges content from multiple divisions of chemistry.

What makes medicinal chemistry an excellent topic for reinforcing introductory chemistry concepts is that almost all students have some exposure to medicinal chemistry topics in their day-to-day lives. Since students already have familiarity with medicinal chemistry and the topic relies on concepts from other branches of chemistry, medicinal chemistry helps students apply and extend their existing chemistry knowledge in an area that seems relevant in society.

DEFINING MEDICINAL CHEMISTRY

The field of medicinal chemistry does not have distinct boundaries, and therefore the content of a medicinal chemistry textbook is debatable. This particular textbook is modeled after medicinal chemistry lectures that are presented at meetings of the American Chemical Society. Indeed, the goal of this textbook is to allow a student to attend and understand a professional medicinal chemistry lecture. A typical medicinal chemistry lecture includes an introduction to a disease, coverage of possible points of intervention for a drug, discovery of weakly active drug hits through screening, selection of drug leads, optimization of a clinical drug candidate, and (often) some details of drug synthesis. These are the central topics of the book and are broadly presented as follows:

- Introduction (three chapters): Selecting a disease for treatment, Food and Drug Administration approval process
- Pharmacodynamics (three chapters): Measuring drug activity against enzymes and receptors
- Pharmacokinetics (two chapters): Drug absorption, distribution, metabolism, elimination
- Drug discovery and synthesis (five chapters): Hit discovery, lead optimization, drug synthesis

In all parts of the text, general concepts that can be applied to many different drugs are favored over specialized ideas with limited scope.

TO THE STUDENT

Students using this textbook are presumed to have a strong foundation in general chemistry and organic chemistry. Only a high school background in biology is required. The text introduces ideas of biochemistry, cell biology, and physiology as needed.

FEATURES OF THE TEXT

The text includes a number of features to help students absorb and retain the material:

- Learning objectives at the opening of each chapter highlight key ideas.
- Case Studies in each chapter demonstrate real-life applications of newly introduced ideas.
- End-of-chapter summaries provide a synopsis of the chapter's material.
- Highlighted keywords are collectively listed and defined in the glossary.
- Sample Calculations throughout each chapter demonstrate the use of equations to analyze quantitative data.
- Conceptual end-of-chapter problems test students' understanding rather than memorization skills.

Acknowledgments

A large number of people have been instrumental in making this textbook become a reality, and I am thankful to them all. A handful of people deserve special mention. First, I thank all the reviewers who have been involved in reading the manuscript. The reviews caught typos, fixed outright errors, and suggested improved methods for presenting material. Because of their efforts, the final text is much better than the original draft. Second, I thank Jessica Moro of Pearson Publishing and Tracy Duff of PreMediaGlobal as well as their colleagues. Both Jessica and Tracy shepherded me as a first time author through the publishing process. My life was much easier because of their guidance. Finally, I thank my wife, Karin. She patiently listened, acted interested, and empathized throughout the many years that this book has gone from being an idea to a final product. I owe her (big time).

Reviewers

Zeynep Alagoz
University of the Sciences, Philadelphia

Peter Bell
Tarleton State University

Michael Berg
Virginia Tech

Adiel Coca
Southern Connecticut State University

Matthew J. DellaVecchia
Palm Beach Atlantic University

Michael Detty
SUNY Buffalo

Richard Fitch
Indiana State University

E. Eugene Gooch
Elon University

Marion Gotz
Whitman College

January Haile
Centre College

Bob Hanson
St. Olaf College

John Hofferberth
Kenyon College

David Hunt
The College of New Jersey

Joanne Kehlbeck
Union College

William Malachowski
Bryn Mawr College

Jennifer Muzyka
Centre College

J. Ty Redd
Southern Utah University

Edward Skibo
Arizona State University

Chapter 1

A Brief History of Drug Discovery

Chapter Outline

The history of medicine through the ages is a rich field. Both ancient and medieval documents note the use of herbs and formulations with biological activities that have since been attributed to specific chemical compounds. How the biological activities of most plants and compounds were discovered is not known, but the findings presumably involved significant speculation. The modern drug discovery process strives to eliminate guesswork and unnecessary risk in developing effective medicines. Coordination of scientific research with regulatory agencies is a vital part of this process. While the current scientific method was nonexistent in the ancient and medieval world, medicines certainly did exist and were in use. Most were extracts from natural sources of irreproducible potency. In the absence of a thorough and critical approval process, early drugs were of questionable benefit and carried considerable hazards. Although the current drug approval process remains far from perfect, the modern drug supply is much safer today than even just 50 years ago.

Learning Goals for Chapter 1

- Appreciate the hazards inherent to unregulated medicine
- Know the origins of the U.S. Food and Drug Administration (FDA)

1.1 Selected Drugs from Early Recorded History

A *drug* is a compound that affects physiological function once it is absorbed into the body. Early humans almost certainly knew of many different drugs and compounds that influenced their health and mood. Evidence of these drugs comes from both archaeological research and examination of historical documents. Very few drugs that were recorded in the ancient and middle ages are still in use today. The following sections provide a few examples of early documented drugs and medicines.

China

The first documented drug treatment for any disease can be traced back to approximately 3000 BCE in China. Emperor Shen Nung, also known as the Yan Emperor, noted that the plant ma huang (*Ephedra vulgaris*) treated a number of conditions, including cough. The active component of ma huang is ephedrine (**1.1**), which was not isolated until 1897 (**Figure 1.1**).[1] Although ephedrine is not used today in the treatment of cough or other cold symptoms, its diastereomer pseudoephedrine (Sudafed, **1.2**) is a common decongestant. Pseudoephedrine has often been extracted from over-the-counter cold medications and illegally converted in methamphetamine (**1.3**), which is known as meth.

India

Turmeric (*Curcuma longa*) is an herb that is native to southern Asia. Turmeric, the spice, is derived from the rhizomes of the turmeric plant. Many people in Asia hold the spice in high regard for treatment of a large range of ailments: cough, liver problems, rheumatism, inflammation, and anorexia. The molecule believed to be most responsible for the benefits of turmeric is curcumin (**1.4**) (**Figure 1.2**). Tumeric spice contains between 2–5% curcumin by weight. Because of its many reported biological effects, curcumin has been extensively studied. Curcumin shows a remarkably diverse range of activity and has been found to prevent cancer, lower cholesterol levels, inhibit blood clotting, suppress diabetes, and stimulate healing. The cancer prevention properties of curcumin have received the most attention. Many people today use turmeric as a dietary supplement for its potentially long-term preventive effects against a variety of cancers.[2]

Middle East

In *Genesis*, a great flood is recounted. During the flood, Noah and his family survived in an ark. After the waters receded and his family was safe, Noah is recounted as planting a vineyard and drinking wine.[3] Some historical scholars place the writing of *Genesis* to around 600 BCE, while a literal reading of the text places the postflood time period at 2000 BCE. Noah's drug of choice, ethanol, was well known to ancient people and continues to be used

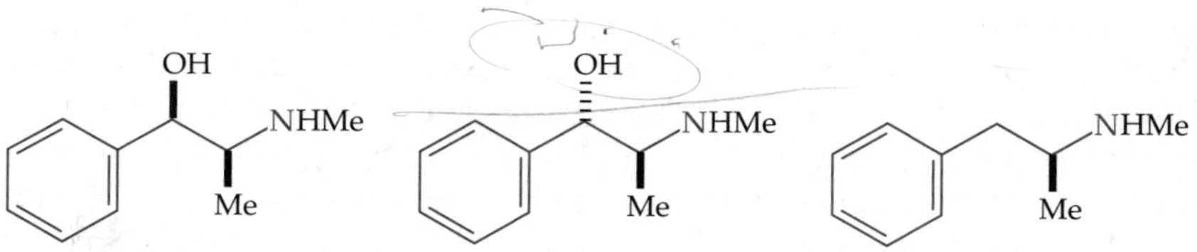

FIGURE 1.1 Ephedrine and analogues

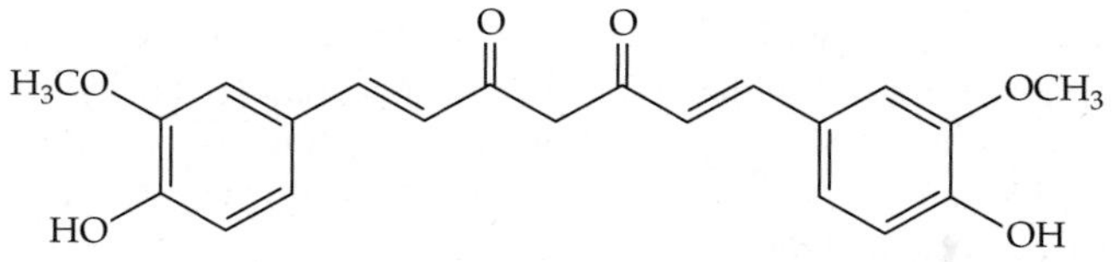

ephedrine
stimulant
1.1

pseudoephedrine
(Sudafed)
decongestant
1.2

methamphetamine
(meth)
stimulant
1.3

curcumin
dietary supplement
1.4

FIGURE 1.2

widely today. The earliest evidence of wine production has been traced to a settlement in northern Iraq. The research site has afforded pottery fragments residues of calcium tartrate, a salt found in grapes. The potsherds have been dated to around 7000 BCE.[4]

Greek, Roman, and Byzantine Periods

Ancient Greece boasts Hippocrates, the father of medicine. Hippocrates (ca. 460–380 BCE) is credited with the development of *humoral medicine*. Humoral medicine links the four elements of the world (air, earth, fire, and water) with the four humors of the body (black bile, blood, phlegm, and yellow bile). The followers of Hippocrates further expanded his theories. In particular, Galen (129–199 CE), who wrote *On the Elements According to Hippocrates*, cemented the dominance of humoral medicine for an additional 1,000 years. Galen's theories were available to the emerging Islamic world, and humoral medicine strongly influenced the writings of Avicenna (980–1037 CE) of Persia, whose *Canon of Medicine* was an authoritative medical text for 500 years.[1]

Although humoral medicine had many followers, a notable exception was Celsus (25 BCE–50 CE), a Roman encyclopedist. Celsus wrote *De Medicina*, which advocates a "reasoned theory of medicine" based on experimentation as opposed to "speculation."[5] While the ideas of Celsus are more consistent with modern medical practice, they failed to displace the mystical appeal of humoral medicine.[1]

A notable pharmaceutical that can be dated to the Greek period is morphine (**1.5**), the active ingredient of opium (**Figure 1.3**). The earliest explicit mention of opium was made by Theophratus (372–287 BCE).[1] Raw opium is obtained by drying the white latex secreted by damaged poppy pods (*Papaver somniferum*). Following refining, opium contains 9–14% morphine.[6] Morphine itself remains widely used as a powerful analgesic. Structural modifications to morphine as seen in heroin (**1.6**) and codeine (**1.7**) lengthen the half-life of the drug in the bloodstream. Compounds with a longer half-life can be administered less frequently without a drop in biological effect in the patient.

1.2 Middle Ages to Modern Times

As society progressed through the Middle Ages to the modern era, newly discovered drugs reflected available technology. The use of crude extracts from natural sources required the least skill and was the major type of drug in use in the Middle Ages. Minerals and inorganic materials became possible with improvements in refining. Eventually, as chemistry became a viable field in the modern era, synthetic drugs became the standard.

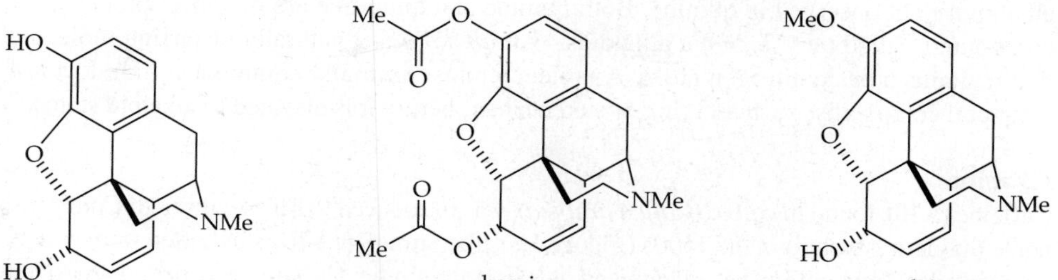

morphine
1.5

heroin
(diacetylmorphine)
1.6

codeine
(3-methylmorphine)
1.7

FIGURE 1.3 Morphine and related opiates

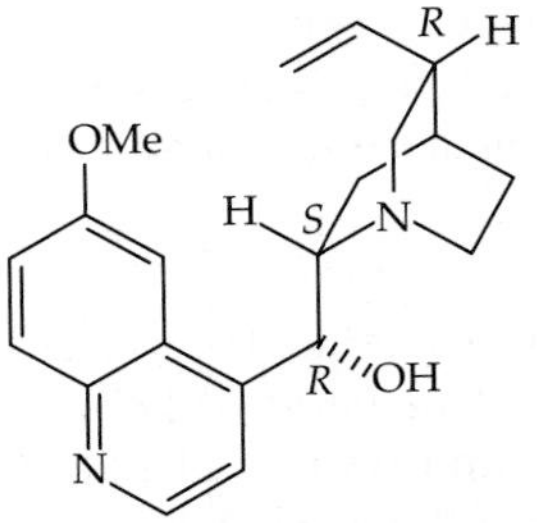

quinine
antimalarial
1.8

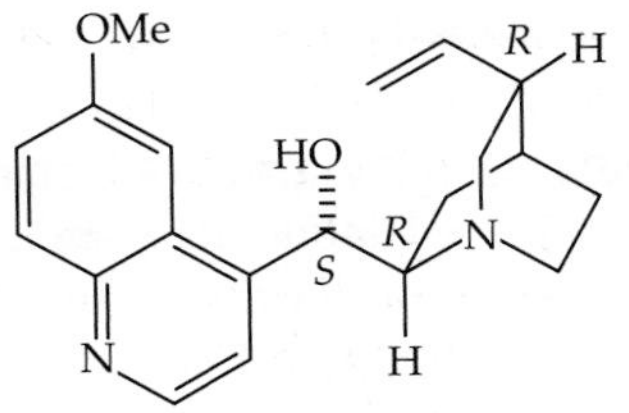

quinidine
antiarrythmic
1.9

FIGURE 1.4 Diastereomeric *Cinchona* alkaloids

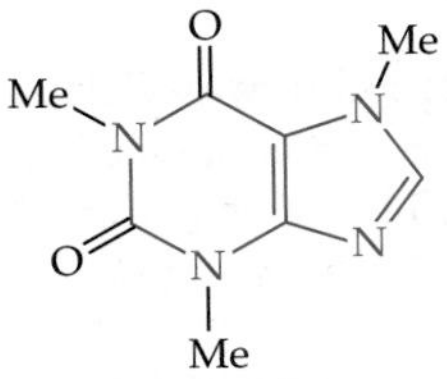

caffeine
stimulant
1.10

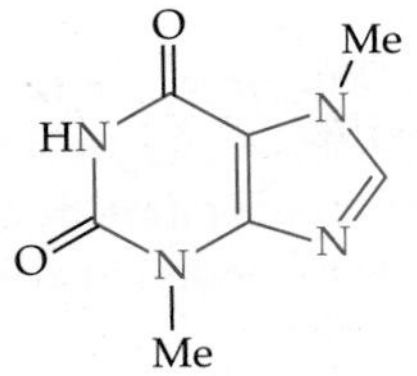

theobromine
diuretic
1.11

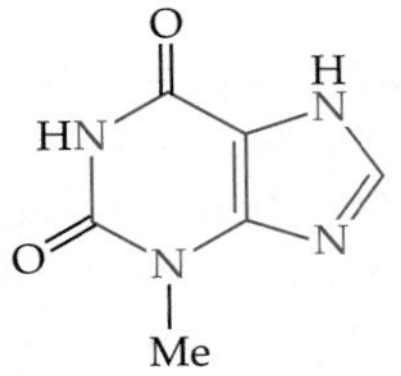

theophylline
asthma
1.12

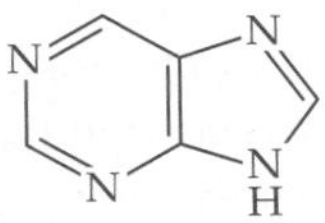

purine
1.13

FIGURE 1.5 Caffeine, related diuretics, and purine

Natural Products

Legitimate discoveries of materials with biological activity in the ancient world were dominated by natural products. This trend continued into the Middle Ages.

Quinine

Quinine (**1.8**) is found in cinchona bark (*Cinchona officinalis*) and is useful as a treatment for malaria and fever (**Figure 1.4**). The first documented use of cinchona bark as a medicine is disputable, but all early accounts originate with the native population of Peru in the seventeenth century. Following their travels to South America, Jesuit priests likely introduced the bark to Europe in the first half of the 1600s.[1] Quinine makes up 0.8–4.0% of the weight of cinchona bark.[6] The compound was isolated in 1820 by Pelletier and Caventou. Also available from cinchona bark is quinidine (**1.9**), a diastereomer of quinine that is less effective in treating malaria. Quinidine, however, is an antiarrythmic agent (stabilizes heart rhythm), an activity not observed in quinine. Both quinine and quinidine are members of a class of compounds called the *Cinchona* alkaloids.[1] An *alkaloid* is a naturally occurring molecule that contains a basic nitrogen atom. A molecule must normally contain a certain level of structural complexity, such as a ring or stereocenter, before it is elevated to alkaloid status.

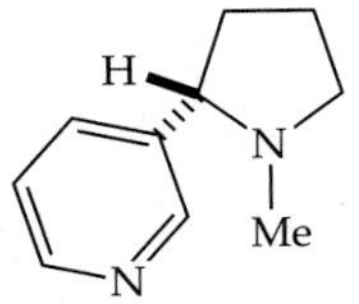

Caffeine

Caffeine (**1.10**), found in coffee (*Coffea arabica*), was introduced to Europe through Constantinople (modern Istanbul) in the 1500s (**Figure 1.5**). The stimulant effects of coffee were widely acknowledged, but coffee was recognized as a useful diuretic. Caffeine was first synthesized by Emil Fischer in 1882. Two related compounds, theobromine (**1.11**) and theophylline (**1.12**), found in cacao seeds (*Theobroma cacao*) and tea (*Camellia sinensi*), respectively, are more potent diuretics than caffeine.[1] All three compounds are based on the purine ring system (**1.13**).

nicotine
stimulant
1.14

FIGURE 1.6

Nicotine

Through tobacco (*Nicotiana tabacum*) from Cuba, Spanish explorers introduced nicotine (**1.14**) to Europe early in the sixteenth century (**Figure 1.6**). While sometimes socially shunned, smoking was believed to have antispasmotic benefits. Nicotine is the active component of tobacco and was isolated in 1828.[1] Nicotine may be found in dried tobacco leaves (*N. tabacum* and *N. rustica*) in amounts of 2–8% as acid salts.[6] The addictive properties of nicotine remain a problem for those who smoke and wish to quit. As a therapeutic category, smoking cessation drugs such as NicoDerm CQ, a transdermal nicotine delivery system for those trying to quit smoking, had collective annual sales approaching US$3 billion in 2008.[7]

Ascorbic Acid (Vitamin C)

Long voyages at sea put sailors at risk for developing scurvy, a condition characterized by a range of symptoms ranging from bleeding gums to extreme fatigue. Both oranges and lemon juice had been recognized to be of value in preventing scurvy by the 1600s, but other treatments of more questionable value were also practiced.[1]

In 1747, John Lind, a Scottish naval physician, performed the first known controlled clinical trial. A clinical trial explicitly involves the testing of a compound on people. Lind tested five possible scurvy remedies—cider, dilute sulfuric acid, vinegar, sea water, and oranges and lemons—on 10 sailors who were suffering from scurvy. Two affected sailors were given a control diet. Only the two sailors who received two oranges and one lemon daily in addition to the control diet recovered. Despite the convincing nature of Lind's work, lemon juice was not part of standard fare for British sailors until 1795. Following the widespread use of lemon juice, the incidence of scurvy essentially dropped to zero in the British navy within a decade.[1]

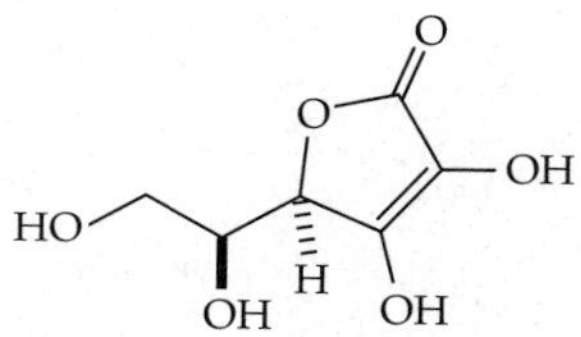

While Lind's clinical trial provided convincing evidence for the ability of oranges and lemons to cure scurvy, the identity of the active curative agent for scurvy remained unknown for well over 100 years. Following approximately 25 years of heightened research into the causes of scurvy, Edmund Hirst of the University of Birmingham published the correct structure of L-ascorbic acid (**1.15**), or vitamin C, in 1933 (**Figure 1.7**). One of Hirst's colleagues, Norman Haworth, known to organic chemistry students because of Haworth projections, shared the Nobel Prize in Chemistry in 1937 for his work on elucidating the biological role of ascorbic acid.[1]

L-ascorbic acid
or vitamin C
1.15

FIGURE 1.7

Advances in Inorganic Chemistry

Modern herbal medications available in health food stores can vary in quality from batch to batch, so one can only imagine the poor quality of plant extracts during the medieval period. Pure organic substances were not available until the mid-1800s, and then only sparingly. In contrast, relatively pure inorganic substances were widely available even in the Middle Ages from advances in mining and smithing. The availability of pure inorganic substances allowed accurate quantification for proper dosing of the patient.

Equally important to advances in chemistry, the field of biology also developed rapidly during the period of the Renaissance. More sophisticated biological techniques and an evolving scientific method allowed for improved observation and documentation of a drug's effects.

Ferrous Sulfate (FeSO$_4$)

Pierre Blaud (1774–1858), a French physician, first noted the medicinal use of ferrous (Fe^{+2}) sulfate (FeSO$_4$) in 1832. Blaud used ferrous sulfate to treat anemia, a condition that was then called chlorosis or "love sickness" and frequently observed in younger women. The standard treatment had been ferric (Fe^{+3}) oxide (Fe$_2$O$_3$). Blaud's protocols soon became the treatment of choice because of the higher solubility of the sulfate over the oxide. Blaud's nephew, who was a pharmacist, marketed "the veritable pills of Doctor Blaud" worldwide.[1]

Zinc Sulfate ($ZnSO_4$)

Zinc sulfate ($ZnSO_4$) was reportedly used by the Dutch physician Sylvius de le Böe (1614–1672) as an emetic (induces vomiting). Thereafter, zinc sulfate was widely used whenever it was necessary to empty a patient's stomach. The rapid onset of zinc sulfate made it more desirable than other treatments. Today, zinc sulfate is sometimes used in small quantities to treat a deficiency of zinc, a required trace element in the body.[1]

Arsenic Anti-Infectives

The development of arsenic derivatives for medicinal purposes bridges inorganic and organic chemistry as well as older and modern pharmaceutical practices. Although these compounds do contain at least one arsenic atom (a metalloid), they are still considered to be organic molecules. The highly covalent carbon-arsenic bond makes these organometallic compounds stable and easy to isolate.

Arsenic trioxide (As_2O_3) was used sporadically for the treatment of various skin conditions throughout the second millennium. The toxicity of arsenic, which interferes with ATP production, prevented widespread acceptance. A promising application of As_2O_3 that emerged in the 1800s was the treatment of trypanosomes, a protozoan infection. Another arsenic derivative, sodium arsanilate or atoxyl (**1.16**), was developed specifically for trypanosome treatment (**Figure 1.8**). In 1902, Paul Ehrlich (1854–1915), a German bacteriologist, developed a related compound with improved antitrypanosomal activity. The new compound was called arsphenamine (**1.17**) or "606," a reference to the compound's activity being discovered in the 606th experiment performed by Ehrlich. Arsphenamine was later found to be effective against *Treponema pallidum*, the bacterium responsible for syphilis.[1] The true structure for **1.17** became a point of controversy in the 1970s when Ehrlich's original structural assignment (**1.18**) was called into question. Not until 2005 was the structure confirmed to be a cyclic oligomer of three or five identical subunits. Structure **1.17** undergoes metabolism in the body to form **1.19**, the actual antiprotizoal species.[8]

FIGURE 1.8 Arsenical antimicrobial drugs developed by Ehrlich

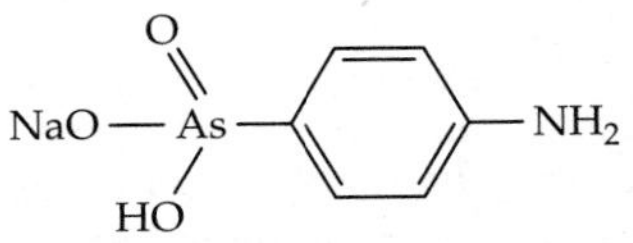

atoxyl
1.16

arsphenamine
(606, Salvarsan)
1.17

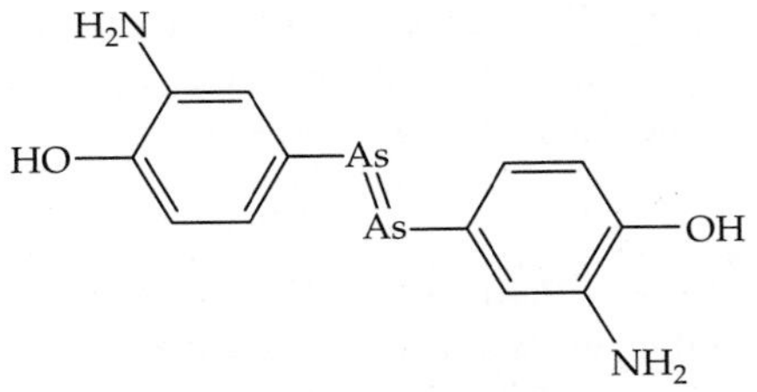

original, incorrect
structural assignment
of arsphenamine
1.18

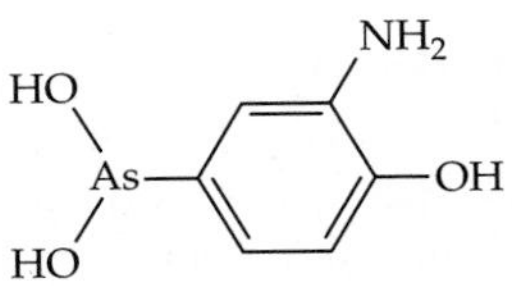

active metabolite
of arsphenamine
1.19

Ehrlich's approach to developing arsphenamine involved extensive screening studies on each newly prepared compound. Information gained from earlier screens helped determine the structure of the next compound to be synthesized. This iterative method is the predominant method for modern lead optimization in medicinal chemistry. Under his method, Ehrlich's antibacterial research continued to progress and ultimately resulted in the discovery of the sulfa drugs.

Ehrlich was keenly aware of the need for safe and effective pharmaceuticals. Among his many other contributions to the field of pharmacology, Ehrlich developed the idea of a *chemotherapeutic index*. Ehrlich defined chemotherapeutic index as the minimum curative dose divided by the maximum tolerated dose (Equation 1.1).

$$\text{chemotherapeutic index} = \frac{\text{minimum curative dose}}{\text{maximum tolerated dose}} \tag{1.1}$$

Ehrlich provided a quantitative measure to accompany the concept that a therapeutic dose must be evaluated relative to a toxic dose. Although Ehrlich's chemotherapeutic index has been displaced by more precise measures of effectiveness and toxicity, the spirit of Equation 1.1 remains a central idea of modern drug discovery. Ehrlich's work in the field of medicine, especially in the area of autoimmune disorders, resulted in his sharing of the Nobel Prize in Medicine in 1908.

Sulfa Drugs

Sulfamidochrysoidine (Prontosil Rubrum, **1.20**), a red dye, was marketed as an antibiotic in the 1930s by the German company I.G. Farbenindustrie, now part of Bayer. Based on his work on the project, Gerhard Domagk was awarded the Nobel Prize in Medicine in 1939 (**Figure 1.9**).[1]

Researchers at I.G. Farbenindustrie had known that some azo dyes, which contain an N=N linkage, have antibacterial properties. Sulfamidochrysoidine, developed in 1932, was particularly effective against staphylococcal and streptococcal infections in mice. Sulfamidochrysoidine was initially named Streptozon by the I.G. Farbenindustrie research team. In 1933, Streptozon was first used in humans and successfully healed a 10-year-old boy who was dying from staphylococcal septicaemia. Physicians in Germany were soon receiving supplies of Streptozon. As Streptozon was found to have broad activity against many bacterial strains, the trade name of sulfamidochrysoidine was changed to Protosil Rubrum. By 1935, sulfamidochrysoidine gained acceptance throughout Europe and the United States. One of the most notable patients to be treated with sulfamidochrysoidine was Domagk's own daughter, who developed a blood infection after pricking her finger with a needle.[1]

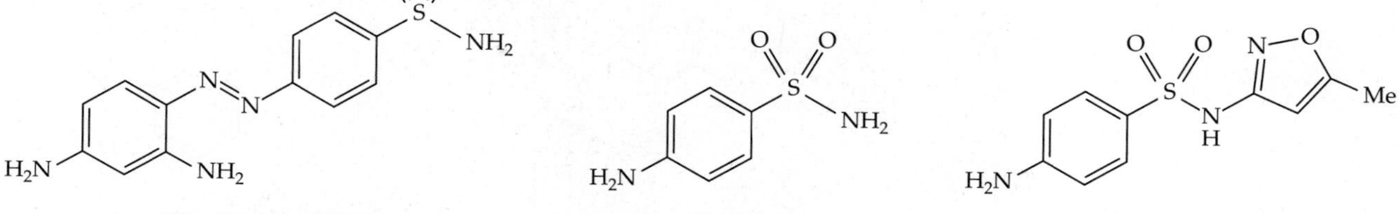

sulfamidochrysoidine
(Streptozon or
Protosil Rubrum)
1.20

sulfanilamide
1.21

sulfamethoxazole
1.22

FIGURE 1.9 Representative sulfa drug antibiotics

Continuing research on sulfamidochrysoidine revealed that sulfanilamide (**1.21**), a metabolite of sulfamidochrysoidine, was the active antibacterial compound (Figure 1.9). Compounds that are chemically altered in the body to form the biologically active drug are called prodrugs.[1]

Sulfanilamide is the parent member of a drug class called sulfa drugs. Sulfa drugs have been extensively investigated, with thousands of analogues having been prepared. Sulfa drugs, while not as potent for certain infections as newer antibiotics, remain in use today. The patents on sulfa drugs have almost all expired, so they are relatively inexpensive compared to other antibiotics on the market. One actively prescribed sulfa drug is sulfamethoxazole (**1.22**) (Figure 1.9). Sulfamethoxaole treats a range of bacterial infections, including many ear infections in children.

1.3 Evolution of the U.S. Food and Drug Administration

Many early drugs, including the sulfa drugs, were used safely and to great benefit; not all were so successful.

Corporate Self-Regulation

Josiah Lilly, son of Colonel Eli Lilly, was a supervisor with Eli Lilly in Indianapolis in the 1880s.[9] One morning, Lilly was faced with a reported problem. A company product, fluidextract thimbleweed, was apparently making people ill. Lilly and a company chemist, Ernest Eberhart, decided to test a sample of the product on themselves to determine its safety. Later that same day, Lilly was found unconscious in his office. Lilly was resuscitated quickly, but Eberhart spent the rest of the day concerned about his own welfare. Fluidextract thimbleweed was quickly removed from Eli Lilly's product list.[10]

Concern for safety was not limited to Eli Lilly. Wallace Abbott started Abbott Alkaloidal Company in 1888. One goal of the company was to produce "dosimetric granules," a reference to the need for consistent dosing in the emerging field of pharmaceuticals.[10] Animal testing was recognized early on as a method of standardizing doses. One of the first animal assays involved fluidextract of ergot, a vasoconstrictor initially used to treat excessive bleeding following childbirth. An active component of the extract is ergotamine (**1.23**), a structural relative of lysergic acid diethylamide (LSD, **1.24**) (**Figure 1.10**). Administration of the extract to roosters resulted in "bluing" of their normally red comb and wattle. Evaluation of both the duration and degree of color change of the comb allowed quantification of the potency of the dose. The following excerpt is taken from a journal

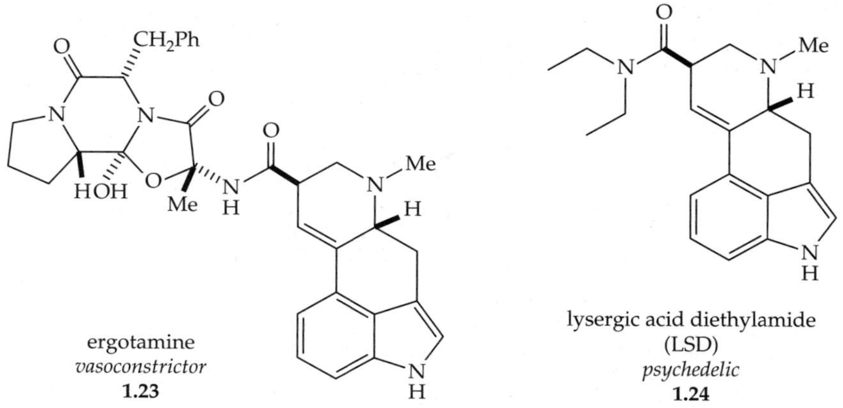

ergotamine
vasoconstrictor
1.23

lysergic acid diethylamide
(LSD)
psychedelic
1.24

FIGURE 1.10

article written by Dr. A. R. L. Dohme.[11] Dohme became president of Sharp & Dohme, which later merged with Merck & Co. to become Merck Sharp & Dohme, or simply Merck as it is known today.

> Experiment I. Black rooster, weight 5 lbs. Before the injection the wattles and comb were red and warm. 5 Cc. fluid extract of best Spanish ergot were injected hypodermically at 10 a.m.
>> At 10:35 a.m., comb bluish but warm; wattles still red and warm.
>> At 11 a.m., comb and wattles blue and cooler.
>> At 11:40 a.m., comb and wattles very blue and much cooler.
>> At 3:30 p.m., comb and wattles still very blue and cool.
> Rooster's bill open, and rooster looking quite sick. Conclusion: Fluid extract of ergot quite active.

Although most early pharmaceutical businesses focused on building a reputable drug industry, some individuals and firms were more interested in turning a quick profit. Safety issues abounded, and government regulation became a necessity to protect the welfare of the general public.

Food, Drug, and Cosmetic Act of 1938

After originally starting as a branch of the U.S. Department of Agriculture, the Food and Drug Administration (FDA) assumed its current title in 1930. At its inception, the FDA held little regulatory power, and the current drug approval process did not exist. In 1937, S. E. Massengill marketed Elixir Sulfanilamide, a new liquid formulation of sulfanilamide with 10% diethylene glycol as a cosolvent. Diethylene glycol, a component of antifreeze, is toxic to the kidneys and liver. Over 100 people died as a result of using the new formulation. Following public outcry, the government passed the Food, Drug, and Cosmetic Act in 1938.[12]

Continuing Challenges and Refinements

The Food, Drug, and Cosmetic Act gave the FDA much larger powers to oversee consumer safety. Perhaps most notably, the FDA now required preapproval of all new drugs. The new drug application (NDA) remains a fixture in the modern drug approval process and precedes the final approval of a drug. Periodic updates to FDA review protocols have attempted to address new problems as they have arisen. Because many problems are difficult to anticipate, new legislation tends to be reactive rather than proactive.

Thalidomide

Thalidomide (**1.25**) was initially released in Germany in 1956 by Chemie Grünenthal (**Figure 1.11**). The drug had been developed as a sedative and also showed antiemetic

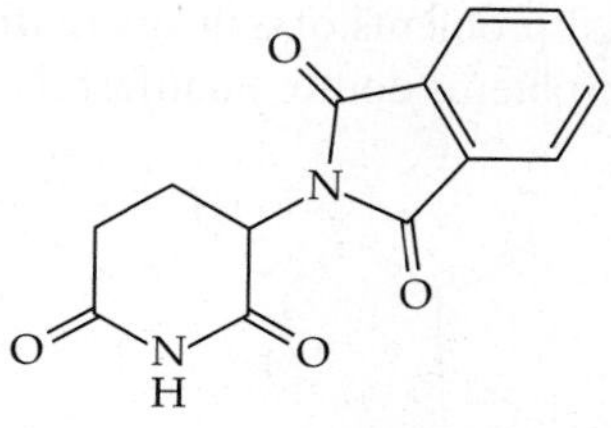

thalidomide
sedative
1.25

FIGURE 1.11

(antinausea) activity. A natural market for thalidomide was pregnant women suffering from morning sickness in their first trimester. Unfortunately, thalidomide was a teratogen and induced birth defects in the fetuses of many of the expectant women who received the drug. Thousands of infants, mostly in Europe and Canada, were delivered with malformed limbs during 1960 and 1961. Tests determined that thalidomide was the cause, and the drug was rapidly removed from the market.[1]

Before the disaster in Europe and Canada, thalidomide had been submitted to the FDA for approval in the United States. The FDA did not consider the safety tests of Chemie Grünenthal to be adequate and withheld approval of thalidomide. The United States was therefore spared the thalidomide disaster. Withholding approval for thalidomide is considered to be one of the triumphs of the FDA in protecting public safety.[1]

The story of thalidomide demonstrates the importance of stereochemistry in pharmaceuticals. Thalidomide contains a stereocenter. Researchers found that the *R*-enantiomer of thalidomide carries the sedative properties of the drug, while the *S*-enantiomer causes birth defects. Formulating thalidomide as a single enantiomer drug is not viable, however, because thalidomide slowly racemizes in the body.[13] Most drugs do not racemize in vivo, and different enantiomers of a compound often give rise to different biological effects within the chiral environment of the body. The incidents surrounding thalidomide raised awareness of the potential impact of enantiomers in the drug industry. Today, the biological effects of enantiomers are tested individually as a matter of standard practice.

The tragic story of thalidomide does have a silver lining. The limb deformities observed in children affected by thalidomide are believed to be caused by impaired bloodflow to the developing appendages in the fetus.[14] The ability of thalidomide to decrease bloodflow makes the drug a potential treatment for tumors that require a steady blood supply to grow. Thalidomide has been approved by the FDA for treatment of multiple myeloma and is being investigated for use against other cancers. Additionally, the sedative properties of thalidomide have been found to relieve the debilitating pain experienced by some leprosy patients.[15]

Fen-Phen

Phentermine (Adipex P, **1.26**), fenfluramine (Pondimin, **1.27**), and its (+)-enantiomer dexfenfluramine (Redux, **1.28**) are antiobesity drugs (**Figure 1.12**). All three engage the body's flight-or-fight response pathways. Part of the response involves decreased appetite. In the mid-1990s, physicians often prescribed a mixture of either fenfluramine or dexfenfluramine with phentermine patients at risk for weight problems. The drug mixture was casually referred to as fen-phen. In 1997, some patients taking fen-phen were found to have damaged heart valves. Initially, the problem was blamed on simultaneous use of the two drugs. Upon further investigation, some patients taking only fenfluramine or dexfenfluramine were also found to have suffered valve damage. By the end of 1997, both fenfluramine and dexfenfluramine had been removed from the market.[16]

The case of fen-phen highlights the complexity of evaluating the safety of a drug. Upon approval, all the potential problems of a drug are difficult to anticipate. This is especially true if physiological problems do not manifest themselves during the animal and

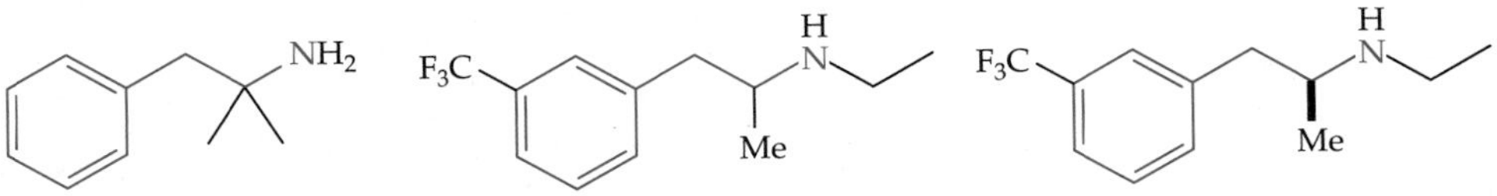

phentermine
(Adipex P)
1.26

fenfluramine
(Pondimin)
1.27

dexfenfluramine
(Redux)
1.28

FIGURE 1.12 Problematic antiobesity drugs

clinical trials. In the case of fenfluramine, clinical testing did not reveal heart irregularities. Valve damage was detectable only through an echocardiogram, an expensive diagnostic test that is not typically included in drug trial monitoring.

Rofecoxib

Cyclooxygenase (COX) is an enzyme that affects some biological responses, including pain and inflammation. COX has two different forms, bearing the creative names COX-1 and COX-2. COX-1 is found in almost every cell in the body; COX-2 is active only at the site of an injury. Common pain relievers like aspirin (**1.29**) and ibuprofen (Advil or Motrin, **1.30**) act by blocking the action of both COX-1 and COX-2 (**Figure 1.13**). Therefore, aspirin and ibuprofen have effects beyond their desired site of action. A problematic side effect is that they cause stomach irritation and even ulcers with long-term use. Researchers rationalized that a drug with selectivity against COX-2 would not have some of the side effects of aspirin and ibuprofen. Investigations into selective COX-2 inhibitors yielded several promising drugs, including rofecoxib (Vioxx, **1.31**), celecoxib (Celebrex, **1.32**), and valdecoxib (Bextra, **1.33**). Between 1997 and 2001, the FDA approved all three compounds.

In 2004, rofecoxib was pulled from the market after long-term studies linked the drug to an increased incidence of heart attacks and strokes. Merck, the marketer of rofecoxib, has since faced allegations that the corporation knew of cardiovascular problems linked to rofecoxib before the long-term studies became public information.[17] This incident has raised issues of ethical information disclosure, sales practices, and the rigor of the FDA review process. In 2005, Pfizer, the marketer of both celecoxib and valdecoxib, voluntarily removed valdecoxib from the market pending further tests on the cardiovascular safety of the drug. Celecoxib, while still available, is now packaged with additional safety warnings.[18]

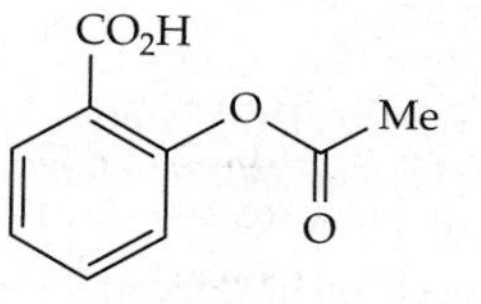

aspirin
1.29

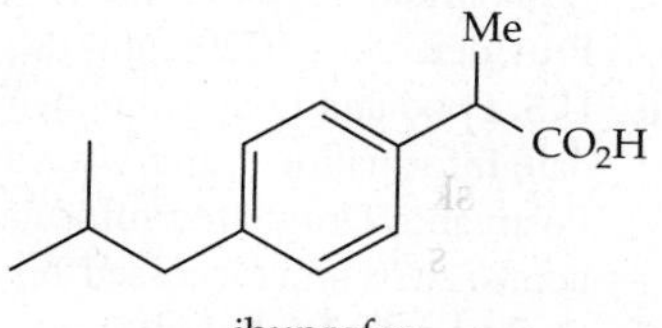

ibuprofen
(Advil, Motrin)
1.30

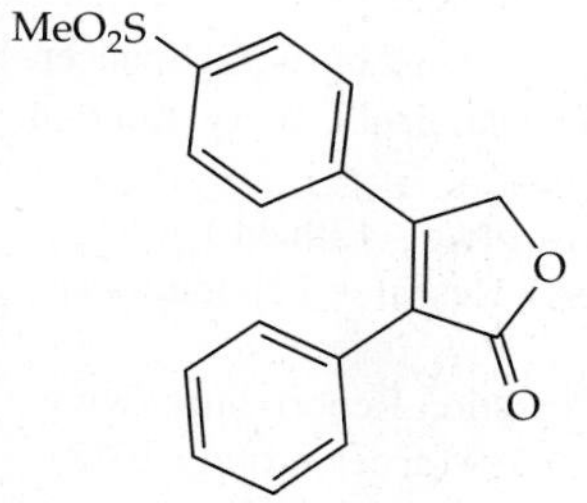

rofecoxib
(Vioxx)
1.31

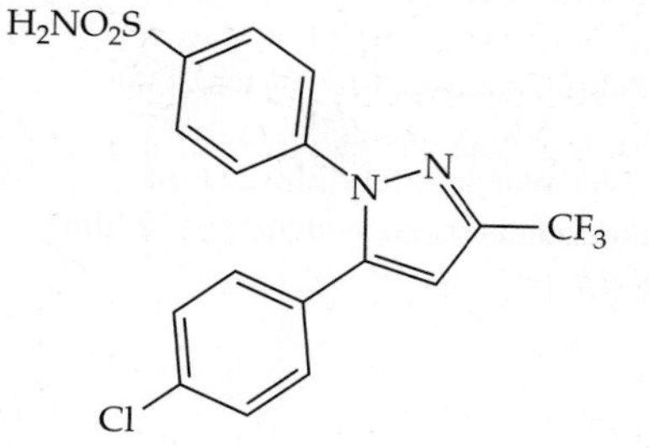

celecoxib
(Celebrex)
1.32

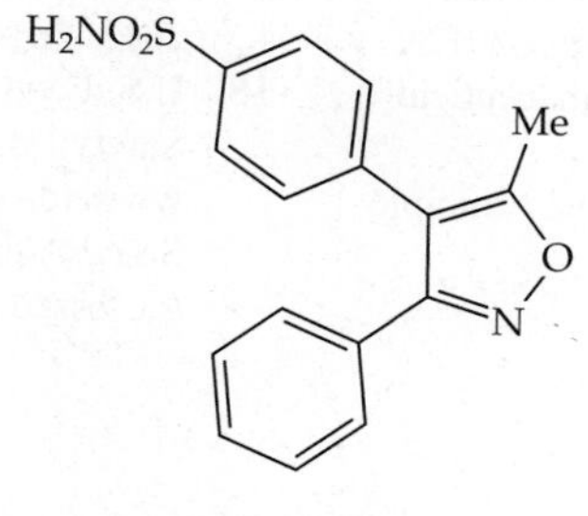

valdecoxib
(Bextra)
1.33

FIGURE 1.13 COX inhibitors used as analgesics

Summary

The earliest medicines were extracts from natural sources, mostly plants. The resulting extracts were of irreproducible purity and unproven effectiveness. Many treatments were surrounded by myths rather than facts. Despite a lack of scientific methodology, early practitioners of healthcare discovered many medicines that were indeed very effective. Regardless, variability in the extracts made accurate dosing difficult, if not impossible. Administered drugs could be either ineffective or overly potent depending on the skill of the extractor and quality of the plant material. Advances in chemistry allowed the development of purification techniques for both inorganic and organic materials. With relatively pure materials in hand, analytically minded physicians could better quantify (and document) patient doses. Continued advances in chemistry allowed the synthesis of new materials. Medicines were no longer limited to what was available from nature. Synthetic chemistry opened medicine to previously unknown compounds with similarly unknown effects, both positive and negative. By the early 1900s, the market swelled with new drugs from various manufacturers with a range of claims of effects. Safety quickly became a problem, and government regulation became a necessity. Today, large regulatory agencies like the FDA in the United States strive to maintain a safe drug supply. For the foreseeable future, however, medicine will continue to be an experimental science, and pharmaceuticals will carry a level of risk.

References

1. Sneader, W. *Drug Prototypes and their Exploitation*. Chichester, UK: Wiley & Sons, 1996.

2. Aggarwal, B. B., Kumar, A., Aggarwal, M. S., & Shishodia, S. Curcumin Derived from Turmeric (*Curcuma longa*): A Spice for All Seasons. In D. Bagchi (ed.) *Phytopharmaceuticals in Cancer Chemoprevention* (Chapter 23). Boca Raton, FL: Chemical Rubber Company, 2005.

3. *Genesis* 9:20-21.

4. Berkowitz, M. World's Earliest Wine. *Archaeol.* 1996, *49*, 26.

5. Celsus, A. C. *De Medicina*. Translated by W.G. Spencer. In E. H. Warmington (ed.), Loeb Classical Library. London: William Heinemann, Ltd. 1935.

6. O'Neil, M. J. (ed.). *The Merck Index*, (13th ed.). Whitehouse Station, NJ: Merck Research Laboratories, 2001.

7. Lead Discovery. Smoking Cessation Report. http://www.leaddiscovery.co.uk/Reports/1389 (accessed July 2012).

8. Lloyd, N. C., Morgan, H. W., Nicholson, B. K., & Ronimus, R. S. The Composition of Ehrlich's Salvarsan: Resolution of a Century-Old Debate. *Angew. Chem. Int. Ed.* 2005, *44*, 941–944.

9. Eli Lilly and Company. History. http://www.lilly.com/about/heritage/Pages/heritage.aspx (accessed July 2012).

10. Stafford, R. O. The Growth of American Pharmaceutical Biology. *BioSci.* 1966, *16*, 675–679.

11. Dohme, A. R. L., & Crawford, A. C. The Active Principle of Ergot. *Am. J. Pharm.* 1902, *74*, 503–507.

12. U.S. Food and Drug Administration. FDA History. http://www.fda.gov/AboutFDA/WhatWeDo/History/Origin/ucm054826.htm (accessed July 2012).

13. Teo, S. K., Colburn, W. A., Tracewell, W. G., Kook, K. A., Stirling, D. I., Jaworsky, M. S., & et al. Clinical Pharmacokinetics of Thalidomide. *Clin. Pharmacokinet.* 2004, *43*, 311–327.

14. Therapontos, C., Erskine, L., Gardner, E. R., Figg, W. D., & Vargesson, N. Thalidomide Induces Limb Defects by Preventing Angiogenic Outgrowth during Early Limb Formation. *Proc. Nat. Acad. Sci.* 2009, *106*, 8573–8578.

15. U.S. Food and Drug Administration. Thalidomide Information. http://www.fda.gov/Drugs/DrugSafety/PostmarketDrugSafetyInformationforPatientsandProviders/ucm107296.htm (accessed July 2012).

16. U.S. Food and Drug Administration. Fen-Phen Safety Update Information. http://www.fda.gov/Drugs/DrugSafety/PostmarketDrugSafetyInformationforPatientsandProviders/ucm072820.htm (accessed July 2012).

17. Curfman, G. D., & Morrissey, S. Expression of Concern: Bambardier et al., "Comparison of Upper Gastrointestinal Toxicity of Rofecoxib and Naproxen in Patients with Rheumatoid Arthritis," *N. Engl. J. Med.* 2005, *353*, 2813–2814.

18. U.S. Food and Drug Administration. MedWatch 2004 Safety Information Alerts. Vioxx Sep 2004. http://www.fda.gov/Safety/MedWatch/SafetyInformation/SafetyAlertsforHumanMedicalProducts/ucm166532.htm (accessed July 2012).

Chapter 2

The Modern Drug Discovery Process

Chapter Outline

The typical journey from concept to market for a drug passes through many groups of people with a variety of responsibilities and skills. Specialty fields in the process include finance, biology, chemistry, law, and medicine. The overall development process is sometimes called *integrated drug discovery* to highlight the cooperative effort of various teams within a corporation. This chapter presents an overview of the drug discovery process. Awareness of the process as a whole is vital for understanding the specific role of medicinal chemistry in the larger picture. Because this chapter touches on many ideas of drug discovery, some terminology is left loosely defined until it can be properly described in later chapters. The glossary at the end of the text should be helpful.

2.1 Market Analysis

From a business standpoint, pharmaceutical companies are no different from any other for-profit business. All sell products to consumers with the intent of generating a profit that will be invested back into the company or distributed to its owners or investors. Just like all businesses, drug companies develop only products that are anticipated to be profitable. The cost of developing a drug and bringing it to market is immense—US$802 million in 2000 according to one estimate.[1] This estimate was revised upward to US$1.2 billion in a follow-up study performed in 2010.[2] Cost estimates vary wildly, but one point is always clear: drug development is very expensive. The profit from each successful drug must be large enough to cover not only the costs of the successful drug but also the expenses of other drugs that were cut in development. For this reason, new drugs are normally developed for diseases and conditions that affect a large number of people. A survey of successful (i.e., profitable) drugs reveals many that treat problems of high blood pressure, high cholesterol, and pain management in arthritis. All affect large segments of the population and treat conditions that require long-term management. A drug for any of these disease conditions will enjoy a large number of patients and subsequent prolonged sales.

Learning Goals for Chapter 2

- Understand the sequence of events required to bring a drug to market

- Know the role of the drug target and how its activity is screened

- Recognize the type of information gained in the various animal and human trials

- Understand the interaction between regulatory agencies and pharmaceutical companies

- Appreciate the overall effort and cost required to develop a new pharmaceutical

Of greatest interest to the pharmaceutical industry are diseases for which no effective treatments exist. The development of a drug for a new category likely has a large potential market. Desirable new categories include drugs for disorders such as Alzheimer's disease, obesity, Parkinson's disease, type 1 and 2 diabetes, and asthma. Creating drugs that are effective in a new category is extremely challenging. The risk of failure is high.

From a humanitarian perspective, a market-driven approach to drug development often leaves unaddressed the diseases of less-developed nations. Nations with a low per capita income are less able to contribute to an economically profitable drug. Some scientists have attempted to address this situation. For example, Dennis Liotta of Emory University spearheaded the creation of iThemba Pharmaceuticals in 2009 in Johannesburg, South Africa. iThemba specifically targets novel treatments for HIV/AIDS, malaria, and tuberculosis, three exceptionally common diseases in Africa.[3]

The following Case Study describes the development of an effective drug for acid reflux, highlights the profitability of such a drug, and demonstrates how quickly competing companies create their own therapies.

CASE STUDY

Treatment of Acid Reflux

Around 1980, the H_2 receptor antagonists emerged as a category creator for the treatment of acid reflux, a condition that causes heartburn. *Receptors* are proteins in the body that regulate biological processes. Receptors are activated and/or deactivated through binding small molecules, or *ligands*. An *antagonist* is a ligand that blocks the activation of a receptor. H_2 Receptor antagonists act by preventing the binding of histamine (**2.1**) to the H_2 receptor (**Figure 2.1**). The H_2 receptor regulates stomach acid secretion. The first H_2 receptor antagonist to reach the market was cimetidine (Tagamet, **2.2**), which shares many key structural features with histamine. Given the prevalence of acid reflux, especially related to the Western diet, cimetidine became an instant blockbuster drug. Ranitidine (Zantac, **2.3**) and famotidine (Pepcid, **2.4**), both with similar structures to cimetidine and the same mode of action, were quickly developed by competing companies to cash in on the large market. Late entries into a market are often called me-too drugs and tend to have lower sales than the first entrant within a drug class. Regardless, because of the size of the market for acid reflux drugs, both ranitidine and famotidine enjoyed strong sales.

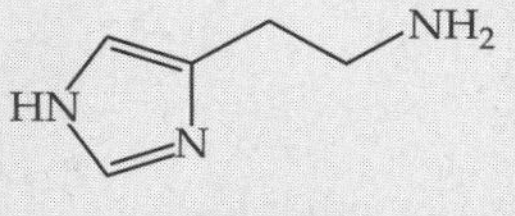

histamine
2.1

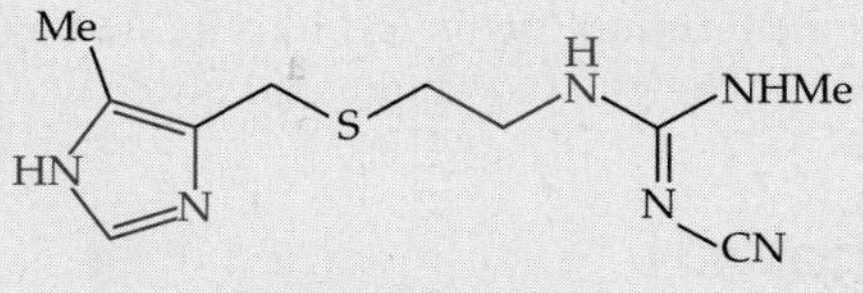

cimetidine
(Tagamet)
2.2

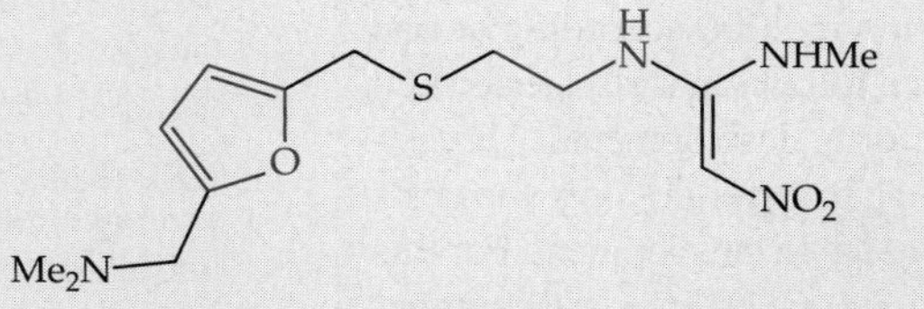

ranitidine
(Zantac)
2.3

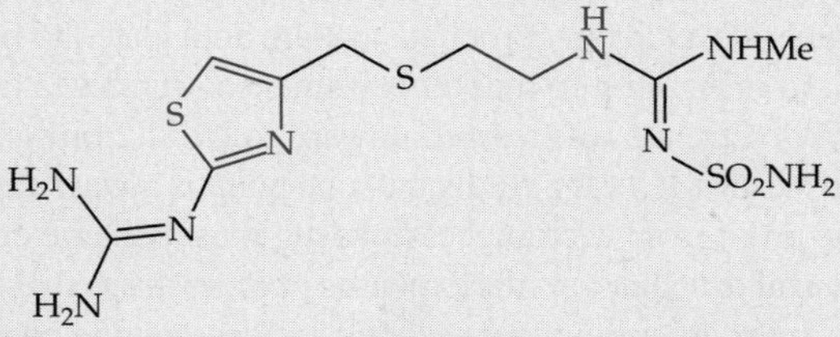

famotidine
(Pepcid)
2.4

FIGURE 2.1 Histamine (**2.1**) and H_2 receptor antagonists (**2.2–2.4**)

Few new drugs or drug classes are category creators. Much more common are new drug classes that have a new mode of action and improve upon an existing drug class. For example, after the H_2 receptor antagonists enjoyed outstanding commercial success for most of the 1980s, a new acid reflux treatment emerged—the proton pump inhibitors, or PPIs. Proton pumps are proteins that help maintain the acidity of the stomach. A PPI blocks the action of the pump and therefore lowers the acid content of the stomach. The PPIs represented a new drug class with advantages over the H_2 receptor antagonists. The acid reflux market quickly shifted to the PPI drugs, including omeprazole (Prilosec, **2.5**) and lansoprazole (Prevacid, **2.6**) (**Figure 2.2**).

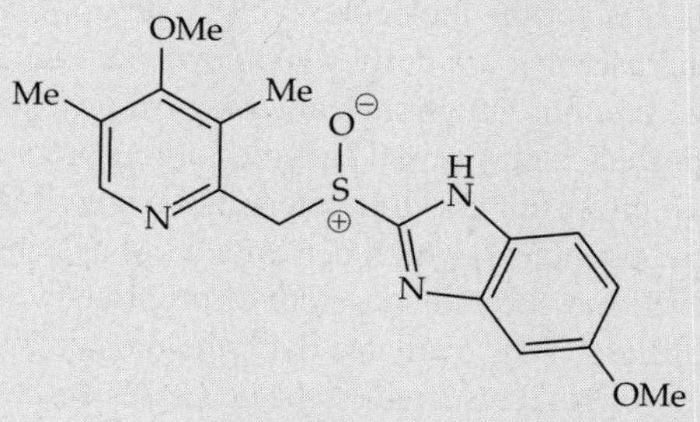

omeprazole
(Prilosec)
2.5

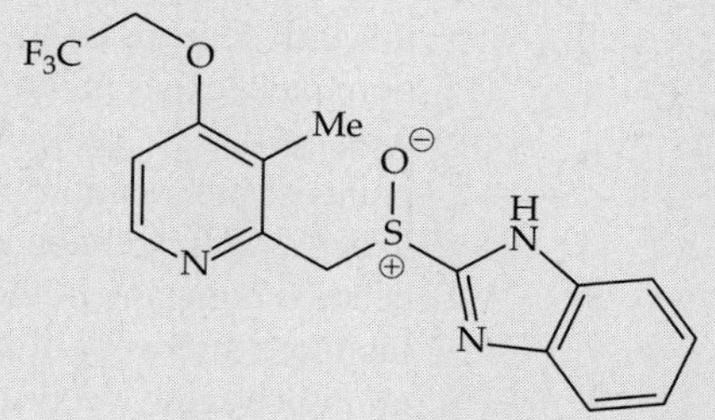

lansoprazole
(Prevacid)
2.6

FIGURE 2.2 Proton pump inhibitors (PPIs)

Some drugs are developed for diseases with small potential markets. In the United States, the Orphan Drug Act of 1983 legislated subsidies for the exploration of drugs that have the potential to help small numbers of patients. A small market is defined as one with less than 200,000 patients in the United States, or any market that is small enough to prevent the company from recouping its research costs through drug sales. Other nations also have similar legislation and definitions. While potentially exploitable and prone to abuse, the Orphan Drug Act has led to the creation of drugs that have a limited market potential and may have otherwise not been developed.

Absent of outside factors, a pharmaceutical company typically addresses diseases affecting large numbers of patients. The size of the market is estimated based on studies of demographic and medical information. Once a disease with a favorable market has been selected, the decision is passed along to the molecular biology team.

2.2 Target Selection

Informed of the specific disease of interest to the pharmaceutical company, the molecular biology team tries to identify a *target* that may be exploited to change the state of the disease. Many different types of targets are known, but the most frequently encountered are enzymes (Chapter 4) and receptors (Chapter 5). Oligonucleotides (Chapter 6) are also occasionally targeted by drugs. The target is the site of action for the drug and controls a biological process related to the diseased state of a patient. Although a company does not need to know the target of a drug, knowing the target and its structure is extremely beneficial to the discovery team and can make for a more successful drug search.

Finding a Target

The classical approach for discovering a drug target is to study the inner workings and signal pathways in a cell. Proteins that play key roles in the processes are candidate targets. The use of genomics to find previously unknown proteins with potential value as targets has become much more commonplace.

SCHEME 2.1 Hypothetical biological pathway

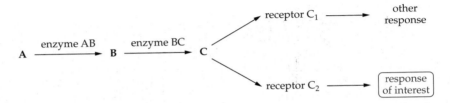

Traditional Methods

Finding a target can be a challenging task for the molecular biology group. The difficulty of this task depends on the target. For diseases that are fairly well known, considerable information likely exists in the literature. The literature might reveal biological processes associated with the disease, and some of those processes may reveal enzymes or receptors as appropriate points of intervention. Targets that are promising for drug intervention are often described as *druggable*. In addition to the literature, previous research performed in-house by the pharmaceutical company in the area of interest may provide a wealth of proprietary information to facilitate discovery of a target. Regardless of the available data, the molecular biology team almost certainly needs to perform new, original research on the processes of the disease.

The molecular biology team tracks the production, transport, and binding of chemical messengers related to the condition of interest. Chemical messengers bind targets along the pathway and cause different effects. Candidates for druggable targets include structures that either manufacture or bind the chemical messengers. Finding a suitable target along this pathway requires determining a point of intervention along the pathway that affects the diseased state without significantly impacting other biological processes. A simple, hypothetical process is shown in **Scheme 2.1**. The entire sequence involves the conversion of **A** to **C** by two enzymes. Once formed, compound **C** binds two different receptors, one (receptor C_2) controls the response of interest, and the other (receptor C_1) affects a process unrelated to the response of interest.

In this example, the preferred target is receptor C_2. Blocking receptor C_2 affects only the response of interest. Another possibility is targeting enzyme BC. This option, however, stops the conversion of **B** to **C** and impacts both receptors C_1 and C_2. A third option, inhibition of enzyme AB, stops production of **C** as well as **B**. If **B** has roles not shown in Scheme 2.1, then those additional pathways will be impacted by targeting enzyme AB. In general, hitting a target toward the end of a pathway (*downstream*) causes fewer side effects than intervening further *upstream*. If the example outlined in Scheme 2.1 were a real scenario, both enzyme BC and receptor C_2 might be investigated as potential targets. Over time, one target would likely emerge as the better option, and research on the second target would be abandoned.

The next two Case Studies demonstrate typical factors that pharmaceutical companies consider when selecting a drug target.

CASE STUDY

ACE Inhibitors and A-II Receptor Blockers[4,5]

High blood pressure, or hypertension, is a condition that affects a large segment of the population. Therefore, drugs that effectively manage hypertension are potentially profitable for a pharmaceutical company. In the 1970s, the principal drugs for treating hypertension were β-blockers. β-blockers shut down β-adrenergic receptors and act as β-adrenergic receptor antagonists. The biological pathways associated with β-blockers are shown in **Scheme 2.2**. The β-adrenergic receptors are triggered by epinephrine (**2.7**) and norepinephrine (**2.8**), and may influence a number of processes including release of renin from the kidneys (**Figure 2.3**). (*Nor* is a prefix used occasionally to denote loss of a methyl group from a nitrogen atom.) Renin, an enzyme, influences vasoconstriction and water retention by converting angiotensinogen into angiotensin I. β-blockers produce side effects including fatigue and depression. In an attempt to avoid the side effects of β-blockers, inhibition of angiotensin-converting enzyme (ACE) was selected as a more downstream target in the blood pressure regulation pathway. The first ACE inhibitor to reach the market was captopril (Capoten,

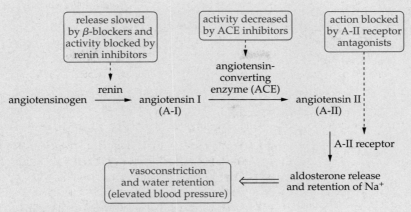

SCHEME 2.2 A hypertension regulation pathway

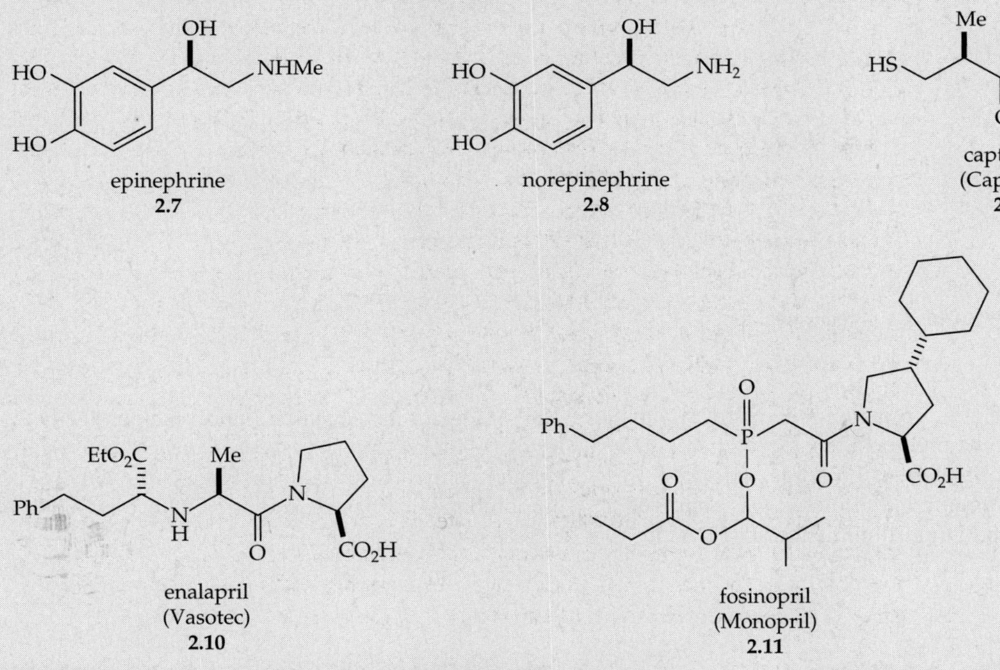

epinephrine
2.7

norepinephrine
2.8

captopril
(Capoten)
2.9

enalapril
(Vasotec)
2.10

fosinopril
(Monopril)
2.11

FIGURE 2.3 Ligands of β-adrenergic receptors and ACE inhibitors

2.9). Later drugs included enalapril (Vasotec, **2.10**) and fosinopril (Monopril, **2.11**). ACE inhibitors quickly displaced β-blockers as the leading antihypertensive drugs in the 1980s.[4]

In addition to converting A-I to A-II, ACE also plays a role in the breakdown of bradykinin. Some patients taking ACE inhibitors develop a dry cough, a symptom that is associated with elevated bradykinin levels. Another class of blood pressure medication, the A-II receptor antagonists, was developed to intervene downstream of ACE and bypass side effects of ACE inhibitors (Scheme 2.2). Examples of A-II receptor antagonists include losartan (Cozaar, **2.12**) and valsartan (Diovan, **2.13**) (**Figure 2.4**). Once A-II receptor antagonists hit the market, ACE inhibitors fell out of favor as the preferred method for treating hypertension. The progression of preferred hypertension drugs from β-blockers to ACE inhibitors to A-II antagonists nicely demonstrates the benefits of intervening as far downstream in a biological process as possible.[5]

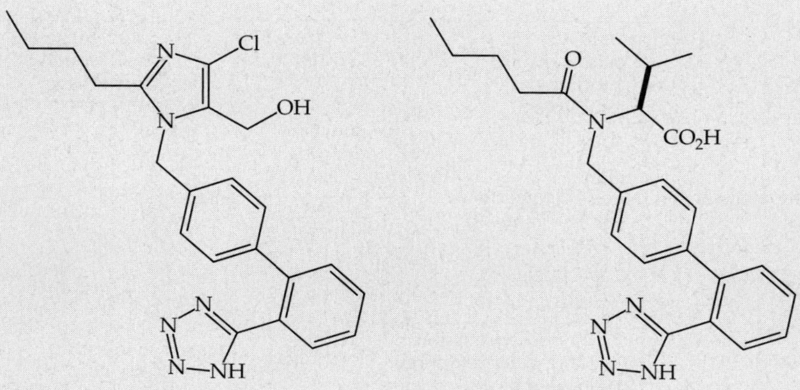

losartan
(Cozaar)
2.12

valsartan
(Diovan)
2.13

FIGURE 2.4 A-II receptor antagonists

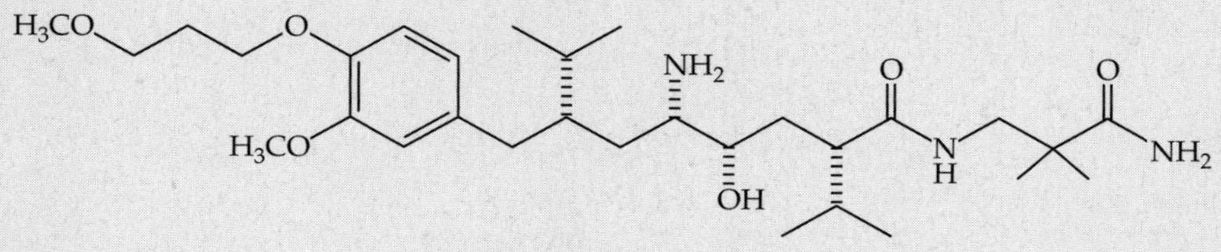

aliskiren
(Tekturna)
2.14

FIGURE 2.5 A renin inhibitor

Despite its being far upsteam in the pathway shown in Scheme 2.2, the enzyme renin has recently emerged as a viable target for hypertension drugs. Aliskiren (Tekturna, **2.14**) is a renin inhibitor and was approved by the U.S. Food and Drug Administration (FDA) in 2007 (**Figure 2.5**). Aliskiren improves upon the activity of β-blockers because its use does not cause a dry cough in patients. To date, aliskiren is the only renin inhibitor that has been approved for treatment of hypertension.

CASE STUDY

HMG-CoA Reductase Inhibitors[6]

Elevated levels of cholesterol in the blood, or hypercholesterolemia, have been linked to heart disease. Cholesterol, a necessary component of all cell membranes, is available to the body by two routes: through the diet and through synthesis by the body itself. When changes to one's diet fail to reduce cholesterol levels, medication that inhibits cholesterol biosynthesis may be necessary. The biosynthetic pathway for cholesterol synthesis starts with acetyl-CoA (**2.15**) (**Scheme 2.3**). Through a series of steps, three acetyl-CoAs are brought together to form 3-hydroxy-3-methylglutaryl-CoA (HMG-CoA, **2.17**). HMG-CoA is reduced to mevalonic acid (**2.18**) by the enzyme HMG-CoA reductase. Mevalonic acid is the precursor of isopentenyl pyrophosphate (**2.19**) and dimethylallyl pyrophosphate (**2.20**), the building blocks of cholesterol (**2.21**) and related steroids. HMG-CoA reductase, which catalyzes the rate-determining step in the biosynthesis of cholesterol, was recognized as an attractive target for drugs treating hypercholesterolemia.

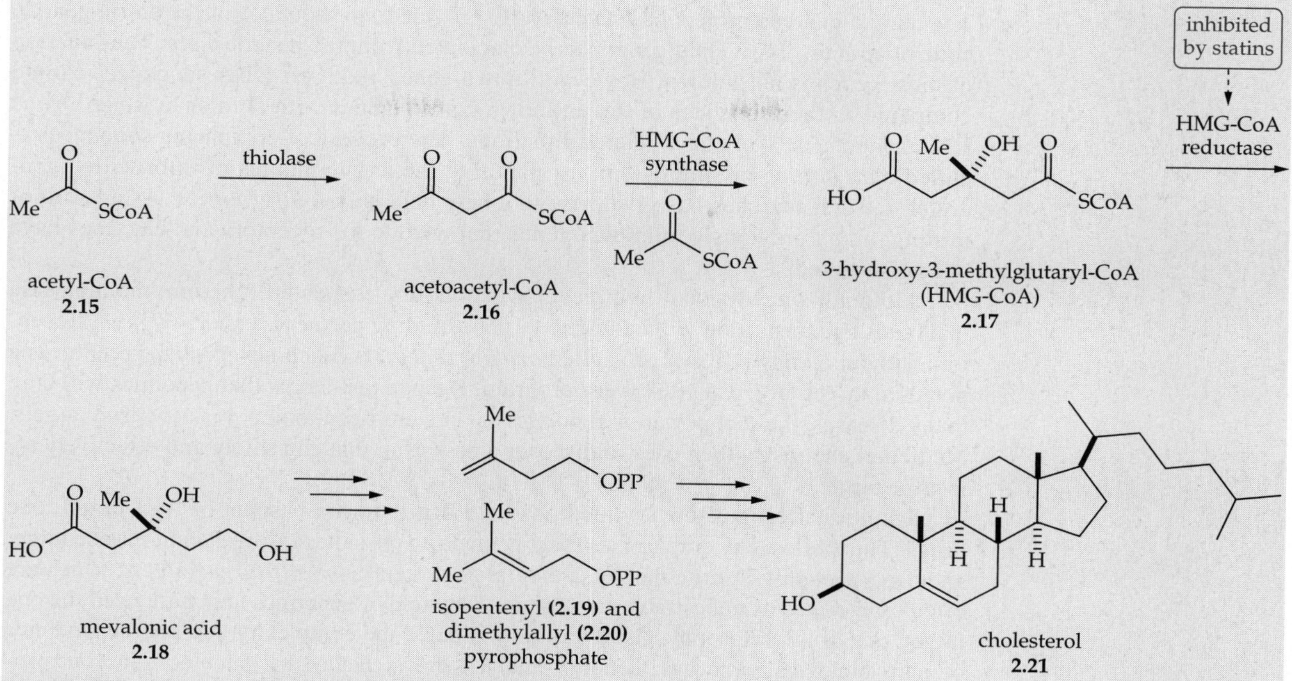

SCHEME 2.3 Biosynthesis of isoprenoid precursors of cholesterol

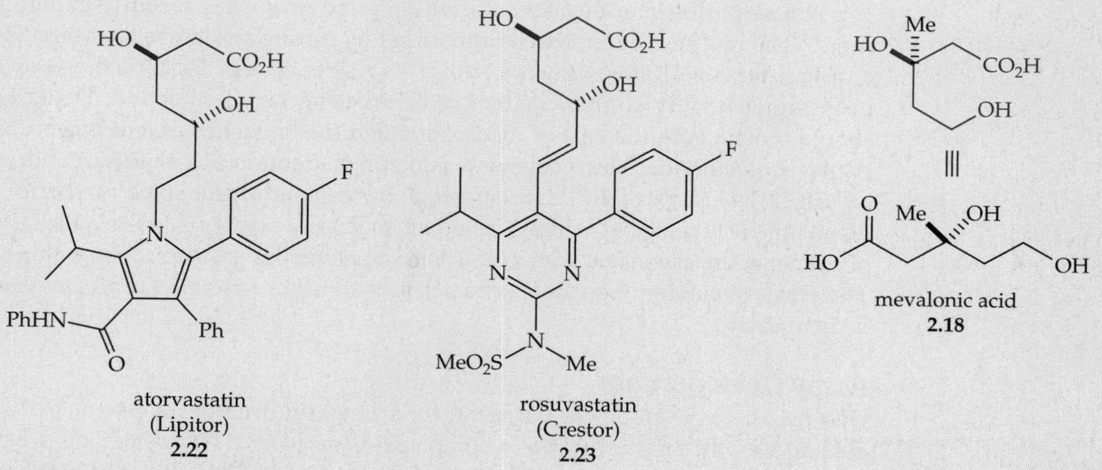

FIGURE 2.6 Two examples of statin drugs

The point of Scheme 2.3 is not to show off some complicated biological transformation. Scheme 2.3 highlights that cholesterol (**2.21**), a vital yet problematic molecule for many individuals, is formed in the body through a series of chemical steps. Each step is performed by an enzyme, and each enzyme is a potential target for a drug discovery program.

Research on the pathway of cholesterol biosynthesis led to the development of a new class of drugs called statins. All statins, including atorvastatin (Lipitor, **2.22**) and rosuvastatin (Crestor, **2.23**), are inhibitors of the enzyme HMG-CoA reductase (**Figure 2.6**). Structurally, the acid side chain found on statin drugs closely resembles mevalonic acid (**2.18**). The side chain plays an important role in the binding of statins to HMG-CoA reductase.

Impact of Genomics and Informatics on Target Discovery

The human genome project has successfully provided the sequence of the entire genetic code of humans. Individual genes can be elucidated from the genetic code. The function of most genes is not known. However, known genes and their DNA sequences can be compared to the sequences of the unknown genes. Genes with similar or *homologous* DNA sequences likely have related functions. The process of comparing sequences is called *data mining* or simply *mining*. Handling the large amounts of information generated in such searches has given rise to a new field called *informatics*. As a result of mining, many previously unknown genes that encode for receptors and enzymes have been discovered.[7]

Although a newly identified receptor may have a suspected function, the molecule that binds and turns it on will be unknown. The binding partner is called a *ligand*. Receptors without a known ligand are called *orphan receptors*. Each new orphan receptor is a potential target for a drug discovery program. Despite the claims that genomics will cure many diseases, the real power of genomics to date has been to uncover new drug targets. Medicinal chemistry then tries to discover a new drug that can safely and effectively affect the target.[8]

Traditional drug discovery methods hold firmly to the concept of "one target, one drug." This philosophy emphasizes that, as much as possible, a drug should hit one target and one target only. A drug that hits other targets often shows *off-target* effects, which are often categorized as undesirable side effects. The field of genomics has challenged the one target, one drug philosophy. Genomics has revealed the complexity and interdependence of many biological pathways. Genomic analysis shows that a drug that blocks just one process may have a diminished effect because the system compensates by enhancing other, alternative pathways.[9]

A new philosophy that favors multitargeted drugs has recently gained momentum.[10] The multitarget approach argues that by hitting several related targets weakly, a drug affects multiple pathways and gives rise to a large biological effect. Because each target is only slightly influenced, less drug may be needed. Using less drug should reduce both the cost of medication and the incidence of side effects. One particular molecule that has received considerable attention as a multitarget drug is curcumin (**2.24**) (**Figure 2.7**). Curcumin, a component of the spice turmeric, shows a broad range of biological activity, including anticancer activity.[11,12] The anticancer effects of curcumin are associated with its influence on numerous growth factors within the cell.[13] The effect of curcumin on any particular growth factor is small, but its aggregate effect is significant.

Assay Development

Quality *assay* development is crucial for successful drug development. An assay is a method by which the activity of a compound is measured. As compounds are tested, or *screened,* in assays, compounds that show activity help steer the medicinal chemistry

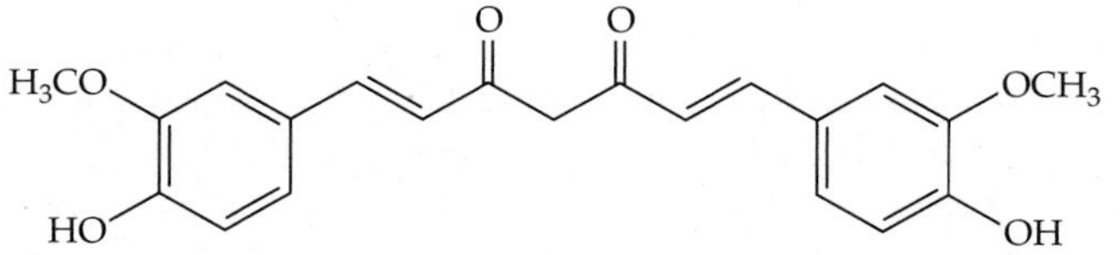

curcumin
possible anticancer
2.24

FIGURE 2.7

team toward designing a new compound with even higher activity. A repetitive feedback loop that alternates synthesis with screening should ultimately afford a highly active compound. For this process to work well, each assay must be reproducible and generate clear, quantifiable results. Developing a drug requires many assays of various types. Some assays give information on binding of a compound to the target, while others reveal how the compound behaves in a biological system. Binding information is related to the *pharmacodynamics,* which can be defined as the action of a drug when it reaches a target. Compound behavior within a biological system falls under *pharmacokinetics,* which covers how a drug is absorbed, distributed, metabolized, and eliminated from an organism. Testing a compound in living organisms is called *in vivo* testing. Assays performed in tubes, dishes, glassware, and plates are called *in vitro* tests.

Biochemical Assays

Biochemical assays normally test either binding to a receptor or the rate of a reaction that is catalyzed by an enzyme. As much as possible, these assays are performed in the absence of cells and other biological material and therefore give information only about the receptor or enzyme (the target) and how it binds the drug being tested. Developing a biochemical assay requires considerable knowledge about the target and its related biological processes. The molecular biology team must be able to prepare or commercially purchase the enzyme or receptor of interest.

By monitoring the consumption of an enzyme substrate and/or the formation of a product, enzyme assays allow the molecular biology team to follow the progress of a reaction. Checking the reaction at regular time intervals gives a full reaction profile, which can be used to determine the kinetic parameters of the reaction. Generally, it is hoped that the compound being tested will inhibit the enzyme. Measuring the rate of the reaction with and without the inhibitor determines the activity of the compound. More details on this process will be covered in Chapter 4.

Receptor binding assays are more complicated, and the screens can take many different approaches. The traditional method involves *tagging* the compound of interest with a radioactive label such as ^{3}H, ^{14}C, ^{32}P, or ^{125}I. The compound is incubated in the presence of the receptor and binds the receptor. Filtration of the mixture removes the receptor as well as any receptor-bound compounds from the assay mixture. Measurement of the radioactivity of the filter paper gives an indication of how much compound bound to the receptor. Over time, binding assays based on detecting a radiolabeled drug have been replaced with fluorescence-based assays. Fluorescence screens tend to be faster, are more reproducible, and avoid generation of hazardous radioactive waste. Receptor binding will be discussed more thoroughly in Chapter 5.

In the initial stages of searching for active compounds, the number of compounds to be screened is often greater than 100,000 and can exceed 500,000. Even 1 million compounds or more may be screened in a single assay.[14] For this reason, biochemical screens must be inexpensive and quick. Ideally, the screen is automated and performed by a robot. Solutions of all necessary reagents and compounds to be tested are loaded into the instrument. The robot, called a *liquid handler,* dispenses reagents and solutions into microtiter plates.

Microtiter plates have a standard size of approximately 3.4" × 5.0". Within the area of the plate are typically 96 wells (8 × 12) for reagents. As technology has improved, 384-well (16 × 24) and 1,536-well (32 × 48) plates have become more common. More wells per plate translates to smaller volumes of reagents and reduced costs. However, very small volumes are difficult to dispense accurately, and the screens can lose reliability. Continued improvements in robotic technology are making the 1,536-well plate the standard format.[15]

Assays are often performed in an aqueous medium with water as the solvent. Most drugs are organic molecules with poor water solubility. To improve solubility or at least evenly disperse the compound throughout the assay well, compounds are commonly dissolved in dimethyl sulfoxide (DMSO). DMSO has several attractive properties. DMSO dissolves organic compounds well, has a low vapor pressure to limit evaporation, and is fairly inert. Furthermore, DMSO freezes around 18 °C. A solution of a compound in DMSO can be easily frozen for long-term storage. Because high concentrations of DMSO may affect the reliability of an assay, compound solutions must be prepared such that, when mixed in the assay, the final DMSO concentration is 1% or less.

For enzyme activity assays, the concentrations of the substrates and/or products can often be determined with a UV-VIS spectrophotometer that is designed to take measurements on solutions while they are still contained within the microtiter plate. These *plate readers* can record data quickly. When a plate reader is used in conjunction with the robotic liquid handlers, large numbers of compounds can be screened in short order. This process is called *high-throughput screening* (HTS). If a screen can be adapted to high-throughput technology, screening costs can be reduced to the range of US$0.10 to $0.50 per well.[14]

Low per well costs allow each compound to be tested in a wider range of concentrations and in duplicate or triplicate. For example, a 384-well plate might be used to test 16 compounds. Each compound may be tested at 12 different concentrations at intervals of 0.5 log units, for a total concentration range of 6 log units, for example 10^{-9} to 10^{-3} mol/L. Performed in duplicate, these tests would fill one 384-well plate ($16 \times 12 \times 2$).[14] Representative data for three compounds ($\times$ strong, $\square$ weak, Δ very weak) are shown in **Figure 2.8**. A more active compound ($\times$) shows activity (percentage effect) at a lower concentration. Screening in duplicate at many different concentrations is normally reserved for compounds that are known or suspected to have significant activity, not in random testing of large libraries.

Cellular Assays

While biochemical assays can provide receptor binding or enzyme inhibition information, they cannot determine how a compound behaves in a more complex cellular environment. Assays involving whole cells are required for this information. Compounds that show activity in a biochemical screen but fail in a cellular assay likely suffer from interference from the cell

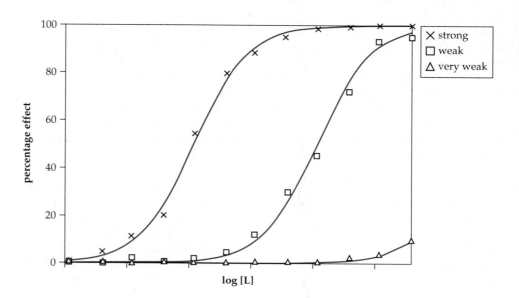

FIGURE 2.8 Representative activity data for compounds from a molecular library

components. Problems can include strong binding of the compound to another protein and difficulty crossing the cell membrane. Cellular assays are less amenable to HTS techniques because the particulate nature of cells interferes with UV-VIS detection methods. Therefore, cell-based assay can be more expensive to perform, costing up to several dollars per well.[14]

Cellular assays are often performed on genetically engineered cells. When a new assay is performed, new cells are grown from an earlier cell culture. While all cultures are ultimately descended from the same original cells, mutations are possible. Therefore, all the cells in each assay are not necessarily identical. A mutant strain may behave differently from the original cell line in the assay. To avoid unreliable results in the screen, the individual performing the assay must be aware of the number of generations, or *passage number*, that the cells being used in the screen are removed from the original culture. Including compounds with known activity as standards in an assay can help identify problems with mutated cell lines. If the standard molecule shows unexpected activity, then the genetic integrity of the cell culture may be called into question.[16]

Though they are more complex, cellular assays are vital to a drug discovery program. Whole cell assays are often used to determine properties not directly related to receptor binding or enzyme interaction. Cellular assays can determine the ability of a compound to cross a cell membrane, which is a key aspect for absorption of a drug in the small intestine. This type of permeability screening generally uses Caco-2 (human colonic adenocarcinoma) cells. Caco-2 cells effectively replicate the permeability of the intestinal lining. Cytotoxicity is another compound property that requires whole cell screening. Furthermore, proteins from blood serum can be added to the cell culture medium to indicate whether the compound being tested binds to serum proteins (see Chapter 7).[17]

Animal Testing

Animal testing was the original method for screening compounds, and the original animal in the screens was a human. For safety reasons, testing in humans gave way to testing in other animals. Rats are a favored species. They are inexpensive to maintain yet large enough to provide workable volumes of blood and tissues for analyses. Unlike developing a biochemical assay, animal screening requires no knowledge of the molecular action of the compound. Activity is observed through symptoms and behaviors exhibited by the animal, such as swelling, body temperature, or blood pressure. Animal testing was most important in the 1950s through the 1970s when molecular biology was making its original advances. Because the technology did not generally exist at that time to determine molecular targets, animal testing was the only option. While animal testing gives large amounts of information on a compound (oral availability, half-life, metabolites, organ toxicity), the screening process tends to be slow, is expensive, and requires sufficient quantities of the compound to dose an entire animal. Animal testing is certainly not a high-throughput method.

Many animal tests have been standardized and are widely used today. Two are the *paw edema test* and the *tail flick test*. In the paw edema test, a rat's paw is injured or irritated to induce swelling. The degree of swelling is measured, and the rat is then treated with an anti-inflammatory. Tracking changes in swelling gives an indication of the effectiveness of the drug. The tail flick test is used with analgesics. The tail of a rat is placed in an uncomfortably warm water bath or on a warm surface. The time it takes for the rat to remove its tail is recorded. Rats that receive an effective analgesic feel less pain and have longer response times than untreated rats in the control group.

Animal tests can be extremely elaborate. Ezetimibe (Zetia, **2.25**), a cholesterol-lowering drug, was developed in the mid-1990s through testing in rats (**Figure 2.9**). The mode of action was known to be inhibition of active transport of cholesterol in the small intestine, but the exact mechanism of transport was unknown. Without an established mode of action, a

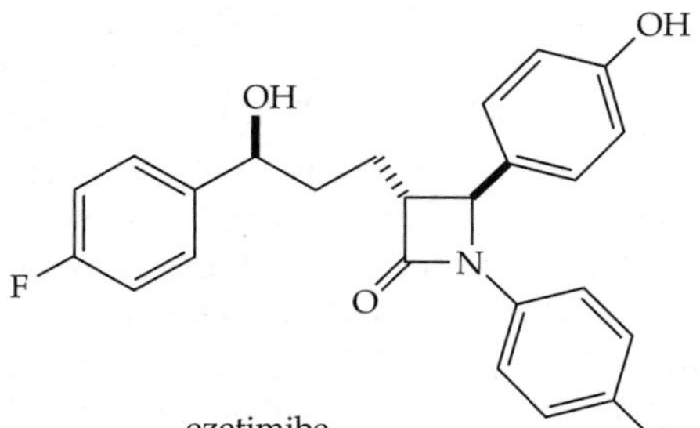

ezetimibe
(Zetia)
inhibits cholesterol absorption
2.25

FIGURE 2.9

target could not be determined and an appropriate biochemical or cellular assay could not be developed. Studying the drug in live animals was the only option. The animal testing for ezetimibe was challenging and ultimately required extensive study of metabolites based on pairs of rats. The first rat was fed a prospective drug. As the first rat metabolized the drug, some of the metabolites were secreted in the first rat's bile. Researchers then removed bile from the first rat and administered the bile to a second rat. If the second rat (receiving bile) showed decreased cholesterol levels, then the compound fed to the first rat must have been metabolized to an active molecule. From these intricate studies, ezetimibe emerged as a commercial drug.[18]

2.3 Lead Discovery

The lead discovery process is thoroughly discussed in Chapter 10, so it is addressed only briefly here. The goal of lead discovery is to find compounds that show modest activity in biological screens. Compounds are selected for screening based on information from the molecular biology team and any x-ray crystallographic data available on the receptor or enzyme of interest. If no information is available, random screening of compounds is used. Screening likely affords compounds of varying degrees of activity. Compounds showing activity beyond a predetermined threshold are called *hits*. The percentage of screened compounds that are active enough to be labeled hits varies widely from one study to another. A desirable hit rate may range from 0.1% to 10%, depending on the target and drug class. Hits are further tested for their ability to cross membranes, cytotoxicity, and metabolism profile. Hits that pass these criteria are called *leads*. The lead discovery process hopefully generates several promising leads because some leads may fail to progress in the lead optimization stage.

2.4 Lead Optimization

Once a hit is elevated to the status of a lead compound, the activity of the lead must be increased. This is accomplished by making structural modifications to the lead compound. The link between modification of a lead's structure and changes in activity is called a *structure-activity relationship* (SAR). As it becomes better understood, the SAR of a lead guides the medicinal chemistry team as it seeks the best methods of increasing the lead's activity. Once the lead has been adequately optimized and shows desirable properties, the lead then graduates to *candidate* status. Standardized safety testing in animals, called *animal trials,* is the next step. A more detailed discussion of the lead optimization process may be found in Chapters 11 and 12.

The areas of lead discovery and lead optimization define the field of *medicinal chemistry*. A medicinal chemist typically possesses formal training in synthetic organic chemistry because medicinal chemists need to continuously make new molecules for testing in some type of assay. Although their background may be in organic chemistry, medicinal chemists must learn the basic ideas of pharmacodynamics and pharmacokinetics so that they may communicate with those who perform the various assays.

2.5 Patent Filing

A pharmaceutical company must be profitable to stay in business, and protecting its work through patents is part of the process. When a compound is protected by a patent, the patent holder may legally exclude others from using that compound for the lifetime of the patent. Patents on compounds are generally filed early in the lead optimization stage.

Patents have a finite lifetime, which is 20 years from the date of filing in most jurisdictions. A typical drug candidate requires seven years or more to reach the market. Therefore, the pharmaceutical company has only a dozen or so years to enjoy sole marketing privileges, recoup its development costs, and turn a profit. Filing a patent too early in the development phase leaves little of the 20-year window for profit on the drug. Filing a patent too late increases the risk of a competitor's possibly coming up with the same idea. The realm of patent law is beyond the scope of this text, although some aspects of generic drugs are covered at the end of this chapter as well as in Chapter 13.

2.6 Animal Trials

Before a drug may be tested on humans, it must first be tested on animals. Tests on humans are called *clinical trials,* and animal trials are often referred to as *preclinical trials.* Preclinical trials differ from the animal tests mentioned earlier in this chapter. The previously discussed animal tests help the drug discovery team determine and optimize the pharmacodynamic and pharmacokinetic behavior of a hit or lead. Preclinical trials, in contrast, are standardized, industrywide tests. The preclinical tests have technical names such as "Segment II Reproductive Study in Rabbits" or "6-Month Toxicity Study in Rats." The specific names suggest the exact nature of each study. Each trial seeks to answer predefined safety questions concerning a drug candidate. Preclinical trials do not address the therapeutic effectiveness of the drug candidate in any way. Preclinical trials examine exclusively safety issues.

Early preclinical trials involve animals including mice, rats, dogs, rabbits, and monkeys. The tests must involve at least one rodent and one nonrodent. The goal of these tests is to determine whether the drug is safe to administer to humans. These are toxicology tests intended to reveal short-term (acute) side effects, and the animals must be given enough of the drug to demonstrate toxic effects. Observed toxic effects help clinical researchers better monitor side effects when the drug candidate is administered to humans. The pharmaceutical company may meet with the FDA to discuss the initial animal tests. The FDA may suggest additional specific animal trials to facilitate the start of trials on humans.

Animal trials continue to be performed after clinical trials in humans have begun. The later studies focus on long-term (chronic) side effects that require extended drug exposure to appear. The long-term animal trials can last up to two years and often overlap with phase I and II trials in humans. Reproductive studies on pregnant rats and rabbits are normally not complete when the drug is first tested on humans. For this reason, subjects involved in early clinical trials are almost always male.

During the animal trials stage, bulk quantities of the drug candidate are necessary. Lead optimization studies may have required only gram quantities for testing, but animal studies may call for a kilogram or more. The exact quantity depends on the potency of the drug, its molecular weight, and the amount needed to achieve toxic levels in the animals. To prepare large amounts of the candidate, process chemists study and modify the synthetic route used by the medicinal chemistry team. Having an efficient, reproducible route for preparing the candidate in high purity is vital for the FDA approval process. Furthermore, the route for making the drug is patented to give the pharmaceutical company additional protection from competitors. Process chemistry is covered in greater detail in Chapter 13.

2.7 Investigational New Drug Application

Before the candidate drug may be tested on humans in the United States, the sponsoring company must file an *investigational new drug* (IND) application with the Center for Drug Evaluation and Research (CDER) of the FDA. This application contains a full summary of in-house screening, animal trials, and information on the synthetic route. The company must also meet with FDA representatives to discuss preclinical data. IND candidates are often called *new chemical* or *molecular entities* (NCE or NME). If the drug seems safe and has promise of being effective, the FDA approves the IND application and phase trials may begin. Based on preclinical data, the FDA may have specific requests for how the phase trials should be conducted.

Sometimes a company seeks approval for use of a drug in the United States that has already been approved in another country. Under such a circumstance, extensive data on human trials is already available, and full phase trials may be dramatically shortened or altogether avoided.

2.8 Phase Trials

Phase trials are vital for determining the safety and effectiveness of a drug. Each type of phase trial answers specific questions about the action of a drug candidate.

Phase I

Phase I clinical trials are the first introduction of the IND into humans. The purpose of phase I trials is to determine the safety and tolerance of the drug. Members, or subjects, of the trials are healthy "volunteers" who are usually paid for their assistance. Subjects are monitored for adverse side effects with increasing doses of the drug. Blood levels are also watched to determine properties such as half-life and metabolism of the drug. The number of subjects normally ranges from 20 to 80.

Drugs being tested in phase I have already been through the necessary animal tests to satisfy the FDA's IND requirements. Animal tests, however, are no guarantee of safety. In March 2006, eight healthy volunteers participated in a phase I study of TGN1412, an antibody. Six volunteers received low intravenous doses (0.1 mg TGN1412 per kg body weight). Two volunteers received a saline placebo. Within approximately an hour, the six volunteers who received TGN1412 began to suffer from a long list of negative effects. Within 16 hours, all six patients were admitted to an intensive care unit with problems in their kidneys, heart, and lungs. All patients eventually recovered but may suffer long-term difficulties. Initial investigations into the failed trial focused on protocol violations and contaminants in the drug. Ultimately, the volunteers' adverse reactions were blamed on biological response pathways that could not be observed in the animal trials. This incident highlights the true research nature of clinical trials.[19,20]

Although the subjects of phase I trials are healthy, it may be possible to gain early clues on the effectiveness of the drug. A drug that treats insomnia should give useful information based on changes in the subjects' sleep patterns. On the other hand, a drug such as an antibiotic would give no activity data during a phase I trial because the subjects would not have an infection. Regardless of the activity of the drug, the phase I trial must establish that the drug is safe before phase II can begin.

Phase II

Phase II marks the first time the drug is used on diseased patients. The patients are closely monitored to determine the effectiveness of the drug as well as proper dosing. While most serious side effects and safety problems should have appeared in phase I, phase II patients are still monitored for unwanted drug effects. Phase II trials involve several hundred patients, and the cost per patient has been estimated to be about US$25,000.[1]

Toward the end of a phase II trial, the sponsoring company often meets with the FDA to present clinical and late animal data collected. Plans for moving forward into phase III trials are also covered. Depending on the data, the FDA may recommend specific clinical tests or additional animal trials to clarify any questions.

Phase III

Phase III trials can involve hundreds to thousands of patients. Because the behavior and safety of the drug is understood from the phase II trial data, some phase III trial protocols are not as intensive, and per patient costs can be somewhat lower than phase II. The larger number of patients allows representation of a range of demographic groups with regard to age, race, and gender. Differences, if any, across groups can be monitored to determine more accurate prescribing and dosing information. Effectiveness and long-term safety are the primary goals of phase III trials. Late in a phase III trial, the pharmaceutical company meets with the FDA to ensure that the data collected to date is adequate for the IND to be fully considered for final approval.

Ideally, phase trials should be conducted in a *controlled, double-blind* fashion. A controlled study includes two pools of patients. Some receive the active drug, while others receive a placebo. In a single-blind study, patients do not know whether they are receiving active drug or the placebo. In a double-blind study, neither the patient nor the healthcare staff knows who is or is not receiving the drug. All drug and placebo samples are encoded. Only clinical staff members who are unassociated with administering the drug and collecting data know which patients are receiving the active drug. Despite being well designed, phase trials can become *unblinded*. A phase III trial of orlistat (Xenical or Alli, **2.26**), a weight loss drug, was possibly unblinded (**Figure 2.10**). Patients who received orlistat experienced significant weight loss. Patients receiving a placebo lost less, if any, weight and were more likely to become discouraged and withdraw from the trial.[21] If a disproportionate number of patients in a trial withdraw from the study, the statistical validity of the trial may be called into question.

Relatively few INDs "die" in a phase III trial, but it certainly does happen. A recent example is torcetrapib (**2.27**), a drug designed to elevate HDL cholesterol levels (**Figure 2.11**). HDL cholesterol is the "good" cholesterol. A phase III trial of torcetrapib was halted when a quarterly study of clinical data revealed elevated mortality rates in patients receiving torcetrapib in comparison to a placebo.[22]

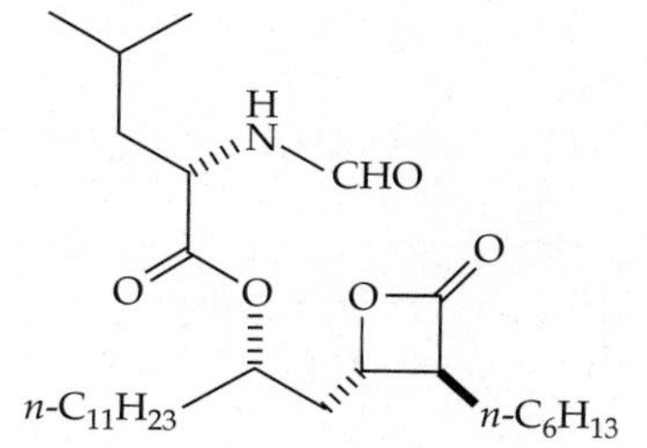

orlistat
(Xenical, Alli)
blocks fat metabolism
2.26

FIGURE 2.10

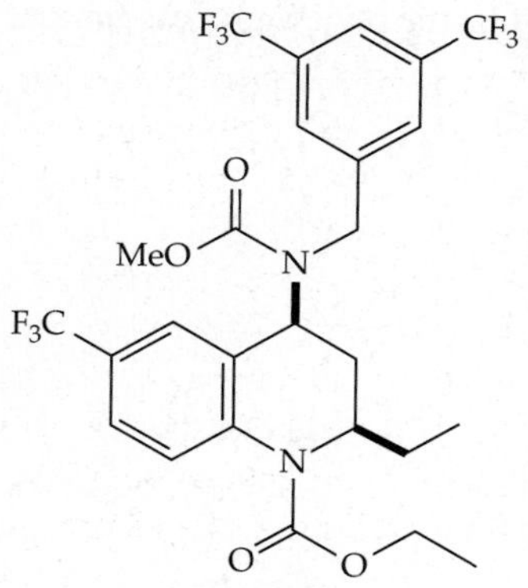

torcetrapib
raises HDL cholesterol
2.27

FIGURE 2.11

2.9 New Drug Application

The final step in FDA approval is for the pharmaceutical company to file a new drug application (NDA). Technically, this step alone is the FDA approval process. All the previously mentioned stages of the IND's development, from synthesis to clinical testing to formulation and package labeling, are data for presentation to and review by the FDA. The application is often called a *dossier,* which conjures an image of a small packet of papers. In reality, the application consists of huge amounts of information. The FDA actually makes available (at a cost) special binders for the submission of NDA and IND data. The binders are color-coded according to content. The amount of data contained in the dossier is staggering.

The dossier is filed during phase III trials. Phase III trials that are included in the dossier are called phase IIIa trials. Phase IIIa trials test for items that are vital for FDA approval. The phase IIIb trials are conducted after NDA filing and cover topics such as cost effectiveness and efficacy in comparison to existing therapies.

In analyzing the full dossier, the FDA may request site visits at the pharmaceutical company and form special review committees to gain additional expertise. If additional information is needed, the sponsor must provide it, possibly requiring design of another phase III protocol or other test. A smooth NDA may be completed within 12 months. Once all FDA requests have been satisfied, the NDA is approved. The sponsoring pharmaceutical company may then market the drug.

2.10 Launch

Once a drug reaches the market, the long-term activity and safety of the drug continue to be monitored. This period of monitoring is sometimes called *phase IV.* Phase IV trials often study changes to the original drug, such as formulation (e.g., gelatin capsule versus tablet). Sometimes, after a drug is released, additional long-term effects may be observed. If the effects are sufficiently undesirable or pose a safety concern, the sponsoring company may voluntarily pull the drug off the market, or approval of the drug may be revoked by the FDA. New effects can be beneficial. For example, raloxifene (Evista, **2.28**) is a drug that was originally intended to treat osteoporosis in postmenopausal women (**Figure 2.12**). Recent studies on the drug indicate that raloxifene may lower the incidence of invasive breast cancer as well. Regardless of new data, raloxifene may not be marketed as a breast cancer drug until it has been approved by the FDA for that specific purpose. Therefore, Eli Lilly, the company that markets raloxifene, would need to file an IND application with the FDA to start the review process for a new use of an already approved drug.

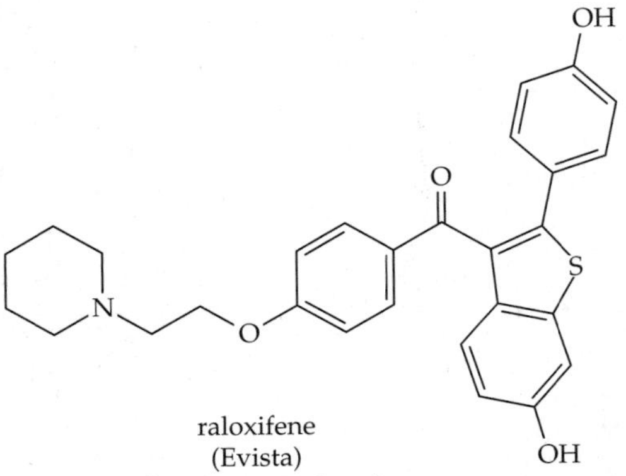

raloxifene
(Evista)
*slows osteoporosis and
antitumor activity*
2.28

FIGURE 2.12

Few drug candidates are actually launched as drugs. Approximately 20% of all drug candidates that reach IND status survive safety and efficacy testing in clinical trials and are eventually approved by the FDA. Success rates fluctuate somewhat across different drug classes. For example, from 1968 to 1992, INDs targeting respiratory illnesses had an approval rate of 12%. Anti-infective INDs were closer to 30%. Specific reasons for failed approval of 319 compounds developed in the United Kingdom from 1964 to 1985 are shown in **Figure 2.13**.[23] More recent data for INDs filed through the FDA break down failures by both the reason for the failure and the phase in which the failure occurred (**Figures 2.14** and **2.15**).[24]

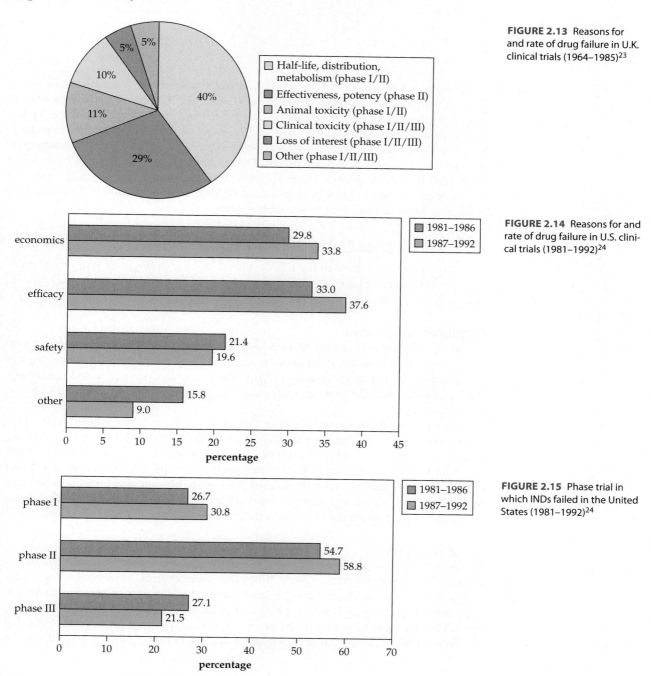

FIGURE 2.13 Reasons for and rate of drug failure in U.K. clinical trials (1964–1985)[23]

Pie chart legend:
- Half-life, distribution, metabolism (phase I/II)
- Effectiveness, potency (phase II)
- Animal toxicity (phase I/II)
- Clinical toxicity (phase I/II/III)
- Loss of interest (phase I/II/III)
- Other (phase I/II/III)

Pie chart values: 40%, 29%, 11%, 10%, 5%, 5%

FIGURE 2.14 Reasons for and rate of drug failure in U.S. clinical trials (1981–1992)[24]

Figure 2.14 (1981–1986 / 1987–1992):
- economics: 29.8 / 33.8
- efficacy: 33.0 / 37.6
- safety: 21.4 / 19.6
- other: 15.8 / 9.0

FIGURE 2.15 Phase trial in which INDs failed in the United States (1981–1992)[24]

Figure 2.15 (1981–1986 / 1987–1992):
- phase I: 26.7 / 30.8
- phase II: 54.7 / 58.8
- phase III: 27.1 / 21.5

FIGURE 2.16 Costs by development stage[1]

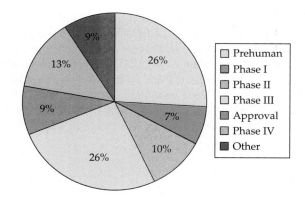

Differences between the U.S. and U.K. studies make direct comparisons difficult, but both reveal that most drugs fail in phase II or earlier. Naturally, the further an IND progresses through the approval process, the greater the investment in that drug by the pharmaceutical company. A failure in phase III is more expensive than a failure in phase I. Based on **Figure 2.16**, a drug that fails in phase II has accumulated a cost of nearly half the total cost of a successful development. The development costs of the 80% or so failed INDs must be recouped by the approximately 20% approved INDs. Keep in mind that the whole pie in Figure 2.16 was estimated at US$800 million in 2000.[1] The investment required to bring just one drug to market is sizable.

2.11 Additional Considerations

Not all drugs proceed through the FDA approval process in the same fashion. Alternative approval processes apply to INDs that treat serious or life-threatening conditions, especially conditions that have no satisfactory alternative treatment.

Accelerated Approval

The FDA can grant *accelerated development/review* status to INDs that fit the right profile. A rationale behind the accelerated review option is that adverse drug effects are less of a concern for critically ill patients. While the accelerated status can shorten the approval timeline, the FDA requires more extensive postapproval testing (phase IV) for an accelerated drug.

Very ill patients may receive a drug that shows promise in late clinical trials. This requires the IND to be classified as a *treatment IND*. Treatment IND status is valuable for critically ill patients who may not be able to wait until the drug receives full NDA approval. Related to a treatment IND is the *parallel track* policy. Specifically developed to help AIDS patients, INDs with parallel track classification may be administered to patients who are not healthy enough to qualify for inclusion in a phase trial. Data from parallel track patients are not included in the clinical trial data.

Generic Drugs

A company's exclusive marketing rights to a particular drug cease when that company's composition of matter patent expires. When the patent expires, anyone may begin marketing that same drug under a generic label. Given the high cost of prescription drugs, generic manufacturers generate competition in the marketplace and can reduce healthcare costs. There is a catch, however. The generic manufacturer must prove to the FDA that its generic product is identical in effectiveness and safety to the branded drug. While the chemical structure of the approved drug and generic drug are identical, differences in manufacturing processes can introduce different impurities that lead to new side effects.

The methods of showing equal effectiveness, or *bioequivalence,* are often called *equivalency testing* and are part of an abbreviated new drug application (ANDA). The tests generally include approximately 25 to 50 healthy volunteers. The sponsoring company must show that its drug is pure and gives the same drug concentration in the blood as the approved drug. One reason generic drugs tend to be less expensive is because the cost of these simplified clinical trials is much, much lower than a standard NDA.

Approval Outside the United States

Up to this point, the discussion has almost exclusively focused on the discovery process in the United States. North America, and more specifically the United States, accounts for a large percentage of the pharmaceutical market (**Figure 2.17**). The importance of the United States is even greater with regard to sales of new drugs (**Figure 2.18**). The pharmaceutical industry is regardless global in scope.[25] Other nations do not fall under the jurisdiction of the FDA, although the drug approval processes of all nations share more similarities than differences. Terminology is fairly consistent across the regulatory agencies. **Table 2.1** shows the names of drug approval agencies in selected countries.

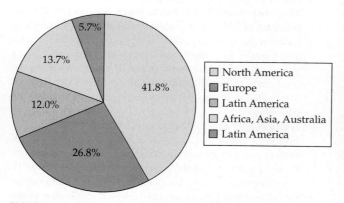

FIGURE 2.17 Pharmaceutical market sales in 2011[25]

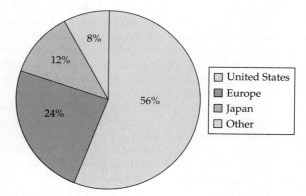

FIGURE 2.18 Pharmacy sales for new medicines launched from 2006 through 2010[25]

TABLE 2.1   Drug regulatory agencies in selected countries

Country/Region	Agency
Canada	Health Products and Food Branch (HPFB)
United Kingdom	Medicines and Healthcare Products Regulatory Agency (MHRA)
France	French Agency for the Safety of Health Products (AFSSAPS)
European Union	European Agency for Evaluation of Medical Products (EMEA) (regulates biotechnology products and cancer, AIDS, diabetes, and neurodegenerative drugs only)
India	Central Drugs Standard Control Organization (CDSCO)
Japan	Pharmaceuticals and Medical Devices Evaluation Center (PMDEC)
China	State Food and Drug Administration (SFDA)

Summary

The complete drug development process, from concept to market, conservatively requires approximately eight years for completion. The business of drug discovery begins with identification of a disease or condition with a sufficient patient base to constitute a market for a new drug. If the disease is not well understood, considerable cellular and molecular biology research may be necessary to discover a point of intervention, or target. A method or assay for testing the activity of the target must be developed for screening potential drugs. To handle possibly hundreds of thousands of molecules for testing, initial screens are often required to be high throughput in their design so that they may be automated with robotic technologies. Once an effective screen method has been developed, medicinal chemists submit large groups of candidate molecules, called libraries, for testing. Molecules that show a promising level of activity are called hits and subjected to additional assays to check metabolic and diffusion properties. The best hits are promoted to the status of lead and iteratively modified and tested. The final outcome is an optimized lead, which should be highly active against the target, both in biochemical and animal tests. Approval to market the drug requires testing in humans through carefully monitored clinical trials. The three phases of clinical trials roughly test for safety, then effectiveness, and finally long-term safety. If the clinical trials proceed as planned, regulatory approval will follow along with marketing of the drug. Even after approval, the safety of the drug is monitored for any long-term effects that could not be uncovered in the initial clinical trials. During the marketing lifetime of a drug, a pharmaceutical company recoups its research & development (R&D) investment in both the successful drug and failed candidates, and ultimately turns a profit.

Questions

1. Phase trials and the FDA approval process often require seven to eight years. Raloxifene (**2.28**) (Figure 2.12) was initially approved to treat osteoporosis. After being on the market for several years, raloxifene was reported to be an effective treatment for certain forms of breast cancer. Estimate how much time would be required for clinical trials and approval of raloxifene as a breast cancer treatment. Justify your answer.

2. The following list provides several problems that a drug candidate might show during the drug discovery process. Indicate the stage(s) in the drug discovery process at which the problem may be discovered. Options include preclinical testing, the three phase trials, and postrelease.
   - Causes extensive liver damage
   - Insufficient efficacy
   - Causes arthritis after several years with chronic use
   - Interacts with certain cardiovascular drugs

3. Assume a company develops five compounds and files an IND for each. One drug dies at the end of phase I, two drugs die at the end of phase II, and one drug dies at the end of phase III. The fifth drug reaches the market. If a drug costs US$800 million to reach the market, use the pie chart in Figure 2.16 to determine the cost of development for the four drugs that failed to complete clinical trials. Assume the "other" and "approval" costs are a factor only for the fifth drug (post–phase III costs). What are the total costs of development for all five drug candidates? These are the losses that must be offset by the sales of the fifth drug.

4. A blockbuster drug is defined as a drug that generates US$1 billion in annual sales. Assume Drug X has 12 years of market exclusivity and has $1 billion in sales for each of those 12 years. Using the guidelines in question 3, over 12 years, how many drugs that fail after phase I would be offset by sales of Drug X? How many drugs that fail after phase III would be offset by sales of Drug X? Assume the profit margin for drugs is 20%, that is, $1 billion in sales equals $200 million in profit.

5. What are some challenging diseases for which an effective drug would be considered a category creator?

References

1. DiMasi, J. A., Hansen, R. W., & Grabowski, H. G. The Price of Innovation: New Estimates of Drug Development Costs. *J. Health Econ.* 2003, *22*, 151–185.

2. Adams, C. P., Branter, V. V. Spending on New Drug Development. *Health Econ.* 2010, *19*, 130–141.

3. iThemba Pharmaceuticals. http://www.ithembapharma.com/newsflash/the-official-launch (accessed January 2012).

4. Saunders, J. Inhibitors of Angiotensin Converting Enzyme as Effective Antihypertensive Agents. In *Top Drugs: Top Synthetic Routes* (Chapter 1). Oxford, UK: Oxford University Press, 2000.

5. Saunders, J. Blockade of Angiotensin-II Receptors. In *Top Drugs: Top Synthetic Routes* (Chapter 2). Oxford, UK: Oxford University Press, 2000.

6. Lehninger, A. L., Nelson, D. L., & Cox, M. M. Lipid Biosynthesis. In *Principles of Biochemistry* (5th ed., Chapter 21). W. H. Freeman, 2008.

7. Hopkins, A. L. & Groom, C. R. The Druggable Genome. *Nat. Rev. Drug Disc.* 2002, *1*, 727–730.

8. Willson, T. M. & Moore, J. T. Genomics versus Orphan Nuclear Receptors: A Half-Time Report. *Mol. Endocrin.* 2002, *16*, 1135–1144.

9. Borisy, A. A., Elliott, P. J., & Hurst, N. W. Systematic Discovery of Multicomponent Therapeutics. *Proc. Natl. Acad. Sci. USA.* 2003, *100,* 7977–7982.

10. Zimmermann, G. R., Lehar, J., & Keith, C. T. Multi-Target Therapeutics: When the Whole Is Greater Than the Sum of the Parts. *Drug Disc. Today.* 2007, *12,* 34–42.

11. Hatcher, H., Planalp, R., Cho, R., Torti, F. M., & Torti, S. V. Curcumin: From Ancient Medicine to Current Clinical Trials. *Cell. Mol. Life Sci.* 2008, *65,* 1631–1652.

12. Goel, A., Jhurani, S., & Aggarwal, B. B. Multi-Targeted Therapy by Curcumin: How Spicy Is It? *Mol. Nutr. Food Res.* 2008, *52,* 1010–1030.

13. Sharma, R. A., Euden, S. A., Platton, S. L., Cooke, D. N., Shafayat, A., Hewith, H. R., et al. Phase I Clinical Trial of Oral Curcumin: Biomarkers of Systemic Activity and Compliance. *Clin. Cancer Res.* 2004, *10,* 6847–6854.

14. Sundberg, S. A. High-Throughput and Ultra-High-Throughput Screening: Solution- and Cell-Based Approaches. *Curr. Opin. Biotech.* 2000, *11,* 47–53.

15. Comley, J. Nanolitre Dispensing: On the Point of Delivery. *Drug Disc. World.* 2002, *3,* 33–46.

16. Jones, J. O. & Diamond, M. I. Design and Implementation of Cell-Based Assays to Model Human Disease. *ACS Chem. Biol.* 2007, *2,* 718–724.

17. Xi, B., Yu, N., Wang, X., Xu, X., & Abassi, Y. The Application of Cell-Based Label-Free Technology in Drug Discovery. *Biotech. J.* 2008, *3,* 484–495.

18. van Heek, M., France, C. F., Compton, D. S., McLeod, R. L., Yumibe, N. P., Alton, K. B., et al. *In vivo* Metabolism-Based Discovery of a Potent Cholesterol Absorption Inhibitor, SCH58235, in the Rat and Rhesus Monkey through the Identification of the Active Metabolites of SCH48461. *J. Pharmacol. Exp. Ther.* 1997, *283,* 157–163.

19. Suntharalingam, G., Perry, M. R., Ward, S., Brett, S. J., Castello-Cortes, A., Brunner, M. D., et al. Brief Report: Cytokine Storm in a Phase 1 Trial of the Anti-CD28 Monoclonal Antibody TGN1412. *N. Engl. J. Med.* 2006, *355,* 1018–1028.

20. Medical and Healthcare Products Regulatory Agency. TGN1412: MHRA Response to Final Report by Independent Expert Working Group on Phase 1 Clinical Trials. http://www.mhra.gov.uk/NewsCentre/Pressreleases/CON2025434 (accessed July 2012).

21. Davidson, M. H., Hauptman, J., DiGirolamo, M., Foreyt, J. P., Halsted, C. H., Heber, D., et al. Weight Control and Risk Factor Reduction in Obese Subjects Treated for 2 Years with Orlistat. *J. Am. Med. Assoc.* 1999, *281,* 235–242.

22. U.S. Food and Drug Administration. Pfizer Stops All Torcetrapib Clinical Trials in Interest of Patient Safety. http://www.fda.gov/NewsEvents/Newsroom/PressAnnouncements/2006/ucm108792.htm (accessed July 2012).

23. Prentis, R. A., Lis, Y. & Walker, S. R. Pharmaceutical Innovation by the Seven UK-Owned Pharmaceutical Companies (1964–1985). *Br. J. Clin. Pharmacol.* 1988, *25,* 387–396.

24. DiMasi, J. A. Risks in New Drug Development: Approval Success Rates for Investigational Drugs. *Clin. Pharmacol. Ther.* 2001, *69,* 297–307.

25. *The Pharmaceutical Industry in Figures: Key Data 2012.* European Federation of Pharmaceutical Industries and Associations.

A Trip through the Body

Chapter Outline

Learning Goals for Chapter 3

- Gain a basic understanding of the composition of the body
- Appreciate how the drug transport pathways of the body are connected
- Know the standard drug administration routes
- Understand how a drug moves through the body from the site of administration to the site of action
- Distinguish between elimination and metabolism as methods of removing a drug from the body

The previous chapter covered how a drug progresses through a pharmaceutical company, that is, how it is conceived, designed, and marketed. Understanding the stages of drug development has value, but a drug's realm of action is within the complex confines of the human body. This chapter qualitatively describes the movement of a drug through the body, a topic that is collectively called pharmacokinetics. The key aspects of pharmacokinetics—absorption, distribution, metabolism, and elimination—are often summarized with the acronym ADME. A T for toxicity is sometimes added to the end to give the acronym ADMET. An effective drug must not only be active against its target (pharmacodynamics) but also move through the body to and from the site of action in a predictable fashion. Upon conclusion of this chapter, the requisite background information will be in place for more sophisticated treatment of target activity, quantitative pharmacokinetics, and finally drug development.

3.1 The Complexity of the Body

The human body is fantastically intricate. Introduction of a drug into such a complex environment can lead to many outcomes, some of which can be readily anticipated and others that cannot. The complexity of the body stands in stark contrast to the relative simplicity of a biochemical assay for activity determination.

What Is the Body?

The body can be defined in a multitude of ways. From a chemical perspective, the simplest analysis is based on elemental composition. **Tables 3.1** and **3.2** show the most abundant

TABLE 3.1 **Major elements in the body[1]**

Element	Mass (kg)	Percentage
Oxygen (O)	43	61.4
Carbon (C)	16	22.9
Hydrogen (H)	7.0	10.0
Nitrogen (N)	1.8	2.6
Calcium (Ca)	1.0	1.4
Phosphorus (P)	0.78	1.1
Potassium (K)	0.14	0.2
Sulfur (S)	0.14	0.2
Sodium (Na)	0.10	0.1
Chlorine (Cl)	0.095	0.1

TABLE 3.2 **Selected minor elements in the body[1]**

Element	Mass (mg)
Magnesium (Mg)	19,000
Iron (Fe)	4,200
Fluorine (F)	2,600
Zinc (Zn)	2,300
Copper (Cu)	72
Iodine (I)	20
Boron (B)	18
Selenium (Se)	15
Chromium (Cr)	14
Manganese (Mn)	12
Molybdenum (Mo)	5
Cobalt (Co)	3

elements found in the body based on a 70 kg person. In drug studies, 70 kg is used as the mass of an average patient. Remarkably, just six elements (O, C, H, N, Ca, and P) are responsible for approximately 99.6% of the body's mass. The remaining elements are mostly electrolytes (K, Na, Cl) or trace metals bound in the active sites of enzymes.[1]

From a molecular perspective, water is by far the predominant compound in the body and accounts for approximately 60% of the body's mass (**Figure 3.1**).[2] Most oxygen and hydrogen atoms in the body are in water molecules. Lipids contain a small amount of oxygen but are primarily carbon and hydrogen. Both nitrogen and phosphorus are found in proteins, DNA, and RNA. Calcium is mostly found in bones, with a small amount also present in blood.

Organization of the Body

The physiology of the body is assembled from the four hierarchical structures listed in order of increasing complexity: cells, tissues, organs, and organ systems. The more complex structures are formed through aggregation of simpler structures.

Cells

The basic unit of a biological system is a cell (**Figure 3.2**). Cells are compartments for chemical reactions. Each cell is enclosed by a lipid bilayer. The surface of the cell is interspersed

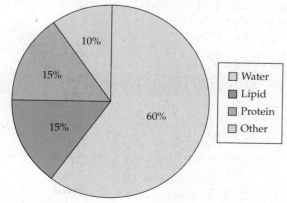

FIGURE 3.1 Content of the body by type of molecule[2]

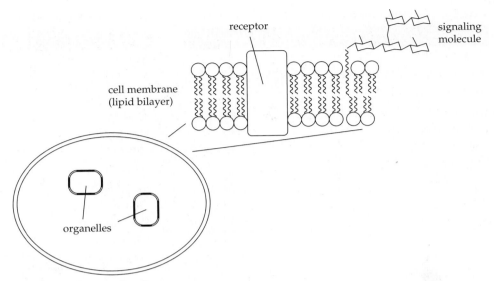

FIGURE 3.2 Features of a cell

with membrane-bound receptors and lipid-anchored signaling molecules. Within the cell are the organelles, cell machinery, enzymes, and structural proteins required for proper cell function. Almost all mammalian cells have a nucleus, the largest organelle in cells.

The human body consists of approximately 10^{15} cells. Cells can have a variety of shapes and sizes according to their specialized functions. A typical cell has a rounded shape and is about 10 μm in diameter. Nerve cells can be very long. The nerve cells that run all the way from the tips of one's toes to the base of the spinal cord are about 1 m in length. Epithelial cells that line capillaries are extremely flat to minimize flow resistance. Bone cells (osteocytes) have a star shape to better interlock with neighboring cells and impart strength and rigidity.[3]

Tissues

Tissues are collections of cells with a common function within an organism. Despite the apparent variety of tissues in the body, tissues are classified into just four types: connective, epithelium, muscle, and nervous (**Table 3.3**).

Of all the tissues in the body, connective tissue is the most broadly defined. The most obvious examples of connective tissue are ligaments, tendons, bone, and cartilage. Both fat and blood are also connective tissues.[3] From a drug standpoint, blood is the most important tissue. Blood serves as the medium by which most drugs travel throughout the body and reach their site of action.

TABLE 3.3 **Types of tissues**

Type	Role	Examples
Connective	Join other tissues	Ligaments, cartilage, bone, blood
Epithelium	Form a protective layer, allow secretion/excretion	Linings and surfaces of organs
Muscle	Allow motion and changes in shape and size	Smooth, cardiac, skeletal muscle
Nervous	Communicate	Nerves, brain

Organs

Organs are one step up the hierarchical ladder from tissues. Organs contain multiple tissue types that work together in a common role in the body. Just like cells and tissues, organs come in a variety of shapes and sizes. Organs can be large (heart, 300 g; both kidneys, 300 g; liver, 1.5 kg). The largest human organ is the skin, which accounts for approximately 15% of the body's mass.[3] Small organs include the teeth and pituitary gland.

Because organs contain several different tissues, they can perform complex tasks. The stomach, for example, has layers of muscle for mechanically breaking down food. The lining of the stomach contains cells that secrete acid and digestive enzymes. Furthermore, the lining is able to absorb water and electrolytes. Another organ, the eye, contains muscle to rotate the eye and allow the eye to focus. Cells in the retina are able to convert light to electrical impulses that are transmitted down the optic nerve to the brain.

Organ Systems

Organ systems are the highest level of organization in the body. A few organ systems are shown in **Table 3.4**. Collectively, organ systems take care of the full needs of the body.

Balance of Life

All cells, tissues, organs, and organ systems work together to maintain proper function of the body. The entire body is both flexible enough to adapt to external environmental changes (e.g. temperature, diet) while still being a delicately balanced machine. Because of the finely tuned nature of the body, even a few milligrams of a drug can cause dramatic changes.

By targeting events on a cellular level, drugs can influence tissues, organs, and/or entire systems. Raloxifene (Evista, **3.1**) is a selective estrogen receptor modulator that is used for the treatment and prevention of osteoporosis in postmenopausal women (**Figure 3.3**). Raloxifene helps maintain bone mineral density in all bone tissues. The primary effect is observed in the larger, load-bearing bones of the skeletal system. However, even the tiny bones of the ear are affected. Similarly, drugs that bind adrenergic receptors cause muscle tissue to contract or relax. Adrenergic drugs impact many systems because almost all systems include some type of muscle tissue. Propranolol (Inderal, **3.2**) is an adrenergic drug that reduces blood pressure by relaxing smooth muscle in blood vessels of the circulatory system. Propranolol also causes undesirable relaxation of smooth muscle in the bronchioles of the lungs (respiratory system). Some drugs do have isolated effects. Warfarin (Coumadin, **3.3**), an anticoagulant, acts only in the blood, and therefore its activity is restricted to the circulatory system.

TABLE 3.4 Selected organ systems

System	Example Organs and Tissues	Function
Circulatory	Heart, blood, blood vessels	Circulate blood and nutrients
Digestive	Esophagus, stomach, liver, colon	Absorb nutrients from food
Nervous	Brain, spinal cord, nerves, sensory organs	Transmit and process information
Skeletal	Bones, tendons, ligaments	Provide structural support
Urinary	Kidneys, bladder, urethra	Balance fluid and remove waste

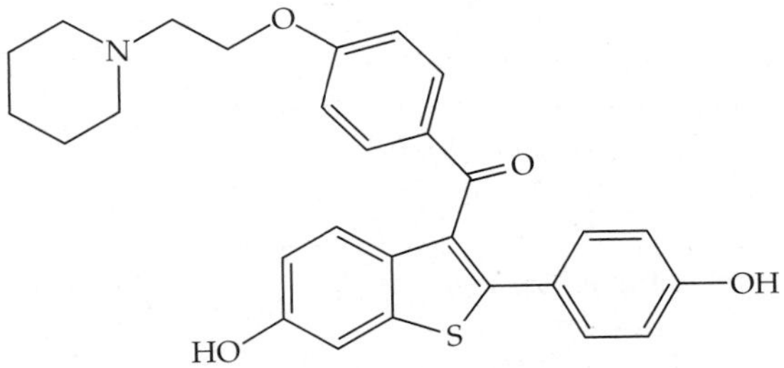

raloxifene
(Evista)
osteoporosis preventative
3.1

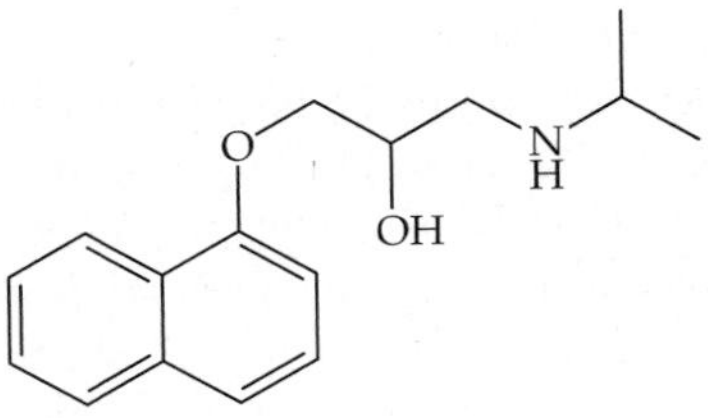

propranolol
(Inderal)
antihypertensive
3.2

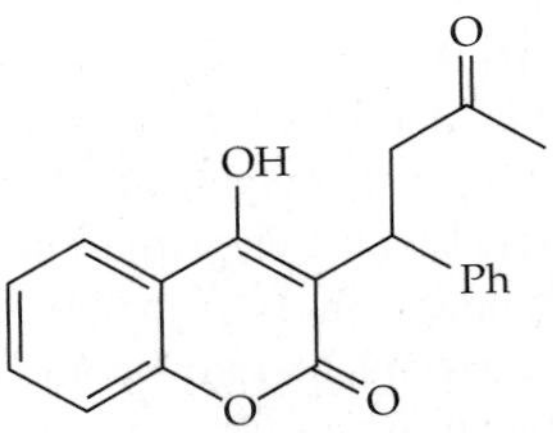

warfarin
(Coumadin)
anticoagulant
3.3

FIGURE 3.3

3.2 Absorption: Drug Entry into the Bloodstream

All drugs must be introduced into the body by some method, normally called the *route of administration* or *route of delivery*. The goal of a delivery method is to place a drug into the bloodstream. From the bloodstream, a drug has access to all points in the body. *Absorption* is the process of a drug's movement from its site of administration to the bloodstream.

Most delivery methods are classified as *enteral*, *parenteral*, or *topical*. Enteral routes involve absorption of the drug through the digestive system. Examples of enteral routes include oral (PO, *per os*, Latin for *through the mouth*) and rectal delivery. Parenterally delivered drugs reach the bloodstream by a route other than the digestive system and include intravenous (IV), intramuscular (IM), and subcutaneous (SC) delivery. Topical drugs are administered directly to their site of action and do not rely on transport through the bloodstream. Examples are auricular (applied to the ear), conjunctival (eye), and cutaneous (skin) delivery. In total, the U.S. Food and Drug Administration (FDA) recognizes over 100 distinct enteral, parenteral, and topical delivery routes.[4]

Oral

Orally delivered drugs provide a characteristic jagged curve relating drug concentration in blood plasma and time (**Figure 3.4**). The low points correspond to the time at which a pill is administered. The concentration rises as the drug is absorbed from the digestive system. As absorption slows, elimination (to be covered later in this chapter) becomes the predominant fate of the drug, and the concentration curve drops until the next dose is ingested. Ideally, the valleys in the curve should not fall below the minimum effective concentration of drug.

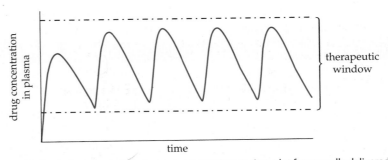

FIGURE 3.4 Representative plasma concentration vs. time plot for an orally delivered drug

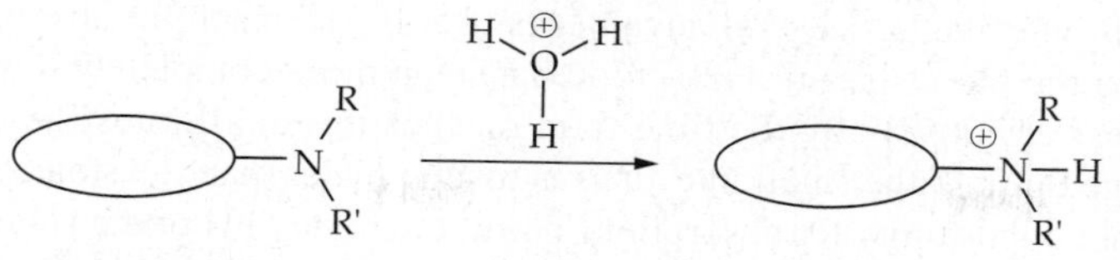

SCHEME 3.1 Protonation of basic drugs

Similarly, the peaks between doses should not exceed toxic levels. Together, the minimum effective concentration and toxic level define the *therapeutic window* of the drug.

Oral is the preferred method of drug delivery. Drugs administered as a tablet, capsule, or gel-cap are the goal of almost all drug discovery programs. Oral drugs are what patients want, and oral drugs are what a drug company tries to provide. Regardless of market pressures for orally available drugs, oral delivery is not the simplest method for a drug to reach the bloodstream. The drug has many hoops—the stomach, small intestine, hepatic portal system, and liver—through which it must jump before entering the general circulatory system.

Stomach

Once swallowed, a drug passes down the esophagus and reaches its first stop, the stomach. The pH of the stomach varies depending on the presence of food and can reach a value as low as 1. The primary role of the stomach is to chemically break down the components of food, especially proteins.

A secondary function of the stomach is to absorb nutrients. Like nutrients, drugs are often absorbed from the stomach. In particular, drugs that are lipophilic in the acidic environment of the stomach tend to be readily absorbed. A classic example is aspirin (**3.4**), or *O*-acetyl salicylic acid (**Figure 3.5**). Aspirin, with a pK_a of about 3.0, is mostly neutral in the stomach. Aspirin is therefore able to cross the lining of the stomach wall and reach the bloodstream. Ethanol is also absorbed in this fashion. In contrast, most marketed drugs contain a basic nitrogen that is protonated at the low pH of the stomach. The resulting charged conjugate acid is less able to cross lipophilic membranes than the drug in its neutral, basic form (**Scheme 3.1**).

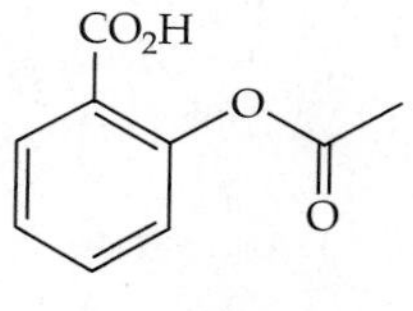

The presence of food in the stomach can greatly influence the rate of drug absorption. Food has many effects. It increases the volume of the stomach, slows the drug's rate of passage into the small intestine, and raises the pH in the stomach. For most drugs, these factors increase the absorption rate from the stomach. Depending on the type of food and the particular drug in question, food can bind the drug and slow its rate of absorption. Because food introduces many variables into a drug's absorption, most early clinical trials are performed on patients with empty stomachs (fasting).

aspirin
3.4

FIGURE 3.5

Small Intestine

The small intestine continues the chemical digestion process started in the stomach with a focus on enzymatic cleavage of fats and sugars. Proteins, already partially broken down in the stomach, are cleaved into individual amino acids. Once different food types have been digested completely, they are ready for absorption. The small intestine consists of three regions: duodenum, jejunum, and ileum (**Table 3.5**).[5]

TABLE 3.5 Regions of the small intestine[5]

Name	Length (m)	pH Range
Duodenum	0.3	5–7
Jejunum	2.5	7–8
Ileum	3.5	7–8

The small intestine has several advantages over the stomach for absorbing nutrients and drugs into the bloodstream. First, food has been more completely digested and retains drugs less effectively by the time food reaches the small intestines. Second, the time of passage through the small intestines is much greater than the stomach. More time equates to more opportunity for absorption. Third, the higher pH of the small intestines is more favorable for the absorption of most drugs relative to the stomach. Fourth and most importantly, the small intestine has a large surface area, which maximizes exposure of the intestine's lining and contents. The small intestine achieves a high surface area through folds (plicae circulares), projections from the intestinal wall (villi), and tiny projections (microvilli) from the epithelial cells lining the wall. Beyond increasing surface area for absorption, the elaborate texture of the intestine's lining also facilitates secretion of digestive enzymes and fluids, and impedes the rate of flow through the small intestine.[3]

As with the stomach, the type of food consumed can influence drug absorption from the small intestine. A particular food that has received considerable interest is grapefruit juice. Grapefruit juice is known to increase the amount of certain drugs that reaches the general circulatory system. In other words, a patient who takes a drug along with grapefruit juice may have a higher concentration of drug in his/her bloodstream than a patient who receives the drug alone. This situation is called the *grapefruit juice effect*.[6] Examples of drugs that are subject to the grapefruit juice effect include felodipine (Plendil, **3.5**), verapamil (Isoptin, **3.6**), midazolam (Versed, **3.7**), and lapatinib (Tykerb, **3.8**) (**Figure 3.6**). Traditionally, the grapefruit juice effect has been viewed

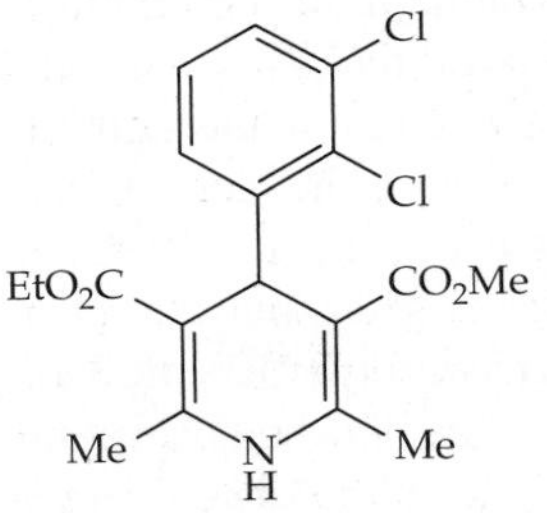

felodipine
(Plendil)
antihypertensive
3.5

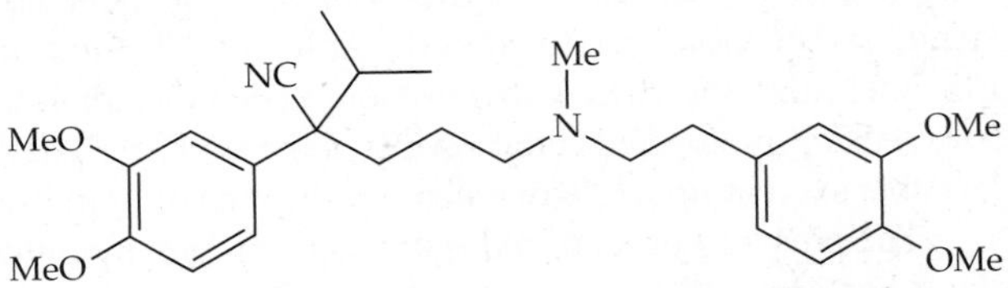

verapamil
(Isoptin)
antihypertensive
3.6

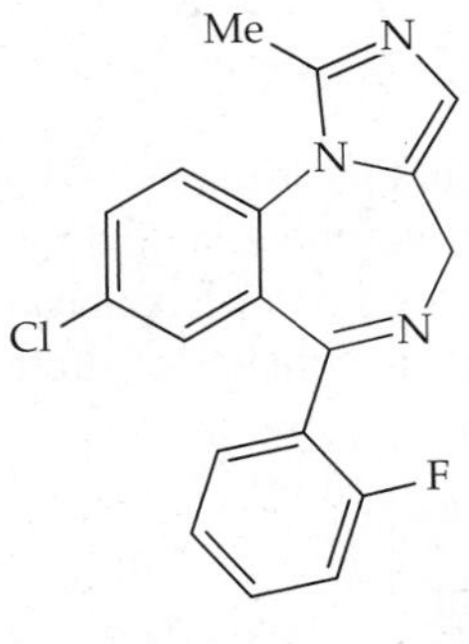

midazolam
(Versed)
sedative
3.7

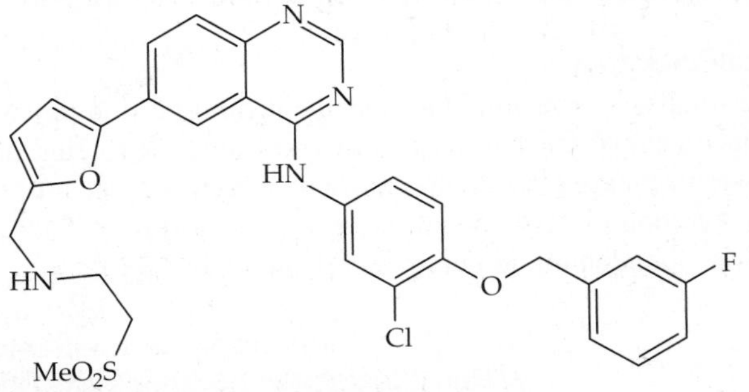

lapatinib
(Tykerb)
anticancer
3.8

FIGURE 3.6 Drugs subject to the grapefruit juice effect

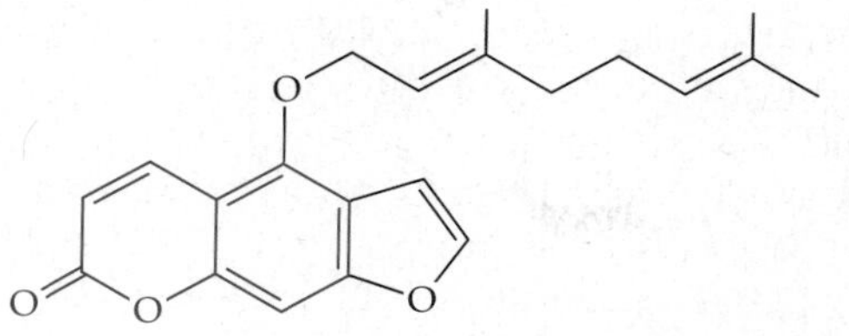

bergamottin
3.9

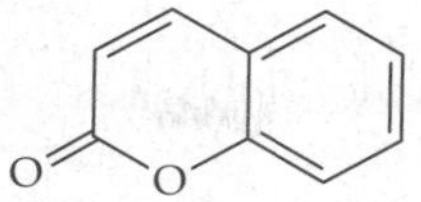

coumarin
3.10

FIGURE 3.7

negatively and as a means by which a patient may receive an overdose of a drug. Recently, the grapefruit juice effect has been suggested as a means of lowering the necessary oral dose of a drug without diminishing the effect of the drug.[6] Proper exploitation of the grapefruit juice effect would require a drug to be administered and tested with grapefruit juice during a later phase trial.

The cause of the grapefruit juice effect on a molecular level has recently been traced to a compound called bergamottin (**3.9**) (**Figure 3.7**), which is a coumarin (**3.10**) derivative. Bergamottin is a potent inhibitor of cytochrome P-450 3A4 (CYP3A4), a metabolic enzyme that breaks down many drugs. CYP3A4 is found in both the liver and the cells that line the small intestine. By inhibiting CYP3A4, bergamottin allows a larger fraction of certain drugs to be absorbed without being first metabolized during the absorption process.[7] All the drugs in Figure 3.6 are significantly metabolized by CYP3A4 and are therefore affected by grapefruit juice. The *first-pass effect* occurs when a drug is metabolized before it even reaches the general circulatory system. The first-pass effect is normally associated with the liver (*hepatic first-pass effect*), but the grapefruit juice effect demonstrates the *intestinal* or *gut first-pass effect*.

Hepatic Portal System and Liver

Drugs that are absorbed from the contents of the stomach and small intestine initially enter the hepatic portal system. The hepatic portal system is a set of blood vessels that collects all the blood from the stomach, small intestine, large intestine, spleen, and pancreas and delivers it to the liver. From the liver, blood is then able to enter the general circulatory system. The liver, therefore, acts as a gatekeeper. Nothing can enter the bloodstream from the digestive system without first passing through the liver. If a person has ingested an unusual foreign substance, called a *xenobiotic*, then the liver and its metabolic enzymes have a chance to destroy or modify the chemical before the substance reaches the rest of the body. The liver cannot break down all foreign chemicals, but it is a remarkably effective defense system for the rest of the body.

Almost all drugs are foreign to the body and therefore are examples of xenobiotics. A drug may be absorbed well from the small intestine only to be nearly completely metabolized by the liver. This *first-pass effect* or first-pass metabolism is a commonly encountered problem in drug development. Drugs prone to first-pass metabolism show low *oral bioavailability*, a value that relates the ability of a drug to be absorbed from the digestive tract, pass through the liver, and reach the general circulatory system. Low bioavailability indicates that the drug is either poorly absorbed from the gastrointestinal tract or subject to high first-pass metabolism. *Bioavailability* is more precisely described and quantified in Chapter 7.

Testing lead compounds and drug candidates for degradation by liver microsomes can inform a discovery team whether a compound suffers from significant first-pass metabolism. Animal testing also provides valuable early information on metabolic susceptibility. The real test, however, comes during clinical trials. Drugs that display an unfavorably

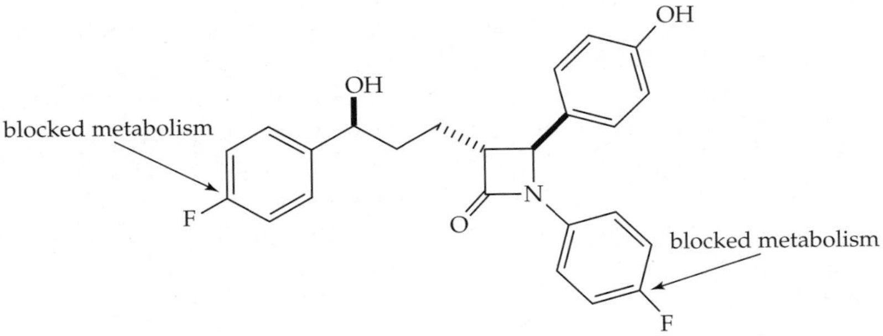

blocked metabolism

blocked metabolism

ezetimibe
(Zetia)
blocks cholesterol absorption
3.11

FIGURE 3.8

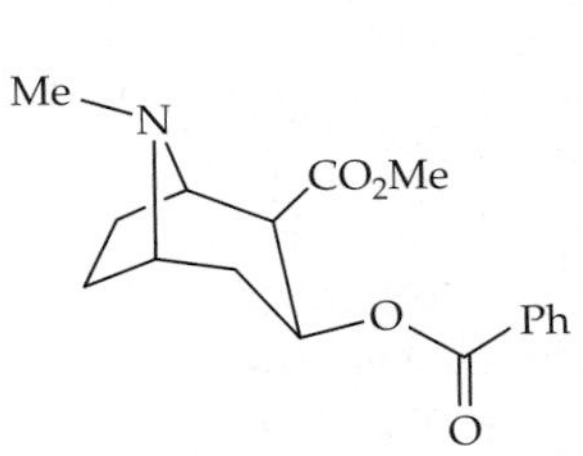

cocaine
3.12

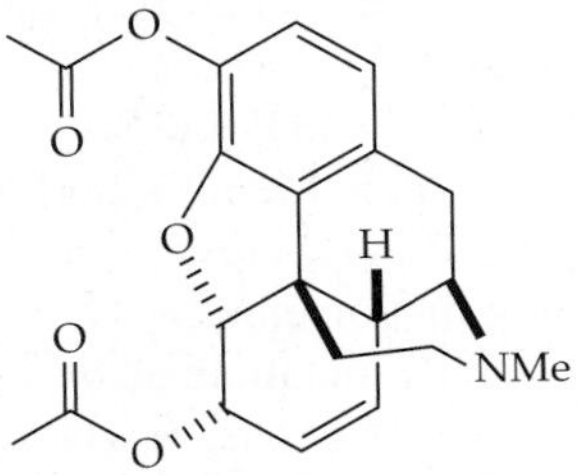

heroin
3.13

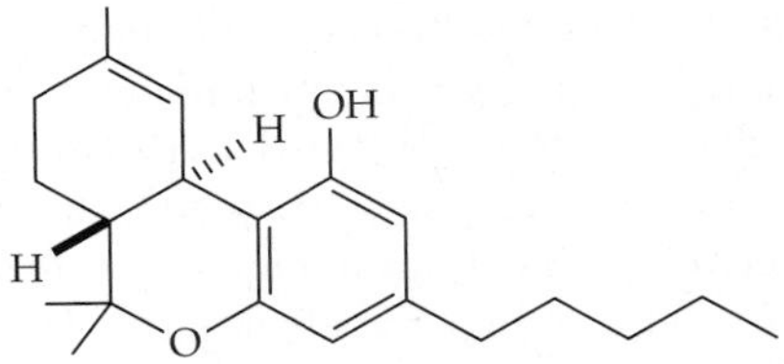

Δ^9-tetrahydrocannabinol
3.14

FIGURE 3.9 Street drugs with poor oral bioavailability

high degree of first-pass metabolism can sometimes be improved by blocking the sites of reaction on the drug. Ezetimibe (Zetia, **3.11**) contains two fluorine atoms that, in part, minimize first-pass metabolism by the liver and increase oral bioavailability (**Figure 3.8**).[8]

Many "street drugs" are not administered in pill form because they have poor oral bioavailability. Examples include cocaine (**3.12**), heroin (**3.13**), and marijuana (active component, Δ^{10}-tetrahydrocannabinol [**3.14**]) (**Figure 3.9**). Many legal prescription drugs also cannot be taken in oral form. Perhaps the most widely used injectable drug is insulin, a 51-peptide hormone that is rapidly broken down in the stomach when taken orally.

If a drug is not extensively metabolized as it passes through the liver, a sufficient amount of the drug will reach the general circulatory system. From the circulatory system, a drug has access to all parts of the body.

Large Intestine

Ideally, an orally administered drug never reaches the large intestine. Instead, the drug is fully absorbed before it passes out of the small intestine. The presence of a drug in a patient's feces is typically a sign of a drug that is not well absorbed in the stomach or small intestine. The drug simply passes through the patient. In addition to the first-pass effect, poor absorption along the digestive tract leads to low bioavailability of a drug.

Injection

Drug injection is a parenteral delivery method and bypasses problems associated with the first-pass effect and absorption through the digestive system. Delivery by injection is common despite several issues: fear of needles, painful injections, and potential infection at the site of injection. For these reasons, injections are normally performed by a health-care professional. Most drugs that are administered only by injection treat severely ill patients. These patients are likely in a hospital or under close supervision of a healthcare worker who can safely administer the drug. An exception is insulin. Insulin is widely used, delivered by injection, and the injections are normally performed by the patient. It is no coincidence that new formulation of insulin not dependent upon injection would be a highly profitable new product for a pharmaceutical company. Inhaled insulin is another dosing option that's been considered but not yet marketed.

Intravenous

Intravenous (IV) delivery has two variations: injection and infusion. In an injection, the entire dose is rapidly placed in a vein via a syringe and needle. The dose is called an IV *bolus*. An IV *infusion*, sometimes called an IV drip, involves administering a drug from an IV bag over a predetermined amount of time. The concentration of the drug solution and IV drip rate determine the eventual drug concentration in the patients blood.

Intravenous delivery is the most direct route into the circulatory system. The entirety of the administered drug reaches the blood because the dose is placed into a vein, and therefore IV routes define 100% bioavailability. Absorption is not an issue.

Intravenous injections deliver the whole dose almost instantaneously, so the effect of the drug is immediate. An infusion requires time to build blood concentrations to a thera-peutic level. Injections and infusions can be used together to provide nearly constant levels of drug in the blood (**Figure 3.10**). An IV injection is used first to serve as a *loading dose* and establish an appropriate drug concentration. An IV infusion then follows with a rate that sustains the drug in the blood and offsets drug lost through normal elimination pro-cesses. An infusion used in this manner is often said to provide a *maintenance dose*. The mathematical relationships that define the curves in Figure 3.10 are covered in Chapter 7.

Intramuscular

In an intramuscular (IM) injection, the drug is injected directly into a large muscle. As with intravenous injections, the IM dose is called a bolus. The drug reaches the blood-stream by diffusing through the surrounding muscle tissue. Because the drug does not im-mediately reach the bloodstream, the drug's effect is delayed. Injections of aqueous drug solutions are absorbed fairly quickly. Drugs can be injected as a solution or suspension in oil to prolong the release time and provide a more sustained effect. Increased bloodflow to the muscle raises the rate of delivery of the drug to the blood. The reservoir of drug in the

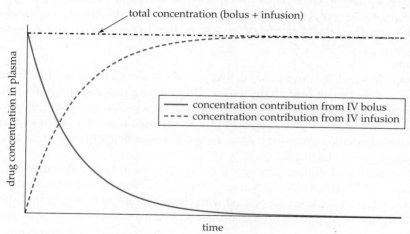

FIGURE 3.10 Plasma concentration vs. time for an IV bolus, IV infusion, and both combined

muscle is sometimes called a *depot*. Bioavailability for IM injections tends to be high, but it is less than 100%. Some of the drug is normally metabolized in the muscle tissue before being distributed to the rest of the body.

Subcutaneous

Subcutaneous administration involves injecting the drug directly under a patient's skin. Subcutaneous and IM injections are similar. Bioavailability is high but lower than administration directly into the bloodstream. The rate of drug release can be somewhat controlled by the formulation of the drug. The pH of the injected solution must be close to physiological pH; otherwise, severe skin irritation can result.

Transdermal

Transdermal delivery involves application of an adhesive patch to the skin. The patch contains a film of the drug that can diffuse through the skin and enter the bloodstream. Absorption through the digestive tract and first-pass metabolism are nonfactors for transdermal drugs, so bioavailability tends to be high. Furthermore, patches are simple for a patient to use. Finally, when patches are well-designed, they can provide a remarkably consistent amount of drug to the bloodstream over a period of time. The plasma concentration versus time curve of a transdermal drug can closely resemble that of an IV infusion. Disadvantages to transdermal delivery include skin irritation at the site of the patch and difficulty for the drug to penetrate the skin.

By design, skin is a protective cover for the body. Skin is approximately 2 mm thick, but the stratum corneum, the primary barrier to drug delivery, is only 10–20 μm thick. The stratum corneum is the outermost layer of the skin. It consists of dried, dead cells that are tightly packed together. The dead cells contain lipids and proteins that collectively form an effective barrier to loss of water and entry of drugs. In general, lipid soluble drugs cross the stratum corneum more easily. Solvent additives such as dimethyl sulfoxide may be included in the patch to help the drug cross into the bloodstream.[9]

While not nearly as common as oral drugs, several transdermal drugs are on the market. Examples include the female contraceptive Ortho Evra (ethinyl estradiol, **3.15**, combined with norelgestromin, **3.16**), the motion sickness drug scopolamine (Scopoderm TTS, **3.17**), nicotine patches (NicoDerm CQ, **3.18**), and the antidepressant selegiline (Emsam Patch, **3.19**) (**Figure 3.11**).

Other Routes

Numerous other, less common routes exist as alternative methods of delivering a drug to the bloodstream. All the routes listed in this section provide drug administration to a part of the body that is heavily perfused with blood vessels and therefore amenable to drug absorption.

Sublingual

Sublingual (Latin, *under the tongue*) delivery involves absorption of a drug through the soft tissues of the mouth beneath the tongue. Drugs administered sublingually reach the bloodstream quickly and completely bypass absorption and first-pass issues of the digestive system. Sublingual tablets are much less compressed than oral tablets to facilitate breakup and dissolution under the tongue. Drugs with a particularly unpleasant taste may not be suitable for sublingual delivery.

Two examples of sublingual drugs include nitroglycerine (**3.20**) and ondansetron (Zofran ODT, **3.21**) (**Figure 3.12**). Nitroglycerine treats chest tightness (angina). Ondansetron is used to treat nausea associated with chemotherapy and anesthesia. Although not formally approved, ondansetron is also prescribed for nausea in women in their first

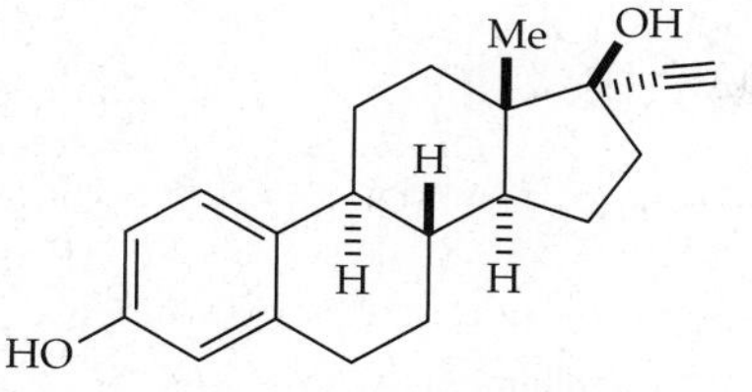

FIGURE 3.11 Drugs administered by a transdermal patch

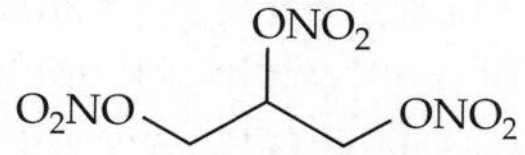

ethinyl estradiol
(Ortho Evra)
contraceptive
3.15

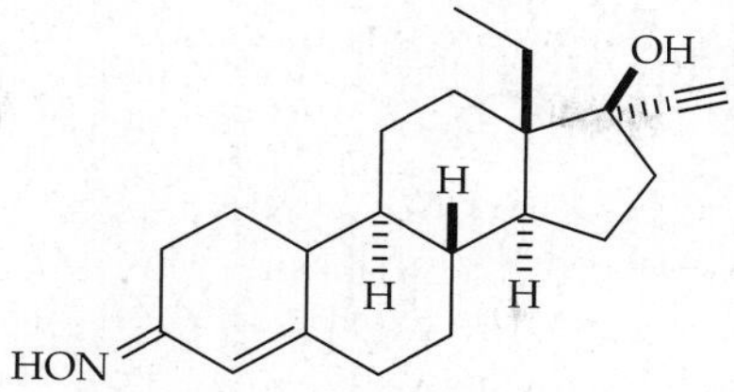

norelgestromin
(Ortho Evra)
contraceptive
3.16

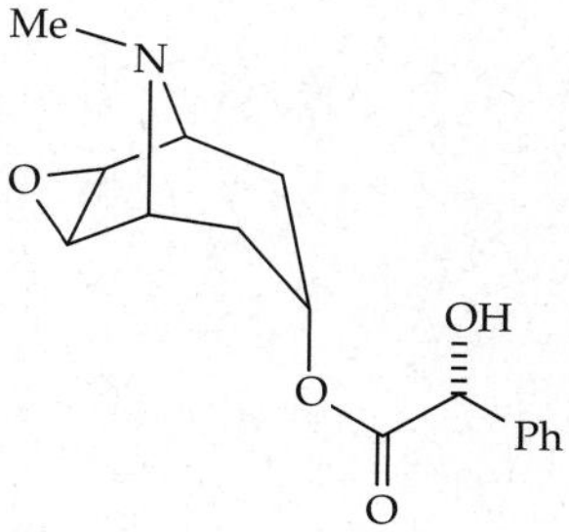

scopolamine
(Scopoderm TTS)
motion sickness
3.17

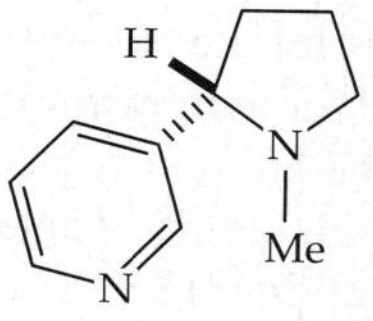

nicotine
(NicoDerm CQ)
smoking cessation
3.18

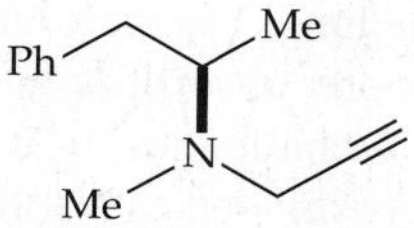

selegiline
(Emsam Patch)
antidepressant
3.19

nitroglycerine
angina
3.20

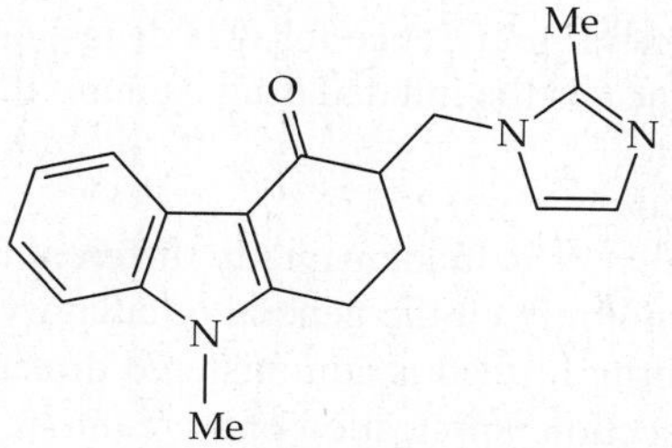

ondansetron
(Zofran ODT)
antinausea
3.21

FIGURE 3.12 Sublingual drugs

trimester of pregnancy. Sublingual delivery is particularly attractive for patients suffering from nausea because emesis empties the stomach contents, including any orally administered antinausea drugs.

Rectal

Rectal suppositories administer drugs to the absorbent environment of the rectum. Although rectal delivery is technically an enteral route, the first-pass effect tends to be considerably lower for rectal drugs relative to traditional oral drugs. For patients who have difficulty swallowing, rectal delivery can be a viable option. Regardless of the advantages of rectal suppositories, patients are generally not receptive to this mode of drug delivery.

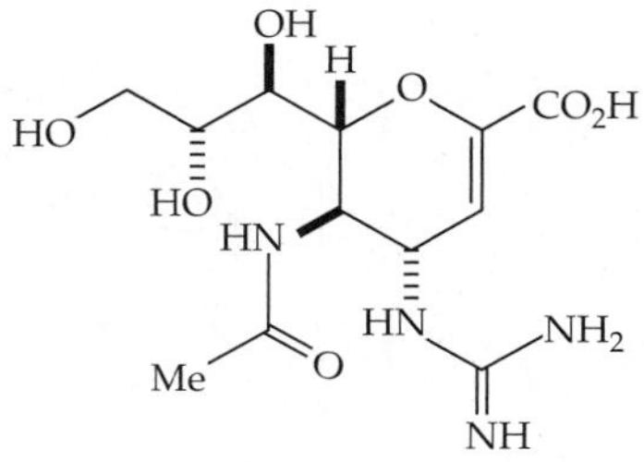

zanamivir
(Relenza)
3.22

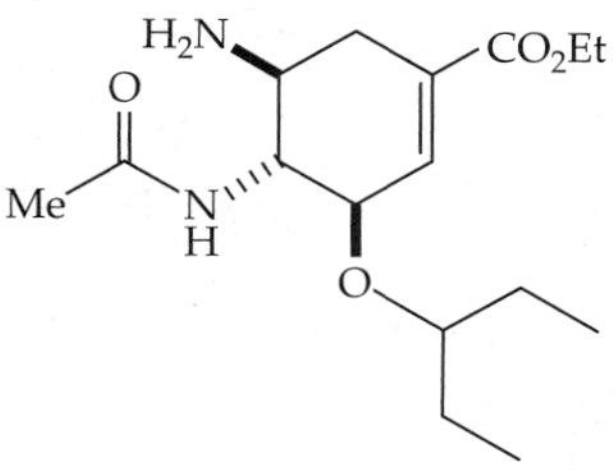

oseltamivir
(Tamiflu)
3.23

FIGURE 3.13 Influenza treatments

Inhalation

The lining of the lungs is rich in blood vessels for the absorption of O_2 and elimination of CO_2. The lungs are therefore a promising site for administration and absorption of a drug. Furthermore, absorption of a drug through the lungs is fast and avoids first-pass effects. Technical challenges, however, can be formidable for inhaled drugs. The drug must be *very* finely divided as a solid (powder) or solution (mist) to be carried deep into the lungs for efficient absorption. Particle sizes of <2 μm are required for the drug to reach the terminal bronchioles and aveoli, where absorption is highest. If not properly administered, a significant amount of drug may end up in the patient's mouth and throat. Another limitation is that the patient's lungs must be healthy and fully functioning.

Zanamivir (Relenza, **3.22**) is used to fight the viruses that cause influenza A and influenza B (**Figure 3.13**). Because its oral bioavailability is low at only 2%, an alternative delivery route, inhalation, is required for zanamivir. As a possible demonstration of the public's general preference for oral drugs, oseltamivir (Tamiflu, **3.23**), an oral influenza drug, far outsells the inhaled drug zanamivir.

Topical

Topical delivery is fundamentally different from enteral and parenteral routes. A topical drug does not rely on the general circulatory system for distribution to the site of action. A topical drug instead is administered directly where it needs to act. Administration can involve injection, application of a cream, or ingestion of a pill while still not technically falling under intravenous, transdermal, or oral delivery, respectively. Injection of a drug directly to its site of action is most common when treating cancerous tumors and is called *intratumoral injection*. A popular topical drug is hydrocortisone (**3.24**), an anti-inflammatory for skin rashes (**Figure 3.14**). Orlistat (Xenical or Alli, **3.25**) prevents pancreatic lipase from

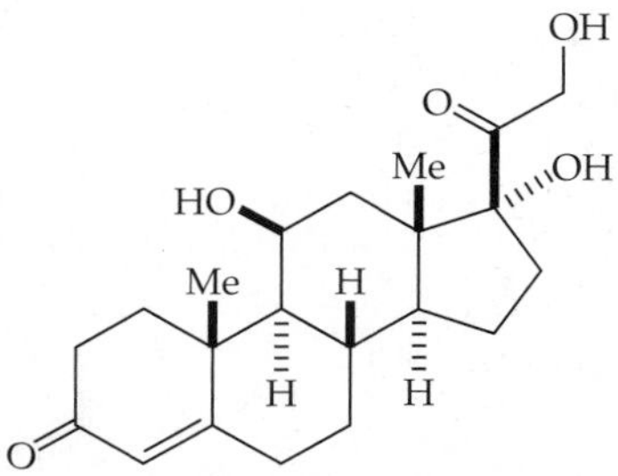

hydrocortisone
anti-inflammatory
3.24

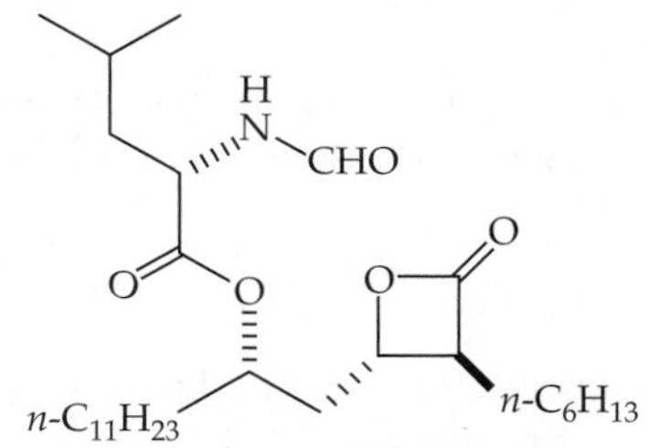

orlistat
(Xenical, Alli)
weight loss
3.25

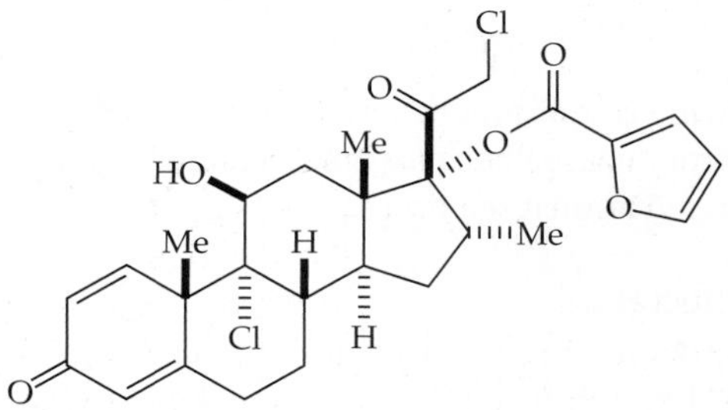

mometasone furoate
(Nasonex)
seasonal allergies
3.26

FIGURE 3.14 Topical drugs

hydrolyzing triglycerides in the intestines. Orlistat is administered as an oral pill but is not absorbed into the bloodstream. Orlistat essentially acts as a topical drug for the contents of the small intestine. Mometasone furoate (Nasonex, **3.26**) is a glucocorticoid steroid that can treat seasonal allergic reactions by topical administration as a nasal spray. The nasal sinuses are the site of action. The drug is virtually undetectable in the blood and therefore is not absorbed into the bloodstream across the sinus membranes.[10]

3.3 Distribution: Drug Transport

Blood is the means by which molecules are transported throughout the body. Blood and the circulatory system in general act as the body's mass transit system. Almost all compounds that enter or leave the body rely on the circulatory system. Aside from topically administered compounds, drugs are designed to be transported throughout the body by the bloodstream. The process of a drug being taken from its site of administration to the body at large is called *distribution*.

Blood

Whole blood consists of water, cells, proteins, and electrolytes (**Figure 3.15**). Cells comprise approximately 45% of total blood volume. The noncellular components of blood are called *plasma*. Plasma is mostly water with some proteins (8% by weight). Plasma proteins may be divided into three categories: serum albumin (60%); the α-, β-, and γ-globulins (36%); and fibrinogens (4%). Blood *serum* is the residual liquid left behind after whole blood is allowed to clot.[5]

Monitoring drug concentration at its site of action is generally impractical, but blood is easy to draw and test for drug content. For this reason, blood is the tissue of choice for determining the amount of drug in the body. This is an example of a transitive relationship: drug concentration in the blood is related to concentration at the site of action which is, in turn, related to the biological effect. Therefore, drug concentration in the blood is related to the biological effect.

Blood is drawn as whole blood, that is, cells and plasma. Drug concentrations are measured in plasma, not whole blood. Whole blood is spun down in a centrifuge to allow separation of plasma from the cellular components. Plasma is more homogeneous than whole blood and therefore easier to analyze. The blood-to-plasma ratio of most drugs falls in a range of 0.5–1.5, and a value of 1 is generally not an unsafe assumption. In other words, the concentration of a drug in whole blood is often assumed to be the same as that

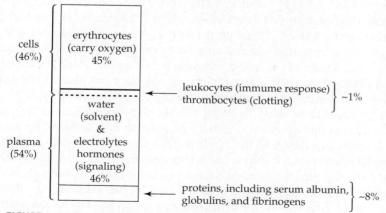

FIGURE 3.15 Composition of whole blood[5]

Acidic drugs bound by albumin

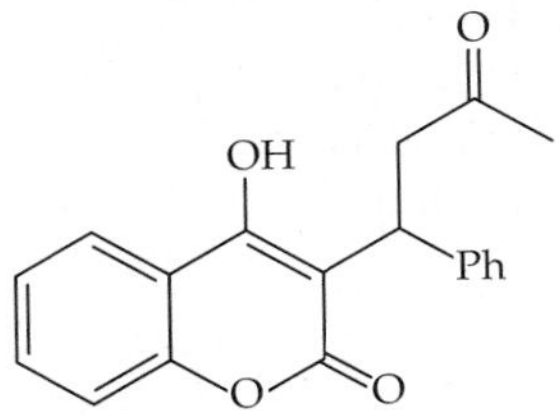

warfarin (Coumadin) *anticoagulant* **3.3** 99.5%	phenylbutazone *anti-inflammatory* **3.27** >95%	ibuprofen (Advil, Midol) *anti-inflammatory* **3.28** >99%

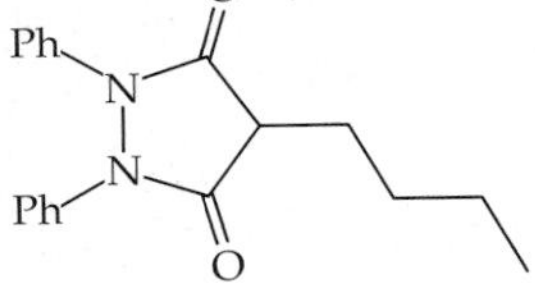

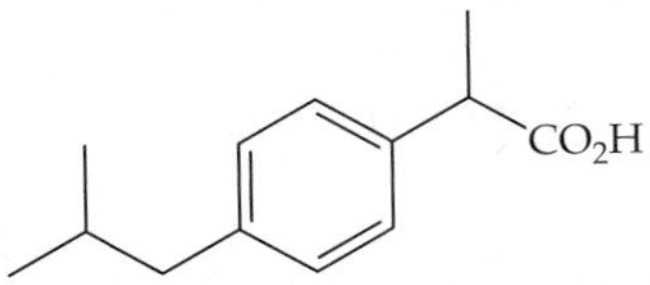

Basic drugs bound by α_1-acid glycoprotein

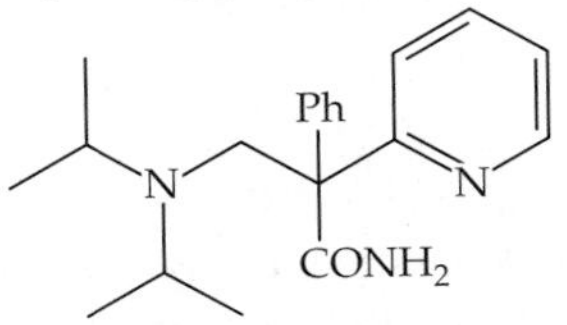

disopyramide (Norpace) *treats irregular heart rhythms* **3.29** ~70%	lidocaine (Xylocaine) *local anesthetic* **3.30** 70%

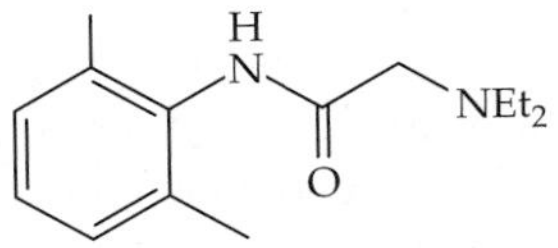

FIGURE 3.16 Plasma protein binding for selected drugs[11]

in plasma. Determining a drug's concentration in whole blood requires breaking open (*lysing*) blood cells so that any drug that has entered the cells is available for detection.

Most drugs target proteins, either enzymes or receptors. Drugs often show affinity for proteins found in plasma. This phenomenon is called *plasma binding* or *serum binding* because the proteins in question are present in both plasma and serum. Albumin tends to bind acidic drugs. α_1-Acid glycoproteins bind basic drugs. Examples of extensively protein-bound drugs are shown in **Figure 3.16** with typical percentage binding at therapeutic doses. All plasma proteins contain sites for the potential binding of lipophilic drugs. Very lipophilic drugs enter the membranes of cells in whole blood and may or may not bind serum proteins.

Serum protein binding by a drug or a drug's entry into cells tends to lengthen the half-life of a drug. Drugs that are protein bound or trapped in cells cannot be metabolized by the liver or filtered by the kidneys. Therefore, the drug stays in the bloodstream longer. Keep in mind that both protein binding and cell entry are reversible processes. As unbound drug in the plasma is metabolized or eliminated, according to Le Chatlier's Principle, cell- or protein-bound drug enters the plasma to restore equilibrium concentrations. The newly unbound drug may then undergo elimination or metabolism. Therefore, protein and cell binding do not permanently keep a drug in the blood. They only slow the rate of metabolism and elimination.

As blood travels throughout the circulatory system away from the heart, the blood vessels branch and narrow until they reach the capillaries. The junctions between the endothelial cells that line capillaries form pores. Plasma can flow through the pores, as can the smaller solutes in plasma.[12] The size limit for the pores corresponds to the plasma proteins, the smallest being albumin with a molecular weight of 67 kDa. Drugs fall well below this size threshold and freely pass through the pores. Outside the capillaries is the *interstitial fluid*, a liquid that bathes tissues and fills the gaps between cells. Once a drug reaches the interstitial fluid, the drug has access to cell surfaces as well as any surface proteins and membrane-bound receptors.

Crossing Membranes

In traveling from the point of administration to the blood and then to the site of action, a drug likely must cross a membrane at least once. Oral drugs must cross the intestinal wall, and drugs binding an intracellular target must cross an additional cell membrane. Membranes may be crossed by several methods. In general, only one method is operative for a given drug.

Membranes themselves consist primarily of lipids (40%), proteins (50%), and cholesterol (10%).[12] Exact composition depends on the type of cell. Despite the large amount of protein present in a membrane, membranes are normally considered in terms of lipids, especially glycerophospholipids. Glycerophospholipids consist of glycerol, a phosphodiester, two fatty acid chains, and often an additional, polar group (**Figure 3.17**). The additional group determines the identity of the glycerophospholipid (**Table 3.6**).

Glycerophospholipids, as well as other lipids, aggregate and align to form a bilayer (**Figure 3.18**). Both faces of the bilayer are polar because of the orientation of the polar

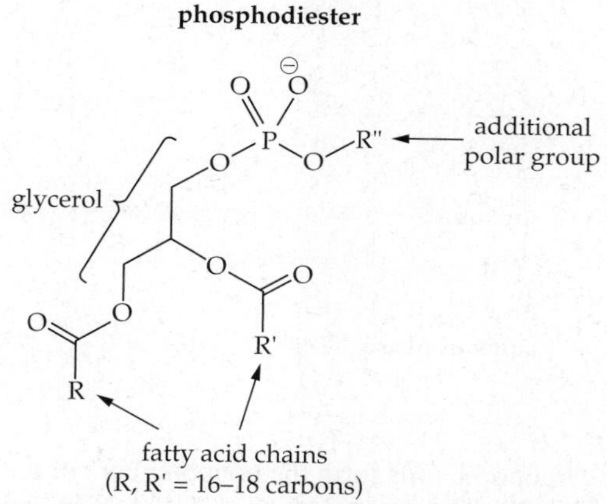

FIGURE 3.17 General structure of a glycerophospholipid

TABLE 3.6 Selected glycerophospholipids[12]

Glycerophospholipid	R″ group	Structure of R″
Phosphatidate	Hydrogen	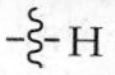 $-H$
Phosphatidylcholine	Choline	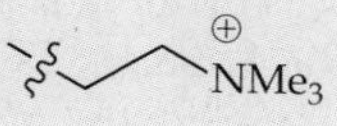 $\overset{\oplus}{N}Me_3$
Phosphatidylethanolamine	Ethanolamine	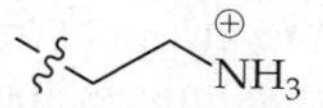 $\overset{\oplus}{N}H_3$
Phosphatidylserine	Serine	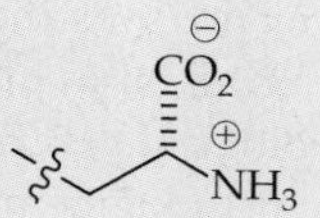 $\overset{\ominus}{C}O_2,\ \overset{\oplus}{N}H_3$
Phosphatidylglycerol	Glycerol	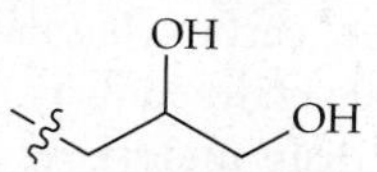 $OH,\ OH$
Phosphatidylinositol	Inositol	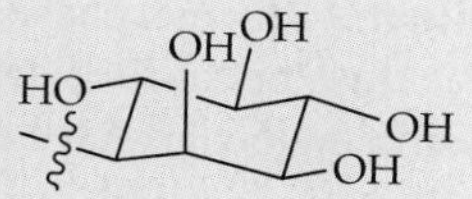 $HO,\ OH,\ OH,\ OH,\ OH$

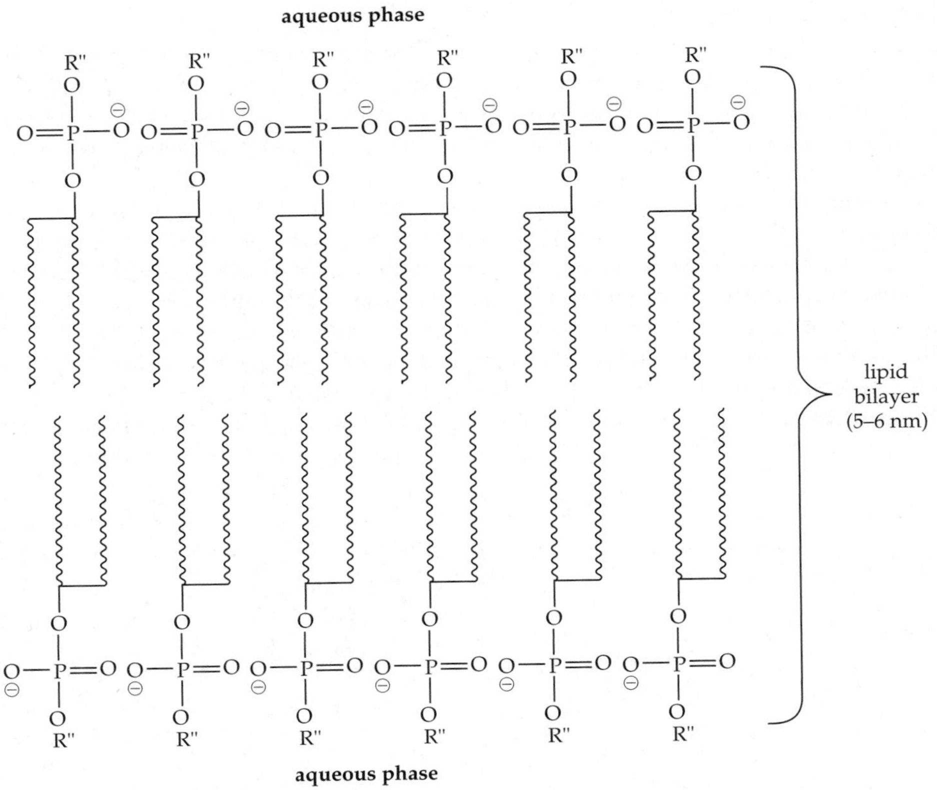

aqueous phase

FIGURE 3.18

groups on the lipids. The lipophilic tails form the nonpolar core of the membrane. In total, a cell membrane has a thickness of 5-6 nm.

Passive Diffusion

Most drugs cross cell membranes by *passive diffusion*. Passive diffusion is simple. Initially the drug is dissolved in an aqueous solution such as digestive fluids in the small intestine or interstitial fluid. After the drug diffuses to the surface of a membrane, the drug *partitions* into the nonpolar interior of the membrane. Diffusion through the membrane exposes the drug to the opposite face of the membrane. The drug then partitions across the interface and enters the aqueous layer on the other side.

Entering the membrane requires the drug to be nonpolar and therefore neutral. This leads to an apparent paradox—how can a drug be polar enough for solubility in plasma, digestive juices, and interstitial fluid while being nonpolar for crossing a membrane? The answer involves incorporation of weak acids and/or bases into the organic drug structure. Weak bases and acids are significantly protonated and deprotonated, respectively, at physiological pH. The resulting charged drug is sufficiently soluble in aqueous solution for transport to a membrane. Fortunately, a cell membrane contains numerous basic (phosphate, carboxylate) and acidic (ammonium) groups for proton transfer and neutralization of the charged drug. When the neutral drug emerges on the other side of the membrane, the drug undergoes another proton transfer at the membrane surface and redissolves into the aqueous phase.

Just because a drug is neutral does not mean it crosses a membrane effectively. Some molecules have a large number of neutral, yet polar, functional groups that prevent

TABLE 3.7   Lipinski's rules[13]

1.	Molecular weight $\leq$ 500 g/mol
2.	Lipophilicity (log P) $\leq$ 5
3.	Sum of hydrogen-bond donors $\leq$ 5
4.	Sum of hydrogen-bond acceptors $\leq$ 10

membrane crossing. One example is glucose with its many OH groups. Similarly, a drug that is very nonpolar tends to enter membranes well and never diffuse back into the cytosol or interstitial fluid.

In 1997, Christopher Lipinski of Pfizer reported simple criteria for predicting whether a drug can readily cross membranes by passive diffusion.[13] Lipinski's criteria are known as *Lipinski's rules* or the *Rule of Five* (**Table 3.7**) and are based on a compound's molecular weight, lipophilicity, and number of hydrogen bond donors and acceptors.

The first rule of Lipinski indicates an ideal molecular weight of 500 g/mol or less. As molecules increase in size, their ability to passively cross a membrane decreases. Larger molecules cause a greater disruption in the supramolecular organization of a cell membrane. The energetic cost of reorganizing the cell membrane to accommodate a molecule is greater for a larger molecule and serves as a barrier to passive diffusion.

The second rule of Lipinski introduces the concept of *lipophilicity*. Lipophilicity describes the solubility of a compound in a nonpolar solvent. Lipophilic compounds dissolve readily in nonpolar solvents such as alkanes, benzene, and fats, while nonlipophilic compounds do not. One method for quantifying lipophilicity is to measure the equilibration of a molecule between two immiscible solvents, one polar and the other nonpolar. The ratio of the concentrations of a drug in each of the two solvents represents a partition coefficient (P) (Equation 3.1).

$$P = \frac{[\text{drug}]_{\text{nonpolar solvent}}}{[\text{drug}]_{\text{polar solvent}}} \tag{3.1}$$

The partition coefficient or its logarithm serves as a quantitative measure of a compound's lipophilicity. For pharmaceuticals, water and 1-octanol are the two solvents used most frequently to measure lipophilicity. The two-phase octanol/water system closely models the cell membrane/interstitial fluid interface. While Lipinski's rules do not specify a minimum lipophilicity, a log P value of 5 or less prevents a drug from effectively hiding within membranes.

The third and fourth Lipinski rules address the maximum number of hydrogen-bond donors and acceptors that a drug should contain. In the body, hydrogen-bond donors and acceptors are sites for strong interactions with water. When a molecule crosses into a membrane, the molecule must shed its shell of water molecules. If the molecule tightly binds water, losing its shell is more difficult energetically. Lipinski therefore suggests a maximal number of hydrogen-bond donors and acceptors.

Passive diffusion is a reversible process that reaches an equilibrium and requires no energy expenditure on the part of the cell. The direction of diffusion is from high concentration to low concentration. If the concentration on both sides is the same, then the *net* rate of diffusion is 0. In rare instances, a drug concentrates to higher levels within a cell or cell compartment, especially lysosomes. Lysosomes are cellular organelles that break down unneeded molecules and structures. The interior of lysosomes have a pH of approximately 4.8, which is much more acidic than blood plasma, typical cytosol, or the contents of the small intestine. In such a relatively acidic environment, a basic drug is almost completely protonated and unable to lose a proton, become neutral, and exit the lysosome by crossing a membrane.

Carrier Proteins

Most drugs that are fully charged or otherwise too polar for passive diffusion cross membranes with assistance from *carrier* or *transport proteins*. Carrier proteins span the membrane and can shuttle small molecules from one side to the other. These proteins are technically catalysts because they accelerate a process (membrane crossing) without being consumed.

Carrier proteins fall under two categories: *equilibrative* and *concentrative*. Equilibrative proteins can be envisioned as a revolving door in the cell membrane. Molecules that fit the door, that is, they properly bind the protein, can flow across the membrane. The direction of net flow is from high concentration to low concentration. The process requires no energy from the cell. In contrast, concentrative carrier proteins are able to pump a drug into or out of a cell. To accomplish this active pumping process, concentrative transport proteins are coupled with ion channels that ultimately rely on hydrolysis of ATP for energy. Because carrier proteins have access to external energy input, they can pump a drug across a membrane in the opposite direction of the concentration gradient.

Carrier proteins have defined roles in a cell, and transporting drugs is not part of their job description. For this reason, a drug discovery team rarely relies on carrier proteins for delivery of a drug to a target. This fact underscores the importance of passive diffusion and straightforward selection criteria such as Lipinski's rules. Regardless, some drugs are indeed transported by carrier proteins. Nucleoside analogues are excellent instructional examples. Nucleosides are the building blocks for DNA and RNA (**Figure 3.19**). Their densely functionalized structure consisting of a (**3.31–3.33**) polar sugar and nucleobase makes nucleosides poor candidates for entering a cell by passive diffusion. To make nucleosides available inside a cell, cells use a host of equilibrative and concentrative nucleoside transporters (ENTs and CNTs, respectively).

Drugs that act as nucleoside analogues mimic the activity of nucleosides while lacking full function. Because nucleoside analogues structurally resemble natural nucleosides, the analogues (1) do not passively diffuse across a membrane and (2) can readily hitch a ride across the membrane by means of an ENT and/or a CNT. Several nucleoside analogues and their corresponding carrier proteins are shown in **Table 3.8**.[14]

Amino acids and short polypeptides are charged and very polar and, like nucleosides, require a carrier protein to cross cell membranes. Valacyclovir (Valtrex, **3.38**), an antiherpes nucleoside analogue, exploits the peptide transport system through incorporation of a valine residue on the 5′-position of acyclovir (Zovirax, **3.39**) (**Figure 3.20**). The valine

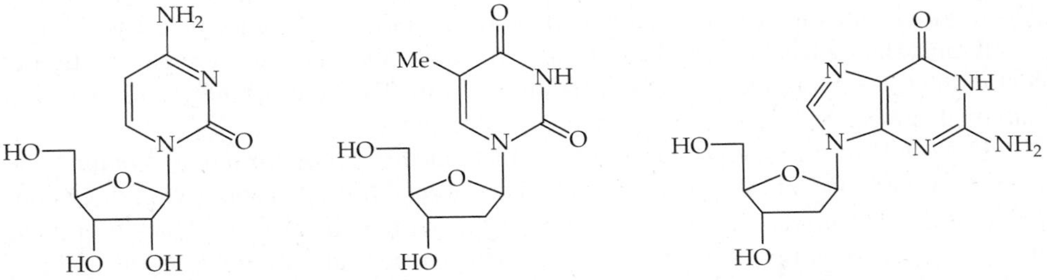

cytidine
3.31

thymidine
3.32

2′-deoxyguanine
3.33

FIGURE 3.19 Representative nucleosides

TABLE 3.8 **Selected antiviral nucleoside analogues and associated nucleoside transport proteins[14]**

Entry	Structure	Names	Transporter
1	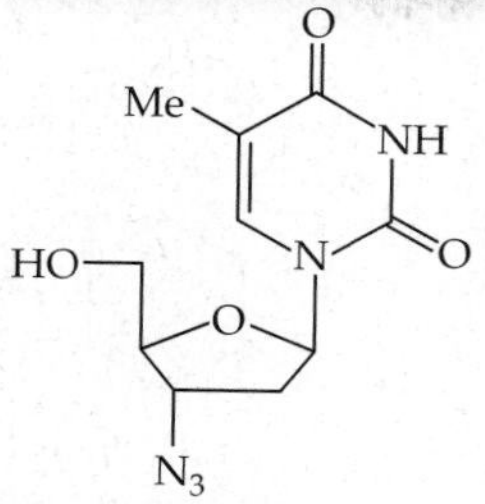	Azidothymidine AZT (Retrovir) **3.34**	CNT1 CNT3 ENT2
2	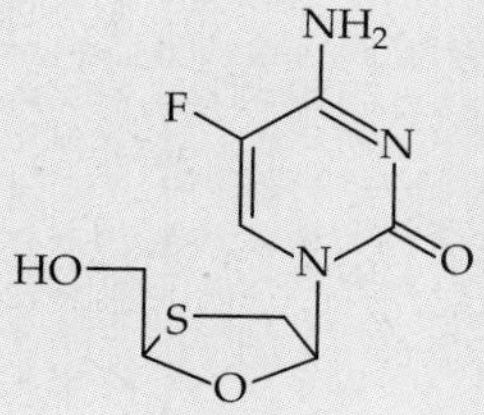	Lamivudine 3TC (Epivir) **3.35**	CNT1
3	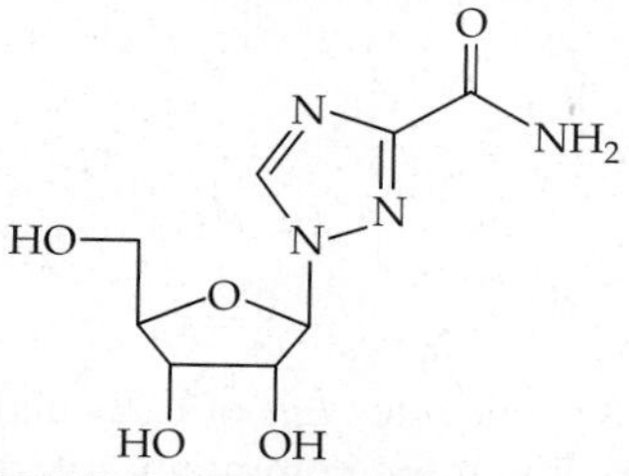	Ribavirin (Rebetol) **3.36**	CNT2 ENT1
4	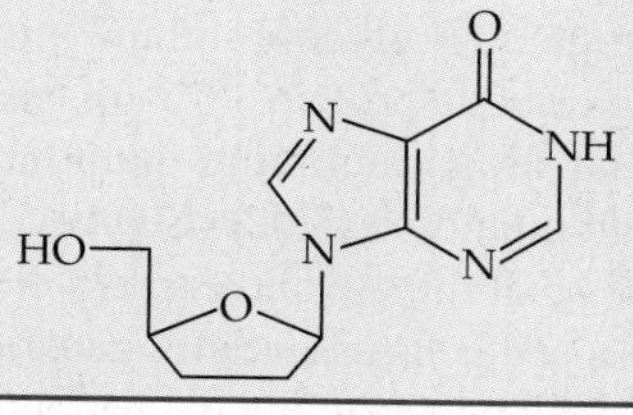	2′,3′-dideoxyinosine ddI (Videx) **3.37**	CNT3 ENT1

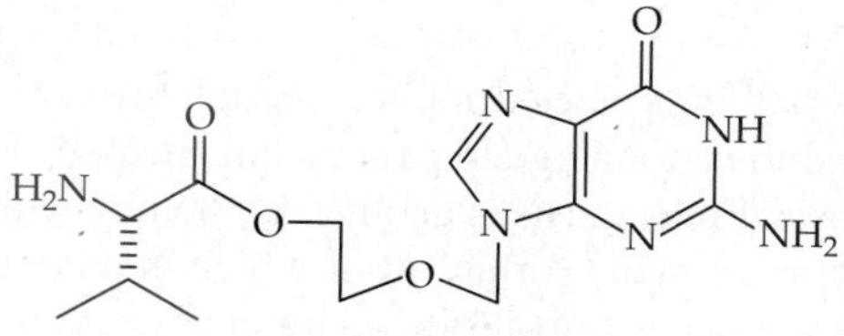

valacyclovir
(Valtrex)
3.38

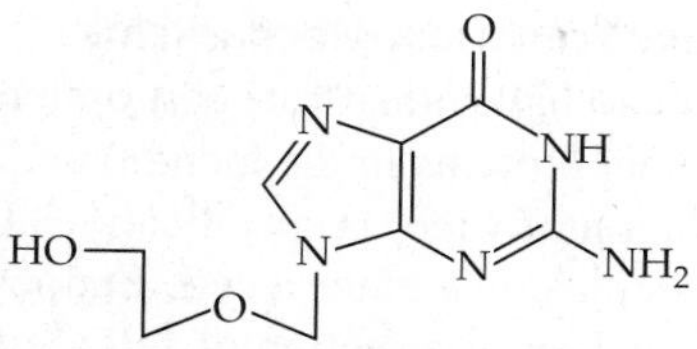

acyclovir
(Zovirax)
3.39

FIGURE 3.20

residue (shown in blue) allows valacyclovir to be recognized and transported out of the small intestine lumen by human oligopeptide transporter 1. The valine group is then removed by hydrolysis to afford the active drug, acyclovir.[15]

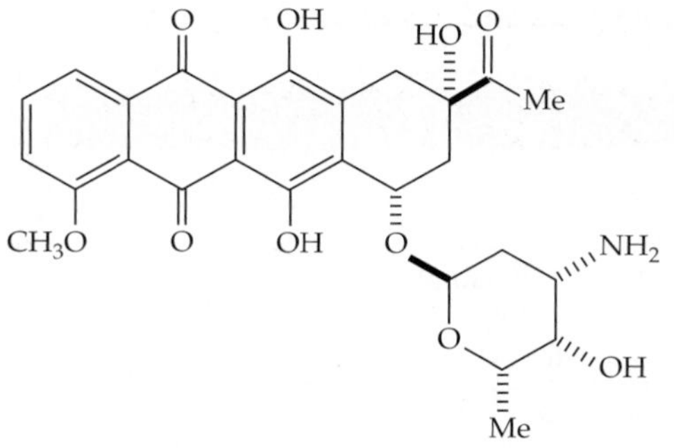

daunomycin
antitumor
3.40

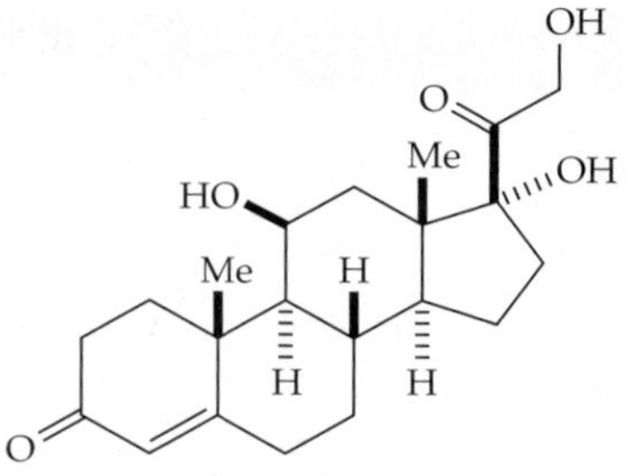

hydrocortisone
anti-inflammatory
3.24

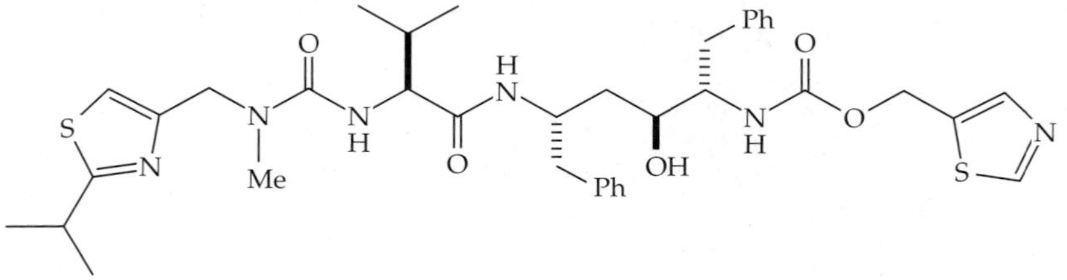

ritonavir
(Norvir)
anti-HIV
3.41

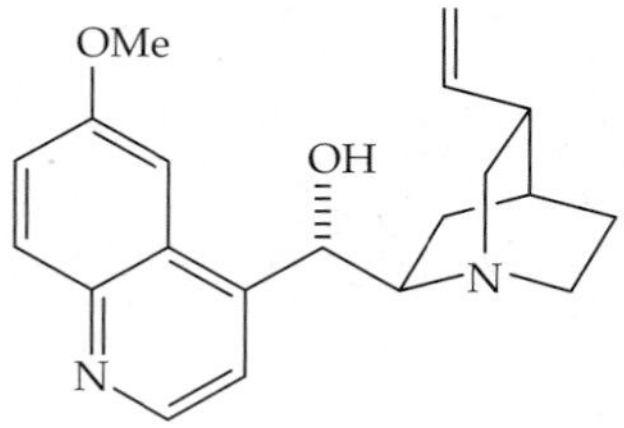

quinidine
antiarrhythmic
3.42

FIGURE 3.21 Drugs subject to efflux by P-glycoprotein

A handful of carrier proteins actively pump molecules *out* of cells and give rise to a phenomenon called *multidrug resistance*, which is more commonly referred to by the acronym *MDR*. Perhaps the most nefarious example is P-glycoprotein (P-gp), a member of a large family of MDR-associated ATP-binding cassette proteins.[16] P-gp has been linked to the efflux of a range of structurally dissimilar drugs including daunomycin (**3.40**), hydrocortisone (**3.24**), ritonavir (Norvir, **3.41**), and quinidine (**3.42**) (**Figure 3.21**). Particularly problematic are bacteria and cancer cell lines that mutate and express elevated levels of resistance-related carrier proteins.[17] These diseased cells essentially generate their own means of removing drugs, the same drugs that are being used to aid the patient.

Other Transport Mechanisms

Cells do have other transport options. Very small molecules can filter through the various channel proteins in the membrane. Drugs tend to be too large to pass by this method. The filtration of water is specifically called osmosis. Endocytosis is another method by which molecules may enter a cell. Endocytosis comes in many forms, all of which involve the indentation of a section of cell membrane that pinches off to form a new internal vesicle in the cell (**Scheme 3.2**). Once the cell breaks down the vesicle, the contents are released

SCHEME 3.2 Endocytosis

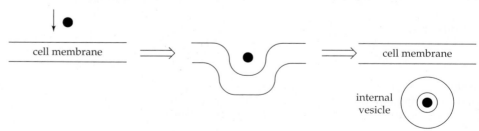

cell membrane

cell membrane

internal
vesicle

TABLE 3.9   Properties of oral drugs marketed from 1983 to 2002[18]

Characteristic or Property	All Drugs (n = 329)	CNS Drugs (n = 74)
Molecular weight (g/mol)	377	310
Percentage PSA	21.0	16.3
OH + NH	1.77	1.50
O + N	6.33	4.32
Hydrogen bond acceptors	3.74	2.12

into the cytoplasm. Endocytosis can be used by a cell to absorb fluid from outside the cell, large molecules such as proteins, and even other cells. Some types of relatively high molecular weight antibiotics enter the cell by endocytosis.

Blood–Brain Barrier

The term *blood–brain barrier* (BBB) refers to the special obstacle that drugs encounter when trying to enter the brain from the circulatory system. The difference between the brain and other tissues and organs is that the capillaries in the brain do not have pores for the free flow of small molecules in the interstitial fluid of the brain. To enter the interstitial fluid, *all* molecules must cross a membrane. This design is a protective measure to defend the brain from unwanted and potentially hazardous xenobiotics. Traditionally, drugs that target the brain or central nervous system (CNS) cross the BBB by passive diffusion. Transport by carrier proteins across the BBB is becoming better understood but remains an area of active research.

A recent study that compared CNS drugs to marketed drugs as a whole revealed significant differences between the two data sets (**Table 3.9**).[18] In all, data from 329 oral drugs marketed from 1983 to 2002 were included. CNS drugs tend to have a lower molecular weight, lower percentage of polar surface area (PSA), fewer hydrogen bond donors (NH and OH), fewer oxygen and nitrogen atoms, fewer hydrogen bond acceptors, and fewer rotatable bonds. *Polar surface area* refers to strongly polarized bonds that are exposed to solvent. PSA tends to interact strongly with polar solvents and therefore decreases the tendency of a molecule to enter a nonpolar membrane.

Given the unique nature of CNS drugs, Lipinski has defined a set of rules customized for CNS drugs based on marketed drugs.[19] Lipinski's revised rules for CNS drugs reduce the recommended maxima for molecular weight, hydrogen-bond donors, and hydrogen-bond acceptors to 400 g/mol, 3, and 7, respectively. The advised lipophilicity limit is unchanged. All trends and data indicate that CNS drugs require properties that allow crossing of a nonpolar membrane by passive diffusion.

3.4 Pharmacodynamics: At the Drug Target

At some point after a drug has been administered, absorbed into the blood, and distributed into other tissues, a drug reaches its site of action, that is, the location of the target to which the drug was designed to bind. A drug begins to have the desired effect on a patient only after it reaches its target. *Pharmacodynamics* is the impact of a drug on its target.

Drugs typically do not travel only to the site of their intended target. Drugs instead are distributed throughout the body to the location of many possible targets.

SCHEME 3.3 Bioactivation of enalapril (**3.43**), a prodrug

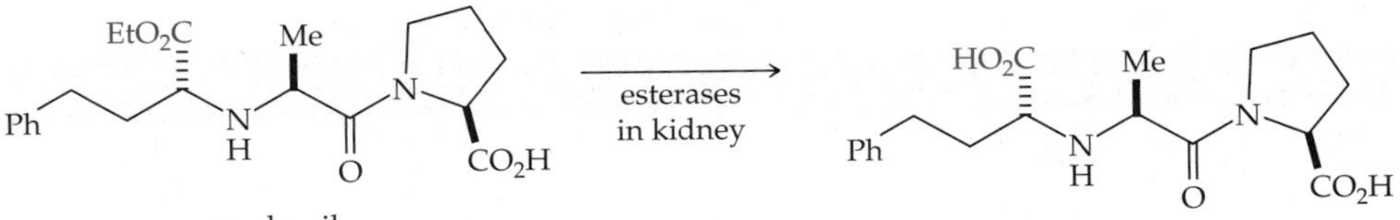

enalapril
(Vasotec)
transported
inactive against target
3.43

enalaprilat
not transported
active against target
3.44

While drugs are designed to interact primarily with just one target, they often interact weakly with multiple targets. With access to multiple targets, drugs frequently show side effects. Not all side effects occur because of undesired, off-target interactions, but many do.

Not all drugs reach their site of action in their original form. Some drugs are administered in a form that is easily absorbed and then modified to an active target-binding form. The active form is not administered to the patient because it is not well absorbed. Compounds that are administered in one form and then activated within the body are called *prodrugs*. An example of a prodrug is enalapril (Vasotec, **3.43**) (**Scheme 3.3**). Enalapril, an antihypertensive peptide, contains an ester group. The presence of this ester allows the drug to be recognized by a transporter protein and absorbed into the bloodstream from the digestive tract.[20] Once in the bloodstream, the ethyl ester of enalapril has reached the site of its target, angiotensin-converting enzyme. Enalapril, however, does not bind angiotensin-converting enzyme. The ethyl ester of enalapril must first be hydrolyzed by enzymes in the kidneys. The hydrolyzed product, enalaprilat (**3.44**), contains a carboxylic acid group that is instrumental for strong binding to angiotensin-converting enzyme. Oral administration of enalapritat, the acid form, is not an option because enalaprilat is not recognized by the transporter protein and does not cross membranes by passive diffusion. Valacyclovir (**3.38**) in Figure 3.19 (see above) is another example of a prodrug.

The discovery and design of drugs in a form that is well absorbed, distributes effectively, and shows sufficient activity falls under the areas of lead discovery and optimization, the topics of Chapters 10 through 12.

While a drug reaches its site of action via the bloodstream, the bloodstream also helps remove a drug from the target. The removal process is called elimination.

3.5 Metabolism and Elimination: Drug Removal

An effective drug must be distributed to its site of action, but it also must be *eliminated* from the body. Elimination encompasses *any* process, typically irreversible, that diminishes the amount of a drug that is distributed throughout the body. Elimination manifests itself in two fashions. The first involves the removal of drug from the body in its original, unchanged form. This specific type of elimination is called either *elimination* or *excretion*. The second form of elimination is called *metabolism*. Metabolism involves the chemical modification of the drug into a different form. Both of these processes reduce the amount and concentration of the drug in the body. Drugs have many possible elimination sites and routes. Generally speaking, the kidney is the primary site of elimination, while the liver is the principle organ for metabolism (**Scheme 3.4**).

Kidneys

The kidneys receive blood through the renal artery (**Scheme 3.5**). Blood in the kidneys is ultimately filtered through nephrons, the operating unit of the kidney. Each kidney

cells
(site of action) ⇌ drug in general circulatory system ⇌ liver ⟶ *metabolism*

kidney ⟶ *elimination*

blood
(general circulatory system) →renal artery→ renal arterioles (nephrons) →glomerular filtration→ **filtered plasma** →cross Bowman's capsule→ **tubular fluid** →collect into ureter→ **urine**

renal blood supply

secretion ‖ reabsorption

renal blood supply

contains approximately a million nephrons. Each nephron passes blood through a spherical collection of capillaries called the *glomerulus*. As with all capillaries, those in the glomerulus allow passage of the plasma with solutes smaller than albumin. Some of the filtered plasma is returned to the renal blood supply, and the rest is received by a collection of epithelial cells called *Bowman's capsule*. The fluid is absorbed and passed into a network of tubules. These tubules are bathed in blood, and both molecules and ions in the filtrate can cross the tubule membrane and re-enter the blood. This process is called *reabsorption*. Molecules and ions in the blood may also enter the tubule in a process called *secretion*. Transport into and out of the tubule is accomplished by filtration, diffusion, and carrier proteins and it depends on the structure of the transported species. The fluid remaining in the tubule system is collected by the ureter and stored in the bladder until it is eliminated from the body. The net effect of the reabsorption and excretion processes in the tubules is to recover water and vital electrolytes from the fluid lost in the glomerulus while allowing waste products to pass on to the bladder.[3]

Drugs in the bloodstream are also affected by tubular reabsorption and secretion. In general, drugs that are readily reabsorbed from tubular fluid are *cleared* from the blood more slowly than drugs that are mostly excreted into the tubule. The concept of *clearance* is related to elimination and is measured, perhaps surprisingly, in terms of bloodflow (mL/min). However, because the rate at which an organ can remove a drug is limited by the rate of bloodflow to that organ, bloodflow is a natural measure for clearance. Clearance throughout the body can be related to elimination, which is quantified in terms of rate constants (inverse time units) or half-life (time units). Again, the mathematical modeling of absorption, distribution, and elimination is covered in Chapter 7.

Tubular reabsorption effectively removes water from the original filtrate collected by the glomerulus. Reabsorption therefore raises the concentration of compounds in the tubular fluid. Filtered drugs present in the tubules are also concentrated, and drug solubility can be an issue. Some drugs, including methotrexate (**3.45**), can actually precipitate within the tubules of the kidney and cause damage (**Figure 3.22**).[21] To prevent crystal-induced acute renal failure, a patient may need to increase fluid intake, which decreases water reabsorption in the kidney.

Many drugs are eliminated in their originally administered chemical state by the kidney and into the urine. A drastic demonstration of this idea can be found with the

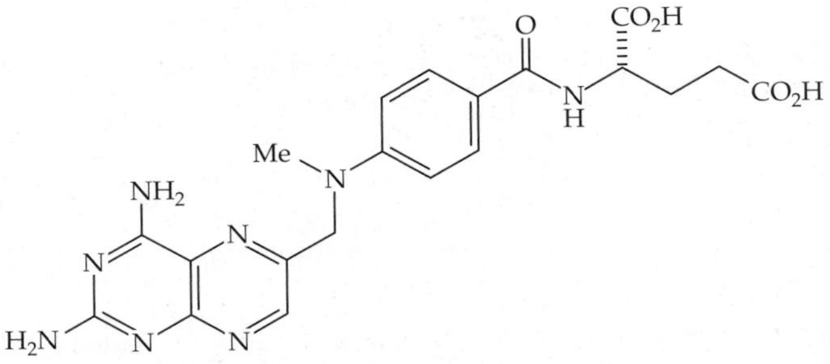

methotrexate
3.45

FIGURE 3.22

β-lactam antibiotic penicillin. Penicillin is eliminated unchanged in the urine. The first human patient to receive penicillin for an infection was initially treated on February 12, 1941 in England. The patient received 200 mg intravenously followed by 100 mg IV doses every 3 hours. The supply of penicillin was severely limited, and available sources of penicillin were exhausted by the third day of treatment. To secure additional drug, scientists extracted penicillin from the patient's urine. By this method, enough penicillin was recovered to continue treatment for three additional days.[22]

Because of the important role of the kidney in eliminating a drug, proper kidney function is vital for a patient to receive a proper dose of a drug. Patients with impaired kidney function are often unable to eliminate a drug effectively. In this situation, the concentration of the drug may be elevated in the blood and ultimately reach toxic levels. In their prescribing information insert, many drugs contain warnings concerning proper kidney function.

Liver

The liver accomplishes elimination primarily by breaking down a drug through oxidations and reductions (metabolism). Several classes of enzymes are involved in metabolism, but the most commonly mentioned class is the cytochrome P-450 family. Liver enzymes can also perform conjugations, that is, reactions that couple a drug to a very polar, water soluble molecule. The products of all these reactions are called *metabolites*. The oxidations, reductions, and conjugations accomplish two tasks. First, the drug is eliminated by being converted into a different, likely inactive structure. Second, the new metabolite is often more water soluble than the original drug. Increased water solubility tends to decrease reabsorption in the kidney and therefore facilitates elimination of the metabolite. The general aspects of metabolism are covered more thoroughly in Chapter 8.

In conjunction with the gall bladder, the liver can eliminate drugs and metabolites (**Scheme 3.6**). The liver produces bile, which is then either stored in the gall bladder or sent directly into the duodenum at the start of the small intestine. Drugs and their metabolites can be included in the bile that leaves the liver. In the small intestine, the drug or metabolite may be reabsorbed and returned to the liver and bloodstream via the hepatic portal systems. This cycle—blood to liver to small intestine and back to blood—can be repeated many times by a drug and greatly prolong the time the drug resides in the body. The recycling process is called *enterohepatic circulation*.

Bacteria in the small intestine facilitate enterohepatic circulation. Many drugs that are excreted into bile by the liver are first conjugated with a polar, solubilizing group. Once the metabolite reaches the small intestine, its increased polarity prevents it from being reabsorbed. Bacteria in the small intestine remove the solubilizing group and thereby restore the original drug. The drug, with its increased lipophilicity, is then absorbed into the hepatic portal system. From that point, the drug passes back through the liver and re-enters general circulation.

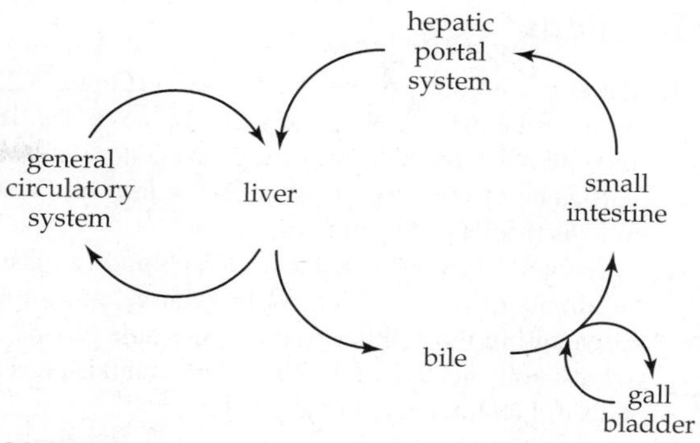

Other Routes

While urine is the primary mode for elimination of a drug, any bodily excretion or secretion represents a possible exit venue. Examples include feces, tears, sweat, breath, hair, and fingernails. One elimination route that is of importance to a specific type of patient is breast milk. Nursing mothers must be very careful when taking medication because many drugs are eliminated, at least partially, through breast milk. A mother may unintentionally serve as a drug administration vehicle for her nursing child.

SCHEME 3.6 Tracing a drug through enterohepatic circulation

One requirement for drug approval is a *mass balance study*.[23] A mass balance study requires administration of a radiolabeled drug and careful analysis of excretions to trace the exit of radioactivity from the organism. Mass balance studies are performed on rats and dogs before humans. A mass balance study focuses less on how the drug is eliminated and more on making sure that the drug is indeed eliminated. Drugs that distribute into tissues and become sequestered can cause serious toxicity problems. Due to the technical challenges of collecting excretions, mass balance studies rarely account for all of an administered dose. In an analysis of recently developed drugs at Pfizer, discovery teams were typically able to account for 80% of an administered dose. The satisfactory recovery level depends on the individual drug, its function, and its elimination pathways.

Summary

Aside from binding strongly to the desired target drugs must be able to reach their site of action in the body. The overall process of a drug's entering a body, distributing via the blood to other tissues, reaching its target, and then exiting via an elimination route defines the field of pharmacokinetics. A drug enters a body and ultimately the bloodstream by its administration route. Intravenous injection is the simplest and most direct route to the bloodstream. The most popular method is oral delivery. Despite its popularity with patients, design of oral drugs poses several challenges. The drug must survive the acidic environment of the stomach, be able to be absorbed along the digestive system, and resist being metabolized by the liver before reaching the general circulatory system. Other delivery routes are possible, but intravenous and oral administration are far and away the most common. Once in the bloodstream, a drug has access to nearly the entire body and can reach its site of action. Drugs that bind receptors on the surface of cells do not need to cross a membrane to reach their targets, but many drugs need to access the interior of a cell. Most drugs cross cell membranes by passive diffusion. Passive diffusion requires the drug to be (or to be able to become) neutral and sufficiently lipophilic to pass through the

nonpolar membrane. Some drugs rely on transport proteins to enter a cell. To exploit a transport protein, a drug must somewhat resemble the natural substrate of the protein. Drugs that bind a target in the brain must be able to cross the blood–brain barrier. Traditionally, drugs affecting the brain have been designed to cross the blood–brain barrier by diffusion. As the transport systems of the brain become more fully understood, actively transported drugs should become more common. While a drug must be able to reach its site of action, a drug must also be able to leave the body. This process, called elimination, is largely performed by the kidneys. Blood is cleared of drug by the kidneys in a filtration process performed by nephrons. Drugs and other wastes are then channeled to the bladder for elimination. The rate of clearance for a drug depends on the bloodflow through the kidney and how the drug behaves while being filtered in the nephron. The liver assists in the elimination process by metabolizing drugs with enzymatic oxidations, reductions, and conjugations. The resulting metabolites tend to be more polar than the original drug. By increasing a molecule's polarity, the liver makes the metabolite more likely to be removed by nephrons in the kidney.

Questions

1. Patients who take methotrexate (**3.45**) (Figure 3.22) must monitor the pH of their urine several times a day. If the pH rises above 7, the patient must receive a dose of sodium bicarbonate (a weak base) either orally or as an IV solution. Why is the pH of the urine important?

2. Alfentanil (**3.a**) is an analgesic. The blood-to-plasma ratio of alfentanil is 0.86. What is the relative concentration of alfentanil in the cellular fraction of whole blood? In other words, how much (concentration) alfentanil is present in the cells of blood relative to the plasma?

alfentanil
3.a

3. Most drugs have a blood-to-plasma ratio of around 1, but exceptions can be found. Epoetin alfa is a protein with a molecular weight of over 18,000 and stimulates the production of red blood cells in anemic patients. Would the blood-to-plasma ratio of epoetin be expected to be around 1? Why or why not?

4. Perhaps surprisingly, the most common element in the body by mass is oxygen, not carbon (Table 3.1). Most of the oxygen is found within water molecules. Use the data in Figure 3.1 to remove the impact of water from Table 3.1. In other words, ignoring water, what is the elemental composition of the body with respect to oxygen, carbon, hydrogen, and nitrogen? (The other elements must be kept in the picture for the calculation to work correctly.)

5. A mass balance study has been performed on laromustine (**3.b**), an anticancer alkylating agent that failed during clinical trials. In male rats after 7 days, the following

lauromustine
3.b

amounts of **3.b** were recovered: 65.6% in urine, 4.9% in cage rinse, 5.5% in feces, and 8.8% in tissues. What is the total mass balance? Is this drug well absorbed? Justify your answer. (See Mao, J., Xu, Y., Wu, D., & Almassain, B. Pharmacokinetics, Mass Balance, and Tissue Distribution of a Novel DNA Alkylating Agent, VNP40101M, in Rats. *Am. Assoc. Pharm. Sci. J.* 2002, 4(4), 46–52.)

6. Two molecular properties that are often watched in the development of orally available drugs are polar surface area and the number hydrogen bond donors and acceptors. Compounds with many hydrogen bond donors and acceptors tend to have a higher polar surface area than compounds with few hydrogen bond donors and acceptors. In this sense, tracking both polar surface area and hydrogen bond donors and acceptors in a drug candidate may seem to be redundant. What functional groups would increase polar surface area without adding to the number of hydrogen bond donors and acceptors?

7. The normal level of protein in urine is <5 mg/dL. Excessive protein in urine is called *proteinuria* and is a sign of kidney problems. Why?

8. Alzheimer's drugs must cross the blood–brain barrier to be effective. Three Alzheimer's drugs are shown in the following figures (**3.c, 3.d, 3.e**). Determine their MW, number of

donepezil
(Aricept)
3.c

galantamine
(Nivalin)
3.d

rivastigmine
(Exelon)
3.e

hydrogen bond donors, and number of hydrogen bond acceptors. Do these drugs satisfy Lipinski's rules for CNS active molecules? If an Alzheimer's drug violated Lipinski's CNS rules (yet still acted as an effective Alzheimer's drug), what might you conclude about the distribution pathways of that drug?

9. The pH of the rectum is between 5.5 and 7.0. Explain why rectal administration may often be favored over oral administration.

10. The standard patient has a mass of 70 kg, and the density of the human body is very close to 1 g/mL. What is the body volume of the standard patient in liters? Based on this volume and using the data in Table 3.1, determine the molar concentration of iron in the body. Assume all the iron in the body is evenly distributed throughout all tissues. Repeat the calculation on selenium, an element involved in certain oxidation-reduction processes in the body. The lesson of this question is that trace elements can be *very* trace indeed.

References

1. D. R. Lide (Ed.), *Handbook of Chemistry and Physics* (83rd ed.). Boca Raton, FL: CRC Press, 2002.
2. Wang, Z.-M., Pierson, R. N. Jr., & Heymsfield, S. B. The Five-Level Model: A New Approach to Organizing Body-Composition Research. *Am. J. Clin. Nutr.* 1992, *56*, 19–28.
3. Tortora, G. J. & Derrickson, B. H. *Principles of Anatomy and Physiology* (11th ed.). New York: Wiley, 2005.
4. U.S. Food and Drug Administration. Route of Administration. http://www.fda.gov/Drugs/Development ApprovalProcess/FormsSubmissionRequirements/ElectronicSubmissions/DataStandardsManualmonographs/ucm071667.htm (accessed July 2012).
5. Widmaier, E. P., Raff, H., & Strang, K. T. *Vander's Human Physiology* (10th ed.). New York: McGraw-Hill, 2005.
6. Ratain, M. J. & Cohen, E. E. The Value Meal: How to Save $1,700 per Month or More on Lapatinib. *J. Clin. Oncol.* 2007, *25*, 3397–3398.
7. Girennavar, B., Poulose, S. M., Jayaprakasha, G. K., Bhat, N. G., & Patil, B. S. Furocoumarins from Grapefruit Juice and Their Effect on Human CYP 3A4 and CYP 1B1 Isoenzymes. *Bioorg. Med. Chem.* 2006, *14*, 2606–2612.
8. Meanwell, N. A. Synopsis of Some Recent Tactical Application of Bioisosteres in Drug Design. *J. Med. Chem.* 2011, *54*, 2529–2591.
9. Margetts, L. & Sawyer, R. Transdermal Drug Delivery: Principles and Opioid Therapy. *Contin. Educ. Anaesth. Crit. Care Pain.* 2007, *7*, 171–176.
10. Patient Information: Nasonex. http://www.spfiles.com/pinasonex.pdf (accessed July 2012).
11. Thummel, K. E. & Shen, D. D. Design and Optimization of Dosage Regimens: Pharmacokinetic Data. In J. G. Hardman & L. E. Limbird (eds.), *Goodman and Gilman's The Pharmacological Basis of Therapeutics* (10th ed., Appendix II). New York: McGraw-Hill, 2001.
12. Alberts, B., Johnson, A., Lewis, J., Raff, M., Roberts, K., & Walter, P. *Molecular Biology of the Cell* (5th ed.). New York: GarlandScience, 2007.
13. Lipinski, C. A., Lombardo, F., Dominy, B. W., & Feeney, P. J. Experimental and Computational Approaches to Estimate Solubility and Permeability in Drug Discovery and Development Settings. *Adv. Drug Del. Rev.* 1997, *23*, 3–25.
14. Pastor-Anglada, M., Cano-Soldado, P., Molina-Arcas, M., Lostao, M. P., Larrayoz, I., Martínez-Picado, J., & Casado, F.J. Cell Entry and Export of Nucleoside Analogues. *Virus Res.* 2004, *107*, 151–164.
15. Landowski, C. P., Sun, D., Foster, D. R., Menon, S. S., Barnett, J. L., Welage, L. S., et al. Gene Expression in the Human Intestine and Correlation with Oral Valacyclovir Pharmacokinetic Parameters. *J. Pharmacol. Exptl. Ther.* 2003, *306*, 778–786.
16. Robert, J. & Jarry, C. Multidrug Resistance Reversal Agents. *J. Med. Chem.* 2003, *46*, 4805–4817.
17. Dey, S., Gunda, S., & Mitra, A. K. Pharmacokinetics of Erythromycin in Rabbit Corneas after Single-Dose Infusion: Role of P-Glycoprotein as a Barrier to in Vivo Ocular Drug Absorption. *J. Pharmacol. Exptl. Ther.* 2004, *311*, 246–256.
18. Leeson, P. D. & Davis, A. M. Time-Related Differences in the Physical Property Profiles of Oral Drugs. *J. Med. Chem.* 2004, *47*, 6338–6348.
19. Pajouhesh, H. & Lenz, G. R. Medicinal Chemical Properties of Successful Central Nervous System Drugs. *NeuroRx* 2005, *2*, 541–553.
20. Swaan, P. W. & Tukker, J. J. Molecular Determinants of Recognition for the Intestinal Peptide Carrier. *J. Pharm. Sci.* 1997, *86*, 596–602.
21. Perazella, M. A. Crystal-Induced Acute Renal Failure. *Am. J. Med.* 1999, *106*, 459–465.
22. Sneader, W. *Drug Prototypes and Their Exploitation.* New York: Wiley, 1996, p. 475.
23. Roffey, S. J., Obach, R. S., Gedge, J. I., & Smith, D. A. What Is the Objective of the Mass Balance Study? A Retrospective Analysis of Data in Animal and Human Excretion Studies Employing Radiolabeled Drugs. *Drug Metab. Rev.* 2007, *39*, 17–43.

Enzymes as Drug Targets

Chapter Outline

Learning Goals for Chapter 4

- Understand the different levels of protein and enzyme structure
- Know how enzyme activity is controlled by the body
- Understand and be able to use the kinetic models of enzymes
- Recognize the different types of enzyme inhibitors
- Appreciate potential complications of inhibiting enzymes with drugs

Enzymes catalyze chemical reactions in the body. Affecting the efficiency of an enzyme can therefore significantly change a biological system. For this reason, enzymes are common targets in a drug discovery program. Enzymes are proteins with generally well defined three-dimensional structures. Furthermore, the catalytic effect of an enzyme upon a reaction can normally be precisely modeled with various kinetic theories. Just as enzyme kinetics is predictable, the manner in which drugs interact with and block enzymes is also predictable. The mode of action and potency of enzyme blocking drugs are easily determined, which makes enzyme inhibitors very attractive drug candidates.

4.1 Introduction to Enzymes

The catalytic action of enzymes arises from their underlying protein structure. Therefore, appreciation of enzymes first requires a working understanding of the different levels of protein structure.

Definition

Enzymes are catalysts, and catalysts have two key qualities. First, catalysts increase the rate of a reaction. Second, catalysts are not consumed in the reaction and therefore do not appear in the overall balanced reaction equation. Catalysts appear throughout everyday life. Most simple hand tools are catalysts. For example, a shovel greatly accelerates the rate at which one can move dirt or gravel. Furthermore, the same shovel can be used day after day and work as efficiently on the last day as on the first. A shovel facilitates a process but is not consumed by the process. Moving dirt does not absolutely require a shovel, but using a shovel certainly beats using one's bare hands.

An example of a simple catalyzed chemical reaction is the hydrolysis of cyclic acetal (**4.1**) to lactol (**4.3**) (**Scheme 4.1**). This reaction is reversible. To a rough approximation, ΔG

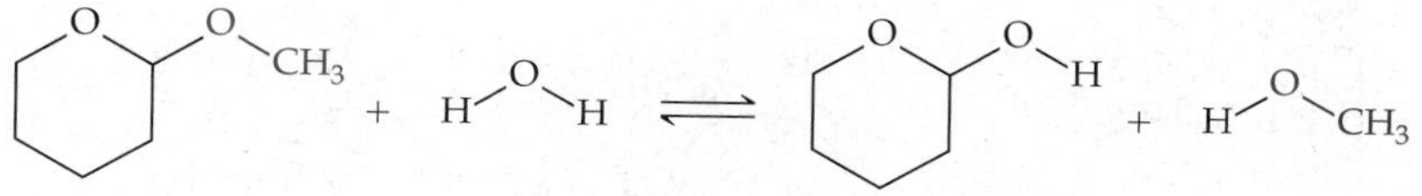

SCHEME 4.1 Hydrolysis of acetal (**4.1**) to lactol (**4.3**)

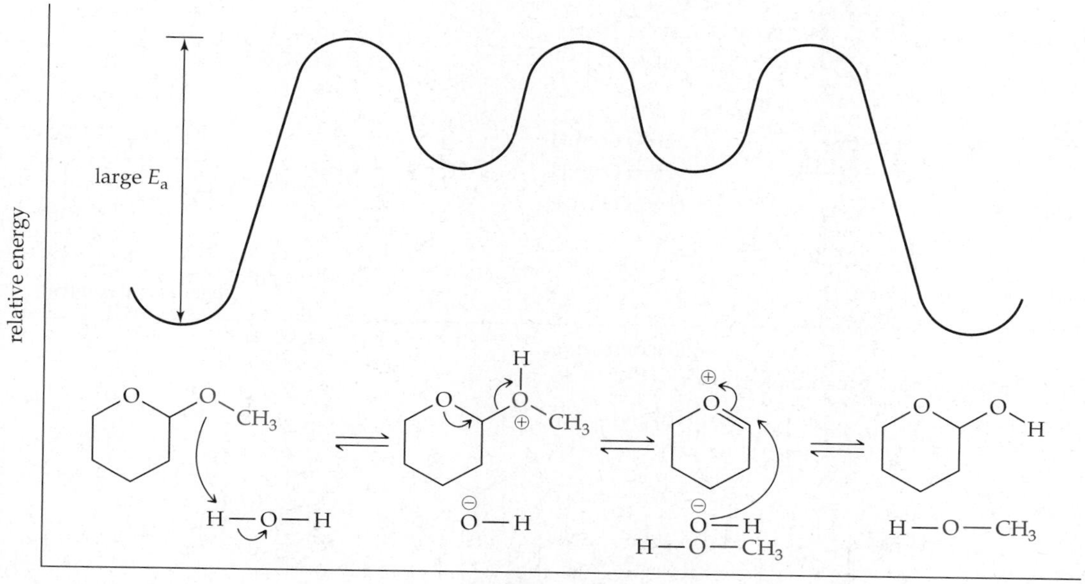

FIGURE 4.1 Uncatalyzed reaction coordinate for hydrolysis of **4.1**

for the overall reaction is 0, and the equilibrium constant is 1. Because the use of a catalyst does not change the starting materials and products of a reaction, ΔG and K for any reaction are unchanged by the use of a catalyst. The catalyst simply allows the reaction to reach equilibrium more quickly.

The intermediates and their relative energies for the *uncatalyzed* hydrolysis of **4.1** are shown in **Figure 4.1**. The first step, a proton transfer, is unfavorable because both the acid (**4.2**) and base (**4.1**) are weak. The first step, therefore, is endothermic, has a high energy of activation, and is very slow. Other uncatalyzed mechanisms are possible, but all involve an energetically unfavorable step at some point along the pathway.

The acid-catalyzed hydrolysis of **4.1** requires a mechanism with four steps instead of three (**Figure 4.2**). Despite the extra step, all the activation energies are relatively small compared to the first step in Figure 4.1. Catalysts typically do add steps to a mechanism, but the overall rate is faster because of reduced activation energies. Although the catalyst may react during the process, it must be re-formed at the end to qualify as a true catalyst. For the reaction in Figure 4.2, the catalyst (boxed) appears as both a starting material and product, so it does not appear in the overall, balanced equation in Scheme 4.1.

Enzymes have all the qualities mentioned previously for an acid catalyst. Enzymes are not consumed in the overall reaction, and they accelerate both the forward and reverse reaction by altering the mechanism and/or lowering the activation energy of the key steps. Often, the product immediately reacts in a subsequent reaction, and the reverse reaction of the enzyme can be ignored.

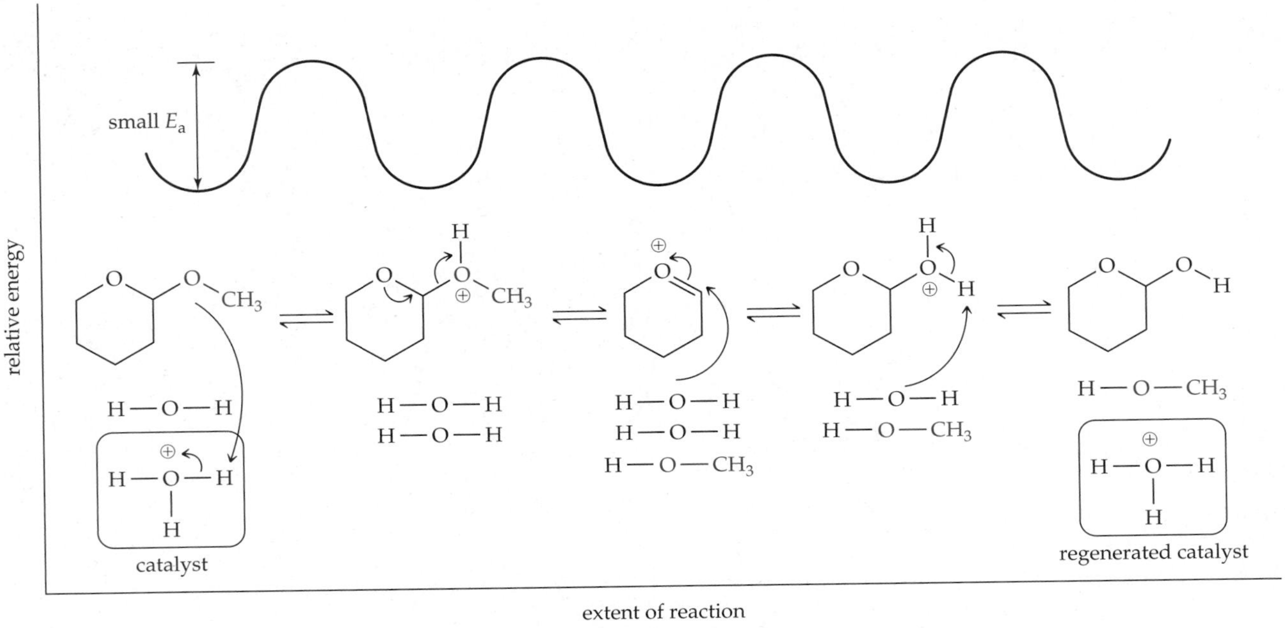

FIGURE 4.2 Acid catalyzed reaction coordinate for hydrolysis of **4.1**

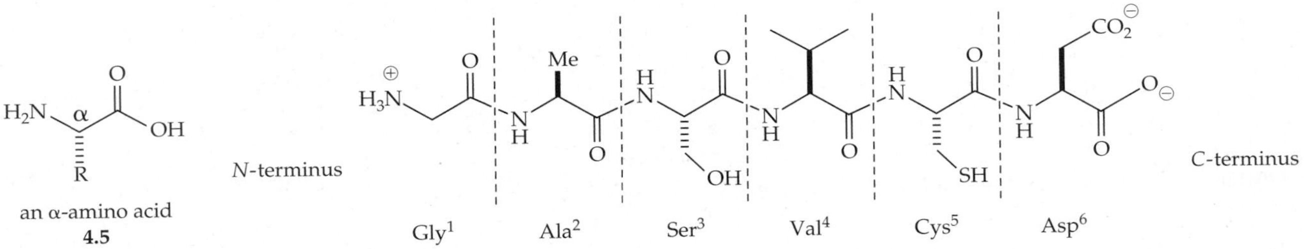

FIGURE 4.3 An α-amino acid and a hexapeptide

Structure

Enzymes are almost exclusively proteins and therefore consist of a chain of amino acids. The number of amino acids in an enzyme varies. An enzyme may contain anywhere from about 100 to over 10,000 individual α-*amino acid* residues. All α-amino acids (**4.5**) consist of a carboxylic acid, an amino group off the α-carbon, and an R-group off the α-carbon (**Figure 4.3**). Amino acids are strung together by formation of an amide bond between the amine nitrogen and carbonyl of the acid. The sequence of amino acids in the chain, or peptide backbone, is called the *primary structure*. Disulfide bonds between cysteine residues are also considered to be part of the primary structure. When a protein is written as an amino acid chain, the primary structure always starts on the left with the *N*-terminus, or amino terminus, and continues to the right before ending with the *C*-terminus, or carboxy terminus (Figure 4.3). Amino acids in the backbone are often numbered, starting with the *N*-terminus. **Table 4.1** lists the 20 standard amino acids with their corresponding abbreviations and R-groups on the α-carbon. Also listed are pK_as of acidic side chains or pK_as of the conjugate acid of basic side chains (as $pK_a{}'$). The different R-groups of amino

TABLE 4.1 Standard amino acids with R-groups[1]

Entry	Name	Symbol	Abbrev.	R-Group	Property	pK_a (pK_a')
1	glycine	G	Gly	H	—	—
2	alanine	A	Ala	CH_3	nonpolar	—
3	valine	V	Val	i-Pr	nonpolar	—
4	leucine	L	Leu	i-Bu	nonpolar	—
5	isoleucine	I	Ile	s-Bu	nonpolar	—
6	phenylalanine	F	Phe	CH_2Ph	nonpolar, aromatic	—
7	proline (full structure shown)	P	Pro	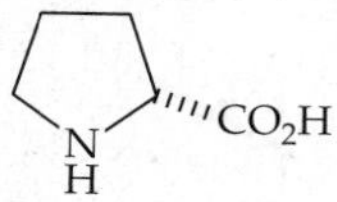	nonpolar	—
8	serine	S	Ser	CH_2OH	polar	—
9	threonine	T	Thr	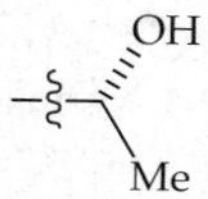	polar	—
10	tyrosine	Y	Tyr	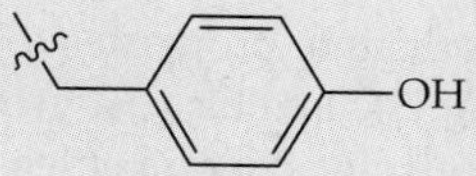	aromatic, polar, weakly acidic	10.1
11	cysteine	C	Cys	CH_2SH	nonpolar, weakly acidic	10.3
12	methionine	M	Met	$CH_2CH_2SCH_3$	nonpolar	—
13	asparagine	N	Asn	CH_2CONH_2	polar	—
14	glutamine	Q	Gln	$CH_2CH_2CONH_2$	polar	—
15	tryptophan	W	Trp	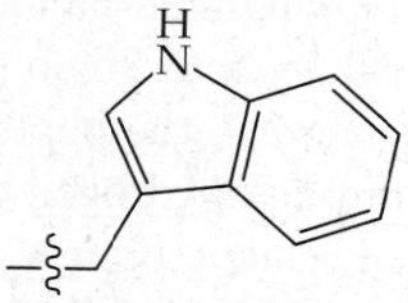	weakly polar, aromatic	—
16	aspartic acid	D	Asp	CH_2CO_2H	polar and acidic	3.7
17	glutamic acid	E	Glu	$CH_2CH_2CO_2H$	polar and acidic	4.3
18	lysine	K	Lys	$CH_2CH_2CH_2CH_2NH_2$	polar and basic	(10.5)
19	arginine	R	Arg	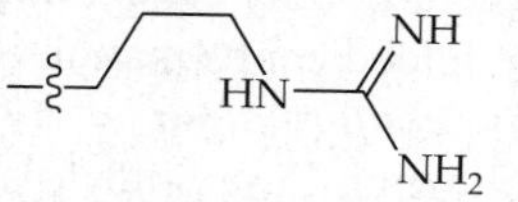	polar and basic	(12.5)
20	histidine	H	His	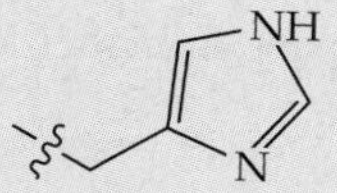	basic	(6.0)

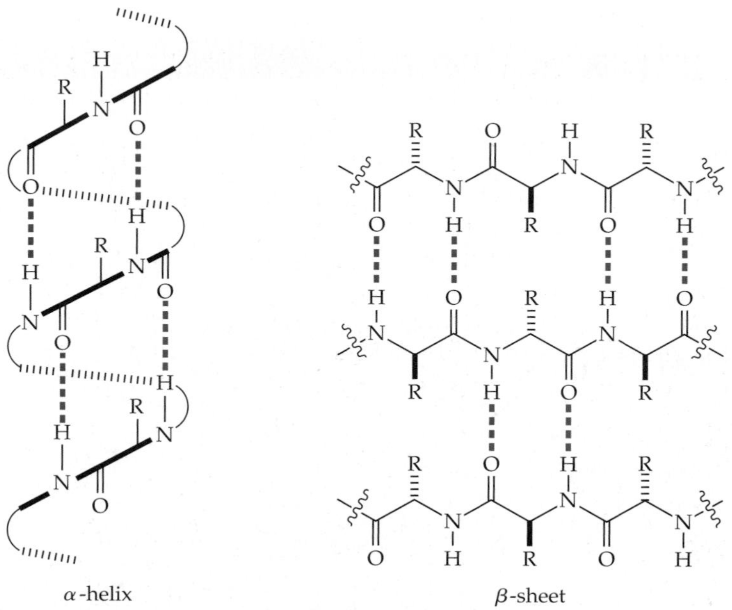

FIGURE 4.4 Hydrogen bonding in α-helices and β-sheets

acids impart distinct properties to a peptide chain. Amino acid R-groups may be polar or non-polar. Acidic and basic R-groups may also be charged at physiological pH. All amino acids except glycine contain at least one stereocenter. Under the traditional Fischer system, amino acids have L-configuration. Under the Cahn–Ingold–Prelog convention, all are (*S*)-configuration at the α-carbon except cysteine.

Portions of an amino acid chain fold in patterns such as *α-helices* and *β-sheets*, both of which are examples of *secondary structure* (**Figure 4.4**). Even a lack of regular secondary structure, called a *random coil*, is itself a type of secondary structure. Secondary structures, aside from random coils, are formed from hydrogen bonds within the peptide backbone. In an α-helix, the backbone winds into a helical coil. The helix is most often right-handed, although left-handed helices are possible. The carbonyl of one amide acts as a hydrogen bond acceptor to an amide N-H farther along the backbone. In a β-sheet, also called a *pleated sheet*, the peptide backbone folds back upon itself with the folds being held together through hydrogen bonding.

The overall relative orientation of the secondary structures of an enzyme determines its three-dimensional shape, or *tertiary structure*. Some enzymes require multiple copies of the same enzyme to function. The individual enzymes cluster into groups of two or more (called dimers, trimers, etc.) and are held together by intermolecular forces. The relative positioning of the separate enzymes in the cluster determines the overall structure, or *quaternary structure*, of the supramolecular complex. While all enzymes have tertiary structure, only clusters of multiple enzyme subunits have quaternary structure. The overall folded conformation of a protein in its active, catalytic form is called the *active* or *native conformation*.

The following Case Study discusses two different roles of α-helices within proteins. The illustration is meant to highlight how seemingly small changes in primary structure can have an impact on higher levels of function for a protein.

CASE STUDY

Use of α-Helices to Cross Cell Membranes[2, 3]

A common structural role for α-helices is to allow a protein to cross a cell membrane. The coil of the helix passes through a lipophilic membrane with an orientation that is nearly perpendicular to the plane of the membrane. Crossing a membrane typically requires a helix formed from 20 to 25 residues.

Human estrone sulfatase is an enzyme that maintains elevated levels of estrogens within certain types of tumor cells. The enzyme is associated with a network of membranes within cells and relies on a pair of transmembrane helices that hold the enzyme to the surface of the membrane. The majority of the over 550–residue protein has a tertiary structure with a roughly spherical shape. A section of the chain loops away from the main structure. This loop contains two α-helices that twice cross a membrane and maintain the correct placement of the enzyme within the cell. α-Helices that hold a protein in place are called *membrane anchors*.[2]

α-Helices that serve as membrane anchors contain amino acids with primarily nonpolar R-groups. The section of human estrone sulfatase that forms one of its transmembrane α-helices is shown in **Figure 4.5**. The figure shows the position of the residue, the identity of the residue (one-letter code), and the residue's property (polar or nonpolar: P or N).[2]

Figure 4.5 indicates that aside from Ser[218] and Thr[228], the center of the helix contains exclusively nonpolar residues. The nonpolar residues allow the helix to favorably interact with the surrounding nonpolar environment of the

lipid bilayer. Polar residues are often found at the ends of a transmembrane helix since the ends interact with the cytoplasm.[2]

While membrane anchors normally contain nonpolar residues, not all transmembrane α-helices act as membrane anchors. Some transmembrane helices have a more complex functional role than serving as a simple tether. Some helices aggregate within a membrane to create a channel. The channel allows the passage of ions across an otherwise impermeable membrane. Many proteins form ion channels, and most use helices to form the breach in the cell membrane.

The ATP-gated ion channels are a family of channel proteins. ATP-gated ion channels form from three separate proteins, each of which contains two transmembrane α-helices. The collective total of six helices form a pore through the membrane and allow the passage of ions. ATP-gated ion channels therefore involve quaternary structure based on the relative orientation of the three protein subunits.[3]

Figure 4.6 shows the amino acid residues found in one of the two transmembrane helices for ATP-gated P2X$_4$ ion channel. In contrast to the membrane anchor in Figure 4.5, the ion channel helix contains far more polar residues and two charged (+) residues. The polar and charged residues are fairly well distributed throughout the helix. The polar residues interact with polar ions as they cross the membrane. The polar residues tend to be oriented on one side of the helix so that the polar R-groups can line the channel

Position	2 1 3	2 1 4	2 1 5	2 1 6	2 1 7	2 1 8	2 1 9	2 2 0	2 2 1	2 2 2	2 2 3	2 2 4	2 2 5	2 2 6	2 2 7	2 2 8	2 2 9	2 3 0	2 3 1	2 3 2	2 3 3	2 3 4	2 3 5	2 3 6
Identity	L	G	V	F	F	S	L	L	F	L	A	A	L	I	L	T	L	F	L	G	F	L	H	Y
Property	N		N	N	N	P	N	N	N	N	N	N	N	N	N	P	N	N	N			N	P	P

FIGURE 4.5 Residues in a transmembrane α-helix of human estrone sulfatase[2]

Position	3 2	3 3	3 4	3 5	3 6	3 7	3 8	3 9	4 0	4 1	4 2	4 3	4 4	4 5	4 6	4 7	4 8	4 9	5 0	5 1	5 2	5 3	5 4	5 5	5 6	5 7
Identity	G	T	L	N	R	F	T	Q	A	L	V	I	A	Y	V	I	G	Y	V	F	V	Y	N	K	G	Y
Property		P	N	P	+	N	P	P	N	N	N	N	N	P	N	N		P	N	N	N	P	P	+		P

FIGURE 4.6 Residues in a transmembrane α-helix of ATP-gated P2X$_4$ ion channel[3]

created by the cooperative alignment of all the helices. Similarly, the parts of the α-helices that do not directly face the channel interact with the nonpolar membrane (**Figure 4.7**). Residues on the nonchannel faces are correspondingly predominantly nonpolar.[3]

Two residues, Arg[36] and Lys[55], in one of the two transmembrane helices of ATP-gated P2X$_4$ ion channel are positively charged. Positively charged amino acids are commonly found toward the end of transmembrane helices. The positive R-groups interact favorably with the negative phosphate surface of the phospholipid bilayer.

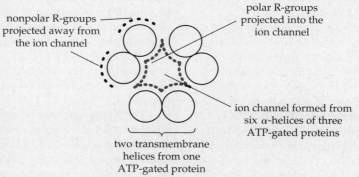

FIGURE 4.7 Depiction of α-helix orientation in ATP-gated ion channels

All enzymes have an *active site*, generally a cavity or pocket in the tertiary structure of the enzyme where chemical reactions can occur. The starting material for an enzyme is called the *substrate*. For a reaction to occur, the substrate must bind in the active site. The binding is based on appropriate intermolecular forces between functional groups on the substrate and the amino acids in the peptide backbone of the enzyme. Because the enzyme is folded, amino acids that are distant according to the primary structure can be close to one another in the tertiary structure. For example, the active site of chymotrypsin, a protease, contains many key residues, including His[57], Ser[195], Gly[193], and Asp[102]. Even though His[57] and Ser[195] are separated by over 100 residues in the backbone, folding of the chain brings them into close proximity within the active site.[4]

There are many methods for approximating the folding of the peptide chain in an enzyme. Computer modeling can be helpful, but x-ray crystallography is the most important. In x-ray crystallography x-rays are scattered as they strike atoms in a crystallized enzyme. The pattern of the scattering can be correlated to the structure of the enzyme. One limitation to x-ray crystallography is that the enzyme must first be crystallized. Amorphous enzyme powders, the typical form of an enzyme, do not scatter light in a sufficiently regular pattern to provide structural information. Another limitation is that the structure corresponds to the solid state of the enzyme. Enzymes are active in aqueous solution, and their aqueous conformations may be different due to interactions with water molecules. Despite these issues, x-ray crystallographic information on an enzyme and its active site is beneficial for developing enzyme inhibitors that block the active site. Sometimes enzymes can be cocrystallized with an inhibitor already bound in the active site. The bound inhibitor is then part of the solved x-ray structure. The drug discovery team may then be able to determine exactly what intermolecular forces are necessary for binding and even how further changes to the inhibitor may alter potency. X-ray crystallography and the determination of protein structure will be covered more thoroughly in Chapter 9. Discussion of types of intermolecular forces may be found in Chapter 10.

Environmental factors can affect the folding of an enzyme. Changes in pH influence the degree of protonation and deprotonation of acidic and basic residues (Table 4.1, entries 16–20) in the peptide chain. As the charge state of an R-group changes with protonation, attractive forces between the residues also change. The overall conformation of the enzyme as well as its activity will be affected, likely in a negative way. Similarly, increases in temperature can give the enzyme enough energy to explore new conformations with decreased catalytic activity. Temperatures that are too high irreversibly destroy the protein in a process called denaturation. Therefore, enzymes generally function well only within tightly defined temperature and pH windows.

Some enzymes require additional chemical species, called *cofactors* or *coenzymes*, to be fully active. Cofactors include metal ions such as Fe^{+2}, Zn^{+2}, and Cu^{+2}. Coenzymes are organic molecules that allow transfer of functional groups or reduction/oxidation reactions. Examples include thiamine pyrophosphate (**4.6**) and pyridoxine 5′-phosphate (**4.7**) (**Figure 4.8**).

Types

Many categorization systems exist for enzymes, and the most commonly encountered method is international Enzyme Commission (EC) numbers (**Table 4.2**). The first number defines the main type of reaction and ranges from one to six. The later numbers further

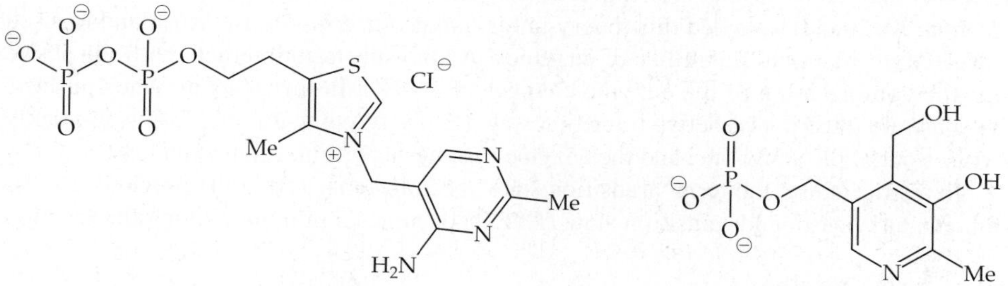

thiamine pyrophosphate
4.6

pyridoxine 5′-phosphate
4.7

FIGURE 4.8 Two coenzymes: thiamine pyrophosphate (**4.6**) and pyridoxine 5′-phosphate (**4.7**)

TABLE 4.2 EC classification of enzymes[4]

First EC Number	Name	Reaction Type
1	Oxidoreductases	Oxidation-reduction
2	Transferases	Functional group transfer
3	Hydrolyases	Hydrolysis of acid and ketone derivatives
4	Lyases	Addition-elimination on C=C, C=N, and C=O bonds
5	Isomerases	Isomerization of alkenes and stereocenters
6	Ligases	Formation-cleavage of C–O, C–S, C–N, and C–C bonds with triphosphate cleavage

describe more specific functions of a given enzyme. For example, porcine pancreatic lipase, a hydrolytic enzyme isolated from pig pancreas, is classified as EC 3.1.1.3.

4.2 Mode of Action

Enzymes catalyze the conversion of a substrate into a product. This conversion involves breaking and forming new chemical bonds. In order to accomplish the conversion, both the enzyme and substrate interact and adopt high-energy, reactive conformations. Conformational flexibility allows enzymes to both perform their catalytic role and be regulated by the body.

Theory

A substrate binds an enzyme at the active site. Substrate-enzyme binding is based on weak intermolecular attractions: contact forces, dipole forces, and hydrogen bonding. Steric effects also play an important role because the substrate must physically fit into the active site. Some enzymes have confined active sites, while others are open and accessible. A restricted active site can lead to high selectivity for a specific substrate. Low specificity can be advantageous for some enzymes, particularly metabolic and digestive enzymes that need to process a broad range of compounds with a variety of structures. Because enzymes are composed of chiral amino acids, enzymes interact differently with stereoisomers, whether diastereomers or enantiomers.

Early enzymatic theory emphasized the importance of high complementarity between an enzyme's active site and the substrate. A closer match was thought to be better. This idea was formally described in Fischer's "lock and key" model. The role of an enzyme (E), however, is not simply to bind the substrate (S) and form an enzyme-substrate complex (ES) but instead to catalyze the conversion of a substrate to a product (P) (**Scheme 4.2**). Haldane, and later Pauling, stated that an enzyme binds the transition state ($TS^{\ddagger}$) of the reaction. Koshland expanded this theory in his *induced fit hypothesis*.[5] Koshland focused on the conformational flexibility of enzymes. As the substrate interacts with the active site, the conformation of the enzyme changes (E $\rightarrow$ E$'$). In turn, the enzyme "pushes" the substrate toward its reactive transition state (E$'$TS‡). As the product forms, it quickly diffuses out of the active site, and the enzyme assumes its original conformation.

In a strict sense, the term "transition state" becomes misused in the previous discussion. An enzyme-bound transition state (E$'$TS‡) occupies a minimum along the reaction

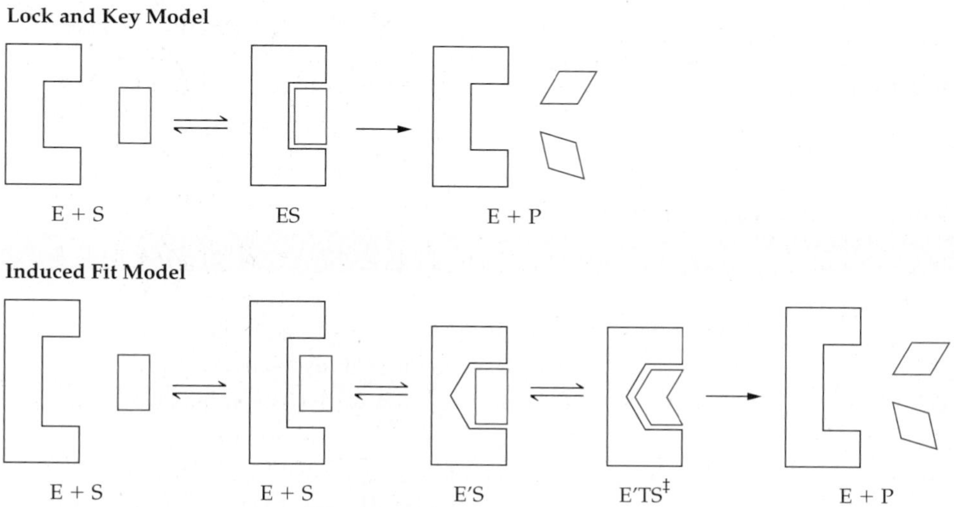

Lock and Key Model

$$E + S \qquad\qquad ES \qquad\qquad E + P$$

Induced Fit Model

$$E + S \qquad E + S \qquad E'S \qquad E'TS^{\ddagger} \qquad E + P$$

SCHEME 4.2 Theories on enzymatic activity

energy coordinate and is therefore an intermediate. The true transition state lies at an energy maximum just before or after $E'TS^{\ddagger}$. Instead of being bogged down by limitations in terminology, it is perhaps better to focus on the big picture: enzymes stabilize key points along the reaction coordinate and therefore reduce activation energies in the process. Reducing the activation energy of the rate-determining step accelerates the overall reaction rate.

Regulation

Because enzymes are involved in so many biological processes, understanding enzyme regulation is directly relevant to medicinal chemistry and drug discovery. Organisms use several different methods for controlling enzyme activity. These same methods are available to medicinal chemists.

Allosteric Control

Some enzymes are regulated to control their activity. The most common form of regulation is *allosteric control*. Allosteric regulation is common in, but not limited to, multienzyme pathways that produce a specific molecule for a cell. To prevent an unwanted build-up of the final product, the product of the pathway often inhibits an enzyme that is upstream in the overall cascade. This process is called *feedback inhibition*. A representation is shown in **Scheme 4.3**. (In this scheme, all enzymes bear the suffix "ase," which is commonly used among enzymes.) The product typically inhibits an enzyme not at the active site but at another position on the enzyme called an *allosteric site*. By binding the allosteric site, the product inhibitor changes the conformation of the enzyme so that it is less able to bind its substrate (**Scheme 4.4**). Binding at the allosteric site is reversible. As the final product of the pathway is depleted in the cell, inhibition by the product will decrease and the synthetic pathway will again become fully active. From a medicinal chemistry perspective, drugs are typically designed to inhibit enzymes by binding the active site. An allosteric site, however, does offer another site for influencing an enzyme's efficiency.[4]

Allosteric binding can manifest itself in many ways, and feedback inhibition is just one example. Entire proteins can bind an allosteric site, and sometimes allosteric binding enhances an enzyme's activity.

allosteric inhibition

$$A \rightleftharpoons \overset{\text{AB-ase}}{} B \rightleftharpoons \overset{\text{BC-ase}}{} C \rightleftharpoons \overset{\text{CD-ase}}{} D$$

SCHEME 4.3 Allosteric inhibition in a multienzyme pathway

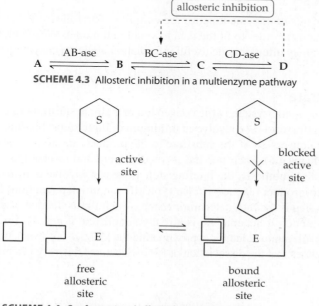

SCHEME 4.4 Conformational effect of allosteric binding

$$\text{chymotrypsinogen} \xrightarrow[\substack{\text{cleave} \\ \text{Arg}^{15}\text{-Ile}^{16}}]{\text{trypsin}} \underset{\substack{\text{π-chymotrypsin} \\ \text{(245 residues)}}}{} \xrightarrow[\substack{\text{remove} \\ \text{Ser}^{14}\text{-Arg}^{15} \\ \text{Thr}^{147}\text{-Asn}^{148}}]{\text{π-chymotrypsin}} \underset{\substack{\text{α-chymotrypsin} \\ \text{(241 residues)}}}{}$$

chymotrypsinogen (245 residues) → π-chymotrypsin (245 residues) → α-chymotrypsin (241 residues)

SCHEME 4.5 Formation of α-chymotrypsin[4]

Covalent Modification

Enzymes are sometimes inactive as simple polypeptides. Addition of another group forms the fully active enzyme. Phosphorylation or glycosylation of an OH or NH in an R-group is typical. Phosphorylations are especially common, and this process is itself mediated by another enzyme, called a *kinase*. Kinases are an important family of enzymes, and kinase inhibition can be a powerful means of impacting cellular function.

Proenzymes

A few enzymes are synthesized in an inactive form by an organism. The inactive enzyme, called a *proenzyme*, must have some sections of the peptide removed to become active. Chymotrypsin is such an enzyme. Produced in the pancreas, chymotrypsin cleaves amide linkages in proteins at aromatic residues. Producing an active digestive enzyme like chymotrypsin in a cell would likely lead to damage or destruction of the pancreatic cell. Chymotrypsin is therefore produced in an inactive form, chymotrypsinogen, a 245-residue polypeptide proenzyme (**Scheme 4.5**). Chymotrypsinogen is secreted into the stomach where trypsin, another digestive enzyme, cleaves chymotrypsinogen between Arg^{15} and Ile^{16} to form π-chymotrypsin. π-Chymotrypsin then autocatalytically loses two dipeptides, Ser^{14}-Arg^{15} and Thr^{147}-Asn^{148}, to afford the fully active enzyme α-chymotrypsin. α-Chymotrypsin contains breaks in its overall peptide backbone, and several disulfide bonds hold together the separate sections of the enzyme.[4]

Enzyme regulation through proenzymes is not a standard target for drug discovery. Direct inhibition of an enzyme through its active site, or sometimes an allosteric site, is much more common.

4.3 Kinetics

Rigorous treatments of enzyme kinetics are complex. The following discussions require several assumptions in order to fit the data to simplified models. While these models have their limitations, they often provide useful information about an enzyme's action on a specific substrate.

Single Substrate

The simplest enzymatic system is the conversion of a single substrate to a single product. Even this straightforward case involves a minimum of three steps: binding of the substrate by the enzyme, conversion of the substrate to the product, and release of the product by the enzyme (**Scheme 4.6**). Each step has its own forward and reverse rate constant. Based on the induced fit hypothesis, the binding step alone can involve multiple distinct steps. The substrate-to-product reaction is also typically a multistep reaction. Kinetically, the most important step is the rate-determining step, which limits the rate of conversion.

While the model in Scheme 4.6 is already simplified, it may be further reduced to **Scheme 4.7** and still retain value for approximating the kinetics of the reaction. This streamlined model assumes that the dissociation of EP in Scheme 4.6 is very fast relative to the rate

$$\text{E} + \text{S} \underset{k_{-1}}{\overset{k_1}{\rightleftharpoons}} \text{ES} \underset{k_{-2}}{\overset{k_2}{\rightleftharpoons}} \text{EP} \underset{k_{-3}}{\overset{k_3}{\rightleftharpoons}} \text{E} + \text{P}$$

SCHEME 4.6 A simplified catalyzed reaction model

$$E + S \underset{k_{-1}}{\overset{k_1}{\rightleftarrows}} ES \xrightarrow{k_2} E + P$$

SCHEME 4.7 A fully simplified catalyzed reaction model

of conversion (Scheme 4.6, $k_3 \gg k_2$). Scheme 4.7 also assumes that the reaction is irreversible (Scheme 4.6). While all enzymes are reversible, the products are often used as quickly as they form. With no available product, which is the substrate for the reverse reaction, an enzyme cannot perform the reverse reaction. In practice, the model in Scheme 4.7 works well when the ES complex forms and dissociates quickly relative to the rate of substrate conversion (k_1 and $k_{-1} \gg k_2$). In other words, forming the product is the rate-determining step.[4]

Michaelis–Menten Equation

Using the model in Scheme 4.7, Michaelis and Menten were able to generate an equation that related the rate of an enzymatic reaction to the concentration of the substrate. The second step is rate determining, so the rate of the reaction (V) as determined by the formation of the product ($d[P]/dt$) is given by Equation 4.1.[4]

$$V = \frac{d[P]}{dt} = k_2[ES] \tag{4.1}$$

Experimentally, values for V are determined at the start of a reaction and correspond to initial rates, V_0 or V at time 0. Testing in this manner ensures that $[P] \approx 0$, and that the reverse reaction may be ignored. In many treatments of enzyme kinetics, the initial rates are explicitly labeled V_0 instead of V.

Unfortunately, [ES] is difficult to measure during a reaction, so the rate equation needs to be written without this term. Michaelis and Menten recognized that [E], unbound enzyme, is equivalent to the total enzyme concentration, $[E_t]$, less [ES] (Equation 4.2). Therefore, the rate of [ES] formation is described in Equation 4.3. The rate of ES consumption can also be described in an equation (Equation 4.4).

$$[E] = [E_t] - [ES] \tag{4.2}$$

$$\text{rate of ES formation} = k_1[E][S] = k_1([E_t] - [ES])[S] \tag{4.3}$$

$$\text{rate of ES consumption} = k_{-1}[ES] + k_2[ES] \tag{4.4}$$

Because the second step of the reaction is relatively slow, the formation of ES quickly establishes a concentration that is maintained early in the reaction by a high [S]. Therefore, early in the reaction, [ES] is essentially constant ($d[ES]/dt = 0$) (Equation 4.5).

$$\frac{d[ES]}{dt} = 0 = \text{rate of formation} - \text{rate of ES consumption} \tag{4.5}$$

This is called a *steady-state approximation* and is expressed mathematically by setting the rate of ES formation equal to the rate of ES consumption (Equations 4.6 and 4.7). After a number of rearrangements, Equation 4.7 can be solved for [ES] (Equation 4.8). The collection of three rate constants is replaced with a single term, K_m, the *Michaelis constant*.

$$\text{rate of ES formation} = \text{rate of ES consumption} \tag{4.6}$$

$$k_1([E_t] - [ES])[S] = k_{-1}[ES] + k_2[ES] \tag{4.7}$$

$$[ES] = \frac{[E_t][S]}{\dfrac{k_2 + k_{-1}}{k_1} + [S]} = \frac{[E_t][S]}{K_m + [S]} \tag{4.8}$$

Substitution of Equation 4.8 into Equation 4.1 gives a new equation (Equation 4.9). According to Equation 4.1, the rate of the reaction increases with rising [ES]. In theory, [ES] reaches a maximum when all enzyme molecules are bound with substrate, meaning [ES] equals [E$_t$]. If [ES] equals [E$_t$], V will achieve V_{max} (Equation 4.10). Combining Equations 4.9 and 4.10 affords Equation 4.11, the *Michaelis–Menten equation*.

$$V = \frac{k_2[E_t][S]}{K_m + [S]} \tag{4.9}$$

$$V_{max} = k_2[E_t] \tag{4.10}$$

$$V = \frac{V_{max}[S]}{K_m + [S]} \tag{4.11}$$

The Michaelis–Menten equation is a reciprocal function of the relationship between a reaction rate and [S] (**Figure 4.9**). The corresponding graph is a partial hyperbola. Graphs of this type are generated by measuring initial rates of reaction with a range of [S]. The data points are fit to a hyperbolic function like Equation 4.11 to determine the key parameters, V_{max} and K_m. V_{max} is the theoretical maximum rate of conversion that a catalyzed reaction approaches as [S] becomes very large. At very high [S], the enzyme is said to be *saturated* with substrate. Every active site contains a substrate molecule, that is, $[E_t] = [ES]$. K_m is equivalent to the substrate concentration required for the reaction to reach half of the maximum rate ($\frac{1}{2}V_{max}$). Conceptually, if reaching V_{max} requires substrate binding to all the enzyme molecules, then reaching $\frac{1}{2}V_{max}$ should require just half of the enzyme molecules to be bound to substrate. The units on K_m are the same as the substrate concentration. V_{max} has units of concentration per unit time.

Mathematically, K_m is determined by three separate rate constants. Under the model in Scheme 4.7, $k_{-1} \gg k_2$, and K_m may be approximated as k_{-1}/k_1 (Equation 4.12). This ratio of rate constants is the same as K_D, the dissociation equilibrium constant of the enzyme-substrate complex. Therefore, K_m relates to the affinity of an enzyme for its substrate. A small K_m indicates a low degree of ES dissociation, implying a high affinity between the enzyme and substrate.

$$K_m = \frac{k_2 + k_{-1}}{k_1} \cong \frac{k_{-1}}{k_1} = K_D = \frac{[E][S]}{[ES]} \tag{4.12}$$

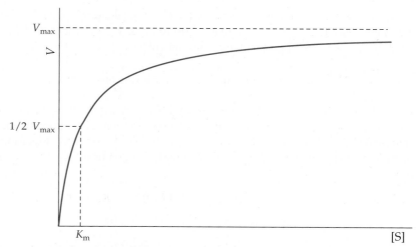

FIGURE 4.9 *V* versus [S] for a catalyzed reaction following Michaelis–Menten kinetics

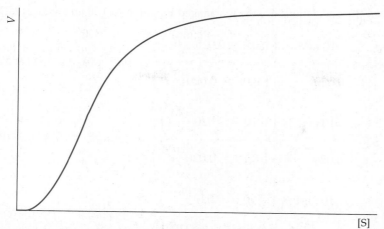

FIGURE 4.10 Sigmoidal relationship between V and [S]

V_{max} has limited utility because it is related to [E$_t$], an experimentally controlled variable. In contrast, k_2 is a property of the enzyme. For the reaction model in Scheme 4.7, k_2 is the rate-determining step in the catalytic process and is frequently labeled k_{cat}. Like k_2, k_{cat} is a first-order rate constant with units of inverse time. k_{cat} is often called the *turnover frequency* (TOF), the maximum number of substrate molecules an enzyme can convert to product per unit time. For an enzyme to achieve a maximum k_{cat}, reaction conditions, namely temperature and pH, must be optimal.

Although the Michaelis–Menten equation places emphasis on V_{max}, published kinetic studies of enzymes generally do not disclose the V_{max} for specific enzymatic reactions. Instead papers report k_{cat}. Deep in the experimental section, a paper will mention the total concentration of the enzyme, [E$_t$], used in the kinetic study. Since $k_2 = k_{cat}$, with both k_{cat} and [E$_t$], one may use Equation 4.10 to estimate V_{max} for the reaction of interest.

Not all enzymes afford a hyperbolic relationship between their rate and substrate concentration. Sigmoidal curves are common and indicate either cooperativity in a multienzyme complex or involvement of an allosteric site on the enzyme (**Figure 4.10**). These types of curves can be fit by introducing exponents on [S] and K_m in Equation 4.11.

Sample Calculation Working with [S] and V

PROBLEM A specific enzymatic reaction acts on a substrate with a K_m of 10 mM and V_{max} of 20 μM/s. How much must [S] increase for V to rise from 5% of V_{max} to 10% of V_{max}? How much must [S] increase for V to rise from 90% of V_{max} to 95% of V_{max}?

SOLUTION To answer this question, we need to calculate four different [S] values. Equation 4.11 lets us do these calculations if we do some rearranging.

$$V = \frac{V_{max}[\text{S}]}{K_m + [\text{S}]} \tag{4.11}$$

For each value of [S], we know K_m and V_{max} (both are constants). We do not directly know V, but we can express V as a percentage of V_{max}. Therefore, we have everything we need to determine all four [S] values. One word of warning: the concentration units on K_m and V_{max} do not match. Either K_m needs to be converted to μM, or V_{max} must be converted to mM/s. We will use mM for concentration ($V_{max} = 0.020$ mM/s).

$$[\text{S}] = \frac{K_m V}{V_{max} - V}$$

All four [S] calculations use the rearranged Equation 4.11 in an identical fashion.

$$[S] = \frac{10 \text{ mM} \times \left(0.05 \times 0.020\frac{\text{mM}}{\text{s}}\right)}{0.020\frac{\text{mM}}{\text{s}} - \left(0.05 \times 0.020\frac{\text{mM}}{\text{s}}\right)} = 0.53 \text{ mM} \qquad [S] \text{ for } 5\% \text{ of } V_{\text{max}}$$

$$[S] = \frac{10 \text{ mM} \times \left(0.10 \times 0.020\frac{\text{mM}}{\text{s}}\right)}{0.020\frac{\text{mM}}{\text{s}} - \left(0.10 \times 0.020\frac{\text{mM}}{\text{s}}\right)} = 1.1 \text{ mM} \qquad [S] \text{ for } 10\% \text{ of } V_{\text{max}}$$

$$[S] = \frac{10 \text{ mM} \times \left(0.90 \times 0.020\frac{\text{mM}}{\text{s}}\right)}{0.020\frac{\text{mM}}{\text{s}} - \left(0.90 \times 0.020\frac{\text{mM}}{\text{s}}\right)} = 90 \text{ mM} \qquad [S] \text{ for } 90\% \text{ of } V_{\text{max}}$$

$$[S] = \frac{10 \text{ mM} \times \left(0.95 \times 0.020\frac{\text{mM}}{\text{s}}\right)}{0.020\frac{\text{mM}}{\text{s}} - \left(0.95 \times 0.020\frac{\text{mM}}{\text{s}}\right)} = 190 \text{ mM} \qquad [S] \text{ for } 95\% \text{ of } V_{\text{max}}$$

The answer is that [S] must increase from 0.53 to 1.1 mM for V to rise from 5 to 10% of V_{max}. To increase V from 90 to 95% of V_{max}, one must raise [S] from 90 to 190 mM. One lesson from these calculations is that as V approaches V_{max}, very large increases in [S] are required for even a small bump in V.

Lineweaver–Burk Equation

Before adequate computer hardware and software were widely available, fitting experimental kinetic data to a curve to determine K_{m} and V_{max} was a significant challenge. Lineweaver and Burk rearranged the Michaelis–Menten equation to form a new linear relationship, the Lineweaver–Burk equation (Equation 4.13).[6]

$$\frac{1}{V} = \frac{K_{\text{m}}}{V_{\text{max}}} \times \frac{1}{[S]} + \frac{1}{V_{\text{max}}} \tag{4.13}$$

Plotting the double reciprocal relationship of $1/V$ against $1/[S]$ gives a line, the equation of which may be determined through a relatively simple linear regression. The slope of the line is $K_{\text{m}}/V_{\text{max}}$. The line may be extrapolated to the y-intercept at a value of $1/V_{\text{max}}$. Furthermore, the theoretical x-intercept occurs at $-1/K_{\text{m}}$ (**Figure 4.11**).

In addition to being easier to fit than the hyperbolic Michaelis–Menten equation, Lineweaver–Burk graphs clearly show differences between types of enzyme inhibitors. This will be discussed in Section 4.5. However, Lineweaver–Burk equations have their own distinct issues. Nonlinear data, possibly indicating cooperative multiunit enzymes or allosteric effects, often seem nearly linear when graphed according to a Lineweaver–Burk equation. Said another way, the Lineweaver–Burk equation forces nonlinear data into a linear relationship. Variations of the Lineweaver–Burk equation that are not double reciprocal relationships include the Eadie–Hofstee equation[7] (V vs. $V/[S]$) (Equation 4.14) and the Hanes–Woolf equation[8] ($[S]/V$ vs. $[S]$) (Equation 4.15). Both are

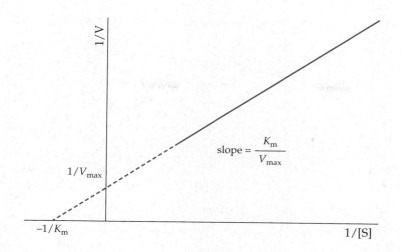

FIGURE 4.11 Lineweaver–Burk graph

considered to be better than the Lineweaver–Burk equation for visually identifying non-linear data.

$$V = -K_m \frac{V}{[S]} + V_{max} \tag{4.14}$$

$$\frac{[S]}{V} = \frac{1}{V_{max}}[S] + \frac{K_m}{V_{max}} \tag{4.15}$$

Sample Calculation   **Understanding Data in a Lineweaver–Burk Plot**

PROBLEM   The following table presents V-[S] data. The units for V are $\mu M/s$, and the units for [S] are mM. Plot the data points first as V versus [S] and then as $1/V$ versus $1/[S]$. Include a trendline with each graph. Where do the data points cluster in each graph?

Entry	[S] (mM)	V ($\mu M/s$)
1	1.1	2.0
2	2.5	4.0
3	4.3	6.0
4	6.7	8.0
5	10	10
6	15	12
7	23	14
8	40	16
9	90	18

SOLUTION   The two graphs, one hyperbolic and the other linear, are shown here. Note that Excel does not have an option for fitting hyperbolic functions. Selecting a logarithmic function gives the best-looking line. In both graphs, the data clusters on one side of the graph: the left. The clustering of data points can undermine the statistical validity of enzyme kinetics data. Ideally, the data points should be spread as uniformly as possible over the range of x- and y-axes.

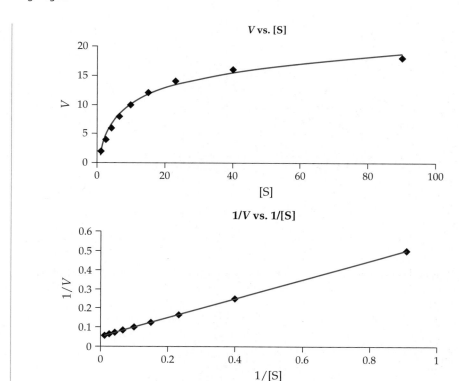

Multiple Substrates

Enzymes often require multiple substrates to complete their catalytic cycle. This may involve combining two compounds into one molecule or transferring atoms or electrons from one substrate to another. The substrates may both bind to an enzyme and react collectively, or each substrate might bind, react, and release sequentially. With two substrates, if both bind to the enzyme, a ternary complex (ES^1S^2) will form (**Scheme 4.8**). The order of substrate addition may be important (ordered) or not (random order). Cases in which the two substrates react sequentially follow a *double-displacement*, or *ping-pong*, mechanism (**Scheme 4.9**). Enzymes requiring more than two substrates have more complicated complexation pathways.

SCHEME 4.8 Formation of a ternary complex in a two substrate process

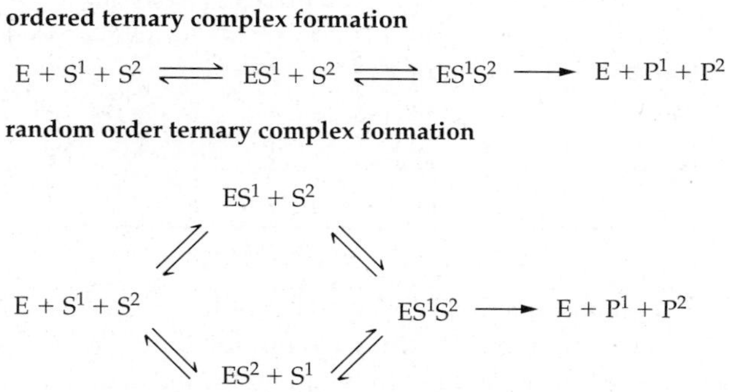

ordered ternary complex formation

$$E + S^1 + S^2 \rightleftharpoons ES^1 + S^2 \rightleftharpoons ES^1S^2 \longrightarrow E + P^1 + P^2$$

random order ternary complex formation

$$\begin{array}{ccc}
 & ES^1 + S^2 & \\
 \nearrow & & \searrow \\
E + S^1 + S^2 & & ES^1S^2 \longrightarrow E + P^1 + P^2 \\
 \searrow & & \nearrow \\
 & ES^2 + S^1 &
\end{array}$$

SCHEME 4.9 Double-displacement or ping-pong mechanism

$$E + S^1 \rightleftharpoons ES^1 \rightleftharpoons E'P^1 \overset{P^1}{\underset{P^1}{\rightleftharpoons}} E' \overset{S^2}{\underset{S^2}{\rightleftharpoons}} E'S^2 \rightleftharpoons EP^2 \rightleftharpoons E + P^2$$

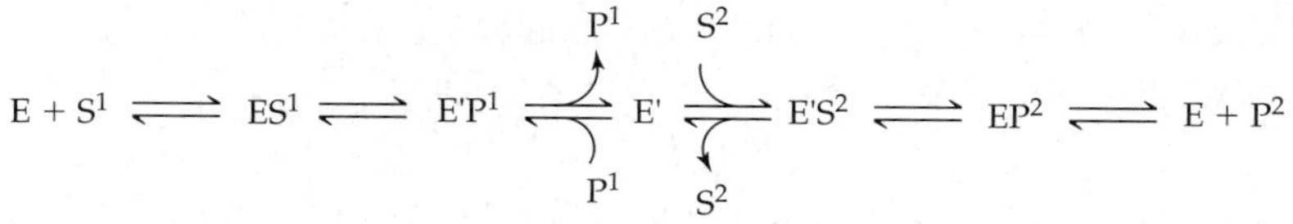

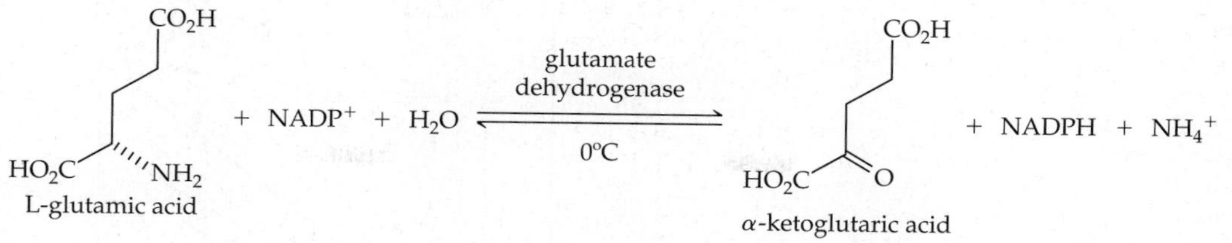

SCHEME 4.10 Substrates and K_m values for glutamate dehydrogenase[9]

| K_m (mM) | 0.68 | 0.11 | | 0.08 | 0.017 | 20 |

Multiple substrate mechanisms follow Michaelis–Menten kinetics. Experiments are performed with constant concentrations of the enzyme and one substrate with variation of the second substrate concentration ($[S^2]$). (Note that the second substrate concentration $[S^2]$ is not the same as a deceptively similar term, the square of the substrate concentration $[S]^2$.) Plotting V against $[S^2]$ gives a hyperbolic curve and allows determination of K_m for the second substrate. The K_m values for all substrates may be found in a similar fashion.

Enzymes that catalyze redox reactions often require a coenzyme such as NAD^+ or $FADH_2$ in addition to a substrate. These are all multiple substrate enzymes. Each substrate and coenzyme will have its own K_m value. The substrates for glutamate dehydrogenase, an enzyme with three substrates in both forward and reverse directions, are shown in **Scheme 4.10** with their K_m values.[9]

4.4 Inhibitors

Inhibitors can act in one of two ways. First, an inhibitor may make act as a *reversible inhibitor*. Reversible inhibitors either decrease k_{cat} and consequently V_{max} (Equation 4.10, $k_2 = k_{cat}$) or decrease an enzyme's affinity for its substrate and therefore raise K_m. A reversible inhibitor can also do both. Second, an inhibitor may be *irreversible*. Irreversible inhibitors decrease V_{max} by destroying or damaging the enzyme, essentially decreasing $[E_t]$ (Equation 4.10). Both reversible and irreversible inhibitors decrease available enzyme function.

Reversible

Reversible inhibitors bind an enzyme through weak, intermolecular forces and establish an equilibrium of being bound or unbound to the enzyme. A *competitive* inhibitor binds at the active site and prevents the substrate from binding. A *noncompetitive* inhibitor binds an allosteric site on the enzyme and prevents conversion of the substrate to product. *Uncompetitive* inhibitors bind the enzyme-substrate complex and make it inactive. All three types of inhibitors show characteristic, distinctive features in a Lineweaver–Burk plot.

Competitive

Competitive, reversible inhibitors bind the active site of the enzyme and therefore block substrate-enzyme interactions. The inhibitor (I) and substrate may not bind simultaneously (**Scheme 4.11**). In something of a chemical love triangle, the enzyme's binding ability is split between two molecules, the substrate and inhibitor. Therefore, the effective affinity of the enzyme for the substrate alone drops. K_m of the substrate will rise accordingly in a system with a competitive inhibitor. (Remember that higher values of K_m correspond to lower affinity.) The K_m of a substrate in the presence of an inhibitor is sometimes written differently, such as K_m^{obs}, to clearly distinguish it from the true, uninhibited K_m value. Because binding of the inhibitor is reversible, addition of enough substrate allows the enzyme to achieve the same V_{max} as the un-inhibited enzyme. These ideas may be summarized by a series of Lineweaver–Burk

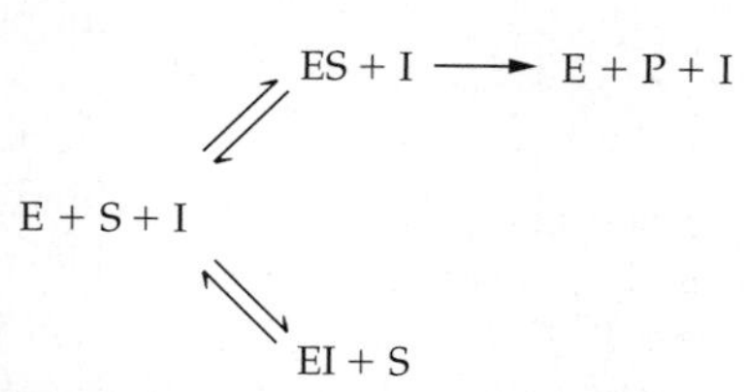

$$ES + I \longrightarrow E + P + I$$

$$E + S + I$$

$$EI + S$$

SCHEME 4.11 Competitive, reversible inhibition

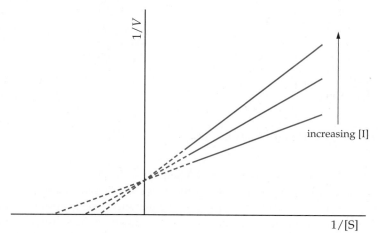

FIGURE 4.12 Effect of a competitive, reversible inhibitor on enzyme kinetics

SCHEME 4.12 Coformycin (**4.11**), a transition state inhibitor of AMP deaminase

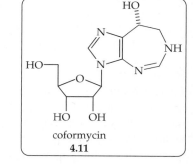

plots with increasing inhibitor concentration (**Figure 4.12**). All the lines have the same y-intercept, indicating identical V_{max} values. Changes in the x-intercepts demonstrate different K_m values. The slope of a line in a Lineweaver–Burk plot is K_m/V_{max}. Because K_m rises with [I] and V_{max} remains constant, lines with more inhibitor have a steeper slope.[4]

Competitive, reversible inhibitors are the most common type of inhibitor developed for pharmaceutical use. If the substrate of an enzyme is known, then a competitive inhibitor will likely somewhat resemble the substrate. The search for an inhibitor will typically start with molecules of similar structure to the substrate. Because enzymes theoretically bind most strongly to a transition state, competitive inhibitors are often designed to resemble a transition state or a high energy intermediate along the reaction coordinate. These types of drugs are called *transition state analogues* or *transition state inhibitors*.

Coformycin (**4.11**) is a transition state analogue that competitively inhibits the enzyme adenosine monophosphate (AMP) deaminase (**Scheme 4.12**). AMP deaminase

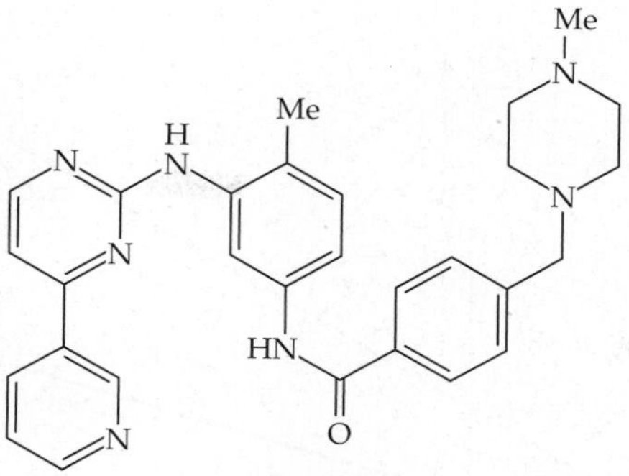

imatinib
(Gleevec)
antitumor
4.12

FIGURE 4.13 Imatinib (**4.12**), a competitive, reversible inhibitor of a tyrosine kinase

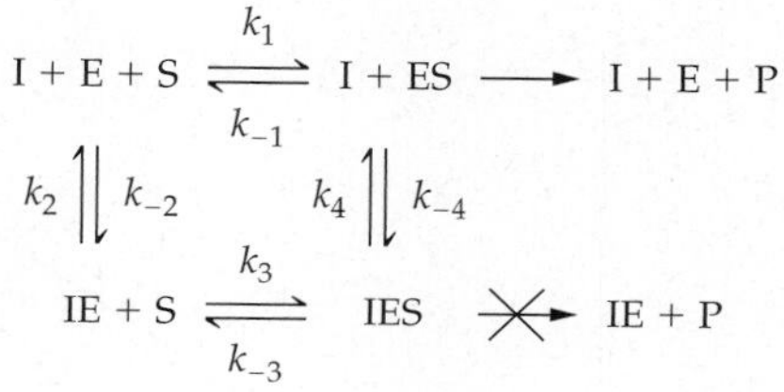

SCHEME 4.13 Pure noncompetitive, reversible inhibition

hydrolyzes AMP (**4.8**) to inosine monophosphate (**4.10**). A likely intermediate along this reaction pathway is structure **4.9**, the product of the addition of water at C6 of the purine ring. Compound **4.9** itself would be a promising inhibitor of AMP deaminase except that it is unstable and readily loses either water or ammonia to re-aromatize the six-membered ring of the purine system. Coformycin contains an extra methylene (CH_2) in the ring and is more stable. Despite the enlarged ring, coformycin still resembles the reactive intermediate closely enough to serve as an effective competitive inhibitor.

Not all competitive inhibitors are obviously based on transition states or intermediates. Imatinib (Gleevec, **4.12**) is a competitive inhibitor of an unregulated tyrosine kinase that is overly active in patients suffering from chronic myelogenous leukemia (**Figure 4.13**). Tyrosine kinases phosphorylate the hydroxyl group of tyrosine residues (see Table 4.1) with ATP. Although imatinib does not have a structure that is clearly analogous to either the 4-hydroxyphenyl group of a tyrosine residue or ATP, imatinib is an effective competitive inhibitor of the enzyme.[10] Although the exact binding of imatinib was not understood when the drug was first approved in 2001, imatinib has since been found to bind near the ATP binding site on the tyrosine kinase.[11]

Noncompetitive

Noncompetitive inhibitors bind an enzyme at an allosteric site and can bind both the free enzyme and the enzyme-substrate complex (**Scheme 4.13**). If the enzyme-substrate complex is bound by the inhibitor, the complex cannot form product. A *pure* noncompetitive inhibitor has no effect on substrate binding ($k_1 = k_3$ and $k_{-1} = k_{-3}$), and likewise, the substrate has no effect on binding of the inhibitor ($k_2 = k_4$ and $k_{-2} = k_{-4}$). Because substrate binding is unaffected by the inhibitor, K_m is constant regardless of amount of inhibitor present (**Figure 4.14**). The inhibitor does, however, decrease V_{max} of the enzyme.[4]

FIGURE 4.14 Effect of a pure noncompetitive, reversible inhibitor on enzyme kinetics

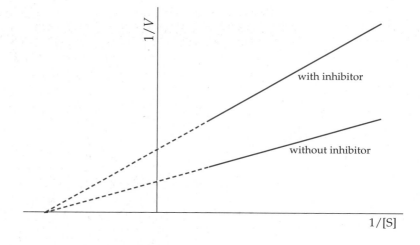

FIGURE 4.15 Effect of a mixed noncompetitive, reversible inhibitor on enzyme kinetics

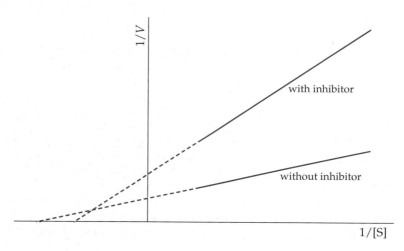

While a pure noncompetitive inhibitor is possible in theory, in practice the binding of the substrate and inhibitor are not independent. None of the rate constants in Scheme 4.13 is necessarily equal to another. If the inhibitor binding the allosteric site does affect substrate binding, the inhibitor is classified as *mixed noncompetitive*, and both V_{max} and K_m will be affected. V_{max} decreases, and K_m may either increase or decrease (**Figure 4.15**).

Rifampin (Rifadin, **4.13**) is a semisynthetic antibiotic that inhibits bacterial RNA polymerase (**Figure 4.16**).[12] Rifampin also shows other activity in vivo—pure, noncompetitive inhibition of aryl hydrocarbon hydroxylase (AHH), a metabolic enzyme. Rifampin's ability to slow metabolism is one of its undesirable side effects. Inhibition of AHH can affect the metabolism of other drugs that are present in the body. Indole (**4.14**) acts as either a pure or mixed noncompetitive inhibitor of α-chymotrypsin depending on the structure of the substrate.[13] The binding of indole to an allosteric site on α-chymotrypsin almost certainly causes changes in regions of the active site. If a substrate interacts with one of these regions in the active site, then that substrate recognizes indole as a mixed noncompetitive inhibitor. If a substrate binds without contacting affected areas in the pocket of the active site, then indole behaves as a pure noncompetitive inhibitor.

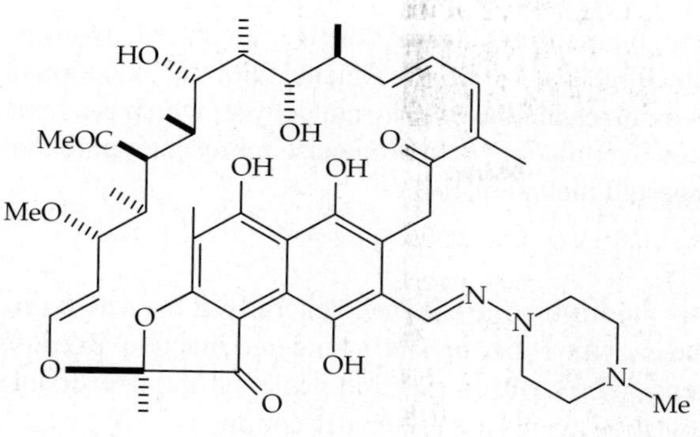

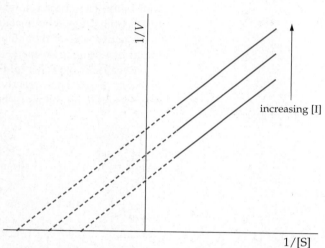

rifampin
(Rifadin)
antibiotic
4.13

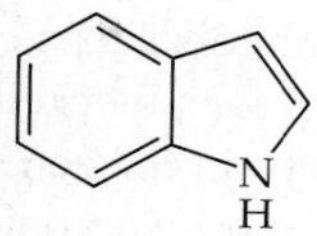

indole
4.14

FIGURE 4.16 Noncompetitive enzyme inhibitors

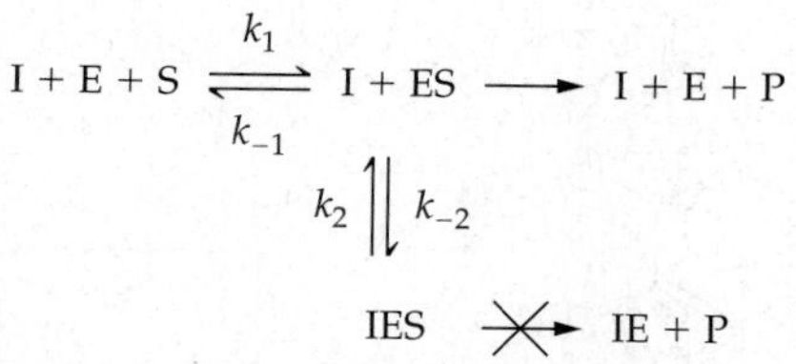

$$I + E + S \underset{k_{-1}}{\overset{k_1}{\rightleftharpoons}} I + ES \longrightarrow I + E + P$$

$$k_2 \big\Updownarrow k_{-2}$$

$$IES \xrightarrow{\quad\times\quad} IE + P$$

SCHEME 4.14 Uncompetitive, reversible inhibition

FIGURE 4.17 Effect of an uncompetitive, reversible inhibitor on enzyme kinetics

Uncompetitive

An uncompetitive inhibitor is much like a noncompetitive inhibitor except that an uncompetitive inhibitor binds only the enzyme-substrate complex (**Scheme 4.14**). The inhibitor-bound ternary complex cannot form product. Uncompetitive inhibitors cause both V_{max} and K_m to decrease by the same factor (**Figure 4.17**). Because the slope of a Lineweaver–Burk plot is K_m/V_{max}, the slope of the line of an inhibited enzyme is unchanged from the uninhibited enzyme.[4]

The effect of an uncompetitive inhibitor on the K_m of a substrate deserves some commentary. A lower K_m implies that an enzyme has been *inhibited* by *increasing* the enzyme's affinity for its substrate. This seemingly counterintuitive statement can be made clearer by examining the equilibria in Scheme 4.14. By binding the enzyme-substrate complex, *the inhibitor forces some free enzyme to bind substrate* in order to maintain equilibrium concentrations of all species in solution. The inhibitor does indeed increase the affinity of the enzyme for its substrate (i.e., decrease K_m).

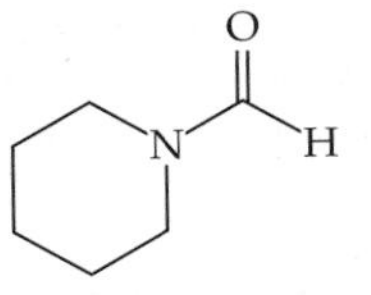

N-formylpiperidine
4.15

FIGURE 4.18 An uncompetitive inhibitor

N-Formylpiperidine (**4.15**) is an uncompetitive inhibitor of liver alcohol dehydrogenase (**Figure 4.18**).[14] Liver alcohol dehydrogenase is often associated with the oxidation of ethanol in the bloodstream, but it also oxidizes methanol to formaldehyde, which is a toxic metabolite. Safe, effective inhibitors of liver alcohol dehydrogenase represent a potential treatment for individuals who have ingested methanol.

Irreversible

Irreversible inhibitors act by covalently modifying the enzyme, generally at the active site. The active site is then blocked, and the enzyme is permanently rendered inactive. Because functional groups in the active site tend to be electron rich and nucleophilic, irreversible inhibitors tend to be electrophiles. Acylation agents are especially common.

Aspirin (**4.16**) is an irreversible inhibitor of cyclooxygenase (COX), an enzyme involved in the synthesis of prostaglandins, a family of hormones (**Scheme 4.15**). The acetyl group in aspirin is transferred to the OH of Ser[530] in the active site of COX-1. The mode of inhibition of COX-2, an isoform of COX-1, by aspirin is similar.[15]

Two other irreversible inhibitors are penicillin G (**4.18**), a β-lactam antibiotic, and orlistat (Xenical or Alli, **4.19**), an antiobesity drug (**Figure 4.19**). Penicillin G inhibits cell wall synthesis in bacteria, while orlistat inhibits the breakdown of fats in the small intestine.[16] Both drugs contain acid derivatives in a strained four-membered ring. Nucleophiles in the active sites of the inhibited enzymes attack the reactive carbonyl groups and open the strained ring in an energetically favorable, irreversible process.

Because irreversible inhibitors tend to be reactive electrophiles, they can be difficult to use as drugs. Their reactivity can translate into instability and/or toxicity. A subclass of irreversible inhibitors is the mechanism-based inhibitor. Mechanism-based inhibitors

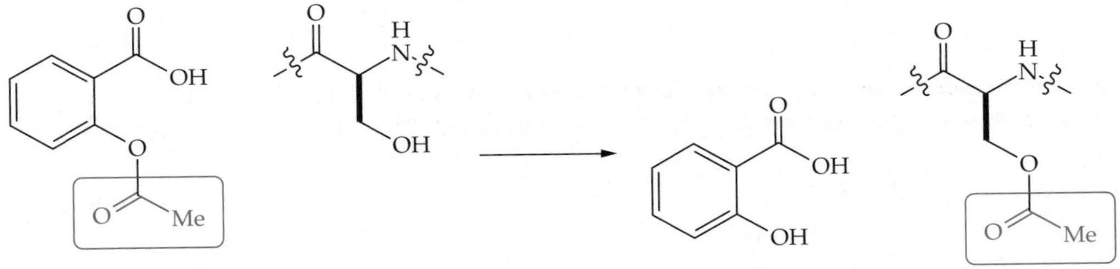

aspirin
analgesic
4.16

Ser[530]

salicylic acid
4.17

SCHEME 4.15 Irreversible inhibition of COX-1 by aspirin

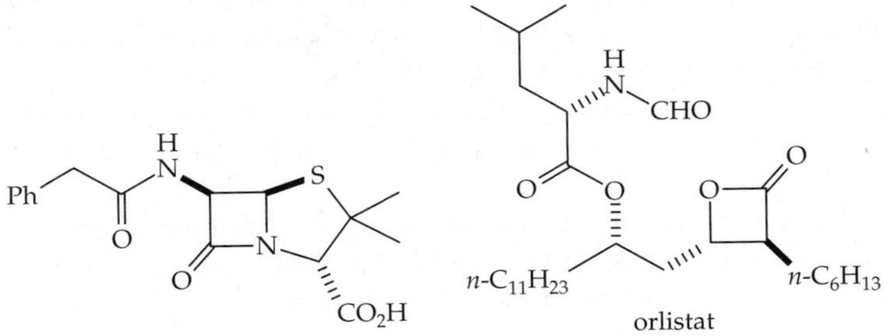

penicillin G
antibiotic
4.18

orlistat
(Xenical, Alli)
weight loss drug
4.19

FIGURE 4.19 Irreversible enzyme inhibitors

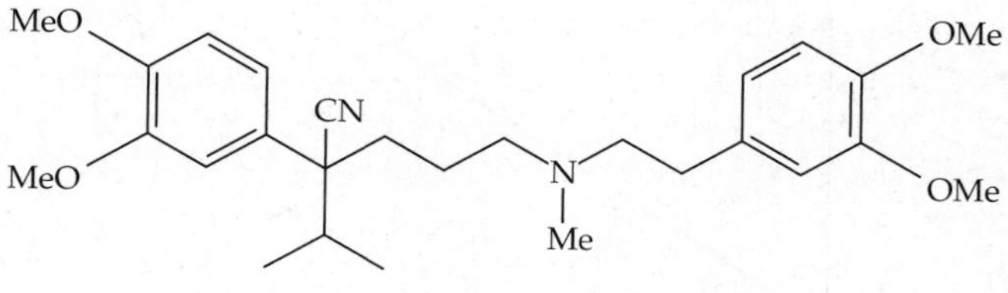

FIGURE 4.20 Verapamil (**4.20**) and a possible nitrosoalkane metabolite (**4.21**)

verapamil
(Calan, Isoptin)
antihypertensive
4.20

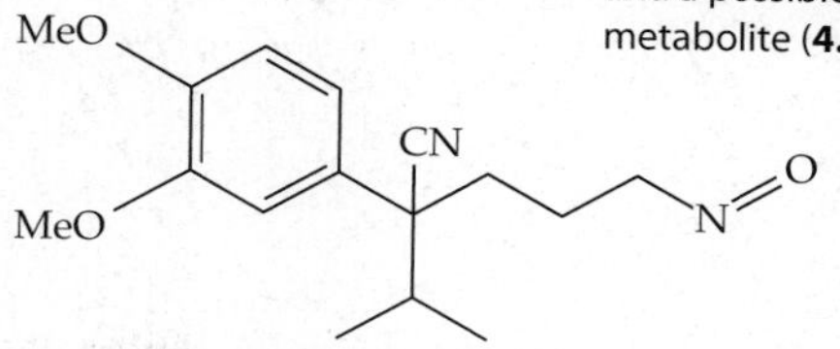

nitrosoalkane
4.21

behave as a substrate in the active site of the enzyme. As the inhibitor reacts in the active site, the inhibitor is converted to a strong electrophile and covalently binds a nucleophilic residue in the active site. The drug becomes fully active only in the active site of its targeted enzyme, so side effects are minimized.

Verapamil (Calan, **4.20**), a drug used to treat high blood pressure, is a mechanism-based inhibitor of cytochrome P-450 3A4 (CYP3A4), one of the many isoforms in the cytochrome P-450 family of metabolic enzymes (**Figure 4.20**).[17] Verapamil undergoes oxidation and dealkylation to form a nitrosoalkane such as **4.21**. Nitroso groups strongly (essentially irreversibly) bind the heme group in CYP3A4 and inhibit the enzyme. As with rifampin (**4.13**) in Figure 4.16, drugs that affect metabolic enzymes can have many undesired consequences.

4.5 Pharmaceutical Concerns

Michaelis–Menten and Lineweaver–Burk plots can help classify an inhibitor, but successful drug development requires the ability to compare the effectiveness of one inhibitor to another based on results from an assay. Enzymes are typically compared based on one of two values: K_i or IC_{50}.

K_i and IC_{50}

K_i is the *inhibition constant* and corresponds to the dissociation equilibrium constant of an enzyme-inhibitor complex. IC_{50} is the concentration of the inhibitor required to decrease the enzyme rate by 50%. Smaller K_i and IC_{50} values are both indicative of a stronger inhibitor.

K_i values are easiest to determine for a competitive, reversible inhibitor because the number of binding possibilities are limited (**Scheme 4.16**, Equation 4.16). K_i for an inhibitor can be determined from Equation 4.17. This equation defines a line generated by plotting the observed K_m (K_m^{obs}) of a substrate against [I] (**Figure 4.21**).[18]

$$EI \overset{K_i}{\rightleftharpoons} E + I$$

SCHEME 4.16 Dissociation of E-I complex

$$K_i = \frac{[E][I]}{[EI]} \tag{4.16}$$

$$K_m^{obs} = \frac{K_m}{K_i}[I] + K_m \tag{4.17}$$

The y-intercept is the original K_m of the substrate with no inhibitor. The theoretical x-intercept is $-K_i$. The slope is K_m/K_i with K_m being the uninhibited K_m (not K_m^{obs}). This type of calculation only works for competitive inhibitors and requires multiple experiments, each performed with a different concentration of inhibitor ([I]).

FIGURE 4.21 K_m against [I] for a competitive, reversible inhibitor

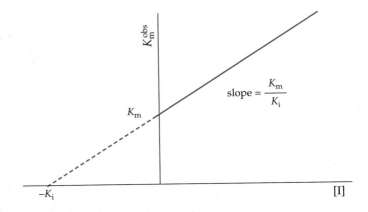

Sample Calculation   Determining a K_i Value from Kinetic Data

PROBLEM The following table presents some K_m^{obs}-[I] data points for a competitive, reversible inhibitor of an enzyme. K_m for the substrate of the uninhibited enzyme is 10 mM. What is the K_i of the inhibitor?

Entry	[I] (mM)	K_m^{obs} (mM)
1	0.0	10
2	1.0	15
3	2.0	20
4	4.0	30

SOLUTION The plot for these points is shown in the following figure with a trendline and its best-fit equation. The graph is a form of Figure 3.18, and the equation we are trying to match is Equation 4.17.

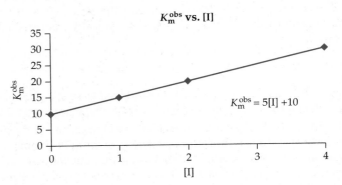

The slope of this line is 5 and corresponds to K_m/K_i. Since K_m is 10 mM, K_i must be 2 mM.

The IC_{50} value for any type of inhibitor can be found by monitoring the rate of an enzyme with varying amounts of inhibitor. Plotting rate against log [I] generates a sigmoidal curve (**Figure 4.22**). The point of inflection of the curve corresponds to 50% of the uninhibited rate and log IC_{50}. Because IC_{50} values can be determined simply with any kind of inhibitor, the effectiveness of most enzymes is measured with an IC_{50} value instead of K_i.

Despite being easier to determine and therefore more amenable for use with high-throughput screening, IC_{50} values do have problems. An IC_{50} value depends on the conditions of the inhibition experiment, namely the concentration of the substrate. If more substrate is used in the experiment, then more inhibitor will be required to suppress the

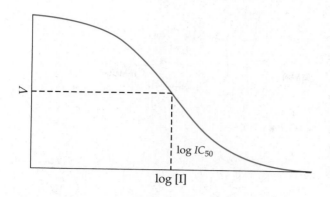

FIGURE 4.22 Determining IC_{50} of an inhibitor

reaction rate and the observed IC_{50} will be lower. For this reason, IC_{50} values for the same enzyme from multiple sources rarely agree with one another. Of course, this problem is less of a concern for a pharmaceutical company performing its own inhibition studies through consistent, internal protocols. Regardless, the literature is full of IC_{50} values that seem contradictory upon first glance. In contrast to IC_{50} value, K_i values are not dependent upon experimental concentrations and may be compared directly. The Cheng–Prussoff relationship was developed in part to allow conversion of IC_{50} values to K_i values and facilitate comparison of inhibition data (Equation 4.18).[19]

$$K_i = \frac{IC_{50}}{\left(1 + \frac{[S]}{K_m}\right)} \tag{4.18}$$

Determination of a K_i value requires one to know the IC_{50} value, the substrate concentration of the experiment, and the K_m of the uninhibited enzyme.

Mutational Resistance to Inhibitors

Enzyme inhibitors are often used to target invading organisms such as bacteria or viruses. Both of these infectious agents have biological processes that are different from those in the host. Unique processes found in an invading organism are ideal targets because a selective inhibitor will have no effect on the host. Unfortunately, because of their rapid rate of division and growth, bacteria and viruses go through many, many generations over the course of an infection. Each generation raises the possibility of a mutant strain. Single point mutations in an enzyme result in changing one amino acid for another in the polypeptide backbone. If the amino acid is involved in recognition and binding of the inhibitor, then a new residue in that position will likely diminish the effectiveness of the inhibitor.

Drug resistance from mutations has been most widely publicized with bacteria. Many antibiotics inhibit enzymes involved in cell wall synthesis. Widespread use of antibiotics combined with frequent failures of patients to finish their prescriptions has led to the formation of strains of bacteria that are resistant to the commonly administered antibiotics, including the penicillins (**4.22**) and cephalosporins (**4.23**) (Figure 4.23). For the most part, these resistant bacteria produce β-lactamase, an enzyme that breaks down the strained β-lactam ring in the drug. Traditionally, methicillin (**4.24**), an antibiotic that is not cleaved by β-lactamase, and related structures have been effective against resistant bacteria. Some bacteria, mostly *Staphylococcus aureus*, have developed mutational resistance to methicillin. These bacteria are called *methicillin-resistant Staphylococcus aureus* (MRSAs), or simply *superbugs*, and are relatively common in hospitals as well as gyms and locker rooms. Indeed, they are so common that methicillin is no longer used to fight persistent infections. Newer antibiotics such as vancomycin (**4.25**) show some success in treating MRSAs, but some bacteria have also developed resistance to vancomycin.

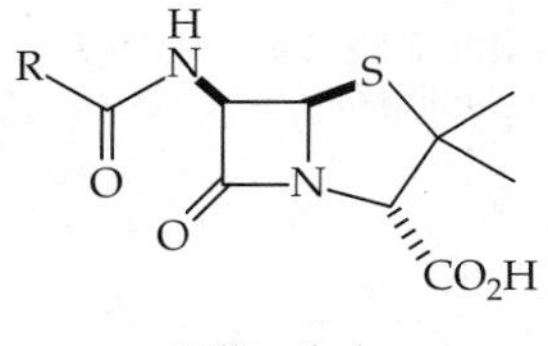

penicillin skeleton
4.22

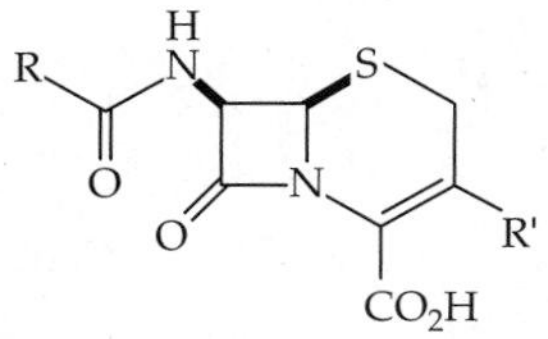

cephalosporin skeleton
4.23

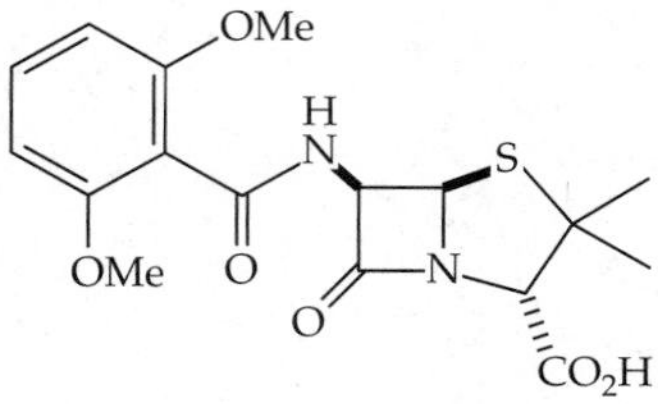

methicillin
4.24

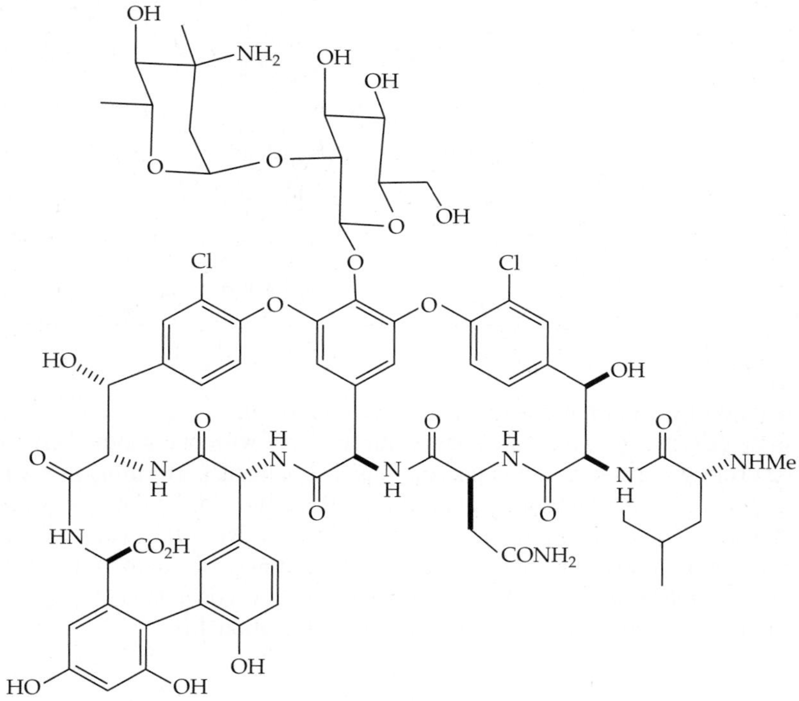

vancomycin
4.25

FIGURE 4.23 Various antibiotic classes and examples

A significant problem in the treatment of human immunodeficiency virus (HIV) is the formation of drug-resistant strains of HIV. Most HIV drugs inhibit viral enzymes, especially HIV reverse transcriptase and HIV protease. The drug-resistant strains of HIV have key mutations in their enzymes to diminish the effectiveness of the antiviral drugs. For this reason, many HIV therapies are a cocktail of several drugs, each of which is potent against different enzyme targets in HIV. The goal is to suppress the viral load in a patient as completely as possible to minimize the possibility of resistant viral mutants. Examples of drug cocktails include Combivir and Trizivir, both products of GlaxoSmith-Kline. Combivir includes two drugs, lamivudine (Epivir, **4.26**) and zidovudine (Retrovir, **4.27**) (**Figure 4.24**). Trizivir adds abacavir (Ziagen, **4.28**). All of these compounds are HIV reverse transcriptase inhibitors. Such drugs are part of HAART, or *highly active antiretroviral treatment*. Several HAART drug combinations have been developed. Enzyme inhibitors that have been more recently included in anti-HIV cocktail drugs include darunavir (Prezista, **4.29**) and raltegravir (Isentress, **4.30**). Darunavir is a protease inhibitor (blocks the cleavage of key proteins), and raltegravir is an integrase inhibitor (blocks the enzyme that incorporates viral DNA into the host's genome).

Because of their rapid rate of cell division, cancer cells are also sometimes able to develop drug resistance. Imatinib (**4.12**) (Figure 4.13) is a kinase inhibitor for the treatment of certain types of leukemia. The x-ray crystal structure of imatinib bound in the active site of the kinase has been solved and shows how mutations in the enzyme interfere with binding and allow drug resistance.[20]

Concentration Effects

The goal of enzyme inhibition is normally to decrease the rate of formation of a product that is ultimately responsible for an undesired effect. As an enzyme is inhibited, the

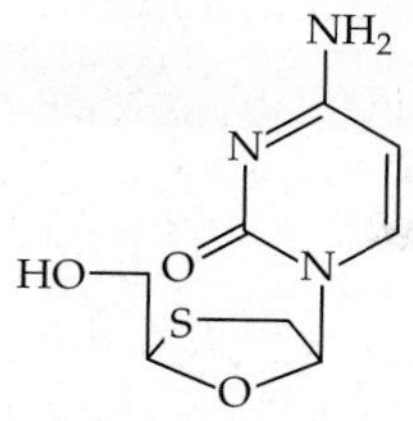

lamivudine
(Epivir)
4.26

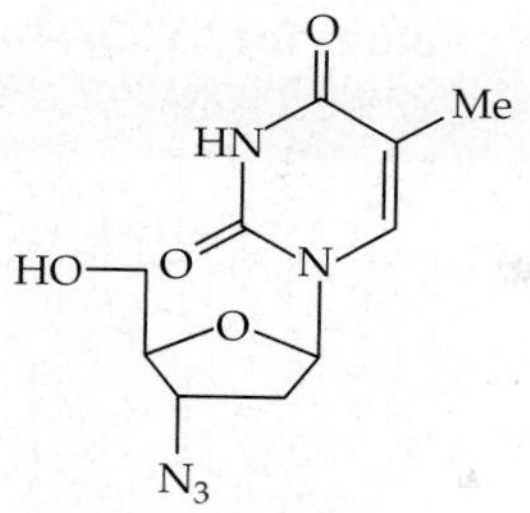

zidovudine
(Retrovir)
4.27

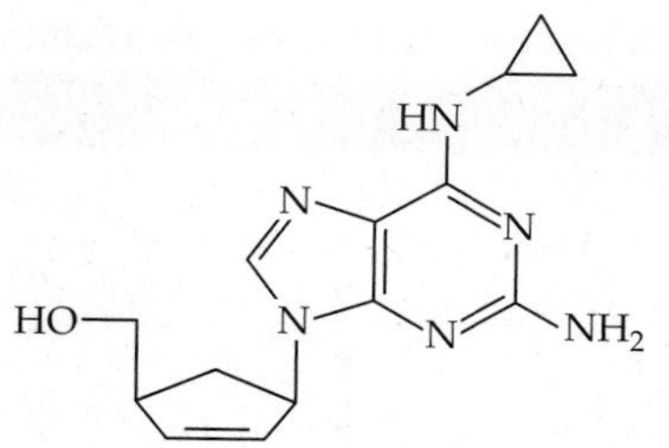

abacavir
(Ziagen)
4.28

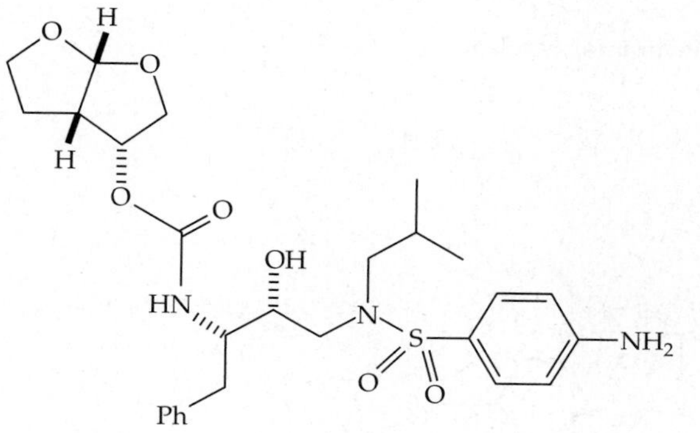

darunavir
(Prezista)
4.29

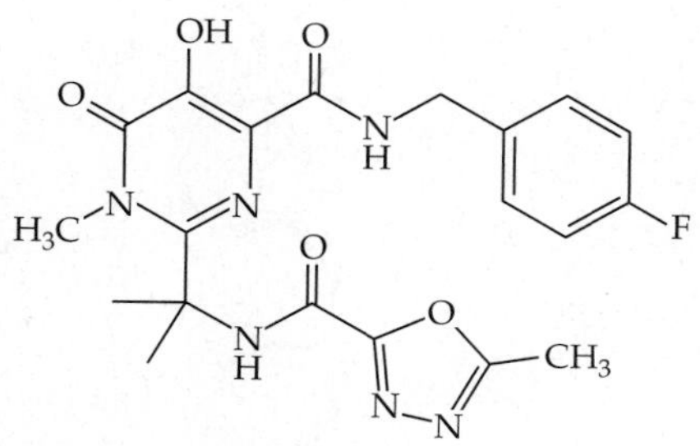

raltegravir
(Isentress)
4.30

FIGURE 4.24 Inhibitors of enzymes vital to the HIV life cycle

substrate of that enzyme will slowly increase in concentration. As [S] builds, the rate of product formation will begin to rise (Figure 3.6). Furthermore, a cell may be able to respond to increases in [S] by producing more enzyme. Therefore, the effect of a successful enzyme inhibitor will be diminished to some degree by increases in the amount of substrate and possibly the enzyme as well.

Metabolism of Drugs

Most drugs are foreign substances, a fact that places them in a category called *xenobiotics*. Other xenobiotics include accidentally ingested toxins such as pesticides. The body must be able to break down and eliminate unanticipated compounds. The cytochrome P-450 enzymes (CYP) found primarily in the liver assume the bulk of the responsibility for breaking down xenobiotics so that they may be eliminated from the body. As foreign materials are absorbed into the bloodstream from the digestive system, they circulate through the liver, which scours the blood to process xenobiotics. In general, metabolites of xenobiotics are less active and are quickly eliminated from the body. Metabolism is therefore a powerful method the body uses to remove foreign substances, including drugs.

CYP is a family of different enzymes. Each variation has a slightly different function, but all are able to accept substrates with a range of structure. Their active sites tend to be broad and shallow so that almost any compound can fit to some degree. **Table 4.3** shows the variety of reactions that can be performed by CYP3A4, a common form of CYP.[21] Despite unrelated substrate structures, the K_m values (binding affinities) are fairly uniform. A more thorough treatment of drug metabolism will be presented in Chapter 8.

TABLE 4.3   **Selected substrates, products, and K_m values for CYP3A4-mediated reactions[21]**

Entry	Substrate	Product	K_m (μM)
1	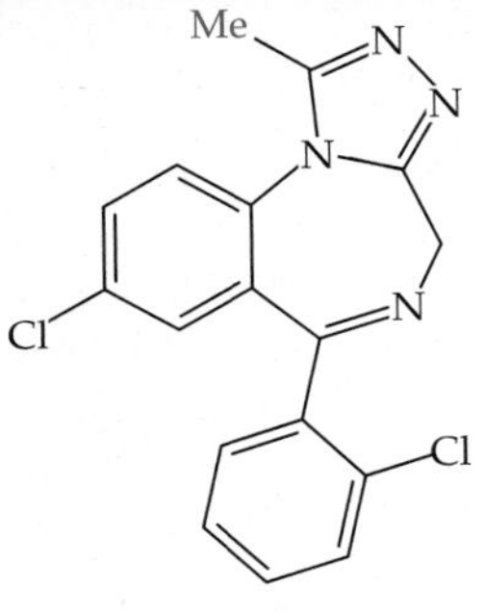 alprazolam (Xanax)	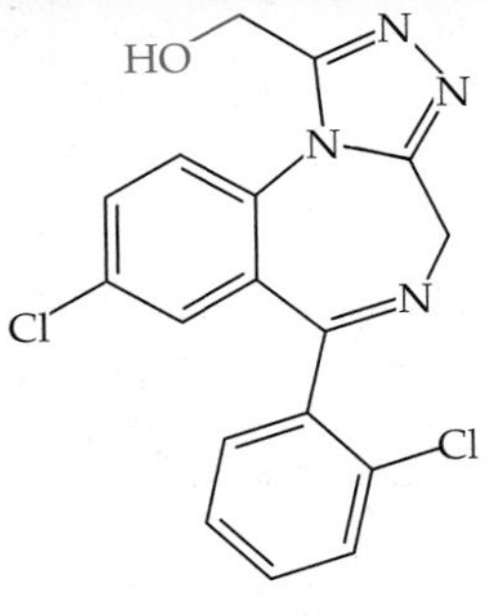 1'-hydroxyalprazolam	55
2	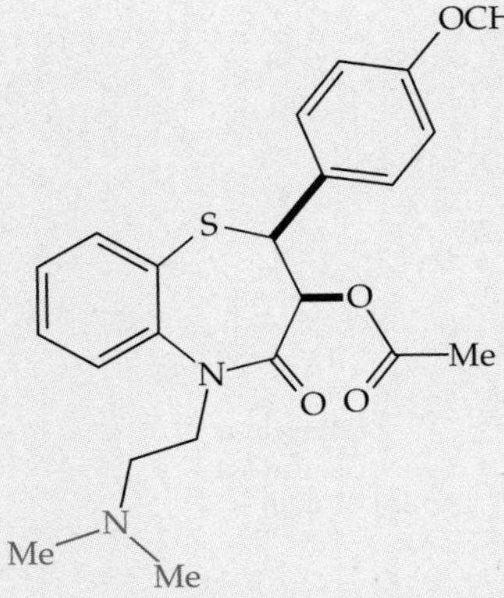 diltiazem (Cardizem)	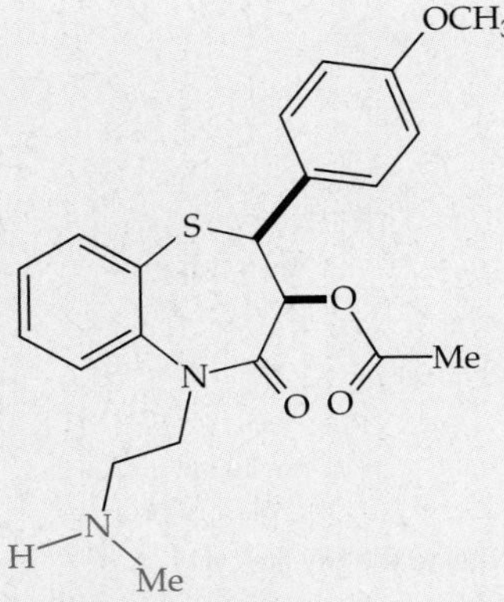 N-desmethyl diltiazem	44
3	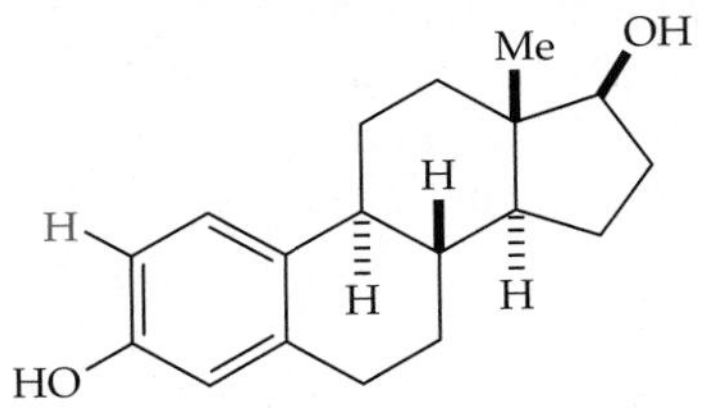 estradiol	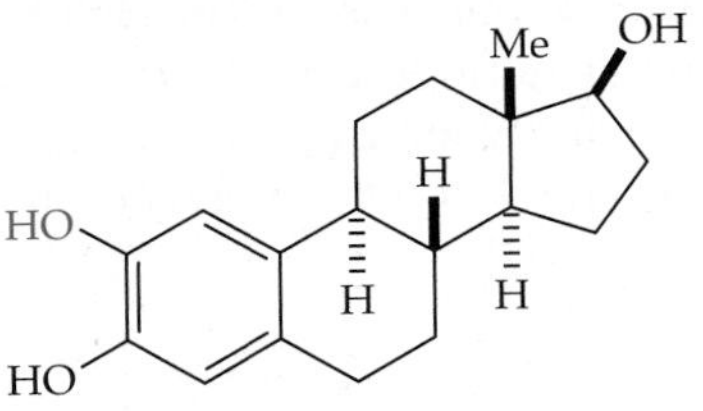 2-hydroxyestradiol	18
4	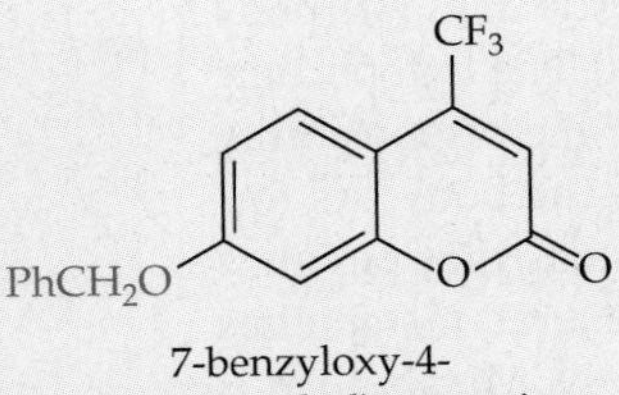 7-benzyloxy-4-(trifluoromethyl)coumarin	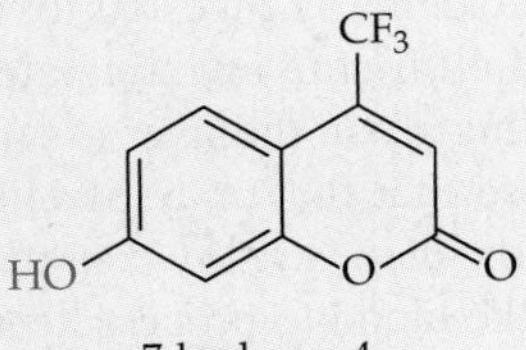 7-hydroxy-4-(trifluoromethyl)coumarin	31
5	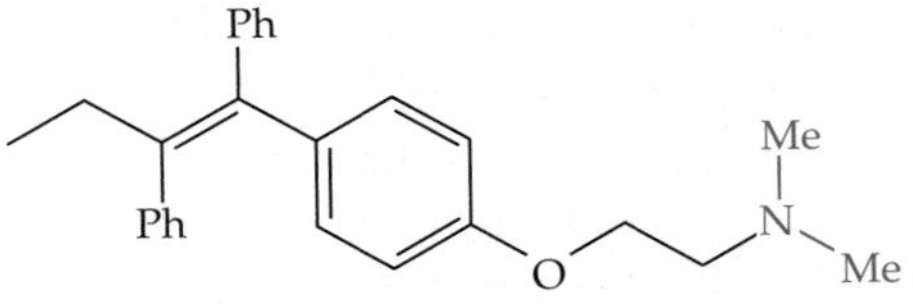 tamoxifen (Nolvadex)	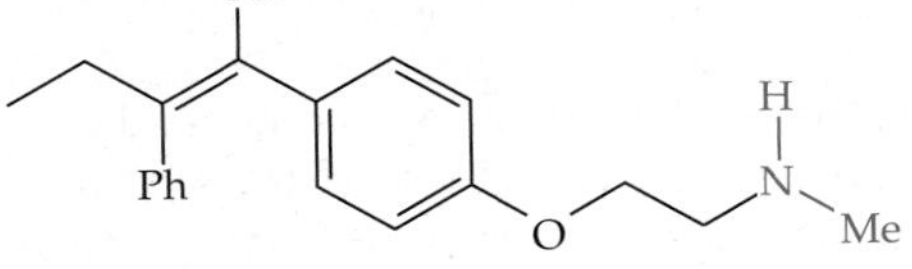 N-desmethyl tamoxifen	50

Summary

Enzymes are the catalysts of the body and facilitate conversion of substrates into products. Without enzymes, necessary reactions would be too slow to sustain life as we know it. Enzymes are proteins assembled from amino acid building blocks. The order and number of amino acid residues represents the primary structure of the enzyme. Localized regions of folding into α-helices or β-sheets determine the secondary structure. Tertiary structure arises from the relative positioning of secondary structures. Some enzymes require multiple proteins to be active. Positioning of the proteins defines the quaternary structure of the fully active complex. Overall folding of enzymes establishes the shape of the active site, a pocket in the enzyme that binds the substrate and other required cofactors needed for the catalyzed reaction. Under the Koshland induced fit hypothesis, substrates bind in the active site and are held together in an orientation that encourages interaction and ultimately reaction. Once formed, the products diffuse out of the active site to make space for new substrate molecules. Properties of an enzyme, such as its affinity for a substrate (K_m) and maximum rate of reaction under a given set of conditions (V_{max}), can be determined through kinetic data. Controlling enzymes through inhibition can take many forms. Some inhibitors bind reversibly to the enzyme active site (competitive), both the enzyme and enzyme-substrate complex (noncompetitive), or just the enzyme-substrate complex (uncompetitive). Other inhibitors covalently and irreversibly bind the enzyme and destroy the enzyme's catalytic ability. The effectiveness of reversible inhibitors is quantified through two parameters: the dissociation equilibrium constant of the enzyme-inhibitor complex (K_i) and the inhibitor concentration required to reduce the rate of enzyme conversion by 50% (IC_{50}). The use of inhibitors to treat diseases can be effective, but some conditions are somewhat problematic. The use of inhibitors to treat bacterial or viral infections as well as many cancers can become less effective over time if genes for the enzyme in the invading organism mutate. Mutations result in changes in the enzyme's shape and decreasing binding of the inhibitor.

Questions

1. Scheme 4.7 includes three separate reactions with three different rate constants. (Be sure to consider reverse reactions, when applicable.) Write out a rate equation for each reaction. Determine the units on each rate constant, k_1, k_{-1}, k_2. Remember that the rate of a reaction always has the units of concentration over time. Use Equation 4.12 and the three rate constants to show that K_m indeed has units of concentration.

2. Figures 4.12, 4.14, and 4.17 are Lineweaver–Burk graphs of competitive, noncompetitive, and uncompetitive reversible inhibitors. Sketch three Michaelis–Menten graphs showing an enzyme with and without one of the three types of inhibitors. Clearly label the important parts of each graph.

3. According to the Michaelis–Menten equation (Equation 4.11), to achieve $\frac{1}{2}V_{max}$ requires $[S] = K_m$. In terms of K_m, what is the value of $[S]$ required to reach 10% of V_{max}? 95% of V_{max}?

4. In introductory geometry classes, the slope of a line is often described as "rise over run" or $\Delta y / \Delta x$. Use the x- and y-intercept values of a Lineweaver–Burk plot to prove that the slope of the line is K_m/V_{max}.

5. (2S,3S)-2,3-Dicarboxyaziridine (**4.c**) is a potent competitive inhibitor ($K_i = 80$ nM) of fumarase, an enzyme that catalyzes the hydration of fumaric acid (**4.a**) to (S)-malic acid (**4.b**). Rationalize how **4.c** might act as a competitive inhibitor of fumarase. Would you expect the enantiomer of the inhibitor to have a higher or lower K_i value? Explain. [Greenhut, J., Umezawa, H. & Rudolph, F. B. Inhibition of Fumarase by S-2,3-Dicarboxyaziridine. *J. Biol. Chem.* 1985, *260*, 6684–6686.]

fumaric acid
4.a

(S)-malic acid
4.b

(2S,3S)-2,3-dicarboxy-
aziridine
4.c

6. Another competitive inhibitor of fumarase is citric acid (**4.d**) with a K_i of 22 μM. Because K_i is an equilibrium constant (Equation 4.16), K_i may be used to calculate the standard free energy of dissociation (ΔG°_{dis}) for the enzyme-inhibitor complex. Use Equation 4.a ($R = 0.00199$ kcal/mol·K, $T = 298$ K) to determine ΔG° for dissociation of fumarase-**4.c** and fumarase-**4.d** complexes (be careful with units). [Teipel, J. W., Hass, G. M. & Hill, R. L. The Substrate Specificity of Fumarase. *J. Biol. Chem.* 1968, *243*, 5684–5694.]

citric acid
4.d

$$\Delta G^\circ_{dis} = -2.3RT \log K_i \qquad (4.a)$$

7. K_i can be used to determine the standard free energy of dissociation (ΔG) of an enzyme-inhibitor complex (see question 6). Free energy is determined by enthalpy and entropy (Equation 4.b). For dissociation, would ΔS be greater or less than 0? Based on Equation 4.b, is the enthalpy of dissociation (ΔH) larger or smaller in magnitude than ΔG? When most medicinal chemists consider binding energies, they focus on contact forces, dipole interactions, and hydrogen bonding. These intermolecular forces are enthalpy contributors.

$$\Delta G = \Delta H - T\Delta S \qquad (4.b)$$

8. Included in the following table are some data points from a hypothetical enzyme kinetics study. Using a spreadsheet program with graphing abilities (such as Excel), generate a Lineweaver–Burk plot of the data points in the table. Determine the best-fit line for the data along with V_{max}, K_m, and r^2 (the square of the correlation coefficient of the line). Does this enzyme follow Michaelis–Menten kinetics? Why or why not?

[S] (mM)	V (s^{-1})
5	4
25	21
50	30
100	35
250	40
500	45
1,000	48

9. Why are mixed noncompetitive inhibitors more common than pure noncompetitive inhibitors?

10. Schemes 4.2 and 4.4 are cartoon-type diagrams of inhibitor-enzyme-substrate interactions. Draw a related diagram that describes how indole (**4.13**) can behave as either a pure or mixed noncompetitive inhibitor for α-chymotrypsin depending on the substrate.

11. Sketch V versus [I] for the data in Figure 4.22.

12. Scheme 4.10 shows the equilibrium for the enzyme glutamate dehydrogenase. Based on the name of the enzyme alone, would you expect this enzyme to drive the reaction to the left or right? Based on the K_m values, does your answer change? Note that the position of the equilibrium for this reaction is not determined by K_m values but the free energy difference between the products and starting materials.

13. This chapter begins with a discussion of how a shovel resembles a catalyst. In what way(s) is a shovel a poor illustration of a catalyst? Can you give a better real-world example of an object that acts like a catalyst?

14. Why do competitive inhibitors of a given enzyme tend to resemble one another? Do noncompetitive inhibitors of a given enzyme necessarily resemble one another? Why or why not?

15. The text mentions that binding an allosteric site in some cases enhances enzyme activity. Describe how this is possible.

References

1. Lehninger, A. L., Nelson, D. L., & Cox, M. M. Amino Acids and Peptides. *Principles of Biochemisty,* 2nd ed., New York: Worth, 1993, Chapter 5.

2. Hernandez-Guzman, F. G., Higashiyama, T., Pangborn, W., Osawa, Y., & Ghosh, D. Structure of Human Estrone Sulfatase Suggests Functional Roles of Membrane Association. *J. Biol. Chem.* 2003, *278,* 22989–22997.

3. Kawate, T., Michel, J. C., Birdsong, W. T., & Gouaux, E. Crystal Structure of the ATP-Gated P2X$_4$ Ion Channel in the Closed State. *Nature.* 2009, *460,* 592–598.

4. Lehninger, A. L., Nelson, D. L., & Cox, M. M. Enzymes. *Principles of Biochemisty,* 2nd ed., New York: Worth, 1993, Chapter 8.

5. Koshland, D. E. Application of a Theory of Enzyme Specificity to Protein Synthesis. *Proc. Natl. Acad. Sci. U.S.A.* 1958, *44,* 98–104.

6. Lineweaver, H., & Burk, D. The Determination of Enzyme Dissociation Constants. *J. Am. Chem. Soc.* 1934, *56,* 6588–6666.

7. Hofstee, B. H. Non-Inverted versus Inverted Plots in Enzyme Kinetics. *Nature.* 1959, *184,* 1296–1298.

8. Hanes, C. S. Studies on Plant Amylases: The Effect of Starch Concentration upon the Velocity of Hydrolysis by the Amylase of Germinated Barley. *Biochem. J.* 1932, *26,* 1406–1421.

9. Bradley, B. A., Colen, A. H., & Fisher, H. F. The Effects of Methanol on the Glutamate Dehydrogenase Reaction at 0 Degrees C. *Biophys. J.* 1979, *25,* 555–561.

10. Deininger, M. W., & Druker, B. J. Specific Targeted Therapy of Chronic Myelogenous Leukemia with Imatinib. *Pharmacol. Rev.* 2003, *55,* 401–423.

11. Weisberg, E., Manley, P. W., Cowan-Jacob, S. W., Hochhaus, A., & Griffin, J. D. Second Generation Inhibitors of Bcr-Abl for the Treatment of Imatinib-Resistant Chronic Myeloid Leukaemia. *Nature Rev.* 2007, *7,* 345–357.

12. Wu, C. L. J., Stohs, S. J. Non-competitive Inhibition of Hepatic and Intestinal Aryl Hydrocarbon Hydroxylase Activities from Rats by Rifampin. *J. Nat. Prod.* 1983, *46,* 108–111.

13. Hein, G. E., & Niemann, C. Steric Course and Specificity of α-Chymotrypsin-Catalyzed Reactions. I. *J. Am. Chem. Soc.* 1962, *84,* 4487–4494.

14. Ramaswamy, S., Scholze, M., & Plapp, B. V. Binding of Formamides to Liver Alcohol Dehydrogenase. *Biochemistry.* 1997, *36,* 3522–3527.

15. Vane, J. R. Inhibition of Prostaglandin Synthesis as a Mechanism of Action for Aspirin-Like Drugs. *Nat. New Biol.* 1971, *231,* 232–235.

16. Lockene, A., Skottova, N., & Olivecrona, G. Interactions of Lipoprotein Lipase with the Active-Site Inhibitor Tetrahydrolipstatin (Orlistat). *Eur. J. Biochem.* 1994, *222,* 395–403.

17. Wang, Y.-H., Jones, D. R., & Hall, S. D. Differential Mechanism-Based Inhibition of CYP3A4 and CYP3A5 by Verapamil. *Drug Metab. Dispos.* 2005, *33,* 664–671.

18. Dixon, M. The Graphical Determination of K_m and K_i. *Biochem. J.* 1972, *129,* 197–202.

19. Burlingham, B. T., & Widlanski, T. S. An Intuitive Look at the Relationship of K_i and IC_{50}: A More General Use of the Dixon Plot. *J. Chem. Ed.* 2003, *80,* 214–218.

20. Schindler, T. Structural Mechanism of STI-571 Inhibition of Abelson Tyrosine Kinase. *Science.* 2000, *289,* 1938–1942.

21. Williams, J. A., Ring, B. J., Cantrell, V. E., Jones, D. R., Eckstein, J., Ruterbories, K., et al. Comparative Metabolic Capabilities of CYP3A4, CYP3A5, and CYP3A7. *Drug. Met. Disp.* 2002, *30,* 883–891.

Receptors as Drug Targets

Chapter Outline

Learning Goals for Chapter 5

- Appreciate the similarities and differences of receptors and enzymes
- Recognize the different structural and functional types of receptors
- Know the types of molecules that bind receptors and their effects
- Understand the different mathematical models related to receptor behavior

Like enzymes, receptors are common targets in drug discovery. Receptors act as switches that can be turned on or off. When turned on, receptors initiate a cascade of events that ultimately produce a biological response. The body contains thousands of different receptors. Despite their diversity, almost all receptors can be classified into a handful of different receptor superfamilies. Receptors bind molecules that either activate or suppress the normal function of the receptor. The impact of receptor binding can be modeled mathematically to allow better understanding of receptor function. Better understanding of receptor function ultimately allows for the design of safer, more effective drugs.

5.1 Receptors

Both enzymes and receptors are proteins, and therefore both biological structures share many fundamental similarities. Regardless their roles within a biological system are distinct from one another.

Similarities to Enzymes

As proteins, both enzymes and receptors possess the same aspects of primary, secondary, tertiary, and quaternary structure. Just as with enzymes, proper folding of a receptor depends on environmental factors, including temperature and pH. The shape of a receptor is crucial because, like enzymes, receptors operate by binding other molecules, called *ligands*. Intermolecular forces and complementary shape determine the strength of reversible binding between a receptor (R) and its ligand (L) (**Scheme 5.1**). The energy of the binding can be quantified by measuring the equilibrium of association (K_A) or dissociation (K_D) (Equation 5.1). As with enzymes, when a receptor binds a ligand, the receptor will undergo conformational changes. These changes affect how the receptor is able to bind additional molecules and induce a response.

$$R + L \underset{K_D}{\overset{K_A}{\rightleftharpoons}} RL$$

SCHEME 5.1 Reversible binding
of a receptor and ligand

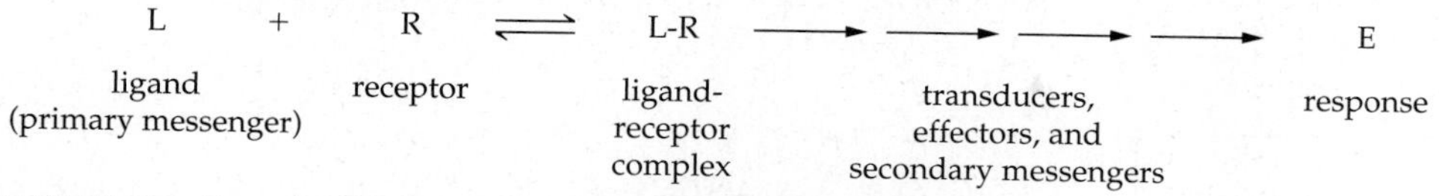

$$L \quad + \quad R \rightleftharpoons \quad L\text{-}R \longrightarrow \longrightarrow \longrightarrow \longrightarrow \quad E$$

ligand   receptor   ligand-   transducers,   response
(primary messenger)   receptor   effectors, and
complex   secondary messengers

SCHEME 5.2 General pathway for response generation from a receptor

$$K_A = \frac{1}{K_D} = \frac{[RL]}{[R][L]} \tag{5.1}$$

The similarities between enzymes and receptors allow both systems to be modeled with many of the same mathematical equations. Most treatments of receptors and enzymes appear to be very different, but the derivations and theories can largely be recycled between the two topics.

Differences from Enzymes

The primary difference between receptors and enzymes lies in what they do. Enzymes convert a substrate to a product. Receptors do not catalyze a reaction or otherwise convert a ligand. Instead, receptors bind a ligand, or *primary messenger* (**Scheme 5.2**). Upon binding a ligand, a receptor changes its conformation to initiate a series of events. These events may involve a number of other agents, including enzymes (*effectors*), binding proteins (*transducers*), and/or other signaling molecules (*secondary messengers*). The number of other players in the pathway depends on the particular receptor. Each ligand can potentially produce many secondary messengers, a phenomenon known as *signal amplification*. Ultimately, the entire process generates an observable biological response (E).[1]

The events initiated from the ligand-receptor complex are more complex than a simple substrate-to-product conversion by an enzyme. For this reason, receptor activity is more difficult to model mathematically than enzymatic activity.

The primary method for determining protein structure is x-ray crystallography. Unfortunately, the receptors most frequently targeted by the pharmaceutical industry are embedded in cell membranes. The cell membrane plays a vital role in determining the overall shape of a membrane-bound receptor. Crystallization is performed in the absence of membrane lipids, so x-ray information is not representative of the true receptor structure. Therefore, reliable structural information on receptors is difficult to obtain.

5.2 Receptor Classification

Receptors can be broadly sorted into *superfamilies* based on their structure and mode of action. The membrane-bound receptor superfamilies are *G-protein–coupled*, *kinase-linked*, and *ion channel–linked* receptors. Superfamilies are divided into families named after the natural, or *endogenous*, ligand that binds all the members of a given family. Unnatural, or *exogenous*, ligands can show selectivity for specific receptors within a family and allow further subdivision.[2]

For example, within the superfamily of G-protein–coupled receptors lies the angiotensin II receptor (AGTR) (**Figure 5.1**). The endogenous ligand for AGTR is, as the name

FIGURE 5.1 Classification of angiotensin II receptors

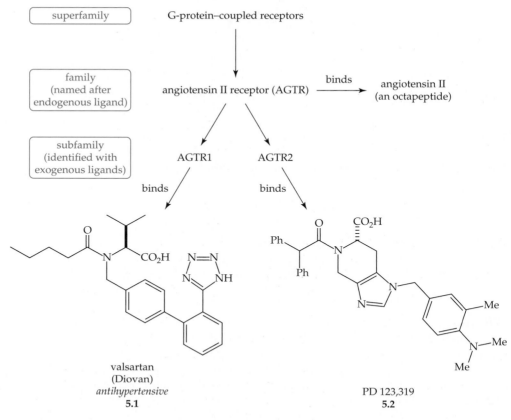

valsartan
(Diovan)
antihypertensive
5.1

PD 123,319
5.2

suggests, angiotensin II, an octapeptide. Two different forms of AGTR have been identified. AGTR1 selectively binds biphenyltetrazoles such as valsartan (Diovan, **5.1**). AGTR2 binds PD 123,319 (**5.2**) as an exogenous ligand.[3] Differences in the binding affinities of valsartan and PD 123,319 for AGTR1 and AGTR2 arise from the differences in structures of AGTR1 and AGTR2. AGTR1 and AGTR2 share a homology of approximately 30%.[4]

All receptors and receptor families cannot be categorized by a one-size-fits-all method. γ-aminobutyric acid (GABA), a neurotransmitter, is the endogenous ligand of the GABA family of receptors. The $GABA_A$ and $GABA_C$ subtypes are of the ion channel–linked superfamily, but $GABA_B$ is of the G-protein–coupled superfamily.[5] Therefore, the GABA family surprisingly spans two superfamilies.

The following subsections describe generally accepted models for the mode of action of the major receptor superfamilies. Conflicting theories and areas of poor understanding are common in research on receptors. Therefore, the following discussions should be regarded as a general guideline only.

Ligand-Gated Ion Channels

Of the membrane-bound receptors, ligand-gated ion channels (LGICs) are the least complicated. LGICs are also called *ionotropic receptors*. LGICs are multiprotein complexes that form a pore in the cell membrane. When a ligand binds the receptor, changes in the receptor conformation cause the pore to open. The opened pore acts as an ion channel. Ions then cross the cell membrane. In some cases, the ions that enter a cell serve as secondary messengers of the signal. Many neurotransmitters act through LGICs.[6] Because the action of ligand binding is so closely associated with the production of secondary

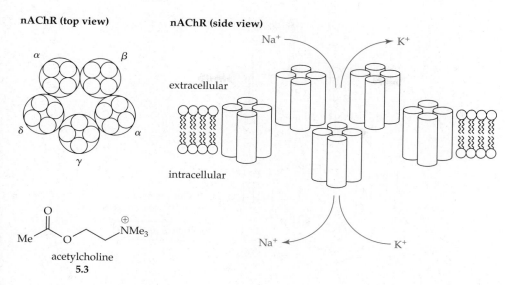

nAChR (top view)

nAChR (side view)

extracellular

intracellular

acetylcholine
5.3

FIGURE 5.2 Structure of nicotinic acetylcholine receptors

messengers, the delay between binding and signal transduction is very short. Therefore, neurotransmitters that act as ligands to LGICs are called *fast neurotransmitters*.

The nicotinic acetylcholine receptor (nAChR) is a thoroughly studied LGIC. Each protein in the nAChR receptor contains four α-helical transmembrane regions that collectively form a rod-like structure (**Figure 5.2**). The rod proteins exist as α-, β-, γ-, and δ-subtypes. In total, the nAChR consists of five rod proteins, two α-proteins, and one each of the β-, γ-, and δ-proteins. In its unbound state, the channel in nAChR is blocked. When two acetylcholines (**5.3**) bind the two α-proteins, the channel opens and allows both Na^+ and K^+ to flow in the direction of the ion gradient. Generally, Na^+ concentrations are higher outside the cell, while K^+ concentrations are higher inside the cell. Different LGICs are selective for different ions. For example, the $GABA_A$ receptor, another LGIC, allows Cl^- to flow into a cell.[7]

G-Protein–Coupled Receptors

G-protein–coupled receptors (GPCRs) consist of a long polypeptide with seven helical regions, each of which crosses the cell membrane (**Figure 5.3**). The result is a protein that, like a strand of thread, stitches back and forth across the membrane. Accordingly, GPCRs are often called *seven transmembrane receptors* (7-TM). The helices are clustered together in a ring-like structure. The *N*-terminus of the protein lies on the extracellular face of the membrane, while the *C*-terminus is intracellular.[8] Understanding of GPCRs was greatly advanced in 2000 when a research group was able to obtain a complete x-ray

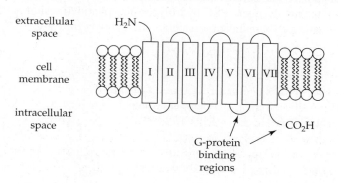

extracellular space

cell membrane

intracellular space

G-protein binding regions

FIGURE 5.3 Model of a typical G-protein–coupled receptor

SCHEME 5.3 G-protein activation by GPCR

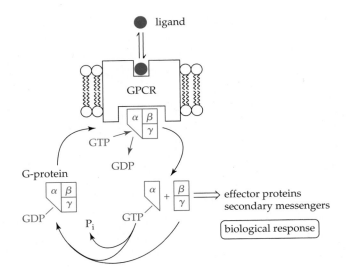

crystal structure of bovine rhodopsin, a GPCR.[9,10] In 2007, an x-ray structure of β_2 adrenoceptor, a human GPCR, was solved.[11]

As their name suggests, GPCRs bind G-proteins. A complete G-protein is a heterotrimer of three separate proteins, consisting of α-, β-, and γ-subunits. The α-subunit binds either guanosine diphosphate (GDP) or triphosphate (GTP). The affinity for GDP/GTP is the origin of the name G-protein. The G-protein heterotrimer binds GPCRs on both the C-terminal tail and the linker between the fifth and sixth transmembrane helices on the intracellular face of the receptor.[12]

Initiation of a signal through a GPCR begins with binding of a ligand on the extracellular side of the receptor (Scheme 5.3). Binding of the ligand changes the conformation of the receptor on the intracellular face. The conformational change causes GTP to replace GDP on the α-subunit of a G-protein that is bound to the GPCR. The GTP-bound G-protein then separates from the GPCR and dissociates into two parts: the GTP-bound α-subunit and a β, γ-heterodimer, both of which are transducers. Either G-protein fragment may subsequently bind the allosteric site of an effector protein (enzyme) and activate or deactivate a catalyzed process. Eventually, the GTP on the α-subunit is hydrolyzed to GDP and phosphate (P_i). The two G-protein fragments then recombine and bind to the GPCR. Binding of a ligand to the G-protein-bound GPCR then restarts the entire cycle.[8,12]

Two of the more studied effector proteins of G-proteins are adenylate cyclase (AC) and phospholipase C (PLC). AC converts adenosine triphosphate (ATP, **5.4**) into 3′,5′-cyclic adenosine monophosphate (cAMP, **5.5**) (Scheme 5.4). cAMP is a secondary messenger that can activate certain kinases (phosphorylation enzymes) and stimulate the breakdown of fats and glycogen. PLC hydrolyzes phosphatidylinositol 3,4-bisphosphate (PIP$_2$, **5.6**) to form two secondary messengers, diacylglycerol (DAG, **5.7**) and inositol

SCHEME 5.4 Formation of cAMP (**5.5**) from ATP (**5.4**) (P = phosphate)

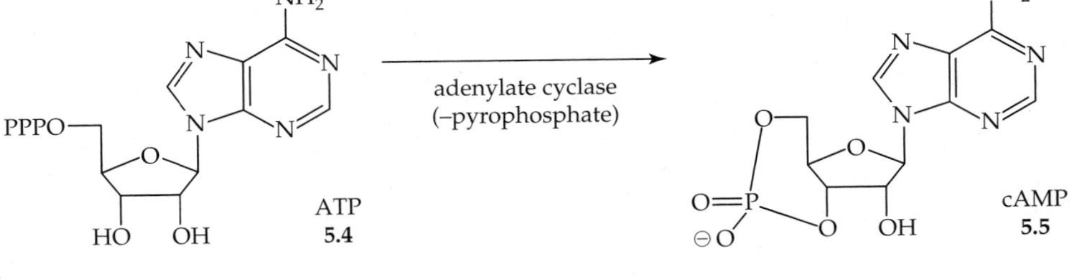

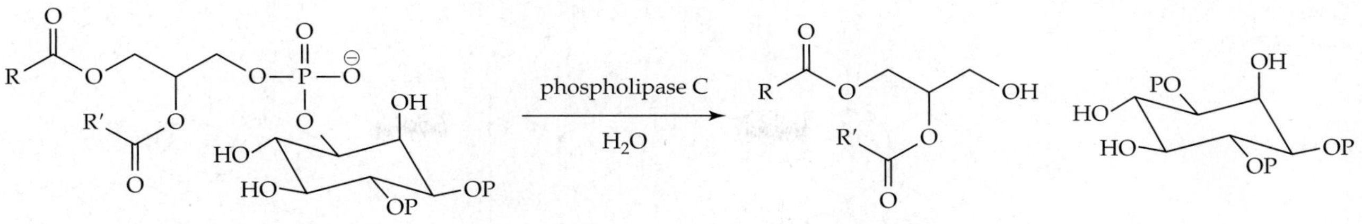

SCHEME 5.5 Formation of DAG (**5.7**) and IP$_3$ (**5.8**) through PLC (P = phosphate)

1,4,5-triphosphate (IP$_3$, **5.8**) (Scheme 5.5). Like cAMP, DAG can activate regulatory kinases. IP$_3$ affects the release of stored Ca^{+2}.[2]

The ligand-binding event in a GPCR is much farther removed from secondary messenger formation than in a LGIC. GPCRs have a longer response time than LGICs. For this reason, neurotransmitters that are ligands for GPCRs are sometimes called *slow neurotransmitters.*

GPCRs are well understood and are frequently targeted in drug discovery programs. In a study from 1996, GPCRs were found to be the target of nearly 50% of all drugs on the market.[13] While that statistic may have changed somewhat over the past 15 years, GPCRs continue to be the most popular drug target.

Tyrosine Kinase–Linked Receptors

Tyrosine kinase-linked receptors (TKLRs) are not as thoroughly explored as the other two membrane-bound receptor superfamilies. The receptor protein in a TKLR consists of three major domains: a ligand-binding domain on the extracellular surface of the membrane, a single α-helix that crosses the membrane, and an intracellular region with tyrosine kinase activity. When a ligand binds a TKLR, conformation changes cause two TKLRs to bind and form a receptor dimer. The intracellular domains phosphorylate the OH groups on tyrosine residues within the peptide chain. This *autophosphorylation* process requires ATPs within the cell to deliver the phosphate groups. The phosphorylated intracellular region then binds proteins with an SH2 domain. The entire supramolecular complex of dimerized TKLR and protein with an SH2 domain then mediates enzymes and generates a response.[14]

SH2 is an abbreviation of *Src homology 2* domain. The protein Src can increase the malignancy of tumor cells. Because of the link between the Src protein and TKLRs, TKLR have become an area of active, intense investigation. One ligand for TKLRs is epidermal growth factor (EGF). EGF regulates cell growth and proliferation. EGF binds the EGF receptor (EGFR), a specific example of a tyrosine kinase-linked receptor. Lapatinib (Tykerb, **5.9**) is a drug used against certain cancers (Figure 5.4). Lapatinib binds EGFR,

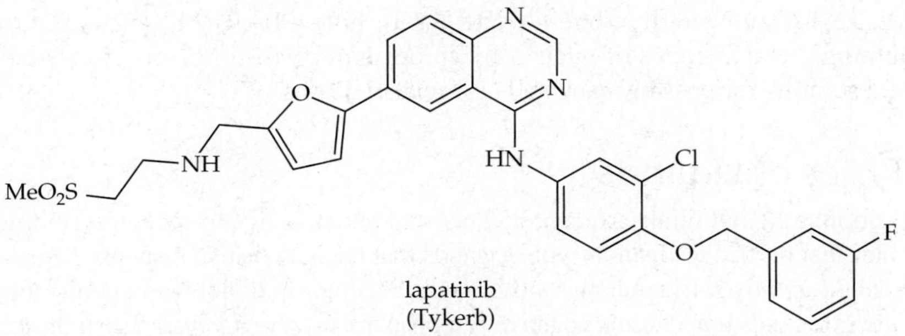

lapatinib
(Tykerb)
antitumor
5.9

FIGURE 5.4

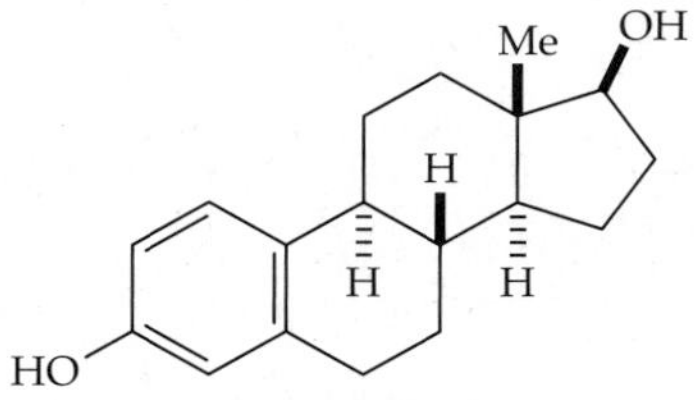

FIGURE 5.5 Ligands for cytoplasmic and nuclear receptors

estradiol
hormone
5.10

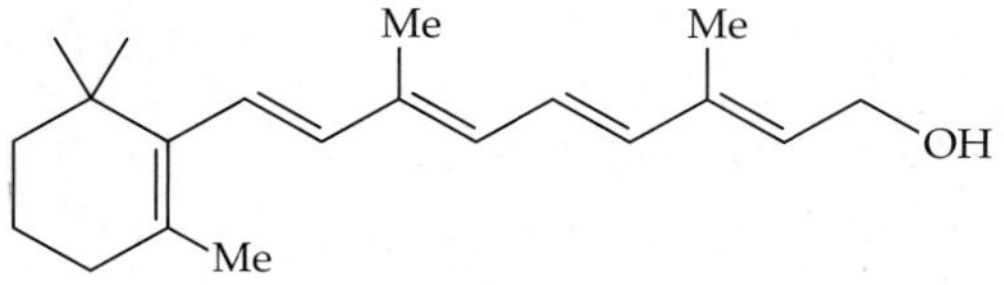

retinol
hormone
5.11

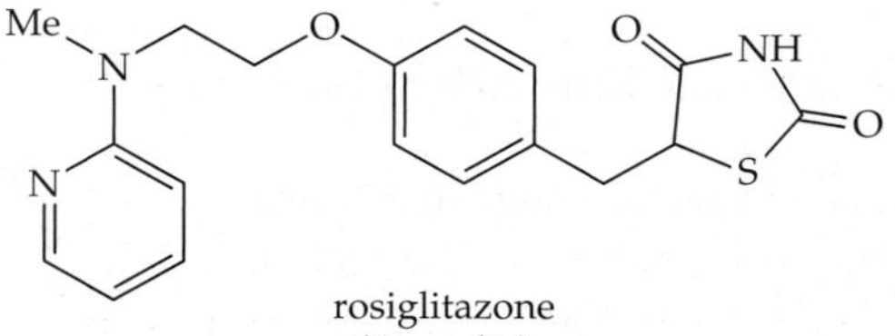

rosiglitazone
(Avandia)
antidiabetic
5.12

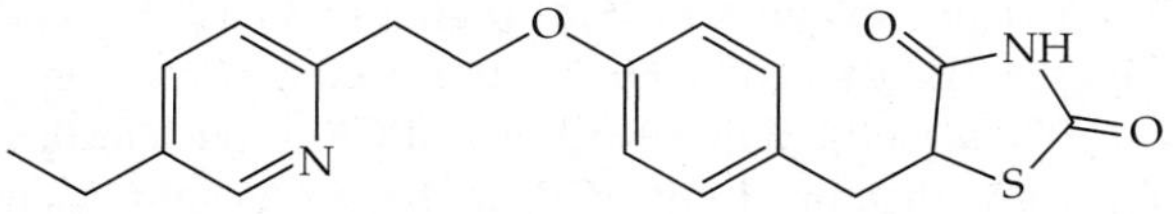

pioglitazone
(Actos)
antidiabetic
5.13

prevents dimerization of the receptor, and therefore prevents the binding of EGFR with any SH2-containing proteins and slows cell growth.[14]

Nuclear Receptors

Not all receptors are associated with the cell membrane. Receptors can be found free in both the cytoplasm and the nucleus. Steroid and thyroid hormones, such as estradiol (**5.10**) and retinol (**5.11**), are examples of ligands for cytoplasmic and nuclear receptors (**Figure 5.5**). The 48 different nuclear receptors found in humans regulate gene expression by binding DNA.[15] All nuclear receptors contain several functional regions, including a *ligand-binding domain* and *DNA-binding domain*. In a series of events, the receptor binds its ligand and becomes able to recognize DNA. Binding typically occurs in the cytoplasm, and the receptor-ligand complex then translocates to the nucleus. Once the receptor-ligand complex binds the appropriate section of DNA, transcription ensues to prepare mRNA, which is then translated to a protein. Nuclear receptors, therefore, control what proteins are synthesized in a cell and are potentially powerful drug targets for affecting cellular function.[16]

One particularly attractive nuclear receptor target is the peroxisome proliferator–activated receptor (PPAR, "pee-PAR"). PPAR influences cellular metabolism and is a common target for treatment of certain forms of diabetes. Examples are rosiglitazone (Avandia, **5.12**) and pioglitazone (Actos, **5.13**), both PPAR-γ ligands (Figure 5.5). Highlighting the challenges of nuclear receptor binding, rosiglitazone has come under increased scrutiny for possibly causing heart-related deaths.[17]

5.3 Types of Ligands

By definition, a ligand binds a receptor. The outcome of a ligand-receptor binding event determines that particular ligand's type. Ligands that elicit a positive response from a receptor are called *agonists*. Ligands that both bind a receptor and block an agonist are called *antagonists*. Ligands that cause a negative response are *inverse agonists*. Ligands are classified through *ligand-response* or *dose-response* curves. These are graphical plots of a response against ligand concentration or, more commonly, the logarithm of ligand concentration.[18]

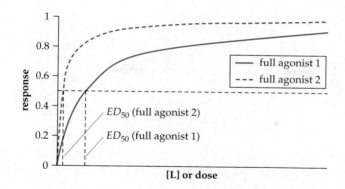

FIGURE 5.6 Full agonist: response vs. [L]

Agonists

Agonists trigger a positive response from a receptor. Agonists are categorized as either *full agonists* or *partial agonists*.

Full Agonists

Full agonists are able to generate a maximal response, defined as 100% or 1.0 response, from a receptor. An example of a full agonist is the endogenous ligand of a receptor. As has already been mentioned, receptors are often named after their endogenous ligand, if it is known. All acetylcholine receptors (AChRs) have acetylcholine (**5.3**) as their endogenous ligand (Figure 5.2), and acetylcholine will invoke a full response from AChRs.

A ligand-response curve for two full agonists is shown in **Figure 5.6**. As the concentration of a ligand increases, the response from a receptor rises and approaches a maximum. The *x*-axis is more often represented in a logarithmic scale (**Figure 5.7**). Figure 5.7 reveals a sigmoidal relationship between response and log [L]. The point of inflection of the curve corresponds to log [L] at 50% maximal response. The ligand concentration required to reach 50% response is called the EC_{50} (*effective concentration*) or ED_{50} (*effective dose*) of that ligand. EC_{50} values are found in data from in vitro studies, for which the exact concentration of the agonist is known. ED_{50} values appear with in vivo studies. For in vivo studies, the dose is known, but the agonist concentration in the affected tissue may not be available. In Figures 4.6 and 4.7, agonist 2 is said to be more *potent* than agonist 1 because agonist 2 requires a smaller concentration/dose to achieve a 50% response.

The term *response* can take a number of different forms. In a biochemical assay, the response may be determined by changes in something quantitative, such as changes in ion concentration, absorbance, or fluorescence. Quantifiable examples of in vivo response include changes in body temperature and blood pressure. With this type of data (*graded response*), the range of responses collected in the study will determine what defines a

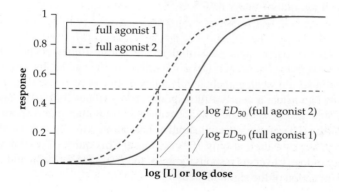

FIGURE 5.7 Full agonist: response vs. log [L]

FIGURE 5.8 Full agonists of certain nAChRs

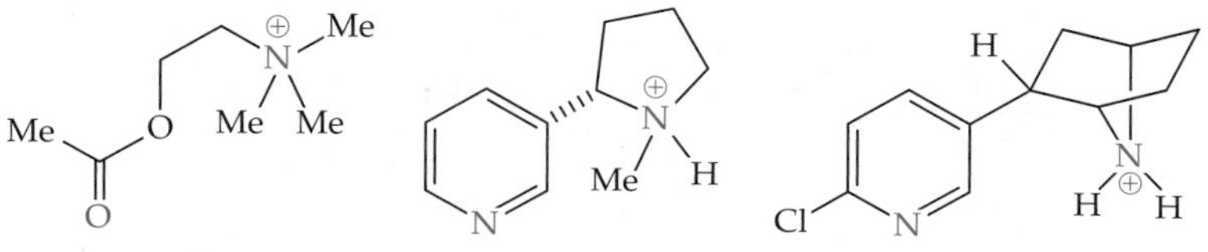

acetylcholine
5.3

(−)-nicotine•H⁺
5.14

(−)-epibatidine•H⁺
5.15

maximal response. Other studies may not afford numerical data. Instead, the response is a yes/no or true/false answer (*quantal response*). In a trial of a sleeping aid, the response may be as simple as answering the question, "Did the patient fall asleep?" In this example, a 50% response makes no sense if defined as 50% sleep. Instead of relating the dose required for 50% response, ED_{50} would define the dose required for 50% of the members of the study to achieve a sleep state. Changing the question to "How much time elapsed before the patient fell asleep?" would afford graded response data, which may or may not be more informative for determining activity.[19]

From a pharmaceutical perspective, a full agonist with a lower ED_{50} is preferable to one with a higher ED_{50}. The lower ED_{50} compound will require a smaller dose to be effective. If less of the drug is administered, then the drug will have a lower probability of interacting with a second target and therefore cause an unwanted, toxic response. The toxicity of a drug can be defined by its TD_{50}, the dose required for 50% toxic response. The ratio of TD_{50} to ED_{50} defines the *therapeutic index* of a drug (Equation 5.2). Ideally, the therapeutic index of a drug should be as large as possible. A high therapeutic index indicates a large window of safety for a drug.

$$\text{therapeutic index} = \frac{TD_{50}}{ED_{50}} \tag{5.2}$$

Full agonists of a given receptor tend to bind the receptor in a similar fashion. Similar binding results in a similar effect: 100% response. Because they bind in a related fashion, full agonists normally have comparable structures in terms of the spatial arrangement of the functional groups that are important for binding. For example, (−)-nicotine (**5.14**) is a full agonist for many of the nicotinic AChRs (nAChRs) (**Figure 5.8**). Nicotine has clear structural parallels with acetylcholine (**5.3**). First, in the binding site on the receptor, the quaternary ammonium ions of **5.3** and **5.14** are attracted to the electron-rich π-face of a tryptophan residue in the receptor protein. Second, the carbonyl oxygen of **5.3** and the pyridine nitrogen of **5.14** serve as hydrogen-bond acceptors with a water molecule that is bound to the receptor. (−)-Epibatidine (**5.15**) shares many structural features of nicotine and acetylcholine and is also a full agonist of many nAChRs.[20]

Partial Agonists

As the name suggests, partial agonists stimulate a response from a receptor. The degree of stimulation, however, does not reach 100% regardless of the partial agonist concentration. Some partial agonists are weak and elicit only a low degree of response (**Figure 5.9**). Others are stronger and can affect a nearly full response. ED_{50} values for partial agonists do not correspond to 50% response on the *y*-axis but instead 50% maximal response possible *by the partial agonist*. Therefore, a partial agonist that can maximally achieve 60% response would have its ED_{50} measured at only 30% response. This quirk of partial agonists highlights the utility of log [L] versus response graphs; the ED_{50} value of a ligand always occurs at the point of inflection of the sigmoidal curve.

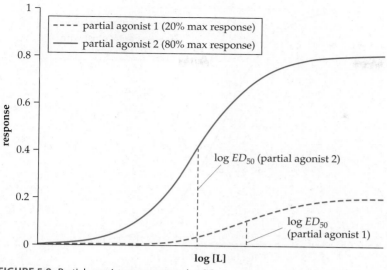

FIGURE 5.10

FIGURE 5.9 Partial agonists: response vs. log [L]

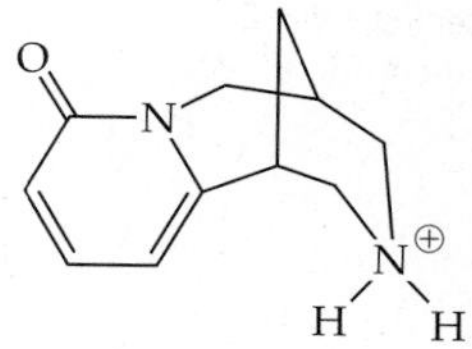

In general, partial agonists with a higher degree of response will be closer structurally to the natural ligand. Very subtle changes in the structure can have a dramatic effect on the maximal response of the partial agonist. Changes in the structure of the receptor can also significantly impact response levels. Within the nAChR family are many members with small differences in one or more of the α-, β-, γ-, and δ-subunits (Figure 5.2). Cytisine (**5.16**) shows partial agonism against the various nAChR members (**Figure 5.10**). The maximal response of cytisine ranges from less than 10% to greater than 90% depending on which nAChR family member is being bound.[21]

When both a partial agonist and a full agonist are present together, the partial agonist *decreases* the response caused by the full agonist. The partial agonist competes with the full agonist for the binding site of the receptor. While the partial agonist can elicit a response, every bound partial agonist is a lost opportunity for binding of a full agonist. The effect of a partial agonist on a receptor already at full response is shown in **Figure 5.11** (solid line). The response decreases only as low as the maximal response of the partial agonist.

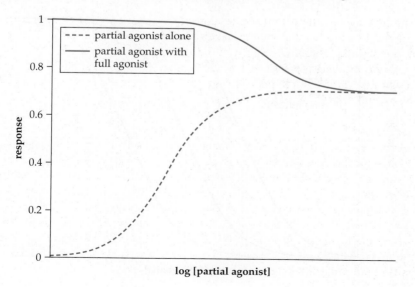

FIGURE 5.11 Effect of partial agonist on a receptor at 100% response

FIGURE 5.12 Effect of
an antagonist on receptor
response

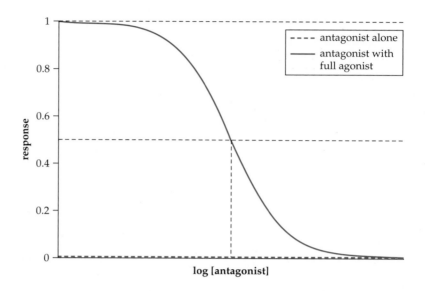

Antagonists

Antagonists have three properties. First, they bind to the receptor just as all ligands do. Second, they do not give rise to a response when they bind (**Figure 5.12**). Third and most important, they reduce the response caused by a full agonist to zero. The antagonist concentration or dose required to decrease the response of a full agonist from 100% to 50% is called the EC_{50}, ED_{50}, or IC_{50} (inhibitory concentration). The IC_{50} value of an antagonist is valid only if the full agonist concentration is held constant throughout the experiment.

Full and partial agonists bind a site on the receptor that causes a response, but antagonists can conceivably bind anywhere on the receptor. Antagonists may directly block the binding position of an agonist. Antagonists may bind to a completely different site on the receptor and cause conformational changes at the agonist binding site. Alternative binding sites for antagonists are called *allosteric sites*, a term borrowed from enzyme inhibitors.

FIGURE 5.13 Competitive
antagonists

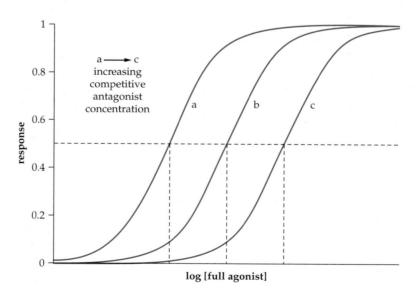

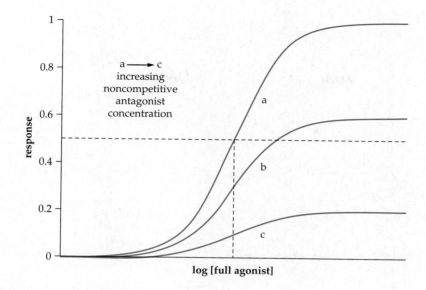

FIGURE 5.14 Noncompetitive antagonists

Antagonists are either *competitive* or *noncompetitive*. Competitive antagonists, which are more common than noncompetitive antagonists, bind reversibly to the receptor through noncovalent intermolecular forces. In the presence of an antagonist, a full agonist can still achieve 100% response, but it will require a greater concentration/dose of the full agonist to do so (**Figure 5.13**). In effect, larger amounts of antagonist raise the ED_{50} value of a full agonist. A noncompetitive antagonist binds the receptor at a different site from the agonist. Noncompetitive antagonist binding makes the receptor incapable of generating a response. No amount of full agonist can counteract a noncompetitive antagonist. In **Figure 5.14**, the ED_{50} value of the full agonist stays constant, but the maximal response decreases as the noncompetitive antagonist concentration is increased.

Because antagonists do not necessarily interact with the receptor at its natural binding site, antagonists can differ structurally from agonists of the receptor. However, based on the methods of drug discovery in the pharmaceutical industry, most drugs that behave as antagonists do share some functional group features of the endogenous ligand or other agonists. Structural similarities are used to achieve binding of the antagonist to the receptor, but the antagonist must be different enough to cause no response. Designing an antagonist that binds without eliciting a response can be a difficult balancing act.

For example, the histamine type 2 (H_2) receptor controls the release of gastric acid. As its name suggests, the endogenous ligand for this receptor is histamine (**5.17**) (**Figure 5.15**). In a search for an H_2 receptor antagonist, N^α-guanylhistamine (**5.18**) was discovered and initially believed to be an effective antagonist. Later research showed that **5.18** is actually a very weak partial agonist for the H_2 receptor. Extending the chain by a single CH_2 unit and modifying the guanidine group to a thiourea afforded compound **5.19**. Compound **5.19** was indeed a true antagonist. Compound **5.19** accomplished two goals. First, it retained enough similarity to **5.17** to strongly bind the H_2 receptor. Second, it was different enough from **5.17** so that it did not elicit a response. This research culminated in the discovery of cimetidine (Tagamet, **5.20**), a very successful drug for the treatment of acid reflux.[22]

Examination of the four H_2 receptor ligands shows that both **5.17** and **5.18** are protonated on the side chain at biological pH, but the side chains of **5.19** and **5.20** are not basic and not protonated. This observation implies that positively charged groups on the side

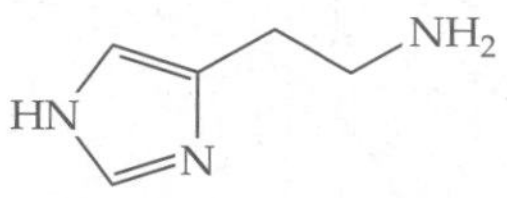

histamine
full agonist
5.17

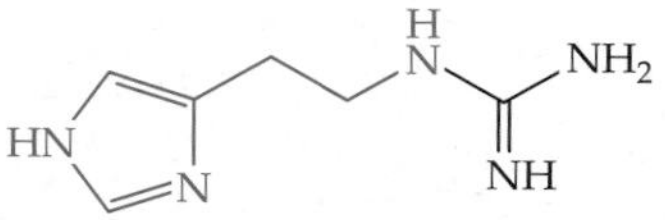

N^α-guanylhistamine
weak partial agonist
5.18

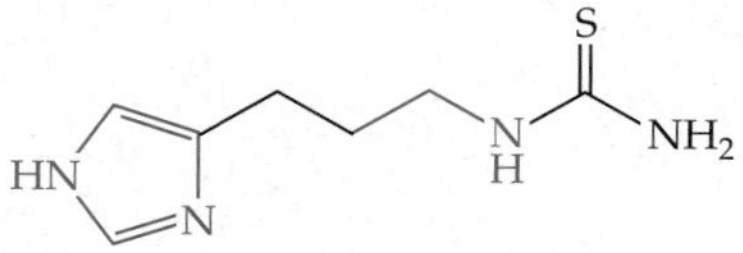

antagonist
5.19

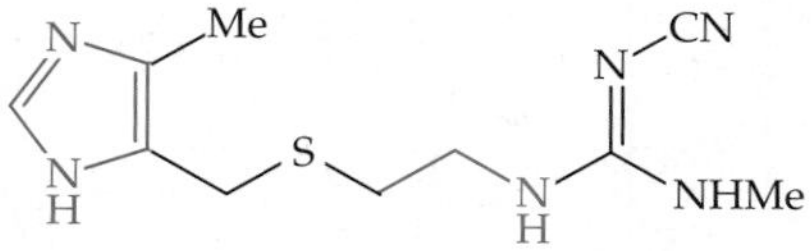

cimetidine
(Tagamet)
antagonist
5.20

FIGURE 5.15 Ligands of the H_2 receptor

chain lead to agonist activity in the molecule.[22] The development of H_2 receptor antagonists is more thoroughly described in Chapter 11.

Inverse Agonists

A third type of ligand, *inverse agonists*, is less frequently encountered. Inverse agonists decrease the response of a receptor that is *constituently active*. A constituently active receptor causes a response in the absence of the receptor's endogenous ligand. In theory, all receptors can be constituently active, but in practice, the level of response is normally too small to be observed. The magnitude of constituent activity can be increased by genetically engineering a cell to express unnaturally high levels of the receptor. In **Figure 5.16**, a hypothetical receptor with constituent activity of 20% of the maximum response is shown. For this receptor, addition of an agonist further increases the response. An antagonist has no effect. An inverse agonist decreases the constituent activity.

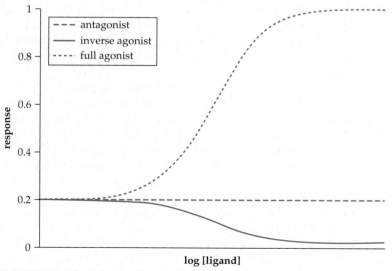

FIGURE 5.16 Effect of various ligands on a constituently active receptor

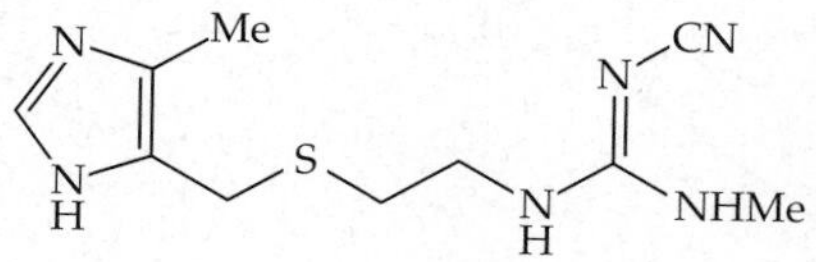

cimetidine
(Tagamet)
antiulcer
5.20

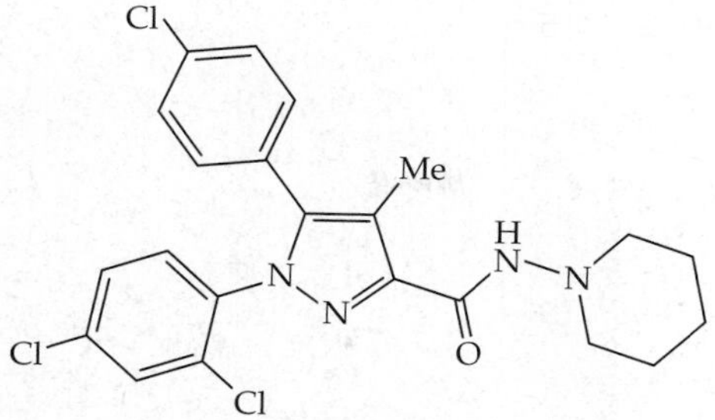

rimonabant
(Acomplia)
appetite suppressant
5.21

FIGURE 5.17 Examples of inverse agonists

Unless a receptor either shows or can be coaxed to show constituent activity, both antagonists and inverse agonists will seem to have the same impact on response: no effect. The previous subsection describes the design of H_2 receptor antagonists. Cimetidine (**5.20**), the drug that arose from the search for H_2 receptor *antagonists*, was later found to act as an *inverse agonist* instead of an antagonist (**Figure 5.17**).[22] Another inverse agonist is rimonabant (**5.21**). Rimonabant is an appetite-suppressing drug that acts on cannabanoid receptor-1, a constituently active GPCR.

5.4 Receptor Theories

Receptor theories strive to explain how the various types of ligands give rise to a response. The variety of receptors and the signaling pathways to which they are linked pose a significant challenge to receptor theory. Receptor response pathways are far more complex than the activity of enzymes, and therefore the modeling of receptors is more complex than the Michaelis–Menten model seen in Chapter 4.

Early receptor theories have some ability to predict receptor behavior, but additional ideas have been incorporated to make the theories more broadly applicable. Despite advances, the expanded theories cannot account for all receptor activities. New and sometimes conflicting models have since been proposed. As a result, the current state of receptor theory is messy at best.

From a drug development perspective, the creation of a universal receptor theory is not important. Medicinal chemists and molecular biologists in a pharmaceutical company must identify which model best fits the intended drug target and allows accurate determination of ligand activity.

Occupancy Theory

Occupancy theory is the predominant receptor theory and is closely related to the enzyme model of Michaelis and Menten.

Clark

Clark developed occupancy theory in the 1920s and 1930s.[23] He built his theory on the premise that a response (E) arises only when a receptor is occupied by a ligand, that is, from a ligand-receptor complex (RL). The response is directly proportional to [RL] (Equation 5.3).

$$E \propto [\mathrm{RL}] \tag{5.3}$$

FIGURE 5.18 Fractional response vs. fractional occupation

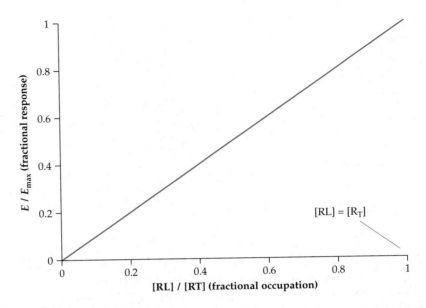

FIGURE 5.18 Fractional response vs. fractional occupation

The receptor itself can be present in one of two states: free (R) or bound in a ligand-receptor complex (RL) (Equation 5.4).

$$[R_t] = [R] + [RL] \tag{5.4}$$

The maximal response (E_{max}) will be reached if all the receptors are bound and no receptors are free (Equation 5.5).

$$E_{max} \propto [R_t] \tag{5.5}$$

Collectively, these statements are summarized in Equation 5.6 and **Figure 5.18**.

$$\frac{E}{E_{max}} = \frac{[RL]}{[R_t]} \tag{5.6}$$

A ligand and receptor exist in an equilibrium (Scheme 5.1) that can be quantified through the *dissociation equilibrium constant*, K_D (Equation 5.7).

$$K_D = \frac{[R][L]}{[RL]} \tag{5.7}$$

Through algebraic substitutions and rearrangements of Equations 5.4, 5.6, and 5.7, Equation 5.8 can be derived.

$$\frac{E}{E_{max}} = \frac{[L]}{K_D + [L]} \tag{5.8}$$

Equation 5.8 bears a striking resemblance to the Michaelis–Menten equation (Equation 5.9) because both equations link a binding event (receptor-ligand or enzyme-substrate) to an outcome (response or conversion).

$$V = \frac{V_{max}[S]}{K_m + [S]} \tag{5.9}$$

By Equation 5.7, when the free receptor concentration, [R], equals the bound receptor concentration, [RL], K_D is equivalent to [L]. Interestingly, when free and bound receptor concentrations are equal ($[RL] = 0.5[R_t]$), the response is 50% (Equations 5.4 and 5.6). In previous discussions, EC_{50} was defined to be [L] required to provide 50% response.

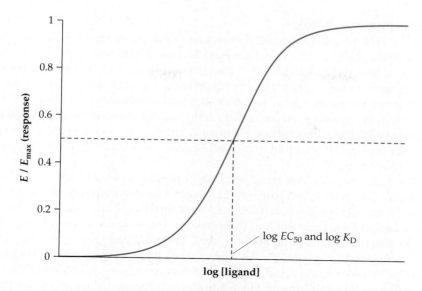

FIGURE 5.19 Equivalence of EC_{50} and K_D

Therefore, based on Clark's occupancy theory, EC_{50} equals K_D. Both occur at the point of inflection of a response versus log [ligand] plot (**Figure 5.19**).

K_D is an important property of a receptor-ligand system. A ligand with a small K_D is said to have a high *affinity* for the receptor. Affinity is related to binding energy, and the standard free energy of binding can be determined from the *association equilibrium constant*, K_A, or $1/K_D$ (Equation 5.10).

$$\Delta G^\circ = -2.3RT \log K_A = +2.3RT \log K_D \tag{5.10}$$

Conveniently, once a response versus log [L] plot has been prepared, the binding energy of a ligand may be directly calculated based on log K_D. Often, binding energies will be calculated to highlight energy gains or costs associated with structural changes in a ligand. This leads directly to the concept called *quantitative structure-activity relationships*, or QSAR. QSAR is more fully discussed in Chapter 12.

| Sample Calculation | The Relationship of K_D, [L], and Response |

PROBLEM   The K_D of a receptor-ligand complex is 30 nM. The ligand is a full agonist. What concentration of ligand is required to achieve 75% response?

SOLUTION   Rearrangement of Equation 5.8 gives of form of the equation that allows [L] to be calculated directly. At this point, the problem is just a matter of substituting the correct values in place of the variables. The concentration of the ligand for 75% response would be 90 nM.

$$\frac{E}{E_{max}} = \frac{[L]}{K_D + [L]} \tag{5.8}$$

$$[L] = \frac{\dfrac{E}{E_{max}} K_D}{1 - \dfrac{E}{E_{max}}}$$

$$[L] = \frac{\dfrac{0.75}{1.0} \times 30 \text{ nM}}{1 - \dfrac{0.75}{1.0}} = \frac{0.75 \times 30 \text{ nM}}{0.25} = 90 \text{ nM}$$

Rosenthal (Scatchard) Plots

K_D values can be predicted through dose-response graphs, or they may be determined from a receptor binding study. Scatchard published a mathematical model for correlating binding data into a K_D value in 1949.[24] Rosenthal later modified Scatchard's equations to provide more useful graphs for interpreting the experimental data.[25] Three things are required for obtaining data for a Rosenthal plot: cell membranes containing the receptor of interest, the ligand of interest, and an antagonist for the receptor. The ligand being tested must be labeled with a radioactive isotope. Tritium, ^3H, is commonly used in place of a regular proton, ^1H, because virtually all ligands contain hydrogen atoms. Labels of ^{32}P and ^{14}C are common as well.

In the experiment, the cell membranes are isolated to serve as a source of the receptor. Some membranes are incubated with a range of concentrations of the radiolabeled ligand (L*), and the rest are incubated with both various concentrations of ligand and a large excess of an antagonist (Ant) (Scheme 5.6). The samples containing only the ligand will contain both free and bound ligand. There are two types of binding for the ligand: binding to the receptor (specific binding) and binding to other cellular components (nonspecific binding). In the samples containing antagonist, specific binding of the ligand to the receptor is blocked. Nonspecific binding is still present. At the end of incubation, all samples are filtered. Ligands bound to the receptor or other cellular structures are caught in the filter, while unbound ligand is washed away. Measuring the radioactivity in the filter allows determination of both nonspecific binding and total binding (specific and nonspecific binding). The difference between these two values is the specific binding of the ligand to the receptor.[24]

The data from the binding study may be plotted to generate a saturation curve (**Figure 5.20**). Saturation curves plot specific binding against [L]. Specific binding is generally measured as the amount of ligand (unit = fmol, 1 femtomole = 10^{-15} mole) bound per amount of membrane protein (unit = mg). Equation 5.11 provides the relationship of a hyperbola in the saturation plot. Maximal specific binding corresponds to B_{max}, and K_D corresponds to the ligand concentration needed to achieve 50% occupancy of the receptors. If the ligand and receptor bind in a 1:1 ratio, then B_{max} also indicates the concentration of receptors in the membrane (fmol/mg).

$$\text{specific binding} = \frac{B_{max}[L]}{K_D + [L]} \tag{5.11}$$

specific and nonspecific binding (total binding)

$$\text{receptors} + \text{other components} \xrightarrow{\text{L*}} \begin{array}{c} \text{receptors-L*} \\ + \\ \text{other components-L*} \\ + \\ \text{unbound L*} \end{array} \xrightarrow{\text{filter}} \begin{array}{c} \text{receptors-L*} \\ + \\ \text{other components-L*} \end{array}$$

nonspecific binding

$$\text{receptors} + \text{other components} \xrightarrow{\text{L* + Ant}} \begin{array}{c} \text{receptors-Ant} \\ + \\ \text{other components-L*} \\ + \\ \text{unbound L*} \end{array} \xrightarrow{\text{filter}} \begin{array}{c} \text{receptors-Ant} \\ + \\ \text{other components-L*} \end{array}$$

SCHEME 5.6 Determination of specific and nonspecific binding through a radiolabeled ligand (L*)

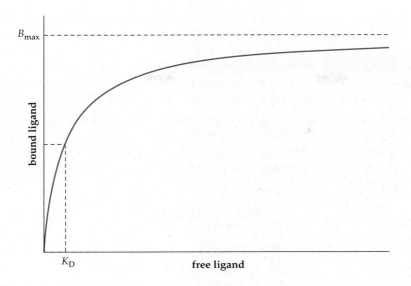

FIGURE 5.20 Representative saturation curve

Binding data is more traditionally presented after the manner of Equation 5.12 in the form of a Rosenthal plot (**Figure 5.21**).[25]

$$\frac{\text{bound}}{\text{free}} = -\frac{1}{K_D}\text{bound} + \frac{B_{max}}{K_D} \qquad (5.12)$$

The y-axis in a Rosenthal plot is the ratio of bound (specific binding, fmol/mg) to free (nM) ligand. The x-axis is simply the bound ligand. The line in a Rosenthal plot has a slope of $-1/K_D$. The units of K_D are concentration, which are the same units assigned to the free ligand, nM in this example. The x-intercept corresponds to B_{max}.

Determining K_D values through binding studies is notoriously difficult and requires excellent technique and proper accounting for background levels of radioactivity. Regardless, it has been a standard method for determining a K_D value, a process casually referred to as "grind and bind." The Rosenthal approach often does not provide the same K_D value as occupancy theory and dose-response curves. This fact, along with the inability of

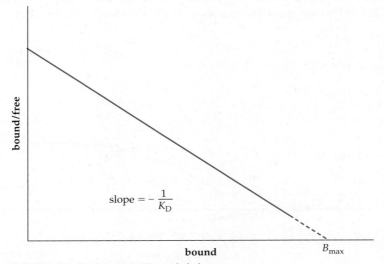

FIGURE 5.21 Representative Rosenthal plot

Clark's simple version of occupancy theory to explain partial agonists, has led researchers to expand and improve the model of occupancy theory.[26]

Sample Calculation   **Being Careful with Units**

PROBLEM   Using Equation 5.12 and the units provided in the discussion preceding Equation 5.12, prove that the units on B_{max} are fmol/g.

SOLUTION   The units on the variables in Equation 5.12 are as follows: bound = fmol/g, free = nM, and K_D = nM. Substitution of these units into Equation 5.12 sets up the solution. The nM concentration unit cancels across all parts of the equation to leave only fmol/g. The units on B_{max} are fmol/g, the number of moles of binding sites per gram of membrane protein.

$$\frac{bound}{free} = -\frac{1}{K_D}bound + \frac{B_{max}}{K_D} \tag{5.12}$$

$$\frac{\frac{fmol}{g}}{nM} = -\frac{1}{nM} \times \frac{fmol}{g} + \frac{B_{max}}{nM}$$

$$\frac{fmol}{g} = -\frac{fmol}{g} + B_{max}$$

$$B_{max} = \frac{fmol}{g}$$

Ariens and Stephenson

In the 1950s, Ariens[27] and Stephenson[28] independently developed modifications to extend Clark's occupancy theory. Ariens introduced the concept of *intrinsic activity*, denoted by the variable e (Equation 5.13).

$$\frac{E}{E_{max}} = \frac{e[L]}{K_D + [L]} \tag{5.13}$$

Intrinsic activity is a variable that accommodates the observation that not all ligands are able to elicit a full response from the receptor. For full agonists, e is simply equal to 1.0 (**Figure 5.22**). In the case of partial agonists, e will be less than 1 and greater than 0. For an

FIGURE 5.22 Intrinsic activity and various types of ligands with same K_D

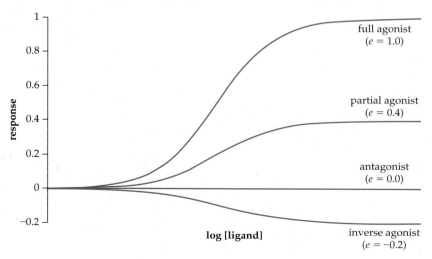

antagonist, e is 0. While the existence of inverse agonists is inconsistent with occupancy theory, by using a negative e value, one can generate a dose-response curve consistent with an inverse agonist. Technically, however, inverse agonists do not cause a negative response. Inverse agonists instead decrease the positive response seen in constituently active receptors. Despite Ariens' advancements, occupancy theory still directly tied ligand binding and response (Figure 5.18). The slope for a response-occupancy plot may be less than 1, but the relationship is still necessarily linear.[26]

Stephenson broke the direct link between binding and response.[28] Stephenson redefined response as a new term, *stimulus* (S) (Equation 5.14).

$$S = \frac{e[\mathrm{L}]}{K_{\mathrm{D}} + [\mathrm{L}]} \qquad (5.14)$$

The ligand causes a stimulus by binding a receptor. That stimulus is then transformed by the cellular machinery into a response. Stephenson described the mathematical relationship between the stimulus and response as the *transducer function* (Equation 5.15). The response is some function, most likely nonlinear, of the stimulus.

$$E = f(S) \qquad (5.15)$$

The transducer function allows occupancy theory to explain ideas that cannot be accommodated by Clark's original theory. One idea is that of *spare receptors*.[28] In some cases, maximal response can be achieved without all the receptors being occupied (**Figure 5.23**). Binding the remaining receptors, called spare receptors, has no impact on the response. Another problem topic for Clark and Ariens is the idea of inverse agonists. Transducer functions can be written to give an initial response (E_0) even in the absence of a stimulus (Equation 5.16). E_0 is consistent with the observed activity of constituently active receptors. If the transducer function gives a negative

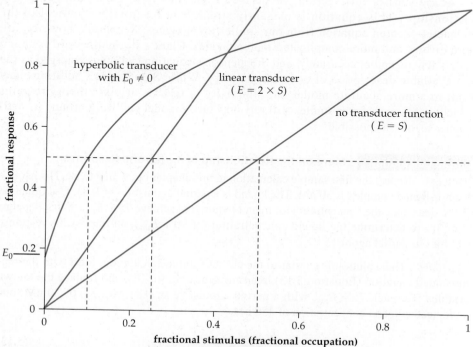

FIGURE 5.23 Response vs. stimulus relationships from transducer functions

value in Equation 5.16, then the response will decrease below E_0, just as would be expected with an inverse agonist.

$$E = E_0 + f(S) \tag{5.16}$$

Stephenson also expanded the intrinsic activity term as the product of the total receptor concentration ($[R_t]$) and *intrinsic efficacy* (ε) (Equation 5.17).

$$e = \varepsilon[R_t] \tag{5.17}$$

Intrinsic efficacy is the efficacy, or activity, per unit receptor. By including the idea of intrinsic efficacy, Stephenson explained how tissues with the same concentrations of a receptor can give rise to different dose-response graphs. The full results of Stephenson's contributions to occupancy theory are summarized in Equation 5.18.[26]

$$E = E_0 + f\!\left(\frac{\varepsilon[R_t][L]}{K_D + [L]}\right) \tag{5.18}$$

Further complicating the receptor picture, receptor concentration is not necessarily constant over time within a given tissue. If a receptor is stimulated for a long period of time by a high ligand concentration, cells in that tissue may react by decreasing the concentration of receptors in a process called *downregulation*. As $[R_t]$ decreases, the response decreases as well. If the original level of response is to be reached, the ligand concentration must be raised. This phenomenon is called *desensitization*. Patients taking certain drugs, such as opioids for pain management, need to elevate their dose over time as they become desensitized to the medication. If those same patients stop taking the drug, the stimulation will cease, and the cells eventually *upregulate* the receptor concentration to the original levels. The patients are again *sensitized* to the drug. If they resume taking the drug, they must start with their initial dosing regimen or else risk an overdose.[26]

At the start of this section, we derived Equation 5.8 to model dose-response relationships. This equation is elegantly simple and essentially identical to the Michaelis–Menten equation from our studies on enzymes. Receptors, however, are more diverse and more complicated than enzymes. Clark's straightforward equation models few receptors accurately, and Stephenson's equation (5.18) has emerged as the best available description of occupancy theory. While Stephenson's additions may result in a more accurate model, the simplicity of Clark's original theory remains attractive. Many receptor studies still rely on Clark's model and work around its deficiencies as best as possible.

Sample Calculation | **More on the Relationship of K_D, [L], and Response**

PROBLEM Repeat the first sample calculation in this chapter but with a twist. The K_D of a receptor-ligand complex is 30 nM. The ligand is a partial agonist ($e = 0.8$). Use Equation 5.13 to determine the concentration of ligand required to achieve 75% E_{max}. What happens if you try to determine the ligand concentration for 90% E_{max}, an impossible response level for our partial agonist?

SOLUTION This solution is similar to the earlier sample calculation. Start with the recommended equation (Equation 5.13), rearrange to solve for [L], and replace the known variables. To reach 75% E_{max} with a partial agonist ($e = 0.8$) with K_D of 30 nM, one would need a ligand concentration of 450 nM.

$$\frac{E}{E_{max}} = \frac{e[L]}{K_D + [L]} \tag{5.13}$$

$$[L] = \frac{\dfrac{E}{E_{max}} K_D}{e - \dfrac{E}{E_{max}}}$$

$$[L] = \frac{\dfrac{0.75}{1.0} \times 30\text{ nM}}{0.8 - \dfrac{0.75}{1.0}} = \frac{0.75 \times 30\text{ nM}}{0.05} = 450\text{ nM}$$

The second part of the problem is just as straightforward. Replace the known variables and determine the concentration of L.

$$[L] = \frac{\dfrac{0.9}{1.0} \times 30\text{ nM}}{0.8 - \dfrac{0.9}{1.0}} = \frac{0.9 \times 30\text{ nM}}{-0.1} = -270\text{ nM}$$

A negative ligand concentration may not make chemical sense, but we must remember that we are asking the impossible here, that is, a partial agonist ($e = 0.8$) to reach 90% maximal response. To make sense of a negative concentration, we must consider that Equation 5.13 is a hyperbolic function. We concern ourselves only with the part of the function that falls within the first quadrant: x-axis (concentration) ≥ 0 and y-axis (response) ≥ 0. The function is defined in the other three quadrants despite not making sense in a biological system. The following graph shows a larger part of the function. Note that the first quadrant alone is the same curve as Figure 5.6, a standard response curve. The full function has two asymptotic lines. One is $x = -K_D$, and the other is $y = e\ (0.8)$.

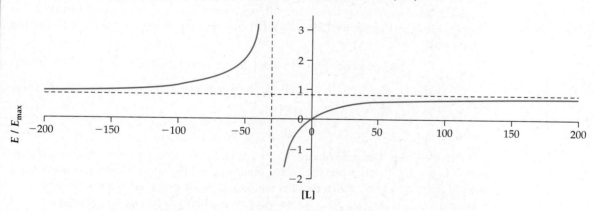

The take-home lesson of this exercise is to be critical of the numbers that formulas give you. Just because you get a number does not mean it is reasonable.

Allosteric Theory

While occupancy theory is far and away the most widely used model for describing dose-response curves, other theories do exist. One example is allosteric theory. At the center of allosteric theory, sometimes called the *two-state model*, is the idea that a receptor can exist in conformations that either cause a response (*relaxed state*) or do not cause a response (*tensed state*).[29] These conformations, represented by T and R, are in equilibrium (Scheme 5.7).

$$R \quad \underset{}{\overset{K}{\rightleftharpoons}} \quad T$$

(response)   (no response)

SCHEME 5.7 Equilibrium of the relaxed and tensed receptor conformations

The binding of a ligand to a receptor shifts the position of the equilibrium. According to allosteric theory, agonists stabilize the relaxed conformation. The equilibrium shifts to the left, and the response increases. Inverse agonists stabilize the tensed conformation and decrease the baseline response. Antagonists bind both conformations equally and do not shift the equilibrium from its original position.

In the absence of a ligand, the receptor will exist to some extent in its relaxed conformation and give rise to a degree of response. By this logic, allosteric theory successfully explains a baseline response of greater than zero as seen with constituently active receptors. This advantage of allosteric theory comes with a cost of complexity. A receptor has essentially an infinite number of conformations, with likely hundreds or even thousands corresponding to important energy minima. The roles of all the important, low-energy conformations must be understood to predict the equilibrium position and response of the receptor. Even for a very simple system, this type of challenge pushes the current understanding of molecular biology and molecular modeling to its absolute limits.

Rate Theory

Another alternative to occupancy theory is rate theory. Rate theory was developed by Paton through examination of receptors that bind stimulants.[30] Paton proposed that a response is caused by the act of binding, not the state of being bound or free (**Scheme 5.8**). This seemingly subtle difference shifts the theory away from K_D and toward k_{on} and k_{off}, the *rate constants* of association and dissociation. Interestingly, at equilibrium, K_D is equal to k_{off}/k_{on} (Equations 5.19–5.21). For this reason, occupancy and rate theory are closely related.

$$\text{rate of association} = \text{rate of dissociation (at equilibrium)} \tag{5.19}$$

$$k_{on}[R][L] = k_{off}[RL] \tag{5.20}$$

$$\frac{[R][L]}{[RL]} = \frac{k_{off}}{k_{on}} = K_D \tag{5.21}$$

An interesting outcome of rate theory is that k_{off} is very important. Because the response is triggered only when the ligand binds, a good ligand (full agonist) must release quickly (high k_{off}) to allow another the binding/response event. Response is directly proportional to k_{off} (Equation 5.22). Under occupancy theory, a full agonist should bind and stay bound. Under rate theory, a full agonist must release the receptor at a high rate.

$$\text{response} \propto k_{off} \tag{5.22}$$

Rate theory does successfully model the behavior of many receptors. It can also be adapted to accommodate partial agonists and antagonists. Determining k_{on} and k_{off} requires more experimental effort than determining K_D.

$$RL \quad \underset{k_{on}}{\overset{k_{off}}{\rightleftharpoons}} \quad R + L$$

SCHEME 5.8

Understanding k_{off}

PROBLEM The value for k_{on} is limited by the rate of diffusion. The diffusion-limited value for k_{on} is $\sim 10^8\ M^{-1}s^{-1}$. Using Equation 5.21, determine the k_{off} of a ligand with a K_D of 100 nM.

SOLUTION Only the right half of Equation 5.21 concerns us. We can rearrange it to directly calculate k_{off}. As always, be careful with the units ($100\ nM = 10^{-7}\ M$). k_{off} turns out to be $10\ s^{-1}$.

$$\frac{k_{off}}{k_{on}} = K_D \tag{5.21}$$

$$k_{off} = K_D k_{on}$$
$$k_{off} = (1 \times 10^{-7}\,M) \times (1 \times 10^8\,M^{-1} \cdot s^{-1}) = 10\ s^{-1}$$

Drug-Target Residence Time

The concept of drug-target residence time is not a formal receptor theory that strives to model the action of various types or receptors. Drug-target residence time is a form of occupancy theory (bound receptor = response) that emphasizes the importance of k_{off}.[31] Remember that occupancy emphasizes the role of the RL complex. Under the simple occupancy model, maximizing [RL] also maximizes response (Equation 5.3). Ligands with a very small K_D, indicating strong RL binding, should be the most potent ligands. Drug discovery groups therefore try to design ligands with strong binding and correspondingly low K_D values.

The idea of drug-target residence time places less emphasis on the *strength* of receptor-ligand interaction and more emphasis on the *duration* of receptor-ligand interaction. The duration of interaction is defined as the *residence time* (τ) of the ligand, which is equal to the inverse of k_{off} (Equation 5.23). A closely related variable is *dissociative half-life* ($t_{1/2}$), also defined by k_{off} (Equation 5.24).[31]

$$\tau = \frac{1}{k_{off}} \tag{5.23}$$

$$t_{1/2} = \frac{0.693}{k_{off}} \tag{5.24}$$

In practice, ligands with a long time of residence are able to become deeply buried in a binding pocket. Such interactions are sometimes associated with either a ligand's becoming entwined in a receptor or a receptor's folding around the ligand. For these reasons, residence time discussions often expand the binding equilibrium picture from the simple model in Scheme 5.1 to the slightly more complex model in Scheme 5.9. which shows binding of the native conformation of the receptor (R) to the ligand. Consistent with ideas from the induced fit model, the receptor then undergoes conformational changes (R*) to more fully interact with the ligand.[31]

Most drug discovery groups emphasize the importance of K_D and maximize the strength of receptor-ligand binding. Strongly bound ligands often also have a long residence time, but the two ideas are not always directly related. In a study of

$$R + L \rightleftharpoons RL \rightleftharpoons R^*L$$

SCHEME 5.9 A residence time receptor-ligand binding model

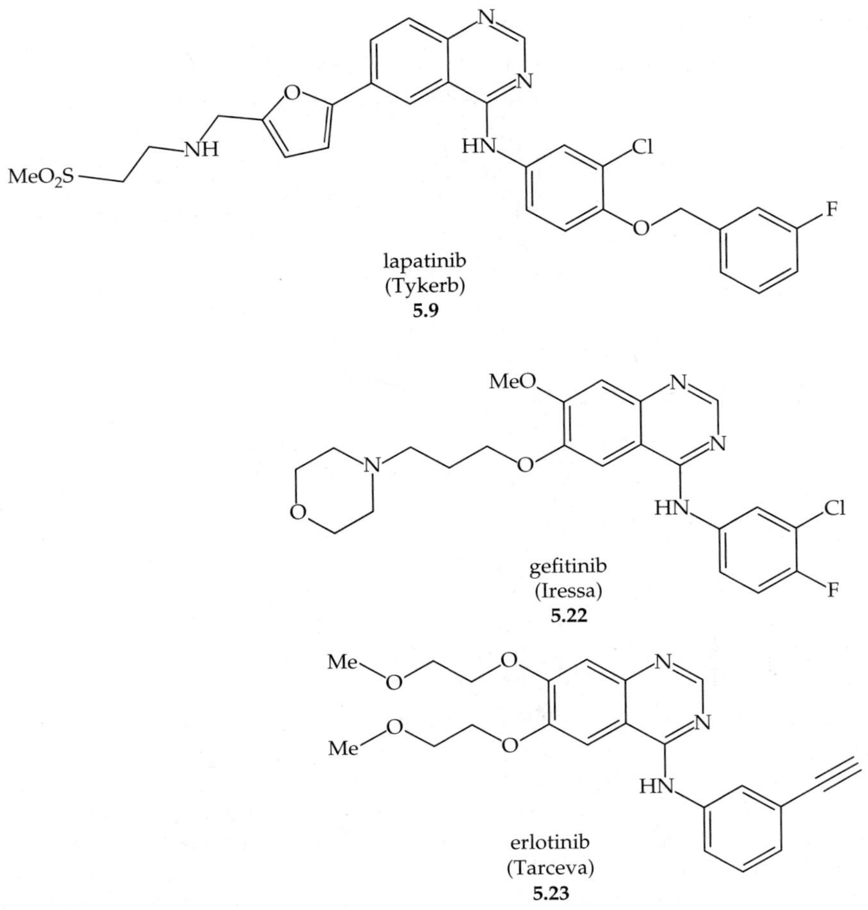

	K_i (nM)	τ (min)
lapatinib (Tykerb) **5.9**	3.0	300
gefitinib (Iressa) **5.22**	0.40	<10
erlotinib (Tarceva) **5.23**	0.70	<10

FIGURE 5.24 EGFR inhibitors with K_i and τ values

EGFR (epidermal growth factor receptor) ligands, a research group compared three drugs: lapatinib (Tykerb, **5.9**), gefitinib (Iressa, **5.22**), and erlotinib (Tarceva, **5.23**) (**Figure 5.24**). Lapatinib was found to be an especially potent inhibitor of the activity of EGFR, a tyrosine-linked kinase receptor. Surprisingly, comparison of the K_i values of the three compounds indicated that lapatinib should be the weakest inhibitor. Further study of the compounds revealed that the residence time of lapatinib on EGFR was at least 30-fold longer than either gefitinib or erlotinib. The long residence time was used to explain the profound inhibition of EGFR that could not be explained by K_i values alone.[32]

A frustrating aspect of receptor theories is that some theories contradict others. Rate theory indicates that a strong ligand has a large value for k_{off} so that ligand-target binding can occur repeatedly. Under the residence time model, a strong ligand has a small value of k_{off} so that the target stays bound for a long time. These two theories are at odds with one another. This apparent contradiction must be viewed with regard to the incredible variety of receptors and ligands found in an organism. Some systems may be modeled best with rate theory, while others are more accurately approximated with the residence time model. Different systems may call for different, even conflicting models.

Summary

Receptors are large proteins or clusters of multiple proteins that initiate biological processes upon binding to an appropriate ligand. Conceptually, both enzymes and receptors share much of the same theory regarding their function. Both control biological pathways. Receptors, however, are not catalysts and do not perform chemical reactions. Receptors are more akin to a switch and often transmit an "on" or "off" signal across a membrane. Receptors are often classified into superfamilies based on their mode of action. The two types that are most often targeted for drug interactions are the ligand-gated ion channels and G-protein–coupled receptors. The ligand-gated ion channels allow ion flow into or out of a cell when a ligand binds the extracellular side of the membrane-bound receptor. G-protein–coupled receptors are also bound by a ligand on their extracellular face. Binding causes conformational changes on the intracellular face to initiate processes inside a cell. Ligands do not always turn on a receptor. Some only partially activate a receptor, and others can turn a receptor off (or prevent it from being turned on). Ligands that bind a receptor and fully initiate a biological response at high concentration are called agonists. Receptors are named after a naturally occurring, or endogenous, molecule present in an organism. The activity of the endogenous ligand defines "full" activity for a ligand. Unnatural, or exogenous, ligands can also be agonists. If an exogenous ligand causes a less than full response at high concentration, it is classified as a partial agonist. Antagonists reduce the response of an activated receptor. Mathematical models have been proposed to describe the actions of agonists and antagonists. The most widely used model is occupancy theory, which determines the response level by the fraction of receptors bound by a ligand. Occupancy theory can effectively model the behavior of most, but not all, receptors. Other receptor theories exist, but they require experimental information about a receptor that can be difficult to obtain.

Questions

1. Thiomuscimol (**5.a**) is a $GABA_A$ agonist. From the following binding data, create a log [L]-response graph and estimate the log K_D (and K_D) values for thiomuscimol. (Binding data estimated from Figure 4A of Ebert, B., Thompson, S. A., Saounatsou, K., McKernan, R., Krogsgaard-Larsen, P., Wafford, K. A. *Mol. Pharmacol.* 1997, *52*, 1150–1156.)

thiomuscimol
5.a

Thiomuscimol (μM)	Response (%)
1.0	0
10	8
30	39
100	78
300	100
1,000	97

2. As a complement to question 1, plot the binding data of thiomuscimol in Michaelis–Menten (response versus [L]) format. Try to create a Lineweaver–Burk plot (1/response versus 1/[L]) and perform a linear regression on the data. What is the problem you encounter while making this graph? Do your best to graph the data. From the best-fit line of the Lineweaver–Burk plot, determine K_D and E_{max}. How well does K_D in this graph match the K_D you determined in question 1? How well does E_{max} match the expected maximum response of 100%?

3. While the equations and mathematics behind the Michaelis–Menten equation and Clark's occupancy theory are similar, the data in an enzyme kinetics and receptor binding study are very different. In the following table are the rate data for a study of invertase, the same enzyme involved in the original work of Michaelis and Menten. Graph V against [S]. Repeat with $1/V$ against $1/[S]$ and perform a linear regression to determine K_m and V_{max} in the spirit of Lineweaver and Burk. Using the calculated V_{max}, create a sigmoidal plot by graphing V/V_{max} against log [S]. This graph is essentially a log [L]-response curve (except with enzymatic data). Can you accurately determine log K_D? What is missing from this data? (Merino, F. D. M. A New Method for Determining the Michaelis Constant. *Biochem. J.* 1974, *143*, 93–95.)

V (time^{-1})	[S] (conc.)
22.0	0.1370
20.5	0.0995
19.0	0.0676
12.5	0.0262
9.0	0.0136
7.0	0.0100
6.0	0.0079

4. Ligand concentration, [L], is generally *assumed* to be the amount of ligand (moles) divided by the volume of the experiment (liters). Specifically, the unbound ligand concentration [L]

is set equal to the total ligand concentration $[L_t]$. In the presence of a receptor, this assumption is incorrect. What is the relationship between [L] and $[L_t]$? (Hint: In what other form does the ligand exist?) Although technically incorrect, why does the relationship $[L] = [L_t]$ work well in most instances?

5. Using Equation 5.12, show that B_{max} equals the x-intercept. Why is B_{max} a theoretical value in Figure 5.21? (Hint: What conditions must be met to reach B_{max}?)

6. Sketch Figure 5.11. Add a third line, a curve for addition of a partial agonist to a mixture with a constant amount of competitive antagonist.

7. Create a table showing the percentage binding of a full agonist to a receptor at response levels of 20, 40, 60, and 80%. Assume a simple Clark occupancy model.

8. What is the percentage response for a full agonist at $[L] = 0.01K_D$? $0.1K_D$? $10K_D$? $100K_D$? What is the percentage response for a partial agonist ($e = 0.35$) at $[L] = 0.01K_D$? $10K_D$?

9. Which equation in this chapter best accounts for the concepts of downregulation and upregulation?

10. Can the allosteric model account for differences between competitive and noncompetitive antagonists? Explain why or why not.

11. A molecule binds a receptor but does not cause a response or affect the response caused by an agonist. How is this possible?

12. A partial agonist can often serve in place of an antagonist. Explain how this statement can be true.

13. Molecules that prevent a response from receptors are called antagonists. Sometimes antagonists for tyrosine kinase–linked receptors are called inhibitors. Why is this appropriate?

14. How might a partial agonist be explained using rate theory?

15. Figure 5.23 shows three lines for three different transducer functions. Sketch a graph of the log [L]-response curve for the $E = S$ line. This should be fairly easy with no surprises. Now, on the same set of axes, sketch the log [L]-response curve for the $E = 2 \times S$ line. If you are uncertain, create an Excel graph to determine the shape of the curve. Discuss the differences.

References

1. Nelson, D. L., & Cox, M. M. Biosignaling. *Lehninger Principles of Biochemistry*, (3rd ed., Chapter 13). New York: Worth, 2000.

2. Alberts, B., Johnson, A., Lewis, J., Raff, M., Roberts, K., & Walter, P. Cell Communication. *Molecular Biology of the Cell*, (4th ed., Chapter 13). New York: Garland Science, 2002.

3. Lo, M., Liu, K. L., Lantelme, P., & Sassard, J. Subtype 2 of Angiotensin II Receptors Controls Pressure-Natriuresis in Rats. *J. Clin. Invest.* 1995, *95*, 1394–1397.

4. Miyata, N., Park, F., Li, X. F., & Cowley, Jr., A. W. Distribution of Angiotensin AT_1 and AT_2 Receptor Subtypes in the Rat Kidney. *Am. J. Physiol. Renal Physiol.* 1999, *277*, F437–F446.

5. Williams, M., Deecher, D. C., & Sullivan, J. P. Drug Receptors. In M. E. Wolff (Ed.), *Burger's Medicinal Chemistry and Drug Discovery*, (5th ed., Chapter 11). New York: Wiley & Sons, 1996.

6. Hucho, F., & Weise, C. Ligand-Gated Ion Channels. *Angew. Chem., Int. Ed.* 2001, *40*, 3100–3116.

7. Ortells, M. O., & Lunt, G. G. Evolutionary History of the Ligand-Gated Ion-Channel Superfamily of Receptors. *Trends Neurosci.* 1995, *18*, 121–127.

8. Ji, T. H., Grossmann, M., & Ji, I. G Protein-Coupled Receptors. *J. Biol. Chem.* 1998, *273*, 17299–17302.

9. Palczewski, K. T., Kumasaka, T., Hori, T., Behnke, C. A., Motoshima, H., Fox, B. A., et al. Crystal Structure of Rhodopsin: A G protein-Coupled Receptor. *Science.* 2000, *289*, 739–745.

10. Stenkamp, R. E., Teller, D. C., & Palczewski, K. Crystal Structure of Rhodopsin: A G-Protein-Coupled Receptor. *ChemBioChem.* 2002, *3*, 963–967.

11. Rasmussen, S. G., Choi, H. J., Rosenbaum, D. M., Kobilka, T. S., Thian, F. S., Edwards, P. C., et al. Crystal Structure of the Human β_2 Adrenergic G-protein-Coupled Receptor. *Nature* 2007, *450*, 383–387.

12. Hamm, H. E. The Many Faces of G Protein Signaling. *J. Biol. Chem.* 1998, *273*, 669–672.

13. Drews, J. Drug Discovery: A Historical Perspective. *Science.* 2000, *287*, 1960–1964.

14. Schlessinger, J. Cell Signaling by Receptor Tyrosine Kinases. *Cell.* 2000, *103*, 211–225.

15. Robinson-Rechavi, M., Garcia, H. E., & Laudet, V. The Nuclear Superfamily. *J. Cell Sci.* 2003, *116*, 585–586.

16. Freedman, L. P. Anatomy of the Steroid Receptor Zinc Finger Region. *Endocr. Rev.* 1992, *13*, 129–145.

17. U.S. Food and Drug Administration. Avandia (Rosiglitazone): Ongoing Review of Cardiovascular Safety. http://www.fda.gov/Safety/MedWatch/SafetyInformation/SafetyAlertsforHumanMedicalProducts/ucm201446.htm (accessed July 2012).

18. Ross, E. M., & Kenakin, T. P. Pharmacodynamics. In J. G. Hardman, & L. E. Limbird (Eds.), *Goodman and Gilman's the Pharmacological Basis of Therapeutics*, (10th ed., Chapter 2). New York: McGraw-Hill, 2001.

19. Cronbach, L. J., Gleser, G. C., & Loewe, S. Quantal and Graded Analysis of Dosage-Effect Relations. *Science.* 1961, *133*, 1924–1925.

20. Chavez-Noriega, L. E., Crona, J. H., Washburn, M. S., Urrutia, A., Elliot, K. J., & Johnson, E. C. Pharmacological Characterization of Recombinant Human Neuronal Nicotinic Acetylcholine Receptors h$\alpha2\beta2$, h$\alpha2\beta4$, h$\alpha3\beta4$, h$\alpha4\beta2$, h$\alpha4\beta4$ and h$\alpha7$ Expressed in *Xenopus* Oocytes. *J. Pharmacol. Exp. Ther.* 1997, *280*, 346–356.

21. Wenger, B. W., Bryant, D. L., Boyd, R. T., & McKay, D. B. Evidence for Spare Nicotinic Acetylcholine Receptors and a β_4 Subunit in Bovine Adrenal Chromaffin Cells: Studies

Using Bromoacetylcholine, Epibatidine, Cytisine and mAb35. *J. Pharmacol. Exp. Ther.* 1997, *281,* 905–913.

22. Ganellin, G. R., & Durant, G. J. Histamine H_2-Receptor Agonists and Antagonists. In M. E. Wolff (Ed.), *Burger's Medicinal Chemistry,* (4th ed., Chapter 48). New York: Wiley & Sons, 1981.

23. Clark, A. J. *The Mode of Action of Drugs on Cells.* London: Edward Arnold, 1933.

24. Scatchard, G. The Attractions of Proteins for Small Molecules and Ions. *Ann. N. Y. Acad. Sci.* 1949, *51,* 660–672.

25. Rosenthal, H. E. A Graphical Method for the Determination and Presentation of Binding Parameters in a Complex System. *Anal. Biochem.* 1967, *20,* 525–532.

26. Rang, H. P. The Receptor Concept: Pharmacology's Big Idea. *Br. J. Pharmacol.* 2006, *147,* S9–S16.

27. Ariens, E. J. Affinity and Intrinsic Activity in the Theory of Competitive Inhibition: Part I. Problems and Theory. *Arch. Int. Pharmacodyn. Ther.* 1954, *99,* 32–49.

28. Stephenson, R. P. A Modification of Receptor Theory. *Br. J. Pharmacol.* 1956, *11,* 379–393.

29. Leff, P. The Two-State Model of Receptor Activation. *Trends Pharmacol. Sci.* 1995, *16,* 89–97.

30. Paton, W. D. M. A Theory of Drug Action Based on the Rate of Drug-Receptor Combination. *Proc. R. Soc. London, Ser. B.* 1961, *154,* 21–69.

31. Copeland, R. A., Pompliano, D. L., & Meek, T. D. Drug-Target Residence Time and Its Implications for Lead Optimization. *Nat. Rev. Drug. Disc.* 2006, *5,* 730–739.

32. Wood, E. R., Truesdale, A. T., McDonald, O. B., Yuan, D., Hassell, A., Dickerson, S. H., Ellis, B., et al. A Unique Structure for Epidermal Growth Factor Receptor Bound GW572016 (lapatinib): Relationships among Protein Conformation, Inhibitor Off-Rate, and Receptor Activity in Tumor Cells. *Cancer Res.* 2004, *64,* 6652–6659.

Chapter 6

Oligonucleotides as Drug Targets

Chapter Outline

Learning Goals for Chapter 6

- Recognize the structural uniqueness of oligonucleotides as drug targets

- Know the different types of DNA and RNA and their roles in the cell

- Understand the general methods of binding oligonucleotides

- Distinguish indirect methods of preventing DNA synthesis and function

Enzymes and receptors, both of which are proteins, are common targets for intervention in the pharmaceutical industry. Being so common, proteins are well understood and to some degree the default target for disease treatment. Oligonucleotides do offer an alternative target. Polynucleic acids encode the proteins that comprise enzymes and receptors. Binding a drug to oligonucleotides can interfere with protein synthesis and greatly impair cellular function.

6.1 The Basics of Nucleic Acids

Oligonucleotides are nucleic acid polymers. Each nucleic acid is assembled from a sugar, a nitrogenous heterocyclic base, and a phosphate. The specific identity of the sugar and base determines the type of nucleic acid, which in turn determines the structure and function of the oligonucleotide.

The Building Blocks of Nucleic Acids

Polymeric nucleic acids—deoxyribonucleic acid (DNA) and ribonucleic acid (RNA)—consist of repeating units. Each unit contains a sugar, base, and phosphate. The sugar is either D-ribose (**6.1**) or 2-deoxy-D-ribose (**6.2**). Both sugars can assume many different stable forms, but nucleic acids are built upon the ribofuranose forms that are shown in **Figure 6.1**. Nucleic acids containing D-ribose make up RNA; nucleic acids containing 2-deoxy-D-ribose form DNA. Haworth projections of the sugars commonly show the ring with substituents vertically up or down. A more proper picture would show an envelope conformation of the furanose ring. For the sake of following convention, the sugars of nucleic acids will be shown as Haworth projections in this text.

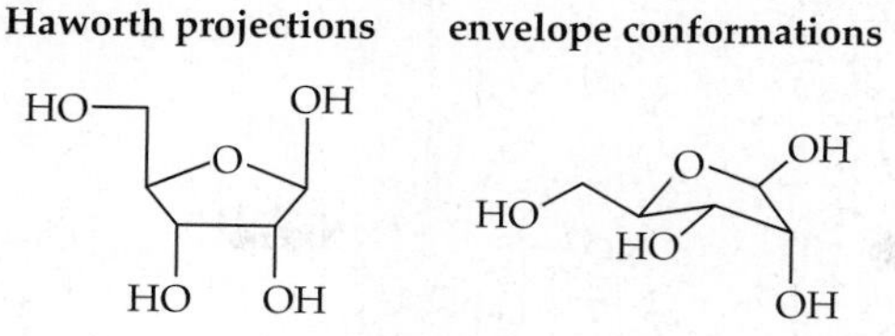

Haworth projections   **envelope conformations**

β-D-ribofuranose
(D-ribose)
6.1

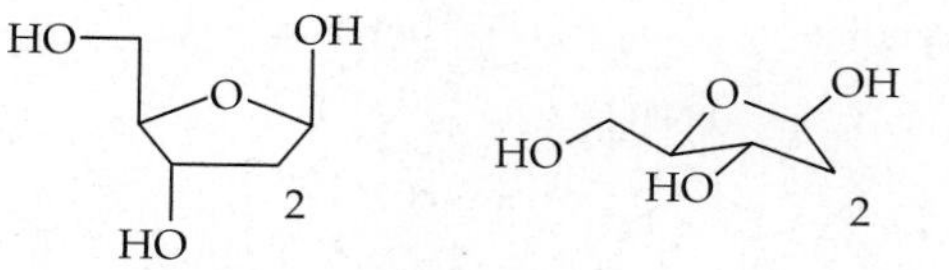

β-D-2-deoxyribofuranose
(2-deoxy-D-ribose)
6.2

FIGURE 6.1 Sugars as found in nucleic acids

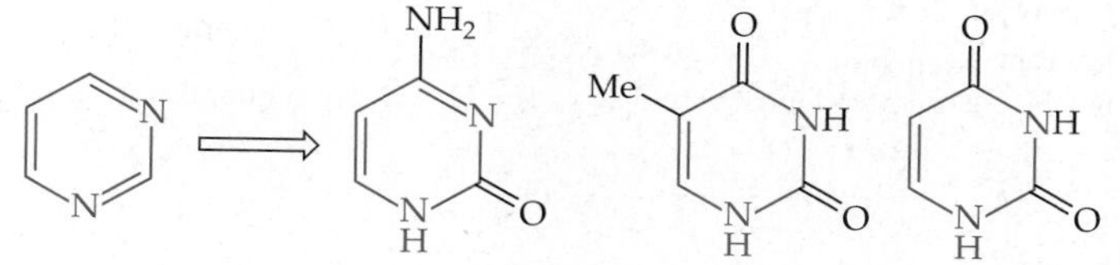

pyrimidine
6.3

cytosine
6.5

thymine
6.6

uracil
6.7

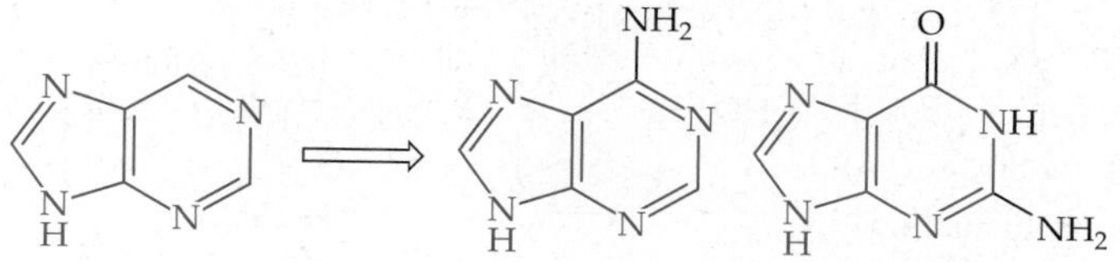

purine
6.4

adenine
6.8

guanine
6.9

FIGURE 6.2 Bases found in nucleic acids

The heterocyclic bases are either monocyclic pyrimidine (**6.3**) or bicyclic purine (**6.4**) derivatives (**Figure 6.2**). The specific bases found in nucleic acids are cytosine (**6.5**), thymine (**6.6**), uracil (**6.7**), adenine (**6.8**), and guanine (**6.9**).

Condensation of one of the two ribose derivatives in Figure 6.1 with a base shown in Figure 6.2 provides a *nucleoside*. Observed combinations with names are shown in **Figure 6.3**. All examples in Figure 6.3 consist of just a base and a sugar. Because the base thymine is only found attached to 2-deoxyribose, the resulting nucleoside is normally called thymidine instead of its full name, 2′-deoxythymidine. Similarly, uracil is only found attached to ribose, not 2′-deoxyribose. To distinguish positions on the base from those on the sugar, positions on the base are noted with regular numbers, while the sugar positions are "primed." Numbering is shown on cytidine and adenosine in Figure 6.3. Analogues of nucleosides can show biological activity, and the analogues are prepared by modifying the base, the sugar, or both. Examples of nucleoside analogues with properly numbered names include 3′-azido-3′-deoxythymidine (Retrovir, **6.10**), 8-aza-2′-deoxyadenosine (**6.11**), and 5-fluorouridine (**6.12**) (**Figure 6.4**). The physiological effects of **6.10** and **6.12** are discussed more fully later in the chapter.

FIGURE 6.3 Common natural nucleosides

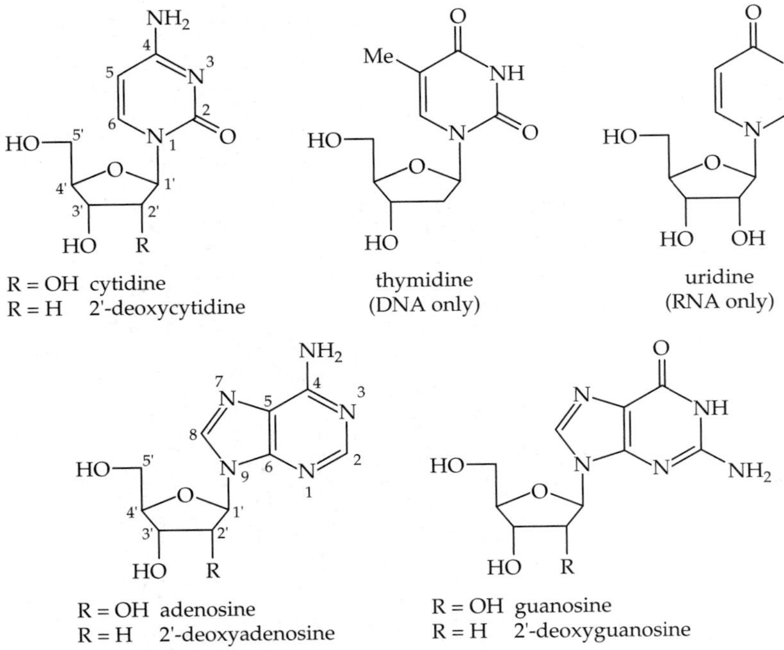

R = OH cytidine
R = H   2'-deoxycytidine

thymidine
(DNA only)

uridine
(RNA only)

R = OH adenosine
R = H   2'-deoxyadenosine

R = OH guanosine
R = H   2'-deoxyguanosine

FIGURE 6.4 Nucleoside analogues

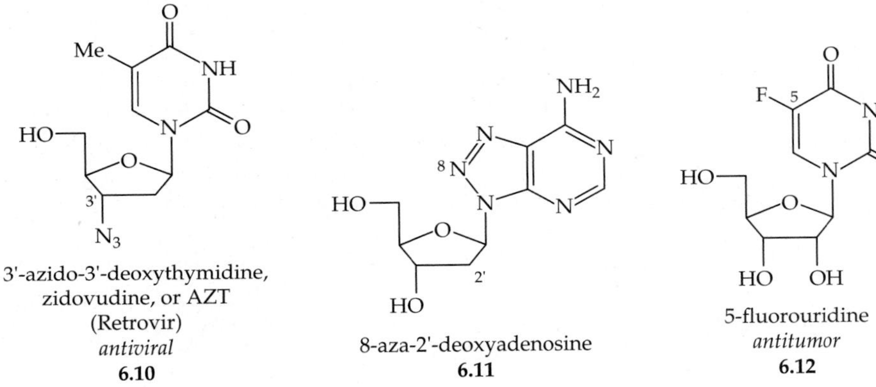

3'-azido-3'-deoxythymidine,
zidovudine, or AZT
(Retrovir)
antiviral
6.10

8-aza-2'-deoxyadenosine
6.11

5-fluorouridine
antitumor
6.12

Nucleotides are nucleosides that have been phosphorylated on the 5'-hydroxyl group. One, two, or three phosphates can be added. Nucleotides are generally named by appending *monophosphate*, *diphosphate*, or *triphosphate* to the end of the name of the corresponding nucleoside, and the names are typically abbreviated. The suffixes *-MP*, *-DP*, and *-TP* correspond to mono-, di-, and triphosphate, respectively. Examples of 2'-deoxyadenosine 5'-diphosphate or ADP (**6.13**) and uridine 5'-monophosphate or UMP (**6.14**) are shown in Figure 6.5.

The phosphate groups are generally depicted as completely deprotonated. A monophosphate ester, such as AMP, has two acidic protons with pK_as of approximately 2 and 7.[1] At pH 7, a significant amount of both the monohydrogen phosphate (**6.15**) and phosphate (**6.16**) will be present. These compounds represent a corresponding acid and conjugate base pair (Scheme 6.1). The acid form is the source of the term *nucleic acid*. Nucleic acids are frequently abbreviated with single letters of *A, C, G, T,* and *U*. 2'-Deoxy forms are denoted as dA, dC, dG, dT, and dU. If the discussion is clearly dealing with DNA, the *d*- prefix may be dropped from the abbreviation. For example, because thymidine is only found in DNA, phosphorylated forms of thymidine are written as TMP, TDP, and TTP instead of dTMP, dTDP, and dTTP.

FIGURE 6.5 Nucleotides

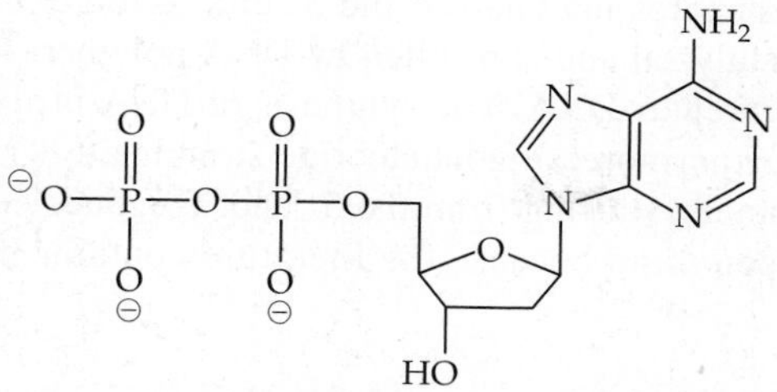

2'-deoxyadenosine 5'-diphosphate
(dADP)
6.13

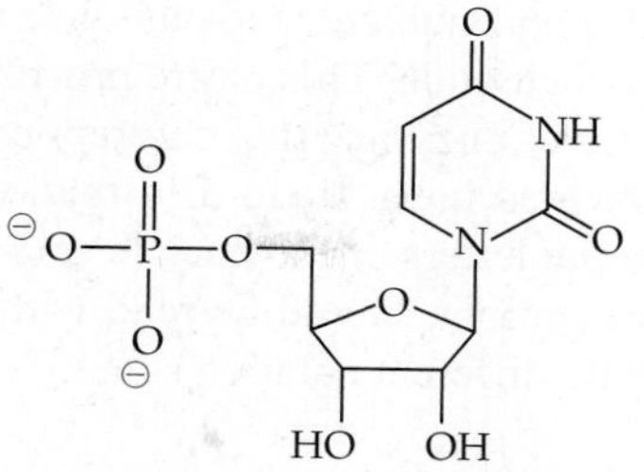

uridine 5'-monophosphate
(UMP)
6.14

SCHEME 6.1 Charge states of nucleotides at pH 7

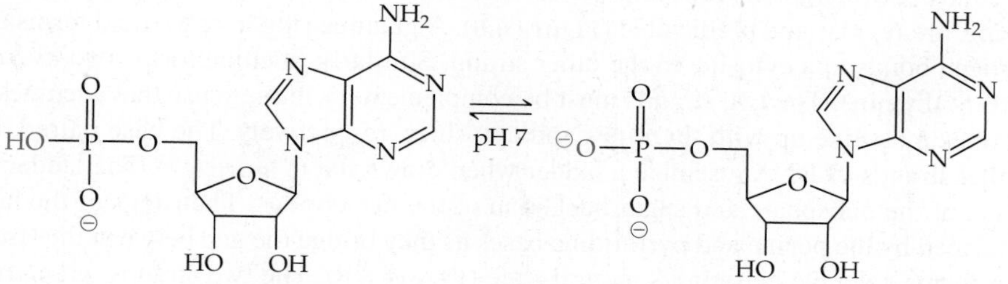

nucleic acid
6.15

conjugate base
6.16

DNA

DNA contains the full genetic blueprint for an organism. Every enzyme, receptor, and structural protein is encoded by DNA in subsections called genes. Collectively, all the genes of an organism comprise its full genome. For humans, the full genome consists of over 3 billion nucleotide pairs.[2]

Synthesis

DNA is a polymer of 2'-deoxynucleic acids (Scheme 6.2). The 3'-hydroxyl group of one nucleic acid reacts with the 5'-triphosphate of another nucleic acid to form a dinucleotide.

2'-deoxyadenosine monophosphate (dAMP)

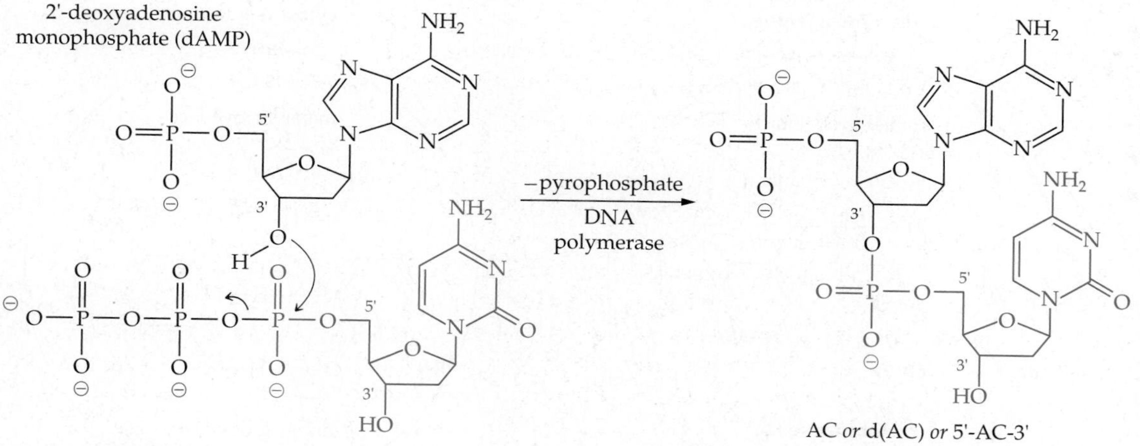

2'-deoxycytidine triphosphate (dCTP)

−pyrophosphate
DNA polymerase

AC or d(AC) or 5'-AC-3'

SCHEME 6.2 Oligonucleotide synthesis by DNA polymerase

Additional nucleic acids as their triphosphates may add to the 3' end of the existing oligonucleotide. This entire process is catalyzed and controlled by DNA polymerases, a family of enzymes that can replicate and repair DNA. The synthesis of DNA is always performed in the 5' to 3' direction.[2] Furthermore, oligonucleotide strands are written with the letters of the nucleic acid monomers, starting from the 5' end. Oligonucleotide abbreviations should be read with an open mind because the literature contains many slightly different notation styles.

Structure and Function

DNA is double stranded, meaning DNA in its full form contains two strands of oligonucleotides. The strands are aligned in an *antiparallel* fashion. The 5' end of one strand lies next to the 3' end of the other strand. The strands interact with one another through hydrogen bonds between the heterocyclic bases. These interactions, called Watson–Crick *base pairing*, are regular and predictable (**Figure 6.6**). A guanine ring in one strand forms three hydrogen bonds to a cytosine in the other strand. Similarly, adenine forms two hydrogen bonds to thymine. The two strands must be complementary in the sense that adenines and guanines must line up with thymines and cytosines, respectively. The base paired, antiparallel strands of DNA resemble a ladder when drawn flat (**Figure 6.7**). The ladder rails consist of the phosphate and sugar backbone of the nucleotides. The rungs of the ladder are formed by the purine and pyrimidine bases as they bridge the gap between the strands. DNA forms a double helix in its natural state (**Figure 6.8**). The two strands wrap around in a right-handed coil. The helix performs a complete turn every 10 nucleotides over a distance of 34 Å. The width of the double helix is approximately 20 Å. Two features created by the winding of the double helix are depressed trenches called the *major* and *minor grooves*. The major groove is wider than the minor groove, but both grooves provide potential sites of interacting with DNA.[3] The double helix in Figure 6.8 is exaggerated to emphasize differences between the two grooves.

RNA

Because of the importance of DNA and the need to keep it as safe and protected as possible, DNA is found only in the restricted environment of the nucleus. Furthermore, it is wound into a compact form to minimize the possibility of damage. When the genetic information of DNA is needed elsewhere in the cell, DNA is partially unwound and

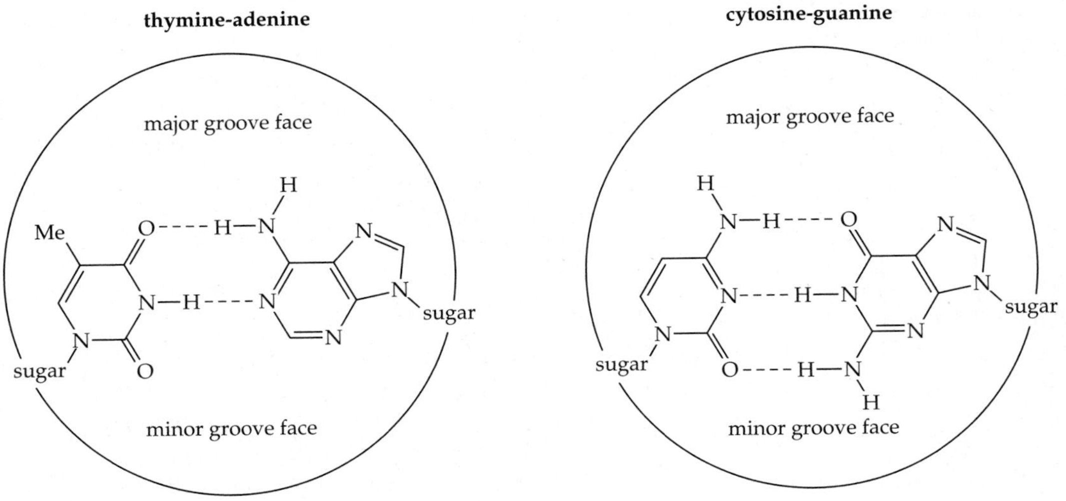

FIGURE 6.6 Watson–Crick base pairing in DNA nucleotides

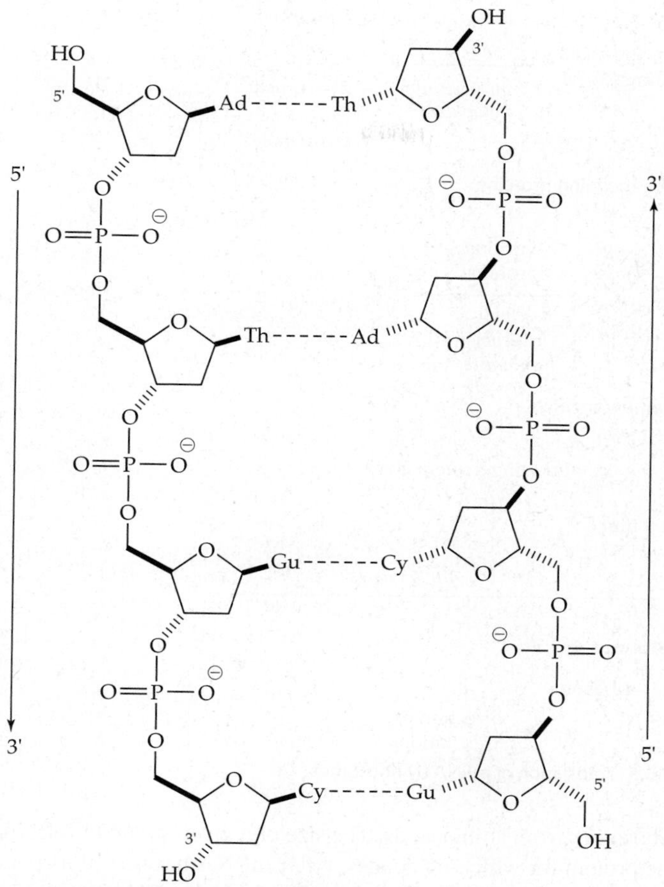

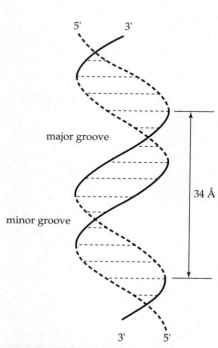

FIGURE 6.7 Rung and ladder structure of DNA

FIGURE 6.8 DNA double helix

transcribed to make RNA by the enzyme RNA polymerase. The original DNA is unchanged by the process. If RNA is somehow damaged, it can always be resynthesized from the original DNA.

Like DNA, RNA can form a double-stranded structure, but an appropriate complementary strand is rarely available. Therefore, RNA is generally single stranded. A single strand of RNA is structurally similar to a single strand of DNA with two exceptions: RNA nucleotides have a 2′-hydroxyl group on the furanose ring, and RNA contains uracil in place of DNA's thymine. Despite structural similarities, RNA and DNA function completely differently. While DNA preserves the genetic code in the nucleus, RNA converts the code into a useable form for the cell. Adapting the information in DNA into a useful form requires many steps, so RNA exists in several distinct forms to perform these various tasks.

mRNA

mRNA, or *messenger RNA*, is a recyclable form of DNA that contains information for the primary sequence of proteins, such as receptors and enzymes. For this reason, mRNA is sometimes called *coding* RNA. Like all RNA, mRNA is prepared by *transcription* of a segment of DNA in the nucleus with the enzyme RNA polymerase (**Scheme 6.3**). The mRNA strand is synthesized from RNA nucleotides. The new single-stranded mRNA then complexes with proteins and is transported into the cytosol. Once in the cytosol, mRNA is

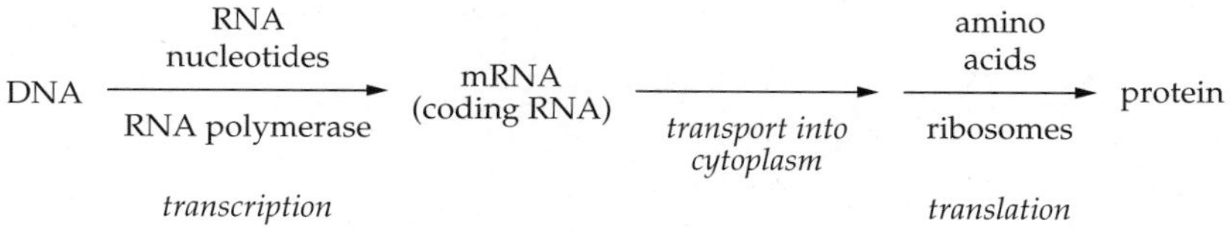

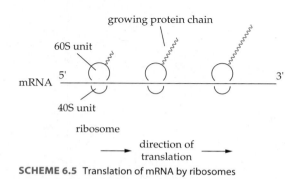

SCHEME 6.3 Function of mRNA for encoding proteins

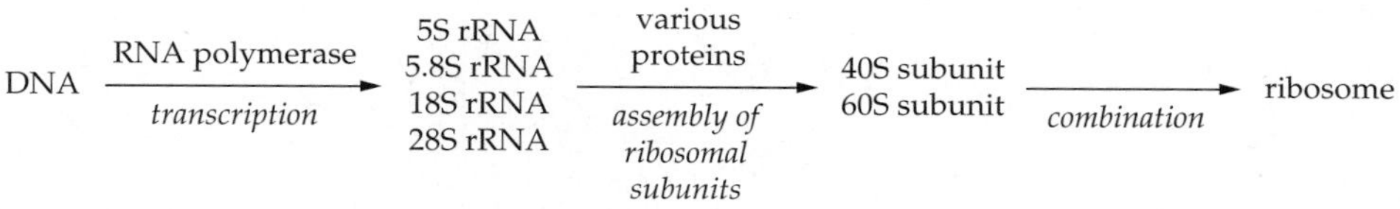

SCHEME 6.4 Assembly of a ribosome from rRNA

SCHEME 6.5 Translation of mRNA by ribosomes

translated into a protein by ribosomes from amino acids. The size of a given strand of mRNA varies based on the size of the protein for which it encodes. After mRNA has served its purpose, a cell will break it down into individual RNA nucleotides for use in other RNA structures.

rRNA

rRNA, or *ribosomal RNA*, provides the catalytic activity of ribosomes. Ribosomes catalyze the synthesis of proteins from amino acids by translating mRNA. rRNA itself does not encode proteins and is therefore an example of *noncoding* RNA. Three segments of rRNA are formed in the nucleus with sizes ranging from just over 100 nucleotides to approximately 5,000 nucleotides (Scheme 6.4). A fourth segment of rRNA is synthesized in the cytosol. The rRNA strands are distinguished with a number followed by an *S*, for Svedberg unit. This notation describes the time of sedimentation of a compound in an ultracentrifuge. The rRNA strands are partially complexed with proteins to form two different subunits, one 40S and another 60S. The two subunits are then combined to form an active ribosome in the cytosol.[2]

Ribosomes contain a channel in which a strand of mRNA binds. During translation, a ribosome slides along a strand of mRNA from the 5′ end to the 3′ end (Scheme 6.5). As the mRNA code is read, the ribosome synthesizes a new protein from the *N*-terminus to the *C*-terminus. The protein is prepared from amino acids that are delivered by another form of RNA called tRNA, which is described in the next subsection. Multiple ribosomes can simultaneously translate a single mRNA strand.

tRNA

tRNA, or *transfer RNA*, is a single strand of RNA with a length of approximately 80 nucleotides (Figure 6.9). Although tRNA is single stranded, tRNA folds in a manner that allows it to base pair to itself. The secondary structure of tRNA, when drawn flat, is often

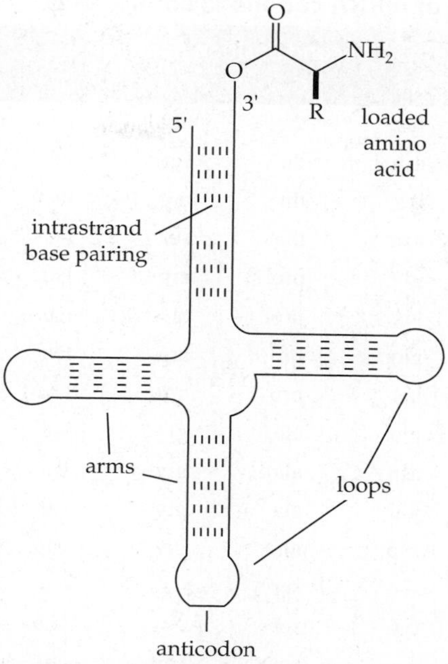

FIGURE 6.9 Structure of tRNA

called the *cloverleaf* structure and is highlighted by arms and loops off a central inter-section point. The role of tRNA is to recognize a mRNA strand that is complexed to a ribosome. Once complexed, tRNA delivers the appropriate amino acid for lengthening a growing protein chain. The required amino acid is attached to the 3′ terminus of tRNA. The *charged* or *loaded* tRNA binds the ribosome-mRNA complex. Transfer of the amino acid to the growing protein depends on a match between a three-nucleotide segment, called a *triplet codon*, in mRNA and the *anticodon* of tRNA. If the mRNA codon and tRNA anticodon are complementary (i.e., able to undergo Watson–Crick base pairing), the amino acid will be transferred. The unloaded tRNA then diffuses from its binding site, and the ribosome slides down the mRNA strand to the next codon. The complex ballet of bind-ing, transfer, and dissociation is remarkably fast. Bacterial ribosomes can extend a protein at an astonishing rate of 20 residues per second.[3]

Table 6.1 shows the relationship between the codon sequence in mRNA and its cor-responding amino acid in the new protein. Because there are 64 (4^3) different anticodon combinations and only 20 encoded amino acids, some different anticodon sequences en-code for the same amino acid. Generally, all the anticodons matching a given amino acid will have the same first two nucleotides. Exceptions are arginine, serine, and isoleucine. For example, the codon for proline will always start with *CC*, but the arginine codon may start with either *AG* or *CG*. The 3′ end of the tRNA anticodon pairs with the 5′ end of the mRNA codon. In other words, the codon and anticodon align and bind in an antiparallel fashion.

siRNA and miRNA

siRNAs, or *short interfering RNAs*, are small, double-stranded segments of RNA with lengths of between 20 and 25 nucleotides. One strand of double-stranded siRNA binds to a complementary strand of mRNA and prevents translation of the mRNA into a protein (**Scheme 6.6**). Therefore, siRNAs are a means of suppressing expression of DNA genes that encode for proteins. Initial binding of a siRNA to mRNA occurs in the nucleus, and siRNA remains bound as the mRNA strand is transported into the cytosol.[4]

TABLE 6.1 Translation of mRNA codons to amino acids[3]

First Position (5′ end)	Second Position (middle)				Third Position (3′ end)
	A	C	G	U	
	lys	thr	arg	ile	A
A	asn	thr	ser	ile	C
	lys	thr	arg	met	G
	asn	thr	ser	ile	U
	gln	pro	arg	leu	A
C	his	pro	arg	leu	C
	gln	pro	arg	leu	G
	his	pro	arg	leu	U
	glu	ala	gly	val	A
G	asp	ala	gly	val	C
	glu	ala	gly	val	G
	asp	ala	gly	val	U
	—	ser	—	leu	A
U	tyr	ser	cys	phe	C
	—	ser	trp	leu	G
	tyr	ser	cys	phe	U

SCHEME 6.6 Translation interference by siRNA

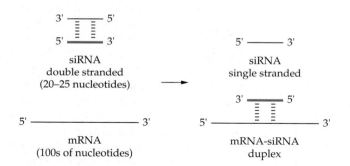

Closely related to siRNAs are miRNAs, or *micro RNAs*. miRNAs are the same size as siRNAs, but miRNAs are single stranded. Like siRNAs, micro RNAs bind mRNAs and interfere with translational processes.[4]

6.2 Common Methods of Oligonucleotide Recognition

Just like any other molecular target, binding a polynucleic acid depends on maximizing available intermolecular forces between the oligonucleotide and drug. The regular, repeating structure of oligonucleotides affords characteristic opportunities for intermolecular attractions.

Base Pairing

Base pairing has been investigated as a tool to allow a drug to recognize a target oligonucleotide. Because DNA is normally double stranded (already paired), RNA is the more common target for a drug intended to imitate or interfere with normal base pairing. For a drug to undergo base pairing to RNA, the drug itself must in essence be a polynucleotide

d(GCGTTTGCTCTTCTTCTTGCG)

fomivirsen
6.17

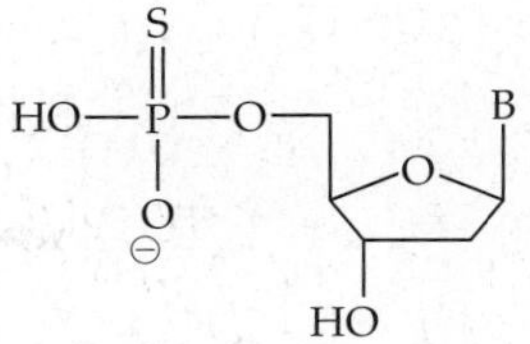

first-generation
nucleotide analogue
6.18

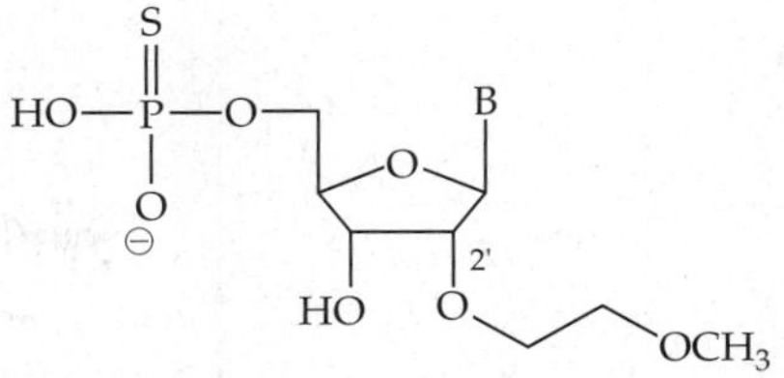

second-generation
nucleotide analogue
6.19

FIGURE 6.10 Fomivirsen (**6.17**) and antisense nucleotide analogues (**6.18** and **6.19**)

with appropriately placed hydrogen bond donors and acceptors. The duplex structure of the RNA-drug complex aligns in an antiparallel fashion. The general technique of base pairing a synthetic oligonucleotide to a strand of RNA has become known as the *antisense* approach. In theory, if a strand of RNA, particularly mRNA, can be selectively complexed into a duplex structure, RNA translation and the corresponding protein synthesis will be compromised. If the blocked protein is an enzyme or receptor, then the physiological state of the cell will be affected. *Gene silencing* is the end result. Antisense drugs function as synthetic versions of miRNAs. Antisense approaches to drug discovery are increasingly called *RNA interference*, or simply *RNAi*.[4]

To date only one antisense drug, fomivirsen (**6.17**), has been approved in the United States and Europe (**Figure 6.10**). Fomivirsen treats cytomegaloviral infections of the eye by binding and inhibiting translation of specific viral mRNA strands. Fomivirsen consists of 21 2′-deoxyribonucleotides with a sulfur in place of an oxygen throughout the phosphodiester backbone. The sulfur substitution improves the chemical stability of fomivirsen in the body. Sulfur analogues of this type, called phosphorothioate deoxynucleotides (**6.18**, B = base), represent a first-generation approach to antisense drugs by Isis Pharmaceuticals, the developer of fomivirsen. Second-generation antisense drugs from Isis are built from RNA phosphorothioate nucleotides with a methoxyethyl cap on the 2′-hydroxyl group (**6.19**). The additional chain further improves metabolic stability of the antisense drug. As of 2012, Isis had many RNAi drugs in clinical trials including two in phase III.[5]

The antisense approach to pharmaceuticals is conceptually attractive and powerful. If a protein target and its sequence are known, then the sequence of the corresponding mRNA will also be largely known. If the exact mRNA sequence can be determined, then a complementary polynucleotide may be prepared to form a duplex. Longer complementary antisense strands give a more stable duplex.

Duplex stability can be measured by *melting temperature* (T_m). T_m corresponds to the temperature at which 50% of the mRNA strands are in duplex form with an antisense strand. Single-stranded and duplex oligonucleotides absorb UV light differently, and UV absorbance at 268 nm (A_{268}) can be used to monitor duplex melting (dissociation) with temperature (**Figure 6.11**). The point of inflection of a graph of UV absorbance against temperature indicates T_m. T_m for short DNA duplexes can be estimated with Equation 6.1, also called the Wallace rule.[6]

$$T_m(°C) = 2(\#A/T) + 4(\#G/C) \qquad (6.1)$$

The terms *#A/T* and *#G/C* refer to the number of A-to-T and G-to-C base pairs, respectively, in the duplex strand. The Wallace rule was designed for duplex strands of 14 to 20 base pairs. The T_m of RNA-RNA and DNA-RNA duplexes may be estimated with Equation 6.2.[7]

$$T_m(°C) = 79.8 + 18.5 \log[M^+] + 58.4(X_G + X_C) + 11.8(X_G + X_C)^2 - \frac{820}{L} \qquad (6.2)$$

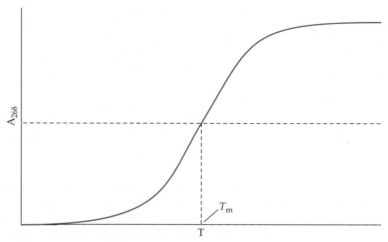

FIGURE 6.11 Duplex oligonucleotide melting graph

The term $(X_G + X_C)$ is the sum of the mole fractions of guanine and cytidine in the antisense strand. The mole fraction of any nucleobase is equal to the number of nucleotides containing that base divided by the total number of nucleotides in the oligonucleotide strand. $[M^+]$ is the molar concentration of monovalent cations. In a typical mammalian cell, $[K^+]$ is 140 mM, and $[Na^+]$ is 10 mM. L is the length of the duplex in base pairs. Based on Equation 6.2 and the assumption that phosphorothioate oligonucleotides behave as regular DNA, fomivirsen (**6.17**) would have a predicted T_m of 59 °C. Equation 6.1 predicts 64 °C. More complex forms of Equation 6.2 with increased accuracy appear regularly in the literature.[8]

Although DNA-RNA duplexes can certainly form, the premise of RNAi is not universally accepted. Bevasiranib, a 21-mer dsRNA that advanced to phase III clinical trials, was suggested to interfere with the translation of mRNA associated with vascular endothelial growth factor and subsequently certain forms of macular degeneration. OPKO Health, Inc., the developer of bevasiranib, proposed that bevasiranib enters cells and binds mRNA. Typically, an oligonucleotide such as bevasiranib with a molecular weight of over 14,000 would be too large to cross the cell membrane. More recent research from 2008 suggests that bevasiranib does *not* enter the cell. Instead, bevasiranib binds toll-like receptor 3 (TLR3) on the surface of cells and initiates an immune response.[9] Coincidentally, one effect of the immune response does slow the progression of macular degeneration. The binding of TLR3 by dsRNA, a common byproduct of viral infection, has been observed by other researchers.[10] OPKO initially refuted the findings of the conflicting research, but bevasiranib was pulled from clinical trials in 2009 because of insufficient efficacy.

Sample Calculation   **Determining a T_m Value**

PROBLEM Show that the T_m of fomivirsen (**6.17**) is indeed 64 °C with Equation 6.1 and 59 °C with Equation 6.2.

SOLUTION Fomivirsen has 10 A/T base pairs and 11 G/C base pairs. For Equation 6.1, the answer is straightforward.

$$T_m(°C) = 2(\#\,A/T) + 4(\#\,G/C) \tag{6.1}$$
$$T_m(°C) = 2(10) + 4(11) = 20 + 44 = 64\ °C$$

Equation 6.2 is not much more difficult, but it certainly has more variables. The mole fraction of G and C is 0.52 (11/21). The length of the strand is 21, and $[M^+]$ is 0.15 M (150 mM). Substitution of the values into Equation 6.2 gives 59 °C.

$$T_m(°C) = 79.8 + 18.5 \log[M^+] + 58.4(X_G + X_C) + 11.8(X_G + X_C)^2 - \frac{820}{L} \quad (6.2)$$

$$T_m(°C) = 79.8 + 18.5 \log 0.15 + 58.4(0.52) + 11.8(0.52)^2 - \frac{820}{21}$$

$$T_m(°C) = 79.8 - 15.2 + 30.4 + 3.2 - 39.0 = 59 °C$$

Charge-Charge Interactions

The phosphate backbone of both DNA and RNA provides a negatively charged template for attracting positively charged species. Drugs that commonly exploit this interaction are the aminoglycoside antibiotics. Examples include neomycin B (**6.20**) and kanamycin A (**6.21**) (**Figure 6.12**). This class of compounds binds rRNA in bacteria to interfere with translation of mRNA into functional proteins.

All these compounds are extensively protonated on their amino groups at biological pH. The 3-aminoalcohol functionality found throughout aminoglycosides can bind rRNA along the phosphate backbone through a combination of charge-charge attraction and hydrogen bonding (**Figure 6.13**).[11] Another binding mode gaining more attention involves hydrogen bonding across the Hoogsteen face of guanine.[12,13] The macrolactide antibiotics such as erythromycin (**6.22**) and clarithromycin (**6.23**) also share a 3-aminoalcohol subunit for interaction with rRNA (**Figure 6.14**).

Intercalation

In duplex DNA, intercalation is the insertion of a relatively flat molecule between adjacent paired bases (**Figure 6.15**). In its ground state, adjacent bases are separated by 3.4 Å, not enough space to accommodate another molecule. Therefore, the double helix must lengthen to create a gap for the intercalating molecule. Increasing the distance between bases distorts the sugar-phosphate backbone and decreases the turn of the helix. Multiple molecules can intercalate into the same strand of DNA, with the limit being approximately one intercalated guest for every two base paired nucleotides. A fully intercalated DNA helix is approximately twice as long and considerably more rigid than the corresponding native DNA structure.[14]

Changes in DNA topology caused by intercalators can impair or prevent DNA replication, a key process for cell division. Therefore, DNA intercalators have the potential

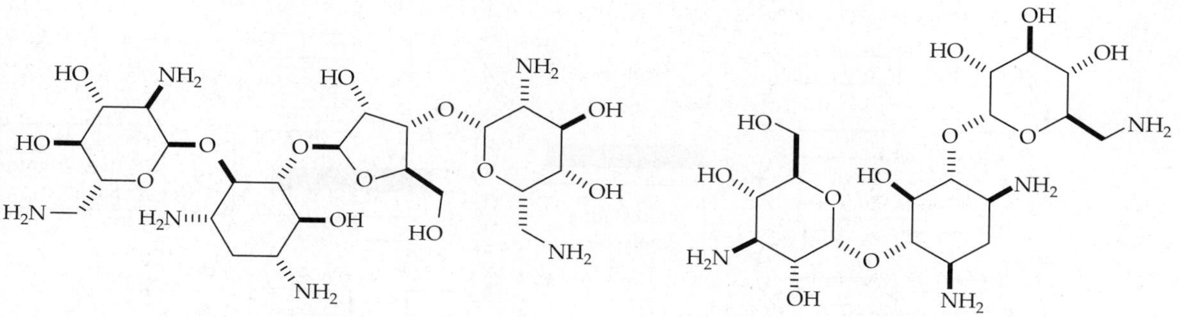

neomycin B
6.20

kanamycin A
6.21

FIGURE 6.12 Aminoglycoside antibiotics

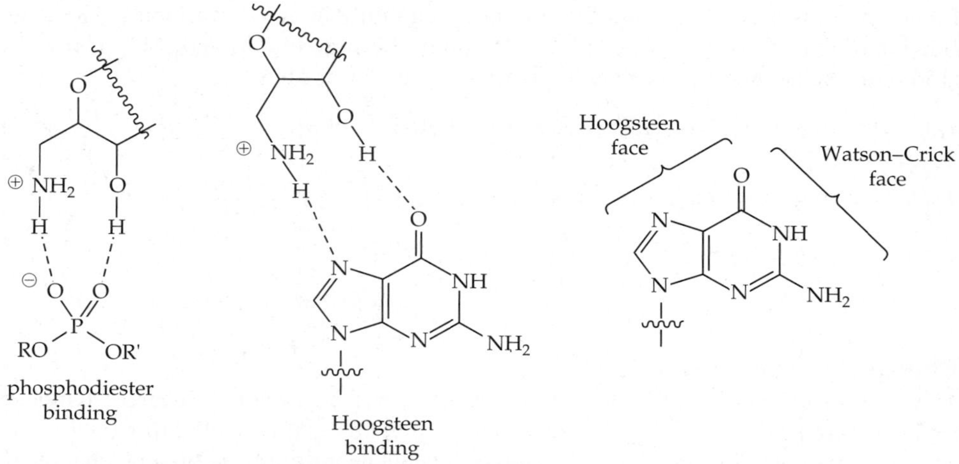

FIGURE 6.13 Aminoglycoside-rRNA modes of interaction

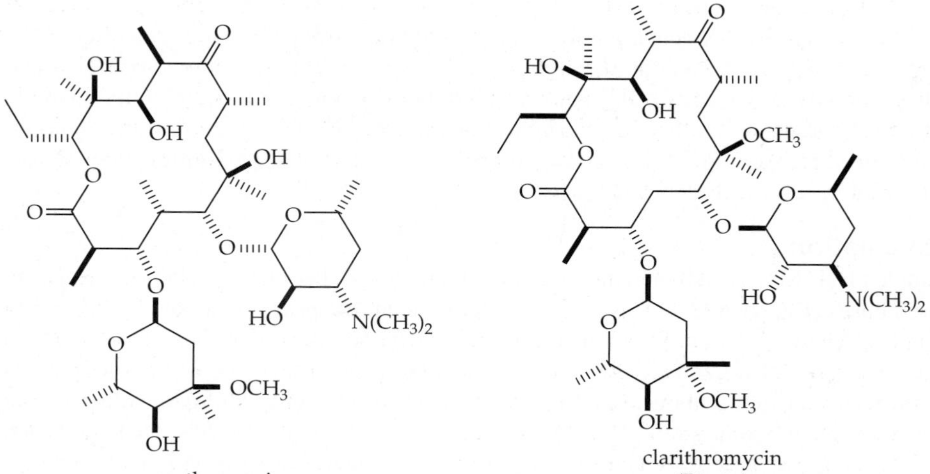

phosphodiester
binding

Hoogsteen
binding

Hoogsteen
face

Watson–Crick
face

erythromycin
6.22

clarithromycin
(Biaxin)
6.23

FIGURE 6.14 Macrolactide antibiotics (**6.22** and **6.23**)

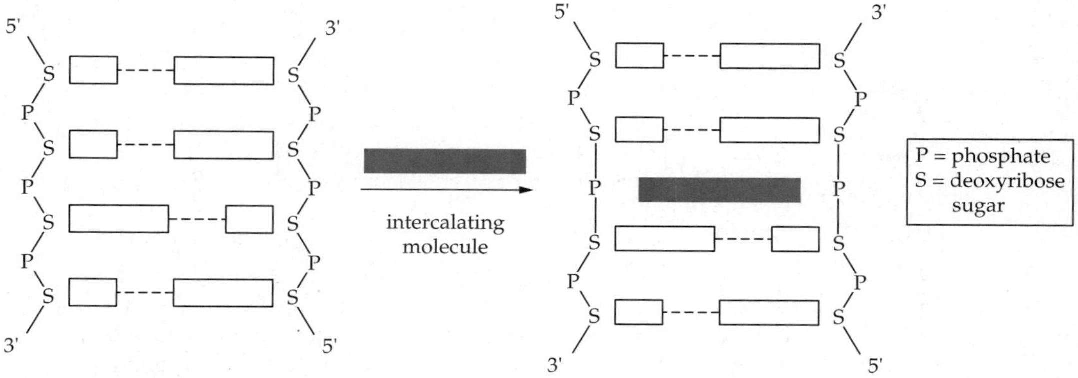

intercalating
molecule

P = phosphate
S = deoxyribose
sugar

FIGURE 6.15 DNA intercalation

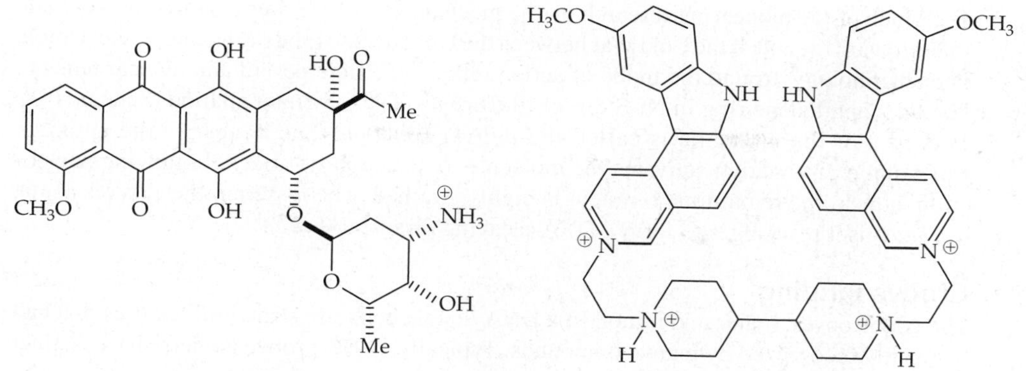

daunomycin
6.24

ditercalinium
6.25

FIGURE 6.16 DNA intercalators (**6.24** and **6.25**)

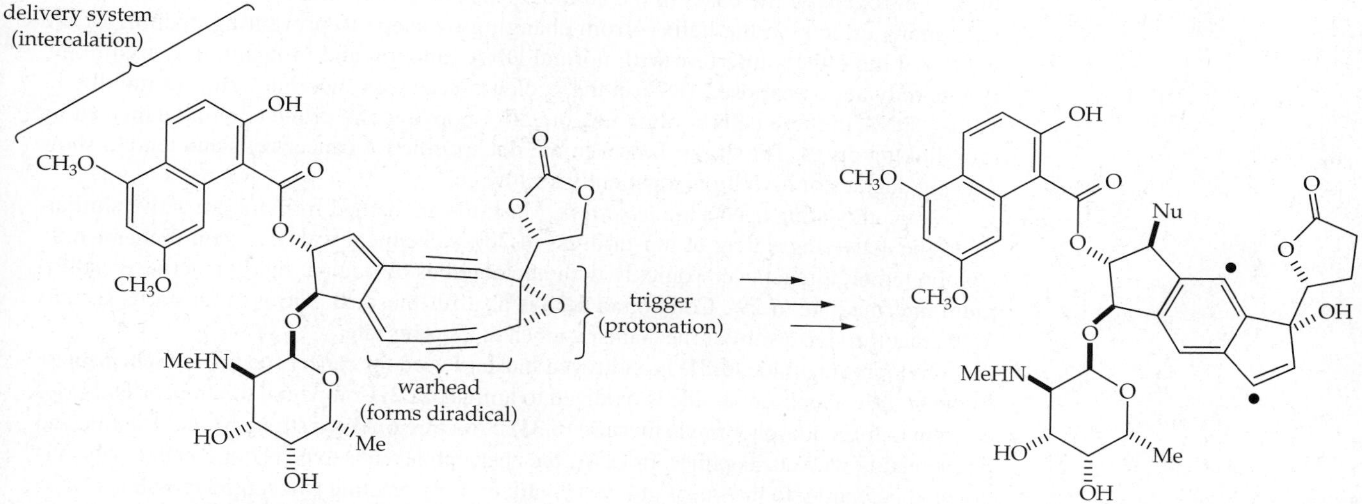

NCS chromophore
6.26

NCS diradical
6.27

SCHEME 6.7 Neocarzinostatin chromophore (**6.26**)

to slow cell division and are particularly prominent in fighting tumors. One chemotherapeutic intercalator is daunomycin (**6.24**) (**Figure 6.16**). The flat, anthraquinone portion of the compound intercalates the DNA helix. Similarly, ditercalinium (**6.25**), an anticancer drug that was found to be too toxic for use, has two flat regions, each of which can intercalate between different bases. Both daunomycin and ditercalinium contain positive charges at biological pH and are attracted to the negatively charged backbone of DNA.

Another anticancer drug that relies on intercalation is the chromophore of neocarzinostatin (**Scheme 6.7**). Neocarzinostatin is a drug consisting of a 1:1 complex of a DNA-cleaving molecule (NCS chromophore, **6.26**) and a small, stabilizing protein. In chemistry discussions, the chromophore alone is often simply called neocarzinostatin (NCS). NCS is an example of an *enediyne* compound. Enediyne compounds contain an alkene and two alkynes in conjugation with each other. All members of this drug class bind DNA, become activated, and then form a diradical intermediate. The diradical causes both strands of a DNA double helix to be cleaved. DNA cleavage of this type cannot be repaired by the cell and leads to cell death.[15]

Much of the pioneering research on the mechanism of NCS was performed in the late 1980s toward the end of the Cold War between the United States and the former Soviet Union. The terminology around the mode of action reflects the time period and nuclear tension. The flat naphthalene ring in NCS directs the drug to DNA by intercalation (Scheme 6.7). This part of the molecule is called the *delivery system*. The *trigger* is the epoxide, which upon protonation activates the molecule to nucleophilic attack. Following nucleophilic attack, the remaining π-system is highly strained. The π-system then serves as the *warhead* as it rearranges to form a DNA-cleaving diradical (**6.27**).[15]

Groove Binding

The two grooves, major and minor, in a DNA double helix are structural features that can be readily recognized by many compounds. Typically, major groove binders share distinct traits from minor groove binders.

Major Groove

Many compounds that act at the major groove of DNA alkylate or otherwise irreversibly bind to nitrogens of the bases in the strands. This alkylation/binding, or *crosslinking*, can have many effects on the helix—from changing its shape to preventing opening of the helix. All the effects interfere with normal DNA function and replication. Crosslinking agents only act on exposed DNA in the nucleus. Because cancerous cells are rapidly dividing, DNA in these cells is often unwound for copying and prone to being damaged by crosslinking drugs. The drugs, however, are not specific for cancer cells and tend to show varying degrees of toxicity to a patient's healthy cells.

Many alkylating agents are *mustards*. Mustards are named from their reactive similarity to the active ingredient of mustard gas (**6.28**), a chemical warfare agent (**Scheme 6.8**). Compound **6.28** reacts very quickly with nucleophiles through a highly reactive episulfonium intermediate (**6.29**). Compound **6.28** is a sulfur mustard. Nitrogen mustards such as **6.30** react through a similar mechanism, albeit at a slower rate.

Cyclophosphamide (**6.31**) is a nitrogen mustard used for cancer treatment (**Scheme 6.9**). In the body, cyclophosphamide is oxidized to aminal **6.32**. Compound **6.32** opens and loses acrolein to form phosphoramide mustard (**6.33**). Structure **6.33** is a strong bis-electrophile and reacts readily with nucleophiles. In DNA, the nucleophile tends to be N7 of guanine, which is oriented outward into the major groove (Figure 6.6). By reacting twice, **6.33** crosslinks DNA either within the same strand (*intrastrand*) or across the double helix (*interstrand*).[16]

Unlike many DNA alkylators, cyclophosphamide shows modest selectivity for cancer cells. Healthy tissues in a patient have high levels of aldehyde dehydrogenase (ALDH). ALDH reduces the potency of cyclophosphamide by oxidizing **6.32** to **6.34**, an inactive compound. Cancerous cells tend to be deficient in ALDH and more prone to damage by cyclophosphamide.

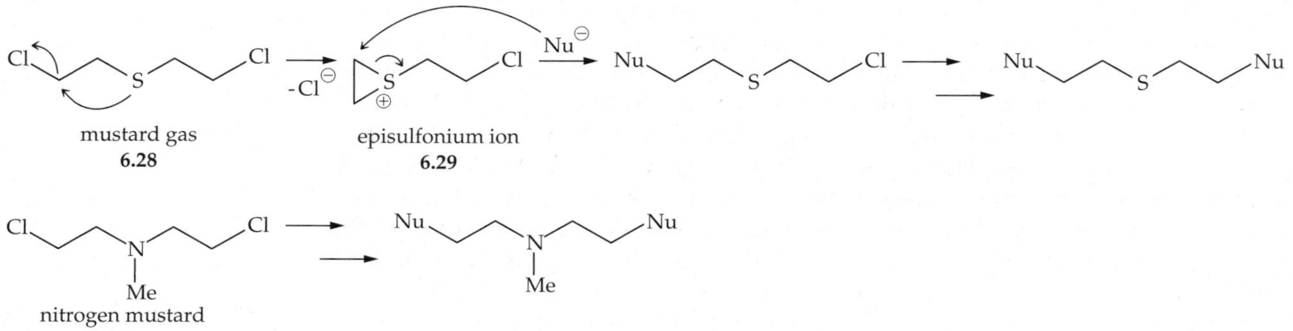

mustard gas
6.28

episulfonium ion
6.29

nitrogen mustard
6.30

SCHEME 6.8 Reactions of mustards with nucleophiles

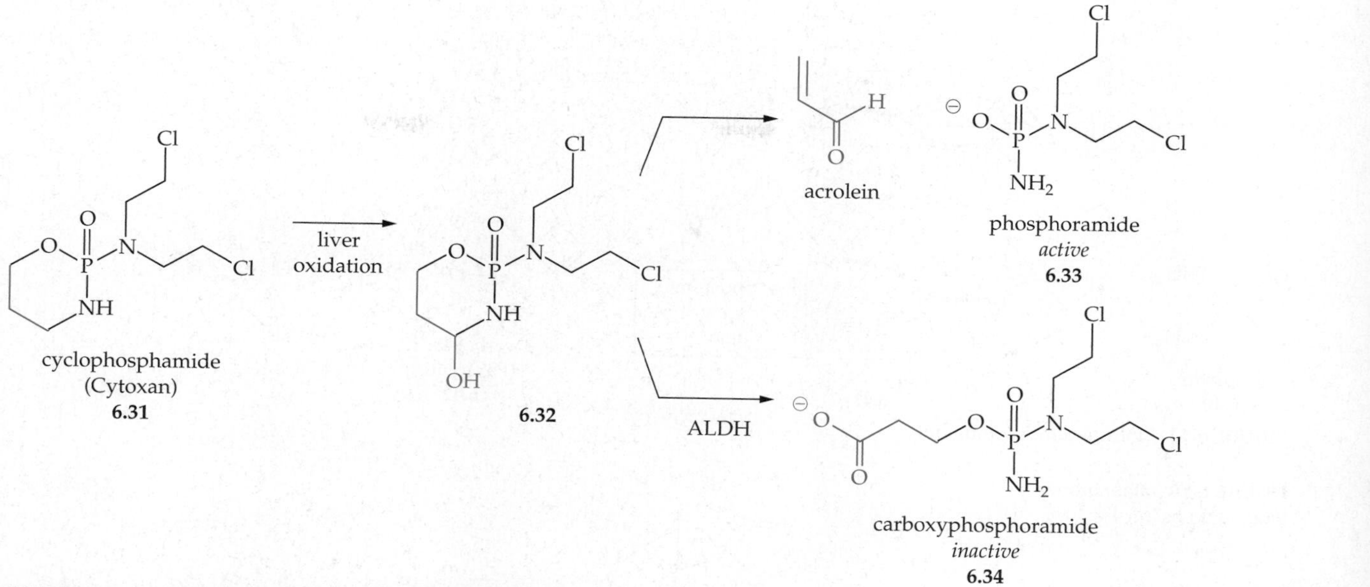

SCHEME 6.9 In vivo activation of cyclophosphamide (**6.31**)

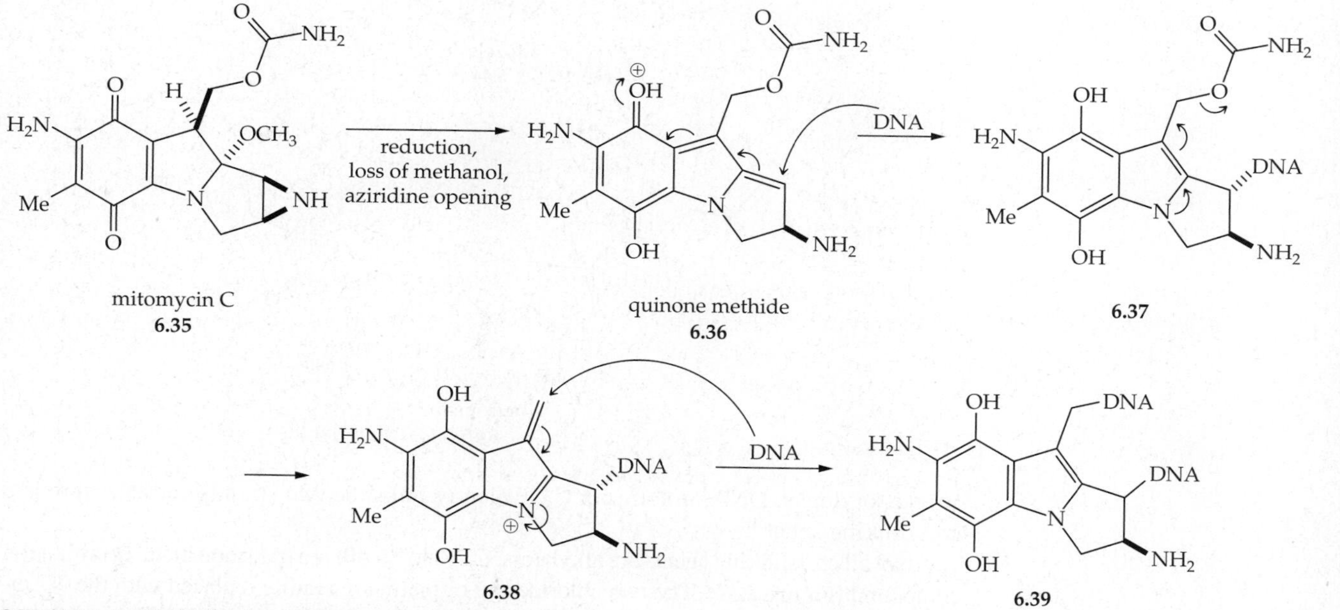

SCHEME 6.10 Mode of action of mitomycin C (**6.35**)

Mitomycin C (**6.35**), a natural product isolated from *Streptomyces lavendulae*, is a nonmustard DNA crosslinking agent (Scheme 6.10). In vivo, the benzoquinone ring of mitomycin C is reduced, methanol is lost, and the aziridine ring opens to form a quinone methide (**6.36**). Quinone methides are a type of molecule that is strongly electrophilic.[17] An NH_2 group at C2 of guanine (the NH_2 group on C2, see Figure 6.3) attacks **6.36** to provide **6.37**. Ionization of **6.37** gives an electrophilic cation (**6.38**), which may be attacked by yet another NH_2 group on guanine. Since the two guanines involved are on separate

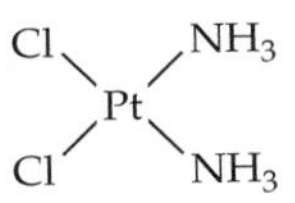

cisplatin
6.40

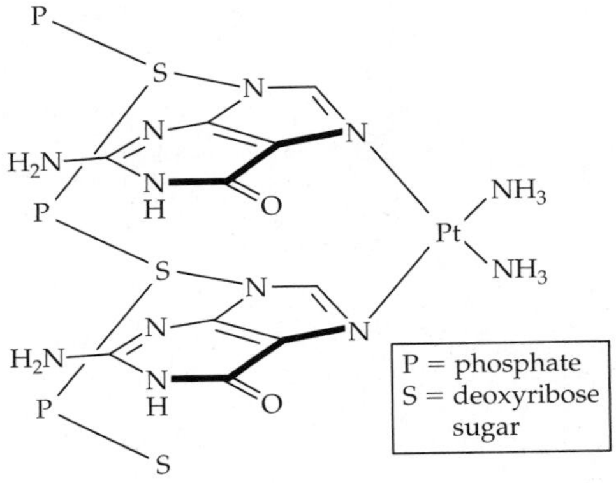

6.41

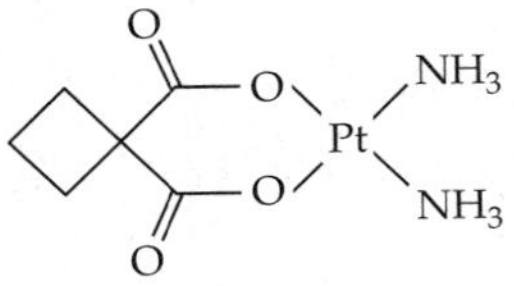

carboplatin
(Paraplatin)
6.42

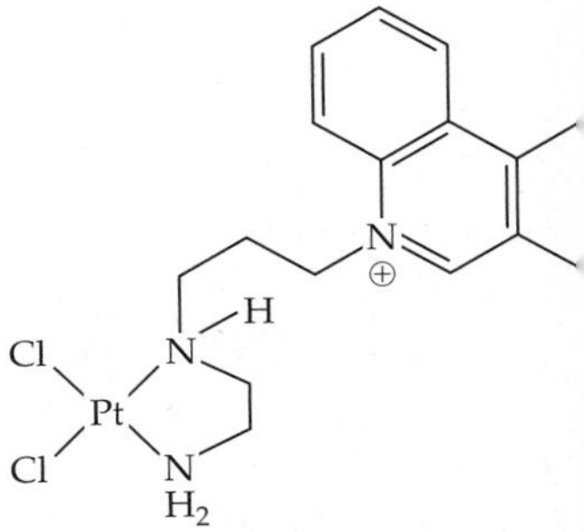

6.43

FIGURE 6.17 Pt(II) antitumor compounds

FIGURE 6.18 Minor groove binding drugs

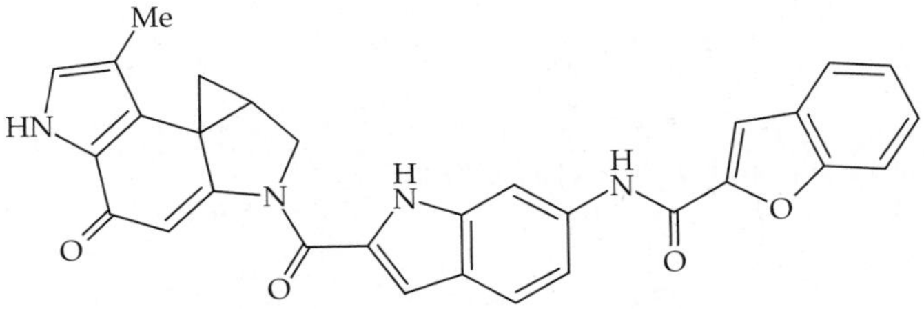

adozelesin
6.44

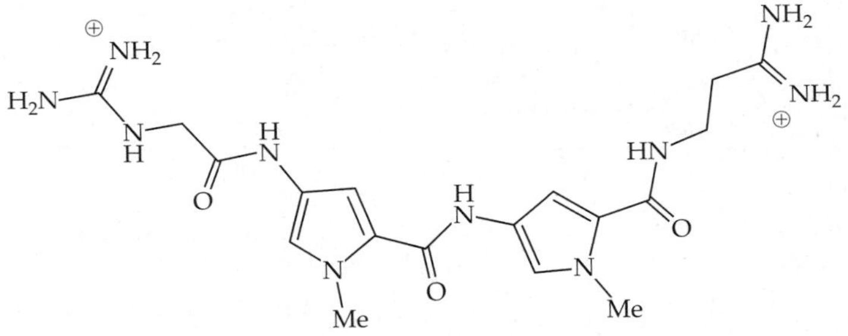

netropsin
6.45

strands of duplex DNA, mitomycin C covalently links the two strands and therefore is a crosslinking agent.[18]

Not all crosslinking agents are alkylators. Cisplatin (**6.40**) is a transition metal coordination compound (**Figure 6.17**). The two chlorides on cisplatin are readily replaced with the N7 nitrogens of two adjacent guanines on the same strand of the double helix (**6.41**). The intrastrand crosslink caused by cisplatin places a kink in the double helix and interferes with DNA replication and transcription.[19] Variations of cisplatin have been developed. One example is carboplatin (Paraplatin, **6.42**). Other research on Pt(II) antitumor drugs has attempted to increase activity by attaching the Pt(II) center to a ligand bearing an intercalating group (**6.43**).[20]

Minor Groove

The minor groove of DNA is narrower than the major groove, and minor groove binding agents tend to be relatively flat structures with a curvature that closely mimics that of DNA. Two examples, adozelesin (**6.44**) and netropsin (**6.45**), both bind strongly in the minor groove (**Figure 6.18**). Adozelesin is a monoalkylator that reacts at N3 of adenine

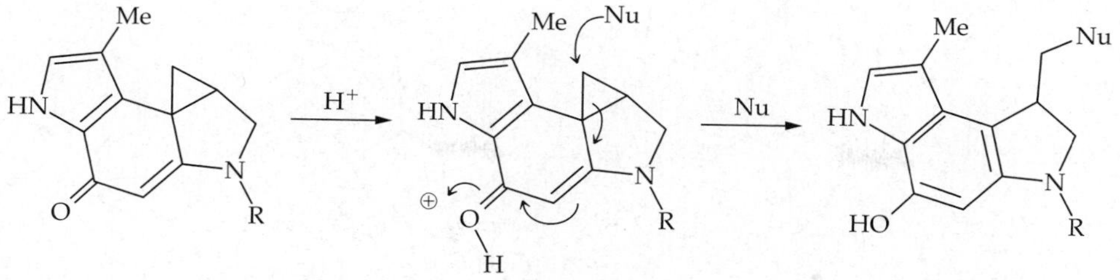

SCHEME 6.11 Alkylation mechanism of adozelesin (**6.44**)

bases in the minor groove. Alkylation involves opening the strained cyclopropane ring of **6.44** (**Scheme 6.11**).[21] Netropsin (**6.45**), a natural product produced by *Streptomyces netropsis*, is both an antimicrobial and antitumor compound. Netropsin is not an alkylator but instead interferes with DNA recognition.[22] Netropsin binds most strongly with adenine-rich sections of DNA, and the positively charged ends likely interact with the phosphate backbone.[23]

6.3 Interference with Nucleic Acid Synthesis and Function

Many compounds do not directly interact with intact DNA but are nonetheless able to affect normal DNA function in a cell. All the following examples are drugs that bind proteins or enzymes.

Anti-HIV Nucleic Acid Analogues

One of the most heavily researched viruses is the human immunodeficiency virus (HIV) as well as its many mutants. HIV carries its genetic material as RNA. Upon infection of a cell, the RNA must be *reverse transcribed* into viral DNA (**Scheme 6.12**). A viral enzyme performs the reverse transcription and is appropriately named *reverse transcriptase* (RT). RT reads single-stranded RNA and synthesizes a corresponding double-stranded segment of DNA. RT is therefore a DNA polymerase, taking up the host cell's own 2'-deoxynucleotide triphosphates to make viral DNA. The new viral double-stranded DNA is then inserted into the host genome by an enzyme called integrase. One approach to fighting HIV is by inhibition of RT with nucleoside/nucleotide analogues, called *nucleoside reverse transcriptase inhibitors* (NRTI). NRTIs can act in one of two ways. First, they can act as competitive inhibitors by binding in the active site of RT. Second, NRTIs have a nonfunctional 3'-position and act as chain terminators of the growing DNA strand. Either mode of action effectively inhibits DNA synthesis by RT.

The first marketed NRTI for HIV was zidovidine (Retrovir, **6.10**) (**Figure 6.19**). Zidovidine is a pyrimidine nucleoside analogue. Substitution of the 3'-hydroxyl with an azido group makes chain elongation of the DNA strand impossible upon zidovidine incorporation. Although nucleo*tides* are the required substrate of RT, most NRTIs are nucleo*sides*. Zidovidine and other NRTIs are transported into a cell without any 5'-phosphates. Once in a cell, the

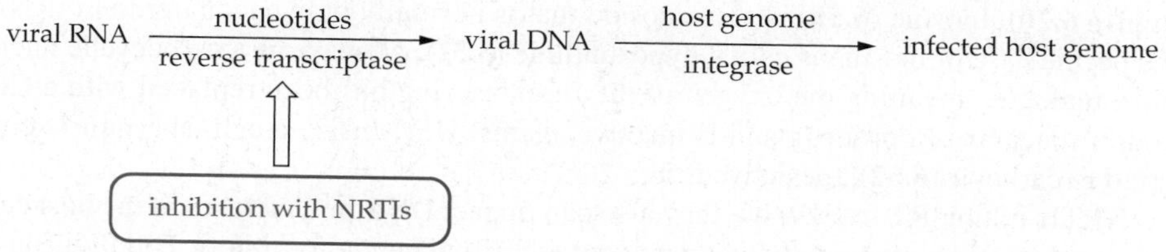

SCHEME 6.12 Activity of inhibition of reverse transcriptase

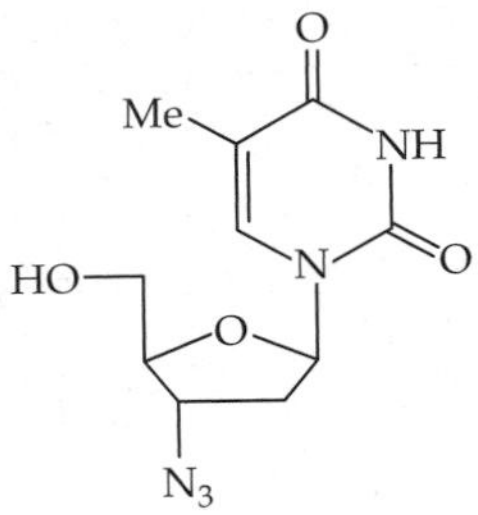

zidovidine or AZT
(Retrovir)
6.10

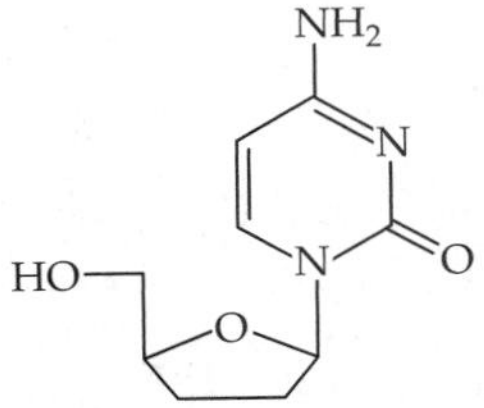

2',3'-dideoxycytidine,
zalcitabine, or ddC
(Hivid)
6.45

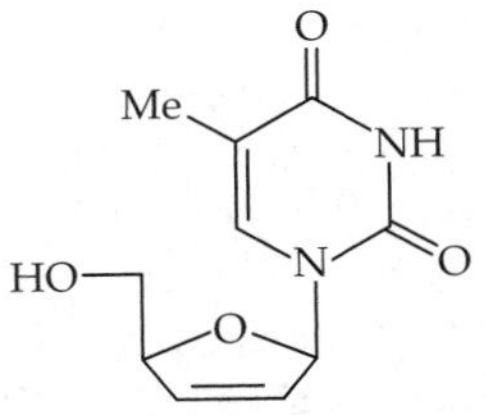

2',3'-didehydro-
3'-deoxythymidine,
stavudine, or d4T
(Zerit)
6.46

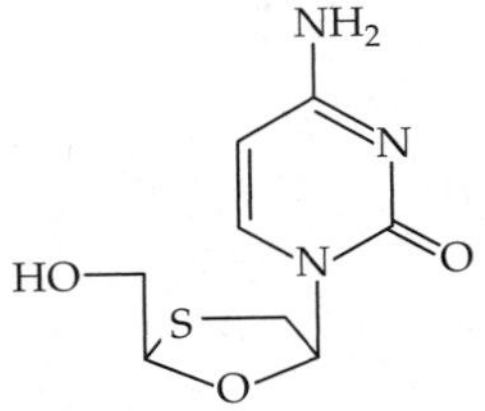

lamivudine or 3TC
(Epivir)
6.47

FIGURE 6.19 Pyrimidine NRTIs[24]

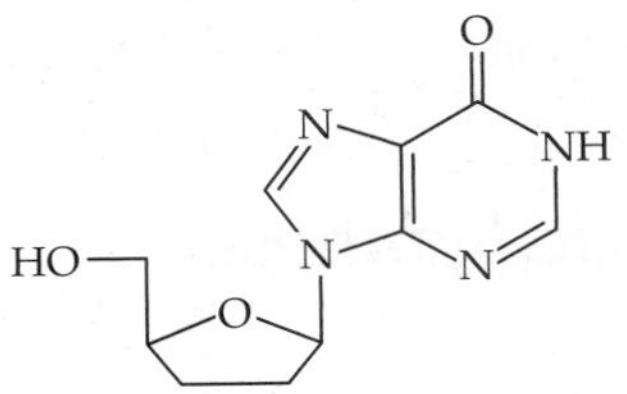

2',3'-dideoxyinosine,
didanosine, or ddI
(Videx)
6.48

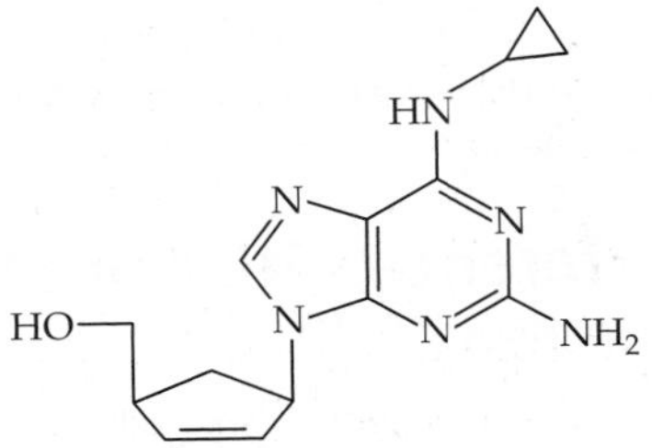

abacavir or ABC
(Ziagen)
6.49

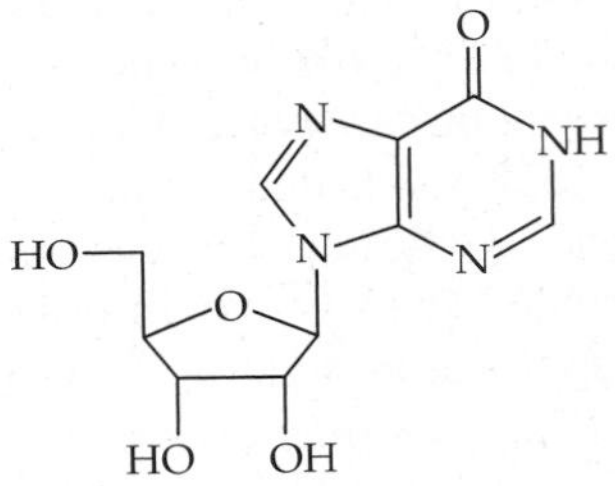

inosine
6.50

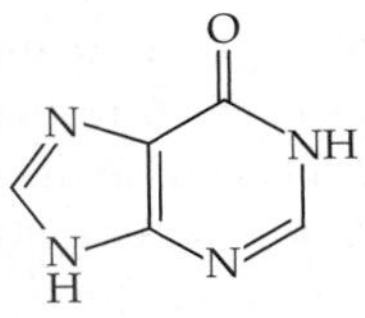

hypoxanthine
6.51

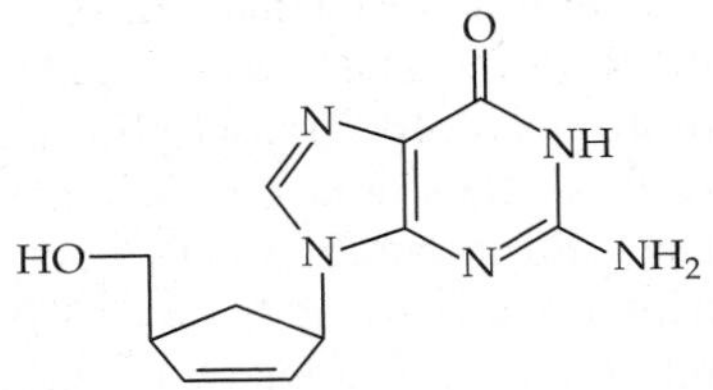

carbovir
6.52

FIGURE 6.20 Purine NRTIs and related compounds[24]

NRTIs are phosphorylated by kinases. Other pyrimidine NRTIs have been approved by the U.S. Food and Drug Administration (FDA) and include 2',3'-dideoxycytidine (Hivid, **6.45**); 2',3'-didehydro-3'-deoxythymidine (Zerit, **6.46**); and lamivudine (Epivir, **6.47**). All are nucleoside analogues that lack a functioning 3'-hydroxyl group. Approved purine NRTIs include 2',3'-dideoxyinosine (Videx, **6.48**) and abacavir (Ziagen, **6.49**) (**Figure 6.20**). Inosine (**6.50**) is a nucleoside that is normally only encountered in tRNA. The purine base of inosine is called hypoxanthine (**6.51**). Abacavir is a carbocyclic nucleoside analogue, meaning the oxygen of the furanose ring has been replaced with a CH_2 group. Abacavir is a prodrug and is inactive against HIV. Inside a cell, abacavir is converted to carbovir (**6.52**), the active drug.

NRTIs inhibit RT in HIV, but they also can impact DNA polymerases in the host cell. Fortunately, DNA polymerases that carry out DNA replication for mitotic cell division in the nucleus resist the effects of NRTIs. A DNA polymerase found in mitochondria, however, is affected by NRTIs. Some side effects of NRTIs, especially lethargy, are attributed to inhibition of DNA replication in mitochondria.

Not all compounds that inhibit HIV RT are nucleoside analogues. Selected approved nonnucleoside RT inhibitors (NNRTIs) (**6.53–6.55**) are shown in **Figure 6.21**. NNRTIs are noncompetitive inhibitors that bind at an allosteric position on RT called the NNRTI pocket.[25] Because NNRTIs do not bind at RT's active site, their structure is not as obviously related to their function as with NRTIs.

Viruses other than HIV may be treated by blocking the action of nucleotide polymerases. Acyclovir (Zovirax, **6.56**) and famciclovir (Famvir, **6.57**) are used against various herpes viruses (**Figure 6.22**). Cidofovir (Vistide, **6.58**) is active against cytomegaloviral retinitis in patients who are infected with HIV. Ribavirin (Rebetol, **6.59**) is approved for use against hepatitis C virus.

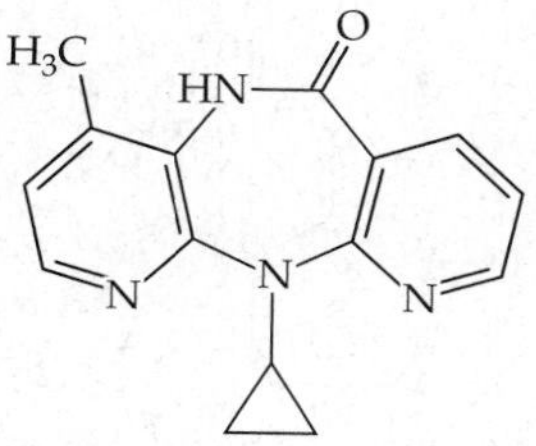

nevirapine
(Viramune)
6.53

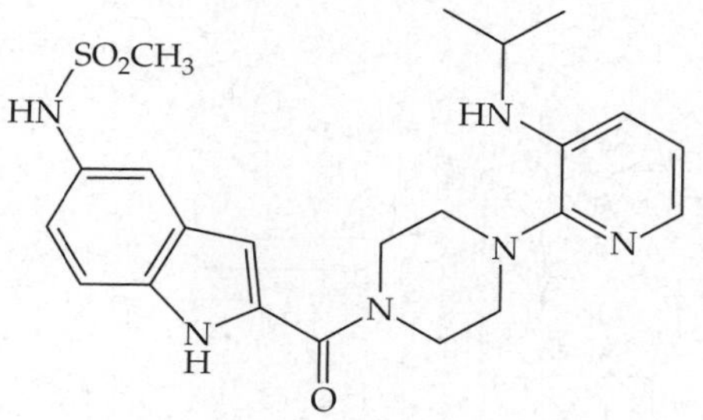

delaviridine
(Rescriptor)
6.54

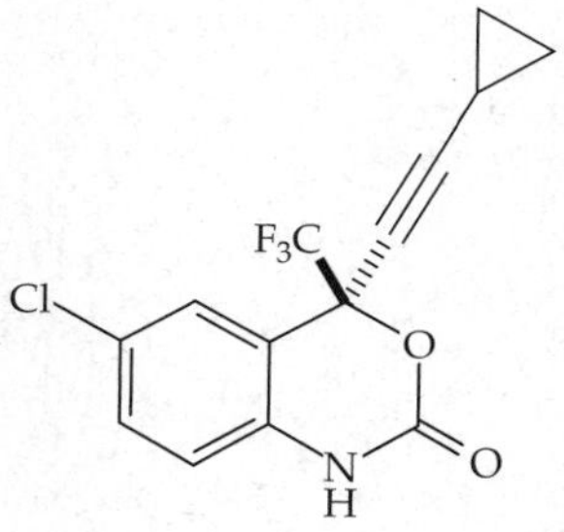

efavirenz
(Sustiva)
6.55

FIGURE 6.21 NNRTIs of reverse transcriptase[24]

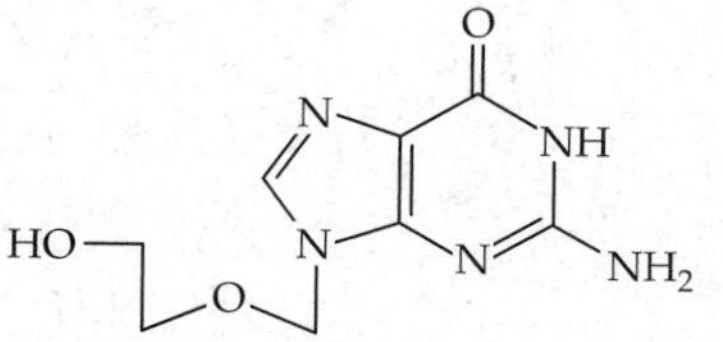

acyclovir
(Zovirax)
6.56

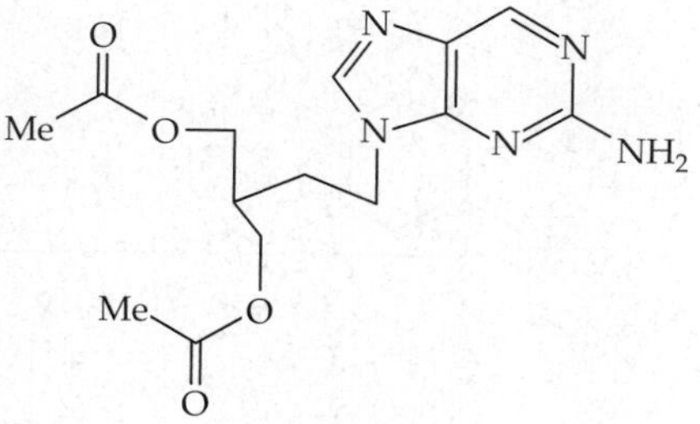

famciclovir
(Famvir)
6.57

FIGURE 6.22 Other nucleoside and nucleotide analogue antivirals

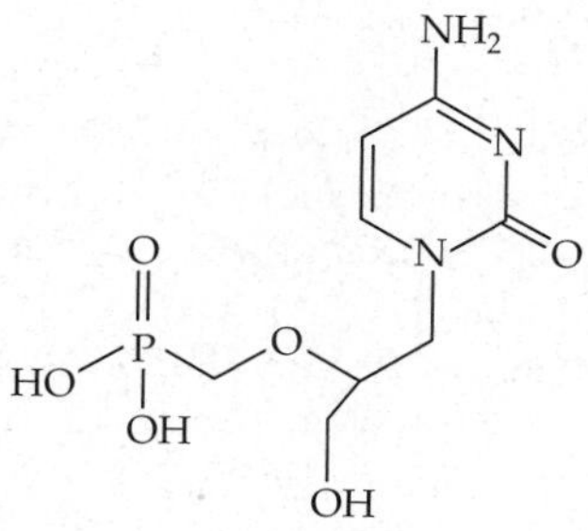

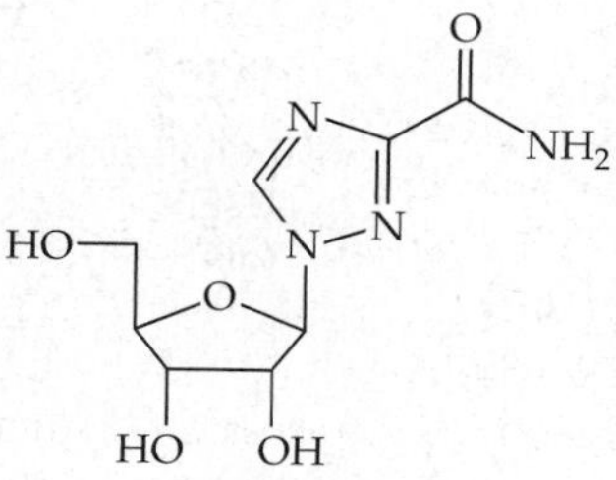

cidofovir
(Vistide)
6.58

ribavirin
(Rebetol)
6.59

Nucleic Acid Antimetabolites

An *antimetabolite* is a compound that is (1) similar to a compound required for normal cellular function and (2) interferes with cellular processes. Based on this definition, many drugs technically qualify as antimetabolites. The term *antimetabolite* is, however, most closely associated with drugs that interfere with the synthesis or action of nucleic acids. Antimetabolites are commonly discussed in the treatment of cancer, viruses, and bacterial infections.

A metabolic pathway that has received considerable attention is the conversion of 2'-deoxyuridine 5'-monophosphate (dUMP, **6.60**) to thymidine 5'-monophosphate (TMP, **6.61**) (Scheme 6.13). Without an adequate supply of TMP, a cell or bacterium cannot create DNA for cell division. Therefore, blocking TMP synthesis is an attractive method for slowing the advancement of certain cancers and bacterial infections. Important molecules in the methylation of dUMP are the various folic acid derivatives: folic acid (FA, **6.62**), dihydrofolic acid (DHF, **6.63**), tetrahydrofolic acid (THF, **6.64**), and N^5, N^{10}-methylene tetrahydrofolic acid (MTHF, **6.65**) (Figure 6.23). These structures

SCHEME 6.13 Methylation of dUMP (**6.60**) to form TMP (**6.61**)

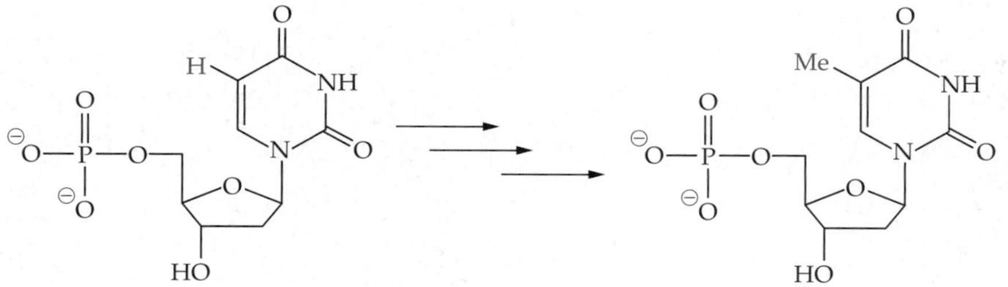

2'-deoxyuridine 5'-monophosphate
(dUMP)
6.60

thymidine 5'-monophosphate
(TMP)
6.61

FIGURE 6.23 Selected folic acid derivatives (**6.62–6.65**)[26]

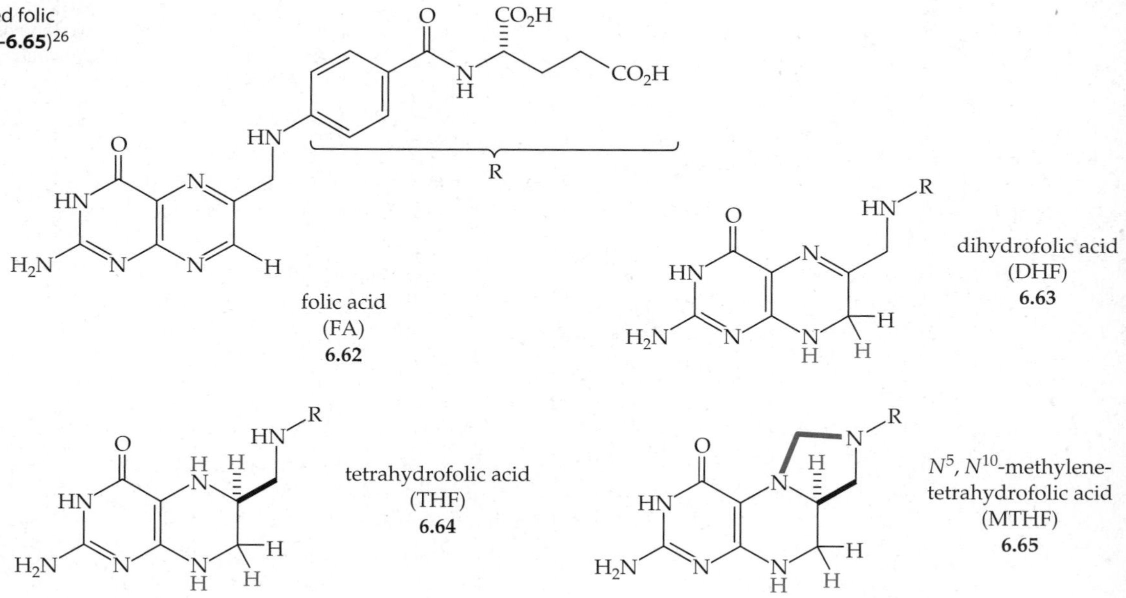

folic acid
(FA)
6.62

dihydrofolic acid
(DHF)
6.63

tetrahydrofolic acid
(THF)
6.64

N^5, N^{10}-methylene-tetrahydrofolic acid
(MTHF)
6.65

$$\text{DHF} + \text{NADPH} + \text{H}^+ \xrightarrow[\text{(DHFR)}]{\substack{\text{dihydrofolate} \\ \text{reductase}}} \text{THF} + \text{NADP}^+$$
6.63   **6.64**

$$\text{THF} + \text{serine} \xrightarrow[\text{(SHMT)}]{\substack{\text{serine hydroxymethyl} \\ \text{transferase}}} \text{MTHF} + \text{glycine}$$
6.64   **6.65**

$$\text{dUMP} + \text{MTHF} \xrightarrow[\text{(TS)}]{\substack{\text{thymidylate} \\ \text{synthase}}} \text{TMP} + \text{DHF}$$
6.60   **6.65**   **6.61**   **6.63**

$$\text{dUMP} + \text{serine} + \text{NADPH} + \text{H}^+ \xrightarrow{\text{overall process}} \text{TMP} + \text{glycine} + \text{NADP}^+$$
6.59   **6.60**

SCHEME 6.14 Overall conversion of dUMP (**6.60**) to TMP (**6.61**)[26]

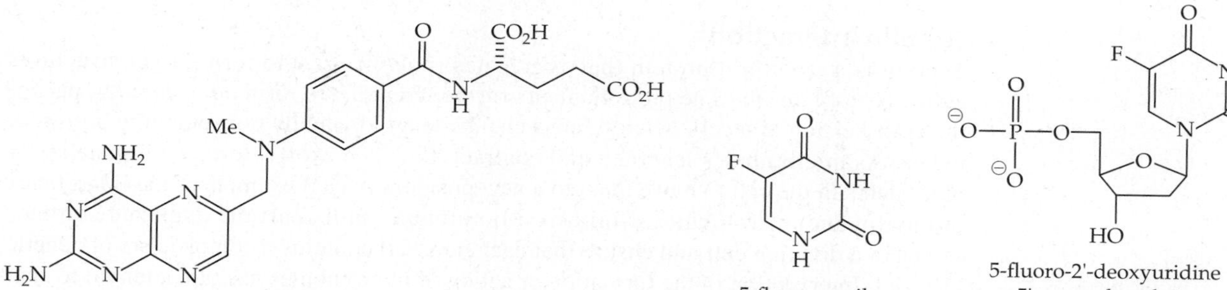

FIGURE 6.24 Antimetabolites in the thymidine synthesis pathway

carry the extra carbon needed to make thymidine. The extra carbon ultimately comes from serine, an amino acid. The overall process is shown in **Scheme 6.14**. Intervention at any one of the three steps shown in Scheme 6.14 will shut down the availability of TMP in a cell.[26]

Many drugs have been developed to intervene in the synthesis of TMP (**Figure 6.24**). For example, methotrexate (**6.66**) inhibits DHFR by blocking the binding site of DHF. Another antimetabolite, 5-fluorouracil (5-FU, **6.67**), is converted in the body to 5-fluoro-2′-deoxyuridine 5′-monophosphate (F-dUMP, **6.68**), a potent inhibitor of TS.[26]

Animals are unable to synthesize folic acid (**6.62**) and must consume adequate quantities in their diets. Plants and bacteria, however, are able to make folic acid. The first step of this synthesis is catalyzed by dihydropteroate synthetase and reacts dihydropteroate diphosphate (**6.69**) and *para*-aminobenzoic acid (PABA, **6.70**) (**Figure 6.25**). Because this pathway is not found in humans, inhibition of the reaction is a method to ultimately stop TMP synthesis in an invading bacterium while not impacting the infected host. The sulfonamides, often called *sulfa drugs*, are a class of antibiotic that exploits the folic acid pathway and inhibits dihydropteroate synthetase. Sulfa drugs bind in the same fashion as PABA and act as competitive inhibitors. The active form of the first sulfa drug is sulfanilamide (**6.71**). Sulfamethoxazole (**6.72**) is a sulfa drug that is widely prescribed today.[26]

dihydropteroate
diphosphate
6.69

para-aminobenzoic acid
(PABA)
6.70

sulfanilamide
6.71

sulfamethoxazole
6.72

Tubulin Interactions

Tubulin is a structural protein that aggregates (polymerizes) to form larger structures called protofilaments. The protofilaments further combine to form *microtubules*, part of the cytoskeleton of a cell. Microtubules change length by adding or removing protofilaments. As microtubules lengthen and contract, they can exert a force on organelles or other items in the cell to move them to a new position. A vital example of the importance of tubulin and microtubules is mitosis. Microtubules pull apart the duplicate chromosomes in a dividing cell and ensure that each new cell contains a complete set of genetic material. Intervention in the formation or action of microtubules has the potential to shut down cell division as well as many other cellular processes.[26]

Vinblastine (**6.73**) is an antimitotic drug that prevents polymerization of tubulin (**Figure 6.26**). When incubated with tubulin, vinblastine complexes in a 1:1 ratio with tubulin proteins. By blocking polymerization, vinblastine prevents microtubule formation and therefore mitosis. In contrast, paclitaxel (Taxol, **6.74**) and epothilone B (**6.75**) stabilize aggregated tubulin. As a result, in the presence of paclitaxel and epothilone B, cells form static bundles of microtubules that are nonfunctional. Vinblastine and paclitaxel are both approved for clinical use against cancer. Ixabepilone (**6.76**), an analogue of epothilone B (**6.75**), has been approved by the FDA for treatment of certain forms of breast cancer. The European Medicines Agency (EMEA) did not approve ixabepilone out of concern over severe side effects.[27]

Vinblastine, paclitaxel, and epothilone B are all natural products. Vinblastine is isolated from the Madagascar periwinkle plant (*Vinca rosea*), paclitaxel from the bark of the Pacific yew tree (*Taxus brevifolia*), and epothilone B as a metabolite of myxobacterium *Sorangium cellulosum*.[28] Complex natural products with promising biological activity such as **6.73–6.75** are attractive challenges for synthetic chemists. Different research groups fiercely compete to publish the first total synthesis of such compounds. The race to paclitaxel was particularly contentious. In the end, the research group of Robert Holton at Florida State University published a synthesis in 1994 in the *Journal of the American Chemical Society*,[29,30] while K. C. Nicolaou of The Scripps Research Institute reported a synthesis in *Nature* in the same year.[31] Other groups were also vying for the first synthesis. In the months leading up to and following the reported syntheses, the propriety of journal reviewing practices and the release of preliminary results to the mainstream news media were topics of discussion in the organic chemistry community.[32]

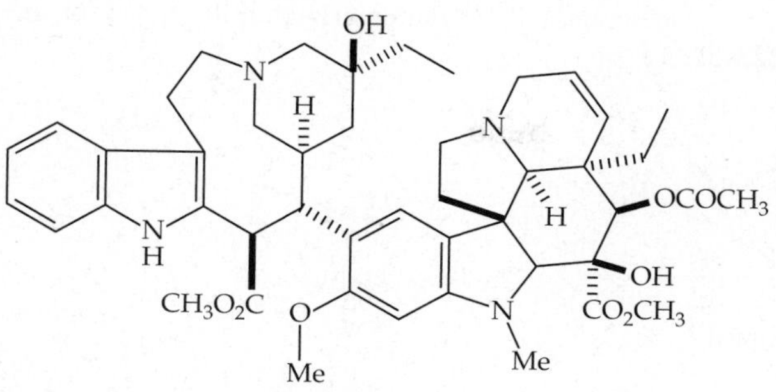

vinblastine
6.73

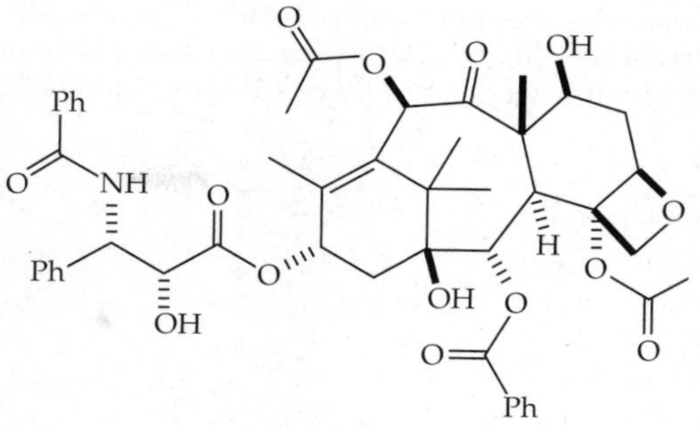

paclitaxel
(Taxol)
6.74

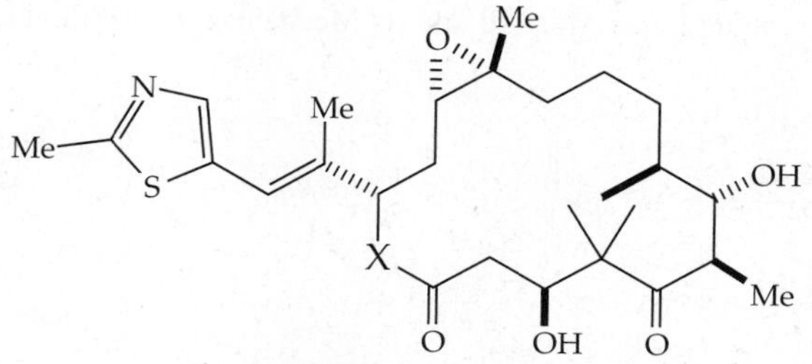

epothilone B (X = O), **6.75**
ixabepilone (X = NH), **6.76**

FIGURE 6.26 Compounds that affect microtubulin polymerization and depolymerization

Summary

Nucleic acids are the building blocks of the genetic code of all living organisms. From only two sugars (ribose and 2-deoxyribose), a phosphate, and five bases (adenine, cytosine, guanine, thymidine, and uracil), virtually all strands of DNA and RNA can be assembled. DNA, in its double-stranded form, resides safely within the nucleus of a cell and consists of genes. Each gene can be transcribed into a segment of single-stranded RNA. RNA then serves in numerous cellular roles—catalysts (rRNA for protein synthesis and ribozymes), regulators (siRNA), information transfer (mRNA), and amino acid shuttling (tRNA). Given the importance of oligonucleotides in biology, DNA and RNA are common targets for drug discovery programs. Therefore, predictable methods of recognizing and binding oligonucleotides are valuable tools for medicinal chemists to understand. The most common means of recognition include traditional Watson–Crick base pairing, charge-charge interaction with the phosphate backbone, intercalation between bases, and binding to the major and minor grooves of helical DNA. Binding DNA can stop transcription and gene expression. Binding RNA typically prevents translation of mRNA into proteins. Some drugs target DNA or RNA synthesis. This is particularly effective for fighting cells that are rapidly dividing or in need of large amounts of genetic material, including cancers, bacterial infections, and viral infections. Approaches to stopping oligonucleotide synthesis include inhibition of polynucleotide polymerases, stopping the synthesis of nucleic acid precursors, and interfering with cell division machinery.

Questions

1. Why would the mole fraction of C and G influence duplex stability (T_m) differently than A and T?

2. What is the effect on T_m of lengthening a duplex strand from 10 base pairs to 15? 15 to 20? 95 to 100? Assume the C:G and A:T ratios are constant throughout all the strands, and use Equation 6.1 for your calculations. Is this trend reasonable? Why or why not?

3. Nucleoside analogues do not always behave as expected. 5-Bromouracil (**6.a**) effectively forms base pairs with guanine nucleotides. This surprising observation has been explained by invoking a different tautomer of **6.a**. Draw a different tautomer of **6.a** and show how it can effectively base pair to a guanine nucleotide. (Tautomerization theory: Topal, M. D., & Fresco, J. R. Base Pairing and

Fidelity in Codon-Anticodon Interaction. *Nature.* 1976, *263*, 289–293. Opposing view: Orozco, M., Hernandez, B., & Luque, F. J. Tautomerization of 1-Methyl Derivatives of Uracil, Thymine, and 5-Bromouracil. Is Tautomerization the Basis for the Mutagenicity of 5-Bromouridine? *J. Phys. Chem. B.* 1998, *102*, 5228–5233.)

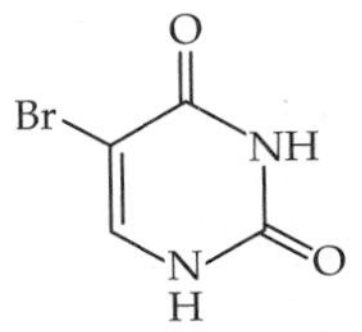

5-bromouracil
6.a

4. When deuterium-labeled mustard **6.b** reacts with a nucleophile (thiosulfate anion, $S_2O_3^{-2}$), a nearly 1:1 mixture of products (**6.c** and **6.d**) is observed. Explain how both products can be formed in approximately equal amounts. (Springer, J. B., Colvin, M. E., Colvin, O. M., & Ludeman, S. M. Isophosphoramide Mustard and Its Mechanism of Bisalkylation. *J. Org. Chem.* 1998, 63, 7218–7222.)

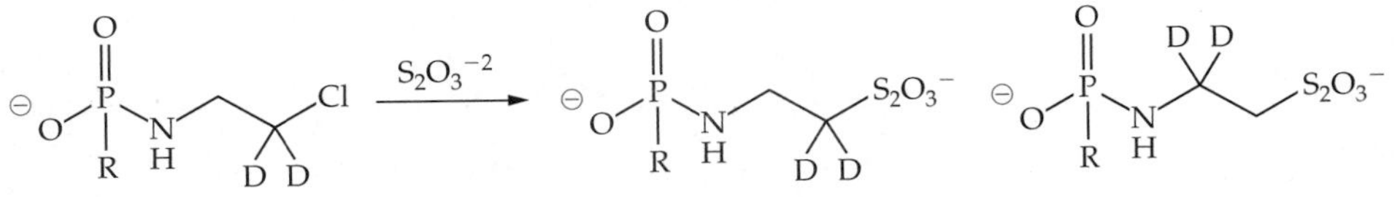

6.b
R = NHCH$_2$CD$_2$Cl
6.c
6.d

5. Bizelesin (**6.e**) is a minor groove binding bis-alkylator. Like almost all DNA alkylators, bizelesin alkylates through a reactive intermediate and not through a simple, one-step S_N2 pathway. Propose an alkylation mechanism for bizelesin. (Hint: Consider adozelesin (**6.44**).) (Pitot, H. C., Reid, J. M., Sloan, J. A., Ames, M. M., Adjei, A. A., Rubin, J., et al. A Phase I Study of Bizelesin [NSC 615291] in Patients with Advanced Solid Tumors. *Clin. Cancer Res.* 2002, *8*, 712–717.)

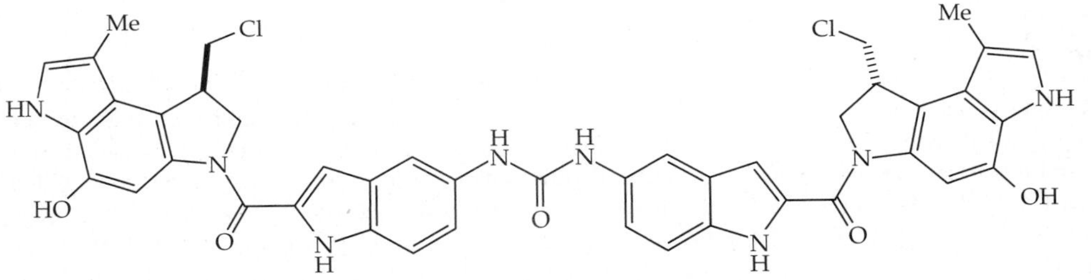

bizelesin
6.e

6. Equation 6.2 fails under two conditions. What are they? Are these conditions likely to be a concern?
7. According to Equation 6.2 and typical cellular conditions, what is the T_m of a duplex strand of DNA as the length approaches infinity? (The ratio of GC base pairs to AT in the human genome is approximately 40:60.)
8. Joining, or *ligating*, DNA involves connecting two separate strands of duplex DNA. The most common approach joins two strands, each with a polynucleotide single-stranded tail called a sticky end. The sticky ends, if complementary, can base pair and hold the strands together until an enzyme, DNA ligase, covalently unifies the phosphate backbone. The following scheme demonstrater the ligation process. A typical sticky end is AATT. What is the T_m of the hybridized yet unligated duplex strand based on Equation 6.1?

9. In the nucleus, DNA is tightly wound about histones, a cylindrical complex of eight proteins. The proteins within histones are rich in lysine and arginine amino acids. Why are these amino acids particularly capable of binding DNA?

10. DNA and RNA often contain many independent binding sites for a given ligand. Therefore, Rosenthal plots (see Chapter 5) are commonly used to determine the number of ligands a saturated strand of DNA or RNA can bind. The binding of netropsin (N) in the minor groove of *Clostridium perfrigens* DNA has been studied and is listed in the following table. D_f represents the concentration of free netropsin (nM), and r is the number of netropsin molecules per DNA base pair. Prepare a Rosenthal plot of the data (r/D_f vs. r) and determine the maximum number of netropsin molecules per DNA base pair. (Wartell, R. M., Larson, J. E., & Wells, R. D. Netropsin: A Specific Probe for A-T Regions of Duplex Deoxyribonucleic Acid. *J. Biol. Chem.* 1974, *249*, 6719–6731.)

Entry	r/D_f (Bound N/Base Pair/Free N)	r (Bound N/Base Pair)
1	$1.9 \cdot 10^{-6}$	0.020
2	$1.8 \cdot 10^{-6}$	0.045
3	$1.5 \cdot 10^{-6}$	0.065
4	$0.8 \cdot 10^{-6}$	0.075
5	$0.4 \cdot 10^{-6}$	0.090

11. In a similar study to Question 10, the binding of netropsin (**6.45**) to DNA was found to increase the T_m of the DNA segment. The following table shows some data for binding netropsin to DNA isolated from *Clostridium perfringens*. Graph the data as a saturation plot (ΔT_m vs. N/nt, molar ratio of netropsin per DNA nucleotide). Estimate the K_D of the DNA-netropsin complex (expressed as the molar ratio of netropsin per DNA nucleotide) based on your graph. (Wartell, R. M., Larson, J. E., & Wells, R. D. The Compatability of Netropsin and Actinomycin Binding to Natural Deoxyribonucleic Acid. *J. Biol. Chem.* 1975, *250*, 2698–2702.)

Entry	$\Delta T_m(°C)$	N/nt
1	4.1	0.01
2	6.8	0.02
3	7.6	0.02
4	10.1	0.04
5	11.5	0.09
6	13.7	0.15
7	14.6	0.20
8	14.4	0.39

12. Repeat Question 11, but graph the data as a linear, double reciprocal plot in the spirit of the Lineweaver–Burk equation (see Chapter 4). Plot $1/\Delta T_m$ vs. $1/(N/nt)$ and perform a linear regression to determine the best-fit line (Equation 4.a). The x-intercept corresponds to the K_D of the DNA-netropsin complex. The K_D value from this method should be more accurate than the estimation in Question 11.

$$\frac{1}{\Delta T_m} = \frac{K_D}{\Delta T_m^{max}} \times \frac{1}{\underset{nt}{N}} + \frac{1}{\Delta T_m^{max}} \qquad (4.a)$$

13. Books sometimes disagree on the preferred tautomer of guanine—**6.f** versus **6.g**. A parallel can be found in 2-pyridone, which also has two likely tautomeric forms, **6.h** and **6.i**. To determine which form of 2-pyridone predominates, the ultraviolet spectra of two isomers, 1-methyl-2-pyridone (**6.j**) and 2-methoxypyridine (**6.k**), were obtained with the observed λ_{max} wavelengths shown. The λ_{max} of 2-pyridone is at 229 nm. Does 2-pyridone likely favor tautomer **6.h** or **6.i**? Which tautomer of guanine is likely favored, **6.f** or **6.g**? (Elguero, J., Marzin, C., Katritzky, A. R., & Linda, P. The Tautomerization of Heterocycles. *Adv. Heterocycl. Chem., Suppl. 1*; New York: Academic Press, 1976.)

tautomers of guanine
6.f   **6.g**

tautomers of pyridone
6.h   **6.i**

1-methyl-2-pyridone
6.j
$\lambda_{max} = 230$ nM

2-methoxypyridine
6.k
$\lambda_{max} = 270$ nM

14. This chapter states that monoesters of phosphates are typically shown with both OH groups of the phosphate deprotonated. The pK_a of the second OH group is

approximately 6.6. The pH of blood is 7.4. Based on the Henderson–Hasselbalch equation (following), determine the ratio of the monobasic form (**6.l**, acid) and dibasic form (**6.m**, conjugate base) of ethyl phosphate at pH 7.4. Is it accurate to represent both OH groups as being deprotonated?

$$pH = pK_a + \log\frac{[\text{conjugate base}]}{[\text{acid}]}$$

Henderson–Hasselbalch equation

monobasic form of
ethyl phosphate
(acid)
6.l

dibasic form
of ehtyl phosphate
(conjugate base)
6.m

15. Antisense drugs target RNA. Why do they not target DNA?

References

1. Kumler, W. D., & Eiler, J. J. The Acid Strength of Mono and Diesters of Phosphoric Acid. The *n*-Alkyl Esters from Methyl to Butyl, the Esters of Biological Importance, and the Natural Guanine Phosphoric Acids. *J. Am. Chem. Soc.* 1943, *65*, 2355–2361.

2. Alberts, B., Johnson, A., Lewis, J., Raff, M., Roberts, K., & Walter, P. How Cells Read the Genome: From DNA to Protein. *Molecular Biology of the Cell* (4th ed., Chapter 6). New York: Garland Science, 2002.

3. Nelson, D. L., & Cox, M. M. Nucleotides and Nucleic Acids. *Lehninger Principles of Biochemistry* (3rd ed., Chapter 10). New York: Worth, 2000.

4. Hannon, G. J., & Rossi, J. J. Unlocking the Potential of the Human Genome with RNA Interference. *Nature.* 2004, *431*, 371–378.

5. Isis Pharmaceuticals Chemistry/Formulations. http://www.isispharm.com/Antisense-Technology/Antisense-Drug-Discovery-Platform/Medicinal-Chemistry.htm (accessed August 2012).

6. Wallace, R. B., Shaffer, J., Murphy, R. F., Bonner, J., Hirose, T., & Itakura, K. Hybridization of Synthetic Oligodeoxyribonucleotides to Phi Chi 174 DNA: The Effect of Single Base Pair Mismatch. *Nucleic Acids Res.* 1979, *6*, 3543–3557.

7. Owczarzy, R., Dunletz, I., Behlke, M. A., Klotz, I. M., & Walder, J. A. Thermodynamic Treatment of Oligonucleotide Duplex-Simplex Equilibria. *Proc. Natl. Acad. Sci. USA.* 2003, *100*, 14840–14845.

8. Owczarzy, R., You, Y., Moreira, B. G., Manthey, J. A., Huang, L., Behlke, M. A., & Walder, J. A. Effects of Sodium Ions on DNA Duplex Oligomers: Improved Predictions of Melting Temperatures. *Biochemistry.* 2004, *43*, 3537–3554.

9. Kleinman, M. E., Yamada, K., Takeda, A., Chandrasekaran, V., Nozaki, M., Baffi, J. Z., et al. Sequence- and Target-Independent Angiogenesis Suppression by siRNA via TLR3. *Nature.* 2008, *452*, 591–597.

10. Sarkar, S. N., Peters, K. L., Elco, C. P., Sakamoto, S., Pai, S., & Sen, G. C. Novel Roles of TLR3 Tyrosine Phosphorylation and PI$_3$ Kinase in Double-Stranded RNA Signaling. *Nat. Struct. Mol. Biol.* 2004, *11*, 1060–1067.

11. Sucheck, S. J., Wong, A. L., Koeller, K. M., Boehr, D. D., Draker, K., Sears, P., et al. Design of Bifunctional Antibiotics That Target Bacterial rRNA and Inhibit Resistance-Causing Enzymes. *J. Am. Chem. Soc.* 2000, *122*, 5230–5231.

12. Hendrix, M., Alper, P. B., Priestley, E. S., & Wong, C.-H. Hydroxyamines as a New Motif for the Molecular Recognition of Phosphodiesters: Implications for Aminoglycoside-RNA Interactions. *Angew. Chem., Int. Ed. Engl.* 1997, *36*, 95–98.

13. Cho, J., Hamasaki, K., & Rando, R. R. The Binding Site of a Specific Aminoglycoside Binding RNA Molecule. *Biochemistry.* 1998, *37*, 4985–4992.

14. Berge, T., Jenkins, N. S., Hopkirk, R. B., Waring, M. J., Edwardson, J. M., & Henderson, R. M. Structural Perturbations in DNA Caused by Bis-Intercalation of Ditercalinium Visualized by Atomic Force Microscopy. *Nucleic Acids Res.* 2002, *30*, 2980–2986.

15. Gredicak, M., & Jeric, I. Enediyne Compounds: New Promises in Anticancer Therapy. *Acta Pharm.* 2007, *57*, 133–150.

16. Fenselau, C. Review of the Metabolism and Mode of Action of Cyclophosphamide. *J. Assoc. Off. Anal. Chem.* 1976, *59*, 1028–1036.

17. Tomasz, M., Chowdary, D., Lipman, R., Shimotakahara, S., Veiro, D., Walker, V., & Verdine, G. L. Reaction of DNA with Chemically or Enzymatically Activated Mitomycin C: Isolation and Structure of the Major Covalent Adduct. *Proc. Natl. Acad. Sci. USA.* 1986, *83*, 6702–6706.

18. Borowski-Borowy, H., Lipman, R., & Tomasz, M. Recognition between Mitomycin C and Specific DNA Sequences for Cross-Link Formation. *Biochemistry* 1990, *29*, 2999–3006.

19. Wing, R. M., Pjura, P., Drew, H. R., & Dickerson, R. E. The Primary Mode of Binding of Cisplatin to a B-DNA Dodecamer: C-C-C-G-A-A-T-T-C-G-C-G. *EMBO J.* 1984, *3*, 1201–1206.

20. Whittaker, J., McFadyen, W. D., Wickham, G., Wakelin, L. P. G., & Murray, V. The Interaction of DNA-Targeted Platinum Phenanthridinium Complexes with DNA. *Nucleic Acids Res.* 1998, *26*, 3933–3939.

21. Liu, J. S., Kuo, S. R., Beerman, T. A., & Melendy, T. Induction of DNA Damage Responses by Adozelesin in S Phase-Specific and Dependent on Active Replication Forks. *Mol. Cancer Ther.* 2003, *2*, 41–47.

22. Wartell, R. M., Larson, J. E., & Wells, R. D. Netropsin: A Specific Probe for A-T Regions of Duplex Deoxyribonucleic Acid. *J. Biol. Chem.* 1974, *249*, 6719–6731.

23. Bruice, T. C., Mei, H.-Y., He, G.-X., & Lopez, V. Rational Design of Substituted Tripyrrole Peptides That Complex with DNA by Both Selective Minor-Groove Binding and Electrostatic Interaction with the Phosphate Backbone. *Proc. Natl. Acad. Sci. USA.* 1992, *89*, 1700–1704.

24. U.S. Food and Drug Administration. FDA Approved Anti-HIV Drugs. http://www.fda.gov/ForConsumers/ByAudience/ForPatientAdvocates/HIVandAIDS Activities/ucm118915.htm (accessed August 2012).

25. Ren, J., Bird, L. E., Chamberlain, P. P., Stewart-Jones, G. B., Stuart, D. I., & Stammers, D. K. Structure of HIV-2 Reverse Transcriptase at 2.35-Å Resolution and the Mechanism of Resistance to Non-Nucleoside Inhibitors. *Proc. Natl. Acad. Sci. USA.* 2002, *99*, 14410–14415.

26. Chabner, B. A., Ryan, D. P., Paz-Ares, L., Garcia-Carbonero, R., & Calabresi, P. Antineoplastic Agents. In J. G. Hardman & L. E. Limbird (Eds.), *Goodman and Gilman's the Pharmacological Basis of Therapeutics* (10th ed., Chapter 52). New York: McGraw-Hill, 2001.

27. Questions and Answers on Recommendation for the Refusal of the Marketing Authorisation for Ixempra. http://www.emea.europa.eu/pdfs/human/opinion/IxempraQ&A_60256908en.pdf (accessed August 2012).

28. Su, D.-S., Meng, D., Bertinato, P., Balog, A., Sorensen, E. J., Danishefsky, S. J., Zheng, Y.-H., et al. Total Synthesis of (π)-Epothilone B: An Extension of the Suzuki Coupling Method and Insights into Structure-Activity Relationships of the Epothilones. *Angew. Chem., Intl. Ed.* 1997, *36*, 757–759.

29. Holton, R. A., Somoza, C., Kim, H. B., Liang, F., Biediger, R. J., Boatman, P. D., et al. First Total Synthesis of Taxol: Part 1: Functionalization of the B Ring. *J. Am. Chem. Soc.* 1994, *116*, 1597–1598.

30. Holton, R. A., Kim, H. B., Somoza, C., Liang, F., Biediger, R. J., Boatman, P. D., et al. First Total Synthesis of Taxol: Part 2: Completion of the C and D Rings. *J. Am. Chem. Soc.* 1994, *116*, 1599–1600.

31. Nicolaou, K. C., Zang, Z., Liu, J. J., Ueno, H., Nantermet, P. G., Guy, R. K., et al. Total Synthesis of Taxol. *Nature.* 1994, *367*, 630–634.

32. Goodman, J., & Walsh, V. *The Story of Taxol: Nature and Politics in the Pursuit of an Anti-Cancer Drug.* Cambridge: Cambridge University Press 2001, pp. 176–182.

Chapter 7

Pharmacokinetics

Chapter Outline

Learning Goals for Chapter 7

- Be able to apply the relevant mathematical models to the various routes of administration
- Know how to derive pharmacokinetic parameters from C_p-time data

While Chapter 3 qualitatively described the pharmacokinetic processes of absorption, distribution, metabolism, and elimination, this chapter focuses on the mathematical side of pharmacokinetics. Despite the complexity of drug passage into, throughout, and out of a biological system, an individual drug's behavior can generally be described by the two principle pharmacokinetic parameters, volume of distribution and clearance. Clearance and volume of distribution are determined through analysis of intravenous bolus drug concentration versus time data. Additional parameters, such as the absorption rate constant and bioavailability, may be determined by subsequent analysis of oral drug data. The pharmacokinetics of a drug will ultimately determine the dosing regimen necessary to maintain effective and safe drug levels in a patient. Regulatory approval of a drug requires a full understanding of a drug's pharmacokinetics. A poorly understood or unpredictable drug is a safety risk and not approvable.

7.1 Intravenous Bolus

The intravenous (IV) bolus is the administration method with the fewest variables and provides a natural starting point for understanding the pharmacokinetic behavior of pharmaceuticals.

C_p vs. Time and Elimination

All pharmacokinetic studies rely on C_p versus time data. C_p is the *plasma concentration* of a drug. For the purposes of our discussion, we will be assuming that the drug concentration in plasma is the same as in whole blood.

The units for C_p can vary. The most common units are in the form of mass/volume, such as $\mu g/mL$ or ng/mL. This chapter exclusively uses C_p with mass/volume units. Very rarely, C_p is encountered in the chemist-familiar moles/volume format like μM or nM.

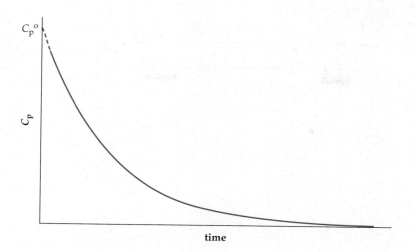

FIGURE 7.1 C_p vs. time for an
IV bolus

The time unit may be minutes, hours (most common), or days. The following statement
will be repeated many times throughout this chapter: *pay careful attention to the units of
variables* before performing any calculations! When dealing with graphs, always check
your axes.

An idealized C_p versus time plot for an IV bolus is shown in **Figure 7.1**. The drug
virtually instantaneously rises to a high level at the time of the injection and then decreases as it is eliminated from the bloodstream by the liver, kidneys, or other organs/
tissues.

The relationship between C_p and time in Figure 7.1 is that of a first-order process and
is expressed in Equation 7.1.[1]

$$C_p = C_p^{\circ}e^{-k_{el}t} \qquad (7.1)$$

C_p is the plasma concentration of the drug at time t. The elimination rate constant is k_{el}, with
units of inverse time. C_p° is the plasma concentration at $t = 0$. C_p° is a special case that
is more theoretical than real. At the time of the injection ($t = 0$), the drug bolus hits the
bloodstream. At this instant, the drug has not mixed with the entire blood supply. The
concentration at the site of injection is very high, but blood in other parts of the body still
has a C_p of 0. For this reason, C_p° cannot be directly measured experimentally but must be
determined by extrapolation of the C_p-time line back to the y-axis.

C_p decreases over time. The rate of elimination is defined by the first derivative of
Equation 7.1 with respect to time (Equation 7.2).

$$\text{rate of elimination} = \frac{dC_p}{dt} = -k_{el}C_p \qquad (7.2)$$

Equation 7.2 highlights the first-order nature of Figure 7.1. The rate of elimination is
directly proportional to C_p. If a patient has twice as much drug in her bloodstream, then
that patient's rate of elimination will be twice as high. Because drug elimination so frequently behaves in a first-order fashion, drugs tend to display a regular *half-life* ($t_{1/2}$). A
half-life is the time interval required for C_p to decrease by a factor of 0.5. The relationship
between $t_{1/2}$ and k_{el} is shown in Equation 7.3.

$$t_{1/2} = \frac{0.693}{k_{el}} \qquad (7.3)$$

FIGURE 7.2 Periodicity of half-life and its effect on C_p

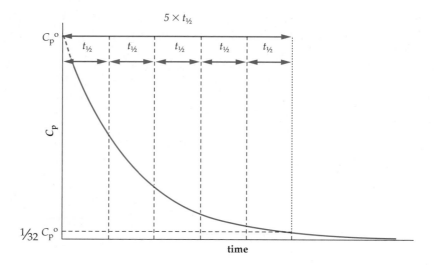

As a rule of thumb, a drug is said to be completely eliminated after a time interval of five half-lives. After five half-lives, the amount of drug in the body is only around 3% of the theoretical C_p^o value (**Figure 7.2**, see the following Sample Calculation). Although 3% residual C_p is not complete elimination in a strict sense, once C_p of a drug falls to such a low level, C_p is below the therapeutic window and is no longer causing a measurable therapeutic effect.

The equations and discussion that have been presented to this point all follow something called the *one-compartment model*. The remaining discussion in this subsection and the subsection on clearance continue to rely on the one-compartment model. Only in the volume of distribution subsection will the curtain be pulled back and the assumptions and failings of the one-compartment model be revealed. Despite the shortcomings of the one-compartment model, it is regardless a valuable means of gaining insight into the behavior of a drug in the body.

Sample Calculation   **Determination of C_p Five Half-Lives after Administration of an IV Bolus**

PROBLEM   What fraction of a drug remains from an IV bolus after five half-lives have elapsed?

SOLUTION   This problem can be solved multiple ways. The simplest method is to consider what happens after one half-life: the C_p drops by a factor of 0.5. In other words, once one half-life has elapsed from the time of an IV bolus, C_p will be equal to $0.5C_p^o$. After two half-lives, C_p drops by a factor of 0.25 (0.5^2). Continuing to five half-lives, we see that C_p has dropped by a factor of 0.0325 (0.5^5). Therefore, after five half-lives, C_p from an IV bolus will be $0.0325C_p$.

Another solution, shown here, relies on Equations 7.1 and 7.3.

$$C_p = C_p^o e^{-k_{el}t} \tag{7.1}$$

$$t_{1/2} = \frac{0.693}{k_{el}} \tag{7.3}$$

By Equation 7.3, five half-lives can be calculated as $3.465/k_{el}$.

$$5 \times t_{1/2} = 5 \times \frac{0.693}{k_{el}} = \frac{3.465}{k_{el}}$$

Equation 7.1 allows one to determine C_p at any time in terms of $C_p{}^o$. Our time value is five half-lives. Substitution of $3.465/k_{el}$ in place of t into Equation 7.1 gives C_p after five half-lives have elapsed.

$$C_p = C_p{}^o e^{-k_{el}t} = C_p{}^o e^{-k_{el} \times \frac{3.465}{k_{el}}} = C_p{}^o e^{-3.465} = 0.0325 C_p{}^o$$

Only 3% residual drug in the bloodstream is small, but it is not zero. In theory, under Equation 7.1, a drug *never* completely eliminates from a patient.

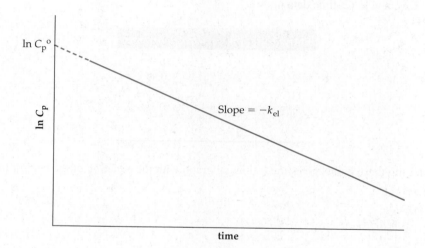

FIGURE 7.3 ln C_p vs. time for an IV bolus

When plotted as ln C_p versus time, a drug that undergoes first-order elimination generates a straight line (**Figure 7.3**). The slope is $-k_{el}$, and the y-intercept is ln $C_p{}^o$. Unlike C_p, ln C_p can be less than 0, and the line for ln C_p versus time can cross the x-axis.

Equation 7.4 summarizes the relationship shown in Figure 7.3.

$$\ln C_p = -k_{el}t + \ln C_p{}^o \tag{7.4}$$

The first derivative of Equation 7.4 with respect to time reveals that the slope is $-k_{el}$ (Equation 7.5).

$$\frac{d(\ln C_p)}{dt} = -k_{el} \tag{7.5}$$

With real C_p-time data (clinical data), the first data point is normally collected approximately 15 minutes after administration. Therefore, the value of ln $C_p{}^o$ is determined by extrapolation of the ln C_p-time line back to t_o rather than as an experimental data point.

Sample Calculation **Calculating Half-Life from C_p-Time Data**

PROBLEM   The following table presents C_p-time data for a drug. Use the data to construct a ln C_p versus time graph. Determine the best-fit line of the graph to calculate the half-life of the drug.

Entry	t (h)	C_p (μg/mL)
1	1	3.46
2	2	2.44
3	3	1.73
4	4	1.22
5	5	0.86

SOLUTION   To generate a ln C_p graph, we need ln C_p data. Taking the natural logarithm of C_p gives new ln C_p-time data points.

Entry	t (h)	ln C_p (μg/mL)
1	1	1.24
2	2	0.89
3	3	0.55
4	4	0.20
5	5	−0.15

Plotting these points gives the following graph with the equation of the best-fit line.

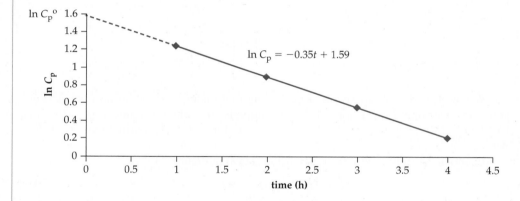

The slope of the line is −0.35 and is equal to $-k_{el}$. If $k_{el} = 0.35$ h^{-1}, then by Equation 7.3, $t_{1/2} = 1.98$ h.

$$t_{1/2} = \frac{0.693}{k_{el}} \tag{7.3}$$

$$t_{1/2} = \frac{0.693}{0.35 \frac{1}{h}} = 1.98 \text{ h}$$

The equation of the best-fit line is in the format $y = mx + b$. The value of the y-intercept, b, is 1.59 and is equivalent to ln C_p^o. This corresponds to a C_p^o value of 4.90 μg/mL.

Clearance

From either a C_p or $\ln C_p$ versus time plot, one feature is immediately clear: the drug concentration drops over time. This process is called elimination and is determined by *clearance* (CL). Clearance is the process of removal of drug from the bloodstream. As was discussed in Chapter 3, clearance occurs primarily either through filtration of a drug by the kidneys (*renal clearance*, CL_R) or metabolism of a drug in the liver by the action of enzymes (*hepatic clearance*, CL_H). Other clearance processes are possible, but CL_R and CL_H normally comprise the large majority of *total clearance* (CL_T or simply CL) (Equation 7.6).

$$CL_T = CL_R + CL_H \qquad (7.6)$$

For a drug that is not eliminated by the kidneys ($CL_R = 0$), the liver dominates clearance and $CL_T \approx CL_H$. Likewise, if a drug is not metabolized by the liver ($CL_H = 0$), the kidneys dominate clearance and $CL_T \approx CL_R$. A drug with a higher clearance is removed more quickly from the bloodstream and displays a shorter half-life. The units on clearance are volume/time.

The kidneys and liver receive blood from the general circulatory system. That blood enters the organ with a certain concentration of drug ($C_p{}^{in}$), the organ removes a fraction of the drug from the incoming blood, and the blood is returned to the circulatory system with a reduced concentration of drug ($C_p{}^{out}$). The fraction of drug removed from the blood as it travels through the organ determines the *extraction ratio* (E) (Equation 7.7).[1]

$$E = \frac{C_p{}^{in} - C_p{}^{out}}{C_p{}^{in}} \qquad (7.7)$$

Since the C_p values in Equation 7.7 are specific to a particular organ, any given extraction ratio is also specific to a particular organ. Furthermore, knowing the extraction ratio of a drug for one organ does not provide insight into the extraction ratio of the same drug for a different organ. Note that valid values for extraction ratio range from 0 to 1, and extraction ratio is a dimensionless number.

Extraction ratio and bloodflow through an organ (Q) collectively determine the clearance of a particular organ (Equation 7.8).

$$CL = Q \times E \qquad (7.8)$$

The units on Q are volume over time, typically mL/min. Bloodflow is occasionally reported on a per mass basis with units of mL/min/kg. Just like extraction ratio, bloodflow in Equation 7.8 is specific to a particular organ. The rate of bloodflow through the liver (Q_H) is 20–25 mL/min/kg, or around 1.5 L/min for a 70-kg patient. The rate of blood filtration through the kidneys (Q_R) is much lower at approximately 2 mL/min/kg. Since bloodflow is tied to a specific organ, Equation 7.8 is appropriate for determining the clearance for individual organs, such as CL_H or CL_R. Since E has no units and Q is in terms of volume over time, Equation 7.8 does provide clearance with the correct units, volume over time.

Calculation of total clearance requires addition of the clearance values of the separate organs, as shown in Equation 7.9, which is an expanded form of Equation 7.6.

$$CL_T = Q_R E_R + Q_H E_H \qquad (7.9)$$

Although the discussion thus far has defined the concept of clearance and presented several equations involving clearance, clearance as a mechanism of removing a drug from the body is likely still unclear. The link between bloodflow and drug

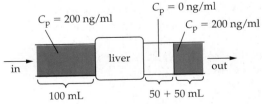

codeine
analgesic
7.1

FIGURE 7.4 Codeine, a drug that is cleared almost exclusively by the liver

removal is admittedly not intuitive. Therefore, an illustration will almost certainly be helpful.

Codeine (**7.1**) is an analgesic that may be administered by an IV bolus (**Figure 7.4**). A typical therapeutic C_p value for codeine is 200 ng/mL, and the hepatic extraction ratio (E_H) for codeine is 0.50. The clearance of codeine (CL_T) is reported to be around 10.7 mL/min/kg, or 750 mL/min, in a 70-kg patient. Codeine is not observed in the urine of patients, so E_R must be 0, and renal clearance (CL_R) is 0 mL/min. As a result, hepatic clearance of codeine (CL_H) is equal to total clearance. Because codeine is cleared only by the liver, codeine provides a simple illustration of the ideas surrounding clearance.

Remember that the liver of a 70-kg patient handles 1.5 L or 1,500 mL of blood per minute. That corresponds to 100 mL every 4.0 seconds. The fate of those 100 mL can be represented in two ways, with each giving different insight into clearance.

The first model is a literal presentation of clearance (**Scheme 7.1**). Based on a clearance value of 750 mL/min for codeine, in 4.0 seconds the liver clears 50 mL of blood and handles a total of 100 mL of blood. If clearance is interpreted literally, 100 mL of blood enter the liver with a concentration of 200 ng/mL, 50 mL (the cleared 50 mL) exit with a concentration of 0 ng/mL, and 50 mL exit at the original concentration of 200 ng/mL. The idea of some blood leaving the liver with no drug and other blood leaving untouched is unrealistic, but this is the literal depiction of clearance. This model aligns with the ideas of bloodflow (Q) and focuses on distinct volumes of blood that are either completely cleared of drug or not yet affected by the extracting organ. The second model of clearance is more in line with extraction ratio (E) (Scheme 7.1). All the blood that enters the liver is acted upon uniformly such that C_p^{out} is equal to the product of E and C_p^{in} (0.5 × 100 ng/mL). While different, both models portray the same net effect on the amount of drug in the bloodsteam. A total of 20 mg (100 mL × 200 ng/mL) enter the liver, and 10 mg (50 mL × 200 ng/mL, or 100 mL × 100 ng/mL) exit.

By watching the effect of multiple passages of blood through the liver, one can see a pattern emerge (**Scheme 7.2**). Again, with codeine as the example, blood with a C_p of 200 ng/mL enters the liver and exits at a concentration of 100 ng/mL. Upon the next pass, C_p drops to 50 ng/mL. The third pass gives a C_p of 25 ng/mL, which is less than the therapeutic concentration of codeine. For each pass, the ratio of C_p^{out} to C_p^{in} is constant at 0.5, the same as the value for the extraction ratio. The magnitude of the decrease, however, is smaller with each pass through the liver. The rate of elimination decreases

literal interpretation of clearance

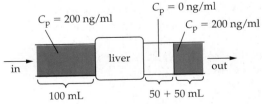

Some blood leaving the liver is completely devoid of drug, and other blood leaving the liver has drug at its original concentration.

actual effect of clearance

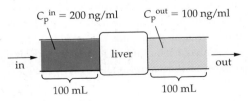

All blood leaving the liver contains drug at a concentration that has been decreased from the original concentration.

SCHEME 7.1 Different representations of clearance

Pass #1 (codeine, 4.0 seconds, Q_H = 25 mL/s, CL_H = 12.5 mL/s, E_H = 0.5)

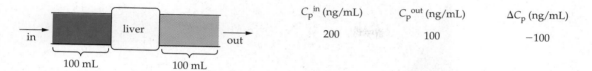

	C_p^{in} (ng/mL)	C_p^{out} (ng/mL)	ΔC_p (ng/mL)
	200	100	−100

Pass #2

	100	50	−50

Pass #3

	50	25	−25

SCHEME 7.2 Multiple passes of blood through the liver

with time. This is entirely consistent with Equation 7.2, which indicates that the rate of elimination is a first-order function of C_p. As C_p drops, so does dC_p/dt.

$$\text{rate of elimination} = \frac{dC_p}{dt} = -k_{el}C_p \qquad (7.2)$$

If the rate of elimination decreases in Scheme 7.2, then what happens to clearance? Clearance is unchanged. For each 4.0-second pass, the liver clears 50 mL (CL_H = 12.5 mL/s) out of the total 100 mL of blood that flows through the organ. Literally, *50% of the blood volume is cleared*, so the actual impact is a decrease in C_p by *50%*. While clearance is constant, the effect of clearance on C_p varies with C_p. Clearance depends on the action of metabolic enzymes on the drug and, at very high drug concentrations, the enzymes can become saturated with substrate. Under these conditions, which are rare, clearance is not constant. Therapeutic concentrations of modern drugs are normally well below the concentrations required to saturate liver enzymes. The tubular secretion and reabsorption processes in the kidneys can also be saturated and affect renal clearance. As with hepatic clearance, variable renal clearance is rare.

Knowing total clearance for a drug is important. Clearance is determined from C_p-time plot data for an IV bolus. C_p-time data for an IV bolus can be used to determine the elimination rate constant of a drug (k_{el}) as well as the hypothetical C_p^o. These two values allow direct calculation of the *area under the curve* (*AUC*) of the C_p-time plot with Equation 7.10 (**Figure 7.5**).[1]

$$AUC = \frac{C_p^o}{k_{el}} \qquad (7.10)$$

Equation 7.10 is obtained by integration of Equation 7.1 and evaluation of the integral over the interval from t_o to t_∞. *AUC* is a measure of the cumulative drug

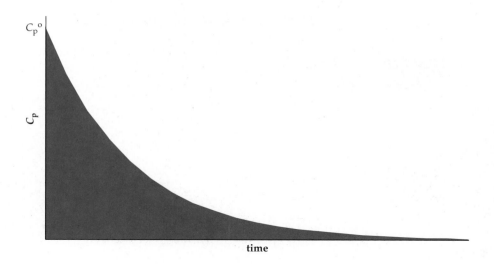

exposure experienced by a patient. Exposure is a function of the length of time the drug is present in the body and the C_p of the drug at various times. The cumulative effect is reasonably determined through an integration. The units of *AUC* are a nonintuitive mass $\cdot$ time/volume, for example $\mu g \cdot h/mL$. Once *AUC* for a bolus has been determined, the dose of the bolus (D_o) may be divided by *AUC* to calculate total clearance through Equation 7.11.[1]

$$CL = \frac{D_o}{AUC} \tag{7.11}$$

The importance of being careful with units when dealing with pharmacokinetic parameters cannot be overstated. The variables used to determine clearance use different standard units for mass and time. For example, the mass unit in C_p tends to be μg or ng, while D_o is most often in mg. Similarly, *CL* uses a time unit of minutes, while the time unit of k_{el} is normally hours. Be vigilant with units.

| Sample Calculation | Determining *CL* from C_p-Time Data (Using *AUC*) |

PROBLEM The following table present some C_p-time data points for an IV bolus dose of 325 mg. What is *CL* for this drug?

Entry	t (h)	C_p (μg/mL)
1	1	3.46
2	2	2.44
3	3	1.73
4	4	1.22
5	5	0.86

SOLUTION These are the same data points as in the previous Sample Calculation, with one extra piece of information: the size (mass) of the dose. From the previous Sample Calculation, we know that k_{el} for this drug is 0.35 h^{-1}. Furthermore, the line we already generated gives the *y*-intercept as 1.59. This value is ln C_p^{o}. Therefore, C_p^{o} is

4.89 μg/mL (inverse natural logarithm of 1.59). Now we have everything we need to use Equation 7.10. *AUC* for this dose is 14.0 μg · h/mL.

$$AUC = \frac{C_p{}^o}{k_{el}} \qquad (7.10)$$

$$AUC = \frac{4.89 \dfrac{\mu g}{mL}}{\dfrac{0.35}{h}} = 14.0 \frac{\mu g \cdot h}{mL}$$

Now Equation 7.11 comes into play. Just be sure to match your units. Convert the dose from mg to μg, and 325 mg becomes 325,000 μg. The initial value for *CL* does not have the standard units, so it must be converted to mL/min.

$$CL = \frac{D_o}{AUC} \qquad (7.11)$$

$$CL = \frac{325,000 \, \mu g}{14.0 \dfrac{\mu g \cdot h}{mL}} = 23,214 \frac{mL}{h} \times \frac{1 \, h}{60 \, min} = 387 \frac{mL}{min}$$

The *CL* value (CL_T) is 387 mL/min.

Volume of Distribution

Along with clearance, volume of distribution determines the pharmacokinetic behavior of a drug. Volume of distribution is an attempt to describe how a drug disperses throughout the various tissues of the body.

Pinning Down Volume of Distribution

Drugs are almost always monitored in vivo by drawing a blood sample from a patient and analyzing the sample for the drug. Drawing blood is quick and convenient, and blood analysis is also accurate and reproducible. The blood, however, is rarely the site of action of a drug. Fortunately, the concentration of a drug in the blood is generally directly proportional to drug concentration in other tissues of the body. Therefore, fluctuations in C_p can serve as a window into concentration changes in the entire body. The argument has been made for the value of determining C_p, but this subsection requires one to understand why C_p alone does not tell the full story of a drug in the body.

Lorazepam (Ativan, **7.2**), an antianxiety drug, nicely demonstrates the weaknesses of C_p (**Figure 7.6**). Following administration of a 2-mg IV bolus, the C_p of lorazepam in a 70-kg patient is approximately 20 ng/mL. For reasons that will be discussed shortly, blood is not sampled until 15 minutes after administration of a bolus. Because lorazepam has a long half-life of 14 hours, one can assume that after just 15 minutes, almost none of the dose has been metabolized or cleared. Essentially the entire 2-mg dose still resides in the patient. Since the dose was administered directly to the blood of the patient, one might *assume* that the drug is in the blood. Based on the C_p of 20 ng/mL (20 μg/L) and the fact that a 70-kg patient has 5 L of blood, the mass of drug contained within the blood of the patient 15 minutes after dosing is 100 μg (20 μg/L × 5 L), or 0.1 mg. So, the patient was dosed 2 mg, but only 0.1 mg resides in the blood. If no drug has been lost (long half-life, low clearance) in 15 minutes, then the remaining 1.9 mg of the original dose must be elsewhere in the body, that is, *not* in the blood. The "still in the blood" assumption was

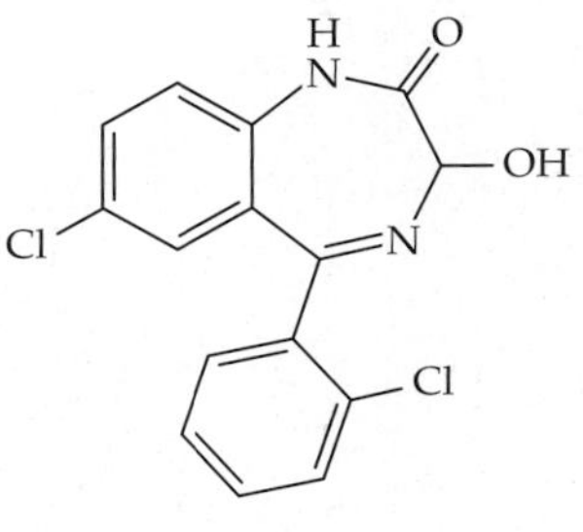

lorazepam
(Ativan)
antianxiety
7.2

FIGURE 7.6

incorrect. Other drugs may behave somewhat differently, but this lorazepam example illustrates a general idea that drugs are not confined to the blood.

Since volume of blood in the body is inadequate for holding the complete lorazepam dose at the observed C_p, what volume of blood would hypothetically be required to contain the full 2-mg dose at a C_p of 20 ng/mL? The required volume is 100 L (2,000 μg ÷ 20 μg/L). This volume—the volume of blood required to contain the drug within a body at an observed concentration—is called the *volume of distribution* (V_d). Volume of distribution is a purely hypothetical value and is sometimes called *apparent volume of distribution* to hammer home the idea that V_d is not a physiologically real number. A human does not have 100 L of blood, but a V_d of 100 L for a drug is completely acceptable.

Conceptually forcing a drug into a hypothetical volume of blood is counterintuitive since drugs clearly reside in other tissues of the body. A compelling reason to envision a drug as being exclusively contained in blood is that clearance occurs almost exclusively on bloodborne compounds. The liver metabolizes compounds in the blood; the kidneys filter the blood. Furthermore, as has already been stated, monitoring C_p by drawing blood is convenient. Therefore, approximating a drug's location as a hypothetical blood volume is at least useful, if not accurate. Because V_d is not a real number, it must be interpreted.

Understanding V_d requires knowledge of the types of fluids into which a drug can distribute. A standard patient (70 kg, or 154 lbs) can be used to demonstrate key ideas. A 70-kg patient has approximately 5.0 L of blood. Because not every patient weighs 70 kg, blood volume is often expressed on a per mass basis as 0.071 L/kg (5 L/70 kg). Multiplication by patient mass gives that patient's approximate blood volume. Whole blood is 54% plasma by volume, so the amount of plasma in a 70-kg patient is 2.7 L. Other sources of fluid in the body include interstitial fluid and lymph (~10 L) and intracellular fluids (~25 L). The *total body water* of a 70-kg male patient is approximately 38 L, or 55% of the body's mass. The value is closer to 50% for females. Total body volume for a 70-kg patient is around 70 L.

Volume of distribution and clearance are both properties of a drug. These two properties determine a drug's elimination rate constant and half-life (Equation 7.12).

$$k_{el} = \frac{CL}{V_d} = \frac{0.693}{t_{1/2}} \tag{7.12}$$

In practice, clinical C_p-time data is used to determine k_{el}, which in turn is used to calculate CL and V_d of a drug. However, physiologically, $t_{1/2}$ and k_{el} are determined by CL and V_d.

Discussion and application of V_d requires definition of compartment models. The simplest model is the *one-compartment model*. Additional compartments can be added as needed with a cost of increased complexity. In the artificial world of paper in a textbook, one can freely force data into any model. When dealing with real drugs and experimental data, one must carefully select the model that best represents the data.

One-Compartment Model

The *one-compartment model* limits distribution of a drug to just one compartment, the *central compartment*.[1] For an IV drug, the drug is administered directly to the central compartment and subsequently eliminates from the same compartment (Scheme 7.3). All the discussion to this point in the chapter falls under the one-compartment model. Despite its weaknesses, which will be described shortly, many drugs can be satisfactorily modeled with one-compartment equations.

SCHEME 7.3 One-compartment model

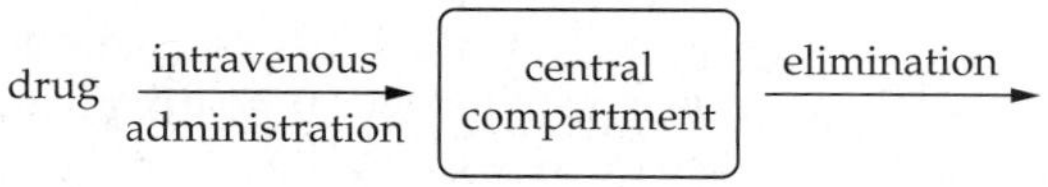

The volume of distribution (V_d) of a drug may be simply calculated under the one-compartment model from C_p-time data. Extrapolation of the ln C_p-time line back to the y-intercept provides the hypothetical $C_p^{\,o}$. As long as the drug mass in the original dose (D_o) is known, Equation 7.13 can calculate V_d.

$$V_d = \frac{D_o}{C_p^{\,o}} \tag{7.13}$$

Equation 7.13 is written to appear as a special case that is true at t_o. This is not true. Equation 7.13 is often rearranged to Equation 7.14.

$$D = C_p V_d \tag{7.14}$$

If V_d is already known, Equation 7.14 allows the calculation of D, the amount of drug present in a patient, at any time from the corresponding C_p value.

Sample Calculation Determining *CL* from C_p-Time Data (Using V_d)

PROBLEM The following table provides some C_p-time data points for an IV bolus dose of 325 mg. What is the *CL* for this drug? The units on C_p are $\mu g/mL$, and time is in hours.

Entry	t (h)	C_p ($\mu g/mL$)
1	1	3.46
2	2	2.44
3	3	1.73
4	4	1.22
5	5	0.86

SOLUTION For the third time, we have the same set of data points. Again, we get the mass of the dose along with the data points. From the second Sample Calculation, we already know that these data points can be used to estimate $C_p^{\,o}$ for the IV bolus as 4.89 $\mu g/mL$. With $C_p^{\,o}$ and the initial dose (D_o), Equation 7.13 can directly calculate V_d. As usual, watch your units. The mass units must be the same. We can convert the dose from 325 mg to 325,000 μg. Although V_d is normally used with units of liters, milliliters will work just fine in the next step.

$$V_d = \frac{D_o}{C_p^{\,o}} \tag{7.13}$$

$$V_d = \frac{325,000 \ \mu g}{4.89 \dfrac{\mu g}{mL}} = 66,462 \ mL \times \frac{1 \ L}{1,000 \ mL} = 66.5 \ L$$

With V_d in hand and k_{el} (from the first Sample Calculation in this chapter), we can directly calculate *CL* from a rearranged Equation 7.12.

$$k_{el} = \frac{CL}{V_d} \tag{7.12}$$

$$CL = k_{el}V_d = \frac{0.35}{h} \times 66,462 \ mL = 23,262 \frac{mL}{h} \times \frac{1 \ h}{60 \ min} = 388 \frac{mL}{min}$$

By this method, *CL* is determined to be 388 mL/min. This is the same number as the previous calculation with some room for rounding errors.

TABLE 7.1   Water and fluid volumes in 70-kg male human[2]

Category	Volume (L)	Volume/Mass (L/kg)
Plasma	2.7	0.039
Whole blood	5.0	0.071
Interstitial fluid	10	0.14
Intracellular fluid	25	0.36
Total body water	38	0.54
Total body volume	70	1.0

The volumes presented in **Table 7.1** can be used to interpret V_d values of moderately distributed drugs. Consider a hypothetical drug with a V_d of 15 L for a 70-kg patient. Since V_d is greater than 5 L, the drug has distributed beyond the plasma alone. The total V_d of 15 L is approximately equal to the volume of both the blood and interstitial fluid in a 70-kg patient. One might conclude that the drug in question does not effectively cross membranes and enter intracellular fluids. If it could, the drug would have a V_d of greater than 15 L, likely in the 35 to 40 L range (blood + interstitial fluid + intracellular fluid). Understanding a drug's V_d can help reveal how deeply the drug is able to penetrate into a body. A drug intended to treat muscle pain from within muscle cells should have a V_d greater than the volume of the blood. Similarly, a V_d of only 5 L (70-kg patient) or 0.07 L/kg, the volume of the blood, would be completely acceptable for an anticoagulant.

A key idea under all compartment models is that the drug must be uniformly distributed throughout the entire compartment. For this reason, the one-compartment model begins to break down with higher V_d values. Consider a drug with a V_d of 200 L for a 70-kg patient. A volume of 200 L is larger than the volume of the patient. The drug cannot be uniformly distributed in the patient because the typical patient (70 L volume) is not physically large enough to hold the drug based on its V_d. (This point is backed up with a Sample Calculation at the end of this discussion.)

Despite its simplicity, very large V_d values can be interpreted sensibly under the one-compartment model. The one-compartment model can often predict the behavior of a large V_d drug reasonably well. Because the blood volume of a 70-kg patient is only 5 L, large V_d values indicate that a drug resides mostly *not* in the blood. A very large V_d value indicates that a drug does *not* reside in body water. Such drugs tend to be lipophilic and are characterized by high log P values. In the primarily aqueous environment of the body, very lipophilic drugs avoid water by *concentrating* within membranes and adipose tissue. A drug that concentrates into adipose tissue is less concentrated in the blood, a polar region, and therefore has a lower C_p and higher V_d. Technically, we have violated the one-compartment model by invoking a second compartment with a concentration that is different from the first. Now is a natural point to move the discussion to the two-compartment model.

Sample Calculation   **Why Large V_d Values Violate the One-Compartment Model**

PROBLEM   A 70-kg patient receives a 350-mg dose of a drug as an IV bolus. The V_d of the drug is 5.0 L/kg. Calculate $C_p^{\,\circ}$ and determine the fraction of the drug contained in the blood.

SOLUTION We will use a form of Equation 7.13. Calculation of $C_p^{\,o}$ is fairly easy and comes out to be 1.0 mg/L.

$$V_d = \frac{D_o}{C_p^{\,o}} \tag{7.13}$$

$$C_p^{\,o} = \frac{D_o}{V_d} = \frac{350 \text{ mg}}{5.0\dfrac{L}{kg} \times 70 \text{ kg}} = 1.0\frac{mg}{L}$$

For a 70-kg patient, the volume of the blood is 5.0 L. The amount (dose) of drug in the blood can be found by using Equation 7.14. The amount of the 350-mg original dose that resides in the blood is only 5.0 mg. Only 1.4% of the drug is contained in the blood.

$$D = C_p V_d \tag{7.14}$$

$$D = 1.0\frac{mg}{L} \times 5.0 \text{ L} = 5.0 \text{ mg}$$

In this particular example, 98.6% of the drug is *not* in the blood.

Let us now look at the rest of the patient's body. A 70-kg patient has a volume of approximately 70 L. We have accounted for the drug in 5 of the 70 L, so how about the other 65 L? Those 65 L contain 345 mg of drug. We can rearrange Equation 7.14 and determine the concentration C of the drug within this 65 L volume. Note that we are assuming that the 345 mg of drug is uniformly distributed.

$$C = \frac{D}{V_d} = \frac{345 \text{ mg}}{65 \text{ L}} = 5.3\frac{mg}{L}$$

Based on the physical confines of the body, the drug is at one concentration in the blood and at a much higher concentration everywhere else. No matter how we might divide the body, we cannot avoid the reality that the drug *must* occupy at least two different compartments with different concentrations. The one-compartment model cannot accommodate this reality.

The conflict for the one-compartment model in this Sample Calculation began with the second sentence of the problem: The V_d of the drug is 5.0 L/kg. Any drug with a V_d that is greater than 1.0 L/kg (the volume/mass of the body) stretches the one-compartment model. Keep in mind that just because the one-compartment model is a simplification, it need not follow reality to be useful for predicting drug behavior.

Two-Compartment Model

The two-compartment model is normally visualized as shown in **Scheme 7.4**. A drug enters the bloodstream (compartment one) and reversibly distributes into peripheral tissues (compartment two). Elimination occurs only from the bloodstream.[1]

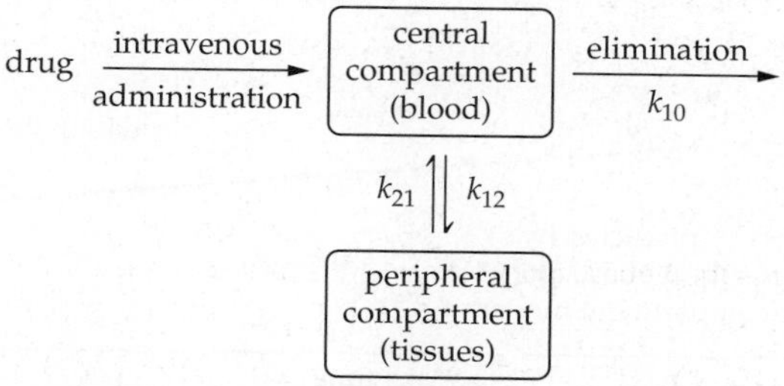

SCHEME 7.4 A common two-compartment model

The early discussion on IV bolus administration was completely based on the one-compartment model. The discussion somewhat avoided covering exactly what was happening to the drug immediately after the injection. Most C_p-time data is not recorded until 15 minutes after injection of the drug. Why? Because interpreting data for the first 15 minutes requires the two-compartment model.

During an IV bolus, the drug is injected and very quickly (within 1 or 2 minutes) is fairly evenly distributed throughout the bloodstream. Even after just 1 minute, some of the dose will have started distributing from the central compartment into the peripheral compartment, and some of the drug will have been lost from the central compartment by elimination. However, the majority of the dose will still be confined to just the central compartment, 5 L in a 70-kg patient if the central compartment is defined as the blood. C_p is therefore very high shortly after the injection because almost all the dose is concentrated within a small volume. Within roughly 15 minutes after administration, distribution of the drug to the peripheral compartment is complete, and the drug is considered to be fully distributed. A drug with a very large V_d value may require more time to fully penetrate all tissues. **Figure 7.7** is a ln C_p versus time plot of these two processes—the *distribution phase* and *elimination phase*—for an IV bolus. Each phase is defined by its own pseudo-linear section in a ln C_p versus time plot. The slope of each line defines an elimination rate constant for the corresponding phase. Because the distribution phase is fast, identification of the distribution phase requires many ln C_p-time data points very early after administration of the IV bolus. Under the one-compartment model, the distribution phase is intentionally missed by not sampling a patient's blood until 15 minutes after dosing.

Mathematically, the C_p-time relationship of a two-component model can be expressed as shown in Equation 7.15, a two-term version of Equation 7.1.

$$C_p = Ae^{-\alpha t} + Be^{-\beta t} \tag{7.15}$$

Equation 7.15 is a simplified form of a more complicated equation in which the terms α and β are written in terms of rate constants in Scheme 7.4. The terms A and B are also a combination of multiple terms and analogous to the $C_p{}^o$ term in Equation 7.1. The slope of

FIGURE 7.7 ln C_p vs. time under the two-compartment model

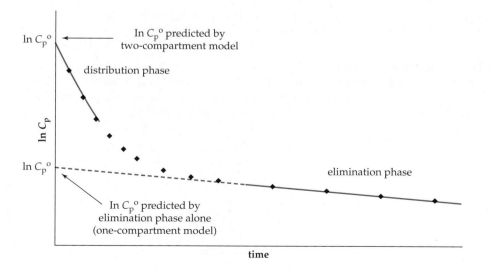

the linear region of the distribution phase is $-\alpha$. The elimination phase has a slope of $-\beta$. Generally, α is much larger than β, and the slope of the distribution phase is steeper than the elimination phase. Because of the magnitude of α, the $Ae^{-\alpha t}$ term approaches 0 a short time ($\sim$15 minutes) after administration. After the distribution phase is complete, Equation 7.15 may be expressed more simply by Equation 7.16, which is the same as Equation 7.1. The two-compartment model simplifies to a one-compartment model. Drugs with a very short $t_{1/2}$ and therefore large value for β may have a less distinct separation between the distribution and elimination phases.

$$C_{\mathrm{p}} \approx Be^{-\beta t} \quad \text{(after distribution phase, } t > 15 \text{ min)} \tag{7.16}$$

While half-life is constant under the one-compartment model, half-life changes with time under the two-compartment model. The two semilinear regions in Figure 7.7 correspond to two elimination constants, that is two different $t_{1/2}$ values according to Equation 7.3. The observed half-life varies because V_{d} is not constant for the drug until the drug has fully distributed.

Variable V_{d} values are troubling because tables of pharmacokinetic data in the literature list single V_{d} values for drugs. The reported V_{d} values correspond to the slowest elimination phase of the drug, the last linear section of the ln C_{p}-time plot. This region, labeled "elimination phase" in Figure 7.7, is more generally called the *terminal elimination phase*. In other words, when pharmacokinetic data is reported, researchers normally present parameters that fit the one-compartment model. Data for other compartment models may be referenced in footnotes, but the one-compartment values take center stage.

Higher compartment models are sometimes needed to match C_{p}-time data for a drug. Inclusion of additional compartments is especially useful for understanding drugs that unexpectedly and rapidly distribute to a high degree into a certain tissue before being slowly released again to the central compartment. Adding compartments and pathways greatly adds complexity to the model but is warranted if it is required to predict or understand the behavior of a drug.

Serum Binding

One factor that exerts significant influence on both CL and V_{d}, and therefore k_{el} and $t_{1/2}$ (Equation 7.12), is the binding of a drug to proteins. The proteins in question are not the drug target but any of the innumerable proteins found in the compartments of distribution. Blood or plasma proteins are most often cited, but protein binding within tissues can also be significant. The binding of proteins in the blood is called *serum* or *plasma binding*.

Serum binding affects both CL and V_{d} for a drug. Drugs that are bound by proteins cannot be metabolized by liver enzymes. Additionally, because serum proteins are too large to be cleared by glomerular filtration in the kidneys, drugs bound to proteins are also too large for removal by renal clearance. Therefore, protein-bound drugs have lower CL values relative to unbound drugs. Serum binding also influences V_{d}. If a drug has a greater affinity for proteins in the blood, the drug will be concentrated in the blood and show a lower V_{d}. Drugs that strongly bind tissue proteins will be drawn out of the blood, and V_{d} will be higher.

Although serum binding slows elimination, all of a drug in the body is still subject to elimination. Protein binding is a dynamic equilibrium. As the concentration of unbound drug is reduced by elimination, bound drug will release to maintain a constant bound-to-unbound drug ratio (**Scheme 7.5**).

$$\underset{\textit{bound}}{\text{protein-drug}} \;\overset{K_{\mathrm{D}}}{\rightleftharpoons}\; \underset{\textit{unbound}}{\text{protein + drug}} \;\xrightarrow{\text{elimination}}$$

SCHEME 7.5 Effect of protein binding upon elimination

C_p hides the significance of serum binding because C_p includes drug that is both bound and unbound by serum proteins. Serum albumin is the predominant protein found in blood. Over the course of investigation of compounds in a drug discovery program, promising compounds are routinely tested for binding to serum albumin. Dialysis methods can determine K_D values for the protein-drug complex.

Advanced treatments of pharmacokinetics introduce mathematical relationships between V_d and serum binding. Serum binding is often noted by variables such as f_u (fraction unbound). Distinctions between drug that is free in the blood (f_u^b) and tissue (f_u^t) are common.

Serum binding can be saturated, and different drugs often compete for the limited available binding sites. An example is warfarin (Coumadin, **7.3**), a potent and potentially dangerous anticoagulant. Warfarin is over 99% bound to serum albumin ($f_u < 1\%$). Two drugs that compete with warfarin for its site of binding are sulfamethoxazole (**7.4**), an antibiotic, and clofibrate (Atromid-S, **7.5**), an antihyperlipidemic (**Figure 7.8**).[3] Patients receiving warfarin should avoid both sulfamethoxazole and clofibrate. If either drug is taken, it will bind albumin and partially displace free warfarin into the blood. Even if the binding of warfarin decreases to only 95% ($f_u = 5\%$), C_p of warfarin will increase dramatically. Uncontrolled hemorrhaging may result if toxic levels of unbound warfarin are exceeded.

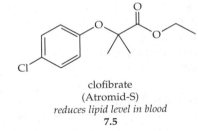

warfarin
(Coumadin)
anticoagulant
7.3

sulfamethoxazole
antibiotic
7.4

clofibrate
(Atromid-S)
reduces lipid level in blood
7.5

FIGURE 7.8 Drugs that competitively bind serum albumin

It is possible to rationally design a compound with minimized serum binding. The following Case Study demonstrates this idea.

CASE STUDY

Minimization of Serum Binding[4]

A cancer research group with Abbott Laboratories and Idun Pharmaceuticals studied compounds that bind and inhibit the activity of Bcl-X_L, a protein associated with B-cell lymphoma. The team developed compound **7.6** as a moderately active inhibitor of Bcl-X_L (**Figure 7.9**). Compound **7.6**, however, lost almost all its inhibitory activity when screened in the presence of human serum. The group determined that **7.6** binds domain III of human serum albumin (HSA-III) with a K_D of 0.1 μM through two key interactions. The 4-fluorophenyl group fits into a hydrophobic pocket, and the ethylene chain of **7.6** fits into a hydrophobic channel. When **7.6** binds Bcl-X_L, both the 4-fluorophenyl group and ethylene chain are exposed to aqueous solvent. The discovery team reasoned that the addition of polar, solublizing R-groups should diminish albumin binding without drastically impacting inhibition of Bcl-X_L. The researchers ultimately developed compound **7.7**, which has a much lower affinity for albumin ($K_D = 94$ μM) and retains Bcl-X_L inhibition in the presence of 10% human serum.

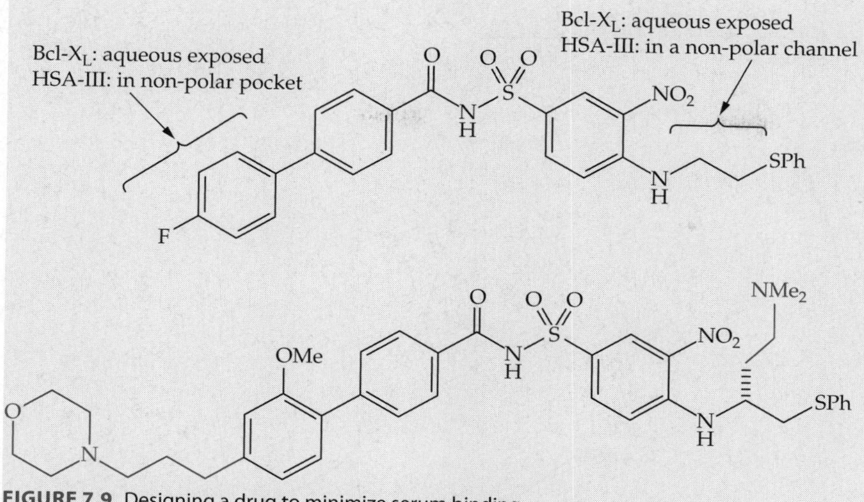

7.6

	K_i
Bcl-X$_L$ alone	36 nM
Bcl-X$_L$ + 10% serum	no inhibition

7.7

	K_i
Bcl-X$_L$ + 10% serum	148 nM

FIGURE 7.9 Designing a drug to minimize serum binding

7.2 Intravenous Infusion

The mathematical modeling of an intravenous infusion builds directly on the ideas introduced in the discussion of the intravenous bolus.

Infusion Rate Constant

An intravenous infusion involves a continuous flow of drug into a patient at a rate defined by the infusion rate constant, R_{inf}, with units of mass/time. Discussions of infusion normally present the infusion rate constant as k_{inf}, which may be confused with a true reaction rate constant. Therefore, this presentation of infusion uses a less ambiguous variable, R_{inf}, for the infusion rate constant.

During an infusion, C_p rises with the start of the infusion. As C_p builds, elimination processes begin to counteract the infusion and slow the rate of increase in C_p. If the infusion continues long enough, C_p approaches a maximum called the *steady-state concentration* (C_p^{ss}) at which the rate of infusion is balanced by the rate of elimination. As soon as the infusion is stopped, C_p is influenced only by elimination and will decrease accordingly. **Figure 7.10** shows C_p versus time data for a representative infusion.

The mathematical relationship between C_p and time for an IV infusion is shown in Equation 7.17.[1]

$$C_p = \frac{R_{inf}}{k_{el}V_d}(1 - e^{-k_{el}t}) \qquad (7.17)$$

This equation assumes a one-compartment model and constant, first-order elimination. Based on Equation 7.17, the e^{-kt} term approaches 0 as t increases, and C_p approaches $R_{inf}/k_{el}V_d$. The value of $R_{inf}/k_{el}V_d$ corresponds to C_p^{ss} (Equation 7.18).

$$C_p^{ss} = \frac{R_{inf}}{k_{el}V_d} \qquad (7.18)$$

In a strict sense, C_p never achieves C_p^{ss}. In practice, infusion of a drug for a period of four $t_{1/2}$ values achieves between 90% and 95% of C_p^{ss}.

An advantage to infusions over an IV bolus is that less drug hits the patient's bloodstream at once. This avoids elevated values of C_p^o, as seen in Figure 7.7, during the

FIGURE 7.10 C_p vs. time for an IV infusion

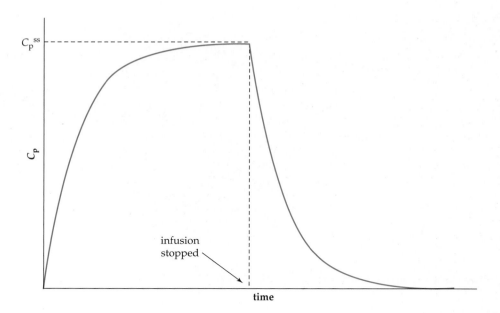

distribution phase of an IV bolus. Infusions are also able to provide a steady C_p level. One disadvantage of infusions is that they require time to reach a therapeutic concentration. Based on Equation 7.18, drugs with a larger k_{el} (shorter $t_{1/2}$) take a shorter time to build by infusion than drugs with smaller k_{el} values. Another disadvantage of infusions is that the patient must have an IV line during the entire time of the infusion.

Sample Calculation   **Calculating C_p^{ss} of an Infusion**

PROBLEM   For a 70-kg patient, the V_d of morphine is 230 L, and CL_T is 1,680 mL/min. Calculate the infusion rate (R_{inf}) of morphine required to sustain a C_p^{ss} of 15 ng/mL in a 70-kg patient.

SOLUTION   The only equation we have that contains C_p^{ss} is Equation 7.18. To use this equation, we need to determine the elimination rate constant (k_{el}) for morphine. Fortunately, with Equation 7.12 we can calculate k_{el} from the two parameters we have been given: V_d and CL. Convert CL to liters to make the units match and switch up the final time units to hours instead of minutes.

$$k_{el} = \frac{CL}{V_d} \tag{7.12}$$

$$k_{el} = \frac{1.68\dfrac{\text{L}}{\text{min}}}{230\,\text{L}} = \frac{0.00730}{\text{min}} \times \frac{60\,\text{min}}{1\,\text{h}} = \frac{0.438}{\text{h}}$$

Now we are ready to calculate R_{inf} with a rearranged Equation 7.18. Remember, all volume and mass units must be consistent. Infusion rates are normally reported as mg/h.

$$C_p^{ss} = \frac{R_{inf}}{k_{el}V_d} \tag{7.18}$$

$$R_{inf} = C_p^{ss}k_{el}V_d$$

$$R_{inf} = 15\frac{\mu\text{g}}{\text{L}} \times 0.438\frac{1}{\text{h}} \times 230\,\text{L} = 1,511\frac{\mu\text{g}}{\text{h}} \times \frac{1\,\text{mg}}{1{,}000\,\mu\text{g}} = 1.5\frac{\text{mg}}{\text{h}}$$

A very slow infusion rate of just 1.5 mg/h provides a C_p^{ss} of 15 ng/mL, which is the minimum analgesic concentration of morphine.

Loading Bolus

Because IV infusions can require considerable time to build C_p to effective levels, an IV bolus may be used to "jump start" C_p. Such dosing is called a *loading bolus*. Through joint use of an IV bolus and infusion, a patient may receive immediate and constant therapeutic effect from a drug. C_p^{total} is equal to the sum of the C_p from the bolus and the infusion (Equations 7.19 and 7.20). Initially, C_p^{total} is dominated by the bolus (**Figure 7.11**). Over time, the drop in the value of C_p from the bolus is countered by a rise in C_p from the infusion. A "perfect" loading bolus, shown in Figure 7.11, places $C_p^{\,o}$ at C_p^{ss} so that C_p^{total} is constant over the entire time of the infusion. Regardless of whether $C_p^{\,o}$ from the bolus is equal to C_p^{ss} of the infusion, C_p^{total} approaches C_p^{ss} over time (**Figure 7.12**).

$$C_p^{total} = C_p^{bolus} + C_p^{inf} \tag{7.19}$$

$$C_p^{total} = C_p^{\,o}e^{-k_{el}t} + \frac{k_{inf}}{k_{el}V_d}(1 - e^{-k_{el}t}) \tag{7.20}$$

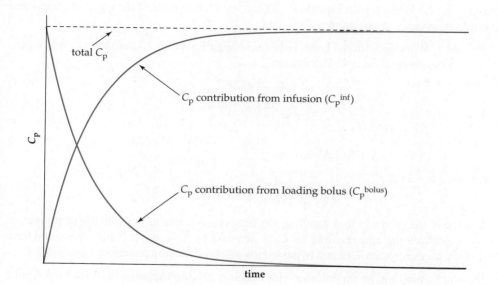

FIGURE 7.11 Use of a loading bolus with an IV infusion

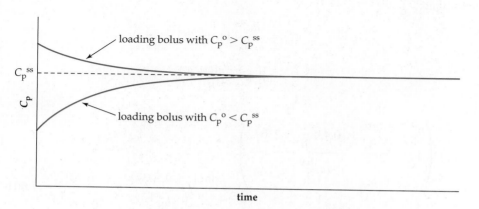

FIGURE 7.12 Infusion with various loading bolus $C_p^{\,o}$ values

---

Sample Calculation   **Calculating a Loading Dose for an Infusion**

PROBLEM   Determine the appropriate dose for a loading bolus that will provide a $C_p^{\,o}$ of 15 ng/mL of morphine in a 70-kg patient. The V_d of morphine is 230 L, and CL_T is

1,680 mL/min. If this bolus is combined with the infusion in the previous Sample Calculation, at what time will the C_p contribution from both the bolus and infusion be the same?

SOLUTION The first part of this question could very well fall under the IV bolus heading (Section 7.1). Equation 7.14 (provided with t_o notation) allows us to calculate the necessary IV bolus dose from the information provided in the question. Keep the units consistent. Remember that ng/mL is equivalent to $\mu g/L$.

$$D_o = C_p^{\,o} V_d \tag{7.14}$$

$$D_o = 15 \frac{\mu g}{L} \times 230\,L = 3.5\,\mu g$$

The required loading dose is 3.5 μg.

There are several ways to approach the second part of the question. (1) Determine the time required for the IV bolus concentration to decrease to 7.5 ng/mL, or half of $C_p^{\,o}$. (2) Determine the time required for the IV infusion concentration to increase to 7.5 ng/mL, or half of $C_p^{\,ss}$. (3) Work with Equation 7.20 and set the two parts of the equation, bolus and infusion, as equal to one another.

Method 1: Working with the bolus requires solving Equation 7.1 for time. k_{el} was calculated in the previous Sample Calculation.

$$C_p = C_p^{\,o} e^{-k_{el}t} \tag{7.1}$$

$$t = \frac{\ln C_p - \ln C_p^{\,o}}{-k_{el}}$$

$$t = \frac{\ln 7.5\frac{\mu g}{L} - \ln 15\frac{\mu g}{L}}{-0.438\frac{1}{h}} = \frac{-0.693}{-0.438\frac{1}{h}} = 1.58\,h$$

If some of the numbers look familiar, the time corresponds to the half-life of morphine. (We calculated the time required for C_p to decrease by 50%, the half-life.) We could have directly used Equation 7.12 and bypassed the algebraic rearrangement of Equation 7.1.

Method 2: Working on the infusion side requires solving Equation 7.17 for time. Units, units, units!

$$C_p = \frac{R_{inf}}{k_{el}V_d}(1 - e^{-k_{el}t})$$

$$t = \frac{\ln\left(1 - \dfrac{C_p k_{el} V_d}{R_{inf}}\right)}{-k_{el}}$$

$$\tag{7.17}$$

$$t = \frac{\ln\left(1 - \dfrac{7.5\frac{\mu g}{L} \times 0.438\frac{1}{h} \times 230\,L}{1{,}500\frac{\mu g}{h}}\right)}{-0.438\frac{1}{h}} = \frac{\ln(1 - 0.5037)}{-0.438\frac{1}{h}} = \frac{-0.701}{-0.438\frac{1}{h}} = 1.60\,h$$

The infusion method should give the same number, and it does within rounding error. The time for the infusion to reach 7.5 ng/mL is 1.60 h.

Method 3: The final approach to determining when the bolus and infusion are contributing equally to C_p is accomplished by setting the two components of Equation 7.20 equal to one another and solving for time.

$$C_p^{\text{total}} = C_p^{\circ} e^{-k_{el}t} + \frac{R_{\text{inf}}}{k_{el} V_d}(1 - e^{-k_{el}t})$$

$$t = \frac{\ln\dfrac{R_{\text{inf}}}{k_{el} V_d} - \ln\left(C_p^{\circ} + \dfrac{R_{\text{inf}}}{k_{el} V_d}\right)}{-k_{el}}$$

$$t = \frac{\ln\dfrac{1{,}500\dfrac{\mu g}{h}}{0.438\dfrac{1}{h} \times 230\,L} - \ln\left(15\dfrac{\mu g}{L} + \dfrac{1{,}500\dfrac{\mu g}{h}}{0.438\dfrac{1}{h} \times 230\,L}\right)}{-0.438\dfrac{1}{h}} \qquad (7.20)$$

$$t = \frac{\ln 14.9\dfrac{\mu g}{L} - \ln 29.9\dfrac{\mu g}{L}}{-0.438\dfrac{1}{h}} = \frac{2.701 - 3.398}{-0.438\dfrac{1}{h}} = 1.59\,h$$

The algebra is definitely messier, but the answer is the same, 1.59 h.

Infusion-Like Administration Routes

Transdermal patches do not fall under the category of intravenous administration. Patches, however, can provide a steady C_p versus time plot with close similarities to that of IV infusions. The example shown in **Figure 7.13** consists of data adapted from the prescribing information insert of Ortho Evra.[5] Ortho Evra is a transdermal version of female contraception. Ortho Evra is administered by applying a patch to an appropriate skin surface. The patch remains in place for a week. A complete cycle of treatment involves three consecutive weeks of wearing a patch followed by a week with no patch. The data in Figure 7.13 correspond to C_p-time points for ethinyl estradiol (**7.8**), one of the two active components in Ortho Evra, for week 3 of cycle 2. At a time of 168 hours (7 days), the patch was removed, and the drug underwent first-order elimination during week 4 of cycle 2. The dip early in the plot corresponds to removal of the week 2 patch and a slight lag in drug ad-

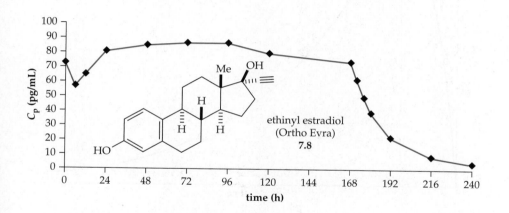

FIGURE 7.13 Transdermal C_p-time data for ethinyl estradiol (**7.8**)[5]

ministration from the newly applied week 3 patch. The C_p^{ss} was determined to be 80 pg/mL. The approximate $t_{1/2}$ was determined to be 17 h ($k_{el} = 0.041$ h^{-1}) from the elimination phase of the plot ($t \geq 168$ hours).

7.3 Oral

Oral administration is the most complicated method of administration covered in this text. Modeling the plasma concentration of an orally administered drug combines both elimination pathways and absorption of a drug after it has been ingested.

Absorption and Elimination Phases

Orally delivered drugs must first be absorbed from the digestive system before they enter the bloodstream, where they are monitored in the form of C_p. Upon entering the bloodstream, the drug will then begin to undergo elimination. C_p versus time curves for oral drugs therefore are dominated by two features: an *absorption phase* and an *elimination phase* (**Figure 7.14**). Between these two phases, the curve reaches a peak (C_p^{max}) at a time of t_{max}.

Equation 7.21 shows the mathematical relationship between C_p versus time in Figure 7.14.

$$C_p = \frac{FD_o}{V_d} \frac{k_{ab}}{(k_{ab} - k_{el})} (e^{-k_{el}t} - e^{-k_{ab}t}) \tag{7.21}$$

The two exponential terms model drug absorption and elimination. The rate of absorption depends on the properties of the drug and is described by k_{ab}, the *absorption rate constant*. The variable F is the *bioavailability* of the drug and a measure of observed drug exposure relative to an experimental maximum. Bioavailability is more thoroughly described in the next subsection.

Valid values of F fall between 0 and 1. The term FD_o/V_d is related to the concentration of drug (same units as C_p) theoretically available to the bloodstream in the absence of elimination. Equation 7.21 assumes first-order elimination and a single compartment model.

FIGURE 7.14 Single-dose oral C_p-time plot

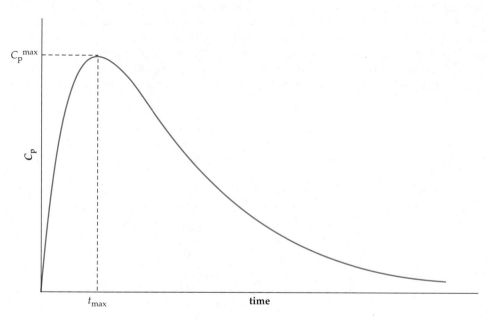

Sample Calculation C_p for Oral Drugs

PROBLEM   An oral drug has a dose of 500 mg, oral bioavailability of 80%, V_d of 70 L, k_{el} of 0.3 h^{-1}, and k_{ab} of 0.8 h^{-1}. The drug has been prescribed to be taken every 4 hours. If the minimum effective concentration of the drug is 1.5 μg/mL, will the patient still have an effective concentration of the drug when the second dose is administered?

SOLUTION   Another way to state this question is whether the first dose is still effective after 4 hours. We need to check C_p at 4 hours. As long as C_p is 1.5 μg/mL or higher, the patient will have therapeutic levels of the drug at the end of the first dose and start of the second dose. This is a simple plug-and-chug problem using Equation 7.21. The hardest part is wading through all the variables.

$$C_p = \frac{FD_o}{V_d} \frac{k_{ab}}{(k_{ab} - k_{el})} (e^{-k_{el}t} - e^{-k_{ab}t}) \tag{7.21}$$

$$C_p = \frac{0.8 \times 500 \text{ mg}}{70 \text{ L}} \frac{0.8\frac{1}{h}}{0.8\frac{1}{h} - 0.3\frac{1}{h}} \left(e^{-0.3\frac{1}{h} \times 4\,h} - e^{-0.8\frac{1}{h} \times 4\,h} \right) = 2.38 \frac{\text{mg}}{\text{L}} = 2.38 \frac{\mu g}{\text{mL}}$$

After 4 hours, the first dose will sustain C_p at therapeutic levels.

Bioavailability

Equation 7.21 contains a number of variables. Experimentally, k_{el} and V_d for a drug can be readily determined from analysis of C_p- or ln C_p-time data from an IV bolus. As long as the administrator knows the amount of drug that he or she gave a patient, then D_o is known as well. That leaves F and k_{ab} as the only unknown terms in Equation 7.21.

Bioavailability (F), or more precisely the *absolute bioavailability*, is a property of a drug, its formulation, and its route of administration. Bioavailability is a measure of drug exposure that results from oral dosing of a drug relative to drug exposure by an IV bolus. As has already been discussed, drug exposure is related to the area under the curve (AUC) of a C_p-time plot.

Intravenous administration is the only method by which an entire drug dose reaches the bloodstream, so the AUC of an IV route (AUC_{IV}) is the benchmark for determining F. Comparison of AUCs from other routes to AUC_{IV} gives F for that particular administration method (Equation 7.22).

$$F_{\text{route}} = \frac{\dfrac{AUC_{\text{route}}}{D_o{}^{\text{route}}}}{\dfrac{AUC_{IV}}{D_o{}^{IV}}} \tag{7.22}$$

Inclusion of the dose sizes in Equation 7.22 is necessary only if the two doses, $D_o{}^{\text{route}}$ and $D_o{}^{IV}$, are not equal. If all variables in Equation 7.21 are already known, AUC for an orally delivered drug can be quickly calculated with Equation 7.23.

$$AUC_{\text{oral}} = \frac{FD_o}{V_d} \frac{k_{ab}}{(k_{ab} - k_{el})} \left(\frac{1}{k_{el}} - \frac{1}{k_{ab}} \right) \tag{7.23}$$

Equation 7.23 is obtained by integration of Equation 7.21 and evaluation of the integral over the interval from t_o to t_∞. Bioavailability is not only an issue for oral drugs. Even for intramuscular delivery, a drug must diffuse through muscle tissue, where the drug may be partially broken down before reaching the bloodstream.

In a clinical setting, AUC_{oral} is determined by trapezoidal estimation of experimental C_p-time data points. From AUC_{oral} and D_o^{oral} and information from an IV bolus, oral bioavailability can be determined through Equation 7.22.

If an oral drug is fully absorbed, then its bioavailability depends on its resistance to first-pass metabolism by the liver. Therefore, the hepatic extraction ratio (E_H) determines the bioavailability of fully absorbed oral drugs (Equation 7.24).

$$F_{oral} = 1 - E_H \qquad (7.24)$$

Drugs with high values of E_H have poor bioavailability. Similarly, by Equation 7.8, high E_H translates into high CL_H, high k_{el}, and short $t_{1/2}$ values. Because of the extensive implications of low bioavailability, drugs with low F values are generally avoided as orally administered drugs. There is not a critical cutoff, but oral drugs with F below 0.2 are uncommon.

Poorly available drugs have more issues than just undesirable pharmacokinetic properties. Safety is also a concern. Despite the clean appearance of the graphs and data used in this textbook, real clinical data can be messy with considerable amounts of scatter. Variability between patients can be large. Only after examining data from many patients can pharmacokinetic parameters be assigned. Regardless, every patient is different, and the parameters for a specific patient will likely not match the average value. How is this relevant to bioavailability? A drug with a good bioavailability ($F = 0.8$) may show a range of F values from 0.85 to 0.75 when used in a range of patients. On a percentage basis, 0.75 and 0.85 are not very different. All patients will absorb a similar amount of drug. A drug with a low bioavailability ($F = 0.1$) may show a range of 0.15 to 0.05 across a patient population. Patients that show an F of 0.15 will reach 1.5 times the C_p of a "typical" patient. Toxic levels could be reached. Patients with an F of 0.05 will receive only half the expected C_p value. The drug may never achieve a therapeutic concentration. This degree of variability is not acceptable and is potentially unsafe. A drug deemed to be unsafe will not be approved.

Aside from problems of first-pass metabolism, drugs can have low bioavailability because they are poorly absorbed from the digestive system. Many factors determine drug absorption. One factor is a drug's time of passage through the digestive system. Absorption of a drug requires time. If a drug passes through the digestive tract too quickly, then some of the drug dose will exit the body through the feces. If necessary, oral drugs may be administered with food to slow rate of drug passage through the digestive system and increase the drug's bioavailability. Another factor in bioavailability is the crystallinity of the drug. Crystallinity affects drug solubility and therefore absorption. The importance of drug crystallinity will be covered more thoroughly in Chapter 13. Related to the issue of polymorphs is the effect of drug formulation. Oral tablets consist of a drug, binding agents, and likely other compounds that fine tune the pharmacokinetic properties of the drug. For proper behavior, the drug must be ground finely enough to raise surface area and its rate of dissolution. Furthermore, when the tablets are formed in a press, excessive pressure must be avoided. It is possible to compress a tablet to such a degree that it fails to break apart in a patient. Such a pellet passes straight through the patient. These tablets have been called bed pan bullets.

Absorption Rate Constant
Once F has been determined for a drug, only one variable in Equation 7.21 remains undiscussed: $k_{ab} \times k_{ab}$ may be found through a $\ln C_p$ versus time plot (**Figure 7.15**). The

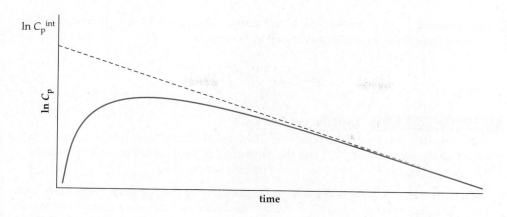

FIGURE 7.15 In C_p vs. time plot for an oral drug

latter part of the graph is the linear first-order elimination phase with a slope of $-k_{el}$. During this phase, the effect from absorption is negligible. Equation 7.21 simplifies to Equation 7.25.

$$C_p = \frac{FD_o}{V_d} \frac{k_{ab}}{(k_{ab} - k_{el})} e^{-k_{el}t} \qquad \text{(elimination phase)} \qquad (7.25)$$

The $\ln C_p$-time version of Equation 7.25 is Equation 7.26.

$$\ln C_p = -k_{el}t + \ln\left(\frac{FD_o}{V_d} \frac{k_{ab}}{(k_{ab} - k_{el})}\right) \qquad \text{(elimination phase)} \qquad (7.26)$$

Remember that Equations 7.25 and 7.26 are valid only for the elimination phase of an oral dose. The y-intercept of the line described in Equation 7.26 ($\ln C_p^{\text{y-int}}$) contains a number of factors. If F has already been determined from AUC data (Equation 7.22), the formula for $C_p^{\text{y-int}}$ (Equation 7.27) can be rearranged to Equation 7.28 for calculation of k_{ab}.

$$C_p^{\text{y-int}} = \frac{FD_o}{V_d} \frac{k_{ab}}{(k_{ab} - k_{el})} \qquad (7.27)$$

$$k_{ab} = \frac{k_{el}}{1 - \left(\dfrac{C_p^{\text{y-int}} V_d}{FD_o}\right)} \qquad (7.28)$$

A particularly significant C_p-time point for an oral drug is C_p^{max} and t_{max}. C_p^{max} occurs during a transition between the absorption and elimination phases. At C_p^{max}, dC_p/dt is 0, and the rate of absorption is balanced by the rate of elimination (Equation 7.29). Through analysis and rearrangement of the first derivative of Equation 7.21 with respect to time, Equation 7.30 can be obtained and used for simple calculation of t_{max}. Substitution of t_{max} into Equation 7.21 then provides C_p^{max}.

$$\frac{dC_p}{dt} = 0 = \text{rate of absorption} - \text{rate of elimination} \qquad \text{(at } t_{\text{max}}) \qquad (7.29)$$

$$t_{\text{max}} = \frac{\ln k_{el} - \ln k_{ab}}{k_{el} - k_{ab}} \qquad (7.30)$$

While t_{max} can be directly calculated with Equation 7.30, t_{max} can also be estimated from a set of clinical C_p-time data points. With both k_{el} and an estimated t_{max}, one may approximate k_{ab} with Equation 7.30. If one knows both k_{ab} and k_{el}, then Equation 7.31 (rearranged from Equation 7.27) can be used to calculate F.

$$F = \frac{C_p^{\text{y-int}} V_d}{D_o} \frac{(k_{ab} - k_{el})}{k_{ab}} \tag{7.31}$$

Sample Calculation   **The Effect of F on C_p^{max}**

PROBLEM   If a drug has 100% oral bioavailability, will an IV bolus and oral dose of the same size have the same C_p^{max}? Use the parameters of the drug in the previous Sample Calculation (except $F = 100\%$) to prove that the answer is no.

SOLUTION   First we need to know the parameters of the previous drug. They were $V_d = 70$ L, $k_{ab} = 0.8$ h^{-1}, and $k_{el} = 0.3$ h^{-1}. Using the same previous dose of 500 mg, we can quickly calculate C_p^{max} for the IV bolus (same as C_p^{o}) with a rearranged Equation 7.13.

$$V_d = \frac{D_o}{C_p^{o}} \tag{7.13}$$

$$C_p^{o} = \frac{D_o}{V_d}$$

$$C_p^{o} = \frac{500 \text{ mg}}{70 \text{ L}} = 7.1 \frac{\text{mg}}{\text{L}}$$

The C_p^{max} for the IV bolus is 7.1 mg/L. This assumes a one-compartment model. If we had enough information to calculate C_p^{max} for a two-compartment model, the value would be higher than 7.1 mg/L (see Figure 7.7).

C_p^{max} for the oral dose occurs at t_{max}, which we can calculate with Equation 7.30.

$$t_{\text{max}} = \frac{\ln k_{el} - \ln k_{ab}}{k_{el} - k_{ab}} \tag{7.30}$$

$$t_{\text{max}} = \frac{\ln 0.3 \frac{1}{h} - \ln 0.8 \frac{1}{h}}{0.3 \frac{1}{h} - 0.8 \frac{1}{h}} = 1.5 \text{ h}$$

Now we are ready to use Equation 7.21 to determine C_p at t_{max}.

$$C_p = \frac{F D_o}{V_d} \frac{k_{ab}}{(k_{ab} - k_{el})} (e^{-k_{el}t} - e^{-k_{ab}t}) \tag{7.21}$$

$$C_p = \frac{1.0 \times 500 \text{ mg}}{70 \text{ L}} \frac{0.8 \frac{1}{h}}{0.8 \frac{1}{h} - 0.3 \frac{1}{h}} \left(e^{-0.3\frac{1}{h} \times 1.5 \text{ h}} - e^{-0.8\frac{1}{h} \times 1.5 \text{ h}} \right) = 3.8 \frac{\text{mg}}{\text{L}}$$

So, C_p^{max} for the IV bolus is 7.1 mg/L, while the oral dose only reaches 3.8 mg/L. The answer is no. Even with 100% oral bioavailability, an oral dose does not achieve the same C_p^{max} as an IV bolus. This statement is true for all oral drugs.

Multiple Oral Doses

Despite its popularity, oral delivery would be worthless if it were unable to sustain consistent C_p levels over time. When spaced properly with a correct dose, orally delivered drugs can

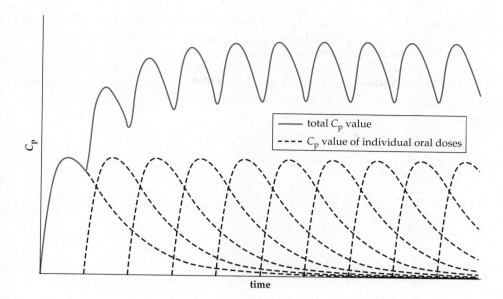

FIGURE 7.16 Construction of a multiple-dose oral C_p vs. time plot

total C_p value
---- C_p value of individual oral doses

indeed maintain a reasonably controlled C_p value. The measured C_p value is actually the sum of the C_p contributions of each individual oral dose (Equation 7.32 and **Figure 7.16**).

$$C_p^{\text{total}} = \sum_{n=1}^{\infty} C_p^{\text{dose } n} \qquad (7.32)$$

Larger, less frequent doses afford a broader range of C_p values. If C_p strays outside the therapeutic window, smaller, more frequent doses provide a narrower gap between the peaks and valleys of the C_p-time graph (**Figure 7.17**). The number of doses required to reach a pseudo C_{ss} range depends on k_{el}, k_{ab}, dose size, and dose frequency.

The goal of most drug discovery programs is to develop a drug in a formulation that allows for once-per-day dosing. Daily dosing is sufficiently frequent so that patients do not forget to take their medication while not so frequent to be an inconvenience.

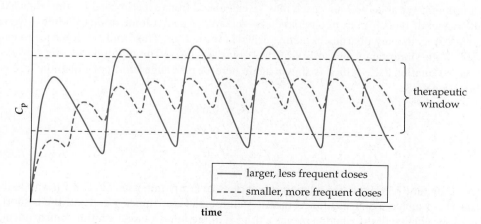

FIGURE 7.17 Effect of dose size and frequency on C_p vs. time for an oral drug

therapeutic window

—— larger, less frequent doses
---- smaller, more frequent doses

| **Sample Calculation** | **Working with Multiple Oral Doses** |

PROBLEM Using the drug in the previous two Sample Calculations, determine C_p at $t = 2$ h after the third 500 mg oral dose. Assume $F = 80\%$.

SOLUTION   This is the same drug as before, so $V_d = 70$ L, $k_{ab} = 0.8$ h^{-1}, and $k_{el} = 0.3$ h^{-1}. Each dose occurs 4 hours apart. The patient will have three doses present. The first was dosed at $t = 0$ h, the second at $t = 4$ h, and the third at $t = 8$ h. So, the C_p of interest is at a total time of 10 h for the first dose, 6 h for the second, and 2 h for the third. For this calculation, we will need to calculate the C_p contribution for each separate dose at its respective time. Add up each contribution for C_p^{total}. This should look something like the following equation below.

$$C_p^{total} = C_p^{1(10 h)} + C_p^{2(6 h)} + C_p^{3(2 h)}$$

Here are the calculations of the individual doses.

$$C_p^1 = \frac{0.8 \times 500 \text{ mg}}{70 \text{ L}} \frac{0.8 \frac{1}{h}}{0.8\frac{1}{h} - 0.3\frac{1}{h}} \left(e^{-0.3\frac{1}{h} \times 10 h} - e^{-0.8\frac{1}{h} \times 10 h} \right) = 0.45 \frac{\text{mg}}{\text{L}}$$

$$C_p^2 = \frac{0.8 \times 500 \text{ mg}}{70 \text{ L}} \frac{0.8 \frac{1}{h}}{0.8\frac{1}{h} - 0.3\frac{1}{h}} \left(e^{-0.3\frac{1}{h} \times 6 h} - e^{-0.8\frac{1}{h} \times 6 h} \right) = 1.44 \frac{\text{mg}}{\text{L}}$$

$$C_p^3 = \frac{0.8 \times 500 \text{ mg}}{70 \text{ L}} \frac{0.8 \frac{1}{h}}{0.8\frac{1}{h} - 0.3\frac{1}{h}} \left(e^{-0.3\frac{1}{h} \times 10 h} - e^{-0.8\frac{1}{h} \times 10 h} \right) = 3.17 \frac{\text{mg}}{\text{L}}$$

The C_p^{total} at 10 h for this dosing regimen is 5.06 mg/L (or 5.06 μg/mL). Note that the contribution from the first dose, C_p^1, is the smallest of the three doses.

Clearance and Volume of Distribution Revisited

The preceding discussion has been intent upon breaking down equations and making sense of different variables and how each may be calculated from experimental C_p-time data. At the outset of this chapter, two parameters—clearance and volume of distribution—were set apart as the key pharmacokinetic variables for a drug. This brief section tries to establish the importance and utility of these two variables. The highlight of this subsection is Equation 7.12, which is shown again here. A rearranged form of Equation 7.12 is Equation 7.33.

$$k_{el} = \frac{CL}{V_d} = \frac{0.693}{t_{1/2}} \tag{7.12}$$

$$CL = \frac{0.693}{t_{1/2}} V_d \tag{7.33}$$

Equation 7.33 shows the relationship of our two key parameters, CL and V_d, with half-life. If a drug has a high clearance, then it is eliminated more quickly and half-life is short. A larger volume of distribution means a drug is diluted into a larger central compartment, and more time is required to clear the drug from a larger volume. A longer time to clear equates to a longer half-life.

One thing that is not obvious in Equation 7.33 is its link to oral bioavailability. The barriers keeping a drug from the blood are absorption and the first-pass effect of the liver. Absorption is often not a factor since the intestines effectively absorb compounds from ingested materials. The liver, with its first-pass effect, is therefore the main determining factor of oral bioavailability.

Hepatic clearance (CL_H) can be crudely approximated from bioavailability (F) according to Equation 7.34.

$$CL_H \approx 25 \frac{mL}{min}(1 - F) \tag{7.34}$$

A high CL_H indicates the liver efficiently removes a drug from incoming blood, so F is low. A low CL_H indicates high F. Hepatic clearance can range in value from 0 to 25 mL/min/kg. With this information, we can begin to put Equation 7.33 to its full use.

Equation 7.33 can be represented as a line of the form $y = mx$ (**Figure 7.18**).[6] The y-axis is CL (CL_H), and the x-axis is V_d. The slope of the line is inversely related to $t_{1/2}$. The dashed horizontal lines on Figure 7.18 represent different oral bioavailability values that are based on the corresponding CL_H values. The three sloped lines represent possible CL_H-V_d values for three different half-lives. Note that this discussion assumes a drug is eliminated exclusively by the kidneys in clearance. This is often a reasonable assumption for many drugs. Figure 7.18 is less useful, however, for understanding drugs that are eliminated to a significant extent by the kidneys in unchanged form.

Before we use Figure 7.18, the units of the plot must be addressed. In Figure 7.18, the volume and time units of the variables included in the graph do not match. Although this chapter preaches careful adherence to matching units, this rule is not a factor for the presentation of Figure 7.18. As will soon be seen, Figure 7.18 is a chart that simply displays CL_H and V_d with their most commonly encountered units. Figure 7.18 is used by plotting CL_H and V_d values for a drug as an individual data point.

Figure 7.18 is helpful to a drug discovery group with regard to the type of changes that might be needed to improve a lead. Late leads, which may have an excellent target binding profile (pharmacodynamics), often need to be optimized with regard to their pharmacokinetics. As a hypothetical example, a late lead might have a V_d of 7 L/kg and CL_H of 5 mL/min/kg. This combination of CL_H and V_d lies on the $t_{1/2} = 16$ h line in Figure 7.18.

If the lead optimization team is seeking an ideal $t_{1/2}$ of 8 h, two options are available. First, the team might try to make the lead less metabolically stable and increase CL_H. This modification would decrease the time of residence of the compound and drop the half-life. A new lead with a V_d of 7 L/kg and CL_H of 10 mL/min/kg would lie on or close to the 8 h $t_{1/2}$ line. Second, the group might try to decrease the V_d of the lead. Confining the lead to a smaller volume while preserving the same CL value will also decrease half-life. A new lead with a V_d of 3 L/kg and CL_H of 7 mL/min/kg would have a $t_{1/2}$ of nearly 8 h.

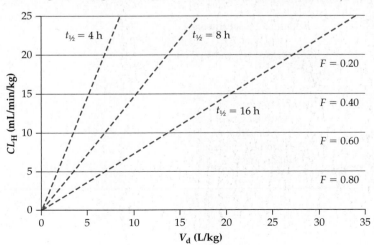

FIGURE 7.18 The CL_H vs. V_d plot

Fast Onset Sertraline Analogues[6]

Sertraline (Zoloft, **7.9**) is an antidepressant that blocks the reuptake of serotonin (**7.10**), a neurotransmitter, from the synaptic junction (**Figure 7.19**). Despite the effectiveness and market success of sertraline, the drug has an unfavorably large V_d of 30 L/kg. Sertraline extensively distributes into tissues and out of the bloodstream. If a drug leaves the blood, it cannot be cleared by the kidneys or liver. Sertraline therefore has a long $t_{1/2}$ (30 h). From the perspective of a patient, drugs with long $t_{1/2}$ values require a long time to reach a consistent, therapeutic concentration. The effect of the drug can also persist long after the last dose. This can be a problem if a drug is causing an adverse reaction in a patient. The patient may stop taking a drug, but that drug will persist in his body for a long time.

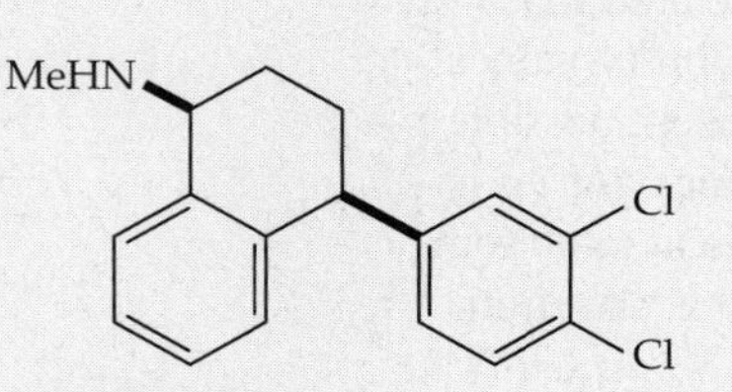

7.9
sertraline
(Zoloft)

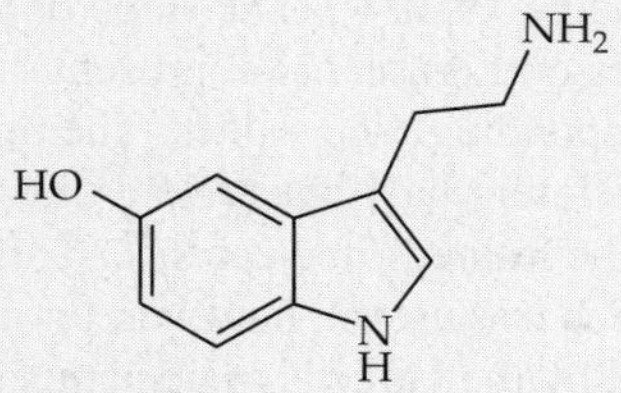

7.10
serotonin

FIGURE 7.19

TABLE 7.2 Sertraline (7.9) and analogues

Entry	Structure	Structural Change (Effect)	$t_{1/2}$ (h)	t_{max} (h)	V_d (L/kg)	IC_{50} (nM)
1	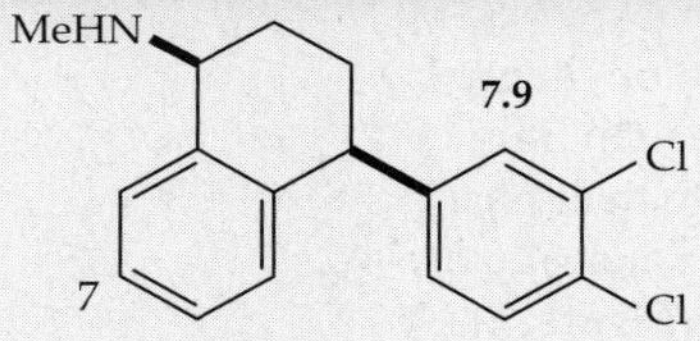 **7.9**	None	30	4–6	30	3
2	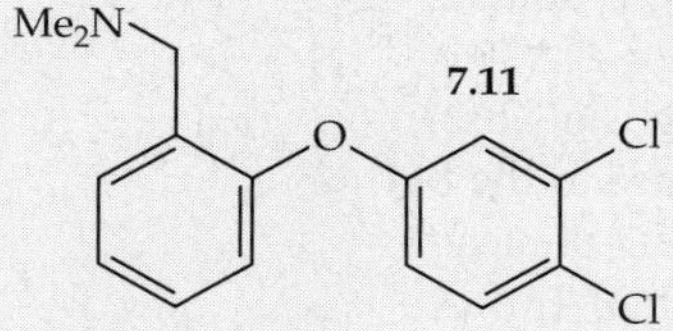 **7.11**	Opened ring and inserted ether (decreased potency)	—	—	—	25
3	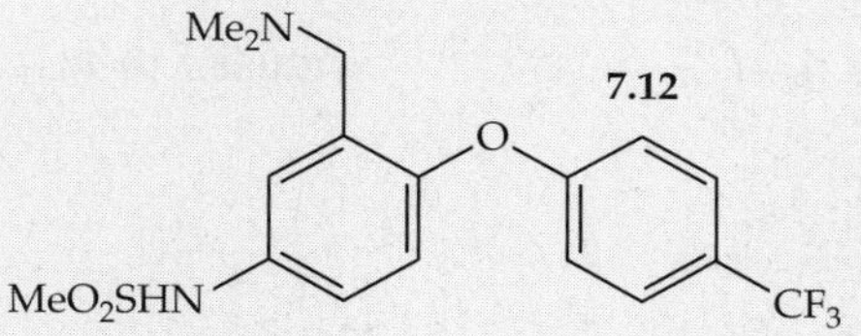 **7.12**	Addition of a substituent at C7 of **7.9** (decreased $t_{1/2}$ and V_d)	5	0.5	2	5
4	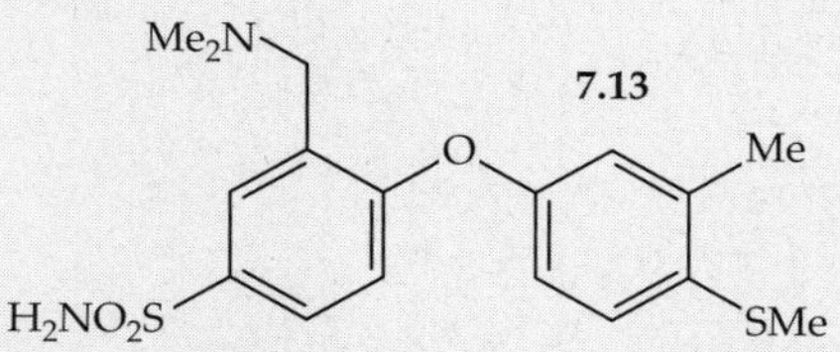 **7.13**	Introduction of a thioether (increased clearance and decreased $t_{1/2}$)	10	<1	1–2	5

Through structural modifications, researchers at Pfizer sought to improve sertraline's pharmacokinetic parameters while maintaining its activity. Changes included opening one of the rings of sertraline (**7.11**) and exploring polar substituents at C7 of the original structure of sertraline (**7.12**). The changes had many effects. The increased polarity of the drug kept the drug more concentrated in the blood (decreased V_d) so that the drug could be more rapidly cleared (decreased $t_{1/2}$) by the kidneys and liver. Introduction of a thioether made the compound (**7.13**) more prone to oxidation by the liver. This raised the hepatic extraction ratio (E_H) to further increase hepatic clearance and decrease $t_{1/2}$. Ultimately, compounds such as **7.13** emerged as promising antidepressants with lower V_d and $t_{1/2}$ and quicker onset of activity (**Table 7.2**).

Decreasing V_d might be accomplished by lowering the lipophilicity of the lead. A more polar lead would be expected to have a higher affinity for the blood and therefore distribute into a smaller volume.

Structural changes to a molecule that already has a high affinity for a target must be made with great care. Late in the drug discovery process, which is often when small adjustments to pharmacokinetic behavior are desired, the discovery team knows the key structural elements required for strong target binding. These elements are the *pharmacophore* of the drug. (Pharmacophores are discussed more thoroughly in Chapter 11.) Structural changes to optimize pharmacokinetics are made away from the pharmacophore of a drug. Regardless, changes that improve pharmacokinetic properties often negatively impact target binding. For this reason, drug discovery groups vigilantly watch the pharmacokinetic properties even as target binding is being optimized. Attempts to fix major pharmacokinetic problems of a potent compound rarely succeed.

7.4 Drug Metabolites

The pharmacokinetic parameters of an administered drug are clearly important for understanding and predicting biological effects. Sometimes drug metabolites form and contribute to the biological effect of the original drug. For example, the analgesic effect of codeine (**7.1**) is attributed primarily to the formation of morphine (**7.14**) as a metabolite of codeine (**Scheme 7.6**). When metabolites show significant activity, the pharmacokinetics of the metabolites must be understood.

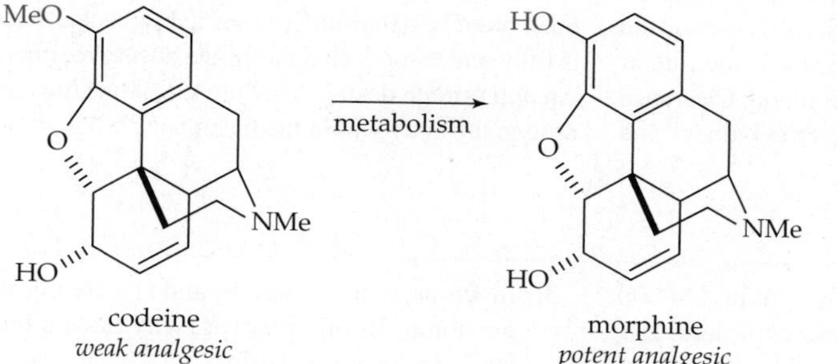

codeine
weak analgesic
7.1

morphine
potent analgesic
7.14

SCHEME 7.6 Metabolism of codeine to morphine

Metabolites often form from drugs that are already circulating in the bloodstream, but they can also be formed in the digestive tract and then absorbed into the bloodstream.

FIGURE 7.20 C_p vs. time for
an IV bolus of a drug and its
subsequent metabolite

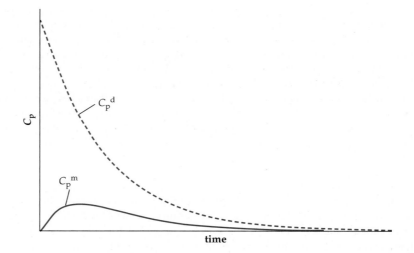

Regardless of where a metabolite forms, C_p-time plots for metabolites often closely resemble **Figure 7.20**. The metabolite's C_p (C_p^m) starts at 0 and rises to a peak. First-order elimination then becomes the predominant process, and C_p^m begins to fall. Overall, the shape of the C_p^m-time curve somewhat resembles C_p-time curve of an oral drug.

The behavior of metabolites may often be modeled reasonably well, but determining the various parameters for metabolites is a challenge. Specialized pharmacokinetic curve-fitting software frequently can tease the parameters from clinical C_p-time data. If necessary, metabolites may be synthesized in lab and then administered directly to animals. From these studies, key parameters such as V_d, CL, F, and k_{ab} may be determined in the absence of the parent drug. Keep in mind that a pharmaceutical company may not administer a metabolite into humans without the metabolite's being classified as an investigational new drug (IND).

Summary

Pharmacokinetic parameters determine the rate of flow of a drug through a biological system. The two basic parameters are volume of distribution and clearance, and both are calculated from intravenous bolus C_p versus time data. Volume of distribution relates the extent to which a drug perfuses into the various tissues of the body to a hypothetical volume of blood. Clearance is the rate at which the body removes a drug from a volume of blood. Clearance is generally a combination of drug elimination by the kidneys and metabolism by the liver. Together, volume of distribution and clearance determine the half-life and elimination rate of a drug. Because most drugs are delivered orally, additional pharmacokinetic parameters, including bioavailability and absorption rate constant, also must often be determined. Once a drug's pharmacokinetic behavior is fully understood, an appropriate dosing regimen can be designed. An appropriate dosing schedule maintains the concentration of the drug in the body within the therapeutic window of the drug.

Questions

1. Infliximab, an antibody with a molecular weight of 144,000 g/mol, is effective in certain autoimmune disorders. If a 350-mg IV dose of infliximab is administered to a 70-kg patient, a C_{max} of 118 μg/mL is observed. What is the V_d of infliximab? What does the V_d indicate about the distribution of infliximab?

2. Compounds with a very small V_d normally have a shorter $t_{1/2}$. The half-life of infliximab is quite long at 228 h (9.5 d). Based on the calculated V_d of infliximab from question 1, speculate why infliximab has such a long half-life.

3. In words, explain why V_d and $t_{1/2}$ are normally directly proportional. In other words, why does a larger V_d normally lead to a longer half-life?

4. If the k_{el} of a drug is much greater than its metabolite, how might one determine the k_{el} of the metabolite? Before answering this question, sketch a superimposed C_p-time plot of both the drug and metabolite.

5. Amiodarone (**7.a**) is used to treat patients with irregular, fast heart rates. The standard maintenance dose is 400 mg/day administered orally, which achieves a C_p of approximately

2.0 $\mu g/mL$. Amiodarone has a V_d of 66 L/kg and a half-life of 53 days. If amiodarone were administered by IV infusion, what would be the infusion rate (R_{inf}) necessary to reach a C_p^{ss} of 2.0 $\mu g/mL$? How long would the infusion require to reach just 1.0 $\mu g/mL$? Do you see a practical problem in using IV infusion to administer amiodarone?

amiodarone
7.a

6. C_p-time data points for a 100-mg IV bolus are given in the following table for a hypothetical drug and a 70-kg patient. Assume a one-compartment model. Determine k_{el}, $t_{1/2}$, and V_d for the drug.

Entry	Time (h)	C_p ($\mu g/mL$)
1	0.25	5.7
2	0.50	4.9
3	0.75	4.3
4	1.00	3.7
5	1.50	2.7
6	2.00	2.0
7	2.50	1.5
8	3.00	1.1
9	4.00	0.6
10	5.00	0.3

7. The same drug as in question 6 was administered as a 100-mg oral dose to the same patient. The C_p-time data is shown in the following table. Determine k_{ab} and then F.

Entry	Time (h)	C_p($\mu g/mL$)
1	0.25	1.7
2	0.50	2.3
3	0.75	2.5
4	1.00	2.4
5	1.50	2.0
6	2.00	1.6
7	2.50	1.2
8	3.00	0.9
9	4.00	0.5
10	5.00	0.2

8. In tables of pharmacokinetic data of drugs, one can readily find information on $t_{1/2}$ and t_{max} for oral drugs. Although k_{ab} may not always be provided, k_{ab} can be determined from $t_{1/2}$ and t_{max}. Ibuprofen (**7.b**) has a half-life of 2 h and t_{max} of 1.6 h. Estimate the k_{ab} of ibuprofen.

ibuprofen
7.b

9. Equation 7.21 and other equations derived from it are not completely general. Under what condition do they fail?

10. If $k_{ab} = \infty$ and $F = 1$, to what does Equation 7.23 reduce? Where have you seen this equation before? Explain in words why this simplification works.

11. Adult males have been estimated to lose 1.0 mg of iron per day. The United States Recommended Daily Allowance for iron is 8.0 mg. What is a reasonable estimate of the bioavailability of dietary iron?

12. Azithromycin (**7.c**) is an antibiotic. The bottom of its therapeutic window for most bacteria has been estimated at 2.0 $\mu g/mL$. Azithromycin is typically dosed orally with 500 mg on day 1 and 250 mg on each of days 2 through 5. The relevant pharmacokinetic parameters of azithromycin are as follows: $k_{el} = 0.0100\ h^{-1}$, $k_{ab} = 1.88\ h^{-1}$, $F = 38\%$, $V_d = 31.1\ L/kg$. Estimate the C_p^{max} for azithromycin over the dosing interval. (To do this problem, you will need to determine the C_p separately for all five doses [at different start times] and add them together.)

azithromycin
7.c

13. The minimum inhibitory concentration (MIC) of azithromycin against many bacteria is 2 $\mu g/mL$. The dosing regimen in question 12 does not reach this threshold concentration in the plasma. Based on azithromycin's V_d of 31.1 L/kg, is it possible for the MIC to be reached within a patient's tissues?

14. Scheme 7.4 shows a diagram of a typical two-compartment model. Diagram a three-compartment model in which the third compartment can only be reached from the peripheral compartment. Diagram a three-compartment model in which the third compartment is in equilibrium with both the central and peripheral compartments.

15. The LC_{50} of a drug in humans is 5 μg/mL. If the drug has a V_d of 10 L/kg, calculate the minimum IV bolus dose necessary to achieve a $C_p^{\,0}$ equal to the LC_{50}. Theoretically, you are calculating the IV LD_{50} of the drug. If the drug causes death very quickly, why is your calculated LD_{50} likely too high?

References

1. Winter, M. E. *Basic Clinical Pharmacokinetics* (4th ed.). Philadelphia: Lippincott Williams & Wilkins, 2003.
2. Hansen, J. T., & Koeppen, B. M. *Netter's Atlas of Human Physiology.* Teterboro, NJ: Icon Learning Systems, 2002.
3. Rolan, P. E. Plasma Protein Binding Displacement Interactions: Why Are They Still Regarded as Clinically Important? *Br. J. Clin. Pharm.* 1994, *37,* 125–128.
4. Wendt, M. D., Shen, W., Kunzer, A., McClellan, W. J., Bruncko, M., Oost, T. K., et al. Discovery and Structure-Activity Relationship of Antagonists of B-Cell Lymphoma 2 Family Proteins with Chemopotentiation Activity In Vitro and In Vivo. *J. Med. Chem.* 2006, *49,* 1165–1181.
5. Ortho-McNeil Pharmaceutical, Inc. *Ortho Evra (Norelgestromin/Ethinyl Estradiol Transdermal System).* Raritan, NJ: Ortho-McNeil, 2001.
6. Middleton, D. The Role of Medicinal Chemistry in Drug Discovery: Controlling the Uncontrollables. Presented at the 230th National Meeting of the American Chemical Society, Washington, DC, August 2005; Paper MEDI 228.

Metabolism

Chapter Outline

Many drugs are not readily eliminated from the body. To facilitate elimination, the body breaks down drugs through one or more chemical reactions. Like most reactions in the body, metabolic reactions are catalyzed by enzymes, the most common being the cytochrome P-450 enzymes in the liver. The products of metabolic reactions are generally more water soluble than the parent drug and therefore more easily cleared from the bloodstream by the kidneys. By studying the metabolism and elimination of a drug, a drug discovery team can better anticipate whether specific groups of patients might need a customized dosing regimen for safety reasons.

8.1 Introduction

The term "metabolism" refers to many separate processes. One process is energy regulation in the body, that is, taking in food and converting it into useful energy. This aspect of metabolism, more specifically called *catabolism*, includes topics such as glycolysis and the citric acid cycle. Another process covered under the umbrella of metabolism is *anabolism*. Anabolism encompasses the use of energy from catabolism to drive the synthesis of proteins, nucleic acids, and other molecules needed for a system to function.[1] Drugs that interfere with anabolic processes are called *antimetabolites* and were briefly discussed in Chapter 6. In the context of drugs, the term "metabolism" describes reactions that enable a system to remove a foreign substance, or *xenobiotic*. Drugs are not metabolized for energy or utility; they are metabolized so that they may be eliminated from an organism as quickly as possible.

When considering metabolism, one must examine the full picture (**Scheme 8.1**). A biological system includes a host of enzymes, some of which recognize, bind, and convert a drug into a metabolite. A single drug will likely be metabolized by multiple enzymes. Some of the reactions will be faster than others. Some reactions may be reversible or may be reversed by a second metabolic enzyme. In turn, each metabolite will be cleared from the body at a different rate. In total, a single drug with its many possible metabolites and competing elimination pathways forms a complex picture. Which metabolite, if any, is most important for elimination of a drug depends on the rate and reversibility of each step.

**Learning Goals
for Chapter 8**

• Recognize the types of metabolic reactions that are common to the functional groups found in drugs

• Understand complicating factors introduced by drug metabolism and the resulting metabolites

• Know how metabolism can be exploited to improve the pharmacokinetics of a potential drug

SCHEME 8.1 Metabolism and elimination pathways for a drug

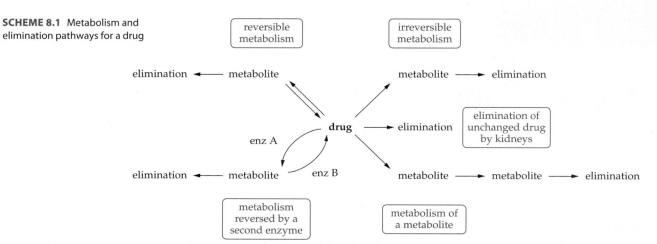

8.2 Metabolic Reactions

Metabolic reactions are divided into two categories: phase I and phase II. Phase I reactions include oxidation, reduction, and hydrolysis. Most phase I reactions fragment the drug into smaller pieces that are more easily eliminated than the starting drug. In contrast, phase II reactions add fragments to the original drug in a process called *conjugation*. The newly appended fragments form a metabolite with increased polarity for improved clearance. Phase I and phase II reactions may initially seem to be in opposition—one process degrades a drug while the other enlarges a drug. Both processes, however, work in concert. The products of phase I reactions contain functional groups such as carboxylic acids, alcohols, and amines. These functional groups may serve as handles for attachment of polar fragments during phase II metabolism. Phase II reactions need not follow phase I reactions. Many drugs are metabolized by only phase I or only phase II processes.

Phase I Reactions

Phase I reactions are typically divided into three categories: oxidation, reduction, and hydrolysis.

Oxidation

Formally, oxidation involves the removal of electrons from a molecule. In a balanced reaction, an oxidation frequently has the net effect of removing hydrogen (H_2) or hydride (H^-) from a molecule. When a molecule or functional group is oxidized by a metabolic enzyme, the oxidation product tends to be more electrophilic and prone to attack by a nucleophile. In a biological system, water is omnipresent and serves as a common nucleophile. Therefore, oxidations are often coupled with a subsequent hydrolysis. These are technically separate steps but are commonly discussed together as one process.

sp³-Hybridized Carbons (Alkyl Groups)

The sp^3-hybridized carbons in an alkyl chain oxidize readily if suitably activated. Activating groups include electronegative heteroatoms (N, O, halogens), aromatic rings, and π-bonds. Oxidation of the carbon requires cleavage of a C–H bond to form a radical or cationic intermediate. Activating groups stabilize the intermediate and thus facilitate the oxidation. It is possible to oxidize unactivated hydrocarbon chains, but the process tends to be slower and less common.

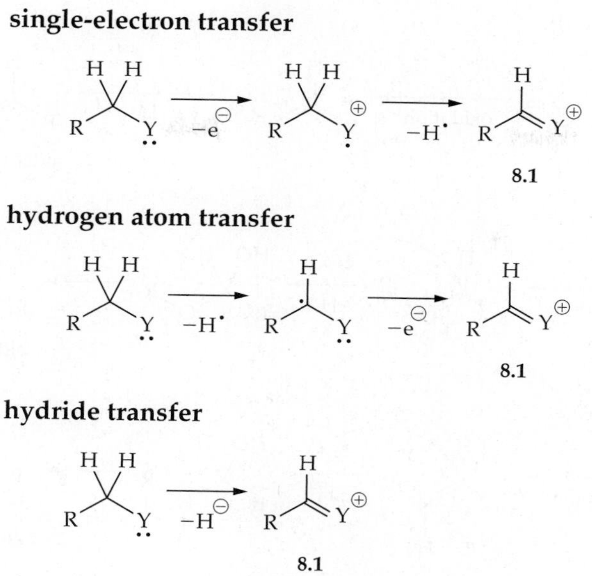

single-electron transfer

8.1

hydrogen atom transfer

8.1

hydride transfer

8.1

SCHEME 8.2 Different mechanisms for oxidative cleavage of alkyl groups

Mechanistically, oxidations fall within one of three categories: single-electron transfer, hydrogen atom transfer, and hydride transfer. The exact mechanism for oxidation of a specific drug depends on the enzyme that is performing the reaction. The cytochrome P-450 (CYP) enzyme family contains an iron-heme core that reacts mostly by hydrogen atom transfer.[2] Monoamine oxidases (MAOs) and flavine-containing monooxygenases (FMOs) involve flavine adenine dinucleotide as a cofactor and often favor a single-electron transfer mechanism.[3] Enzymes such as liver alcohol dehydrogenase use nicotinamide adenine dinucleotide and follow a hydridic process.[3] Knowing the exact mechanism of oxidation is less important than being able to predict the likely structures of metabolites.

When heteroatom-substituted alkyl groups (Y = electronegative heteroatom) are oxidized, all three mechanisms lead to a common, cationic intermediate, **8.1** (Scheme 8.2). Water either attacks or deprotonates intermediate **8.1** depending on the identity of Y (Scheme 8.3). If water attacks as a nucleophile, an alkyl group will be lost from the parent drug. This process is known as an *oxidative dealkylation*. Two specific dealkylations of drugs are shown in **Scheme 8.4**.

Metabolism often occurs through multiple steps, that is, metabolism of metabolites. The lactam metabolite (**8.9**) of thioridazine (Melleril, **8.6**) may appear to be unrelated to the processes in Schemes 8.2 and 8.3 (**Scheme 8.5**). Metabolite **8.9**, however, arises first from oxidation adjacent to nitrogen to give an iminium ion intermediate **8.7**. Reaction with water gives a carbinol (**8.8**) which may be oxidized like any other alcohol to a carbonyl. In this case, the carbonyl is part of a cyclic amide, or a lactam (**8.9**).[4] Similarly, the carboxylic acid metabolite (**8.13**) of losartan (Cozaar, **8.10**) arises initially from the aldehyde metabolite (**8.11**). Reversible hydration of the aldehyde forms a new alcohol that may be oxidized to a carboxylic acid (**8.13**). When considering drug metabolism, almost all drugs have very many *possible* metabolites. In practice, only a small number of metabolites will be important for a drug's elimination.

sp²-Hybridized Carbons (Arenes and Alkenes)

sp²-**Hybridized Carbons (Arenes and Alkenes)** Hydroxylations of aromatic rings are common in metabolic pathways of drugs. The oxidations are performed almost exclusively by CYP enzymes. The reaction involves formation of an unstable arene oxide

SCHEME 8.3 Oxidations of specific alkylated heteroatoms

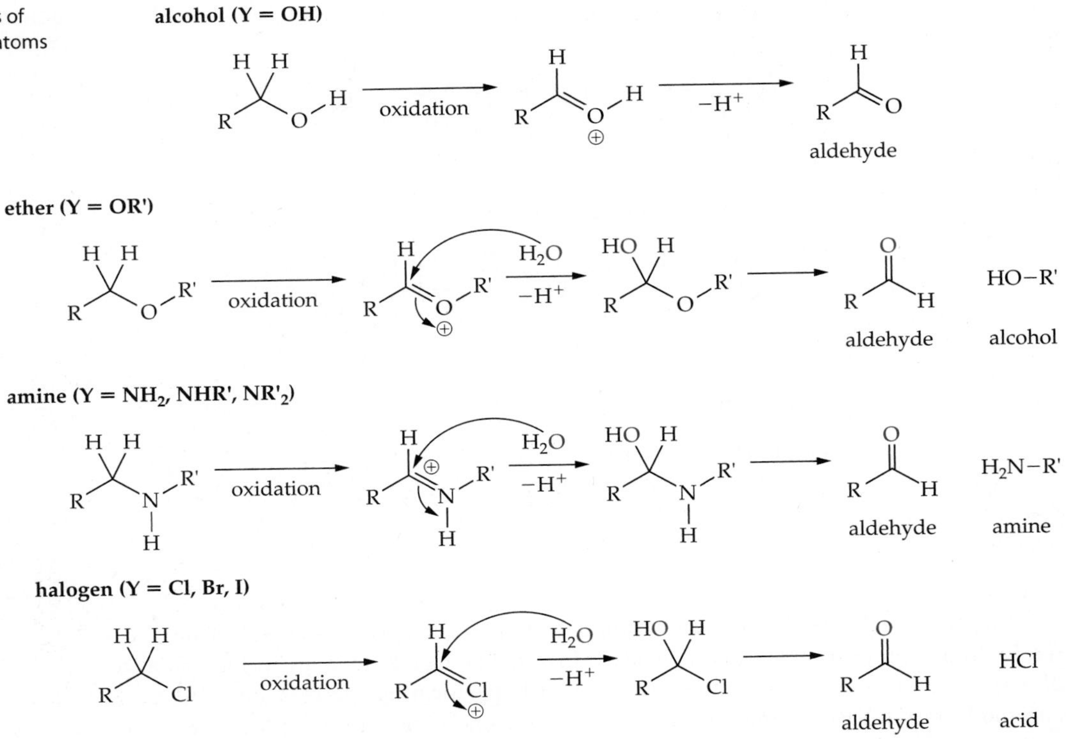
alcohol (Y = OH)
oxidation
$-H^+$
aldehyde
ether (Y = OR')
oxidation
H_2O
$-H^+$
aldehyde
HO–R'
alcohol
amine (Y = NH_2, NHR', NR'_2)
oxidation
H_2O
$-H^+$
aldehyde
H_2N–R'
amine
halogen (Y = Cl, Br, I)
oxidation
H_2O
$-H^+$
aldehyde
HCl
acid

SCHEME 8.4 Dealkylated metabolites of commercial drugs

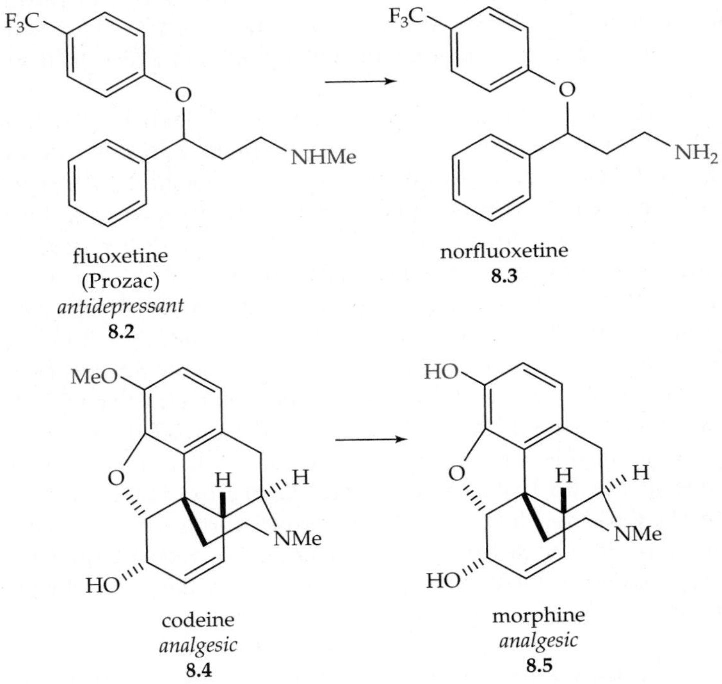
fluoxetine
(Prozac)
antidepressant
8.2
norfluoxetine
8.3
codeine
analgesic
8.4
morphine
analgesic
8.5

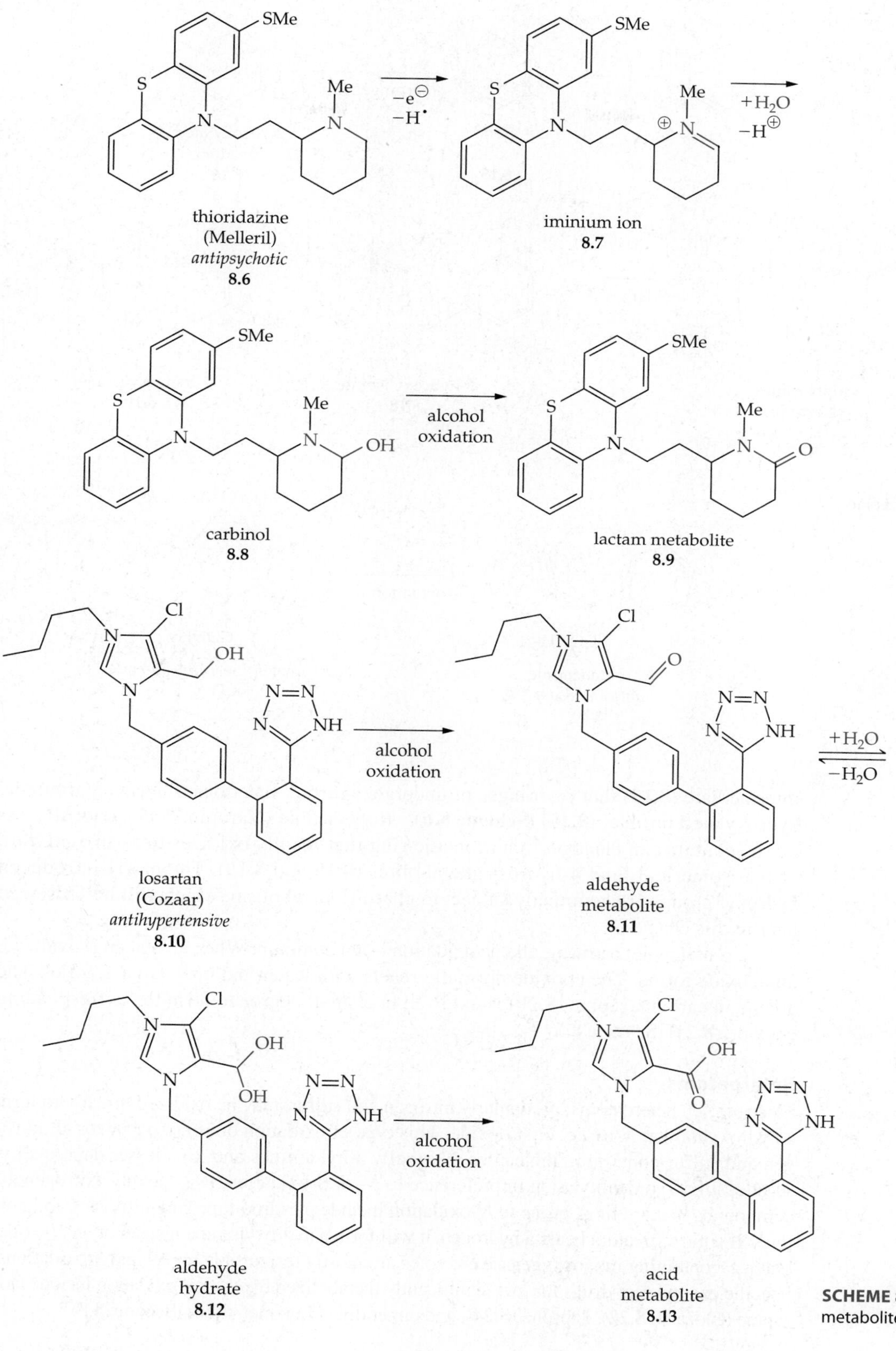

SCHEME 8.5 Metabolites of metabolites

SCHEME 8.6 Arene metabolism

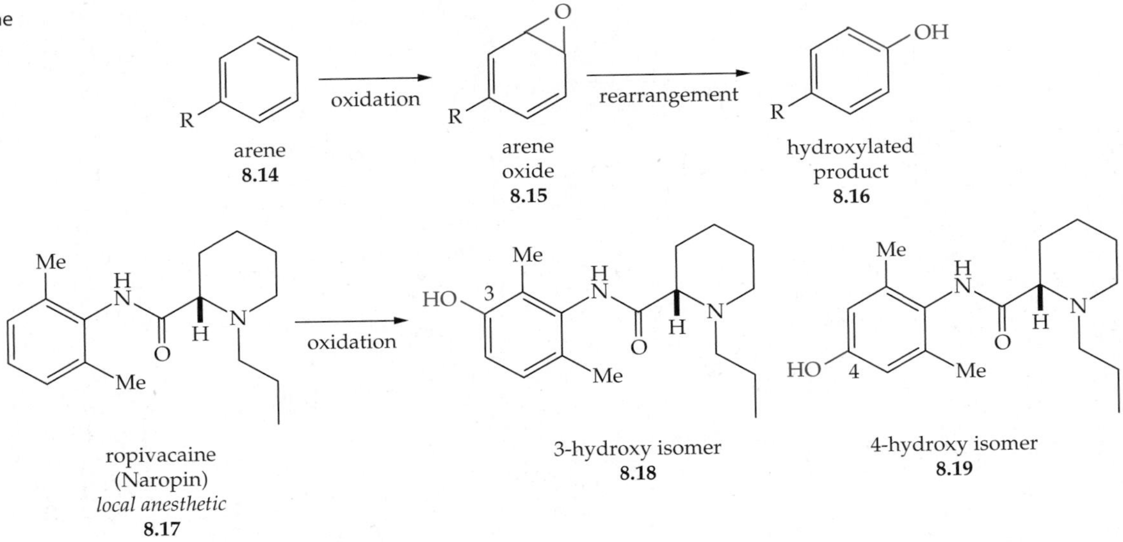

arene
8.14

oxidation

arene
oxide
8.15

rearrangement

hydroxylated
product
8.16

ropivacaine
(Naropin)
local anesthetic
8.17

oxidation

3-hydroxy isomer
8.18

4-hydroxy isomer
8.19

SCHEME 8.7 Carbamazepine epoxidation

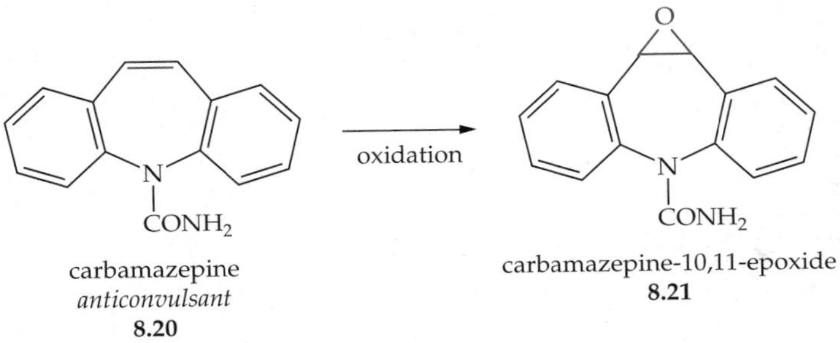

carbamazepine
anticonvulsant
8.20

oxidation

carbamazepine-10,11-epoxide
8.21

intermediate (**8.15**) that rearranges or undergoes further reactions to form an aromatic, hydroxylated product (**8.16**) (**Scheme 8.6**).[5] Ropivacaine (Naropin, **8.17**), a local anesthetic, contains an electron-rich aromatic ring that readily oxidizes to form a mixture of the isomeric 3- and 4-hydroxy metabolites (**8.18** and **8.19**). The newly introduced hydroxyl groups are commonly subject to phase II metabolism, which will be addressed later in this chapter.

As a metabolic pathway, alkene oxidation is not common. When an alkene is oxidized, an epoxide forms. The epoxide normally reacts with water and opens to a 1,2-diol. The alkene in carbamazepine (**8.20**) is oxidized to a small degree to form the corresponding epoxide (**8.21**) (**Scheme 8.7**).

Heteroatoms

Electron-rich heteroatoms, particularly nitrogen and sulfur, may be oxidized directly to form a hydroxylamine, *N*-oxide, or sulfoxide. This type of oxidation does not occur for all nitrogen and sulfur-containing molecules. Normally alkyl amines and, to a lesser degree alkyl sulfides, undergo dealkylation in preference to *N*- or *S*-oxidation, respectively. Nitrogenous compounds that are most prone to *N*-oxidation include pyridine-type ring nitrogens and anilines. If a nitrogen atom bears a hydrogen, it will form a hydroxylamine instead of an *N*-oxide. Flavine-containing monooxygenase enzymes are mostly responsible for *N*- and *S*-oxidations. Specific examples of drugs that are significantly metabolized by *N*- or *S*-oxidation include clozapine (Clozaril, **8.22**), dapsone (**8.24**), and cimetidine (Tagamet, **8.26**) (**Scheme 8.8**).[6]

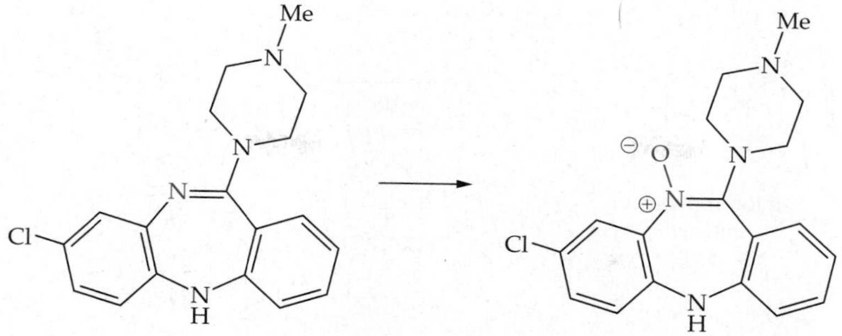

SCHEME 8.8 Examples of heteroatom oxidations in metabolism

clozapine
(Clozaril)
antipsychotic
8.22

clozapine *N*-oxide
8.23

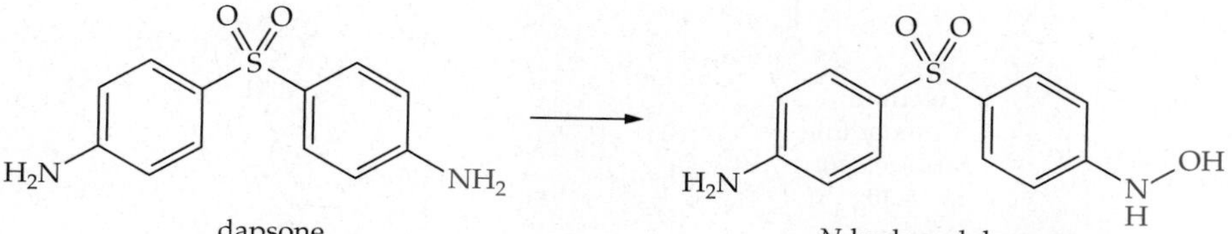

dapsone
leprosy
8.24

N-hydroxyl dapsone
8.25

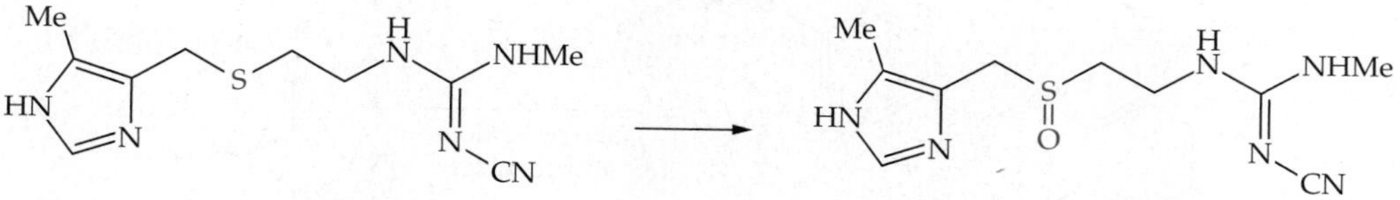

cimetidine
(Tagamet)
acid reflux
8.26

cimetidine sulfoxide
8.27

Reduction

Metabolism is almost always an oxidative process. Reductive metabolism is much more limited. Functional groups that are reduced are, naturally, in a higher oxidation state. The more common examples include nitro groups, which are reduced to amines, and ketones, which reduce to alcohols. Chloramphenicol (**8.28**), an antibiotic that has fallen out of favor because of serious side effects, contains a nitro group that is reduced to the corresponding amine (**8.29**) (Scheme 8.9).[7] Warfarin (Coumadin, **8.30**), an anticlotting agent, is at least partially metabolized by reduction of its ketone to an alcohol.

A rarely encountered functional group is an azo linkage, which consists of an nitrogen-nitrogen double bond. Although almost never found in a commercial drug, one particular azo group and its metabolic reduction resulted in the discovery of an entire class of antibiotics and has saved countless lives. The story was already partially described in Chapter 1. Sulfamidochrysoidine (**8.32**) is a dye with activity against streptococcal bacteria (Scheme 8.10). The drug was marketed as Protosil Rubrum, a name derived from the red color of the dye. Later research proved that the azo linkage in **8.32** is reduced and cleaved in the body to afford sulfanilamide (**8.33**). Furthermore, the observed activity of sulfamidochrysoidine is actually attributable to metabolite **8.33**. From this research rose an entire class of drugs, the sulfonamide antibiotics or simply "sulfa drugs," which remain in use today.

SCHEME 8.9 Examples of reductions in metabolism

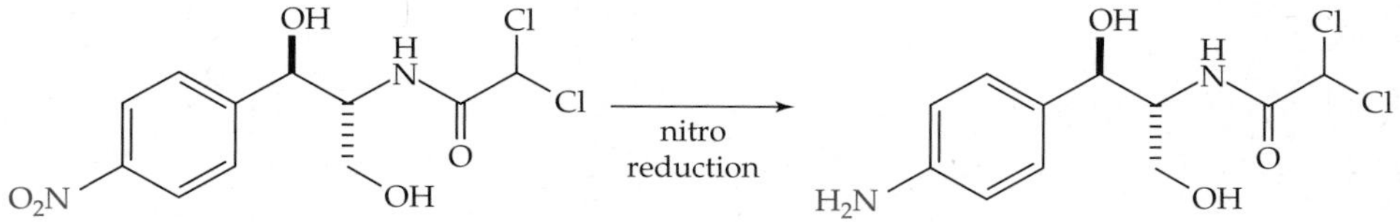

chloramphenicol
antibiotic
8.28

nitro
reduction

8.29

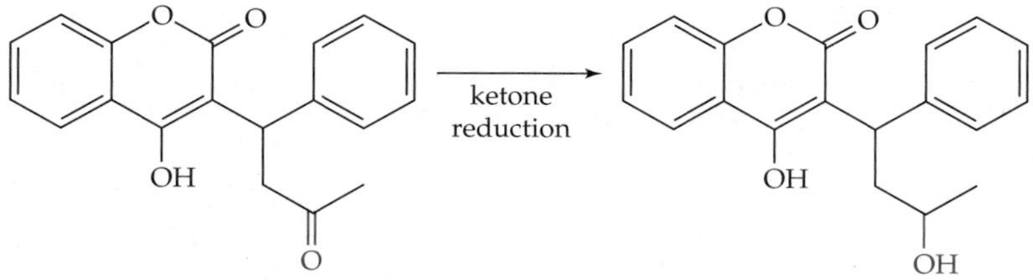

warfarin
(Coumadin)
anticoagulant
8.30

ketone
reduction

8.31

SCHEME 8.10 Reduction of an azo linkage

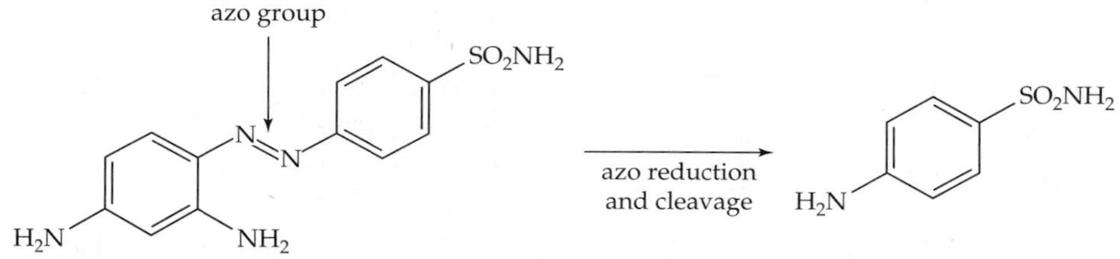

azo group

sulfamidochrysoidine
(Protosil Rubrum)
antibiotic
8.32

azo reduction
and cleavage

sulfanilamide
8.33

SCHEME 8.11 Common hydrolysis reactions

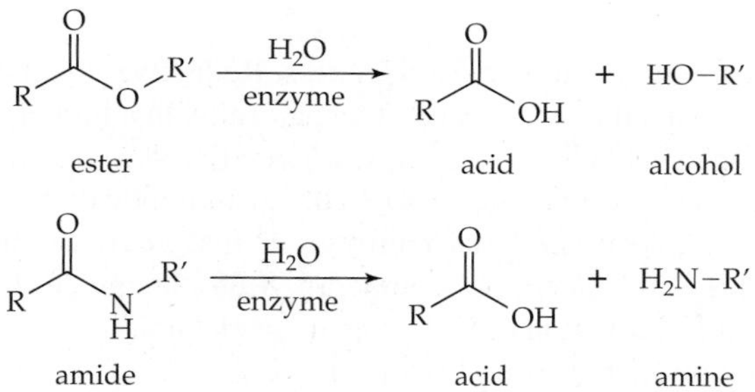

ester

H_2O
enzyme

acid

+ HO−R′

alcohol

amide

H_2O
enzyme

acid

+ H_2N−R′

amine

Drugs that are administered in an inactive form and are metabolized to a structure with the desired biological activity are called *prodrugs*. Prodrugs are highlighted in the final section of this chapter.

Hydrolysis

The last form of phase I metabolism is hydrolysis. Hydrolysis reactions generally involve acid derivatives. Specifically, esters and amides are hydrolyzed to carboxylic acids and an

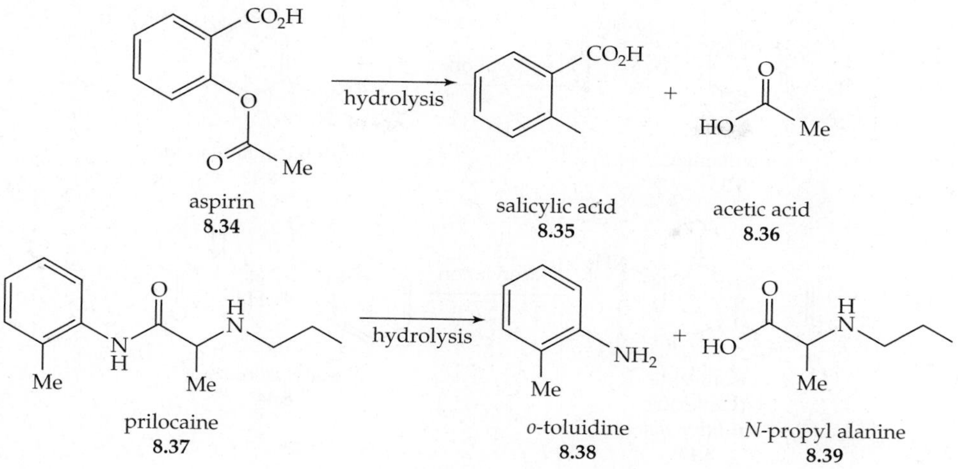

SCHEME 8.12 Hydrolysis reactions of specific drugs

aspirin
8.34

salicylic acid
8.35

acetic acid
8.36

prilocaine
8.37

o-toluidine
8.38

N-propyl alanine
8.39

alcohol or amine, respectively (**Scheme 8.11**). The products of the hydrolysis are generally more water soluble than the original compound. A less common hydrolysis reaction is the ring-opening of an epoxide with water.

Despite the presence of water throughout the body, hydrolysis reactions of esters and amides require enzymes to proceed at an appreciable rate. Numerous enzymes throughout the body carry out hydrolysis reactions. The enzymes, both esterases and amidases, are found in the digestive system, individual cells, and plasma. The exact site of hydrolysis of a specific drug depends on the drug's structure and functionality. The ester in aspirin (**8.34**) is primarily hydrolyzed by pseudocholinesterase in plasma to form salicylic acid (**8.35**) and acetic acid (**8.36**) (**Scheme 8.12**).[8] Prilocaine (**8.37**), a topical anesthetic, undergoes amide hydrolysis in the liver and kidneys.

Phase II Reactions

While phase I metabolism tends to break apart a drug, phase II metabolic reactions create molecules that are larger than the parent drug by *conjugating* polar molecules with the drug. Phase I and II reactions work together because many drugs do not have the right functionality to undergo conjugation reactions. In this situation, a drug first undergoes phase I metabolism to introduce an alcohol, acid, or amine group. These groups are then excellent reactive sites for attaching a separate molecule by phase II reaction. Ultimately, the goal of both processes is the same—get the drug out of the body.

Acetylation

Amines, especially aromatic amines (**8.40**), can be acetyled by acetyl CoA in phase II metabolism (**Scheme 8.13**). Amine acetylation actually does not increase polarity in the interest of improving the rate of elimination. Instead, acetylation decreases the toxicity of anilines, which can be highly reactive in the body. The toxicity of anilines will be more fully discussed in Section 8.2 of this chapter. Sulfanilamide (**8.33**) undergoes significant acetylation during its metabolism (**Scheme 8.14**). Isoniazide (Laniazid, **8.43**), a hydrazide-containing drug, is *N*-acetylated on its NH_2 group.

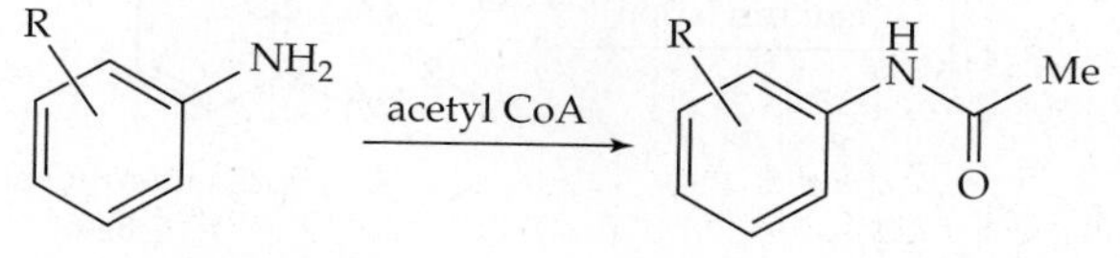

8.40

8.41

SCHEME 8.13 Acetylation of an aromatic amine

SCHEME 8.14 *N*-Acetylation of sulfanilamide and isoniazide

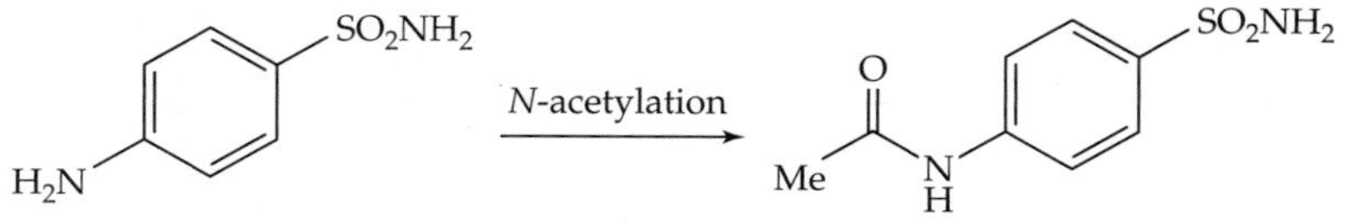

sulfanilamide
8.33

N-acetyl sulfanilamide
8.42

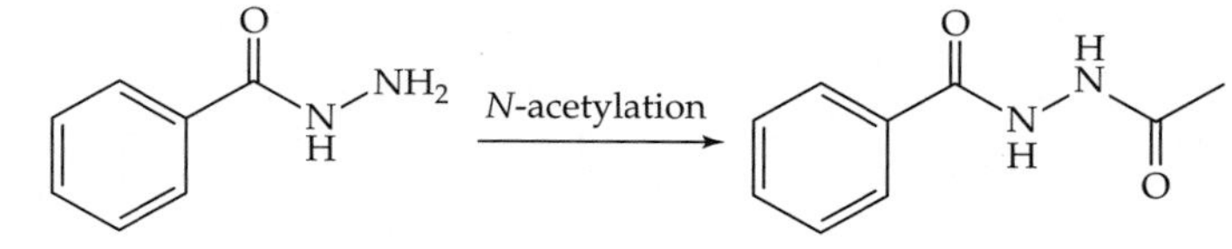

isoniazide
(Laniazid)
antitubercular
8.43

N-acetyl isoniazide
8.44

Sulfonylation

Sulfonylation as a conjugation pathway is most important for drugs containing a phenol. The sulfate ester product is very acidic (pK_a −3.0) and is deprotonated to its anionic conjugate base in water. Therefore, the conjugated drug is considerably more polar and water soluble than the original phenol. Prenalterol (**8.45**), a compound that has been tested in humans but not approved by the U.S. Food and Drug Administraton (FDA), accelerates the heart rate and is metabolized to a large degree by sulfonylation (**Scheme 8.15**).

Glucuronidation

A major phase II metabolism pathway is the conjugation of a compound with uridine diphosphate glucuronic acid (UDP-glucuronic acid) (**8.47**) (**Scheme 8.16**). Glucuronidation products are called *glucuronides* (**8.48**). Drugs bearing alcohols and carboxylic acids are often candidates for conjugation with glucuronic acid. One metabolic pathway for sulindac (Clinoril, **8.49**) involves phase I oxidation of the sulfoxide to a sulfone (**8.50**) followed by phase II conjugation to form a glucuronide (**8.51**).

Drugs are not the only compounds that undergo phase I and phase II metabolism. Waste products from the body are also commonly metabolized during the course of their elimination. One example is bilirubin (**8.52**) (**Scheme 8.17**). Bilirubin arises primarily through the breakdown of hemoglobin from red blood cells. Despite containing two carboxylic acids, bilirubin is lipophilic and poorly excreted by the kidneys. In the liver, bilirubin is conjugated with UDP-glucuronic acid. The resulting glucuronide (**8.53**) is water soluble and eliminated both by the kidneys and into bile through the gall bladder. Bilirubin and its glucuronide have a yellow color and contribute to the coloration of both urine and feces. If bilirubin is not properly metabolized and eliminated, it will accumulate in fatty tissues and organs, including the skin. Yellowing of the skin is a characteristic of a

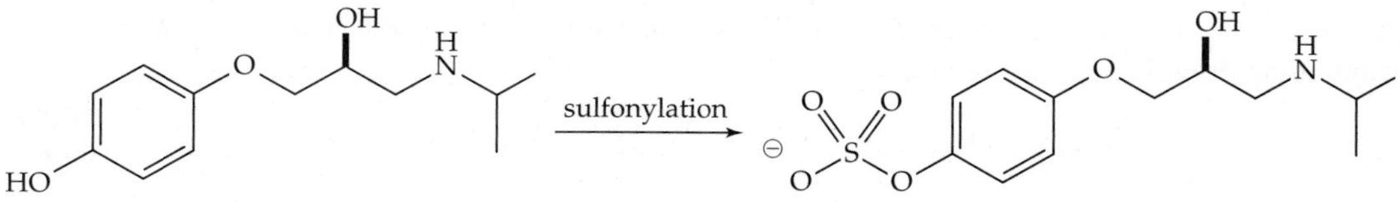

prenalterol
8.45

prenalterol-*O*-sulfate
8.46

SCHEME 8.15 *O*-Sulfonylation of prenalterol

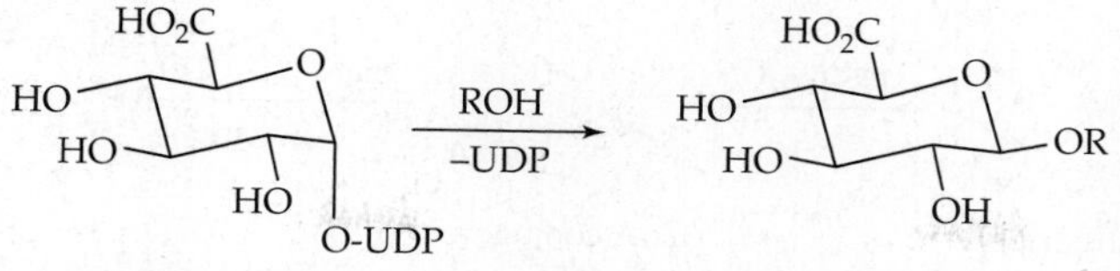

UDP-glucuronic acid
8.47

a glucuronide
8.48

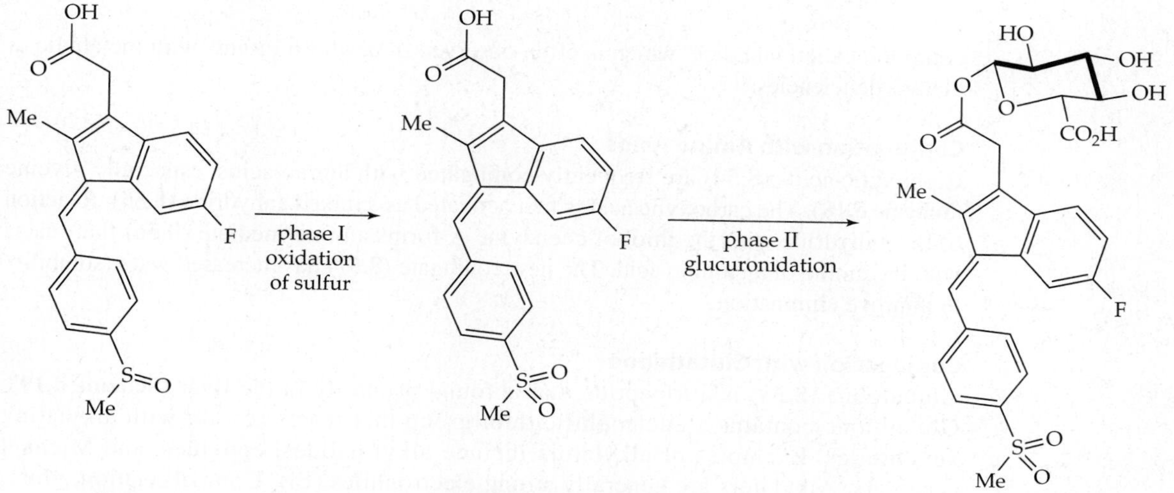

sulindac
(Clinoril)
antiinflammatory
8.49

sulfone
8.50

glucuronide
8.51

SCHEME 8.16 Glucuronidation

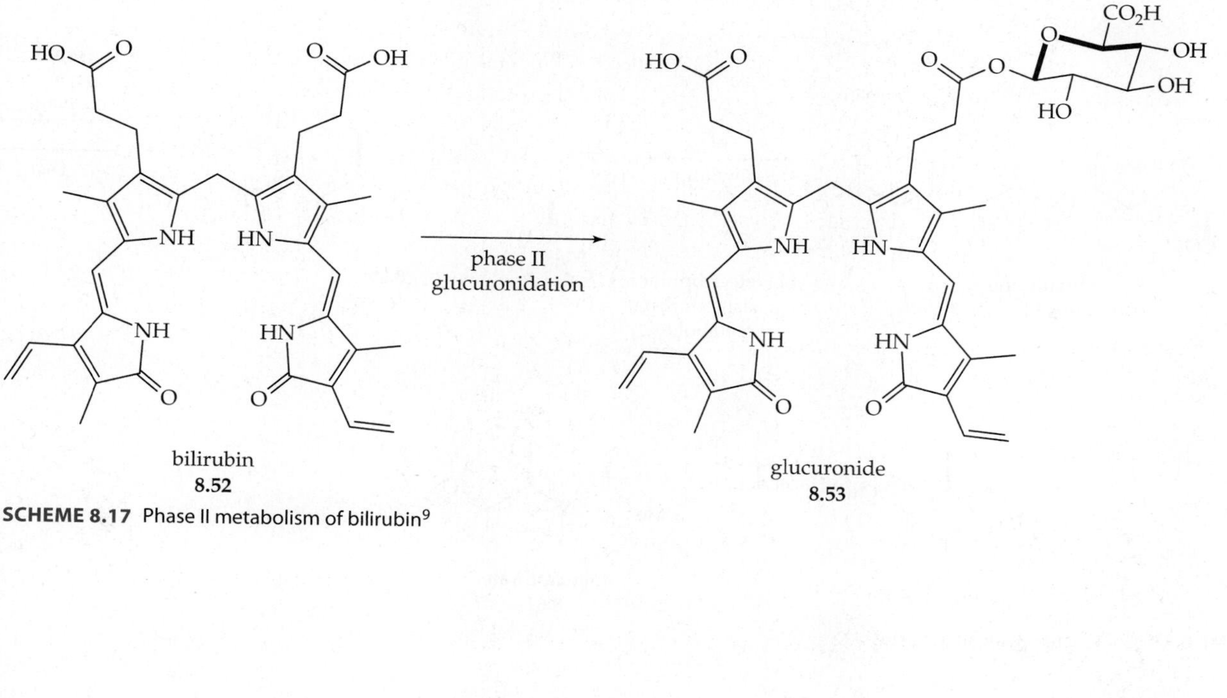

bilirubin
8.52

glucuronide
8.53

SCHEME 8.17 Phase II metabolism of bilirubin[9]

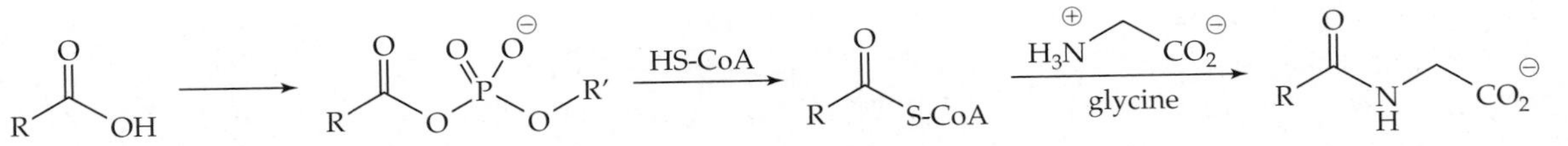

acid **8.54** — mixed anhydride **8.55** — CoA conjugate **8.56** — conjugate **8.57**

SCHEME 8.18 Conjugation with amino acids

condition called jaundice, which is often observed in newborn infants with metabolic or dietary deficiencies.[9]

Conjugation with Amino Acids

Carboxylic acids (**8.54**) are frequently conjugated with amino acids, especially glycine (**Scheme 8.18**). The carboxylic acid is first activated as a mixed anhydride (**8.55**). Reaction of the anhydride with the thiol of coenzyme A forms an intermediate (**8.56**) that reacts with the amine of an amino acid. The new conjugate (**8.57**) has increased water solubility to improve elimination.

Conjugation with Glutathione

Glutathione (**8.58**) is a tripeptide that is found primarily in the liver (**Scheme 8.19**). Glutathione contains a nucleophilic thiol group that reacts readily with alkylating xenobiotics. Examples of alkylators include alkyl halides, epoxides, and Michael acceptors. Alkylators are generally strong electrophiles (El). Upon alkylation, glutathione is systematically broken down and acetylated to form the final conjugate, called a mercapturate (**8.61**). Orally administered drugs rarely contain strongly electrophilic functional groups. Therefore, if a mercapturate is implicated in the metabolism of a drug, the mercapturate is normally a conjugate of an electrophilic phase I metabolite, not the original drug.

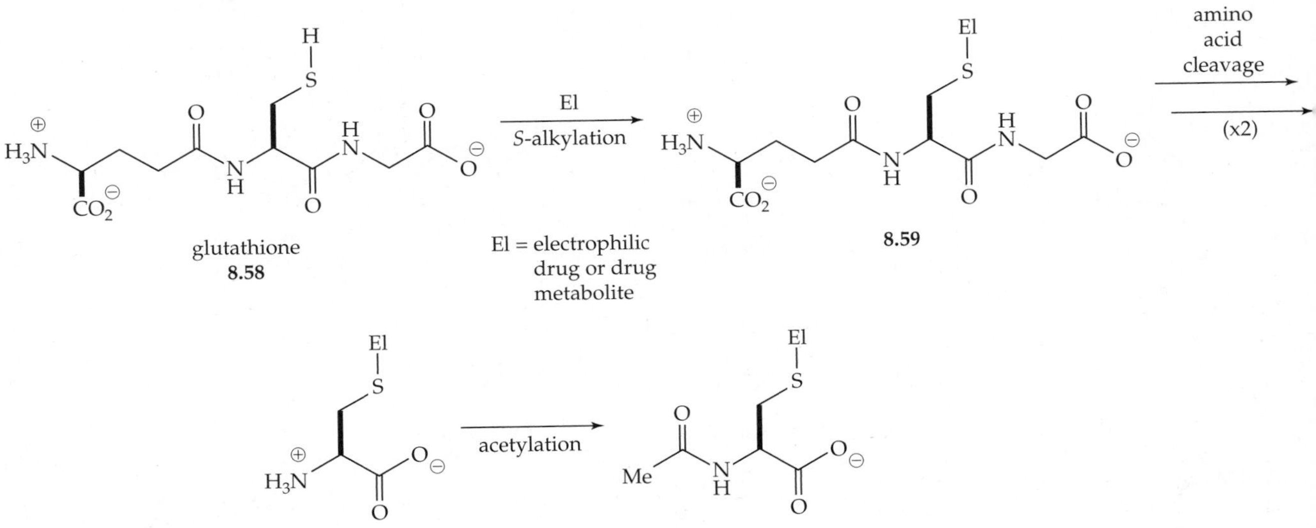

glutathione **8.58** — **8.59** — **8.60** — mercapturate **8.61**

SCHEME 8.19 Conjugation with amino acids

8.3 Metabolism Issues

Metabolism greatly complicates the drug discovery process. Factors that must be considered include the biological activity of new metabolites, potential interference on metabolism by a drug, and variations in a drug's metabolism across different patient populations.

Metabolite Activity

The activity of each metabolite formed in the body must be understood and classified into one of three categories:

- No activity
- Activity against the same target as the original drug
- Activity against a target different from the original drug

Inactive Metabolites

Inactive metabolites are the simplest possibility and desirable for a commercial drug. In this scenario, the drug is metabolized and simply eliminated. Loss of activity can occur for pharmacodynamic and/or pharmacokinetic reasons. If the key parts of a drug that bind the target are significantly altered by metabolism, then the metabolite will lose its affinity for the intended target and its biological activity. If metabolism radically changes how the drug distributes in the body, the metabolite might not be able to reach the target of the drug.

The primary metabolite of sertraline (Zoloft, **8.62**), an antidepressant, arises from oxidative dealkylation of the amine group by CYP enyzmes in the liver (**Figure 8.1**). The metabolite, *N*-desmethylsertraline (**8.63**), is essentially inactive and is readily eliminated by the kidneys. Other metabolites of sertraline include **8.64** and **8.65**, both of which involve modification of the amine. Neither is an effective inhibitor of the serotonin transporter, the target of sertraline.[10]

Metabolites with On-Target Activity

Many metabolites retain activity from the parent drug, that is, *on-target activity*. Selected examples of marketed drugs, the major metabolite of each, and the percentage activity of the metabolite are shown in **Table 8.1**.

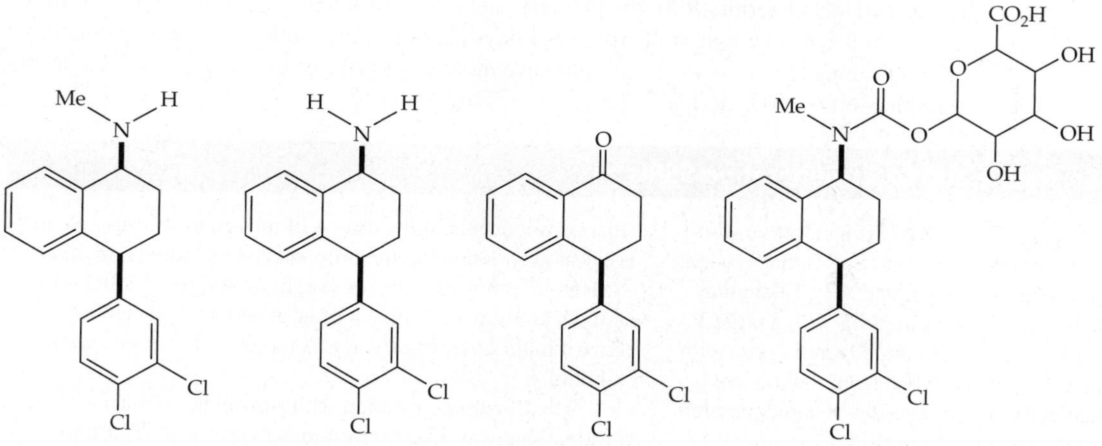

sertraline (Zoloft) *antidepressant* **8.62**

N-desmethylsertraline **8.63**

sertraline ketone **8.64**

sertraline *N*-carbamoyl glucuronide **8.65**

FIGURE 8.1 Sertaline and its inactive metabolites

TABLE 8.1   Drugs with metabolites of similar activity[11]

Entry	Drug	Drug Name	Metabolite	Activity (%
1	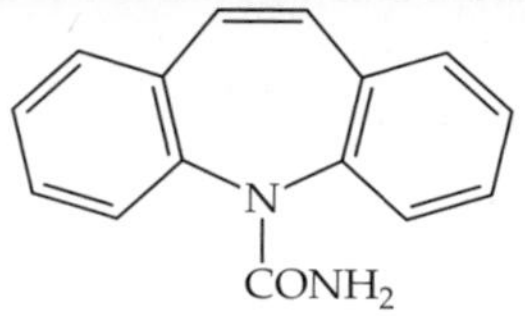	carbamazepine *anticonvulsant* **8.23**	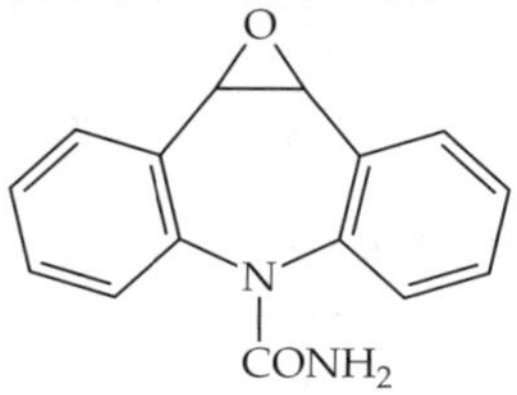	~ 100
2	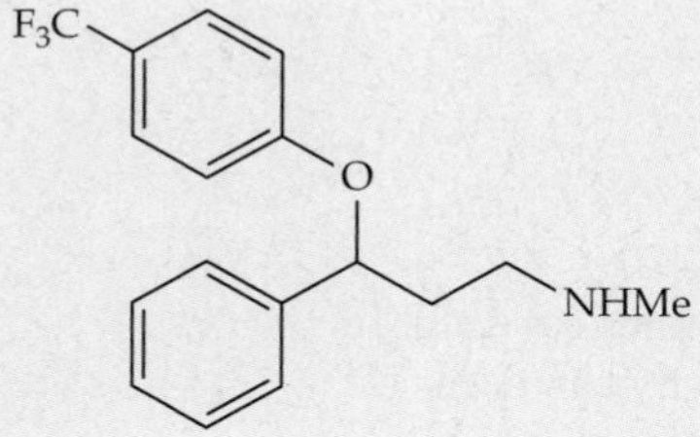	fluoxetine (Prozac) *antidepressant* **8.2**	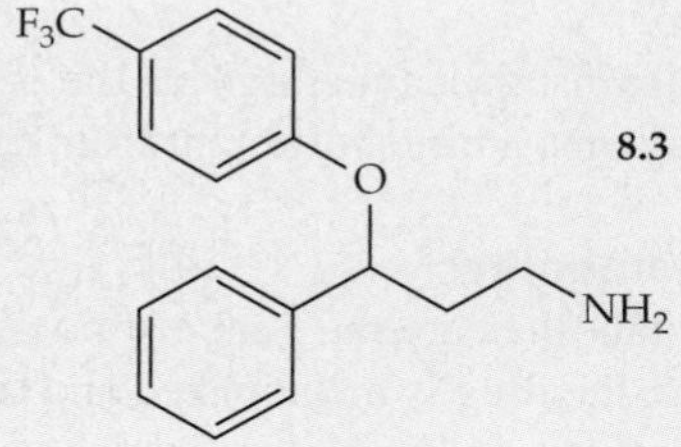 **8.3**	~ 100
3	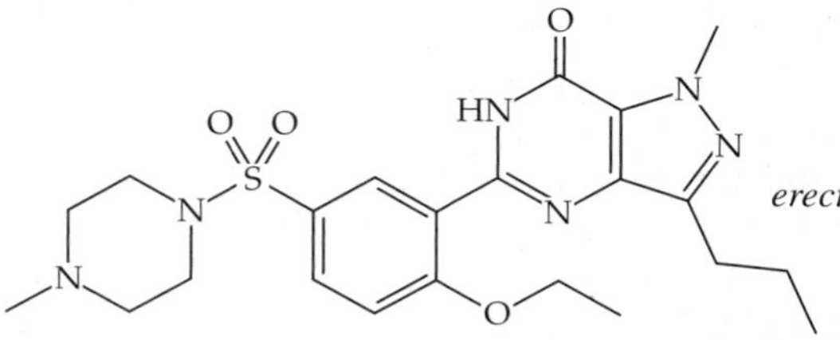	sildenafil (Viagra) *erectile disfunction* **8.66**	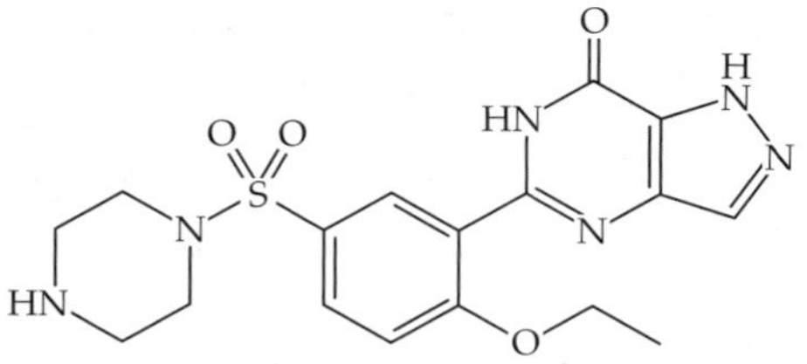	50

Most metabolites are more polar than the original drug and, therefore, eliminate relatively quickly. In other words, the metabolite has a shorter half-life than its drug. Under these conditions, the original drug is responsible for most of the observed biological activity. If the metabolite is active and has a longer half-life than the original drug, then the metabolite will be mostly responsible for activity once steady-state concentrations have been achieved. An example is fluoxetine (Prozac, **8.2**). Fluoxetine has a half-life of 2.2 days. Norfluoxetine (**8.3**), the primary metabolite of fluoxetine, is equipotent to fluoxetine and has a very long half-life of 6.4 days. In one study, patients receiving fluoxetine over a prolonged period were found to have more norfluoxetine (130 μg/L) in their serum than fluoxetine (109 μg/L).[12]

CASE STUDY

Sertraline and Later Generation SSRIs[10]

Sertraline (Zoloft, **8.62**) is a selective serotonin reuptake inhibitor (SSRI) that is used for the treatment of depression (**Figure 8.2**). Like many SSRIs, sertraline is a very lipophilic drug with a large V_d and correspondingly long $t_{1/2}$. Orally delivered drugs with these properties are readily absorbed from the digestive system and immediately collect into the first encountered fatty organ, the liver. The drug then slowly distributes from the liver. As a result, SSRIs require a long time to achieve steady-state concentrations in the body. For the patient, this means that the drug will only slowly take full effect. Similarly, if a patient quits taking the drug,

therapeutic levels of the drug will remain as the drug clears very slowly. Knowing the problems of sertraline and other SSRIs, researchers at Pfizer sought to design an SSRI with low V_d and short $t_{1/2}$. Such a drug would be expected to have a rapid onset of efficacy. A target $t_{1/2}$ value was set at 8 hours.

The Pfizer team started with sertraline, a Pfizer product that was already well understood with regard to which functional groups were required for high activity. Initial efforts focused on the benzofused ring of sertraline. Substitution of a polar sulfonamide group was hoped to decrease both V_d and $t_{1/2}$. Compound **8.67** showed excellent serotonin reuptake inhibition ($IC_{50} = 1$ nM) and an estimated $t_{1/2}$ of 8 h based on animal model predictions (Figure 8.2). In humans, however, **8.67** underwent phase I metabolism to form a demethylated compound with a very long $t_{1/2}$ of 240 h. Attention shifted to opening the aliphatic ring of sertraline. Simple removal of two CH_2 groups from sertraline and linking the aromatic rings with an ether provided **8.68**, a compound with decreased activity ($IC_{50} = 25$ nM) yet a much lower V_d.

Continued research on ring-opened structures ultimately afforded **8.69**. *In vitro* **8.69** had excellent predicted properties: $IC_{50} = 5$ nM, $V_d = 2$ L/kg, and $t_{1/2} = 5$ h (**Figure 8.3**). *In vivo*, however, compound **8.69** was demethylated to **8.70**, an active metabolite with a long

half-life. Variations on **8.69** suffered other problems. Replacing the NMe_2 group with an azetidine (four-membered nitrogen-containing ring) led to an unstable compound (**8.71**). Incorporation of a pyrrolidine ring (**8.72**) introduced stereochemical complications that were not attractive.

Because metabolism of the amino group was causing problems, the discovery team decided to direct metabolism to another position of the molecule. Thioethers are known to undergo rapid oxidation, and the lower half of the molecule was modified to include an SMe group (**8.73**) (**Figure 8.4**). Compound **8.73** did show excellent properties with good activity ($IC_{50} = 5$ nM) and reasonable pharmacokinetics ($V_d = 1.5$ L/kg, $t_{1/2} = 10$ h). Furthermore,

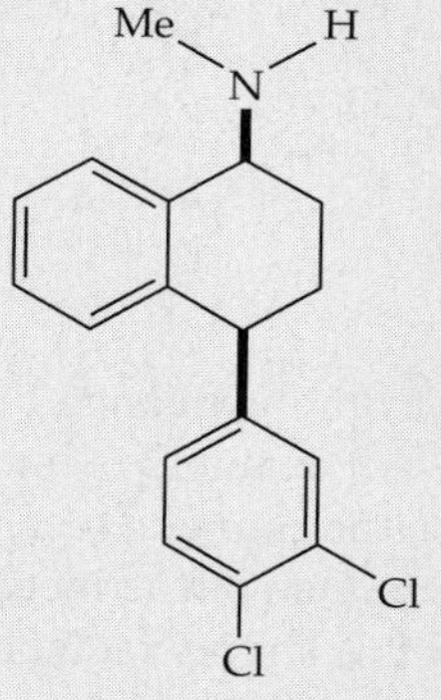

	sertraline **8.62**	**8.67**	**8.68**
IC_{50} (nM)	3	1	25
V_d (L/kg)	1,900	—	100
$t_{1/2}$ (h)	30	8 (calc.)	—

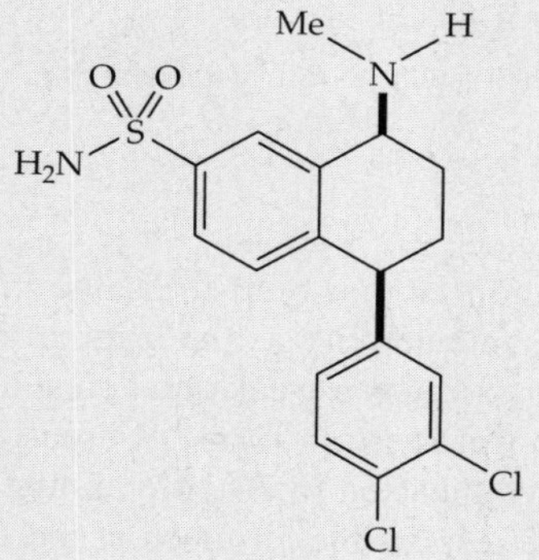

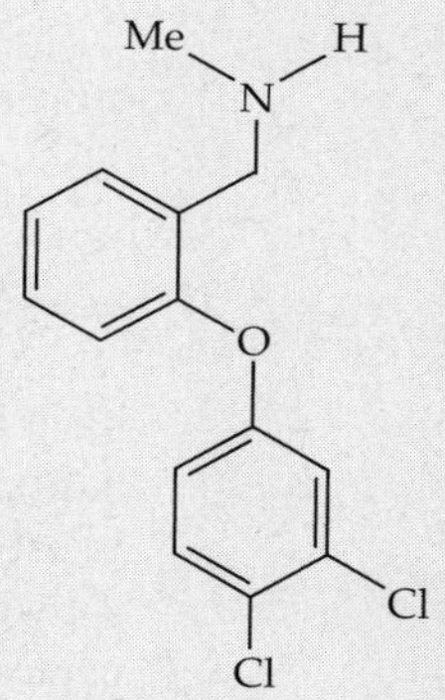

FIGURE 8.2 Sertaline and analogues

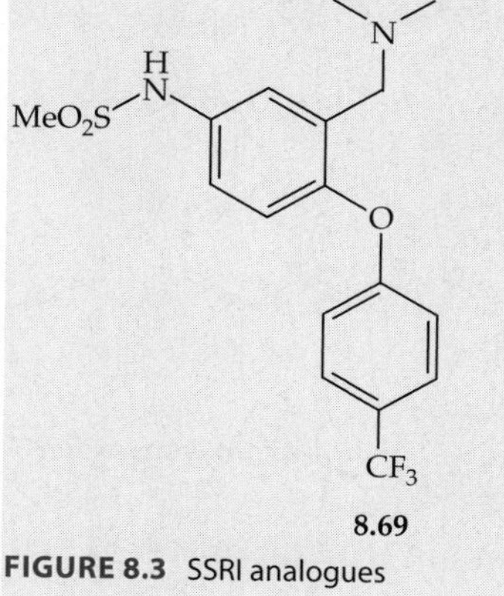

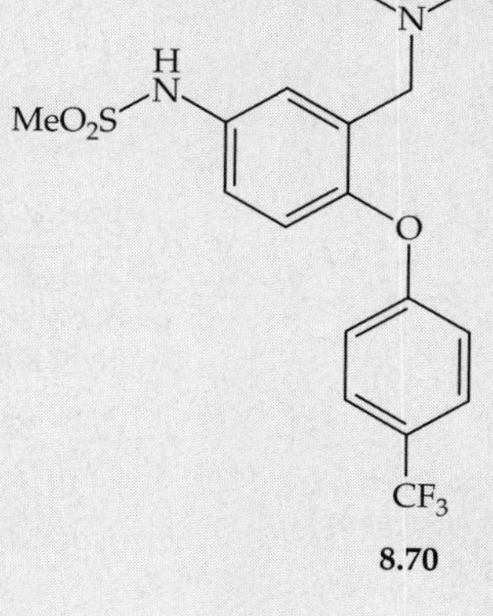

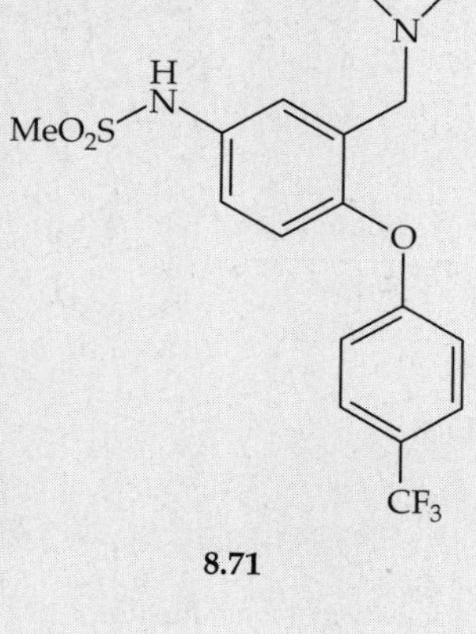

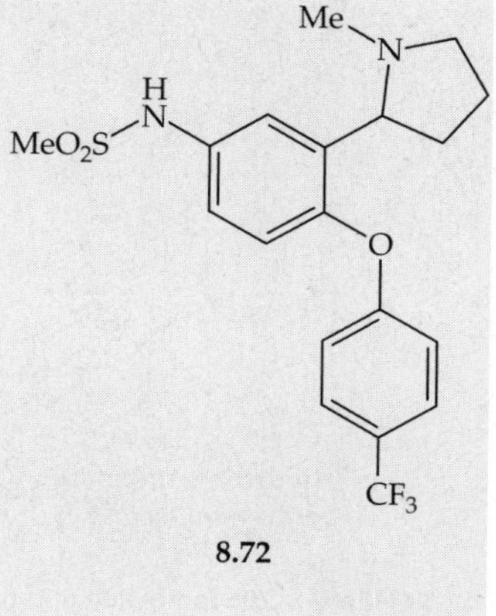

FIGURE 8.3 SSRI analogues

8.73 formed a single, inactive metabolite through phase I oxidation of the thioether to a sulfoxide (**8.74**).

Although compound **8.73** was never approved as a marketable drug, its development highlights complications that drug metabolism can introduce into the drug discovery process. When metabolites show the same activity as the original drug, then the metabolite is also a part of the full activity picture of the drug.

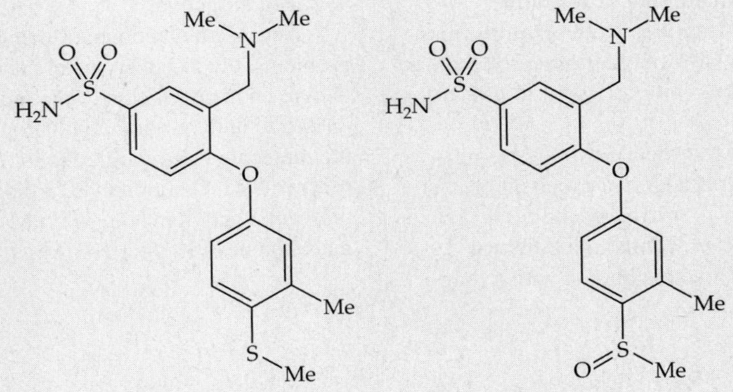

FIGURE 8.4 SSRI candidate with metabolite

Metabolites that bind the same target as the unchanged drug do not necessary have the same activity as the original drug. An example is tamoxifen (**8.75**) (**Scheme 8.20**). Tamoxifen is a weak antagonist against the estrogen receptor in breast tissue and is used to treat certain forms of breast cancer. One of the metabolites of tamoxifen is 4-hydroxytamoxifen (**8.76**). Metabolite **8.76** is a strong antagonist for the estrogen receptor and responsible for much of the observed activity of tamoxifen. A less prominent metabolic pathway for tamoxifen involves the cleavage of the aminoethyl chain to give compound **8.77**. This metabolite also binds the estrogen receptor but insteads acts as an agonist. This type of activity reversal is rare and can occur only with drugs that bind receptors, not enzymes.[13]

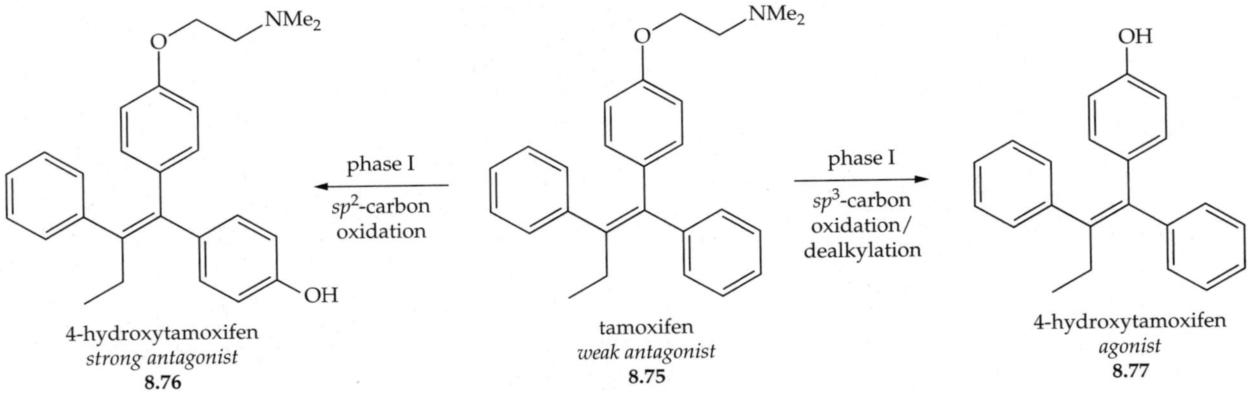

SCHEME 8.20 Tamoxifen and the activity of its metabolites[13]

Metabolites with Off-Target Activity

Metabolites often show activity that is dissimilar to the original drug. This activity, called *off-target activity*, involves interaction of a metabolite with other receptors or enzymes. Off-target activity, while common, is difficult to anticipate and is a major focus of clinical trials. Patients in a clinical trial answer a long list of questions to determine whether they are experiencing any side effects, major or minor. Of course, not all side effects arise from metabolites. Often, the drug itself is responsible for the off-target effects. Side effects from off-target activity of metabolites pose a serious problem for a drug and jeopardize regulatory approval of a drug candidate. The severity of a side effect is judged relative to the condition treated by a drug. A cancer drug would be allowed more severe side effects than a drug that alleviates seasonal allergy symptoms.

The liver is the primary metabolic organ of the body. For this reason, the liver is where most toxic metabolites are formed and where toxic metabolites do most of their damage. Although the liver is equipped to handle toxic metabolites, its capabilities are finite and can be overwhelmed. The following Case Studies describe the types of toxicity issues that are observed in drugs.

CASE STUDY

Acetaminophen[14]

Acetaminophen (Tylenol, **8.78**) is an over-the-counter analgesic (**Scheme 8.21**). Acetaminophen is primarily metabolized in the liver by two phase II processes, sulfonylation and glucuronidation. Both phase II reactions involve the phenol group of acetaminophen.

A small percentage of acetaminophen is metabolized by phase I oxidation with *N*-hydroxylation of the amide (**8.81**) (**Scheme 8.22**). Loss of water affords *N*-acetyl-*p*-benzoquinone imine (**8.82**). Benzoquinones are electron-deficient and strongly electrophilic, a trait of many toxic metabolites. Fortunately, **8.82** forms in the liver, which contains a high concentration (5 mM) of glutathione (**8.58**, G-SH). Glutathione, a strong nucleophile because of its thiol group (SH), reacts with the quinone imine by a conjugate addition. The resulting intermediate (**8.83**) tautomerizes to **8.84**, a stable, water soluble metabolite that is readily eliminated. At high doses, however, large quantities of the benzoquinone imine (**8.82**) form. Excess **8.82** consumes glutathione until the liver is left unprotected, and the remaining benzoquinone imine **8.82** begins to damage the liver.

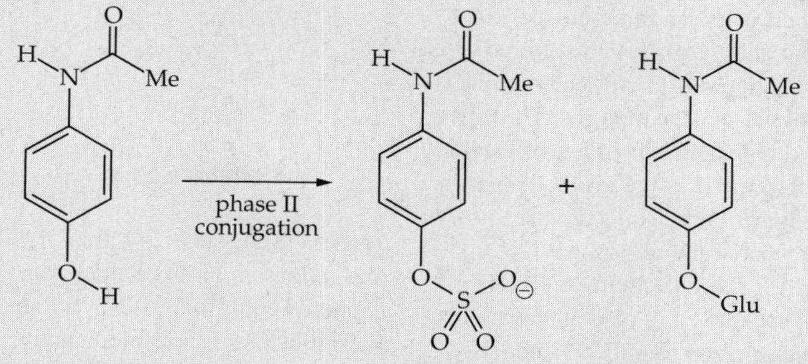

acetaminophen (Tylenol) *analgesic* **8.78**

acetaminophen sulfate **8.79**

acetaminophen glucuronide **8.80**

SCHEME 8.21 Phase II metabolism of acetaminophen

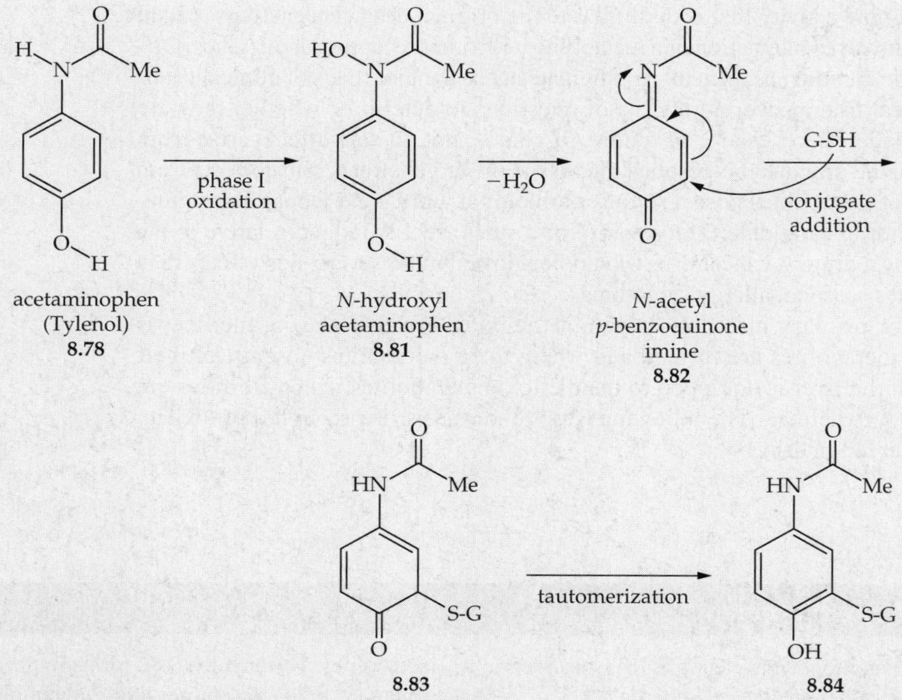

SCHEME 8.22 Toxic metabolites from acetaminophen

CASE STUDY

MPPP[15,16]

In the early 1980s, a 42-year-old heroin addict in northern California was treated for symptoms that resembled advanced Parkinson's disease. The patient, however, had shown no symptoms only a week earlier. Upon further investigation, physicians found the patient had taken a synthetic form of heroin over a three-day period. Within two days, the patient and his girlfriend suffered from late-stage Parkinson's symptoms. Five other heroin users were soon found with the same symptoms. Through an investigation involving the National Institutes of Health (NIH), researchers found impurities in the synthetic heroin that caused rapid progression of Parkinson's disease.

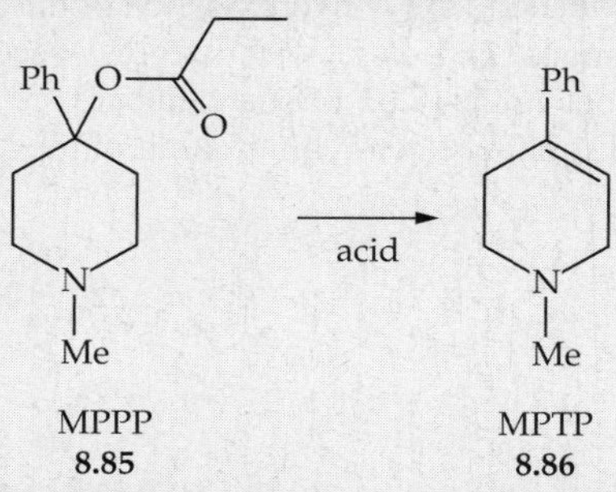

The drug in question was 1-methyl-4-phenyl-4-propionoxypiperidine (MPPP, **8.85**) (**Scheme 8.23**). During its synthesis, if MPPP is exposed to acid at elevated temperatures, an elimination occurs to afford 1-methyl-4-phenyl-1,2,5,6-tetrahydropyridine (MPTP, **8.86**) as an impurity. MPTP is the key impurity responsible for Parkinson's symptoms.

MPTP is oxidized in the liver to 1-methyl-4-phenyl-2,3-dihydropyridininum (MPDP$^+$, **8.87**) in a phase I

SCHEME 8.23 Formation of MPTP from MPPP

process (**Scheme 8.24**). Reaction of MPDP$^+$ with water and oxidation by liver aldehyde oxidase affords 1-methyl-4-phenyl-5,6-dihydro-2-pyridone (MPTP lactam, **8.89**). MPTP lactam is a stable, inactive compound that is eliminated by the kidneys. Unfortunately, MPTP (**8.86**) is not only metabolized in the liver. MPTP is very lipophilic and readily crosses the blood–brain barrier. In the brain, MPTP is oxidized to MPDP$^+$. Liver aldehyde oxidase, the enzyme that converts MPDP$^+$ to MPTP lactam, is

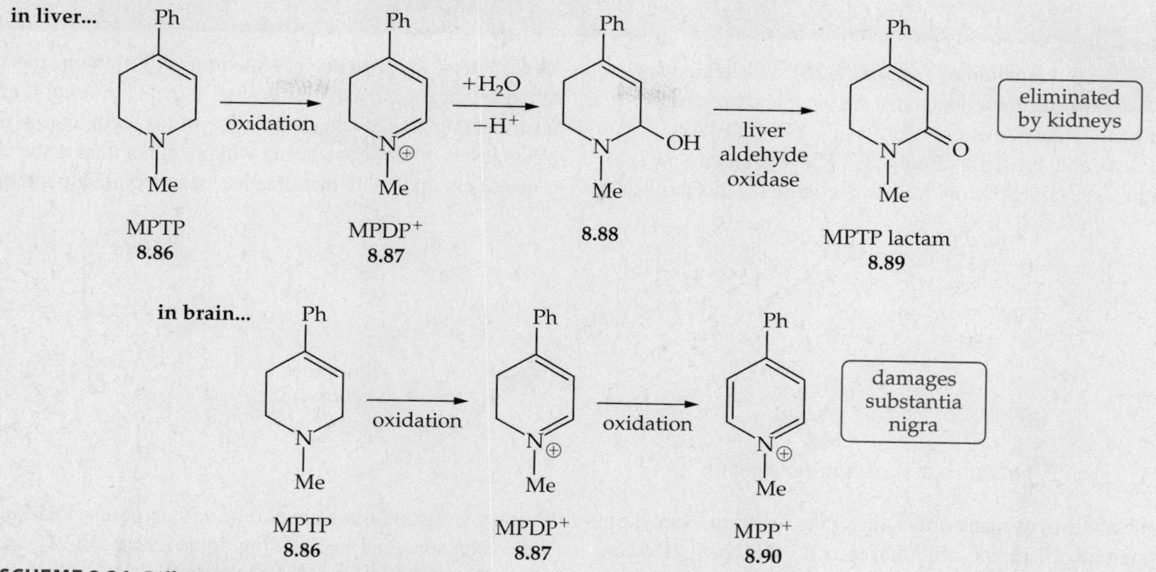

SCHEME 8.24 Different metabolic pathways of MPTP

not present in the brain. Instead, MPDP$^+$ is oxidized to 1-methyl-4-phenylpyridinium (MPP$^+$, **8.90**). Both MPDP$^+$ and MPP$^+$ are polar, charged species that cannot cross the blood–brain barrier. Therefore, once they form in the brain, MPDP$^+$ and MPP$^+$ are stuck. MPP$^+$ severely damages the substantia nigra, the same part of the brain that is damaged in Parkinson's patients.

MPPP is a "street drug" and not approved by the FDA. If MPPP were to be submitted for regulatory approval, the Parkinson's-related issues of MPPP would likely have been discovered early in animal trials. Regardless, the story of MPPP demonstrates the complexity of anticipating the possible metabolic pathways of drugs.

Inhibition of Metabolism by Drugs

Metabolic enzymes degrade compounds that are foreign to an organism. Because the structure of a foreign substance cannot be reasonably well anticipated, metabolic enzymes must have a fairly open active site that can accommodate a range of substrates. As a result, varied xenobiotics fit in the active site and can be metabolized. Just as an open active site can accept structurally diverse substrates, an open active site may be inhibited by molecules of highly varied structures. Inhibition of a metabolic enzyme is of great concern to drug discovery and is particularly important for patients who take more than one drug. Significant problems can arise if one drug or its metabolites inhibit the metabolism of a second drug.

Early in the drug discovery process when hits are being evaluated as potential leads, drug companies screen hits for their ability to inhibit CYP enzymes as well as monoamine oxidase and flavine-containing monooxidase systems. The CYP enzymes deserve special comment. Many proteins are found with multiple, similar variations. These variations are called *isoforms*. Of all the CYP enzymes, CYP3A4 is the most important and responsible for the metabolism of many drugs. A hit that strongly inhibits any metabolic enzyme, especially CYP3A4, is downgraded relative to other hits that are weak inhibitors or cause no inhibition. Leads are also screened for metabolic inhibition as they are developed.

Cimetidine[17]

Cimetidine (Tagamet, **8.26**), a histamine H_2-receptor antagonist for ulcer and heartburn treatment, inhibits several isoforms of CYP: CYP1A2, CYP2D6, and, most importantly, CYP3A4 (**Figure 8.5**). If a patient takes cimetidine as well as a drug that is metabolized by CYP3A4, the presence of cimetidine will slow the metabolism, clearance, and elimination of the other drug. If an unsuspecting patient takes a full dose of the second drug, then blood levels of the second drug will be higher than in the absence of cimetidine. Cimetidine has been reported to inhibit

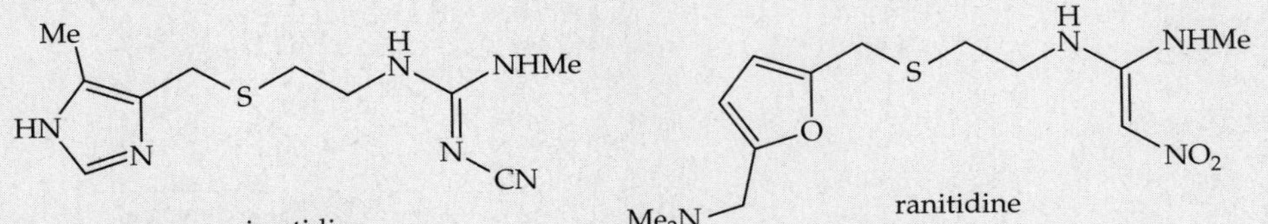

cimetidine
(Tagamet)
8.26

ranitidine
(Zantac)
8.91

FIGURE 8.5 H_2-receptor antagonists

the metabolism of many drugs, including warfarin, phenyltoin, propranolol, nifedipine, chlordiazepoxide, diazepam, lidocaine, theophylline, and metronidazole. These are all common drugs for conditions such as pain, depression, and elevated blood pressure. When cimetidine first reached the market in 1979, it represented a revolutionary treatment for heartburn and ulcers. However, patients with heartburn often have diet issues that cause hypertension and require blood pressure medication.

Problems of combining cimetidine with hypertension drugs were soon apparent. In 1981, ranitidine (Zantac, **8.91**), another H_2-receptor antagonist, reached the market. Ranitidine also inhibits metabolism, but only CYP2D6. Isoform 2D6 is less crucial in the metabolism of common drugs. Ranitidine took much market share from cimetidine because of cimetidine's relatively large number of drug interactions.

St. John's Wort[18,19]

Drug interactions are not limited to regulated drugs. St. John's Wort is an herb that is sold as a dietary supplement for the treatment of depression. Dietary supplements are regulated as a food in the United States and avoid the scrutiny of drugs. Supplements in the European Union are covered under the Food Supplements Directive, which is also less restrictive than drug regulation. Despite the fact that St. John's Wort is not classified as a drug, it does contain chemicals with pharmacological activity. Two of the active ingredients of St. John's Wort are believed to be hyperforin (**8.92**) and hypericin (**8.93**) (**Figure 8.6**).

Hyperforin has been shown to increase the activity of CYP3A4 and CYP2C9 by binding nuclear receptors that regulate gene expression of certain CYP-encoding genes.[18] As a result, St. John's Wort can have the opposite effect as cimetidine. Patients who take St. John's Wort in conjunction with their prescribed medicines may find that their

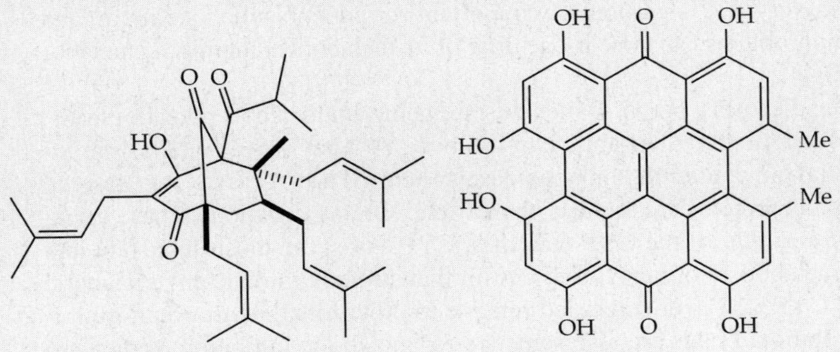

hyperforin
8.92

hypericin
8.93

FIGURE 8.6 Active compounds in St. John's Wort

medications are not effective because they are metabolized and eliminated faster than expected. Indinavir (Crixivan, **8.94**) was singled out by the FDA for concern on problematic interactions with St. John's Wort (**Figure 8.7**). Indinavir inhibits HIV protease and is metabolized by CYP enzymes. If an AIDS patient is taking both St. John's Wort and indinavir, the C_p of indinavir will likely be too low for effective management of an HIV infection.[19]

A general problem with dietary supplements is that they are not viewed as drugs. Of course, legally, they are indeed *not* drugs. Supplements, however, do contain chemicals with biological activity that may interact with prescription drugs. Because patients do not consider supplements as drugs, they may be less likely to mention them to their physicians. Furthermore, because supplements are not regulated as drugs, they are more likely to have undocumented drug interactions.

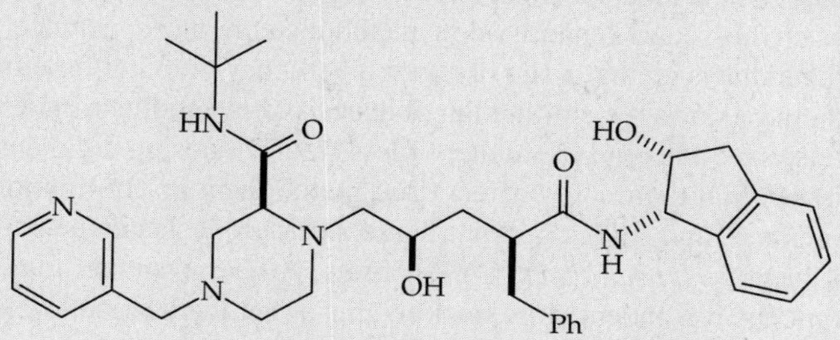

indinavir
(Crixivan)
HIV protease inhibitor
8.94

FIGURE 8.7

Population Variations

Not all patients are the same. Patients differ in their age, gender, genetics, and health. All these factors can play a role in drug metabolism. Metabolism rate is evaluated through the basic pharamacokinetic parameters of V_d, CL, k_{el}, and $t_{1/2}$ (Chapter 7). Of course, to show variability in metabolism, a drug must first be metabolized. Some drugs are eliminated unchanged.

Age

A patient's age can have significant impact on how a drug is metabolized. The baseline rate of metabolism is defined by adult patients who are otherwise healthy. In general, older patients have reduced metabolic function. In contrast, children and possibly adolescents have accelerated metabolism. Infants generally lack full liver function, and drugs commonly have longer $t_{1/2}$ values in newborns than adults. Age differences for a drug are normally found during phase II and phase III clinical trials, when the drug is introduced in a larger range of patients.

Alprazolam (Xanax, **8.95**) and lamivudine (Epivir, **8.96**) both show half-lives that vary with the age of the patient (**Figure 8.8**). Alprazolam, a drug for anxiety, has a $t_{1/2}$ of

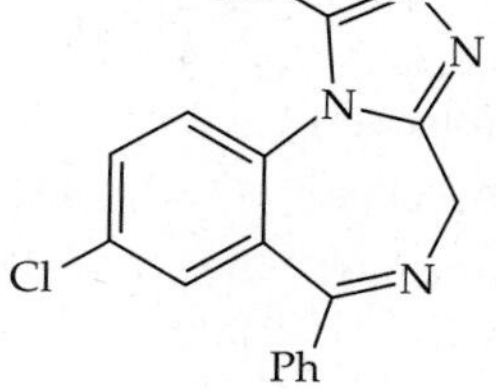

alprazolam
(Xanax)
anti-anxiety
8.95

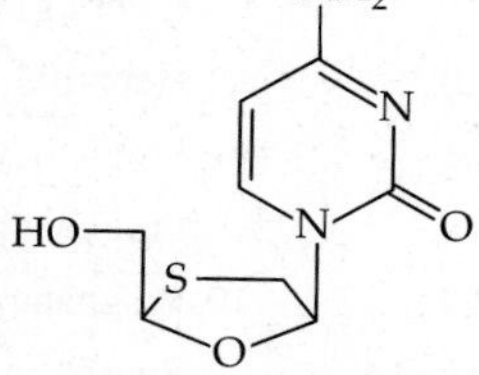

lamivudine
(Epivir)
*HIV reverse
transcriptase inhibitor*
8.96

FIGURE 8.8

11.2 h in adults and 16.3 h in geriatric patients. Lamivudine, an anti-HIV drug, has a half-life of only 2.0 h in pediatric patients (ages 4 months to 14 years). In adults with similar blood sampling, lamivudine has a $t_{1/2}$ of 3.7 h.[11]

If a drug's half-life varies according to the age of the patient, then the drug's dosing may need to be adjusted for each age group. However, drugs are normally developed with a wide therapeutic window so that all populations can receive the same dosing regimen.

Health

The health of a patient can have a dramatic impact on how effectively he or she eliminates a drug. The liver is the most important organ for metabolism. Therefore, liver conditions, such as cirrhosis and hepatitis, slow metabolism, reduce clearance, and lengthen half-life. While the kidneys are not a site of metabolism, they are vital for elimination. Any form of kidney damage slows the elimination of drugs and metabolites. Patients with diabetes or kidney infections often require smaller or less frequent dosing regimens. Other conditions or lifestyle issues that commonly affect drug metabolism or elimination include obesity, high blood pressure, and smoking. Smoking tobacco products can be particularly problematic because it increases, or *induces*, CYP activity. A heavy smoker may metabolize some drugs very quickly. If a patient quits smoking, his or her medication levels may need to be decreased.

CASE STUDY

Theophylline[11]

Theophylline (**8.97**) opens bronchial pathways in the lungs through a variety of mechanisms (**Figure 8.9**). The effect of age and patient health on the half-life of theophylline has been well documented. Selected data is shown in **Table 8.2**. The adult age category $(t_{1/2} = 8.7$ h) is bolded because it is the benchmark category against which all other groups are compared.

Several interesting and representative trends are apparent in the data of Table 8.2. In terms of age, infants tend to have poorly developed metabolic function, but children have very high metabolic rates. Metabolism slows in adulthood and continues to decrease in geriatric patients. With regard to health problems, impaired liver function has the most dramatic effect on the half-life of theophylline relative to all the listed conditions. The impact of pregnancy is modest until the fetus is large enough to put an appreciable burden upon its mother. Finally, thyroid hormones increase metabolism. Elevated or depressed thyroid activity often decreases or increases half-life, respectively.

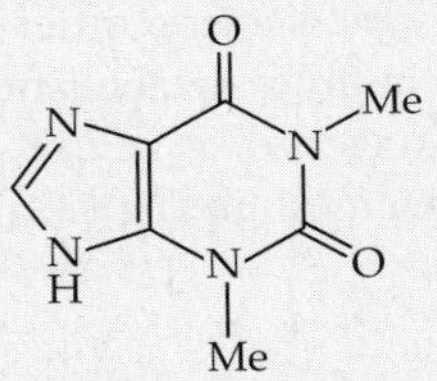

theophylline
anti-asthma
8.97

FIGURE 8.9

TABLE 8.2   Patient age and health effects on the half-life of theophylline

Age Effects		Health Effects	
Age	$t_{1/2}$ (h)	Condition	$t_{1/2}$ (h)
Infants (1–2 days)	25.7	Liver cirrhosis	32.0
Infants (3–30 weeks)	11.0	Hepatitis	19.2
Children (1–4 years)	3.4	Pregnancy (1st trimester)	8.5
Children (6–17 years)	3.7	Pregnancy (2nd trimester)	8.8
Adults (18–60 years)	**8.7**	Pregnancy (3rd trimester)	13.0
Elderly (>60 years)	9.8	Sepsis	18.8
		Hypothyroid	11.6
		Hyperthyroid	4.5

Gender

Gender would seem to be an obvious and important difference among patients, but men and women have essentially the same metabolic activity when compared on a per mass basis. One very significant difference between men and women is that women can be pregnant. As was shown in Table 8.2, pregnancy can significantly impact a patient's metabolism.

Genetics

Of all the factors that affect metabolism, genetics is the most problematic. A physician can gauge a patient's health through questioning and an exam. Age and gender may be readily determined. Genetics, however, are not so easily discerned. For the purposes of this chapter, the term "genetics" refers to either mutations in a patient's metabolic enzymes or the absence of specific enzymes in a patient's genome. The study of genetic variations and how they impact drug action has emerged as its own field, called *pharmacogenetics*.[20]

Humans have approximately 50 active CYP enzymes. The enzymes are divided into families, subfamilies, and isoforms by a number, letter, and another number, respectively. One isoform, CYP3A4, is responsible for nearly half of all drug metabolism in the body. CYP2D6 and CYP2C9 are two other commonly encountered metabolic enzymes. Many others are also active. Some of the less common enzymes are absent from the genome of certain patients. As a result, those patients metabolize some drugs at a very different rate relative to the general population.[20]

One example of a drug that shows genetic variability in its metabolism is terfenadine (Seldane, **8.98**) (**Scheme 8.25**). Both CYP3A4 and CYP2D6 metabolize terfenadine into fexofenadine (Allegra, **8.99**) through a relatively uncommon phase I oxidation of an unactivated carbon. Approximately 10% of the population has genetic variations in CYP2D6 that render the enzyme inactive toward terfenadine. These patients are classified as poor metabolizers. Other phenotypes include intermediate, extensive, and ultrarapid metabolizers. In poor metabolizers, CYP3A4 assumes full responsibility for terfenadine metabolism. If a poor metabolizer on terfenadine also takes a drug that inhibits CYP3A4, terfenadine will be slowly metabolized and reach higher C_p levels. Terfenadine can cause arrhythmias in patients. Indeed, terfenadine was withdrawn from the market because of its risk of arrhythmias. It was replaced by its metabolite, fexofenadine.[21]

Isoniazid (**8.43**) is another drug with variable metabolism (Scheme 8.14). Metabolism involves *N*-acetylation of the terminal nitrogen of isoniazid, and the reaction is catalyzed by arylamine *N*-acetyltransferase (NAT2).[22] The two known isoforms of NAT2 have different rates of acylation of isoniazid. Depending on their form of NAT2, patients are either fast or slow inactivators. The variations can largely be drawn along racial lines. Patients of Inuit or Japanese ancestry are mostly fast inactivators. Scandinavians, Ashkenazi Jews, and North African Caucasians tend to be slow inactivators. The population of the United

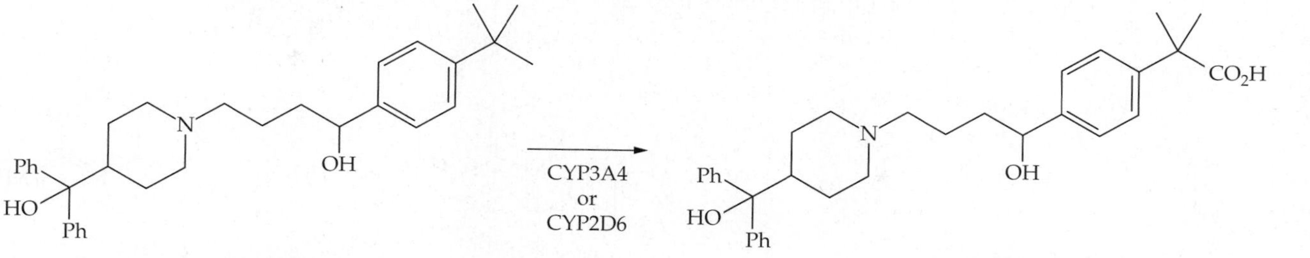

SCHEME 8.25 Metabolism of terfenadine

terfenadine
(Seldane)
seasonal allergies
8.98

CYP3A4
or
CYP2D6

fexofenadine
(Allegra)
seasonal allergies
8.99

States is an approximate 50/50 mixture of fast and slow inactivators. In fast inactivators, the $t_{1/2}$ of isoniazid is 70 minutes on average compared to 2–5 h for slow inactivators. Because isoniazid is a comparatively safe drug with a wide therapeutic window, all patients can be given the larger dose required for fast inactivators.[23]

Warfarin (**8.30**) shows variable metabolism based on genetics (Figure 8.9). CYP2C9 plays a large role in warfarin metabolism. Approximately 10% of the population has variant CYP2C9 genes that express less active forms of CYP2C9. Between 10 and 20% of whites and less than 5% of blacks and Asians have the less effective forms of CYP2C9. Most patients require 2–10 mg per day of warfarin to maintain therapeutic levels. Patients with poorly active CYP2C9 require only 1.5 mg or less. Because the therapeutic window of warfarin is narrow, and warfarin can quickly reach toxic levels, patients begin with a low-dose regimen of warfarin. If the patient's blood levels of warfarin are too low, the patient likely has an active form of CYP2C9 and will be stepped up to a higher dose.[11]

In the course of evaluating hits, a drug discovery team screens compounds against different isoforms of CYP enzymes. If a hit is metabolized by a CYP form known to be highly variable, then that hit may show genetic variations across population groups. Genetic variations complicate dosing regimens, so the hit will be downgraded relative to others.

8.4 Prodrugs

Not all drugs are active in their administered form. These drugs undergo reactions in the body to afford metabolites that are ultimately responsible for the observed biological activity. Drugs that fit this profile are called *prodrugs*. Pharmaceutical companies may develop a prodrug if the original drug is active but poorly available. The prodrug incorporates additional structure or functionality to help it reach its site of action. Once in the body, the prodrug is metabolized into the active drug form (**Scheme 8.26**).

Angiotensin-converting enzyme (ACE) inhibitors are a class of antihypertensives that are almost all prodrugs. Enalaprilat (**8.100**) is one of the original ACE inhibitors (**Figure 8.10**). The ethyl ester of enalaprilat (**8.101**) increases the compound's bioavailability ($F = 60\%$). In the body esterases cleave, the ester to form enalaprilat (**8.100**), the active ACE inhibitor. Enalapril is not orally available ($F = 0\%$), but it is an effective drug if injected. Similarly, valacyclovir (Valtrex, **8.102**), an antiviral, is a prodrug for acyclovir (Zovirax, **8.103**). Attachment of a valine residue to the pseudo 5′-OH group of acyclovir allows valacyclovir to exploit active transport systems in the body. The valine is later cleaved by esterases to afford acyclovir. Inclusion of valine in the molecular structure raises the bioavailability from 10–20% (acyclovir) to 55% (valacyclovir).[11]

Cyclophosphamide (Cytoxan, **8.104**) is an anticancer prodrug that is oxidized by the CYP2B enzyme family in the liver (**Scheme 8.27**) and was first introduced back in Chapter 6. Oxidation occurs adjacent to the ring nitrogen to afford 4-hydroxycyclophosphamide (**8.105**), which is transported throughout the body by the circulatory system. Compound

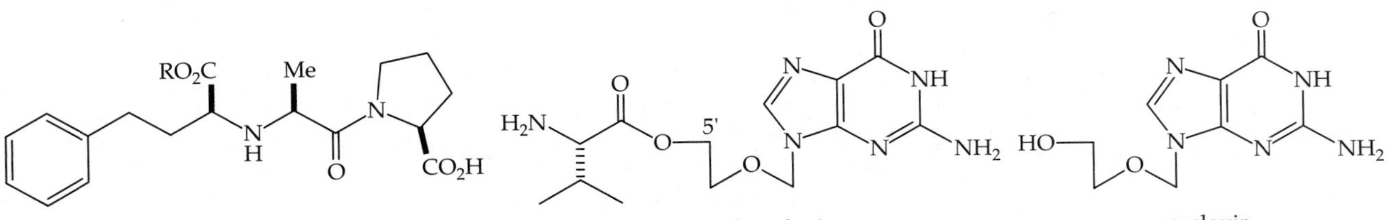

$$\text{prodrug (inactive)} \xrightarrow[\text{enzymes}]{} \text{drug (active)}$$

SCHEME 8.26 Prodrug design

R = H, enalaprilat, **8.100**
(Vasotec)
antihypertensive
R = Et, enalapril, **8.101**

valacyclovir
(valtrex)
antiherpes
8.102

acyclovir
(Zovirax)
antiHIV
8.103

FIGURE 8.10 Prodrugs and their active metabolites

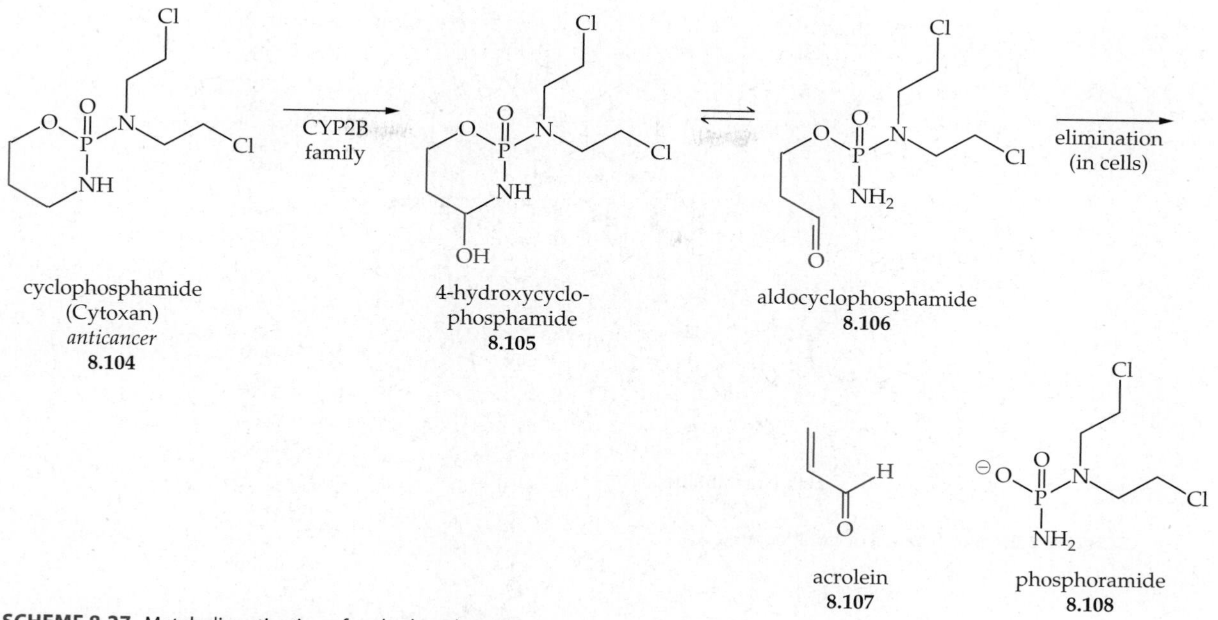

SCHEME 8.27 Metabolic activation of cyclophosphamide

8.105 is in equilibrium with aldocyclophosphamide (**8.106**). In tumor cells, **8.106** undergoes an elimination to yield acrolein (**8.107**) and phosphoramide (**8.108**). Phosphoramide is the active metabolite. The low oral bioavailability of phosphoramide necessitates the use of the more orally available cyclophosphamide ($F = 75\%$).[24]

Prodrugs may appear to be an excellent method for increasing the oral bioavailability of a drug. Indeed, marketable drugs include many examples of successful prodrugs. However, prodrugs are not favored as a general solution. Prodrugs introduce another layer of complexity into the overall picture of a drug. A more desirable solution is the introduction of one or more metabolically stable groups to a nonpharmacophore region of the molecule. The groups may need to be polar or nonpolar, depending on the situation.

The following Case Study demonstrates the challenges that administration of a prodrug can pose to a drug discovery program.

CASE STUDY

Clopidogrel and Prasugrel[25,26]

Clopidogrel (Plavix, **8.109**) acts as an antagonist for receptors that trigger the activation of platelets and therefore affects the coagulation of blood (**Scheme 8.28**). Clopidogrel, in its administered form, is inactive. The drug is activated by metabolism, primarily through the action of CYP2C19. CYP2C19 oxidizes the thiophene ring to form compound **8.110**. The alcohol **8.110** undergoes further reactions, the thiophene ring opens, and compound **8.111** results. This product is the active antagonist that arises from ingestion of clopidogrel.[25]

Clopidogrel has been found to be ineffective for certain segments of the population. Between 2% and 14% of the population in the United States has been estimated to have a variation of CYP2C19 that renders the enzyme less active toward the oxidation of clopidogrel. In this subgroup of patients, clopidogrel is not effectively converted into the active metabolite (**8.111**), and therefore use of clopidogrel has little impact on platelet activity. For this reason, in 2010 the FDA ruled that the prescribing information insert for clopidogrel must carry a *black box warning* advising patients of the possible ineffectiveness of the drug. A black box warning is a means of notifying patients that a drug can cause severe or life-threatening consequences. In the case of clopidogrel, the consequence is that the drug may be ineffective, allow the formation of clots, and subsequently result in a stroke or heart attack.[26]

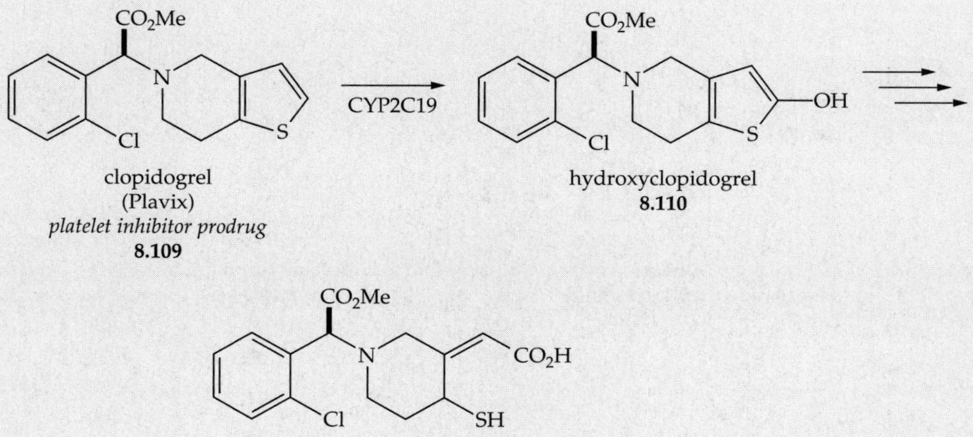

clopidogrel
(Plavix)
platelet inhibitor prodrug
8.109

hydroxyclopidogrel
8.110

active metabolite
8.111

SCHEME 8.28 Bioactivation of clopidogrel

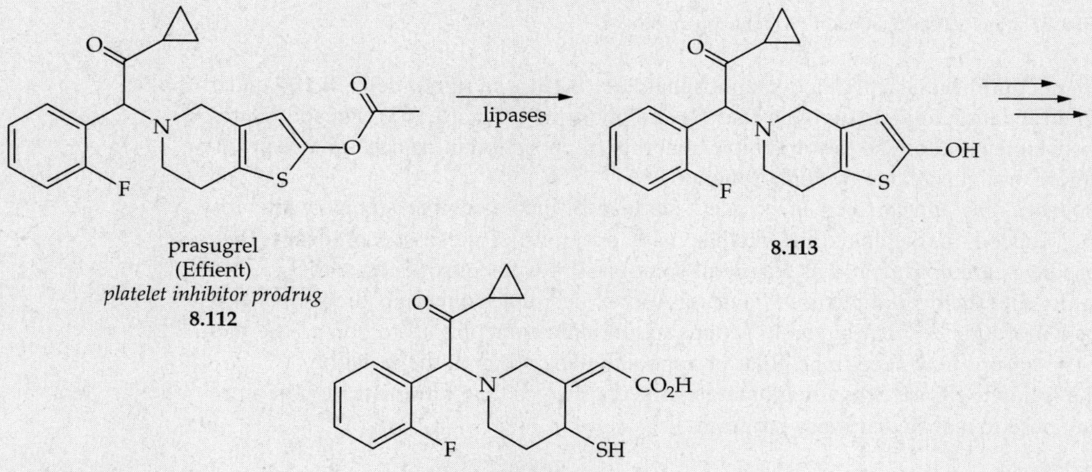

prasugrel
(Effient)
platelet inhibitor prodrug
8.112

8.113

active metabolite
8.114

SCHEME 8.29 Bioactivation of prasugrel

Another platelet inhibitor prodrug is prasugrel (Effient, **8.112**) (**Scheme 8.29**). Prasugrel has a structure that is strikingly similar to that of clopidogrel. The key difference is that the thiophene ring carbon in prasugrel is already in the same oxidation state as the corresponding carbon in hydroxyclopidogrel (**8.110**). Instead of bearing a hydroxyl group, the thiophene ring bears an acetoxy group. In the body, the ester of the acetoxy group is hydrolyzed by lipases to reveal a hydroxyl group (**8.113**). Compound **8.113** then undergoes multiple reactions to form **8.114**, the biologically active form of prasugrel.[25]

Because prasugrel is activated by lipases that show essentially no reaction variability across different populations, prasugrel displays a more uniform level of activity than CYP2C19-dependent clopidogrel. The more predictable behavior of prasugrel allows it to avoid the undesirable black box warning carried by clopidogrel.

Summary

In tandem with elimination, metabolism is the means by which the body clears xenobiotics from its system. Most metabolites are more water soluble than the original compound and therefore more readily eliminated by the kidneys. Metabolic reactions are catalyzed by enzymes and divided into two categories: phase I and phase II. Phase I reactions include oxidations, reductions, and hydrolyses. Oxidations, the most common phase I reactions, cleave alkyl groups from heteroatoms, introduce hydroxyl groups on aromatic rings, and form oxides of amines and sulfides. Phase II reactions attach very polar groups to compounds to increase water solubility. Polarizing groups include sulfate, glucuronic acid, glutathione, and amino acids. Ideally, a drug's metabolites should be inactive and simply eliminate. In reality, metabolites often display either the same activity as the original drug or completely different activity. Metabolites with different activity frequently are responsible for side effects, including toxic and adverse reactions. One of the most problematic effects of drugs and their metabolites is that they can sometimes inhibit metabolic enzymes. Metabolism inhibition often causes other drugs in a patient's system to have extended half-lives and build to toxic levels. Metabolism of a specific drug is not uniform across all patients. A patient's age, health, and genetics all impact how a patient clears a drug. Age and health effects are readily discovered during clinical trials of the drug. Genetic factors are more elusive and are often not fully understood until after a drug is launched. In some instances, metabolic reactions can be exploited to form a biologically active metabolite from an inactive drug, called a prodrug. Prodrugs are useful when the active species has poor pharmacokinetic properties. Because they introduce an additional layer of complexity in the development process, prodrugs are not the favored method of drug development. Prodrugs can, however, be a workable solution for a compound with good activity (pharmacodynamics) yet poor availability (pharmacokinetics).

Questions

1. The following figures represent dextromethorphan (**8.a**) and several of its known metabolites: dextrorphan (**8.b**) (primary metabolite), dextrorphan *O*-sulfate (**8.c**), and 3-hydroxymorphinan (**8.d**). List the different reactions (in their likely order) that dextromethorphan underwent to form each metabolite.

dextromethorphan
8.a

dextrorphan
8.b

dextrorphan
O-sulfate
8.c

3-hydroxymorphinan
8.d

2. Turn to Appendix A and pick five different drugs to demonstrate the following phase I metabolic processes: oxidation (dealkylation, arene/alkene, heteroatom), reduction, and hydrolysis. Be sure that your original drugs contain functional groups that will allow you to highlight the metabolic reaction.

3. Repeat question 2 but select five different drugs to demonstrate the following phase II processes: acetylation, sulfonylation, glucuronidation, conjugation with an amino acid, and conjugation with glutathione.

4. Many drugs show genetic differences in their metabolism. Almost universally, the population segment with a slower rate of metabolism also excretes a higher percentage of the unchanged drug in the urine. Explain why this observation is not surprising.

5. Bicalutamide (**8.e**) is a drug used to treat certain forms of prostate cancer. The drug is administered in racemic form, and the two enantiomers behave differently in the body. The inactive *S*-enantiomer undergoes rapid glucuronidation. The active *R*-enantiomer is slowly oxidized. Draw likely structures of both metabolites.

bicalutamide
8.e

6. The most important organs for drug clearance are the kidneys and liver. Table 8.2 shows how a patient's age and different health conditions affect the half-life of theophylline. Based on Table 8.2, which organ is more important for the clearance of theophylline: the kidneys or liver?

7. Scheme 8.1 indicates that some metabolic reactions are reversible. Give at least two examples of metabolic reactions, phase I or II, that should be reversible.

8. Metabolism typically increases a molecule's polarity, and compounds with a higher polarity generally undergo renal elimination more quickly (have a shorter half-life). Why does increasing polarity normally decrease half-life? Answer this question with respect to V_d. Equation 7.12 may be helpful.

9. Metabolism typically increases a molecule's polarity, and compounds with a higher polarity generally undergo renal elimination more quickly (have a shorter half-life). Why does increasing polarity normally decrease half-life? Answer this question with respect to *CL*. Equations 7.9 and 7.12 may be helpful.

10. Liver damage is especially relevant for orally delivered drugs. Why?

References

1. Gibson, G. G., & Skett, P. *Introduction to Drug Metabolism* (2nd ed.). London: Blackie Academic & Professional, 1994.

2. Bhakta, M. N., & Wilmalasena, K. A Mechanistic Comparison between Cytochrome P_{450}- and Chloroperoxidase-Catalyzed *N*-Dealkylation of *N,N*-Dialkyl Anilines. *Eur. J. Org. Chem.* 2005, 4801–4805.

3. Van Houten, K. A., Kim, J. M., Bogdan, M. A., Ferri, D. C., & Mariano, P. S. A New Strategy for the Design of Monoamine Oxidase Inactivators: Exploratory Studies with Tertiary Allylic and Propargylic Amino Alcohols. *J. Am. Chem. Soc.* 1998, *120*, 5864–5872.

4. Lin, G., Chu, K.-W., Damani, L. A., & Hawes, E. M. Identification of lactams as in vitro metabolites of piperidine-type phenothiazine antipsychotic drugs. *J. Pharmacol. Biomed. Anal.* 1996, *14*, 727–738.

5. Guroff, G., Daly, J. W., Jerina, D. M., Renson, J., Witkop, B., & Udenfriend, S. Hydroxylation-Induced Migration: The NIH Shift: Recent Experiments Reveal an Unexpected and General Result of Enzymatic Hydroxylation of Aromatic Compounds. *Science.* 1967, *157*, 1524–1530.

6. Fleming, C. M., Branch, R. A., Wilkinson, G. R., & Guengerich, F. P. Human Liver Microsomal N-hydroxylation of Dapsone by Cytochrome P-4503A4. *Mol. Pharmacol.* 1992, *41*, 975–980.

7. Holt, D. E., & Bajoria, R. The Role of Nitro-Reduction and Nitric Oxide in the Toxicity of Chloramphenicol. *Hum. Exp. Toxicol.* 1999, *18*, 111–118.

8. Adebayo, G. I., Williams, J., & Healy, S. Aspirin Esterase Activity: Evidence for Skewed Distribution in Healthy Volunteers. *Eur. J. Intern. Med.* 2007, *18*, 299–303.

9. Fevery, J., Leroy, P., Van de Vijver, M., & Heirwegh, K. P. M. Structures of Bilirubin Conjugates Synthesized in vitro from Bilirubin and Uridine Diphosphate Glucuronic Acid, Uridine Diphosphate Glucose or Uridine Diphosphate Xylose by Preparations from Rat Liver. *Biochem. J.* 1972, *129*, 635–644.

10. Middleton, D. The Role of the Medicinal Chemist in Drug Discovery: Controlling the Controllables. 230th National Meeting of the American Chemical Society, 2005 August 28–September 1, Washington, DC.

11. Murray, L., Sr. (Ed.). *Physician's Desk Reference* (58th ed.) Montvale, NJ: Thomson PDR, 2004.

12. Orsulak, P. J., Kenney, J. T., Debus, J. R., Crowley, G., & Wittman, P. D. Determination of the Antidepressant Fluoxetine and Its Metabolite Norfluoxetine in Serum by Reversed-Phase HPLC, with Ultraviolet Detection. *Clin. Chem.* 1988, *34*, 1875–1878.

13. Fan, P. W., & Bolton, J. L. Bioactivation of Tamoxifen to Metabolite E Quinone Methide: Reaction with Glutathione and DNA. *Drug Metab. Disp.* 2001, *29*, 891–896.

14. Klaassen, C. D. Principles of Toxicology and Treatment of Poisoning. In J. G. Hardman & L. E. Limbird (Eds.), *Goodman and Gilman's The Pharmacological Basis of Therapeutics* (10th ed., Chapter 4). New York: McGraw-Hill, 2001.

15. Wallis, C. Surprising Clue to Parkinson's. *Time,* April 8, 1985, pp. 61–62.

16. Staack, R. F., & Maurer, H. H. Metabolism of Designer Drugs of Abuse. *Curr. Drug Metab.* 2005, *6*, 259–274.

17. Ganellin, G. R., & Durant, G. J. Histamine H_2-Receptor Agonists and Antagonists. In M. E. Wolff (Ed.), *Burger's Medicinal Chemistry* (4th ed., Chapter 48). New York: Wiley & Sons, 1981.

18. Kliewer, S. A., Goodwin, B., & Willson, T. M. The Nuclear Pregnane X Receptor: A Key Regulator of Xenobiotic Metabolism. *Endocr. Rev.* 2002, *23*, 687–702.

19. Henney, J. E. Risk of Drug Interactions with St. John's Wort. *J. Am. Med. Assn.* 2000, *283*, 1679.

20. Phillips, K. A., Veenstra, D. L., Oren, E., Lee, J. K., & Sadee, W. Potential Role of Pharmacogenomics in Reducing Adverse Drug Reactions. *J. Am. Med. Assn.* 2001, *286*, 2270–2279.

21. Jones, B. C., Hyland, R., Ackland, M., Tyman, C. A., & Smith, D. A. Interaction of Terfenadine and its Primary Metabolites with Cytochrome P450 2D6. *Drug Met. Disp.* 1998, *26*, 875–882.

22. Wormhoudt, L. W., Commandeur, J. N., & Vermeulen, N. P. Genetic Polymorphisms of Human *N*-acetyltransferase, Cytochrome P450, Glutathione-S-transferase, and Epoxide Hydrolase Enzymes: Relevance to Xenobiotic Metabolism and Toxicity. *Crit. Rev. Toxicol.* 1999, *29*, 59–124.

23. Petri, W. A., Jr. Antimicrobial Agents. In J. G. Hardman & L. E. Limbird (Eds.), *Goodman and Gilman's The Pharmacological Basis of Therapeutics* (10th ed., Chapter 48). New York: McGraw-Hill, 2001.

24. Chabner, B. A., Ryan, D. P., Luiz, P.-A., Garcia-Carbonero, R., & Calabresi, P. Antineoplastic Agents. In J. G. Hardman & L. E. Limbird (Eds.), *Goodman and Gilman's The Pharmacological Basis of Therapeutics* (10th ed., Chapter 52). New York: McGraw-Hill, 2001.

25. Brandt, J. T., Close, S. L., Iturria, S. J., Payne, C. D., Farid, N. A., Ernest, C. S. II, et al. Common Polymorphisms of CYP2C19 and CYP2C9 Affect the Pharmacokinetic and Pharmacodynamic Response to Clopidogrel but Not Prasugrel. *J. Thromb. Haemost.* 2007, *5*, 2429–2436.

26. U.S. Food and Drug Administration (FDA). FDA Announces New Boxed Warning on Plavix. http://www.fda.gov/NewsEvents/Newsroom/PressAnnouncements/ucm204253.htm (accessed April 2012).

Chapter 9

Molecular Structure and Diversity

Chapter Outline

9.1 Determining Target Structure
- Literature
- X-Ray Crystallography
- NMR Spectroscopy
- Molecular Modeling

9.2 Complementarity between a Target and Drug
- Intermolecular Forces
- Molecular Shape
- Drug Pharmacophore

9.3 Searching for Drugs
- Diversity and Molecular Space
- Privileged Structures

9.4 Combinatorial Chemistry
- Parallel Synthesis
- Split Synthesis
- Advantages and Disadvantages

Learning Goals for Chapter 9

- Understand how knowing a target's structure assists the drug discovery process
- Identify the complementary roles of x-ray crystallography, NMR, and modeling for assigning a target's structure
- Know the intermolecular forces most relevant for target-drug binding
- Understand the structural importance of a drug's pharmacophore
- Appreciate the structural diversity of drug-sized molecules
- Distinguish between different combinatorial chemistry approaches

Molecular structure is a topic that has been absent in this text so far. Activity has been the main discussion point—activity of enzymes, receptors, and oligonucleotides as well as pharmacokinetic activity of a molecule in a living organism. This chapter introduces the ideas behind molecular structure. Structural determination of enzymes and receptors, the common targets of drugs, is covered first. X-ray crystallography is the most important tool for determining a protein's structure, but nuclear magnetic resonance (NMR) spectroscopy and molecular modeling are also useful. The structure of a target in turn drives the structure of possible drugs. Most drugs interact with targets through intermolecular forces: hydrogen bonding, ionic bonding, dipole forces, and contact forces. Ideally, a drug contains appropriate functionality positioned to maximize intermolecular forces between the drug and target. While the process of designing a drug is easy to put into words, knowing how to start is challenging. The number of molecules that may be considered as possible drugs is essentially infinite. To sample the endless potential structural diversity of biologically active molecules, pharmaceutical companies synthetically prepare large collections of molecules. By screening these collections for activity against targets, drug companies search for compounds that might be modified into viable drugs. Once a structure has been linked with biological activity, the drug discovery group will closely follow the relationship of structure and activity in the search for a new drug.

9.1 Determining Target Structure

Drugs bind targets. The structure of a drug and its target are therefore closely related. If one understands the structure of the target, then important elements of the drug can generally be inferred. For this reason, the discovery team seeks out as much information on

214

the target as possible. While this section on target structure often plays a prominent role in both lead discovery (Chapter 10) and lead optimization (Chapters 11 and 12), learning about the target structure typically occurs much earlier.

This section deals exclusively with protein targets: enzymes and receptors. Is not the structure of oligonucleotides also important? It certainly is. The difference between proteins and oligonucleotides, especially DNA, is that proteins are far less predictable in terms of structure. An essentially universal structure like the double helix of DNA has no parallel in the world of proteins. Considerable research effort is expended on determining the three-dimensional structure of target proteins.

Literature

The medicinal chemistry team is not in a vacuum when determining a target's structure. First, the team has the benefit of the research performed by the molecular biology group in analyzing the target and preparing any assays. Second, the chemical, biological, and medical literature may afford some useful clues for the target. If the target is known in the literature, other researchers may have reported data on the target structure and perhaps the binding of small molecules to the target. Even weak binding data can be valuable information. Of course, results in the literature are not all of the same quality. The discovery team will need to evaluate the reliability of the information. Because of the novel and cutting edge nature of drug research, worthwhile data from the literature will likely be sparse.

X-Ray Crystallography

X-ray crystallography is the most important method for determining protein structure. X-ray crystallography is a technique by which an x-ray beam is scattered upon contact with a crystallized protein. The pattern of scattering can be used to determine the relative position of the atoms in the crystal. The quality of an x-ray crystal structure is described in terms of its *resolution*. Resolution is listed in units of angstroms (Å). A high resolution for an x-ray structure might be 0.5–2.0 Å with a lower number indicating better resolution. Once the data has been processed, the positions of nonhydrogen atoms appear as spheres in three-dimensional space in an image called an *electron density map*. The electron density map shows atoms as spheres.

Figure 9.1 shows the impact of resolution on the readability of an electron density map. Image (a) has a resolution of 0.62 Å. Atoms are nearly distinct and separate from other atoms. Image (b) has a resolution of 1.2 Å. Atoms are still clear but somewhat larger. As the resolution drops to 2.0 Å (c), the spheres of the individual nuclei become unclear. The feature that now dominates the electron density map is the peptide backbone. Once the resolution drops to 3.0 Å (d) and 4.0 Å (e), single atom branches off the backbone are nearly imperceptible. At 5.0 Å (f), folds of the backbone can mistakenly appear to merge together.

Solving an x-ray structure requires correctly assigning atoms in the protein to match the electron density map. Before solving a structure, an x-ray crystallographer already has the sequence of the target protein from the molecular biology team. The x-ray crystallographer begins to solve the structure by aligning the most obvious residues, such as the aromatic rings, into the electron density map. Less obvious residues then fall into place as possibilities become more limited. If an electron density map has a low resolution (6–8 Å), then only general features such as α-helices or β-sheets may be clear. The confident placement of individual residues is not possible.

A protein will generally have one or more random coil regions in its overall structure Random coils, as the name suggests, do not have a regular, folded structure and undergo

(a) 0.62 Å resolution

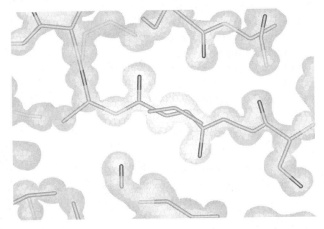

(b) 1.2 Å resolution

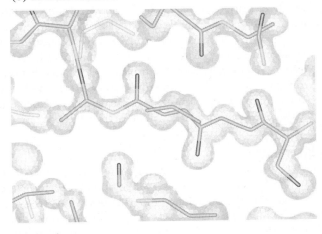

(c) 2.0 Å resolution

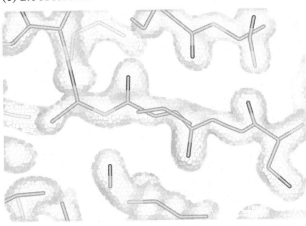

(d) 3.0 Å resolution

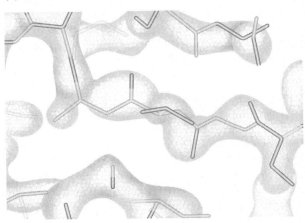

(e) 4.0 Å resolution

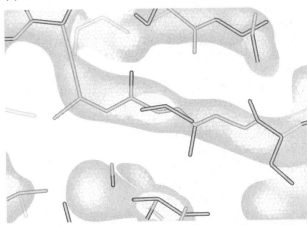

(f) 5.0 Å resolution

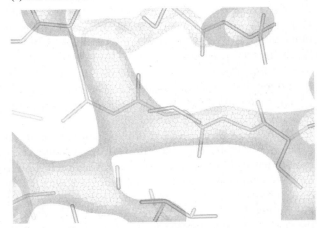

FIGURE 9.1 Electron density maps of varying resolution. (Used by permission of Dr. Paul Emsley, Department of Biochemistry, University of Oxford.)

rapid, continuous conformational changes. These regions appear indistinct even in an otherwise high resolution x-ray structure. Fortunately, binding and active sites generally occupy ordered protein regions that are amenable to characterization by x-ray crystallography.

Knowing the structure and shape of the binding site of a target allows the discovery team to make informed decisions about molecules that will bind to the target. Is the binding site deep or shallow? Are the amino acid residues in the binding site polar, nonpolar, or charged? The properties of the binding site determine the optimal, complementary properties needed in a drug that will interact with the binding site. Often molecules can be cocrystallized with an enzyme target. The x-ray structure of the complex shows both how a compound fills the binding pocket and precisely where a compound may be modified to better fill the active site.

NMR Spectroscopy

The value of a quality x-ray structure to the discovery team cannot be overestimated. However, an x-ray structure is not always available. In this situation, the structure of a protein target can sometimes be estimated, at least partially, by NMR. NMR-elucidated structures have one clear advantage over x-ray structures. NMR data is obtained on a protein that is dissolved in an aqueous solution, just as proteins exist in a living organism. X-ray data is based on proteins in the solid state, so the effect of solvent is less well documented in the resulting structure.

Proteins that are to be studied by NMR must be specially prepared. Natural proteins almost exclusively contain ^{12}C and ^{14}N, neither of which is an NMR-active isotope. For analysis, the protein must be enriched with ^{13}C and ^{15}N. Isotopically-enriched proteins can be isolated from bacteria that are grown in the presence of ^{13}C-enriched glucose and $^{15}NH_4Cl$. The resulting proteins are then handed off to the NMR experts.

Most protein NMR data falls into two categories: dihedral angles and interatomic distances. A computer model that accommodates the observed angles and distances is then constructed. More angles and distances from the NMR data provide more constraints for the structural model. Exactly how the angles and distances are determined by NMR is beyond the scope of this text. Suffice it to say that true experts in the field of NMR, sometimes called NMR jocks, are adept at coaxing a remarkable amount of information from two- and three-dimensional NMR techniques. As more powerful magnets with greater resolving power become available, NMR specialists will be able to gain useful information on larger, more complex proteins.

The NMR data do not form a structure alone. The NMR data are used to guide and refine a molecular model.

Molecular Modeling

The primary sequence of the target may be sufficient information for the discovery team to piece together a workable model of the target. Proteins with 40% or more overlap in their primary sequences are said to be *homologous* and often fold in a similar fashion. Therefore, if a protein with both high homology to the target and known folding can be found, the folding of the target may be inferred by analogy. In a computer simulation, the target is aligned with the literature protein as closely as possible. NMR data, if available, provides restrictions for further assistance. The structural energy of the aligned, restricted target is then minimized through molecular mechanics calculations. The final modeled structure represents an estimate of the true folding of the target protein. The entire process is called *homology* or *comparative modeling*.[1]

Experimental protein folding data based on both x-ray crystallographic and NMR results are available online through the Protein Data Bank (PDB, http://www.rcsb.org/pdb).

FIGURE 9.2 Growth of experimental structures included in the PDB

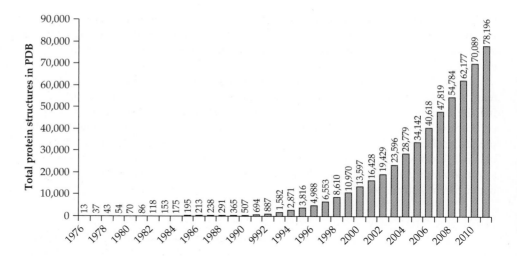

Figure 9.2 shows the historical growth of the number of structures included in the database. The number of PDB structures covers only approximately 2% of known proteins. Regardless of this "sequence-structure gap," the proteins included in the database contain sufficient diversity in their structures and sequences to allow reasonable estimation of folding for many protein targets. Homology modeling is more useful for studying enzymes than receptors, particularly the membrane-bound receptors, which have fewer examples in the PDB. The value of homology modeling will continue to increase as both the PDB grows in content and the field of molecular modeling improves.

The following Case Study shows an example of how comparative modeling can readily lead to the development of a lead structure.

CASE STUDY

Protease Inhibition in Coronaviruses[2,3]

AG7088 (**9.1**) was discovered using homology modeling. AG7088 is a weak inhibitor of an important protease for the activity of a coronavirus that was responsible for the 2003 outbreak of severe acute respiratory syndrome (SARS) (**Figure 9.3**). In the search for how to fight the SARS virus, researchers focused on a protease called M^{pro}, which is common to all coronaviruses with high sequence homology. A peptide inhibitor was known for M^{pro} of another coronavirus, transmittable gastroenteritis virus. The peptide inhibitor was unattractive

AG7088
9.1

AG7122
9.2

FIGURE 9.3 Early hits for inhibition of SARS-coronavirus M^{pro} [2,3]

because small proteins have poor oral availability. Further research showed similarities between the x-ray crystal structure of the peptide-bound M^{pro} and the AG7088-bound protease of a completely unrelated virus. Because AG7088 inhibits a viral protease and the inhibited complex resembles M^{pro} of transmittable gastroenteritis virus, AG7088 was proposed to be a possible inhibitor of M^{pro} for SARS coronavirus.[2] Indeed, AG7088 showed activity against SARS coronavirus M^{pro}. Related structures, such as AG7122 (**9.2**), have shown to be even more potent.[3]

The research on SARS-coronavirus M^{pro} is actually an example of double homology modeling. Insight into SARS coronavirus M^{pro} was gained by analogy to transmittable gastroenteritis virus M^{pro}, which was in turn better understood by analogy to the protease of an unrelated virus.

Determining protein structures can be a highly speculative process. Just because a structure has been assigned does not mean it is accurate. One simple check for validity of the assigned structure involves validating the dihedral angles of the backbone of the target protein.[4] The conformation of each residue in a protein backbone can be defined by three different dihedral angles: φ, ψ, and ω (**Figure 9.4**). The angles range in value from -180 to $+180°$ and are defined by the atoms in the backbone: C$'$ (carbonyl carbon), C$^\alpha$ (carbon adjacent to the carbonyl), and N. Of the three angles, ω is the least useful. Conjugation of the nitrogen lone pair into the π-system of the carbonyl essentially locks ω, the dihedral angle between adjacent C$^\alpha$ atoms, at 0 or 180°, normally 180°. Different secondary structures show preferred angle ranges for φ and ψ. In an α-helix, both φ and ψ have values near $-50°$. Within a β-sheet, φ tends to fall in a range of -50 to $-160°$, while ψ lies between $+100$ and $+180°$. Graphically, angles φ and ψ for all the residues in a protein can be plotted with the x-axis as φ and the y-axis as ψ. This is called a Ramachandran plot (**Figure 9.5**). For a given protein, the data points naturally cluster into core regions defined by α-helices, both right- and left-handed, and β-sheet secondary structures. Outliers from these regions do occur, but they are almost always small in number. If the Ramachandran plot of a proposed protein structure shows too many points outside the α-helix and β-sheet regions, the structure may need to be reexamined.[4]

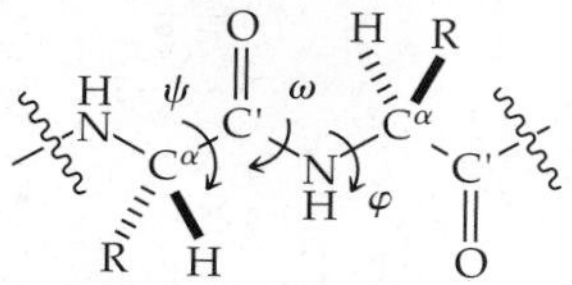

FIGURE 9.4 Dihedral angles defined in a polypeptide backbone

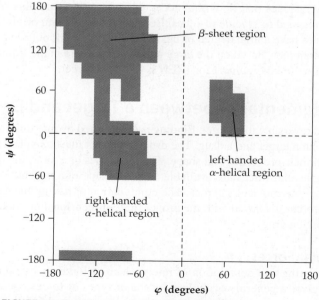

FIGURE 9.5 Typical regions in a Ramachandran plot

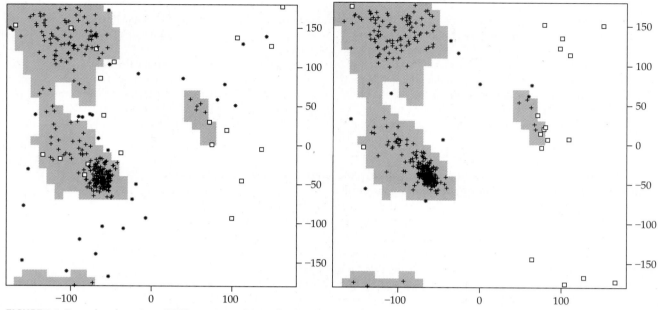

FIGURE 9.6 Ramachandran plots of PDB structures 1ZEN and 1B57 (chain A)

Despite advances in x-ray crystallographic techniques and computer modeling, incorrectly assigned protein structures are encountered in the literature. Inevitably, mistakes have also been included in the PDB archives. The Ramachandran plots for two archived x-ray structures of fructose 1,6-bisphosphate aldolase are shown in **Figure 9.6**. Both plots were generated by the Electron Density Server at Uppsala University (http://eds.bmc. uu.se/eds/ramachan.html).[5] The plot for PDB entry 1ZEN[6] shows considerably more outliers (9.8%) than 1B57[7] (2.9%). For a high-resolution x-ray structure (resolution ≤2.0 Å), outliers should comprise ≤5% of the plotted residues. Outliers are denoted with an asterisk (*), while residues that lie within accepted core regions are plus signs (+). Open squares (□) correspond to glycine residues. Because the side chain of glycine is a hydrogen atom, glycines have a high degree of flexibility and often fall outside the core regions of a Ramachandran plot. Based on the large number of outliers in the Ramachandran plot of 1ZEN, the structural assignment for 1ZEN is likely incorrect.

9.2 Complementarity between a Target and Drug

Most drugs bind reversibly to a target. Binding is determined by maximizing intermolecular forces between a target and a drug. The drug and target must complement one another in terms of both their properties and shape. The properties of a molecule determine what types of intermolecular forces are available, and the shape of a molecule determines the placement of the intermolecular forces. The combination of having the correct available intermolecular forces displayed with the proper three-dimensional orientation defines the *pharmacophore* of a drug.

Intermolecular Forces

The following listing and description of intermolecular forces is by no means exhaustive. It merely gives a general vocabulary for major types of forces encountered in medicinal chemistry discussions. The approximated energies associated with the various intermolecular forces in this section are net binding energies—the difference between a molecule being dissolved in water and being bound by a target through the intermolecular force in question. The highest end of the listed energy range arises from an intermolecular force that is thoroughly buried in the pocket of a binding site. Binding energies for interactions that are exposed to solvent tend to be weaker.[8]

Hydrogen Bonding

Hydrogen bonding is a strong intermolecular force. New examples of hydrogen bonding interactions are frequently discovered, but as a practical matter, this discussion is restricted to systems with O–H and N–H groups as the hydrogen donor, and O and N atoms as the hydrogen acceptor. These are by far the most commonly encountered examples in biological systems and medicinal chemistry. The strength of a hydrogen bond varies in the range of 0 to 3 kcal/mol.[8] If one of the O or N atoms is charged, such as between a carboxylate and an alcohol, then the hydrogen bond tends to be relatively strong. If all atoms are neutral, like between an alcohol O–H and the oxygen of water, the interaction is weaker and only worth up to 1.5 kcal/mol.[8]

Hydrogen bonds are unique among the intermolecular forces because hydrogen bonds involve orbital overlap. The σ^*-orbital of the X–H bond of the hydrogen donor should ideally align with the axis of the lone pair of the hydrogen acceptor (Y). The hydrogen bond will be strongest if the X–H–Y angle is 180°. Deviations from 180° will diminish the strength of the interaction. An example is the interaction between an imidazolium ion (**9.3**) and acetate ion (**9.4**) (**Figure 9.7**). Proper alignment of these two species gives a strong hydrogen bond (Case A). If the imidazolium ion is rotated 45°, the N–H–O angle is 135°, and the strength of the hydrogen bond decreases (Case B). In this particular example, regardless of the N–H–O angle, the imidazolium ion is attracted to the acetate ion through an electrostatic ion-ion force, which is the topic of the next subsection.

Protein targets contain many possible sites for hydrogen bonding interactions. Each residue in the peptide backbone affords an N–H group as a hydrogen bond donor and a carbonyl as a hydrogen bond acceptor. Furthermore, over half of the standard amino acids (Chapter 4, Table 4.1) contain a hydrogen bond donor and/or an acceptor. Maximizing the formation of hydrogen bonds between a target and drug makes binding more favorable and increases the potency of the drug against its target.

Ionic Forces

Intermolecular ionic forces involve at least one molecule with a formal charge. An ion may be attracted to either another ion of opposite charge (ion-ion) or the end of a nearby dipole with opposite charge (ion-dipole). Energetically, ionic attractions are strong and fall in the range of 0 to 4 kcal/mol.[8] The stronger interactions are found between carboxylate anions and ammonium ions. The interactions are a combination of both an electrostatic attraction and a hydrogen bond.

Case A N—H—O = 180° (stronger H-bond)

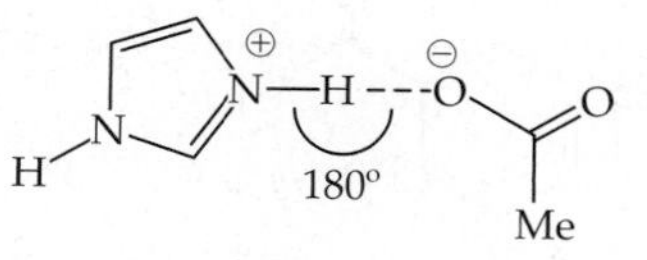

Case B N—H—O = 135° (weaker H-bond)

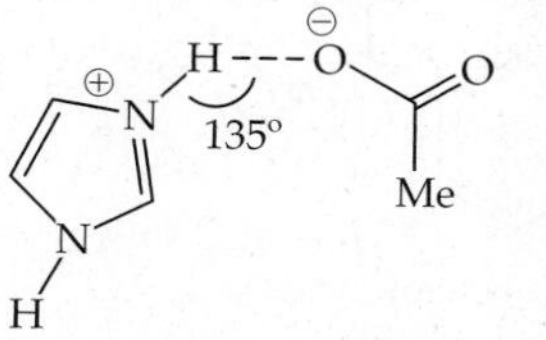

imidazolium ion **9.3**   acetate ion **9.4**

imidazolium ion **9.3**   acetate ion **9.4**

FIGURE 9.7 Effect of relative alignment on hydrogen bonding

SCHEME 9.1 pH effects on anions and ionic forces

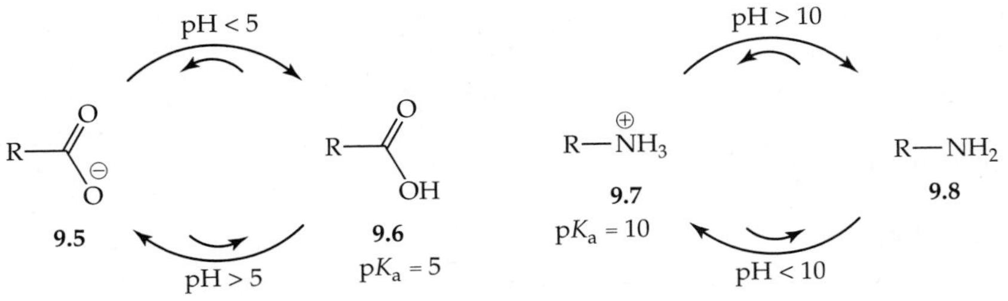

Ionic forces can be very pH dependent. For example, carboxylates (**9.5**) are in equilibrium with their conjugate acid (**9.6**), a carboxylic acid (**Scheme 9.1**). Carboxylic acids have a pK_a of approximately 5. If the pH is above 5, the carboxylate form predominates. Below pH 5, the acid is the major form. Therefore, below pH 5, the ability of this functional group to participate in ionic intermolecular forces is greatly diminished. Cations are also subject to pH effects. Especially common in pharmaceuticals are protonated amines (**9.7**). Alkyl ammonium ions have a pK_a of approximately 10.

Organs and tissues in the body can have very different pHs. The pH of blood is 7.4. The stomach can be as low as 1, while the small intestine is slightly basic at about 8. Functional groups, especially carboxylic acids, can lose or gain a charge as they move from one part of the body to another. If the solution pH and pK_a of the acid are known, then the relative concentrations of the acid and its conjugate base can be calculated through the Henderson–Hasselbalch equation (Equation 9.1).

$$pH = pK_a + \log\frac{[\text{conj. base}]}{[\text{acid}]} \tag{9.1}$$

Five amino acids contain side chains that are significantly charged at physiological pH (Table 4.1). The acidic residues are aspartic acid and glutamic acid. The basic residues are lysine, arginine, and histidine. The presence of any of these residues in the binding pocket of a target is an opportunity for a drug to bind through an ionic interaction. Metals bound within the active site also provide opportunities for ionic bonds.

| Sample Calculation | **Determining the Ratio of a Conjugate Base to Its Acid at a Given pH** |

PROBLEM   Determine the ratio of benzoate (conjugate base) to benzoic acid (acid) in an aqueous solution at pH 7.0. The pK_a of benzoic acid is 4.2.

SOLUTION   The acid–base equilibrium for this reaction is shown in the following figure. Determining the relative amounts of the acid and conjugate base requires direct substitutions in the Henderson–Hasselbalch equation (Equation 9.1). The pH is 7.0, and the pK_a is 4.2. The equation evaluates to reveal that the ratio of the conjugate base to the acid is 630:1. As a check on this result, since pH 7.0 is more basic than the pK_a of benzoic acid, the conjugate base should predominate.

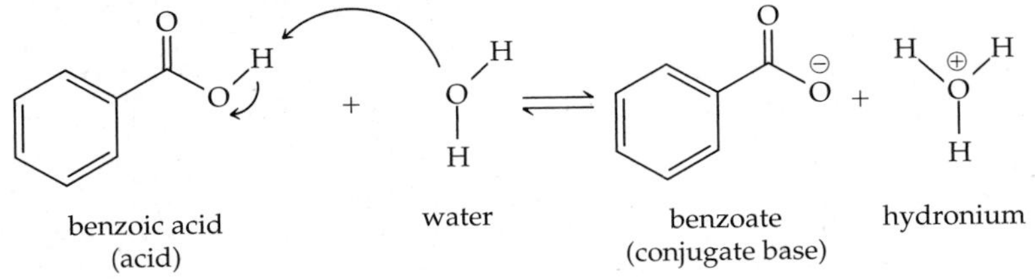

benzoic acid   water   benzoate   hydronium
(acid)   (conjugate base)

$$pH = pK_a + \log \frac{[\text{conj. base}]}{[\text{acid}]} \qquad (9.1)$$

$$7.0 = 4.2 + \log \frac{[\text{conj. base}]}{[\text{acid}]}$$

$$2.8 = \log \frac{[\text{conj. base}]}{[\text{acid}]}$$

$$10^{2.8} = 630 = \frac{[\text{conj. base}]}{[\text{acid}]}$$

Contact Forces and the Hydrophobic Effect

Contact forces, often called London dispersion forces, are a specific type of van der Waals force. Contact forces arise from an instantaneous unequal distribution of electron density in a molecule. This results in a temporary dipole that can induce a temporary dipole in a neighboring molecule. The formation of these complementary, temporary dipoles generates a weak attractive force (**Scheme 9.2**). Because contact forces are classified as being weak, it may be tempting to disregard the role of contact forces in drug-target binding. Contact forces, however, are important for drug binding.

Drug binding is normally described in terms of a measure of activity, such as IC_{50}, K_i, or EC_{50}. As was shown in Chapters 4 and 5 on enzyme and receptor targets, all the activity terms are closely related to the dissociation equilibrium constant (K_D). The equilibrium constant is in turn a reflection of the standard free energy of binding (ΔG°_{bind}) of the drug-target complex (Equation 9.2).

$$\Delta G^\circ_{bind} = -2.3RT \log \frac{1}{K_D} = +2.3RT \log K_D \qquad (9.2)$$

The free energy of binding is a combination of both the enthalpy and entropy of binding (Equation 9.3).

$$\Delta G = \Delta H - T\Delta S \qquad (9.3)$$

In the case of contact forces, the magnitude of the enthalpy term is relatively small. The entropic contribution, however, can be significant. When a nonpolar compound is in an aqueous solution, the water molecules form a highly ordered solvent shell around the nonpolar portions of the compound (**Scheme 9.3**). This phenomenon is called the *hydrophobic effect*. Once the compound buries itself into a binding site on a target, some solvating water molecules will

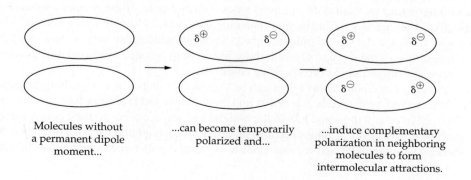

Molecules without a permanent dipole moment...

...can become temporarily polarized and...

...induce complementary polarization in neighboring molecules to form intermolecular attractions.

SCHEME 9.2 Contact forces in molecules with strong, permanent dipoles

SCHEME 9.3 Drug-target binding and the hydrophobic effect

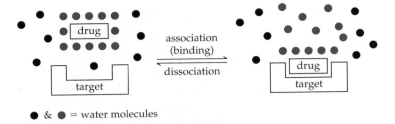

association (binding)

dissociation

● & ● = water molecules

be released to the solution. Water molecules that are free in solution have more freedom of motion than water molecules held within a solvent shell. In Scheme 9.3 the drug is surrounded by a number of water molecules (shown in blue) that define an ordered solvent shell. Once the drug binds the target, most of the water molecules are freed from the drug. The binding of a nonpolar molecule into a lipophilic pocket therefore increases the disorder ($\Delta S > 0$) of the overall system. In other words, the binding of a compound to a target is often entropically favored. This statement may seem counterintuitive since binding of a drug to its target at first glance would seem to decrease, not increase, entropy. If the effect of solvent is included in the analysis, the binding of a hydrophobic drug to a target increases the entropy of a system.

Since the binding of a lipophilic drug to a target is entropically favored and the impact of the hydrophobic effect is a function of the exposed surface area of the drug, the energetic contribution to binding by contact forces and the hydrophobic effect can be expressed in terms of the surface area ($Å^2$) of drug that is buried during binding. In practice, the approximate binding energy of a lipophilic drug is -0.03 kcal/mol/$Å^2$.[9]

An often-cited threshold activity value (IC_{50}, K_i, or EC_{50}) for a drug is 10 nM. By Equation 9.2, this activity value corresponds to a $\Delta G°_{bind}$ of -11 kcal/mol. Achieving -11 kcal/mol of binding energy would require the burying of between 350 and 400 $Å^2$ surface area of a lipophilic drug.[9] Often, knowing the potential binding energy of a specific group is more useful than the energy as a function of surface area. A single CH_2 group in a binding pocket provides approximately 0.8 kcal/mol of binding energy. An entire phenyl ring may be worth around 2.0 kcal/mol.[8]

Dipole Forces

The discussion of dipole forces, another type of van der Waal force, is nearly identical to that of contact forces. The primary difference is that the dipoles are permanent and have a larger magnitude, so the attractive forces are larger. Molecules that interact by dipole forces contain electronegative atoms that generate polarized bonds to carbon. Entire polarized molecules or polarized subregions of molecules line up in antiparallel fashion so that the negative end of one molecule's dipole can align with the positive end of another. Both interacting molecules do not need a permanent dipole. If the second molecule is nonpolar, then the dipole in the first molecule can induce a dipole in the second. The strength of dipole-dipole and induced dipole-dipole attractions can vary widely based on the magnitude of the dipoles. Overall, these interactions are still weak and only perhaps 0.1 kcal/mol larger than a corresponding contact force.[8]

The effect of dipole-dipole forces can be seen in the trend of increasing boiling points in 1-butene (**9.9**), propionaldehyde (**9.10**), and propionitrile (**9.11**) (**Figure 9.8**). These compounds have nearly the same molecular weight and surface area, so the contact forces should be nearly identical. The structures do differ in the magnitude of polarization of the π-bond. The more polarized π-bond leads to stronger dipole forces and a higher boiling point (bp).

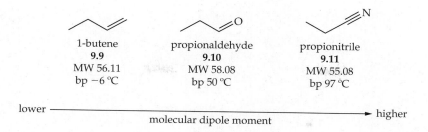

1-butene
9.9
MW 56.11
bp −6 °C

propionaldehyde
9.10
MW 58.08
bp 50 °C

propionitrile
9.11
MW 55.08
bp 97 °C

FIGURE 9.8 Effect of dipole-dipole forces on boiling points

lower ←———————————————————————→ higher
molecular dipole moment

Molecular Shape

A drug interacts with a target through the intermolecular forces described previously. For strong binding, the functional groups on the drug must be positioned in a manner that aligns with the functionality of the target. Said more simply, the shape of a drug is important.

Discussions on molecular shape revolve around two main topics: conformation and stereochemical configuration. Drugs typically have few bonds that can undergo free rotation. Their structures are normally purposefully locked into place through the strategic incorporation of rings and/or π-bonds. Rings and π-bonds keep the structure rigid and hold the drug's shape as close to its ideal as possible. Stereocenters also influence the three-dimensional geometry of the drug and impact the ability of a drug to fully interact with its target.

In practice, both the conformation and configuration of stereocenters of a drug are often studied together when the shape of a molecule is being idealized for target binding. An excellent example of this type of problem is the study of cholinergic activity of acetylcholine.

CASE STUDY

Conformational and Stereochemical Effects on Ligands of Acetylcholine Receptors[10]

Acetylcholine (**9.12**) binds both types of cholinergic receptors: muscarinic and nicotinic (**Figure 9.9**). The names of these receptor types were based on muscarine (**9.13**) and nicotine (**9.14**), selective agonists for each receptor. A muscarinic response is characterized by nausea, salivation, and tearing. Nicotinic responses are noted by an acceleration of the heart rate. Acetylcholine's ability to elicit a response from both subtypes of receptors may imply that different conformations of acetylcholine may be responsible for binding to each receptor.

Quantifying the conformation of acetylcholine requires measurement of the dihedral angle of the ester and ammonium functional groups across the CH_2CH_2 chain (**Figure 9.10**). In comparison to butane, acetylcholine's low energy conformations can be labeled *anti* and *gauche*. High-energy conformations include the eclipsed conformations with dihedral angles of 0° (**9.18**) and 120° (**9.16** and **9.20**). Historically, the relative orientations of the ester and ammonium group at 0° and 120° have been given the

names *synplanar* and *anticlinal*, respectively. The dihedral angles are assigned based on the relative position of the front carbon to the back carbon. The magnitude of rotation, clockwise ($\Theta > 0°$) or counterclockwise ($\Theta < 0°$), of the front carbon required to eclipse the oxygen and nitrogen determines the dihedral angle. Some of the conformations are mirror images of one another. In the chiral environment of an enzyme or receptor, conformers that are mirror images may bind differently. Only six different conformations are shown in Figure 9.10, but an infinite number of intermediate possibilities exist. Testing all is not necessary or possible. The goal is to test a representative sampling.

An obvious means of testing two dihedral angles, 0° and 180°, is to make *Z*- and *E*-alkene analogues (**9.21 and 9.22**) with the alkene restricting rotation about the C–C bond (**Figure 9.11**). Unfortunately, analogues **9.21** and **9.22** are not chemically stable. In general, desirable analogues in a medicinal chemistry study are limited to compounds that can be prepared and purified and are

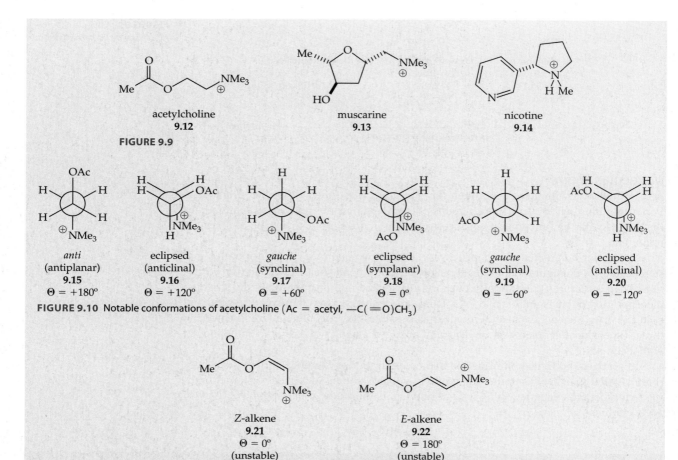

FIGURE 9.9

acetylcholine
9.12

muscarine
9.13

nicotine
9.14

anti
(antiplanar)
9.15
Θ = +180°

eclipsed
(anticlinal)
9.16
Θ = +120°

gauche
(synclinal)
9.17
Θ = +60°

eclipsed
(synplanar)
9.18
Θ = 0°

gauche
(synclinal)
9.19
Θ = −60°

eclipsed
(anticlinal)
9.20
Θ = −120°

FIGURE 9.10 Notable conformations of acetylcholine (Ac = acetyl, —C(=O)CH$_3$)

Z-alkene
9.21
Θ = 0°
(unstable)

E-alkene
9.22
Θ = 180°
(unstable)

FIGURE 9.11 Unstable alkene analogues of acetylcholine

(1S,2S)-*trans*
9.23
Θ = 60°

(1S,2S)-*trans*
9.24
Θ = 180°

trans diaxial decalin
weakly muscarinic
9.25
Θ = 180°
(racemic)

FIGURE 9.12 Conformationally-restrained analogues of acetylcholine

also stable. Sometimes this is a significant problem in a project. For the case of acetylcholine, because the desired alkenes are not viable, rings may instead be used to fix the orientation of the ester and ammonium groups.

The mostly commonly encountered ring size is a six-membered ring. Fusing a six-membered ring to

acetylcholine does not immediately address the conformational flexibility problem. Multiple diastereomers are possible, and each diastereomer can exist as two enantiomers. Furthermore, each new compound can undergo a chair flip. For example, the (1S,2S)-*trans* isomer (**9.23**) has a dihedral angle of 60° (**Figure 9.12**). The higher energy ring flip conformer (**9.24**) has a dihedral angle of 180°. If **9.23** shows strong muscarinic or nicotinic activity, either chair, **9.23** or **9.24**, may be responsible for the activity. One way to determine which conformer gives a response would be to restrict the ability of the cyclohexane ring of **9.24** to flip. Adding a second ring to make a decalin system (**9.25**) accomplishes this goal. All these analogues, **9.23**, **9.24**, and **9.25**, have been prepared. All are essentially inactive. The locked *trans* diaxial isomer (**9.25**) shows very weak muscarinic activity, 0.06% relative to acetylcholine. In the cases of **9.23** through **9.25**, it is possible that the newly introduced ring blocks receptor binding even though one or more of the conformations is correct for optimal interaction with the receptor.

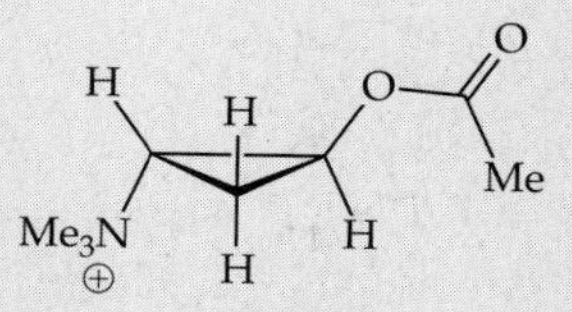

A ring system that is able to restrict conformational flexibility while minimizing additional bulk is cyclopropane. As a strained system, its angles differ from the standard 60° and 180° possibilities with cyclohexane. The (+)-*trans* cyclopropane analogue (**9.26**) does show muscarinic activity comparable to acetylcholine (**Figure 9.13**). The dihedral angle between the ester and ammonium group is approximately 145°. The (−)-*trans* enantiomer and both

(+)-*trans*
muscarinic
9.26
Θ = 145°

FIGURE 9.13 Rigid acetylcholine analogue

enantiomers of the *cis* diastereomer show poor muscarinic activity. None of the cyclopropane analogues elicits an appreciable nicotinic response.

The structures from **9.15** to **9.26** collectively show the factors that must be considered when one tries to determine the optimal three-dimensional orientations of functional groups for drug binding. While searching for ideal orientations can be a challenging process, learning the relative positioning of groups for maximum biological activity is valuable information.

While alkene isomers **9.21** and **9.22** cannot be prepared to probe the geometry of acetylcholine binding (Figure 9.11), alkenes are normally easy to prepare and do have a dramatic impact on activity. In a study of analogues of the antipsychotic drug chlorpromazine (Thorazine, **9.27**), the *Z*-alkene isomer (**9.28**) was found to be equipotent to chlorpromazine (**Figure 9.14**). The *E*-isomer (**9.29**) is inactive.[11] These results give insight into the active conformation of the flexible aminopropyl side chain of **9.27**.

Exploring conformational flexibility and stereochemistry are two basic tools for exploring structure. Conformational probing is almost always informative, but the impact of stereochemistry is not so predictably important. For example, ephedrine (**9.30**) and pseudoephedrine (Sudafed, **9.31**) are diastereomers, but they both elicit similar biological responses (**Figure 9.15**). Esomeprazole (Nexium, **9.33**) is the single enantiomer form of omeprazole (Prilosec, **9.32**), a racemic prodrug commonly used for acid reflux disease. Both enantiomers of **9.32** have essentially the same activity. Indeed, within the cells lining the stomach, both enantiomers of omeprazole are converted to the same achiral compound, which is responsible for the biological effect. The *S*-enantiomer (**9.33**), however, is less subject to the first-pass effect and therefore has a higher bioavailability. Therefore,

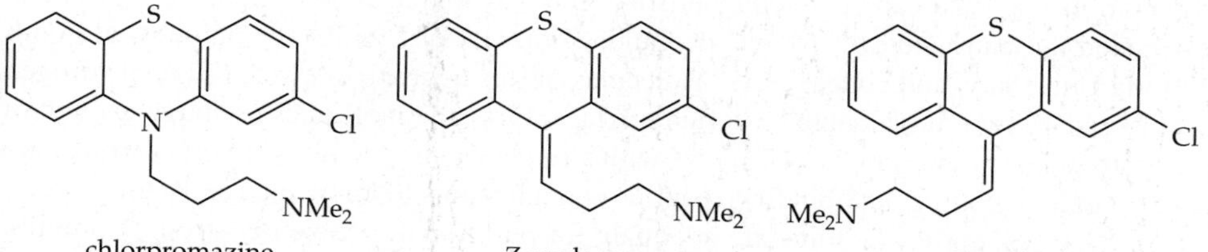

chlorpromazine
(Thorazine)
antipsychotic
9.27

Z-analogue
active
9.28

E-analogue
inactive
9.29

FIGURE 9.14 Chlorpromazine and restrained analogues[11]

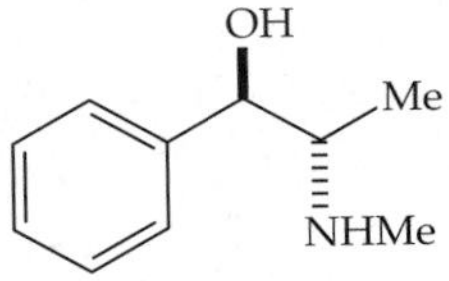

ephedrine
stimulant
9.30

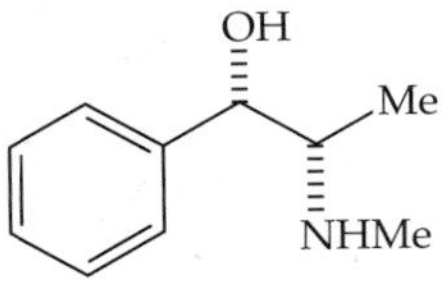

pseudoephedrine
(Sudafed)
decongestant
9.31

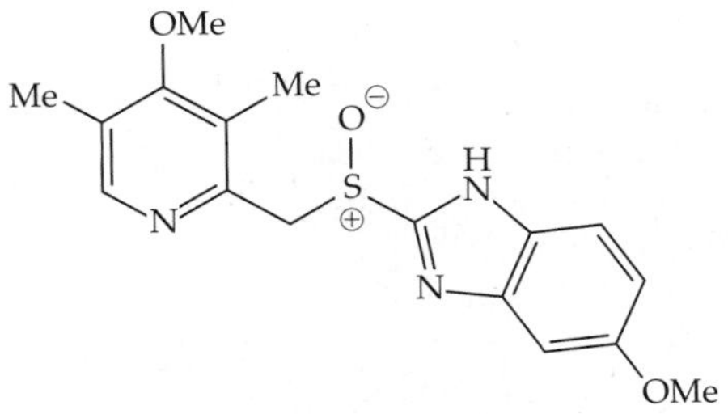

omeprazole
(Prilosec)
anti-acid reflux
9.32

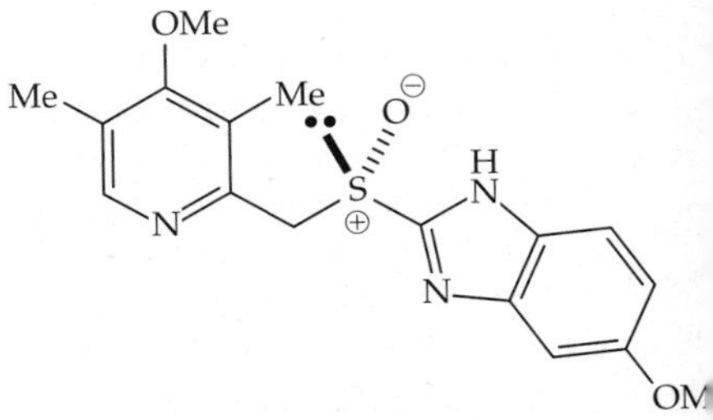

esomeprazole
(Nexium)
anti-acid reflux
9.33

FIGURE 9.15 Active compounds with stereocenters

although the target is not affected by the absolute stereochemistry of the enantiomers of omeprazole, the enzymes that metabolize **9.32** do prefer to break down the *R*-enantiomer over the *S*-enantiomer.[12]

Drug Pharmacophore

The *pharmacophore* of a biologically active compound is the minimal portion of a molecule that is required to provide biological activity. The idea of a pharmacophore connects all the concepts in this chapter thus far. The pharmacophore is the spatial arrangement of key functional groups, and the key functional groups are dictated by the complementary structure of the target. For this reason, knowing the structure of the target is extremely helpful in the development of a new drug. In the drug discovery process, the pharmacophore is normally elucidated late in lead discovery stage (Chapter 10) or early in the lead optimization stage (Chapter 11). Pharmacophores are discussed more thoroughly in the first chapter on lead optimization, Chapter 11.

9.3 Searching for Drugs

This chapter has covered the value of knowing the structure of a target and a target's relevance to the structure of a drug. The topic of searching for a drug still remains. How big is this task? This section covers the challenge of searching for a drug in a broad, almost philosophical manner. The discussion closes with an introduction to compound libraries, one of the most important tools in drug discovery.

Diversity and Molecular Space

Molecular diversity is a key concept in the process of drug discovery. The number of different molecules that can be assembled from 30 nonhydrogen atoms (C, N, O, S) has been estimated to be around 10^{63} by Wayne Guida, formerly of Ciba-Geiby and now at South Florida University, and colleagues.[13] Molecules of this approximate size (30 nonhydrogen atoms) satisfy two fundamental criteria for drugs. First, the molecules are large enough to generate sufficient intermolecular forces with a target for drug activity. In other words, activity requires binding, and binding requires enough atoms to allow intermolecular forces. Thirty C, N, O, and/or S atoms are adequate for building a sufficiently strongly binding drug. Second, a molecule with 30 nonhydrogen atoms is small enough to be absorbed from the digestive tract through passive diffusion and transported in the body in a predictable fashion. An example of a common molecule that contains close to 30 nonhydrogen atoms

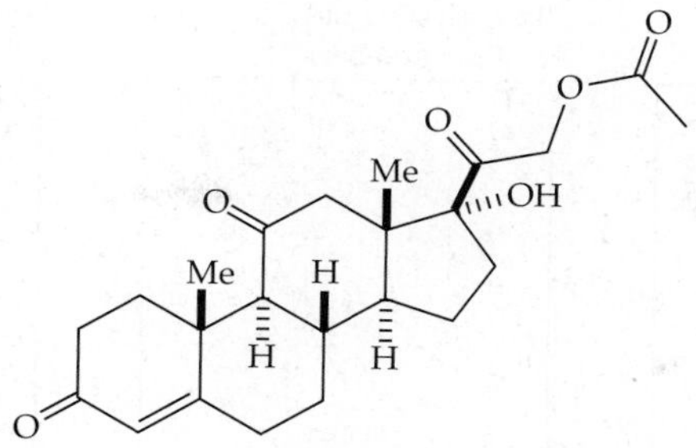

FIGURE 9.16 Cortisone acetate, a molecule similar in size to many pharmaceuticals

cortisone acetate
anti-inflammatory
$C_{23}H_{30}O_6$
MW 402.48
9.34

is cortisone acetate (**9.34**), an anti-inflammatory with a molecular formula of $C_{23}H_{30}O_6$ and a molecular weight of 402 g/mol (**Figure 9.16**). Note that there is nothing magical about 30 nonhydrogen atoms. This is simply *representative* of the size of drugs that are administered as oral tablets or capsules.

The value 10^{63} demonstrates that the number of possible molecules that may be considered as drugs is seemingly infinite. As of 2011, the number of organic and inorganic substances in the Chemical Abstract Services (CAS) Registry was approximately 6×10^{7}.[14] The CAS Registry includes substances that have been published in the literature and patents. Therefore, the number of molecules that have been reported in the chemical literature is miniscule relative to the possible number of drug-sized molecules (10^{63}). Furthermore, if each of the 10^{63} molecules has an average molecular weight of 400, just one molecule of each member of the entire collection would have a mass of 7×10^{38} kg. That value far exceeds the collective mass of all the objects in our solar system ($\sim 2 \times 10^{30}$ kg).

The accuracy of the value 10^{63} is certainly questionable. Regardless, the true number—whatever it may be—is inconceivably large. How does one think about these innumerable molecules and sift through them for biologically active structures? One way to make this number more manageable is to visualize the set of 10^{63} molecules as a two-dimensional surface in an Euler diagram (**Figure 9.17**). This diagram defines a *molecular space*: a region of chemical variation limited by the number of nonhydrogen atoms in a molecule. This space contains regions of molecules with no activity against a target (*dead molecules*), molecules with modest activity and/or properties (*hits*), molecules with improved activity and/or properties (*leads*), as well as molecules with high activity and ideal properties (*drugs*). Most of the space is dead, with the hit spaces, lead spaces, and drug spaces each occupying a successively smaller area. Not all hit spaces contain a lead space.

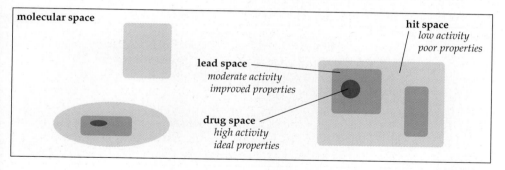

molecular space

hit space
low activity
poor properties

lead space
moderate activity
improved properties

drug space
high activity
ideal properties

FIGURE 9.17 Pockets of activity against a target within a defined molecular space

FIGURE 9.18 Outcomes of screening libraries with (a) low diversity, many members; (b) high diversity, few members; and (c) high diversity, many members

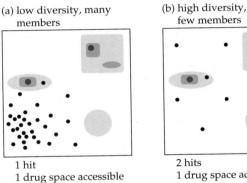

(a) low diversity, many members

1 hit
1 drug space accessible

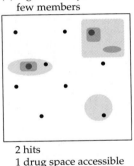

(b) high diversity, few members

2 hits
1 drug space accessible

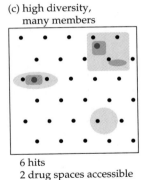

(c) high diversity, many members

6 hits
2 drug spaces accessible

Not all lead spaces contain areas of activity that qualify as drug spaces. Finding a molecule within one of the drug spaces is the goal of a drug discovery program.[13]

When searching for compounds with activity against a target, the medicinal chemistry team often screens the target against a diverse collection, or *library*, of potential drug molecules. This process is called *random screening*. The screened molecular library hopefully includes members that fall within one or more of the hit spaces that also contain leads and drugs. The successful discovery of a pool of hits in the early stages of drug development requires a compound library that is both large in number and diverse in structure. The diversity of the library makes certain that the molecular space is uniformly sampled. A large number of compounds ensures that many of the different hit spaces are represented in the library. A library with too few compounds or a nondiverse collection will not produce as many hits and may not result in discovery of a new drug (**Figure 9.18**).

Hits discovered through screening are prioritized, and the most promising hits are promoted to lead status. Through a series of structural changes, a lead can hopefully be "walked through" molecular space until it occupies a drug space (**Figure 9.19**). The walking of a lead to a drug occurs during the lead optimization process and is discussed further in Chapter 11.

Privileged Structures

Despite the perceived value of diversity, a small number of molecular classes have repeatedly been found to show activity against a range of different target proteins. Examples of such molecular classes, called *privileged structures*, include 2-aminothiazoles (**9.35**), 1,2-diarylcyclopentanes (**9.38**), and 3-aminoalkyl-2-arylindoles (**9.41**) (**Figure 9.20**). Compounds containing these molecular scaffolds are active against many different enzymes and receptors.[15]

Certain areas of molecular space, such as regions containing privileged structures, are rich in hits. The inverse argument—molecular space regions without privileged structures

FIGURE 9.19 Walking a molecule from hit space to drug space by six sequential structural changes

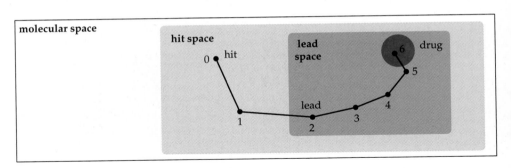

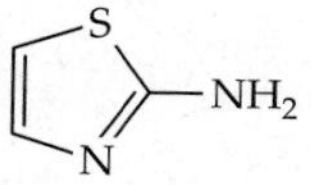
2-aminothiazole
9.35

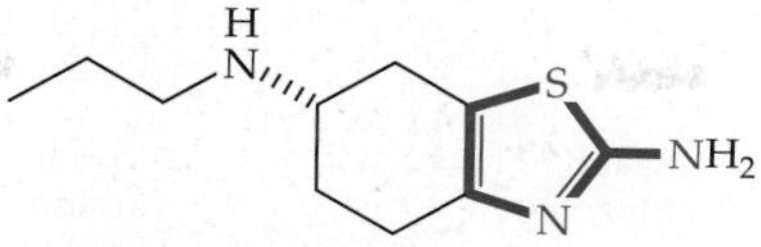
dopamine D2 and D3
receptor agonist
9.36

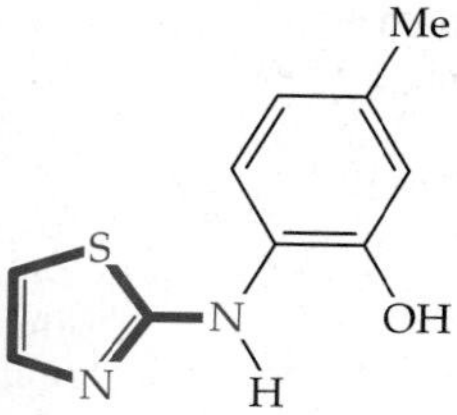
cyclooxygenase and
5-lipoxygenase inhibitor
9.37

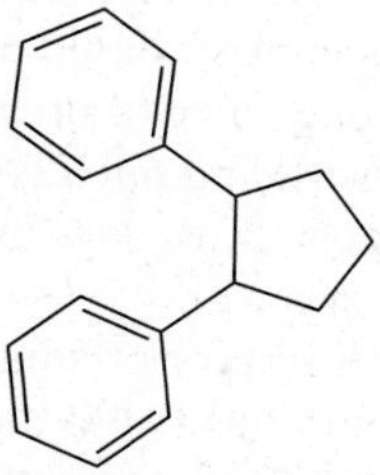
1,2-diarylcyclopentane
9.38

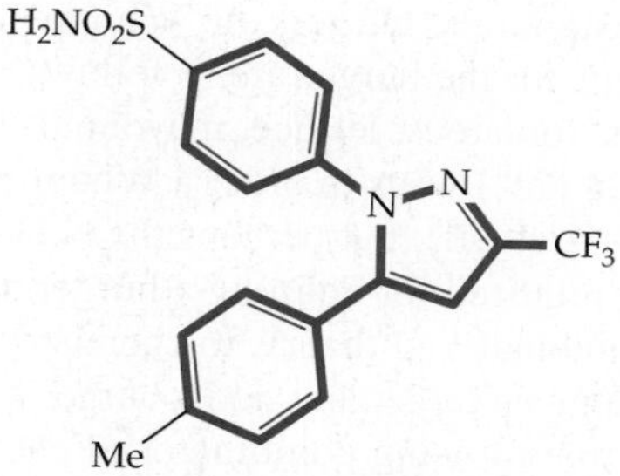
cyclooxygenase-2 inhibitor
(celecoxib, Celebrex)
anti-inflammatory
9.39

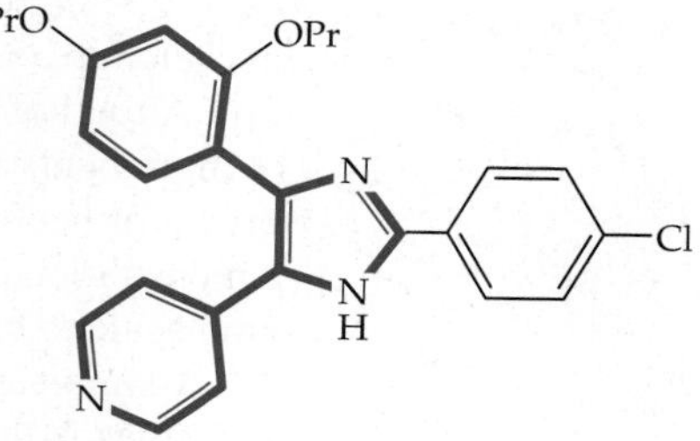
glucagon receptor
antagonist
9.40

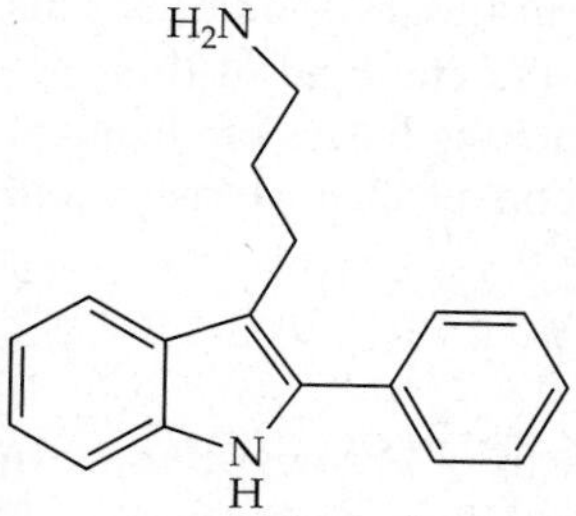
3-aminoalkyl-
2-arylindole
9.41

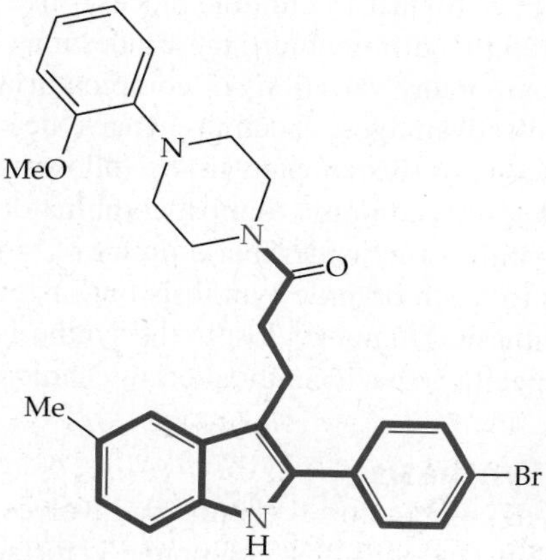
neurokinin
receptor antagonist
9.42

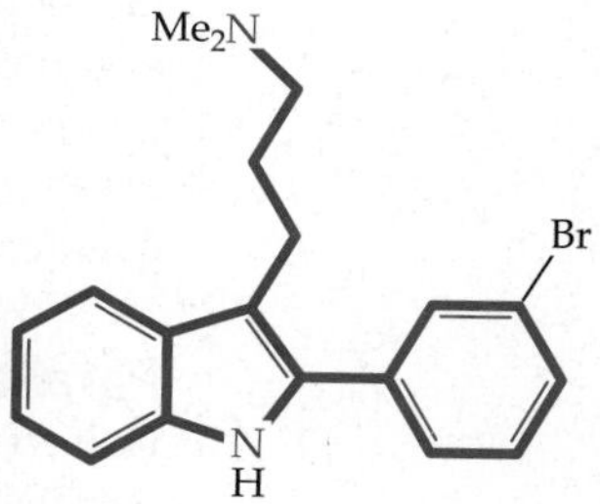
5-hydroxytryptamine receptor
(5-HT_6 and 5-HT_7) antagonist
9.43

FIGURE 9.20 Privileged structures (**9.35**, **9.38**, and **9.41**) with specific examples

are not rich in hits—is not necessarily true. Regardless, some pockets of molecular space are almost certainly relatively dead in terms of hits, leads, and drugs.[16]

9.4 Combinatorial Chemistry

To match the incredible diversity of molecular space, pharmaceutical companies often rely on *combinatorial chemistry*. Combinatorial chemistry, or less formally *combi chem* or even just *combi*, is a technique for synthesizing large numbers of different molecules

SCHEME 9.4 A simple synthetic scheme for combinatorial chemistry

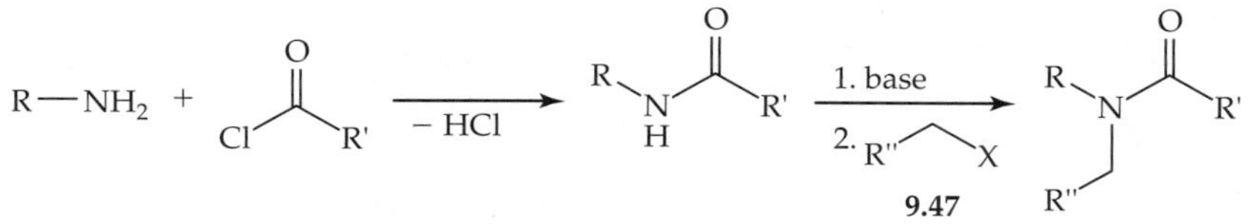

primary amines **9.44**   acid chlorides **9.45**   secondary amides **9.46**   tertiary amides **9.48**

from a relatively small set of starting materials, often called *building blocks*. Applications of combinatorial theory can be found throughout everyday life. In the 1970s, Burger King marketed its signature burger, the Whopper, as being available 1,024 different ways. The ingredients for the burger are as follows: sesame seed crown, beef patty, pickles, ketchup, onions, tomatoes, lettuce, mayonnaise, bun heel, cheese, bacon, and mustard. A total of 1,024 (2^{10}) ways to order a Whopper is not an obvious number based on 12 ingredients ($2^{12} = 4,096$), but perhaps the beef patty and at least one piece of bread (crown or heel) are required for defining a burger. Regardless, the advertising pitch is a demonstration of combinatorial theory. In chemistry, combinatorial theory is used to find a molecule to bind a given target, not find a burger to satisfy a given customer.

A simple example of using combinatorial chemistry to prepare a library of small molecules is the synthesis of secondary amides (**9.46**) from primary amines (**9.44**) and acid chlorides (**9.45**) (**Scheme 9.4**). Amides of type **9.46** have two points of variation: the R-group of the amine and the R'-group from the acid chloride. A third point of variation may be introduced by N-alkylation with an alkyl halide (**9.47**) to form a tertiary amide (**9.48**). With just 15 building blocks—five amines, five acid chlorides, and five alkyl halides—125 (5^3) different tertiary amides may be prepared.

There are many variations of combinatorial chemistry, and all have distinct advantages and disadvantages. A comprehensive description of the different techniques is beyond the scope of this chapter. In the following sections, a few examples will highlight the main features common to most combinatorial approaches to library synthesis. This section contains a considerable amount of synthetic chemistry. The goal of this section is not to teach organic synthesis but instead to demonstrate the basics of chemical library synthesis. Do not get lost in the synthesis! Focus instead on the characteristics and qualities specific to each combinatorial technique.

Parallel Synthesis

One form of combinatorial chemistry involves *parallel synthesis* or *array synthesis*. In parallel synthesis, each new compound is formed in its own miniature reactor or well. If a new library is to contain 500 members, synthesis of the library will require 500 wells in the final step. For this reason, parallel synthesis is associated with the phrase *one compound–one well*.

Managing a large number of individual reactions can be cumbersome. Reactions are often automated with a robotic liquid handler to minimize the tedium. The liquid handler does exactly what its name implies; it dispenses solutions of reagents into racks of wells. Multiple racks are placed on the deck of the liquid handler. Some racks hold wells for reactions, and other racks hold solvents or solutions of reagents. The positions of all the reaction wells and solutions are programmed into a computer. The computer then directs a robotic arm. The robotic arm is suspended over the deck where it takes up and dispenses solvents and reagents as needed.

The racks can include a number of features, including the ability to heat and cool the reaction as well as to exclude air and moisture. Furthermore, a liquid handler can also

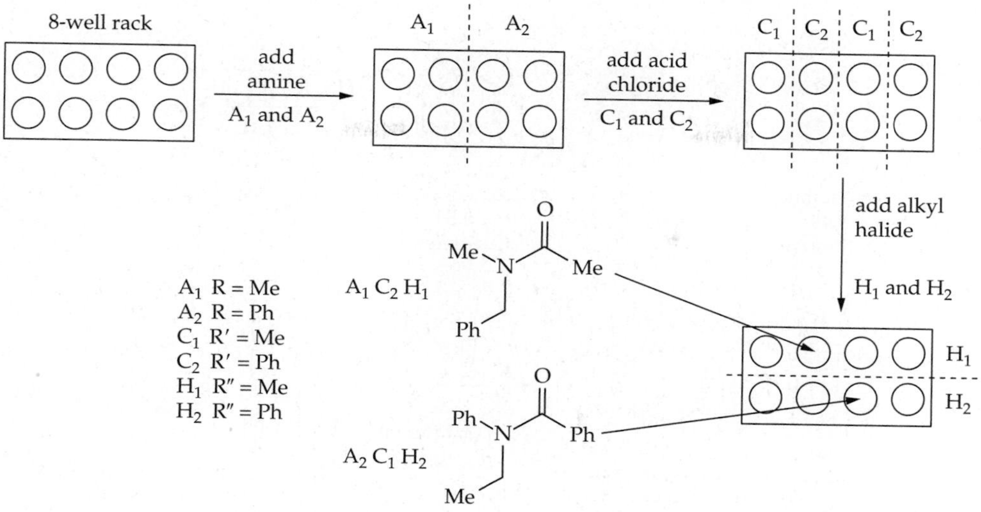

SCHEME 9.5 Parallel synthesis of an eight-member library

A₁ R = Me
A₂ R = Ph
C₁ R′ = Me
C₂ R′ = Ph
H₁ R″ = Me
H₂ R″ = Ph

$A_1 R = Me$
$A_2 R = Ph$
$C_1 R' = Me$
$C_2 R' = Ph$
$H_1 R'' = Me$
$H_2 R'' = Ph$

$A_1 C_2 H_1$

$A_2 C_1 H_2$

remove liquids. With some creativity, a liquid handler can perform almost any needed operation: solvent addition and removal, mixing, and filtration. At the end of all the steps of the synthesis, each well will contain a new library compound.

In Scheme 9.4, a three-component synthesis using an amine, acid chloride, and alkyl halide is shown. If this process were started with two amines (A_1 and A_2), two acid chlorides (C_1 and C_2), and two alkyl halides (H_1 and H_2), eight (2^3) new amides may be formed (**Scheme 9.5**). Such a synthesis would require eight wells. The reagents would be dispensed in a pattern to generate all possible combinations. At the end of the synthesis, the exact position and contents of each well would be known. Therefore, parallel synthesis is often referred to as a *spatially addressable* method. Any particular compound may be accessed directly and immediately.

Having a crude reaction product in a well is different from having the same compound in isolated or purified form. Isolation and purification of reaction products has been estimated to consume over 50% of a synthetic chemist's laboratory time. In other words, isolation and purification require more time than does synthesizing a molecule. Therefore, streamlining synthesis is less critical than streamlining isolation and purification. A properly equipped liquid handler can perform liquid-liquid extractions for automated work-ups. Filtration of crude products through a cartridge of silica gel can serve as a simple form of column chromatography. While extraction and filtration through silica gel may not afford an analytically pure product, the material will likely be sufficiently pure for use in a compound library to screen for activity.

Regardless of how the products are isolated and purified, all parallel syntheses hold to the concepts of one compound–one well and spatial-addressability. Differences between various parallel syntheses are discernable based on whether the synthesis is performed in a solution-phase or solid-phase manner.

Parallel Synthesis: Solution Phase

The vast majority of organic methodology involves solution-phase chemistry. Reagents involved in a reaction are typically dissolved in the same liquid phase. Collisions between the reagent molecules result in bond-breaking and bond-forming reactions.

A simple example of a solution-phase parallel synthesis from Liu and colleagues is shown in **Scheme 9.6**, which depicts a three-step preparation of a 24-member library of 5-aminobenzimidazoles (**9.54**).[17] All manipulations of the reactions were performed

SCHEME 9.6 Solution-phase parallel synthesis of 24 5-aminobenzimidazoles[17]

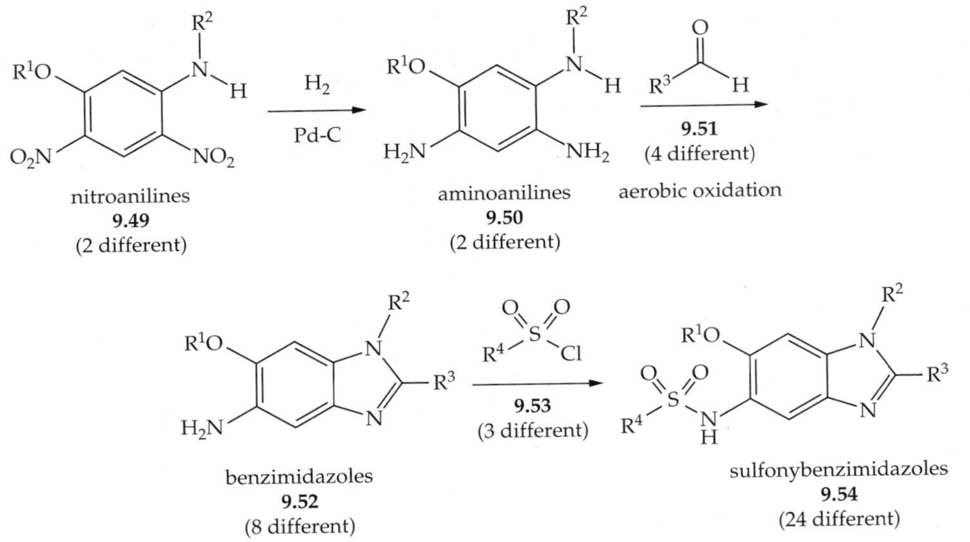

nitroanilines
9.49
(2 different)

aminoanilines
9.50
(2 different)

aerobic oxidation

9.51
(4 different)

benzimidazoles
9.52
(8 different)

9.53
(3 different)

sulfonybenzimidazoles
9.54
(24 different)

by a robotic liquid handler. The synthesis started with eight wells. Four wells contained a solution of one 5-alkoxy-2,4-dinitroaniline (**9.49**, R^1 = methyl, R^2 = 1-methyl-3-phenylpropyl), and four wells contained a solution of another 5-alkoxy-2,4-dinitroaniline (**9.49**, R^1 = ethyl, R^2 = 2-methylcyclohexyl). All wells were subjected to H_2/Pd-C in a mixture of 1,4-dioxane and ethanol to reduce the nitro groups to amines of type **9.50**. Pd-C, an insoluble solid catalyst, was removed by filtering the completed reactions into eight fresh wells. Excess amounts of four aldehydes (**9.51**) were distributed into the eight wells. Condensation and air-oxidation afforded solutions of benzimidazoles **9.52**. Excess aldehyde reagent was quenched, and the eight wells were again filtered and then evaporated to dryness. The residues were dissolved in dichloromethane and washed with aqueous $NaHCO_3$. Solutions of the eight different benzimidazoles were each divided equally into three wells. To these wells were added solutions of three different sulfonyl chlorides (**9.53**). Destruction of excess sulfonyl chloride followed by filtration and concentration afforded 24 5-aminobenzimidazoles (**9.54**). Automated analysis indicated that the products were over 92% pure on average. Only two of the products were less than 80% pure. This level of purity is adequate for reliable screening for biological activity. Although in this example only 24 new compounds are formed, more elaborate solution-phase parallel syntheses have generated hundreds and thousands of compounds.[16]

Aside from the fact that the reactions in Scheme 9.6 were performed in small wells instead of traditional flasks and glassware, the chemical reactions are no different from the types of reactions performed in a traditional introductory organic chemistry class. The reagents are mixed into a common solvent so that they may react with one another in solution.

Parallel Synthesis: Solid Phase

Solid-phase synthesis gained recognition with Merrifield's pioneering resin-bound polypeptide synthesis methodology.[18] In solid-phase synthesis, the reagent of interest is attached to a macroscopic *bead* made of a polymeric resin—essentially, a piece of plastic. The beads are approximately the size of coarse grains of sand. Dangling off the beads are side chains, called *linkers* or *tethers*, to which the molecule of interest is attached (**Scheme 9.7**). If the linker is long enough, the molecule will freely interact with solvent even though it is actually tied to a relatively massive, insoluble bead. Reactions can be

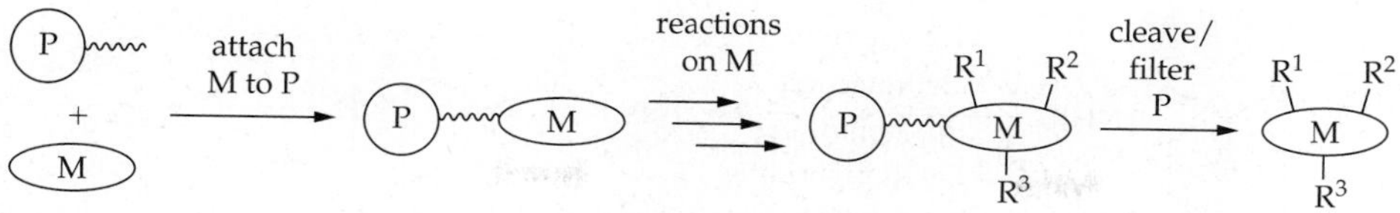

SCHEME 9.7 General set-up of a solid-phase synthesis

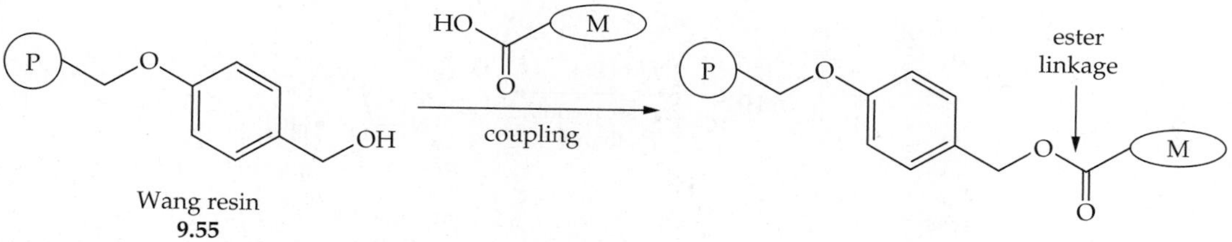

P = polymeric resin with linker
M = molecule of interest

SCHEME 9.8 Loading a Wang resin

Wang resin
9.55

ester linkage

performed on the molecule as desired. Because the ultimate goal is to prepare a molecule for inclusion in a library, the transformed molecule will eventually be *cleaved* from the bead by destroying the linker and then removing the detached bead by filtration.

The stability of the linker is vital to the success of solid-phase synthesis. The linker is the lifeline of the attached molecule. If the linker is cleaved unexpectedly, the molecule will be washed off the bead and lost. Therefore, the linker must be chemically robust enough to withstand various reaction conditions and only "break" during the cleavage conditions. Considerable research efforts have been expended on determining optimal resin-linker structures. A typical example can be found in a popular resin, the Wang resin (**Scheme 9.8**). In the Wang resin (**9.55**), a polystyrene bead (P) is connected to a linker with an OH on the end. Molecules (M) may then be attached to the resin by coupling a carboxylic acid on the molecule to the OH of the linker. Cleavage of the connecting ester through a hydrolysis reaction liberates the molecule from the resin. Esters are cleaved under strongly acidic and basic conditions, so these conditions must be avoided in the preparation of libraries on a Wang resin.

Once the molecule is attached to the resin, performing reactions on the molecule is straightforward. The beads may be free in a reaction well or suspended from above by a *pin* (**Figure 9.21**). If the beads are free, reagents and solvent will be added into the well with a liquid handler. Once the reaction is complete, draining the liquid from the bottom of the well will remove spent reagents and solvent. Additional reagents can then be added for the next reaction. If the beads are attached to pins, the pins may be dipped into reagent-filled wells. After a reaction is complete, the pins can be raised and dipped into another well for the next reaction.

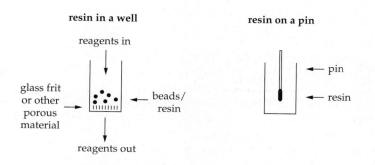

resin in a well

reagents in

glass frit or other porous material

beads/ resin

reagents out

resin on a pin

pin

resin

FIGURE 9.21 Common well formats for solid-phase parallel synthesis

SCHEME 9.9 Solid-phase
approach to hydantoins (**9.60**)[19]

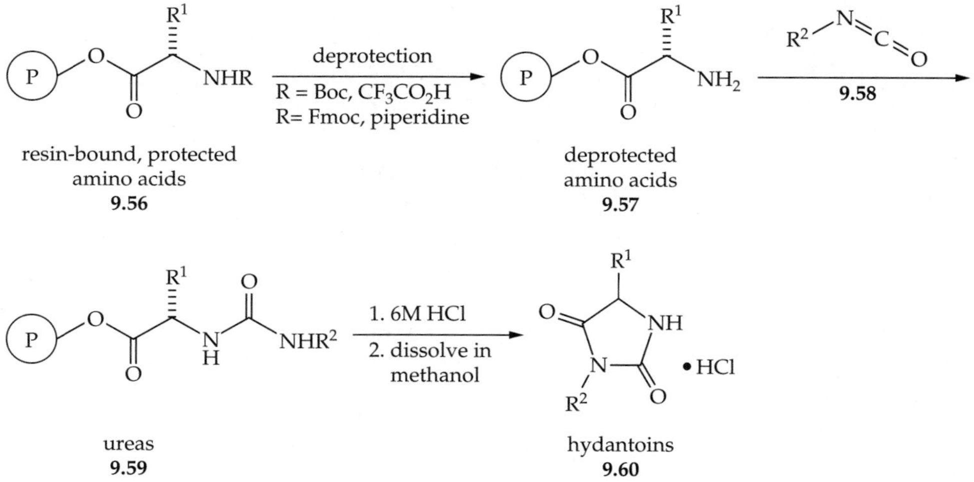

resin-bound, protected
amino acids
9.56

deprotected
amino acids
9.57

ureas
9.59

hydantoins
9.60

The first examples of solid-phase synthesis applied to pharmaceutical library synthesis appeared in the early 1990s. One early report targeted hydantoins (**9.60**), a class of anticonvulsants (**Scheme 9.9**).[19] The synthesis started with eight different protected amino acids (**9.56**, R = Boc or Fmoc) attached to a resin. In a variation on attaching the resin to a pin, the resin was placed inside hollow, porous pins. The porous pins allowed the solutions in the reaction wells to freely bathe the beads while keeping the beads in a defined volume. The pins were dipped in wells with appropriate reagents to remove the protecting group. The resulting deprotected, resin-bound amino acids (**9.57**) were then each dipped in wells containing one of five isocyanates (**9.58**) to provide ureas (**9.59**). The ureas were dipped into 6M HCl to both cleave the compounds from the resin and cyclize the ureas to hydantoins as their HCl salts (**9.60**). The highly acidic cleavage cyclization conditions racemize stereochemistry of the α-carbon of the amino acid. Because the salts were insoluble in aqueous HCl, the compounds remained within the hollow pins. Dipping the pins in wells containing methanol dissolved the salts and allowed them to diffuse out of the pins and into their respective wells. Evaporation of the methanol afforded a library of 39 of the expected 40 hydantoins as their HCl salt in yields ranging from 4 to 81%. Purity of the products was confirmed by TLC, MS, and ^{1}H NMR.[19]

Another early solid-phase synthesis focused on preparing 1,4-benzodiazepine-2, 5-diones, a structure common to many different classes of drugs.[20] The route started with preparation of a custom resin-linker (**9.61**) (**Scheme 9.10**). The resulting resin was divided into 10 equal portions. Each portion was mixed with a different α-amino acid in the presence of NaBH(OAc)$_3$ to form amine **9.62** (**Scheme 9.11**). Each of the 10 different batches of amine **9.62** was in turn evenly divided into 12 portions for a new total of 120 individual batches of resin. Each batch was loaded into a tube with a fritted bottom. The

SCHEME 9.10 Preparation of
resin-linker for a 1,4-benzodi-
azepine-2,5-dione library[20]

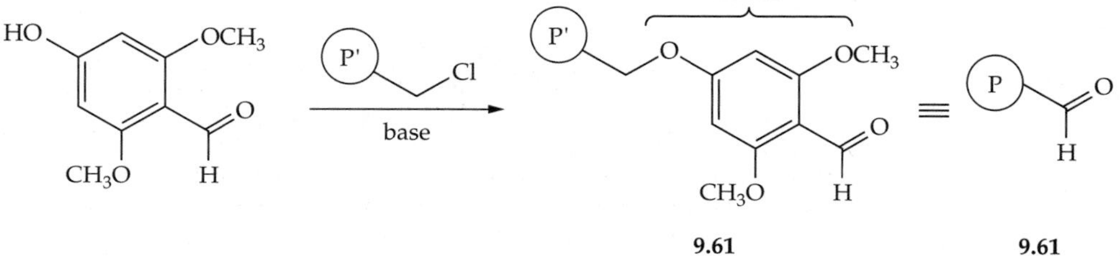

9.61

9.61

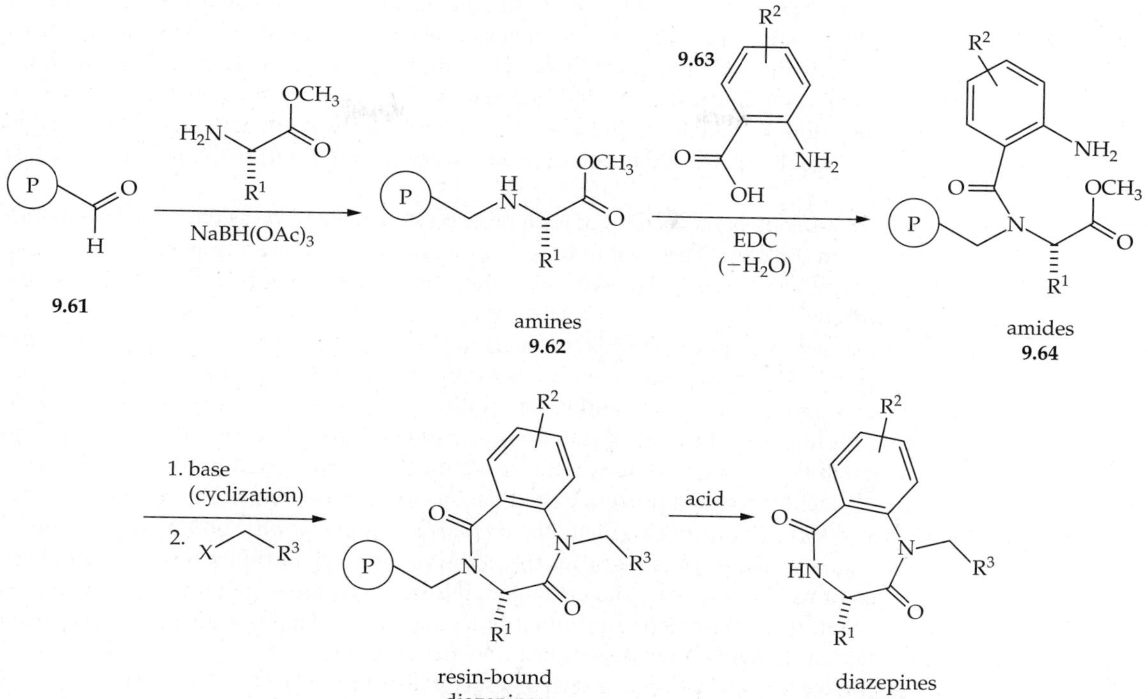

SCHEME 9.11 Solid-phase parallel synthesis of 1,4-benzodiazepine-2,5-diones[20]

amines were then reacted with 12 different anthranilic acids (**9.63**) with EDC (1-ethyl-3-[3-(dimethylamino)propyl] carbodiimide hydrochloride), a water-soluble coupling reagent. The resin of each of the 120 amide products (**9.64**) was further subdivided into 11 lots, each of which was placed into a separate tube with a fritted bottom. The tubes, serving as hollow pins, were dipped into a dilute solution of a base to close the ring of the 1,4-benzodiazepine-2,5-dione. Dipping the pins into an alkylating agent generated the resin-bound product (**9.65**). Finally, each pin was immersed in its own well containing an acid solution, causing cleavage of the linker and formation of the final compounds (**9.66**). This synthesis required 1,320 wells ($10 \times 12 \times 11$) to hold the final products. Because 9 of the 10 amino acids contained stereocenters ($R^1 \neq H$) that could be racemized during the reaction, 1,188 of the 1,320 wells likely contained a mixture of enantiomers. Therefore, under the right conditions, a library of 2,508 compounds was theoretically formed. The purity of the products was randomly checked by HPLC and ^1H NMR.[20]

Note that the syntheses outlined in Schemes 9.9 and 9.11 are consistent with the practice of parallel synthesis. Each final product is formed in its own well.

Split Synthesis

Split synthesis is a completely different approach to library generation from parallel synthesis. Furthermore, split synthesis can be performed only on a resin; solution phase is not an option. Split synthesis involves dividing, or splitting, a pile of loaded resin into batches for reaction with a set of building blocks. After the reaction, all the resin is pooled back together, thoroughly mixed, and then resplit for reaction with the next set of building blocks. Based on this protocol, split synthesis is often also called *mix-split* or *pool-split* synthesis.

An important idea for split synthesis is that each well must contain a large number of individual resin beads. The large number of beads is important for two reasons. One, each well must include a representative sampling of beads from the previous reactions. Two, each reaction must form enough reacted beads to provide samples for each well in the next reaction. Since the mixing and splitting processes are not perfect, the use of large numbers of beads ensures that each desired library member will be prepared on at least one bead.

The number of wells needed for each reaction step is the same as the number of building blocks in the step. This is a reduction in workload compared to parallel synthesis. Once the synthesis is complete, the molecules are cleaved from the resin into wells, one well per bead.

At the end of a split synthesis, because the beads have been pooled and mixed, the exact identity of a molecule on a given bead is unknown. Likewise, the identity and structure of compounds in wells is unknown. Split synthesis is not a spatially addressable method. Fortunately, the exact structure does not need to be known unless a compound shows activity in a screen. If active, the structure of the compound in the well will need to be elucidated through a process called *deconvolution*. Deconvolution is generally accomplished through one of two methods: *recursive deconvolution*[21] or *binary encoding*.[22]

In recursive deconvolution, after the addition of a new building block, a sample of that batch of resin is set aside and cataloged. Furthermore, after the final step, the resin is not repooled, so the last step for all the library members remains known. The cataloged resins and knowledge of the last step facilitate deconvolution.[21]

Recursive deconvolution is perhaps best explained through a simple eight compound library prepared by three steps, addition of *A* or *B*, then *1* or *2*, and finally *a* or *b* (**Scheme 9.12**). In total, such a library has four resin samples. The resin samples are cataloged from intermediate steps—two samples each from the *A/B* step and the *1/2* step. Assume the library gives one hit in a screen. The screener of the library knows only the last step leading to the hit. In this case, the hit is *A1b*. The screener would know only the last step that produced hit, so the hit would be *??b*, with *??* being the unknown building blocks from steps one and two.[21]

In the example shown in Scheme 9.12, recursive deconvolution starts with the cataloged resins from the second step (**Scheme 9.13**). The resins would be either P-*?1* or P-*?2* based on whether the second step added *1* or *2*. Again, only the last step that formed a cataloged resin (second step, in this case) is known. Because the hit came from the batch that reacted with *b* in the final step, the cataloged resins would both be reacted with *b*. Separate reaction of the two sets of cataloged resins with *b* followed by cleavage from the resin gives two mixtures, *?1b* and *?2b*. Screening would reveal activity only in the mixture of *?1b*, indicating that building block *1* in the second step is giving activity instead of *2*. The

SCHEME 9.12 Split synthesis and screening of an eight compound library

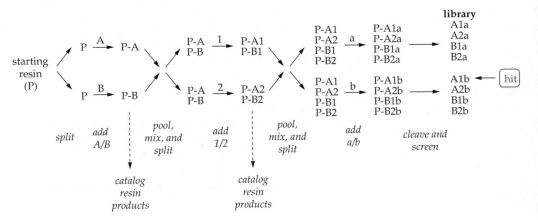

SCHEME 9.13 Recursive deconvolution

One, test cataloged resin from second step,

and then

test cataloged resin from first step.

cataloged resin from second step

P-?1 ——————→ ?1b ←—— active
1. add b
2. cleave
3. screen
P-?2 ——————→ ?2b

cataloged resin from first step

P-A ——————→ A1b ←—— active
1. add 1
2. add b
3. cleave
4. screen
P-B ——————→ B1b

first pair of cataloged resins, **P-A** and **P-B**, would then be reacted with **1** and then **b**. One of the products, **A1b**, is active, and deconvolution of the hit is complete. (Read the previous two paragraphs again [and again]—recursive deconvolution is not a difficult idea, but something about working backward in an abstract discussion tremendously increases the degree of difficulty.)

The alternative to recursive deconvolution is deconvolution by binary encoding. Binary encoding requires the attachment of additional molecules, called *tags*, directly to the resin. After a building block is attached to the resin-bound molecule, one or more corresponding tags are also attached to the resin. If the cleaved molecule shows activity in an assay, the tags will be cleaved separately from the resin. Analysis of the identity of the cleaved tags reveals what the structure of the hit must be. Binary encoding requires each bead to include two linkers, one for the library molecule and another for the tags. The presence of two tethers can complicate a library encoded with tags. Both linkers must be stable throughout the synthesis and yet individually cleavable at the end of the process.[22]

As with recursive deconvolution, binary encoding is most easily explained with a simple example. A 16-member library may be constructed by combining four building blocks of one type (*A,B,C,D*) with four of another type (*1,2,3,4*) (**Scheme 9.14**). In binary encoding, four bits are required to encode for all 16 members. Each binary digit corresponds to a unique tag. Four hypothetical tags might be t_w, t_x, t_y, and t_z. After the first building block is added, a combination of t_w and t_x (one - t_w, the other - t_x, both -$t_w t_x$, or neither) is also added to the resin. Addition of the tag occurs in a separate step from the building block. After the second synthetic step, a combination of t_y and t_z is added (t_y, t_z, $t_y t_z$, or neither). At the end of the synthesis, the library molecule and tags (if any) are cleaved from the resin. The tags and molecule are cleaved by different reactions so that the library compound can be removed without cleaving the tag. If a library member is found to be active, analysis of the cleaved tags reveals the structure of the hit. For library

first step	tag(s)
A	none
B	t_w
C	t_x
D	t_w and t_x

second step	tag(s)
1	none
2	t_y
3	t_z
4	t_y and t_z

resin-bound and tagged C4

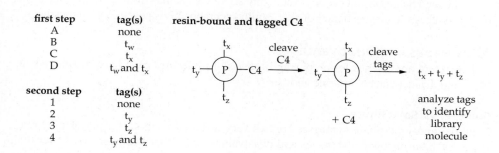

SCHEME 9.14 Binary encoding

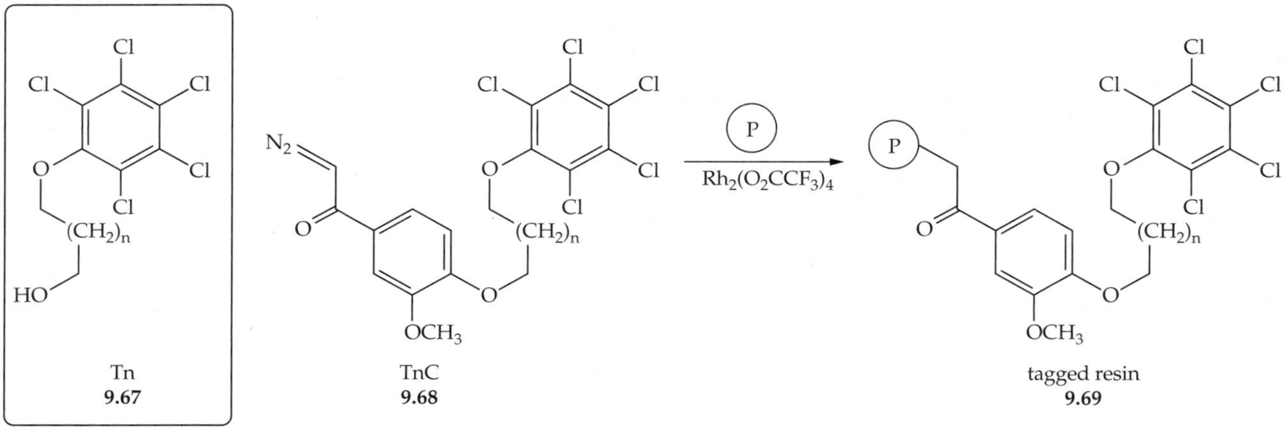

SCHEME 9.15 Still's Tn binary encoding tags[23]

member $C4$, tags t_x, t_y, and t_z would be detected as having been cleaved from the bead. For $A2$, only t_y would be present.

W. Clark Still of Columbia University has developed a widely used type of encoding tag called Tn (**9.67**) in which n stands for a number of carbons in a linker.[23] The tags start in the form of TnC (**9.68**), which attaches to the resin through a reaction with catalytic $Rh_2(O_2CCF_3)_4$ (**Scheme 9.15**). Exposure to ceric ammonium nitrate (CAN) and silylation of the Tn alcohols provides the tags in a form for analysis by electron capture gas chromatography. Electron capture gas chromatography is an extremely sensitive analytical technique, so the amount of tag on the resin, which is miniscule, can still be detected. The tags vary by the length of the linker (n). Changes in linker length give tags with characteristic retention times in GC analysis. The tags can also be varied by replacing Cl with F on the aromatic ring.[23]

Synthesis of a massive, 2.18-million member library through split methods has been described by Stuart Schreiber of Harvard University.[24] Only a few details of the library are presented here. For a combinatorial library, the members show a high level of diversity. The library started with attachment of shikimic acid to a resin (**9.70**) (**Scheme 9.16**). An esterification followed by an intramolecular nitrone cycloaddition formed a complicated core (**9.72**) with a number of possible sites of variation for the library. Some of the reaction sites are indicated. Upon completion of the synthesis, the linker was cleaved through a photochemical reaction. Some portions of the library were encoded with Still's Tn tags, while other portions were not encoded. Three representative library members (**9.73**–**9.75**) are shown. In a screening of a 456-compound sublibrary for activation of transforming growth factor-β-responsive reporter gene, compound **9.74** showed modest activity ($EC_{50} = 50 \ \mu M$). The sublibrary was not encoded with tags, and all hits were identified through recursive deconvolution.[24]

Although Schreiber did not discover highly active compounds from his library, the power of split synthesis to make a library of over 2 million compounds is nonetheless impressive.

Advantages and Disadvantages
The different approaches for preparing libraries of molecules each have distinct advantages and disadvantages over one another.

Solution vs. Solid Phase
Solution- and solid-phase syntheses are two very different methods for performing reactions. Each has its own advantages and disadvantages. Neither is universally better than the other, although both techniques have ardent supporters. A more balanced view is that

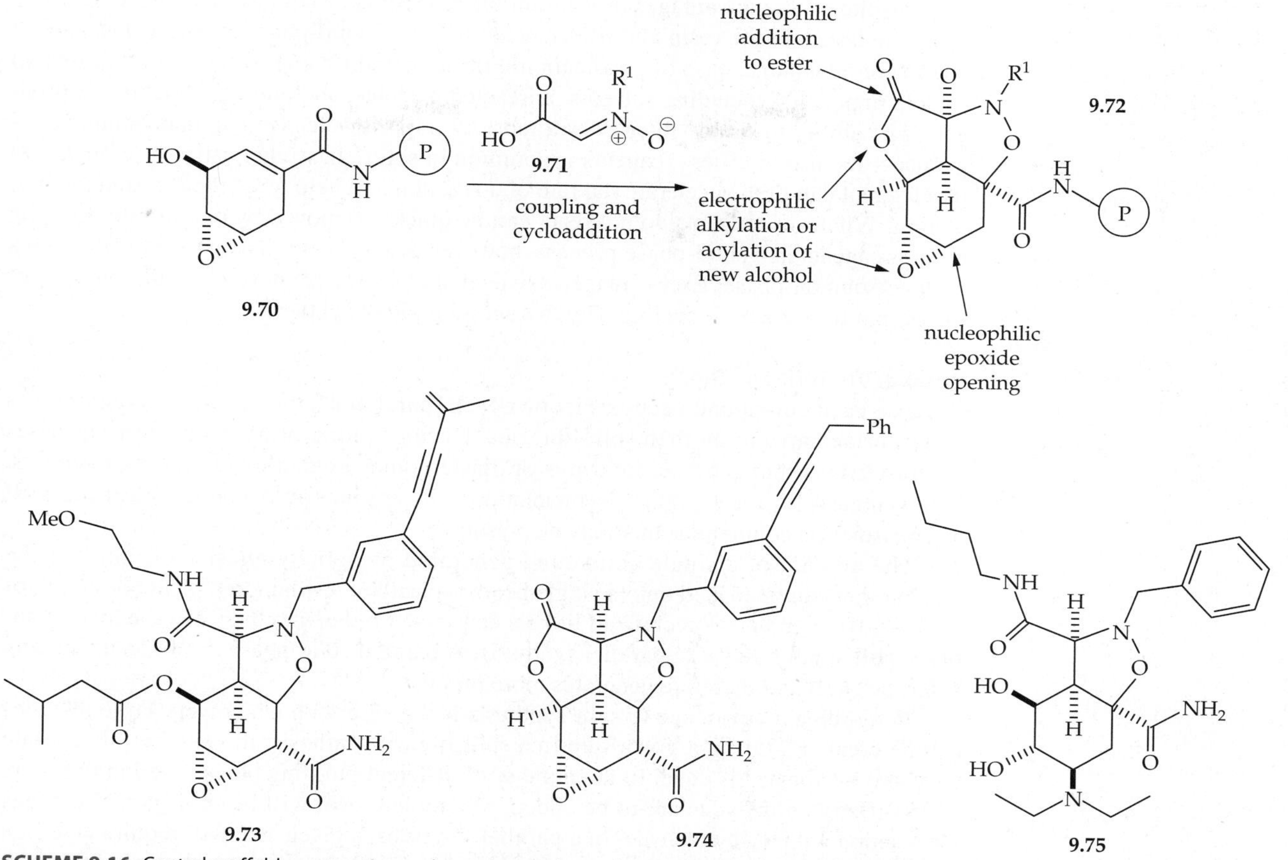

SCHEME 9.16 Central scaffold construction and sample members of a 2.18-million compound library[24]

both solution- and solid-phase synthesis are tools that can be used together to facilitate the drug discovery process.

A major advantage of solution-phase synthesis is that it is the de facto standard. Over 150 years of chemical literature outlines solution-phase methodology. Furthermore, all chemists are initially trained as students in solution-phase chemistry. Laboratories are all equipped with solution-phase equipment. The advantage of dominance will probably decrease slightly over time as more solid-phase chemistry is reported in the literature.

Solution-phase chemistry is simpler than solid-phase. Solid-phase synthesis has two extra variables: the resin and linker. Unwanted reactions on either can cause major problems. The reactivity of the resin, normally polystyrene, is typically very low and not a major concern. Linkers, however, are designed to be broken, so they must have some level of reactivity. The cleavage conditions must be strictly avoided during all the synthetic steps. Because photochemical reactions are uncommon in mainstream synthesis, photochemically labile linkers are attractive in solid-phase synthesis.

If a library compound shows activity, the compound will need to be resynthesized for confirmation testing. If the library was prepared through solution-phase chemistry, hits can be synthesized by scaling up the original procedure. Reactions do not always give the same yield when performed on a larger scale, but modifications will likely be minor. In contrast, a solid-phase synthesis rarely translates well to the solution phase. Resynthesis of hits may require the entire route to be redesigned from scratch.

Despite all the advantages of solution-phase chemistry (literature precedent, familiarity, no issues of the resin and linker, scale-up, etc.), solid-phase synthesis has at least one major advantage: ease of purification. Filtration and rinsing of the resin removes all soluble impurities, including solvents, unreacted reagents, and nonresin-bound side products. The ability to simply filter away almost all impurities leads to an interesting consequence—the use of excess reagents is common in solid-phase chemistry. A large excess of a reagent can both accelerate the rate of a reaction and help force the reaction to completion. Whatever reagents do not react can be quickly removed by one filtration. Using excess reagents in a solid-phase process, however, is a common barrier to transitioning a route to solution-phase. Excess reagents, which are trivial to remove in a solid-phase synthesis, can be a troublesome impurity in a solution-phase synthesis.

Parallel vs. Split Synthesis

Because each compound occupies its own well, parallel synthesis libraries require less effort for deconvolution than split libraries. Deconvolution in parallel synthesis arises from reactions that generate mixtures of diastereomers, enantiomers, or regioisomers. Split synthesis libraries require deconvolution to determine the synthetic pathway as well as any isomeric compounds that may be present.

The amount of a single compound generated by split synthesis is limited to the amount that can be loaded onto a single bead, typically a few hundred picomoles (1 picomole $= 10^{-12}$ moles). Specialized linkers can increase the amount,[25] but the final quantity is still small relative to parallel synthesis. A parallel solid-phase synthesis uses many beads per well and therefore generates more product.

A significant advantage to split synthesis is the ease with which very large libraries can be created. During a given step in a split library synthesis, the number of separate reactions performed is equal to the number of different building blocks used in that step. If 25 different aldehydes are to be added, the pooled resin will be split into 25 batches for reaction with each aldehyde. In a parallel synthesis, a given step will require at least a number of reactions (and wells) equivalent to the number of products at the end of the step. Minimizing the number of individual reactions dramatically reduces the work required for the library synthesis and makes large libraries more accessible. In the simplicity of a split synthesis, the locations of specific products are lost. Everything is together in a mass of beads. Time-intensive deconvolution procedures are the cost of the simplicity of split synthesis.

Summary

A valuable piece of information for the drug discovery group is the structure of the drug target. Structural information is normally derived from x-ray crystallographic data but also may be derived from NMR and molecular modeling data. A proposed target structure must be *validated* through various methods such as a Ramachandran plot. A correct target structure allows the drug discovery team to make educated guesses on the molecular structure of the desired drug that might best complement the target. Elements in the drug under greatest consideration include functionality that maximizes available intermolecular forces and the three-dimensional orientation of the drug to match the shape of the target. Information on the target is valuable because it can help exclude potential molecules from consideration as drug candidates. The number of possible compounds that may be tested for biological activity is essentially endless, so information that helps narrow the search can save significant time and resources. Regardless of whether the drug discovery team is assisted in its search with a target structure, the group will screen compound libraries for molecules with activity against the target. Increased diversity in the library raises the chance of finding many hits through screening. One relatively recent approach for improving the breadth of a library is combinatorial chemistry. Combinatorial chemistry can generate tens of thousands of different molecules by varying side chains about a common molecular core, or scaffold. Combinatorial libraries are prepared through various techniques, each with advantages and disadvantages. The goal of all the methods is the same: to facilitate the exploration of molecular space for biologically active molecules.

Questions

1. Find a polynomial regression applet on the web to fit the PDB structure data in Figure 9.2. (A good one is http://www.arachnoid.com/polysolve/index.html.) Select a high enough order to give a reasonable fit ($r^2 \geq 0.99$). Based on the best-fit equation, how many structures would you predict to be in the PDB in 2015? 2020?

2. The format of the PDB code is as follows: The first character must be a numerical digit. The last three characters may be either a digit or a letter. The code is not case dependent. With your equation from question 1, in what year will the PDB run out of four-character codes for individual structures?

3. Use a Ramachandran plot generator (http://eds.bmc.uu.se/eds/ramachan.html) to compare three different PDB structures for human hemoglobin: 2HHB, 3HHB, and 4HHB. Does any one of the three stand out as better or worse than the others?

4. Draw resonance forms of an amino acid that explain why the dihedral angle ω is almost always either 0 or 180°.

5. All amino acids are predictable in a Ramachandran plot, except for glycine. Why is glycine so unique in its behavior compared to other amino acids?

6. Ionic bonding depends on the charge state of a compound. Organic chemists tend to draw functional groups in a neutral form regardless of their charge state at biological pH (7.4). Salicylic acid (**9.a**), shown in its neutral form, is the biologically active form of aspirin. The pK_a of the acid in salicylic acid is 3.0. Which of the following is least likely? The carboxylic acid of salicylic acid interacting with a target through an ionic bond, as a hydrogen bond acceptor, or as a hydrogen bond donor? Calculate the ratio of the acid to conjugate base with Equation 9.1 to justify your answer.

salicylic acid
active metabolite of aspirin
analgesic and anti-inflammatory
9.a

7. The phenol OH of salicyclic acid has a pK_a of 13.6. At physiological pH, which of the following is least likely? The phenol of salicylic acid interacting with a target through an ionic bond, as a hydrogen bond acceptor, or as a hydrogen bond donor? Calculate the ratio of the acid to conjugate base with Equation 9.1 to justify your answer.

8. The binding of dopamine (**9.b**) to the dopamine receptor has been intensely studied. *trans*-2-(3,4-methylenedioxyphenyl)cyclopropylamine (**9.c**) is an agonist for the dopamine receptor, while the *cis* isomer is inactive. In reference to Figure 9.10, what relative relationship of the amino and aryl group does **9.c** replicate (antiplanar, synclinal, etc.)? Draw a Newman projection of dopamine that mimics the structure of **9.c**. Is this a low-energy conformation of dopamine? (Triggle, D. J. Adrenergics: Catecholamines and Related Agents. In M. E. Wolf (Ed.), *Burger's Medicinal Chemistry* (4th ed.,) Chapter 41.) New York: Wiley & Sons, 1981.

dopamine
neurotransmitter
9.b

trans-2-(3,4-methylenedioxyphenyl)-cyclopropylamine
9.c

9. Of the top 10 drugs of 2010 based on sales in the United States, nine are administered orally. Those nine drugs (**9.d** through **9.l**) are shown in the following figure. Based on this small group of drugs, is the selection of 30 carbon, nitrogen, oxygen, and sulfur atoms as a "typical drug size" by Guida and colleagues a reasonable number?

esomeprazole
(Nexium)
anti–acid reflux
9.d

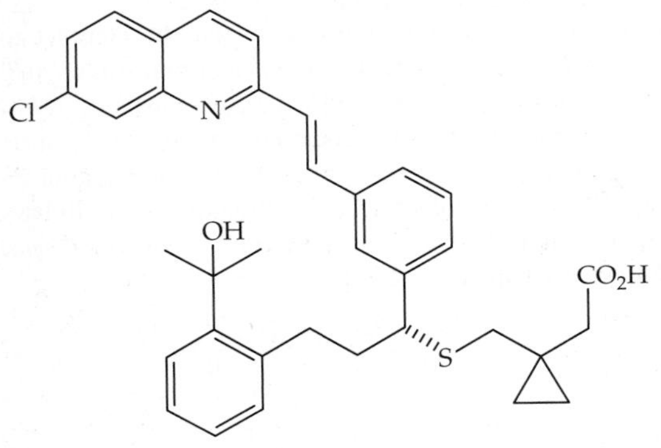

montelukast
(Singulair)
seasonal allergy relief
9.e

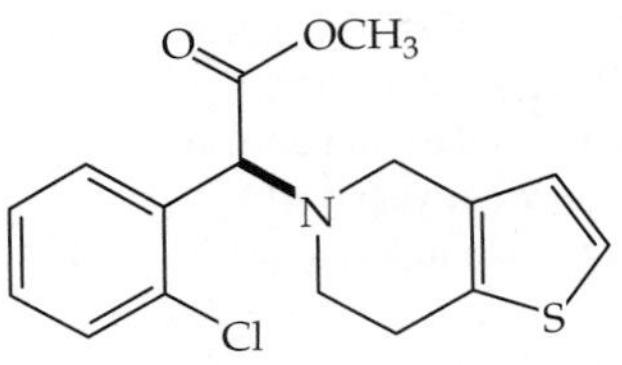

clopidogrel
(Plavix)
antiplatelet agent
9.f

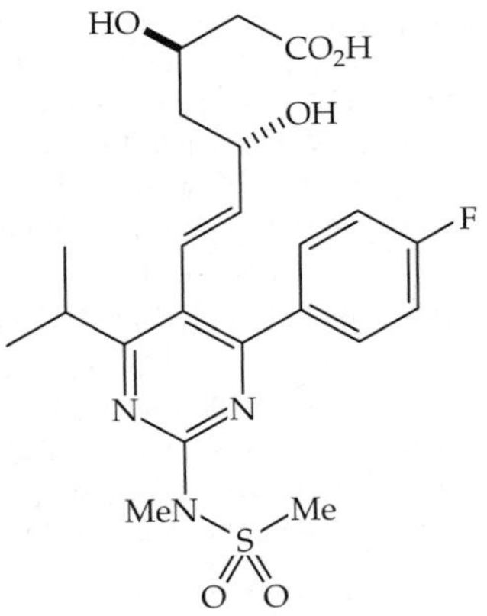

rosuvastatin
(Crestor)
anticholesterol
9.g

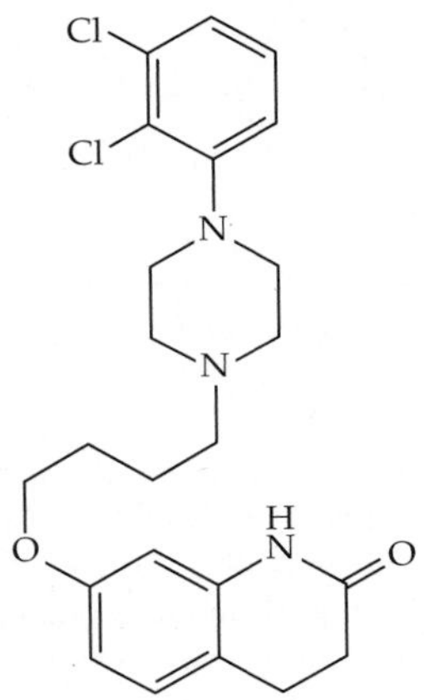

aripiprazole
(Abilify)
antidepressant
9.h

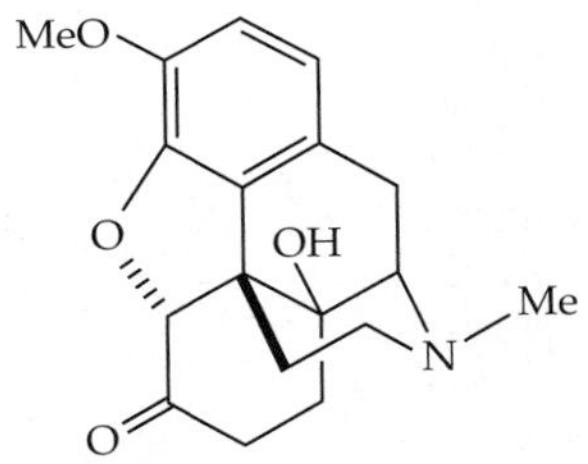

oxycodone
(active component of OxyContin)
analgesic
9.i

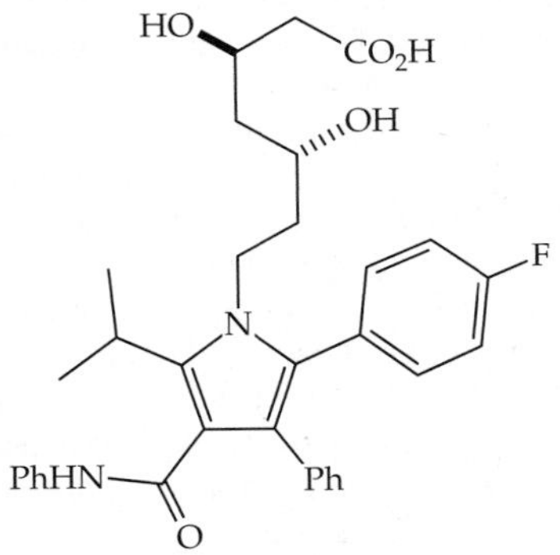

atorvastatin
(Lipitor)
anticholesterol
9.j

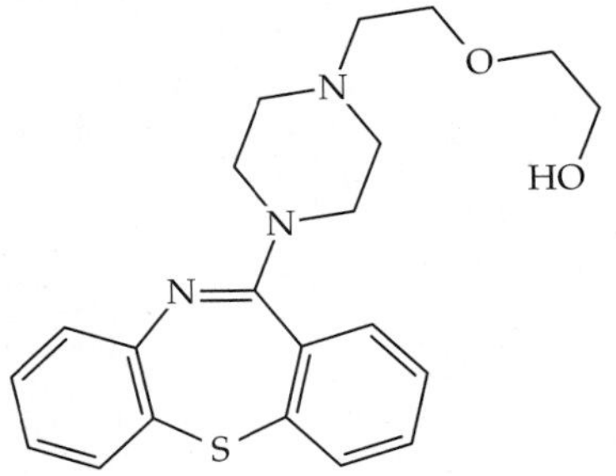

quetiapine
(Seroquel)
antipsychotic
9.k

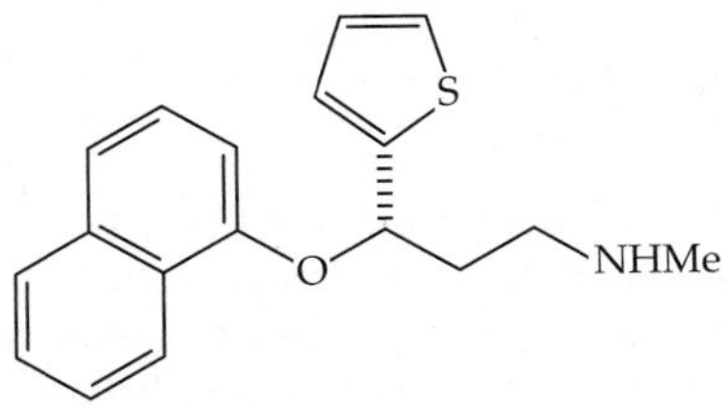

duloxetine
(Cymbalta)
antidepressant
9.l

10. During screening of a molecular library, 0.1% of the tested compounds may show hit level activity. If the screened compounds are representative of molecular space as a whole, how many hits would be found if all drug space (10^{63}) were to be tested? Does this indicate that the number of possible drugs for any given target is large or small?

11. The Ugi reaction is one of the first multicomponent reactions. The reaction combines an aldehyde, acid, amine, and isocyanide to form a new product (**9.m**). The process is efficient, clean, and capable of quickly preparing a large number of compounds. For this reason, reactions like the Ugi reaction are sought for library synthesis. Most major chemical suppliers maintain lists of structures sorted by functional groups. Sigma-Aldrich lists approximately 600 aldehydes, 800 amines, 1,400 acids, and 30 isocyanides in its catalog. Of these compounds, 80% of the acids and aldehydes, 25% of the amines, and all the isocyanides are appropriate for the Ugi reaction. How large a library of Ugi products could be formed from the Aldrich building blocks? (Ugi, I. The α-Addition of Immonium Ions and Anions to Isonitriles Accompanied by Secondary Reactions. *Angew. Chem., Int. Ed. Engl.* 1962, *1*, 8–21. Sigma-Aldrich Co., Organic Building Blocks, http://www.sigmaaldrich.com/chemistry/chemistry-products.html?TablePage=16273490 (accessed August 2011).)

12. Reductants are a common tool in synthetic chemistry. Unfortunately, reducing agents are not always compatible with resins used in solid-phase synthesis. What would be the problem with using $NaBH_4$ or $LiAlH_4$ with a molecule attached to a Wang resin?

13. Propose a mechanism for cleaving the resin in Scheme 9.11 (**9.65** $\rightarrow$ **9.66**). The full structure of **9.61** in Scheme 9.10 will be helpful.

14. In Scheme 9.14, how would one differentiate between compounds D3 and B2?

15. A 27-member library was prepared by split synthesis in three steps. Each step involved three different building blocks: A, B, C; 1, 2, 3; and a, b, c. The library afforded one hit. In a recursive deconvolution of the library, how many compounds would need to be made and tested to determine the identity of the hit? For a hypothetical three-step library of steps involving x, y, and z building blocks, how many compound syntheses would be required for recursive deconvolution? How many compounds would be required to deconvolve a four-step library of m, n, o, and p building blocks?

9.m

References

1. Hillisch, A., Pineda, L. F., & Hilgenfeld, R. Utility of Homology Models in the Drug Discovery Process. *Drug Disc. Today* 2004, *9*, 659–669.

2. Anand, K., Ziebuhr, J., Wadhwani, P., Mesters, J. R., & Hilgenfeld, R. Coronavirus Main Proteinase (3CL^pro) Structure: Basis for Design of Anti-SARS Drugs. *Science* 2003, *300*, 1763–1767.

3. Fuhrman, S. A., Matthews, D. A., Patick, A. K., & Rejto, P. A. Methods and Compositions for Inhibiting SARS-Related Coronavirus Replication by Using Inhibitors of SARS-Related Coronavirus 3C Proteinase. World Patent WO2004093860, November 11, 2004.

4. Ramachandran, G. N., Rmakrishnan, C., & Sasisekharan, V. Stereochemistry of Polypeptide Chain Configurations. *J. Mol. Biol.* 1963, *7*, 95–99.

5. Kleywegt, G. J., Harris, M. R., Zou, J. Y., Taylor, T. C., Wahlby, A., & Jones, T. A. The Uppsala Electron-Density Server. *Acta Crystallogr. D.* 2004, *60*, 2240–2249.

6. Cooper, S. J., Leonard, G. A., McSweeney, S. M., Thompson, A. W., Naismith, J. H., Qamar, S., et al. The Crystal Structure of a Class II Fructose-1,6-Bisphosphate Aldolase Shows a Novel Binuclear Metal-Binding Active Site Embedded in a Familiar Fold. *Structure* 1996, *4*, 1303–1315.

7. Hall, D. R., Leonard, G. A., Reed, C. D., Watt, C. I., Berry, A., & Hunter, W. N. The Crystal Structure of *Escherichia coli* Class II Fructose-1,6-Bisphosphate Aldolase in Complex with Phosphoglycolohydroxamate Reveals Details of Mechanism and Specificity. *J. Mol. Biol.* 1999, *287*, 383–394.

8. Stewart, K. Kinase Inhibitor Design: Introduction and Examples from KDR, CHK, and PIM Kinases. Presented at Medicinal Biochemistry Symposium, University of North Carolina at Greensboro, Greensboro, NC, April 2009.

9. Hopkins, A. L. Pharmacological Space. In C. G. Wermuth (Ed.), *The Practice of Medicinal Chemistry* (3rd ed., Chapter 25). New York: Academic Press, 2008.

10. Cannon, J. G. Cholinergics. In M. E. Wolff (Ed.), *Burger's Medicinal Chemistry* (4th ed., Chapter 43). New York: Wiley & Sons, 1981.

11. Kaiser, C., & Setler, P. E. Antipsychotic Agents. In M. E. Wolff (Ed.), *Burger's Medicinal Chemistry* (4th ed., Chapter 56). New York: Wiley & Sons, 1981.

12. Andersson, T., Hassan-Alin, M., Hasselgren, G., Röhss, K., & Weidolf, L. Pharmacokinetic Studies with Esomeprazaole, the (*S*)-isomer of Omeprazole. *Clin. Pharmacokinet.* 2001, *40*, 411–426.

13. Bohacek, R. S., McMartin, C., & Guida, W. C. The Art and Practice of Structure-Based Drug Design: A Molecular Modeling Perspective. *Med. Res. Rev.* 1996, *16*, 3–50.

14. Chemical Abstracts Service. CAS Registry Keeps Pace with Rapid Growth of Chemical Research, Registers 60 Millionth Substance. http://www.cas.org/newsevents/releases/60millionth052011.html (accessed September 2012).

15. Müller, G. Medicinal Chemistry of Target Family–Directed Masterkeys. *Drug Disc. Today* 2003, *8*, 681–691.

16. Patterson, D. E., Cramer, R. D., Ferguson, A. M., Clark, R. D., & Weinberger, L. E. Neighborhood Behavior: A Useful Concept for Validation of "Molecular Diversity" Descriptors. *J. Med. Chem.* 1996, *39*, 3049–3059.

17. Li, L., Lui, G., Wang, Z., Yuan, Y., Zhang, C., Tian, H., et al. Multistep Parallel Synthesis of Substituted 5-Amino-benzimidazoles in Solution Phase. *J. Comb. Chem.* 2004, *6*, 811–821.

18. Merrifield, R. B. Solid Phase Peptide Synthesis: I: The Synthesis of a Tetrapeptide. *J. Am. Chem. Soc.* 1963, *85*, 2149–2154.

19. DeWitt, S. H., Kiely, J. S., Stankovic, C. J., Schroeder, M. C., Cody, D. M. R., & Pavia, M. R. "Diversomers": An Approach to Nonpeptide, Nonoligomeric Chemical Diversity. *Proc. Natl. Acad. Sci. USA.* 1993, *90*, 6909–6913.

20. Boojamra, C. G., Burow, K. M., Thompson, L. A., & Ellman, J. A. Solid-Phase Synthesis of 1,4-Benzodiazepine-2,5-Ones: Library Preparation and Demonstration of Synthesis Generality. *J. Org. Chem.* 1997, *62*, 1240–1256.

21. Erb, E., Janda, K. D., & Brenner, S. Recursive Deconvolution of Combinatorial Chemical Libraries. *Proc. Natl. Acad. Sci. USA.* 1994, *91*, 11422–11426.

22. Ohlmeyer, M. H. J., Swanson, R. N., Dillard, L. W., Reader, J. C., Asouline, G., Kobayashi, R., et al. Complex Synthetic Chemical Libraries Indexed with Molecular Tags. *Proc. Natl. Acad. Sci. USA.* 1993, *90*, 10922–10926.

23. Nestler, H. P., Bartlett, P. A., & Still, W. C. A General Method for Molecular Tagging of Encoded Combinatorial Chemistry Libraries. *J. Org. Chem.* 1994, *59*, 4723–4724.

24. Tan, D. S., Foley, M. A., Stockwell, B. R., Shair, M. D., & Schreiber, S. L. Synthesis and Preliminary Evaluation of a Library of Polycyclic Small Molecules for Use in Chemical Genetic Assays. *J. Am. Chem. Soc.* 1999, *121*, 9073–9087.

25. Fromont, C., & Bradley, M. High-Loading Resin Beads for Solid Phase Synthesis Using Triple Branching Symmetrical Dendrimers. *Chem. Commun.* 2000, 283–284.

Lead Discovery

Chapter Outline

Once a target, normally an enzyme or receptor, has been established and an assay for activity has been developed, the medicinal chemistry team must discover, find, and make compounds that interact with the target. Through the screening process, some compounds emerge with sufficient activity to warrant further investigation. The active compounds are then examined against a number of criteria, including complexity and anticipated pharmacokinetic behavior. Compounds that satisfy the selection criteria are called leads and advanced for further optimization of activity, selectivity, and biological behavior. Occasionally, leads are found through other methods, such as serendipity or clinical observations. This chapter describes techniques of discovering active compounds through screening and selecting the most promising compounds as leads. The overall process is collectively known as lead discovery.

Learning Goals for Chapter 10

- Appreciate the role that screening of chemical libraries plays in drug discovery
- Distinguish between the different library screening approaches
- Differentiate hits and leads
- Know the properties that allow some hits to be advanced as leads over others
- Understand the non-screening approaches that can produce leads

10.1 Approaches to Searching for Hits

The most common tool for discovering hits is library screening. The library may consist of traditional compounds with potentially high activity molecules, smaller fragments of less activity, or even virtual molecules tested through molecular modeling simulations.

Traditional Library Screening

The goal of screening of a library, in whole or in part, is to discover compounds with modest activity against a target. The active compounds discovered through a screen are called *hits*. The threshold for activity varies based on the target, but hit-level activity is typically 1 μM or lower. Targets are normally enzymes or receptors, so the term *activity* refers to an IC_{50} or EC_{50} value.

General Aspects

Compounds are normally stored as a stock solution to be dispensed as needed by robotic equipment. Dimethylsulfoxide (DMSO) is a preferred solvent for several reasons. First, in low concentrations, DMSO is well tolerated in most assays. Second, the low melting point of DMSO (18 °C) allows samples to be easily frozen. Compounds that are stored in the solid state are less prone to decompose. Third, DMSO is less volatile than most organic solvents. Decreased volatility minimizes solvent evaporation, so concentrations remain nearly constant over prolonged storage. Maintaining a known concentration is vital. The activity of each screened molecule is related by a concentration-effect relationship. If the concentration of a stock solution is not accurate, then any subsequent assessment of activity will also be incorrect.[1]

The purity of compounds in the library is an important factor. Reaction side products (e.g., triphenylphosphine from a Wittig reaction) should be completely removed from a compound in a library. Similarly, compounds should be free of any solvents used in their synthesis. Unfortunately, not all impurities are easily removed. Impurities can sometimes interfere with an assay and lead to inaccurate results.[2]

Some reactions afford mixtures of products. Mixtures include diastereomers, such as *endo* and *exo* products (**10.1** and **10.2**) of a Diels-Alder cycloaddition, and regioisomers, such as *ortho* and *para* products (**10.3** and **10.4**) from an electrophilic aromatic substitution (**Scheme 10.1**). Even a reaction that forms products as subtly similar as enantiomers is technically a mixture of products. Isomeric mixtures violate the spirit of "one compound, one well" in combinatorial chemistry. Isomeric mixtures, however, are often unavoidable and therefore tolerated in compound libraries. Mixtures are also tolerated in libraries of compounds that have been derived from natural sources. Examples include extracts from finely ground vegetation and microbial broths.

Recall from Chapter 2 that high-throughput screens involve some type of biochemical test for binding of the molecule of interest to a target. The screens are performed in 96-, 384-, or 1,536-well microtiter plates. As a library is screened, small amounts of each member of the library will be placed in each well and tested for activity. If each library member is a single compound, a well will contain a single compound or a mixture of compounds. Regardless of whether a library consists of single compounds or mixtures, promising hits in an assay must have their activity confirmed. Confirmation testing involves identifying the library member, synthesizing that molecule, and repeating the

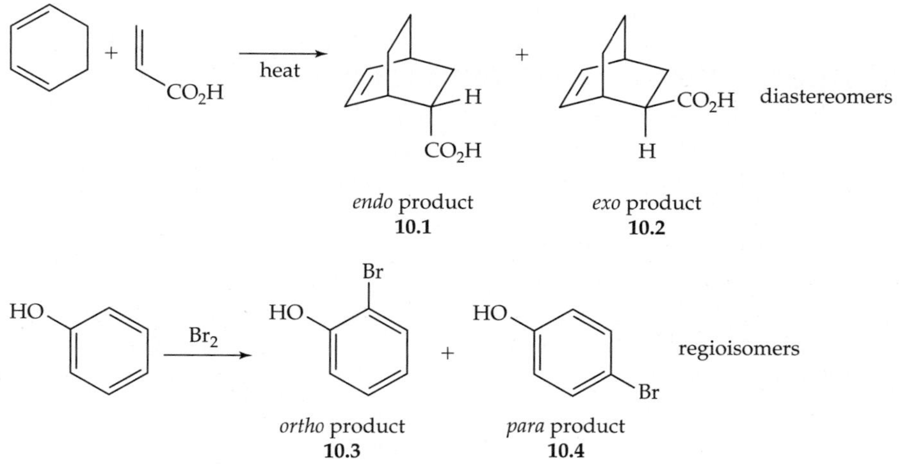

endo product
10.1

exo product
10.2

ortho product
10.3

para product
10.4

SCHEME 10.1 Isomeric products often found in libraries

assay. This process ensures that pure material of a known structure is used in the screen. The testing of pure material also ensures that the activity data is as accurate as possible. If the hit contains a mixture of compounds, each component of the mixture may need to be synthesized to determine which is responsible for the observed activity in the assay.

All screening processes generate some irreproducible data, either *false positives* or *false negatives*. False positives are inactive compounds that erroneously give a positive result in an assay. False negatives correspond to compounds that are active in an assay yet mistakenly show up as inactive or weakly active. False positives are preferred to false negatives in a screen. A false positive will be quickly identified through follow-up confirmation testing. Repeated testing will correctly show the false positive as inactive. In contrast, inactive compounds, including false negatives, are almost never retested. Therefore, the true activity of a false negative will likely never be discovered, and a promising hit, and possibly a drug, may be completely missed.

An important idea of a compound library is that it can be used over and over again in different assays. A pharmaceutical company may screen its library against a receptor one week and an enzyme the next week. A high-throughput screen (HTS) requires little sample. Even a few milligrams of a molecule, when diluted into DMSO, will provide enough stock solution for many assays. The entire library is not normally tested in each assay. Preliminary information, such as x-ray or NMR data, on the target may allow some compounds to be eliminated from the screen based on physical properties (polarity, charge, size) or chemical structure. If less information is known about a target, the odds of the entire library being screened become greater. The cost of an inexpensive high-throughput screen may be as low as US$0.10 per well, so screening a large library of a million compounds would be $100,000 or more. This may be a small price to pay for information that affords hits that can be optimized into a marketable drug, but a marketable drug is not a sure outcome of every large-scale screening.

In summary, compound libraries are a vital resource in a productive drug discovery program. Molecular biological information on a target can advance a drug search only so far. Random screening of either an entire or partial library, then, hopefully provides promising hits for the lead optimization stage.

In-House Libraries

As medicinal chemists synthesize molecules during their day-to-day research, small samples of new compounds are submitted for inclusion in the company's compound library. Over the course of years and decades of research, a compound library steadily grows. A library will reflect the areas of research that have contributed to the collection. A company that has historically been strong in researching β-lactam antibiotics (**10.5** and **10.6**) would have a very different library from a company with strength in estrogen receptor binding compounds (**10.7** and **10.8**) (**Figure 10.1**). Once combinatorial chemistry became recognized in the 1990s as a method for making large numbers of molecules, most pharmaceutical companies hired teams of chemists to create collections of molecules to augment the company's existing library. Samples from natural sources may also be included in a library.

When pharmaceutical companies merge, their libraries merge as well. Through a library, the purchasing company gains a tangible chemical record of the research of the acquired firm. In 1995, Glaxo acquired Wellcome. An area of strength for Wellcome was antiviral research. Wellcome's products at the time included acyclovir (**10.9**) and zidovudine (AZT, **10.10**), nucleoside analogues with activity against the herpes simplex and human immunodeficiency viruses, respectively (**Figure 10.2**). Today, Glaxo, now operating under the name GlaxoSmithKline (GSK), still maintains a strong presence in treatments

FIGURE 10.1 β-Lactam antibiotics (**10.5** and **10.6**) and estrogens (**10.7** and **10.8**)

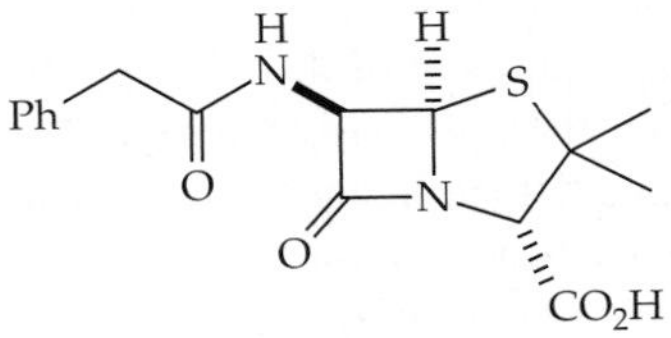

penicillin G
antibiotic
10.5

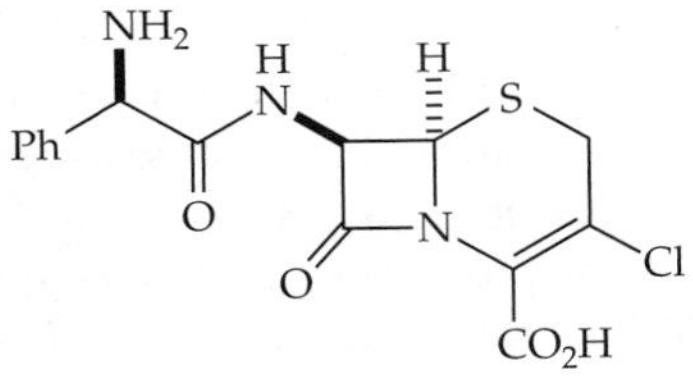

cefaclor
(Ceclor)
antibiotic
10.6

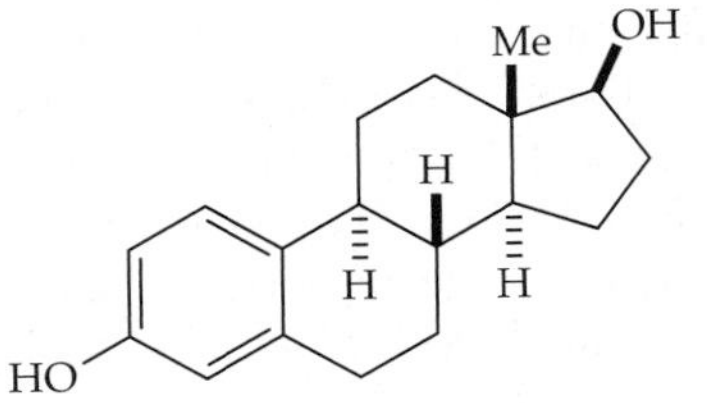

estradiol
hormone
10.7

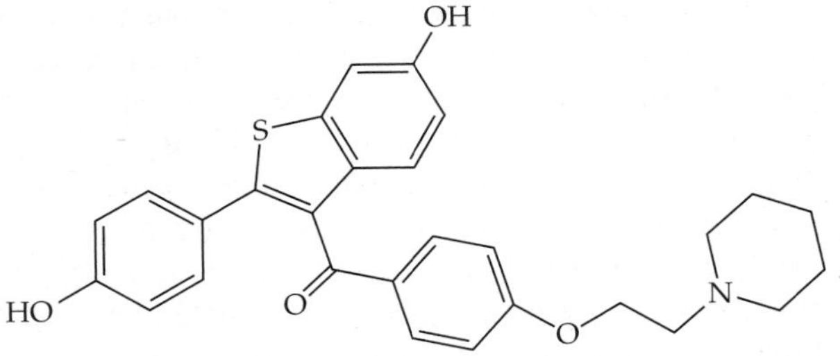

raloxifene
(Evista)
estrogen receptor ligand
10.8

FIGURE 10.2 Antiviral compounds

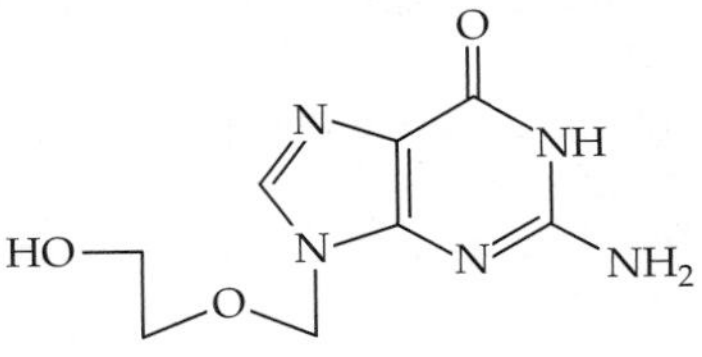

acyclovir
(Zirovax)
10.9

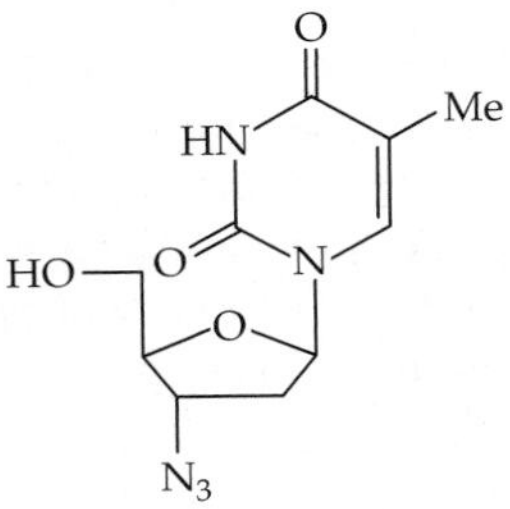

zidovudine
(Retrovir)
10.10

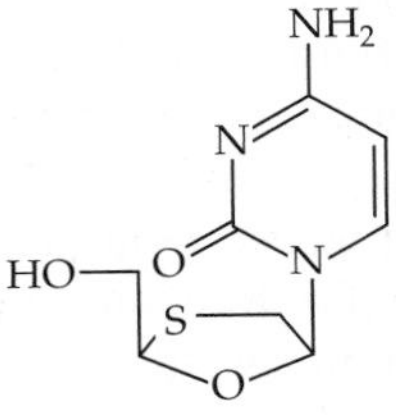

lamivudine
(Epivir)
10.11

for viral infections. One recently developed drug is lamivudine (**10.11**), an anti–human immunodeficiency virus nucleoside analogue.

Compound libraries may be bought and sold individually. After the dissolution of the Soviet Union in 1991, laboratories that had been formerly well funded by the Soviet government suddenly became essentially broke. As a means of generating funds, some research groups began to sell portions of their in-stock compounds. The samples were readily purchased by Western companies, including the pharmaceutical industry. The value of the compounds depends on the novelty of the structures and their purity.

Specs, founded in 1987 and based in Delft, The Netherlands, purchases compounds from all over the world, mostly from academic laboratories. These compounds are added to Specs' existing library and in turn sold to interested companies. A company can search the Specs library and purchase promising compounds or the entire collection. Those compounds become the outright property of the purchasing company. If Specs has a sufficient amount of a given compound, the company will sell samples of each compound many times over. The amount sold for each compound may be only 0.5 to 1.0 mg.

Out-Sourced Libraries

Just as a library can be purchased, a library can also be rented. The owner of the library typically enters into an agreement with a drug company. The drug company pays to access and test the compound in a screen. If the compound eventually results in the discovery of a new drug, the owner of the library may receive a bonus. One company built on this type of business model was Pharmacopeia, Inc., of Cranbury, New Jersey.

Pharmacopeia was founded in the early 1990s by W. Clark Still of Columbia University and Michael H. Wigler of Cold Spring Harbor Laboratory. Still and Wigler were early pioneers in the development of combinatorial chemistry for pharmaceutical development purposes.[3] As of 2007, Pharmacopeia claimed to have used the Still and Wigler techniques to prepare a library of over 7.5 million compounds—a massive number that is far greater than the library of a typical major drug company. With a library of this size, Pharmacopeia entered licensing agreements with drug companies. In each partnership, Pharmacopeia brought a large library for discovering hits and an ability to make additional compounds as needed for lead optimization. In turn, the pharmaceutical company provided expertise in screening compounds, performing clinical trials, and marketing a drug. If a compound provided by Pharmacopeia became a marketed drug, Pharmacopeia shared in the revenues from the drug's sales. Over the years, as its own resources grew, Pharmacopeia shifted its business plan more toward developing drugs independently, without the involvement of an outside pharmaceutical company. In 2008, Pharmacopeia was purchased by Ligand Pharmaceuticals of La Jolla, California. Ligand now owns the full chemical library of Pharmacopeia.

Mycosynthetix of Hillsborough, North Carolina, out-sources a large library of fungal broths. The screening of products from fungi is attractive because the number of different species of fungi is astonishingly large and essentially unexplored. Any biological activity discovered through screening compounds from fungi is almost certainly previously unknown and therefore more easily protected through patents.

Fragment-Based Screening

Fragment libraries are no different than traditional compound libraries except molecules in a fragment library are smaller. Fragments have a molecular weight of only 120 to 250 g/mol. Limiting molecular weight dramatically decreases diversity in the library. Far fewer molecules are required to sample the molecular space of a 250 MW library than a 400 or 500 MW one. Smaller molecules have fewer potential sites for intermolecular binding than larger molecules. Therefore, small molecules rarely bind as strongly as larger compounds. For example, a hit from a typical combinatorial library may show activity (K_D, IC_{50}, K_i, EC_{50}) at concentrations of 1 μM or lower. In contrast, a hit in a fragment library may be selected with activities of around 1 mM, which is a 1,000-fold difference in activity. Remember that a larger IC_{50} value implies weaker enzyme-inhibitor binding.[4,5]

By itself, a single fragment with 1 mM binding is not very interesting. However, if multiple fragments are known to bind near the same site on a target, then the fragments can sometimes be connected to form a single strongly binding hit (**Scheme 10.2**). The key to discovering hits through fragment-based screening requires two steps. First binding fragments must be discovered. Second, the fragments must be properly connected and rescreened to discover a hit. Proper connection of the fragments can be a challenge. The tether between the fragments must be the correct length and placed appropriately. Successful examples of fragment-based hit discovery involve targets with two or more active site pockets, each of which can accommodate a fragment-sized group. Because

SCHEME 10.2 Hit development through fragment-based screening

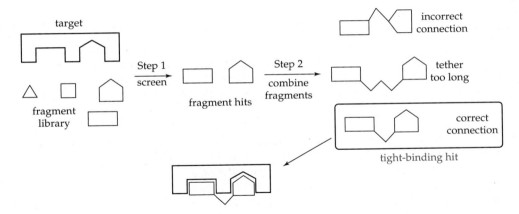

fragment connection requires a spatial understanding of how a fragment and target interact, fragment-based screening methods need to have a three-dimensional model of the target. Visual models arise from NMR, x-ray crystallography, and molecular modeling data which were discussed in the opening section of Chapter 9. In x-ray crystallography, x-ray structures of fragments bound to a target provide both the site and position of binding. With quality structural information to guide the drug discovery group, determining the ideal linker length and position is a much easier task.[4,5]

Recent research performed in the laboratory of George Whitesides at Harvard University suggests that the only poor choice for a linker is one that is too short to span the distance between fragment binding sites. Linkers that are longer than necessary simply fold upon themselves to bring the fragments closer for binding a target. This research implies that initial linkers to tether fragments should be longer rather than shorter. The optimal length can be determined in a subsequent study.[6]

The following three Case Studies demonstrate the use of fragment-based screening to discover leads.

CASE STUDY

Inhibition of Stromelysin[7,8]

Stromelysin is a zinc-dependent protease that is responsible for breaking down and re-forming connective tissues, including collagen. Dysfunctional activity of stromelysin and related enzymes is associated with arthritis and tumor activity. An effective inhibitor of stromelysin may serve as a treatment for these conditions. With these facts in mind, a discovery team at Abbott used NMR to screen a fragment library for stromelysin binding. The technique of using NMR to discover hits from fragment libraries is often called *SAR by NMR*. SAR, or *structure-activity relationship*, is a term normally associated with the lead optimization process, not lead discovery. However, as has already been mentioned in this chapter, lead discovery and optimization are not completely independent processes.

Screening by NMR requires a ^{15}N-labeled target, stromelysin in this case, that has a well-characterized two-dimensional ^{15}N-^{1}H spectrum. In a labeled protein, each amino acid has a ^{15}N-^{1}H bond on the backbone of the protein. Each ^{15}N-^{1}H bond gives a separate signal in the NMR spectrum. If the spectrum is characterized, then the signals corresponding to amino acids in the active site are located. Addition of a fragment, if it binds the target, will cause changes in the spectrum. If the spectral changes occur with signals known to be associated with the active site, then the fragment likely binds at the active site. Once two active site binding fragments are identified, linkers can be designed to connect them.

Before the Abbott study, two important facts about stromelysin were already known. First, the Zn^{2+} ion in the active site binds weakly to hydroxamic acids (**Figure 10.3**). So, acetohydroxamic acid (**10.12**) was chosen as the first fragment with a modest K_D of 17 mM. Second, the active site of stromelysin was known to

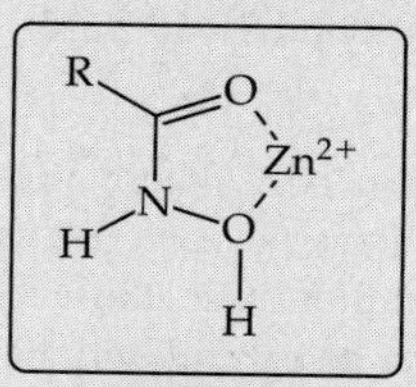

binding of hydroxamic acids to Zn^{2+}

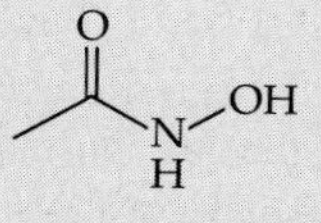

acetohydroxamic acid
MW 75
10.12
$K_D = 17$ mM

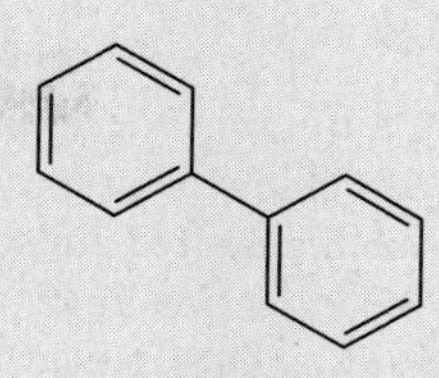

biphenyl
10.13

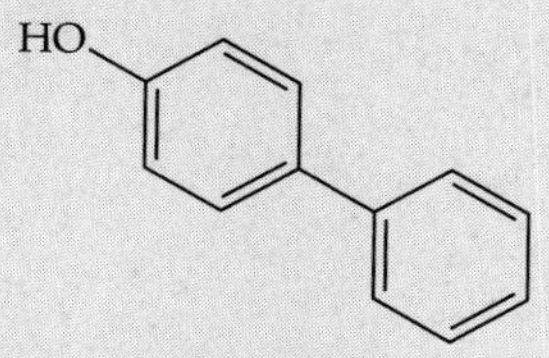

4-hydroxybiphenyl
MW 170
10.14
$K_D = 0.28$ mM

FIGURE 10.3 Fragments with modest binding to stromelysin

contain a hydrophobic binding site. A fragment library of biphenyl (**10.13**) and related compounds was selected to test binding in the hydrophobic pocket. 4-Hydroxybiphenyl (**10.14**) emerged as a promising fragment with a K_D of 0.28 mM.

Based on additional NMR information, the active site of stromelysin was modeled, and the relative orientations of fragments **10.12** and **10.14** were approximated in the binding pocket. Synthesis of a series of compounds (**10.15–10.18**), a combination of both fragments with different linker lengths, generated a set of micromolar inhibitors (**Figure 10.4**). The most potent was **10.16**. Note that both **10.12** and **10.14** are fragment hits with low molecular weights of 75 and 170. Compound **10.16**, as a combination of two fragments, shows an IC_{50} activity that is typical for a traditional hit. Structure **10.19** was subsequently found to be a strongly binding compound with an IC_{50} of 15 nM (0.015 μM).

The effect of combining fragments can be examined energetically. K_D values are equilibrium constants and related to the standard free energy of association/binding ($\Delta G^\circ_{\text{bind}}$) by Equation 10.1.

$$\Delta G^\circ_{\text{bind}} = -2.3RT \log K_A = -2.3RT \log \frac{1}{K_D} = +2.3RT \log K_D \quad (10.1)$$

The standard free energy of binding for **10.12** with a K_D of 17 mM is −2.4 kcal/mol. Compound **10.14**, with a K_D of 0.28 mM, has a binding energy of −4.9 kcal/mol. Bringing both fragments together with an appropriate tether should provide a binding energy of −7.3 kcal/mol (−2.4 + −4.9). Instead, the energy of binding for **10.16** is −8.9 kcal/mol, 1.6 kcal/mol lower than expected. The difference is likely attributable to a combination of two factors: weak binding by the carbons in the tether and/or changes to the fragments by adding the tether. For example, capping the OH of **10.14** forms an ether. The nonpolar ether of **10.16** may complement the binding site better than the polar phenol of **10.14**.

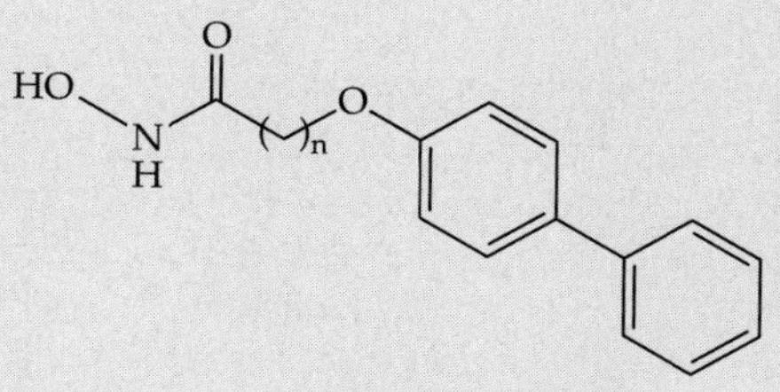

	n	MW	IC_{50} (μM)
10.15	1	243	3.9
10.16	2	257	0.31
10.17	3	271	110
10.18	4	285	100

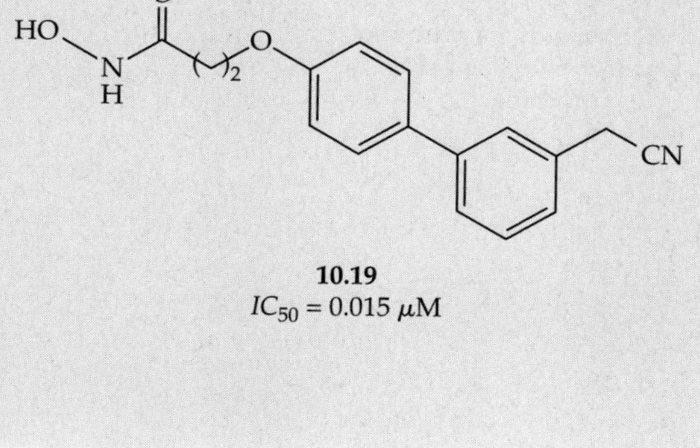

10.19
$IC_{50} = 0.015$ μM

FIGURE 10.4 Inhibitors of stromelysin

| **Sample Calculation** | **Determining a Binding Energy from an IC_{50} Value** |

PROBLEM Determine the standard free energy of binding of compound **10.19** based on its IC_{50} of 0.015 μM.

SOLUTION Binding energies are simple to determine with Equation 10.1. The hardest part may be determining the proper value of R, the gas constant, to use in the equation. The value of R depends on the desired units for $\Delta G°$. We have been using kcal/mol. For these units, we need to use $R = 0.00199$ kcal/mol $\cdot$ K. We will use $T = 298$ K. As always, be careful with the units on K_D and IC_{50} values. Always convert them to molarity. Therefore, in place of 0.015 μM, we need 1.5×10^{-8} M. With all the details handled, the calculation is fairly straightforward. Keep in mind that logarithmic operations are unitless, so the molarity units on IC_{50} disappear in the calculation. The $\Delta G°_{bind}$ calculates as -10.7 kcal/mol.

$$\Delta G°_{bind} = +2.3RT \log K_D \qquad (10.1)$$

$$\Delta G°_{bind} = 2.3 \times 0.00199 \frac{\text{kcal}}{\text{mol} \cdot \text{K}} \times 298 \text{ K} \times \log\left(1.5 \times 10^{-8} \frac{\text{mol}}{\text{L}}\right)$$

$$\Delta G°_{bind} = -10.7 \frac{\text{kcal}}{\text{mol}}$$

CASE STUDY

Inhibition of Cyclin-Dependent Kinase 2[9]

Researchers have been able to allow the target itself to select an ideal hit from a library of fragments and fragment products. This technique is called *dynamic combinatorial chemistry*. Fragments with complementary reactivity react in the presence of a target (**Scheme 10.3**). Either the target binds optimal fragments and causes them to link in situ, or the target selects the best product of a fragment reaction. In either case, the tightest-binding product occupies the active site of the target. Careful analysis of the electron-density map of the target cocrystallized with the product reveals the x-ray structure of the product. Naturally, a limitation of this method is that the target must be able to be crystallized.

An inhibitor of cyclin-dependent kinase 2, an enzyme implicated in a number of human cancers, has been developed through a simple application of dynamic

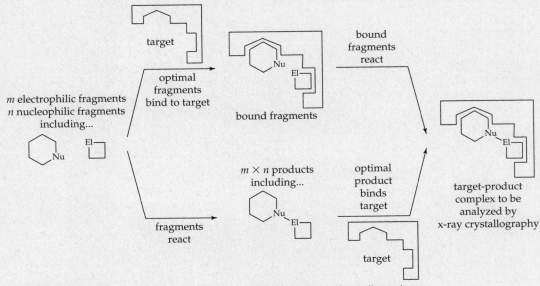

SCHEME 10.3 Application of dynamic combinatorial chemistry with x-ray crystallography

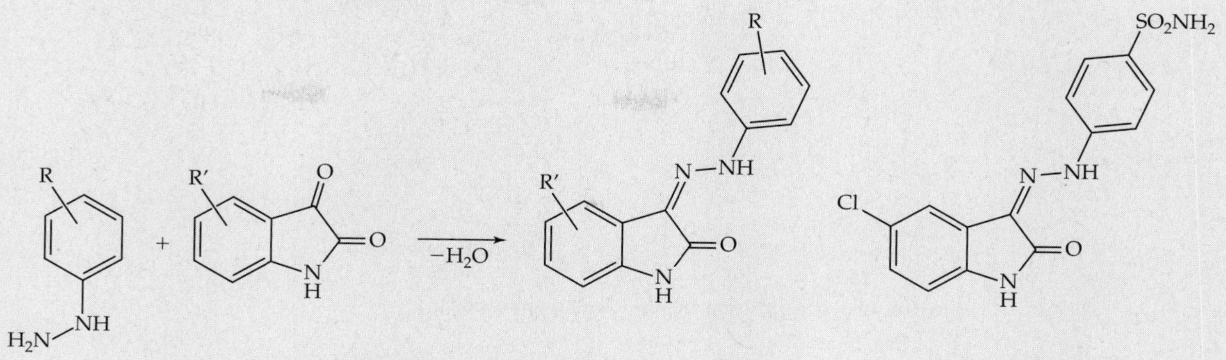

SCHEME 10.4 DCC-based discovery of a CDK2 inhibitor

hydrazine
10.20

oxindole
10.21

condensation
product
10.22

10.23
$IC_{50} = 30$ nM

combinatorial chemistry with x-ray crystallographic analysis. Kinase crystals were added to a solution of a fragment library of six arylhydrazines (**10.20**) and five oxindoles (**10.21**) (**Scheme 10.4**). A total of 30 condensation products (**10.22**) were possible, and all 30 had been shown to form under the conditions of the study. Furthermore, compounds of type **10.22** were previously found to bind the kinase, so the x-ray study was guaranteed to afford

potent hits. The kinase crystals were isolated from the mixture and analyzed to reveal binding of one compound (**10.23**) in the active site.

Although the study of cyclin-dependent kinase 2 was a proof-of-principle example with compounds already known to show activity, the technique of dynamic combinatorial chemistry is a viable method for screening for hits from a fragment library.

CASE STUDY

Inhibition of Acetylcholine Esterase[10]

Mass spectrometry can also be used to screen a library of reactive fragments. The concept is similar to the x-ray crystallography Case Study, except the fragments must not be able to react in the absence of the target. The target binds fragments in close proximity to allow connection of the fragments with a reactive tether. Analysis of the solution by mass spectrometry reveals any high molecular weight compounds that are products of fragment combination.

K. Barry Sharpless and colleagues at the Scripps Research Institute searched for inhibitors of acetylcholine esterase (AChE). AChE is known to bind both tacrine (**10.24**) and ethidium bromide (**10.25**) (**Figure 10.5**). The idea of the Sharpless group was to attach reactive tethers to fragments containing tacrine and ethidium. In the presence of AChE, the fragments would bind the enzyme. Tacrine and ethidium systems bearing optimal tethers would then covalently link in the active site of AChE to form a tight-binding inhibitor of AChE

from the two fragments. The high molecular weight inhibitor could then be detected by mass spectrometry.

The reaction chosen to connect the tethers in situ was the Huisgen azide-alkyne cycloaddition (**Scheme 10.5**). The Huisgen cycloaddition forms 1,2,3-triazoles as a nearly 1:1 mixture of regioisomers (**10.28** and **10.29**). The reaction is slow at room temperature. However, if the azide and alkyne are positioned ideally, such as when bound in close proximity by AChE, then the reaction occurs at room temperature.

The fragment library consisted of tacrine and ethidium derivatives with tethers of variable lengths to both azides and alkynes (**Figure 10.6**). In total, the library included 36 different tacrine-ethidium combinations. Each combination could form two regioisomeric triazoles, so a grand total of 72 different products was possible.

Each of the possible 36 azide-alkyne combinations was reacted separately in the presence of AChE. Only one reaction

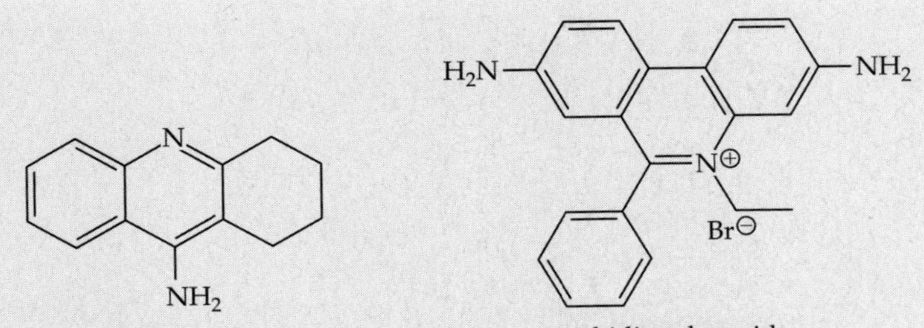

tacrine
10.24

ethidium bromide
10.25

FIGURE 10.5 Acetylcholine esterase-binding compounds

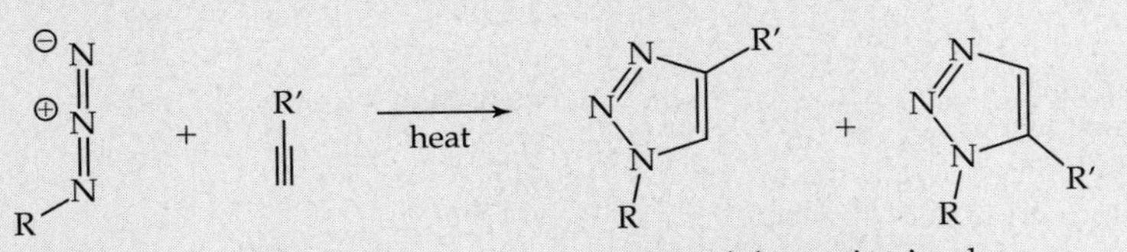

azide
10.26

alkyne
10.27

regioisomeric triazoles
10.28 **10.29**

SCHEME 10.5 Cycloaddition of azides with alkynes to form triazoles

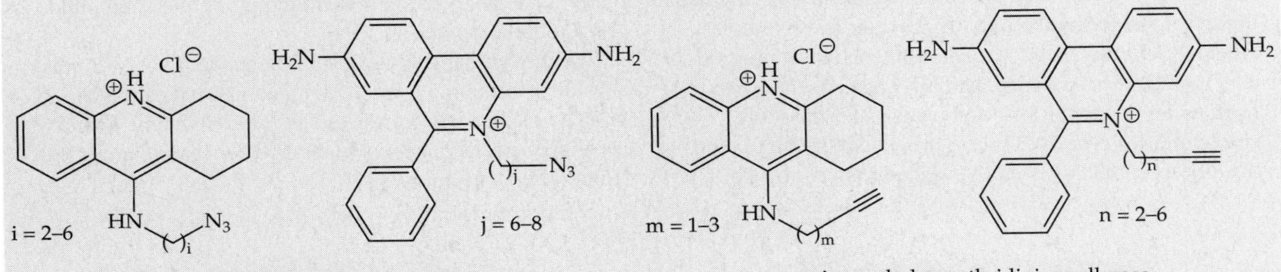

tancrine and phenanthridinium azides

tancrine and phenanthridinium alkynes

FIGURE 10.6 Azide-alkyne fragment library

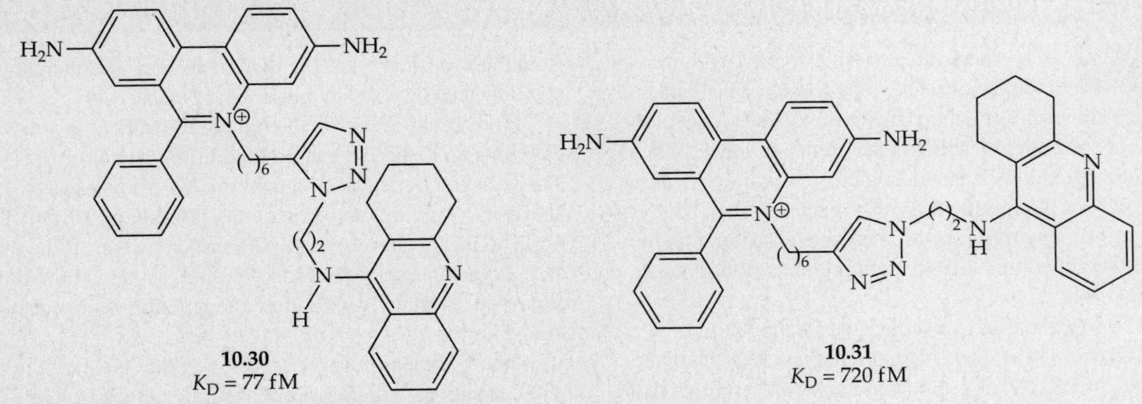

10.30
$K_D = 77$ fM

10.31
$K_D = 720$ fM

FIGURE 10.7 Azide-alkyne fragment library hits

afforded a detectable amount of triazole product: a molecule (**10.30**) with a molecular weight of 661 by mass spectrometry (**Figure 10.7**). The compound was found to be a single regioisomer. Both **10.30** and its regioisomer (**10.31**) could be prepared in the absence of AChE. Independent testing of each regioisomer confirmed the K_D of **10.30** as an astonishing 77 fM (femtomolar, 10^{-15} M) and revealed weaker, yet still remarkably strong, binding of **10.31** with a K_D of 720 fM.

Virtual Screening

Virtual screening, sometimes called *in silico screening*, is a relatively new approach to library testing. In a virtual screen, computerized molecular models of both the target and library member are aligned to determine potential complementary intermolecular interactions. Molecules with a high level of complementarity, indicative of potentially strong binding, are flagged for synthesis and testing. A virtual screen requires sufficient knowledge about the target protein structure, likely from x-ray or NMR data.[11] Virtual screening does *not* require an existing compound library. Any molecule imaginable can be modeled in a computer and screened. Of course, the virtual library should consist of realistically synthesizable compounds.[12] Selected compounds are also normally filtered for those with desired structural elements.

Virtual screening faces a number of challenges. Molecular conformations of both the target and library member are an issue. The target protein will have many low-energy conformations. Some virtual screening methods attempt to accommodate flexibility of the target protein, as complicated as it may be. All in silico approaches try to account for flexibility of the library member, which is a challenge because even small molecules can have many low-energy conformations. Other factors include tautomers, pH-dependent ionizations, and stereochemistry. Properly handling all these variables is not trivial and quickly complicates the modeling process. Once starting conformations of the target and library members have been established, each library member is virtually brought into contact with the target to determine the likelihood of binding. The process, called *docking*, follows the induced-fit model in which the interacting molecules influence each other's conformations until a minimum energy is reached.[11]

After a compound has been docked to the target, the binding energy is estimated in a process called *scoring*. Standard intermolecular forces, contact forces, dipole interactions, and hydrogen bonding are approximated and totaled. Current scoring methods produce many *false positives*. False positives are compounds with high predicted binding that show little or no activity in validation testing. Often, the number of false positives can be reduced by using several different scoring systems to calculate binding. This approach is called *consensus scoring*. Compounds with high predicted activities from more than one method are selected for further investigation.[11]

High-scoring compounds are then prepared and screened in a biochemical assay. "Preparing" requires either synthesizing or purchasing a sample of the compound. Sample purchase is normally much faster than synthesis. For this reason, compounds in a virtual screen are often limited to structures that can be purchased from any number of commercial suppliers. Databases of available molecules have been prepared for the sole purpose of assisting virtual screening.

The following three Case Studies highlight the use of virtual screening to find hits.

CASE STUDY

Inhibition of Protein Kinase B[13]

Protein kinase B uses ATP to phosphorylate other enzymes. Controlling protein kinase B is a method of influencing many cellular processes, including cell division in cancer cells. A virtual screen for protein kinase B inhibitors started with an x-ray structure of protein kinase B bound to AMP-PNP (**10.32**), an unreactive analog of ATP, and a peptide substrate (**Figure 10.8**). A collection of 50,000 compounds from Chembridge (San Diego, California) was docked into the active site of protein kinase B and scored by an algorithm called Drugscore (BioSolvIT, Sankt Augustin, Germany). The top 4,000 compounds were then scored by two other methods, called Goldscore and Chemscore. The lists of the top 700 compounds from both methods contained 200 common compounds. Visual inspection of these

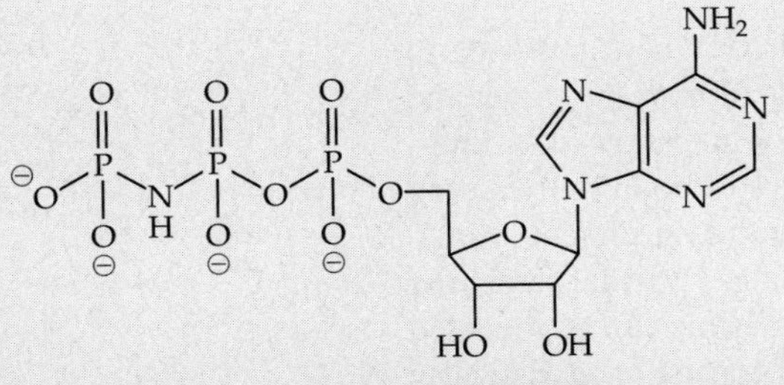

AMP-PNP
10.32

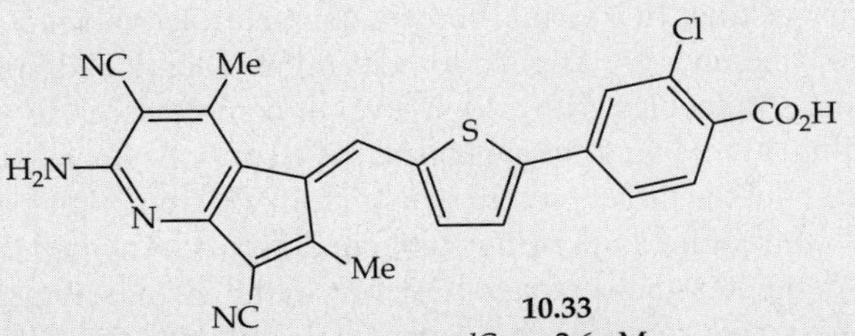

10.33
$IC_{50} = 2.6\ \mu M$

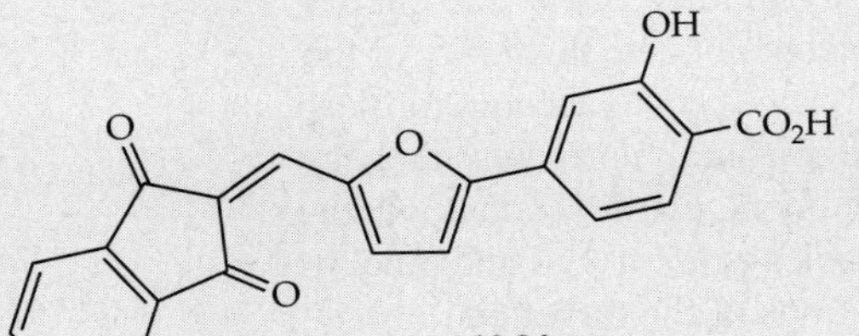

10.34
$IC_{50} = 25\ \mu M$

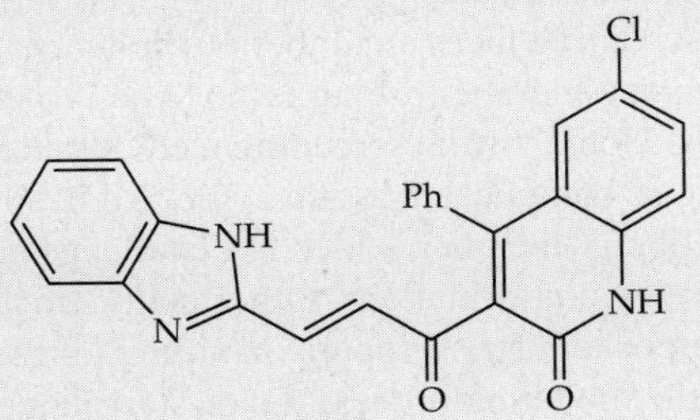

10.35
$IC_{50} = 4.5\ \mu M$

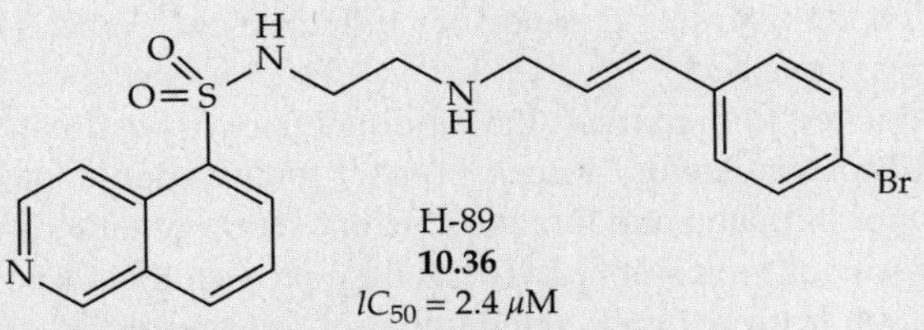

H-89
10.36
$IC_{50} = 2.4\ \mu M$

FIGURE 10.8 AMP-PNP (**10.32**) and inhibitors of protein kinase B (**10.33–10.36**)

200 docked structures allowed 100 to be eliminated because of unfavorable interactions. The remaining 100 compounds were screened against protein kinase B. Three compounds (**10.33–10.35**) emerged with hit-level activity. Experimental activity was comparable to H-89 (**10.36**), a commonly cited protein kinase B inhibitor.

CASE STUDY

Inhibition of Sir2 Type 2[14]

Sir2 type 2 is a regulator enzyme that deacylates tubulin and may be involved in aging and cancer. A virtual screen for an inhibitor started with a molecular model built from a high-resolution x-ray structure of the regulatory enzyme. In a molecular dynamics simulation, the model was allowed to undergo conformational changes. Various low-energy conformers were weighted, averaged, and minimized to afford a new molecular model. A handful of residues in the 389-peptide protein fell outside the allowed regions of a Ramachandran plot. None of the outlier residues was within the active site of the enzyme, so the new minimized model was deemed to be adequate for virtual screening.

Two known Sir2 type 2 inhibitors, sirtinol (**10.37**) and A3 (**10.38**), were docked into the modeled active site (**Figure 10.9**). Studying both complexes revealed the positions of likely important hydrophobic and hydrogen bonding interactions within the active site. A search of a library of approximately 50,000 compounds available from Maybridge (Trevillett, England) was searched for screening candidates. In total, 44 compounds satisfied the selection criteria, largely based on the structure of the modeled

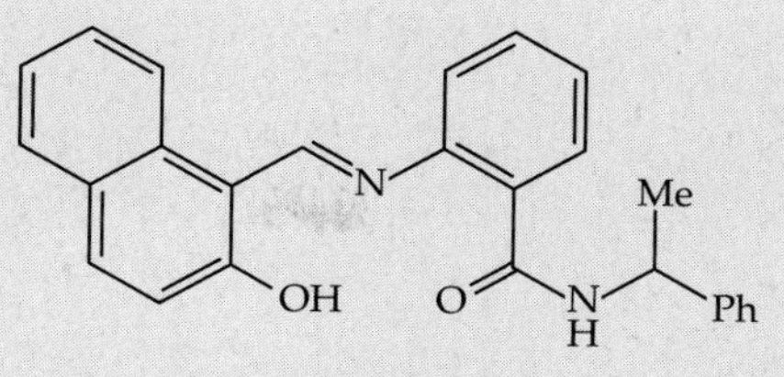

sirtinol
10.37
$IC_{50} = 38\,\mu M$

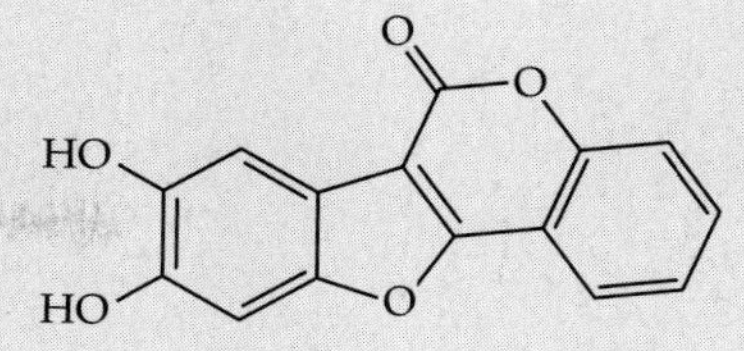

A3
10.38
$IC_{50} = 45\,\mu M$

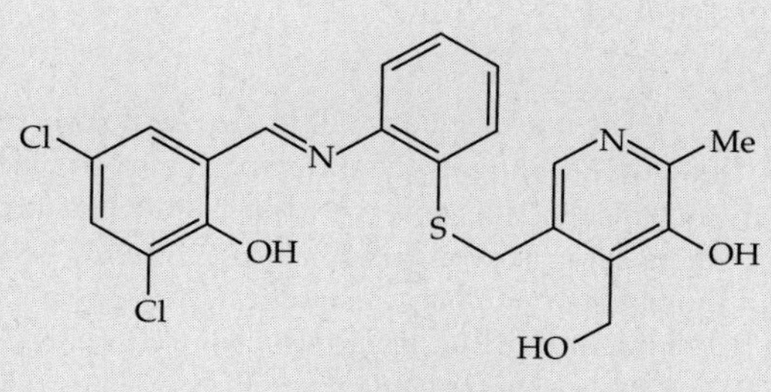

10.39
$IC_{50} = 74\,\mu M$

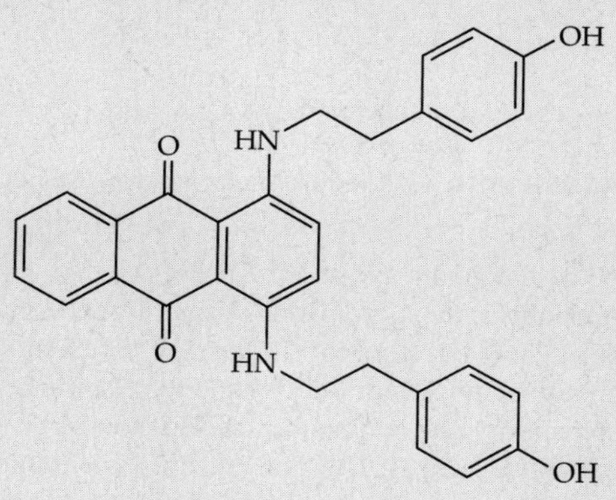

10.40
$IC_{50} = 57\,\mu M$

FIGURE 10.9 Inhibitors of Sir2 type 2 regulatory enzyme

enzyme. Docking of all 44 compounds revealed 15 that were able to interact fully with the active site. Screening of the 15 compounds against the regulatory enzyme indicated promising activity for **10.39** and **10.40**. Interestingly, all attempts to score the binding energy through software analysis of the docked inhibitors failed to distinguish active and inactive structures. Therefore, all docked structures were inspected visually for reasonable interactions.

CASE STUDY

C-C Chemokine Receptor Type 5 Agonists[15]

C-C chemokine receptor type 5 is a receptor that has been implicated in the early stages of HIV infection. Antagonists of this receptor may theoretically block viral infection of a cell. New antagonists were pursued through a virtual screen. A computer model of the receptor, a G-protein–coupled receptor, was constructed based on constraints from an x-ray structure of bovine rhodopsin, another G-protein–coupled receptor. A handful of C-C chemokine receptor type 5 antagonists have been reported in the literature. The binding pocket of the known antagonists had been elucidated through site-directed mutagenesis studies and shown to consist of two distinct hydrophobic regions spanning an anionic glutamic acid residue. Docking each antagonist into the binding site, minimizing the complex, and averaging the structures provided the final model of the receptor.

Initial tests were performed on a pool of 1,000 compounds, including seven known antagonists. Two scoring methods were able to place half the known antagonists into the top 5% of all docked compounds. This exercise validated the model of the C-C chemokine receptor type 5 binding pocket as well as the scoring method for complex evaluation.

The virtual screen was performed on nearly 50,000 compounds that were taken from a library of 1.6 million. Selection for docking required a structure to have a nonpolar aromatic ring and basic amine for matching the binding

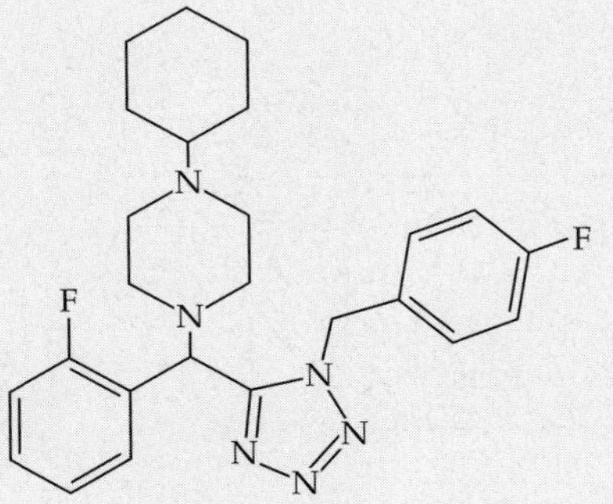

FIGURE 10.10 C-C chemokine receptor type 5 binding compounds

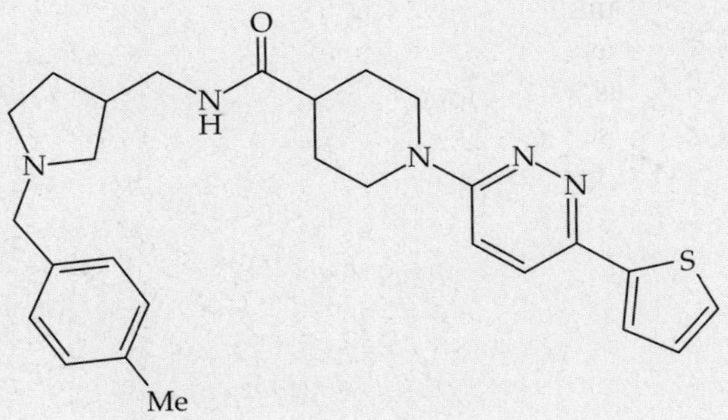

pocket. A repetitive docking and scoring process indicated 81 compounds as most promising. Visual inspection reduced the final candidate pool to 77. Through biochemical screening, seven compounds were found to effectively bind the receptor. Only one compound, **10.41**, showed antagonist activity (**Figure 10.10**). The other six, including **10.42**, were agonists or partial agonists.

The unexpected result in the chemokine receptor example highlights the difference between binding an enzyme and a receptor. A nonsubstrate that strongly binds an enzyme active site will almost certainly be an effective competitive inhibitor. In contrast, a compound that blocks the binding site of an agonist can itself either elicit a response or shut down the receptor.

10.2 Filtering Hits to Leads

If a screening process is successful, a number of hits will be identified. The number of hits varies depending on the target. It could be as high as 5–10% or as small as 0.1% of the tested library. The cutoff for the required activity of a hit is somewhat arbitrary. The discovery group may select an activity level based on other known active compounds. The threshold may also be based on the performance of the entire library. For example, the discovery group may count all compounds that are two or three standard deviations more active than the average of the full library. For a library with normally distributed activity, a cutoff of two standard deviations would give a hit rate of 2.1%. A cutoff of three standard deviations would give a hit rate of 0.1%. Based on a representative hit rate of 1%, screening a library of 100,000 compounds would generate 1,000 hits. This is too many compounds to follow up each hit individually, so the number of hits needs to be reduced, or *filtered*, to reach a more manageable figure.

Pharmacodynamics and Pharmacokinetics

The most obvious filter would be to select the most potent hits. The threshold for activity of a hit may be 1 to 10 μM. Setting the limit at 100 nM would quickly reduce the number of hits. This approach has its problems.

Activity in a biochemical assay is strictly a measure of how a molecule interacts with a target, that is, pharmacodynamics. Since advancements in biochemical assays have made them the norm, pharmaceutical companies have continually watched compounds with excellent pharmacodynamics fail in clinical trials because of poor pharmacokinetics. Drug companies have now learned to emphasize both pharmacodynamics and pharmacokinetics throughout the lead discovery process. Instead of prioritizing hits based on binding (pharmacodynamics) with a simple activity threshold, initial hits are also screened for a preliminary pharmacokinetic behavior. Pharmacokinetic properties of a hit in humans can be estimated with cellular assays as well as animal testing. It is important to be able to estimate properties in humans because U.S. Food and Drug Administration (FDA) approval for testing in humans will not have been obtained for hits from a library screen.

Another selection criterion for hits is the structural complexity of the hit. A hit that is advanced in a discovery program must be modified to increase its binding to a suitable level. If the hit has a complex structure that is difficult to prepare, synthesis of derivatives of the hit will require much time and slow the entire discovery process. Very complex hits are therefore often less attractive for promotion.

Biological Assays

Preliminary pharmacokinetic behavior can be tested through a number of whole cell assays. Most commercially successful drugs are administered orally, meaning the drug must be able to enter the bloodstream by crossing membranes in the intestines. The most common membrane permeability assay is performed by monitoring the absorption and secretion of a compound by colon carcinoma cells (Caco-2). Diffusion across Caco-2 cell membranes is considered to be a valid model for molecular transport in the small intestines.[16]

Drugs are mostly metabolized by liver enzymes, especially the cytochrome P-450 enzyme family. The ability for cytochrome P-450 enzymes to metabolize a hit is tested with liver microsomes. Liver microsomes consist primarily of endoplasmic reticulum that contains metabolic enzymes. Hits are individually incubated in the presence of the liver microsomes. Monitoring changes in concentrations provides a sense of the rate of metabolism of each hit. Liver microsomes are also used to determine whether the hit inhibits metabolic processes. Hits that inhibit liver metabolism are shunned.[17]

Acceptable hits do not need to show ideal behavior, but problem compounds will be removed from consideration. If all the hits fail initial pharmacokinetic screening, several options are possible. First, the search for hits could start over with screening of a new library. Second, the threshold for selection of hits could be lowered to enlarge the pool of hits, some of which may pass the permeability and metabolic screens. Third, the criteria for passing the Caco-2 and microsome screens may be softened to allow some hits to pass.[18]

Lipinski's Rules and Related Indices

Permeability and liver microsome screens are not high throughput. To save time, researchers have sought simple methods for eliminating compounds that will be poor lead candidates. A common method involves calculated indices. The first and most widely recognized index-based filter was reported by Lipinski in 1997.[19] This filter is called *Lipinski's rules* or the *Rule of 5* (**Table 10.1**).

Lipinski's rules are designed to predict oral availability of compounds that passively diffuse across membranes. Lipinski's rules are based on observations of a database of approximately 2,500 drugs or compounds studied in clinical trials. In general, the compounds could be described structurally with limits on their molecular weight, number of hydrogen-bond donors

TABLE 10.1   Lipinski's rules[19]

1. Molecular weight $\leq$ 500
2. Lipophilicity (log P or clog P) $\leq$ 5
3. Sum of hydrogen-bond donors $\leq$ 5
4. Sum of hydrogen-bond acceptors $\leq$ 10

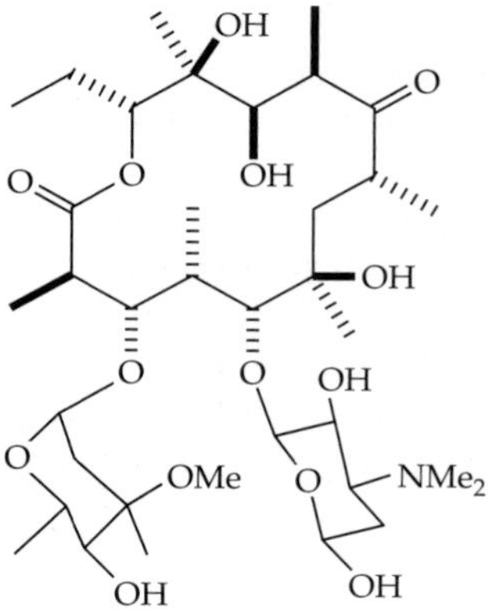

FIGURE 10.11 Erythromycin, a drug that violates Lipinski's rules

erythromycin
10.43

734 g/mol
6 H-bond donors
15 H-bond acceptors

and acceptors, and lipophilicity (log P or clog P). Log P is an experimental measure of lipophilicity that will be fully described in Chapter 12. A higher log P value indicates lower water solubility. The form clog P (pronounced "see log P") is a computer-estimated version of log P. A compound that violates any of Lipinski's rules may not be absorbed well when orally administered. Drugs that cross membranes by active, facilitated, or other means of transport fall beyond Lipinski's rules. Exceptions include the macrolide antibiotics such as erythromycin (**10.43**) (**Figure 10.11**). Methods by which molecules cross membranes are covered more thoroughly in Chapter 3.[19]

Over time, Lipinski's rules have been criticized as inappropriate for the evaluation of hits and leads. Hits and leads have weaker binding energies than final drugs. The process of optimization increases a lead's binding energy with multiple, successive structural modifications. These modifications typically increase a molecule's functionality and subsequently raise the molecular weight of the lead. A hit or lead with a molecular weight of 480 may slip under the Lipinski molecular weight requirement, but after going though the optimization process, the molecule may balloon to a molecular weight of 600 or higher. With this logic, Lipinski's rules are perhaps too permissive to be useful as a filter for hits and leads.

In 1999, Teague distinguished between *lead-like* and *drug-like* hits, and combinatorial library collections (**Table 10.2**).[20] Lead-like hits are characterized as having lower molecular weights (<350), activity (>0.1 μM), and clog P values (<3). The lower values give lead-like compounds room to grow into an optimized, high-affinity drug that still satisfies Lipinski's rules. Drug-like hits have higher molecular weights (>350) and clog P values (>3) but still modest affinity (~0.1 μM). The definition of lead-like has since been used as a preliminary filter for selecting more promising hits from a screen. Sometimes, simple selection criteria such as Lipinski's rules or lead-like properties are applied to a library before the initial screen is even performed.

A recently reported tool for hit evaluation and prioritization is *ligand lipophilicity efficiency* (*LLE*).[21] *LLE* is calculated as the difference between the negative logarithm of a hit's binding affinity, such as $-\log IC_{50}$ or $-\log K_D$, and the logarithm of a hit's partition coefficient, such as log P or clog P (Equation 10.2).

$$LLE = -\log IC_{50} - \log P \qquad (10.2)$$

TABLE 10.2 Lead-like and drug-like compounds[20]

Lead-Like	Drug-Like
1. Activity >0.1 μM	**1.** Activity >0.1 μM
2. MW <350	**2.** MW >350
3. Clog P < 3	**3.** Clog P >3

Higher *LLE* values are considered to be better. Consider what this equation says: "*LLE* equals activity less lipophilicity." Without this equation, one might be tempted to say that two hits with the same activity are equally attractive to a drug discovery team. Based on Equation 10.2, the less lipophilic hit (lower log *P*) has a higher *LLE* value and would be more attractive as a hit. Although this may not seem to be an earth-shattering conclusion, Equation 10.2 does show the trade-off between activity and lipophilicity when prioritizing hits.[22] Equation 10.2 quantitatively relates some of the central ideas behind the guidelines of Lipinski and Teague. An underlying assumption in Equation 10.2 is that growth of a lead into a drug will increase the compound's lipophilicity. A good hit or lead should therefore start with a lower lipophilicity so that the log *P* of the final drug will not surpass Lipinski's magic value of 5.

Other attempts to refine or improve Lipinski's rule set have appeared in the literature. One notable factor for consideration is the number of rotatable bonds in a hit. Increased molecular flexibility can reduce the ability of a molecule to cross a membrane. The maximum number of rotatable bonds has been suggested as 10. The polar surface area, often abbreviated as PSA, of a molecule is another important factor. Polar surface area is tightly correlated to the number of hydrogen-bond donors and acceptors contained in a molecule. A maximum polar surface area of 140 Å^2 or the equivalent of 12 hydrogen-bond donors/acceptors has been suggested. This is in line with Lipinski's rules.[22]

Lipinski terms and related indices exclusively predict oral bioavailability. None addresses metabolism concerns. While the formation of unwanted metabolites is difficult to predict, several functional groups have become recognized as common sources of problems. Examples include quinones and hydroquinones (**10.44**), aryl nitro groups (**10.45**), primary aryl amines (**10.46**), and Michael acceptors (**10.47**) (**Figure 10.12**). Quinones and Michael acceptors are strong electrophiles that tend to react quickly with and deplete glutathione stores in the liver. Aryl nitro compounds are reduced in the body to aryl amines, which are oxidized to electrophilic species with the same problems as quinones and Michael acceptors. Because the liver enzymes perform a large fraction of a body's metabolism of xenobiotics, the liver is the most commonly damaged organ when metabolites are toxic.[23]

Final Concerns for Promotion of a Hit to a Lead

Only a small number of hits remain after various selection criteria have been applied to the initial hit pool. The surviving hits, sometimes called *compounds of interest* or similar, receive additional scrutiny. Each remaining hit undergoes a handful of structural modifications. Modified hits are often called *analogues*. The analogues allow the discovery team to gain a preliminary understanding of the impact of structural changes on the activity of the hit against its target. If similar targets are known and available, filtered hits are tested for selectivity against the desired target and undesired related targets. These selectivity comparisons can predict the likelihood of side effects. Some in vivo testing may be performed in animals, especially rats. The in vivo tests provide a more accurate and reliable picture of

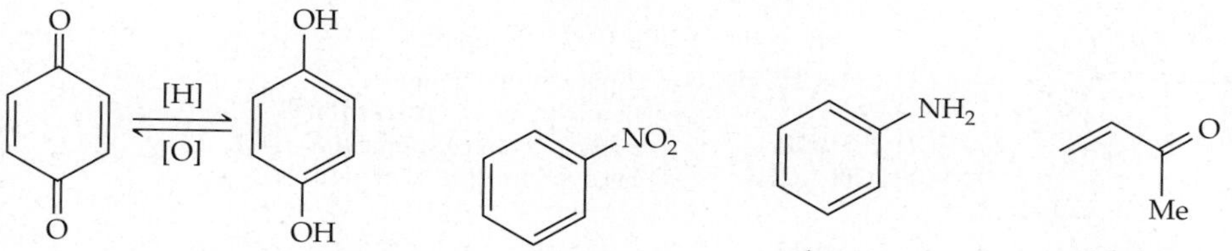

quinone hydroquinone **10.44**

aryl nitro **10.45**

primary aryl amine **10.46**

Michael acceptor **10.47**

FIGURE 10.12 Common problematic functional groups in drug candidates

a compound's pharmacokinetic profile. Finally, patent searches are performed to determine the patentability of the hit and later leads that might arise. If a compound cannot be patented, then that compound will certainly not be advanced as a lead.

The outcome of all these selection steps is hopefully one or more leads. Final leads may differ in structure somewhat from their original respective hits. Early structural modifications hopefully generate analogues of higher activity. While hits are often selected at an activity level of 1 μM (K_D, IC_{50}, etc.), structural changes may provide a lead with activity at concentrations of 0.1 μM (100 nM). Ultimately, potency will typically be improved down to the 1–10 nM level during the lead optimization stage.

10.3 Special Cases

Not all leads are discovered through a high-throughput screen of a chemical library. Two of the more commonly mentioned alternative methods include serendipitous discovery and optimization of side effects from an existing drug or drug candidate. The important role of natural products in drug discovery also deserves extra attention.

Serendipity

Some scientific discoveries, including drugs, are labeled as serendipitous. The word *serendipity* was derived by English author Horace Walpole from a fairy tale, *The Three Princes of Serendip*. Serendip was an old name for Sri Lanka.[24] According to Walpole, the princes ". . . were always making discoveries, by accidents and sagacity, of things they were not in quest of. . . ." Whether serendipitous drug discoveries are accidental is arguable, but the discoveries are unexpected. The sagacity required for such discoveries must be credited in part to the powers of observation of the researchers.

The following two Case Studies are commonly cited examples of serendipitous drug discoveries. Both show the clear role of observation by the researchers.

CASE STUDY

Penicillin[25]

In the summer 1928, Alexander Fleming, an English bacteriologist, went on vacation toward the end of July. A number of Petri dishes inoculated with staphylococci had been left on the bench top. When Fleming returned on August 3, all of the plates except one contained an even distribution of bacterial colonies. The different plate contained a mold, about which no bacterial colonies were growing. It was later found that the mold, a strain of *Penicillium notatum*, produced a family of compounds called penicillins (**10.48**), which killed nearby bacterial colonies (**Figure 10.13**).

After considerable investigation, a series of events had to occur for Fleming to make his observation. First, Fleming left his Petri dishes on the bench top instead of in an incubator. Second, research on the floor below Fleming's focused on the effect of mold on allergies. The source of Fleming's mold, a particularly uncommon strain, was present in the downstairs laboratory. Third, the first days of Fleming's vacation were cold. The cool laboratory allowed the mold to grow and infuse the surrounding culture medium with penicillin. Fourth, the latter vacation days were warm, providing conditions ideal

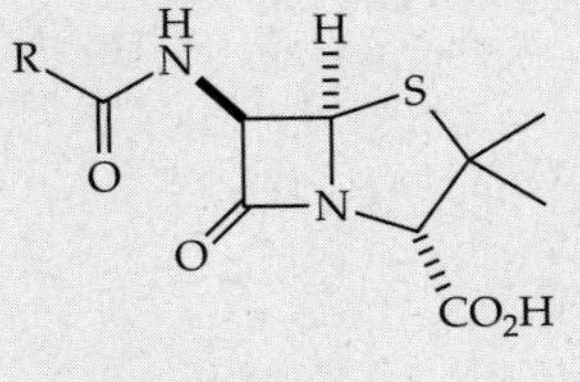

R	name
$CH_2CH=CHCH_2CH_3$	2-pentenylpenicillin
CH_2Ph	benzylpenicillin (penicillin G)
$CH_2C_6H_4OH$	p-hydroxybenzylpenicillin
$CH_2(CH_2)_5CH_3$	heptylpenicillin

10.48

FIGURE 10.13 Selected penicillin antibiotics

for bacterial growth to reveal a zone of inhibition around the mold.

Despite Fleming's recognition of the surprising failure of bacterial growth in the presence of a mold, difficulties in isolating somewhat pure samples of penicillin from "mold juice" prevented proper investigation of the benefits of penicillin. In 1939, Howard Florey, an Australian bacteriologist working at Oxford, took an interest in penicillin. With the assistance of adequate funding and resources, relatively pure samples of penicillin were isolated from mold broth. These samples allowed proper, controlled testing of penicillin. A particularly important experiment was performed in May 1940. Eight mice were infected with streptococci. Four received various treatments of penicillin. The four untreated mice died within 24 hours. One of the treated mice died two days later, and all three other treated mice survived. Based on this success and others, research advanced quickly. By early 1941, penicillin was tested in humans. Until the end of World War II, almost all penicillin produced was used to treat injured Allied soldiers.

CASE STUDY

Chlordiazepoxide[26]

In early 1955, a research group headed by Leo Sternbach at Hoffmann-La Roche was investigating tranquilizers. After several months of interesting yet disappointing research, the focus of the research group shifted to other problems. In April 1957, a laboratory cleanup revealed a few compounds, including chlordiazepoxide (Librium, **10.49**), which had never been submitted for testing (**Figure 10.14**). Instead of throwing out a small number of untested compounds from a long-abandoned project, the compounds were dutifully screened. Upon testing in mice and cats, compound **10.49** showed remarkable activity and low toxicity. Indeed, chlordiazepoxide was sufficiently active to enter clinical trials without further optimization. In 1960, chlordiazepoxide was marketed under the trade name Librium. Librium was the first of the benzodiazepine drugs, a class that continues to be widely researched. Benzodiazepines have since become recognized as a privileged structure. Following Librium, another benzodiazepine tranquilizer, diazepam (Valium, **10.50**), was first marketed in 1963.

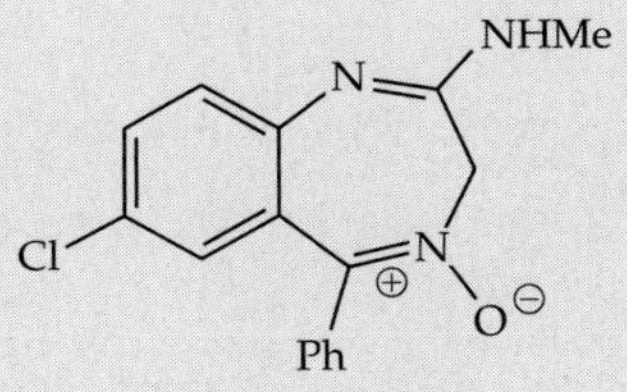

chlordiazepoxide
(Librium)
10.49

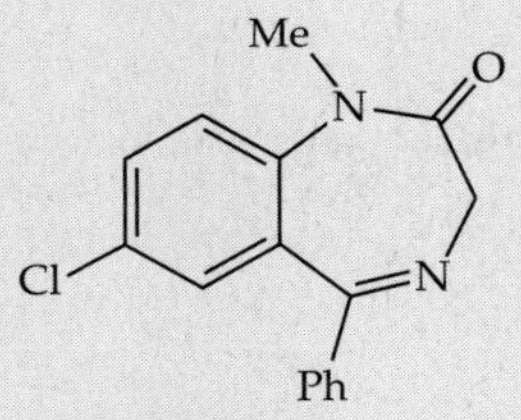

diazepam
(Valium)
10.50

FIGURE 10.14 Benzodiazepine tranquilizers

Clinical Observations

Virtually all drugs have side effects. Side effects are carefully tracked during clinical trials and include any undesired effects of the drug. Side effects can include almost anything—drowsiness, elevated heart rate, diarrhea, loss of sense of taste, weight gain, and so on. Side effects often can arise for many reasons, including binding to more than one target (lack of specificity) and metabolites that bind additional targets.

In theory, a drug with poor target specificity can be modified to enhance the side effect and decrease the original effect. This process is called selective optimization of side activities (SOSA). A significant advantage to SOSA is that the molecules, now considered hits for their secondary activity, have already been studied in clinical trials. The oral availability and safety will already be well established. A disadvantage is that the hit will likely be very drug-like (high molecular weight). Optimization of binding to a different receptor will only increase the molecular weight further. Also, if the hit was developed by another company, patentability of analogues of the hit may be a challenge.[27]

The following two brief Case Studies demonstrate how side effects can be enhanced to afford drugs.

CASE STUDY

Viloxazine[28]

In the 1970s, the standard treatment for hypertension (high blood pressure) included β-blockers and diuretics, such as propranolol (Inderal, **10. 51**) and hydrochlorothiazide (**10.52**), respectively (**Figure 10.15**). β-Blockers possess many other different activities, including sedative and anticonvulsant effects. To enhance some of the secondary effects, the central core, or pharmacophore, of β-blockers (**10.53**) was modified to

form a new ring (**10.54**) (**Scheme 10.6**). A thorough study of derivatives of **10.54** yielded the antidepressant viloxazine (Emovit, **10.55**).

The research on viloxazine implies that different conformations of β-blockers are responsible for their poor target selectivity. Restriction of the flexibility by a standard method, tying the molecule with a ring, boosts selectivity for one of the targets.

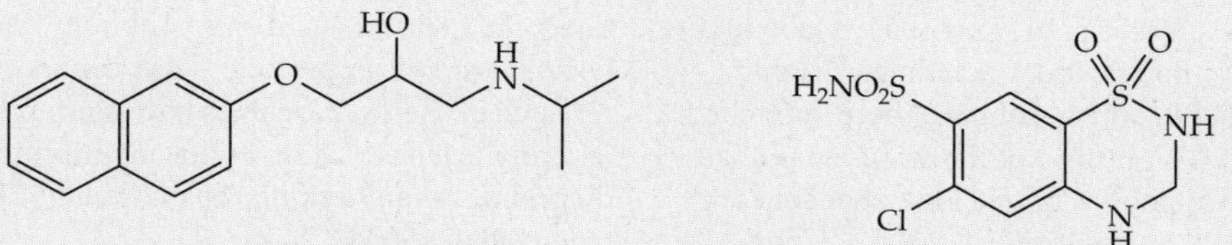

propranolol
(Inderal)
antihypertensive
10.51

hydrochlorothiazide
diuretic
10.52

FIGURE 10.15 Early treatments for hypertension: propranolol and hydrochlorothiazide

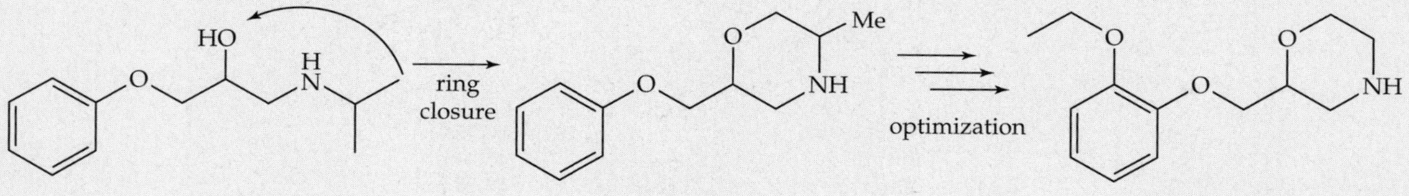

pharmacophore
of β-blockers
10.53

10.54

viloxazine
(Emovit)
antidepressant
10.55

SCHEME 10.6 Development of viloxazine

CASE STUDY

Sildenafil[29]

In the late 1980s, a Pfizer research team in England developed a promising drug for treating angina, a condition characterized by tightness in

the chest because of limited bloodflow to the heart. The compound (**10.56**), known at the time as UK-92,480, entered clinical trials in healthy volunteers, the equivalent of

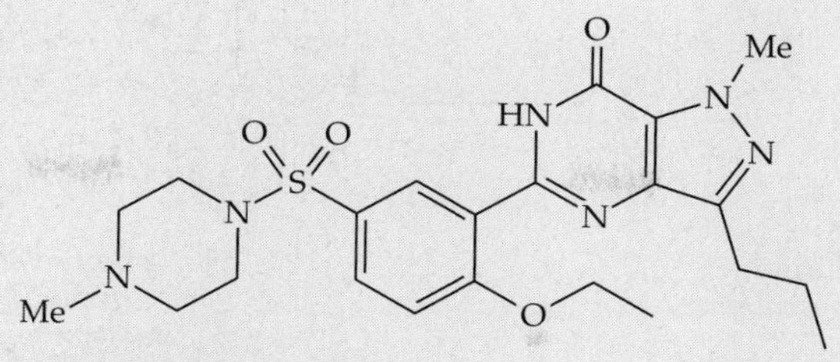

FIGURE 10.16

sildenafil
(Viagra)
erectile dysfunction
10.56

Phase I trials in the United States (**Figure 10.16**). Although UK-92,480 appeared to be safe, it had little impact on blood pressure, heart rate, and cardiac output. In a 1992 study, some volunteers receiving multiple doses reported an increased incidence of penile erections. Preliminary data was not overwhelming, but Pfizer spent another two years investigating the potential market for an erectile dysfunction drug. A study involving 300 patients in 1994 and 1995 showed excellent results for UK-92,480, which had been renamed sildenafil. The trials were so successful that, when the trials were complete, patients were reluctant to turn in their unused medication. Some patients even falsely claimed to have flushed leftover pills down the toilet in order to keep the unused medication. Sildenafil was approved by the FDA in early 1997 under the trade name Viagra. Despite competition in the marketplace from other erectile dysfunction drugs, Viagra has been a blockbuster for Pfizer, with a peak in sales of nearly US$2 billion worldwide in 2008.

Natural Products

In the early 1990s, the philosophy underlying combinatorial chemistry began to strongly influence mainstream drug discovery. As was covered in Chapter 9, the number of molecules that may be considered as potential drugs is essentially unlimited. To get a handle on molecular diversity, chemists developed techniques to synthesize huge numbers of new molecules. The resulting libraries were seen to be the perfect drug discovery tool.

Since the dawning of the combinatorial age in drug discovery, the major pharmaceutical companies have become less productive with regard to discovering drugs for new targets. This drop in creating new drugs has been called the *innovation gap*. Blame for the innovation gap has fallen on several possible culprits, including higher safety standards for new drugs and a flawed research and development structure for drug companies. Overreliance on combinatorial chemistry has also received scrutiny. Opponents of the current utilization of combinatorial chemistry say that libraries should be developed around compounds with a high probability of activity. Natural products are one possible class of molecule that shows a range of biological activity.

Many researchers feel that natural products have been improperly pushed to the background in drug discovery. Evidence in favor of this position can be found in studies by David Newman of the National Cancer Institute in Fredericksburg, Maryland. Newman periodically publishes an updated review of the origins of newly developed drugs.[30] From January 1981 to June 2006, 1,184 new drugs and drug candidates were reported. Approximately 70% of these compounds were either natural products, derived from natural products, or inspired by natural products. Examples of natural products and respective synthetic drugs they have inspired include ephedrine (**10.57**) and propranolol (Inderal, **10.51**), HMG-CoA (**10.58**) and rosuvastatin (Crestor, **10.59**), and 2'-deoxyguanosine (**10.60**) and acyclovir (Zovir, **10.61**) (**Figure 10.17**). With such a large percentage of recent drugs being closely tied to natural products, natural product supporters argue against wholly synthetic libraries, which are the norm in combinatorial chemistry.[30]

Proponents of combinatorial chemistry counter that combinatorial methods have been on the pharmaceutical scene only since the mid-1990s. Since bringing a drug through the

FIGURE 10.17 Natural products and inspired pharmaceuticals[30]

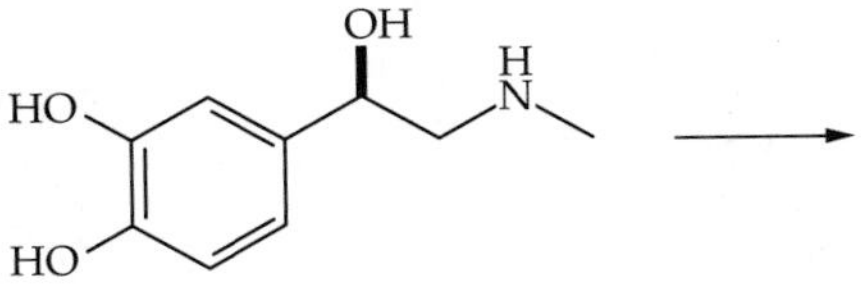

ephedrine
neurotransmitter
10.57

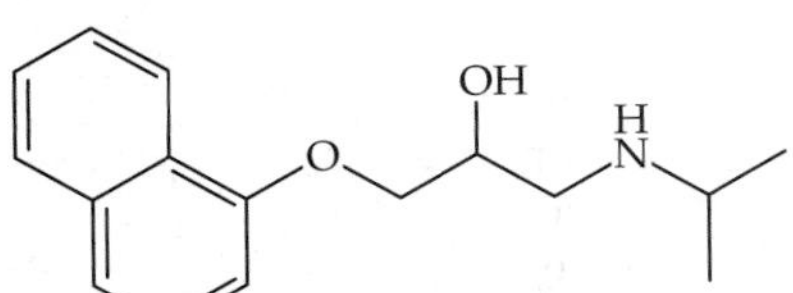

propranolol
(Inderal)
antihypertensive
10.51

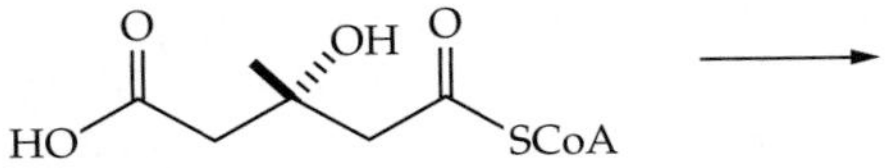

HMG-CoA
metabolite
10.58

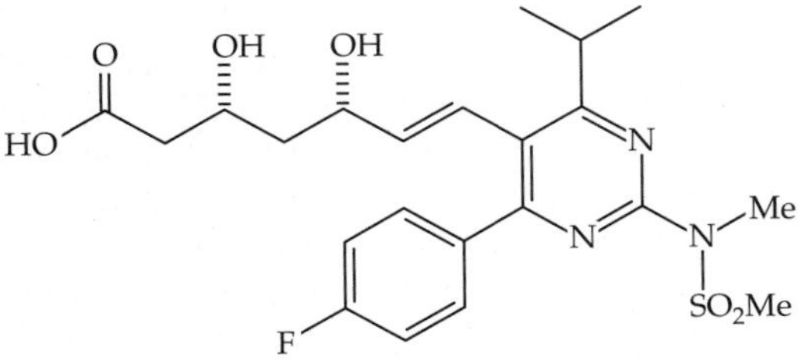

rosuvastatin
(Crestor)
anticholesterol
10.59

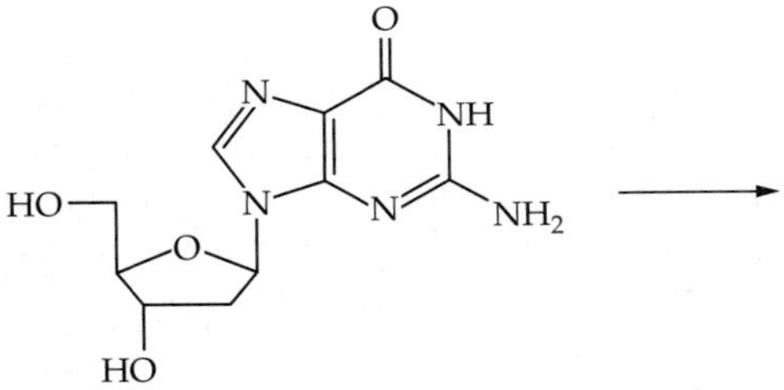

2'-deoxyguanosine
nucleoside
10.60

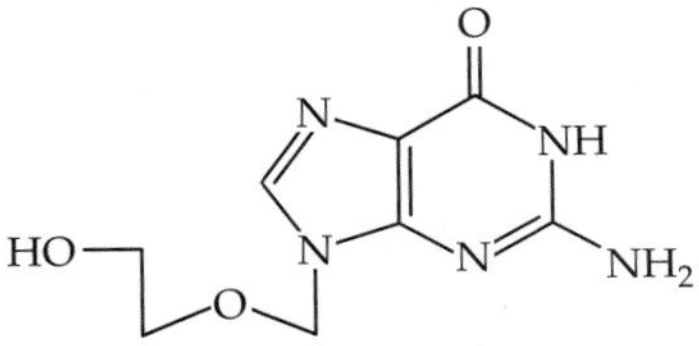

acyclovir
(Zovir)
antiviral
10.61

entire development process requires eight or more years, it is still too early to bring judgment upon combinatorial chemistry. Regardless, natural products have proven their value, and their role should not be diminished. It is not surprising that molecules synthesized by biological systems to interact with targets are also valuable for discovering leads. To this end, some researchers advocate libraries of isolated natural products as an improved tool for discovering hits and leads.[31]

Summary

Lead discovery, the search for compounds with activity against a target, begins with screening molecules. Regardless of the available information on the target and any known ligands, screening normally starts with a compound library that may contain a million or more compounds. Screening may involve the entire library or just some portion of the collection. Active compounds are called hits. A second approach to discovering hits includes the screening of smaller fragment libraries. Active fragments are linked to increased potency. A third screening approach to finding hits involves using computer models to estimate binding energies between a target and ligand. The activity of predicted hits is then confirmed through a traditional assay. The hits, regardless of how they are discovered, are then checked for factors such as their ability to be transported within the body, ease of synthesis, and even patentability. Hits that survive these criteria are promoted to lead status and subjected to further structural optimization. Exceptions to the described lead discovery process are uncommon but do occur. Leads and even hits have been discovered through serendipity. Leads can also arise from observation of desirable side effects of other drugs.

Questions

1. The following list provides three reactions. The product of each reaction will be a member of a library that will be screened for activity against various targets. Assuming the library is being designed under the "one compound, one well" philosophy, predict which reaction(s) will result in the formation of essentially one product from the provided starting material.

   a. E2 elimination of HBr with a base

   b. OsO_4-catalyzed dihydroxylation of an alkene

   c. Diels–Alder cycloaddition of a diene with acrylonitrile

2. The following two compounds are both natural products with promising biological activity. What might be a problem with promoting these compounds to lead status in a drug discovery program?

   galantamine
   (Nivalin)
   anti-Alzheimer's
   10.a

platencin
antibiotic
10.b

3. Like any measurement, activity data from screening a library has many sources of error. Some possible problems are given in the following list. For each problem, indicate whether introduced error would cause the compound activity to be overreported (potential false positive) or underreported (potential false negative). Another option is that the effect of the problem cannot be predicted. Note that the following problems are issues only with the library. The assay itself will also contain sources of error.
   • Evaporation of the stock solution of a library member
   • Degradation of the compound during long-term storage
   • Random error in dispensing of small volumes
   • Impure sample in the library

4. In Chapter 9, the number of possible molecules containing 30 large atoms (C, N, O, and/or S atoms) was estimated to be 10^{63}. A library containing just one molecule of each such compound would have an approximate mass of 10^{39} kg (impossible to achieve on Earth). In the spirit of advocating a fragment-based approach to screening, researchers have calculated that the number of reasonable structures with a MW of 160 or less (up to 11 large atoms taken from C, N, O, S, Si, P, and the halogens) as 1.39×10^7. The average MW was calculated as 153. What would be the mass of a single-molecule sample library of these compounds? Is this a feasible number? (Fink, T., Bruggesser, H., & Reymond, J.-L. Virtual Exploration of the Small-Molecule Chemical Universe below 160 Daltons. *Angew. Chem. Int. Ed.* 2005, *44*, 1504–1508.)

5. ΔG°_{bind} always has a negative value for the compounds encountered in this text (Equation 10.1). What would have to be true about K_D for ΔG°_{bind} to be positive? Why do we not encounter drugs, leads, or hits with a positive ΔG°_{bind}?

6. Nine of the 10 top best selling drugs for 2010 in the United States are shown here. Determine which of the compounds, if any, violate Lipinski's rules (Table 10.1). Log *P* values for each compound may be found online at www.drugbank.ca.

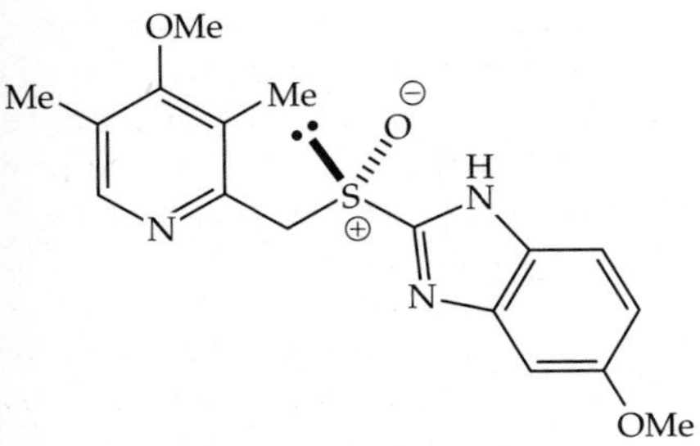

esomeprazole
(Nexium)
anti–acid reflux
10.c

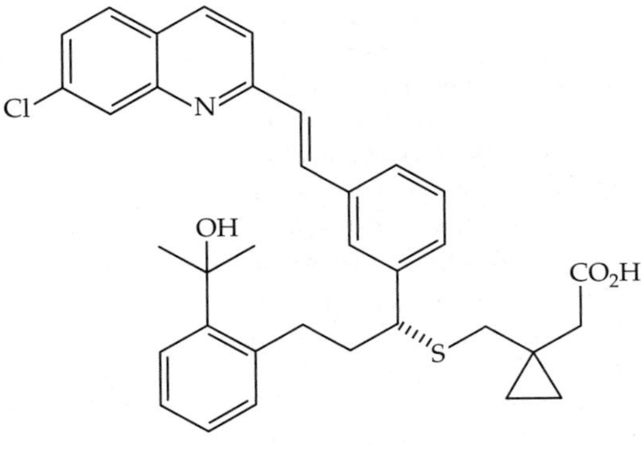

montelukast
(Singulair)
seasonal allergy relief
10.d

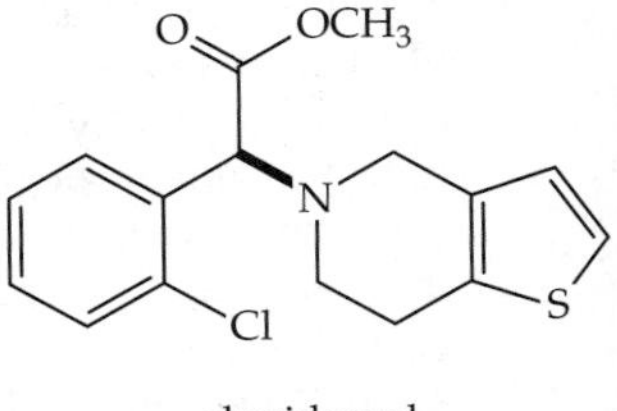

clopidogrel
(Plavix)
antiplatelet agent
10.e

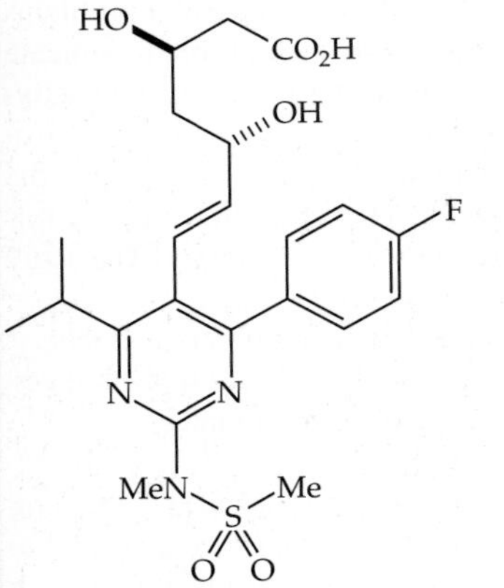

rosuvastatin
(Crestor)
anticholesterol
10.f

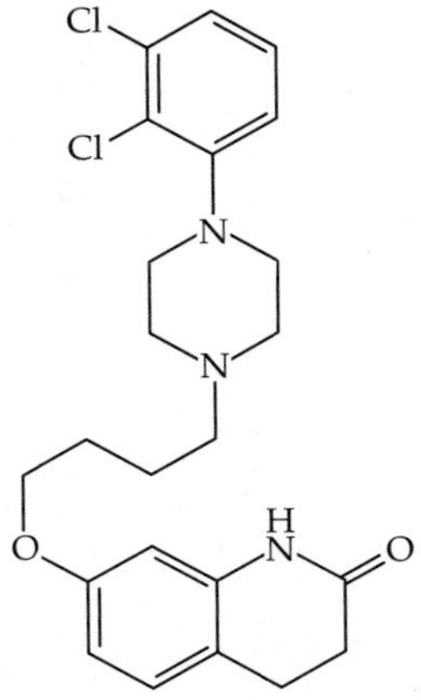

aripiprazole
(Abilify)
antidepressant
10.g

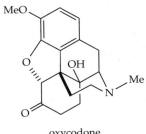

oxycodone
(active component of OxyContin)
analgesic
10.h

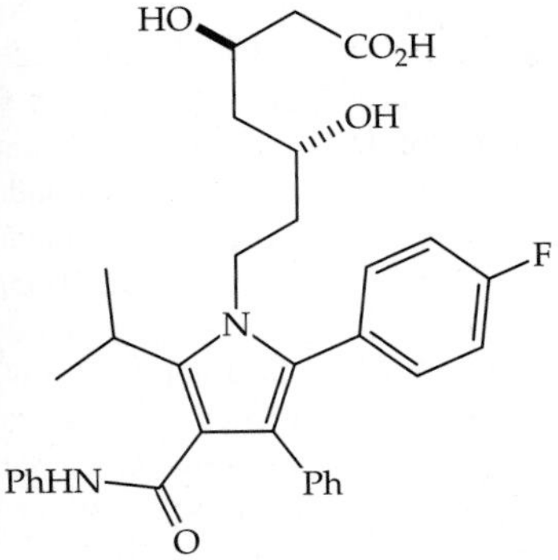

atorvastatin
(Lipitor)
anticholesterol
10.i

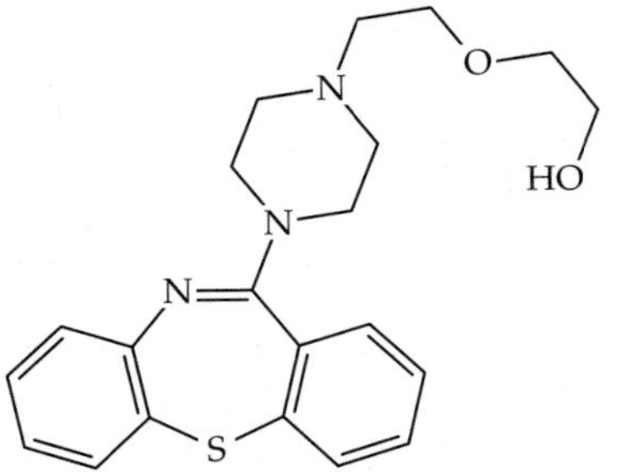

quetiapine
(Seroquel)
antipsychotic
10.j

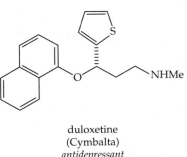

duloxetine
(Cymbalta)
antidepressant
10.k

7. Of the top 10 best selling drugs of 2010, GlaxoSmithKline's Advair, an inhaled asthma medication, was omitted from question 6. Why was this drug left out of the Lipinski analysis of question 6?

8. In the Case Study on stromelysin, two fragments were linked with a CH_2CH_2 tether to form compound **10.16** (Figure 10.4). Compound **10.16** had a binding energy that was 1.6 kcal/mol stronger than expected. Two possible explanations were provided for the surprisingly strong binding. Based on the intermolecular forces discussion in Chapter 9, which explanation is more plausible?

9. The following diagrams show four hits from an assay with their activities (ED_{50}) and lipophilicities (clog P). Which of the four hits is the most attractive based on its ligand lipophilicity efficiency?

10.l
$K_D = 1.1 \times 10^{-7} M$
clog $P = 2.97$

10.m
$K_D = 3.2 \times 10^{-6} M$
clog $P = 0.08$

10.n
$K_D = 7.3 \times 10^{-7} M$
clog $P = 2.76$

10.o
$K_D = 9.5 \times 10^{-8} M$
clog $P = 3.35$

10. Chapter 6 introduced the idea of privileged structures. Despite the wealth of activity found in regions of molecular space with privileged structures, can you think of a reason pharmaceutical companies might avoid privileged structures when searching for hits leading to new drugs?

11. Similar in intent to ligand lipophilicity efficiency is ligand efficiency. Ligand efficiency is defined as the binding energy of a molecule divided by the number of nonhydrogen atoms (n) in the molecule. A higher ligand efficiency indicates a molecule is maximizing its binding from each atom in the structure. The equation for ligand efficiency (LE) is shown in Equation 10.a. Calculate the ligand efficiency for the four compounds shown in question 9 ($R = 0.00199 \text{ kcal/mol} \cdot K; T = 298 \text{ K}$). The units on K_D must be in mol/L. (Hopkins, A. L., Groom, C. R., & Alex, A. Ligand Efficiency: A Useful Metric for Lead Selection. *Drug Disc. Today.* 2004, *9*, 430–431.)

$$LE = \frac{\Delta G^\circ_{bind}}{n} = \frac{2.3RT \log K_D}{n} \quad (10.a)$$

12. Interpreting the ligand efficiency values seen in question 11 requires one to know what a characteristic ligand efficiency value is for a drug. If a typical drug has 30 nonhydrogen atoms and a K_D of 10 nM, then what is the ligand efficiency of a typical drug?

13. What is the ligand lipophilicity efficiency of a typical drug if its K_D is 10 nM and clog P is 5?

14. Turn back to Chapter 9 and examine structure **9.74** in Scheme 9.15. This compound is mentioned as a hit from a library screen. Based on the information in this chapter, comment on the possibility of advancing **9.74** as a lead.

15. Why might high values for polar surface area affect how a molecule crosses a membrane?

References

1. Gordon, E. M., Barrett, R. W., Dower, W. J., Fodor, S. P. A., & Gallop, M. A. Applications of Combinatorial Technologies to Drug Discovery: 2. Combinatorial Organic Synthesis, Library Screening Strategies, and Future Directions. *J. Med. Chem.* 1994, *37*, 1385–1401.

2. Schafffrath, M., von Roedern, E., Hamley, P., & Stilz, H. U. High-Throughput Purification of Single Compounds and Libraries. *J. Comb. Chem.* 2005, *7*, 546–553.

3. Ohlmeyer, M. H., Swanson, R. N., Dillard, L. W., Reader, J. C., Asouline, G., Kobayashi, R., et al. Complex Synthetic Chemical Libraries Indexed with Molecular Tags. *Proc. Natl. Acad. Sci.* 1993, *90*, 10922–10926.

4. Erlandson, D. A., McDowell, R. S., & O'Brien, T. Fragment-Based Drug Discovery. *J. Med. Chem.* 2004, *47*, 3463–3482.

5. Carr, R. A. E., Congreve, M., Murray, C. W., & Rees, D. C. Fragment-Based Lead Discovery: Leads by Design. *Drug Disc. Today.* 2005, *10*, 987–992.

6. Krishnamurthy, V. M., Semetey, V., Bracher, P. J., Shen, N., & Whitesides, G. M. Dependence of Effective Molarity on Linker Length for an Intramolecular Protein-Ligand System. *J. Am. Chem. Soc.* 2007, *129*, 1312–1320.

7. Shuker, S. B., Hajduk, P. J., Meadows, R. P., & Fesik, S. W. Discovering High-Affinity Ligands for Proteins: SAR by NMR. *Science.* 1996, *274*, 1531–1534.

8. Hajduk, P. J., Sheppard, G., Nettesheim, D. G., Olejniczak, E. T., Shuker, S. B., Meadows, R. P., et al. Discovery of Potent Nonpeptide Inhibitors of Stromelysin Using SAR by NMR. *J. Am. Chem. Soc.* 1997, *119*, 5818–5827.

9. Congreve, M. S., Davis, D. J., Devine, L., Granata, C., O'Reilly, M., Wyatt, P. G., & Jhoti, H. Detection of Ligands from a Dynamic Combinatorial Library by X-ray Crystallography. *Angew. Chem. Int. Ed.* 2003, *42*, 4479–4482.

10. Lewis, W. G., Green, L. G., Grynszpan, F., Radić, Z., Carlier, P. R., Taylor, P., et al. Click Chemistry in Situ: Acetylcholinesterase as a Reaction Vessel for the Selective Assembly of a Femtomolar Inhibitor from an Array of Building Blocks. *Angew. Chem. Int. Ed.* 2002, *41*, 1053–1057.

11. Lyne, P. D. Structure-Based Virtual Screening: An Overview. *Drug Disc. Today.* 2002, *7*, 1047–1055.

12. Irwin, J. J., & Shoichet, B. K. ZINC: A Free Database of Commercially Available Compounds for Virtual Screening. *J. Chem. Inf. Model.* 2005, *45*, 177–182.

13. Forino, M., Jung, D., Easton, J. B., Houghton, P. J., & Pellecchia, M. Virtual Docking Approaches to Protein Kinase B Inhibition. *J. Med. Chem.* 2005, *48*, 2278–2281.

14. Tervo, A. J., Kyrylenko, S., Niskanen, P., Salminen, A., Leppänen, J., Nyrönen, T. H., et al. An In Silico Approach to Discovering Novel Inhibitors of Human Sirtuin Type 2. *J. Med. Chem.* 2004, *47*, 6292–6298.

15. Kellenberger, E., Springael, J.-Y., Permentier, M., Hachet-Haas, M., Galzi, J.-L., & Rognan, D. Identification of Nonpeptide CCR5 Receptor Agonists by Structure-Based Virtual Screening. *J. Med. Chem.* 2007, *50*, 1294–1303.

16. Cole, C. J., & Pfund, W. P. Permeability Assays. *Mod. Drug Disc.* 2000, *3*, 73–76.

17. Li, A. P. Preclinical in Vitro Screening Assays for Drug-Like Properties. *Drug Disc. Today.* 2005, *2*, 179–185.

18. Wunberg, T., Hendrix, M., Hillisch, A., Lobell, M., Meier, H., Schmeck, C., et al. Improving the Hit-to-Lead Process: Data-Driven Assessment of Drug-Like and Lead-Like Screening Hits. *Drug Disc. Today.* 2006, *11*, 175–180.

19. Lipinski, C. A., Lombardo, F., Dominy, B. W., & Feeney, P. J. Experimental and Computational Approaches to Estimate Solubility and Permeability in Drug Discovery and Development Settings. *Adv. Drug Del. Rev.* 1997, *23*, 3–25.

20. Teague, S. J., Davis, A. M., Leeson, P. D., & Oprea, T. The Design of Leadlike Combinatorial Libraries. *Angew. Chem. Int. Ed.* 1999, *38*, 3743–3748.

21. Veber, D. F., Johnson, S. R., Cheng, H.-Y., Smith, B. R., Ward, K. W., & Kopple, K. D. Molecular Properties That Influence the Oral Bioavailability of Drug Candidates. *J. Med. Chem.* 2002, *45*, 2615–2623.

22. Edwards, M. P., & Price, D. A. Role of Physiochemical Properties and Ligand Lipophilicity Efficiency in Addressing Drug Safety Risks. *Annu. Rep. Med. Chem.* 2010, *45*, 381–391.

23. Nassar, A.-E. F., Kamel, A. M., & Clarimont, C. Improving the Decision-Making Process in Structural Modification of Drug Candidates: Reducing Toxicity. *Drug Disc. Today.* 2004, *9*, 1055–1064.

24. *The American Heritage Dictionary of the English Language* (4th ed.). New York: Houghton Mifflin, 2000.

25. Sneader, W. *Drug Prototypes and their Exploitation.* New York: Wiley, 1996, pp. 469–478.

26. Sternbach, L. H. The Benzodiazepine Story. *J. Med. Chem.* 1979, *22*, 1–7.

27. Wermuth, C. Selective Optimization of Side Activities: Another Way for Drug Discovery. *J. Med. Chem.* 2004, *47*, 1303–1314.

28. Greenwood, D. T., Mallion, K. B., Todd, A. H., & Turner, R. W. 2-Aryloxymethyl-2,3,5,6-tetrahydro-1,4-oxazines: A New Class of Antidepressants. *J. Med. Chem.* 1975, *18*, 573–577.

29. Kling, J. From Hypertension to Angina to Viagra. *Mod. Drug Disc.* 1998, *1*(2), 31–38.

30. Newman, D. J., & Cragg, G. M. Natural Products as Sources of New Drugs over the Last 25 Years. *J. Nat. Prod.* 2007, *70*, 461–477.

31. Quinn, R. J., Carroll, A. R., Pham, N. B., Baron, P., Palframan, M. E., Suraweera, L., et al. Developing a Drug-Like Natural Product Library. *J. Nat. Prod.* 2008, *71*, 464–468.

Chapter 11

Lead Optimization: Traditional Methods

Chapter Outline

Along with lead discovery, lead optimization is the primary responsibility of a medicinal chemist. This chapter describes classic bench chemistry methods for modifying an existing structure both to increase activity and to improve the biological stability and transport of a structure. These two tasks, increasing activity and improving in vivo properties, are the primary challenges of lead optimization. If successful, the outcome of lead optimization will likely be a promising drug candidate.

11.1 Pharmacophore Determination

The *pharmacophore* of a lead compound is the minimal portion of the molecule required for significant biological activity. The concept of a pharmacophore was introduced in Chapter 9 but will receive its full due in this section. Part of the lead optimization process is determining the pharmacophore. To elucidate the pharmacophore, researchers prepare analogues of the lead with portions of the molecule removed. Testing of these stripped-down leads reveals whether the missing pieces result in a significantly lower binding. Parts of the molecule that, when removed, cause a steep drop in binding are considered to be vital for binding and therefore part of the lead's pharmacophore. Establishing the pharmacophore allows the medicinal chemistry team to know what parts of the lead are most responsible for binding and activity.

At this stage of the drug discovery process, the discovery team focuses on increasing activity of the molecule. Since early assays are primarily in vitro binding studies, the term *activity* refers to pharmacodynamics. Just as important as pharmacodynamics are pharmacokinetics. As the activity of a lead is progressively improved, the medicinal chemistry group also monitors pharmacokinetic properties of the lead. If pharmacokinetic properties of the lead need to be modified, structural changes will be made on the nonpharmacophore regions of the lead. The following two case studies demonstrate the utility of understanding the pharmacophore of a series of compounds.

Learning Goals for Chapter 11

- Know how to determine the pharmacophore of a drug
- Be able to apply standard structure-activity relationship techniques to increase activity
- Understand the use and logic behind isosteres and bioisosteres
- Appreciate the challenges of optimizing the activity of a peptide lead

The Pharmacophore of Morphine[1]

A classic example in pharmacophore determination is the study of morphine (11.1) (Figure 11.1). Morphine is a powerful painkiller (analgesic) with a number of unwanted effects, including dependence and sedation. Morphine is a natural product and may be considered to be a lead compound in a search for other analgesics. Through many extensive studies, the pharmacophore of morphine (11.2) has been found to consist of (1) a phenyl ring next to (2) a quaternary carbon bearing (3) a tertiary aminoethyl chain. This substitution pattern is sometimes called the *morphine rule*. From this information, literally thousands of morphine analogues

have been prepared and tested. One notable example is meperidine (11.3). This structure contains all the crucial groups to fulfill the morphine rule and retains 10–12% of the potency of morphine. The simplicity of the compound minimizes production costs. Fortunately, because morphine is extremely potent, the lower activity of meperidine is still sufficient for its effective use in pain relief. Another interesting morphine analogue is methadone (11.4). Unlike most morphine-based compounds, methadone is orally available and does not cause euphoric sensations. For these reasons, methadone is used to manage addiction to morphine and related compounds.

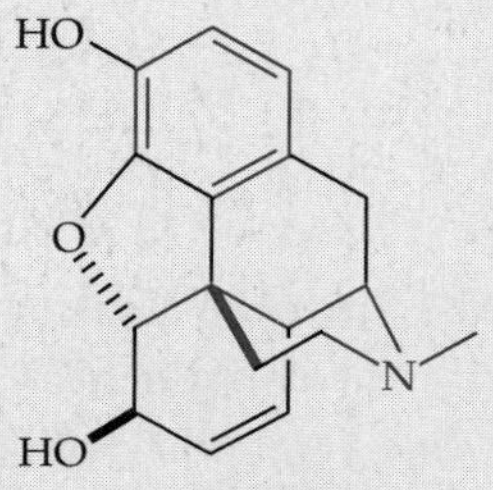

morphine
analgesic
11.1

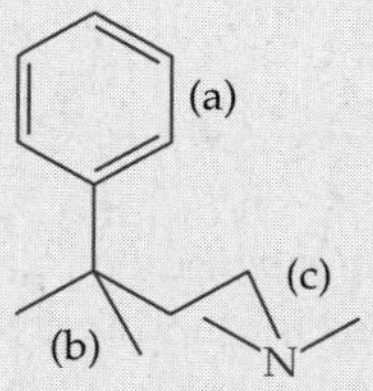

morphine
pharmacophore
11.2

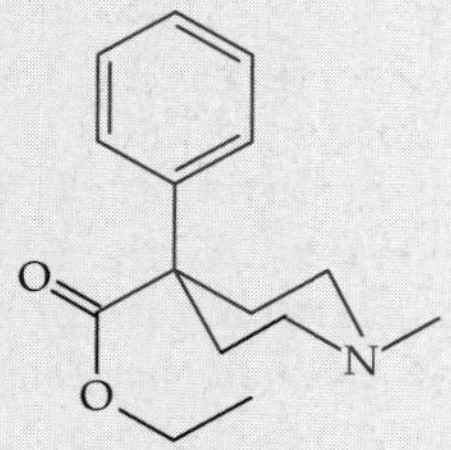

meperidine
(Demerol)
analgesic
11.3

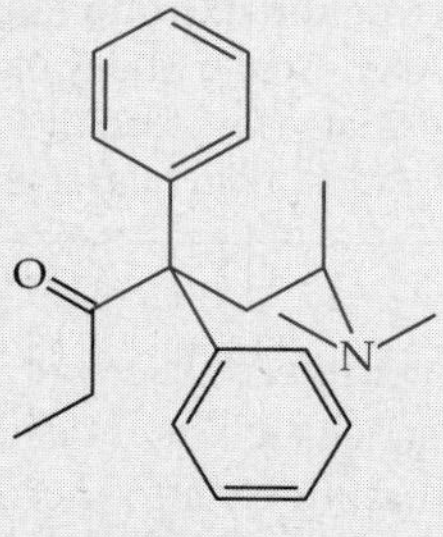

methadone
*analgesic and
addiction treatment*
11.4

FIGURE 11.1 Morphine and compounds with similar activity

The Pharmacophore of Migrastatin, a Promising Anticancer Compound[2,3]

Sometimes the search for a lead's pharmacophore can provide surprising results. (+)-Migrastatin (11.5) was isolated from a bacterial broth in 2000 and found to have hit-level activity ($IC_{50} = 29\ \mu M$) against certain cancers (Figure 11.2). Because of its novel structure, migrastatin drew considerable interest from cancer research groups. The group of Samuel Danishefsky of the Sloan-Kettering Institute for Cancer Research and Columbia University prepared simplified analogues of migrastatin to determine its pharmacophore. Surprisingly, when the large imide side chain was removed from the central ring of migrastatin, the potency of the new structure (11.6) increased by over three orders of magnitude ($IC_{50} = 22\ nM$). Replacement

of the imide chain of 11.5 with an isopropyl group (11.7) decreased activity ($IC_{50} = 227\ \mu M$). Reduction of one of the π-bonds as well as replacement of the ring of oxygen of 11.6 with a nitrogen (11.8) or carbon (11.9) afforded compounds that had less activity than 11.6 but were still much improved over migrastatin. While the research of Danishefsky did not strictly define the pharmacophore of migrastatin, the work shows the type of discoveries that occur as a hit or lead is initially probed for the origins of its activity.[2]

Based on information about the five compounds in Figure 11.2, the migrastatin core (11.6) is the most potent and would appear to be the logical choice to move forward as the lead compound. However, in the presence of mouse

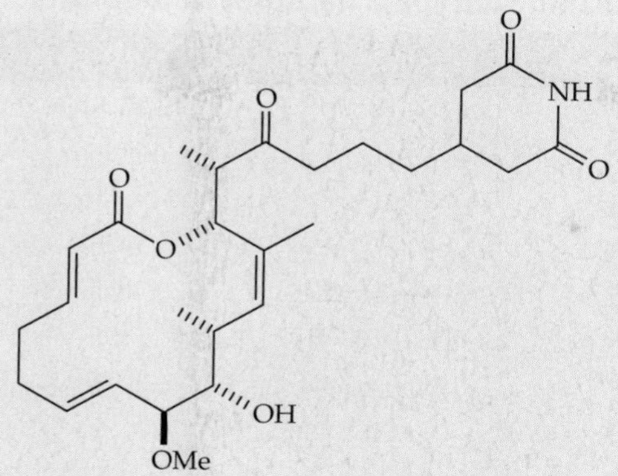

(+)-migrastatin
IC_{50} = 29 μM
stable in plasma
11.5

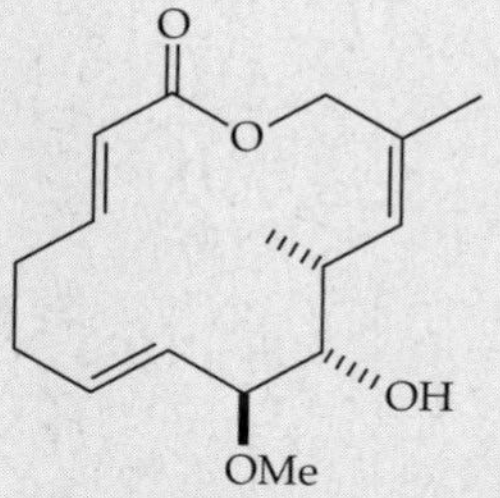

migrastatin core
IC_{50} = 22 nM
$t_{1/2}$ = 20 min
11.6

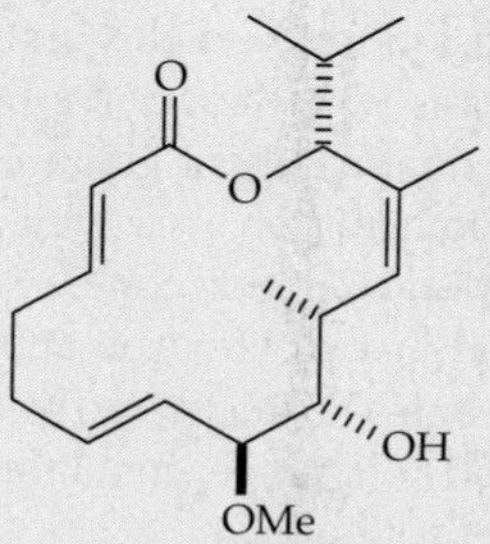

(*R*)-isopropyl-
migrastatin
IC_{50} = 227 μM
stable in plasma
11.7

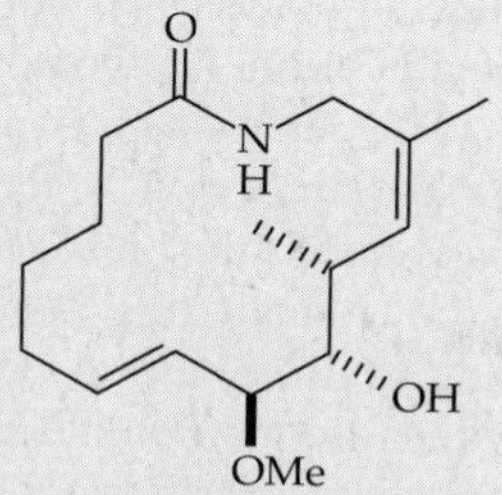

macrolactam
IC_{50} = 255 nM
stable in plasma
11.8

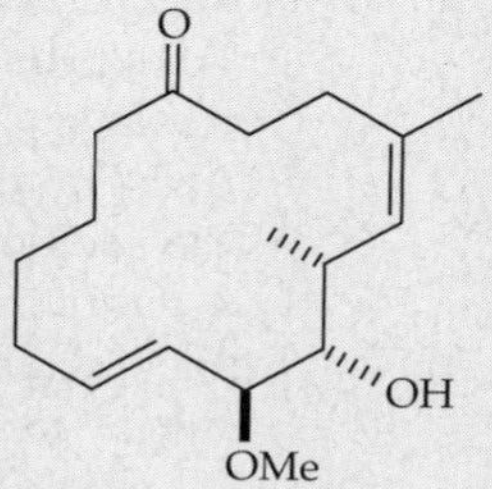

macroketone
IC_{50} = 100 nM
stable in plasma
11.9

FIGURE 11.2 (+)-Migrastatin and analogues[2]

blood plasma, the ester of the core molecule (**11.6**) has a half-life of only 20 minutes while the ketone (**11.9**) is stable for over 60 minutes.[2] Active research in migrastatins currently focuses on the optimization of the less potent ketone (**11.9**) with its better pharmacokinetic properties instead of the more potent but less stable core (**11.6**).[3]

11.2 Functional Group Replacements

Once the pharmacophore of a lead has been established, the medicinal chemistry group begins to fine-tune the lead through modifications around the pharmacophore. This process of making modifications allows the medicinal chemistry, or *med chem*, group to learn the structure-activity relationships (SAR) of the lead. Statements such as "We performed SAR on the lead" are common. One of the first experimenters to use SAR to develop a drug was Ehrlich in his work in the early 1900s on arsenical drugs for syphilis treatment (see Chapter 1). The sulfa drugs of the 1930s and 1940s are another early class that demonstrated the power of SAR in drug development. SAR requires a lead that can be readily synthesized with multiple positions for functionalization and modification. Several different techniques fall under the umbrella of SAR.

TABLE 11.1   Activity of Galanin Subtype-3 Receptor Antagonists[4]

Structure	Entry	R^1	R^2	GAL$_3$ K_i (nM)
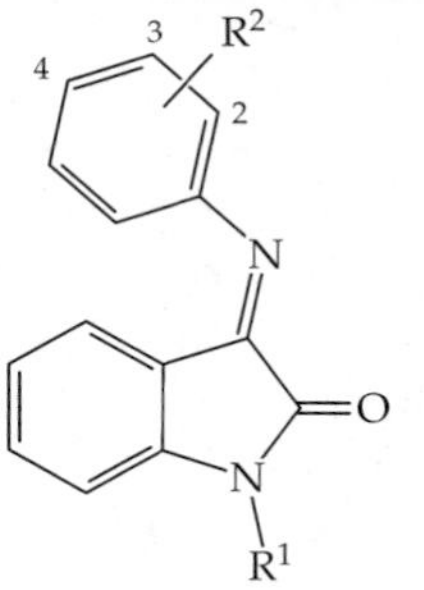	1	H	2,3-dichloro	437
	2	H	4-chloro	850
	3	H	4-methoxy	>10,000
	4	H	3-CF$_3$	596
	5	CH$_2$–CH=CH$_2$	2,3-dichloro	150
	6	CH$_2$–C≡CH	3-CF$_3$	52
	7	1-ethylpropyl	3-CF$_3$	89
	8	Ph	3-CF$_3$	17

11.10

Functional group replacement is the most common form of SAR, and it involves substitution of one functional group with another. Most journal articles in the field of medicinal chemistry include large tables of modified leads with their corresponding activities. For example, the data in **Table 11.1** come from a small study of galanin subtype-3 receptor antagonists (**11.10**).[4] Galanin is a G-protein–coupled receptor with a variety of functions. In this study, the parent ring system, a 3-imino-2-indolinone, is modified through the addition of groups in two positions, on the indolinone nitrogen and in the 2-, 3-, and 4-positions ring on the aromatic ring. Aside from entry 3 (inactive), the tested compounds show activity ranging from 850 nM to 17 nM, a 50-fold range. In a larger study, substitution effects on the other aromatic would likely be explored.

Early in an SAR study, substitutions made to a lead are selected according to a number of factors, including ease of preparation and availability of reagents. Subsequent changes are guided by screening of the initial analogues. Logic in this feedback loop can be simplistic. For example, a methyl group might be added to a lead (**11.11**) to afford a new analogue (**11.12**) (**Scheme 11.1**). If the additional methyl boosts activity, then perhaps a hydrophobic pocket is being filled. Lengthening the chain to an ethyl (**11.13**) or adding a second methyl nearby (**11.14**) would be a logical next action. If the first methyl caused a decrease in activity, then perhaps no open space is available in that part of the binding pocket. Replacing the new methyl with a small group, such as fluorine (**11.15**), could probe for electronic effects. Replacing the ring carbon with a nitrogen atom (**11.16**) could

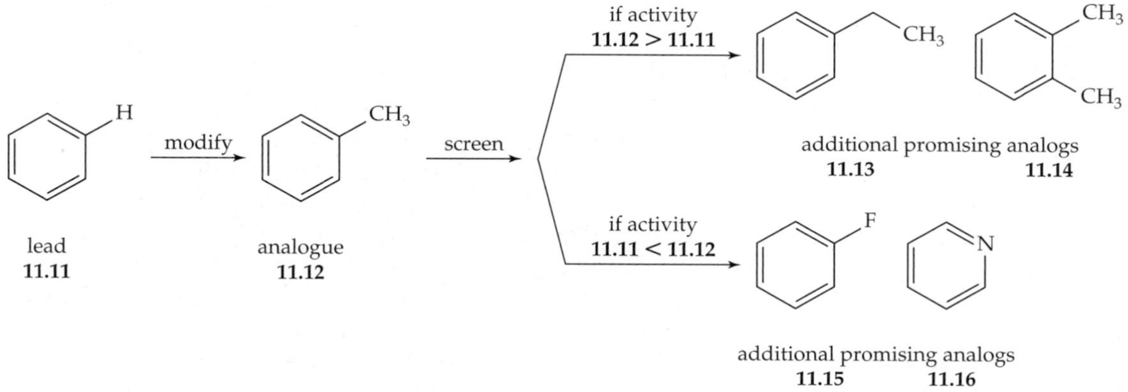

SCHEME 11.1 Logic applied in traditional SAR

search for a potential hydrogen bond donor in the binding pocket. Based on this type of logic, as many positions as possible on the lead are substituted and tested.

Structural information on the target binding site can be immensely helpful in guiding the med chem team with regard to which substituents might be most beneficial. The structural information often comes in the form of x-ray, NMR, and modeling data (see Chapter 9). If the binding pocket is relatively small, then smaller R-groups might be selected preferentially as substituents. A pocket lined with polar groups would steer the med chem team toward adding ionized substituents or groups that can act as hydrogen bond acceptors/donors. All changes are made with the desire to increase the number of attractive intermolecular forces between the lead and target. Increasing the intermolecular forces raises the magnitude of the binding energy (ΔG_{bind}) and decreases activity measures $(K_D, IC_{50}, K_i,$ etc.).

In working with larger molecules, the lead optimization team normally divides the lead into logical sections or regions (**Scheme 11.2**). Each section is then modified separately to determine the best substituent for that one section of the molecule. After the optimal substituents for all the individual sections have been determined, the best substituents are combined within the same analogue to afford a highly optimized lead. This "divide and conquer" approach can minimize the synthetic workload. For a lead with two halves (**11.17**) (left and right rings), testing four different R-groups on each ring requires the synthesis of eight (2×4) analogues (**11.18** and **11.19**). A ninth compound must be prepared to bring together the two optimized halves (**11.20**). In contrast, making all possible combinations would require the synthesis of 16 (4×4) analogues. The time savings of making nine versus 16 compounds is not huge, but most leads contain three or more regions and many possible substituents for each region. A greater number of regions and substituents makes the divide and conquer method more attractive.

Different regions of a molecule sometimes assume interesting names that are related to compass points or geographic locations. Some key functional groups in the development of ezetimibe (**11.21**), a cholesterol-lowering drug, were given characteristic and descriptive names based on the placement of key functional groups (**Figure 11.3**).[5]

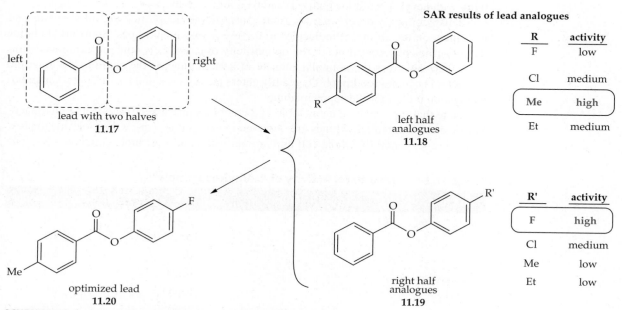

SAR results of lead analogues

R	activity
F	low
Cl	medium
Me	high
Et	medium

left half analogues
11.18

R'	activity
F	high
Cl	medium
Me	low
Et	low

right half analogues
11.19

lead with two halves
11.17

optimized lead
11.20

SCHEME 11.2 Optimization by independently modifying sections of the lead

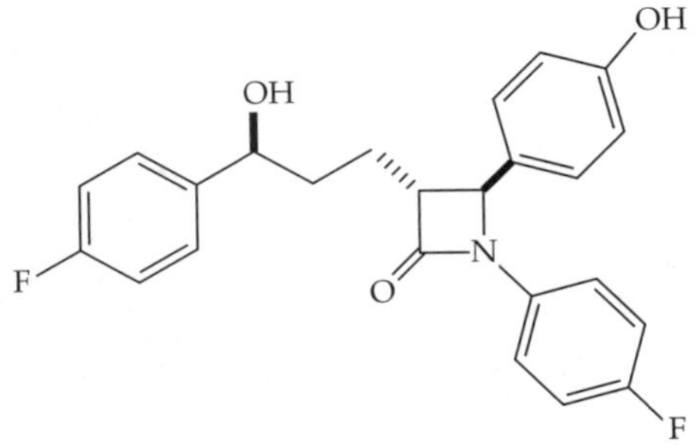

ezetimibe
(Zetia)
anticholesterol
11.21

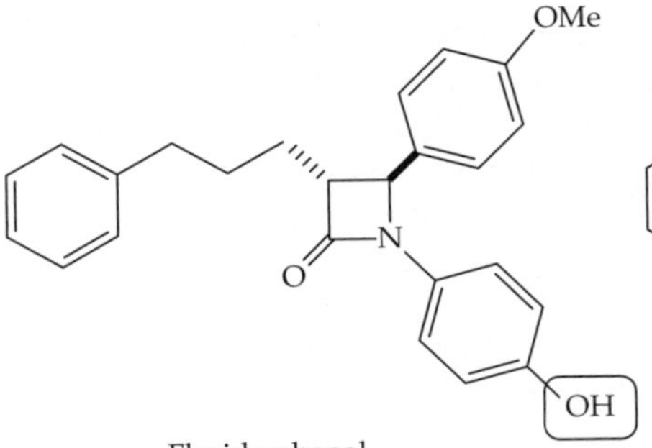

Florida phenol
11.22

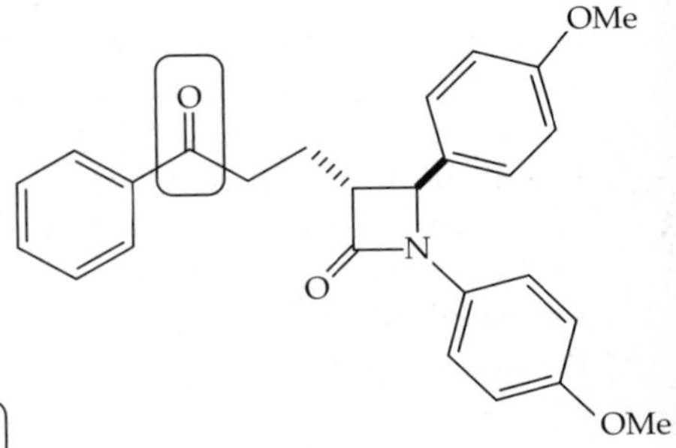

Western ketone
11.23

FIGURE 11.3 Named optimized leads[5]

11.3 Alkyl Group Manipulation

Changes to alkyl groups through elongating, branching, or forming rings can have a significant impact on a lead's activity. Topics that are not covered here but that are highly applicable to SAR are the use of rings and alkenes to restrict conformational flexibility and explore the impact of stereochemistry on biological activity. These ideas were discussed in Chapter 9 under the umbrella of molecular shape.

Chain Homologation

Homologous compounds differ by a single carbon, which is normally a single CH_2 unit. Extending the length of a chain or expanding a ring in one-carbon increments can reveal trends in biological activity. As carbons are added, activity generally increases up to a point and then begins to fall. A representative example is the antibacterial activity of 4-alkylresorcinols (**11.24**) (**Table 11.2**).[6] The substituted resorcinol analogues peak in activity with hexyl substitution (entry 4) and then drop rapidly as the chain is elongated.

This commonly observed trend is attributed to the lipophilicity of alkyl chains and its impact on the ability of molecules to diffuse across membranes. Short alkyl chains do not sufficiently contribute to the lipophilicity of a molecule, and the compound cannot effectively cross the nonpolar interior of a membrane. If the chain is too long, the compound becomes too lipophilic, readily enters membranes, and never leaves. Chain homologation probes for the happy medium.

Note that the activity data in Table 11.2 combines pharmacodynamic and pharmacokinetic data. Lipophilicity trends are observed in assays that require the molecule to cross a membrane, generally whole cell and animal tests, not most biochemical assays. The

TABLE 11.2 Antibacterial Activity of 4-Alkylresorcinols[6]

Structure	Entry	R	Relative Activity
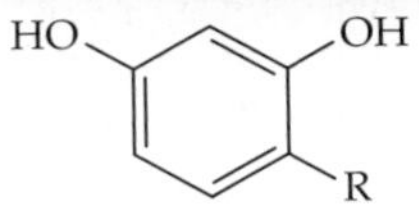 4-alkylresorcinols **11.24**	1	propyl	1.0
	2	butyl	4.2
	3	pentyl	6.6
	4	hexyl	10.2
	5	heptyl	6.0
	6	octyl	0.0

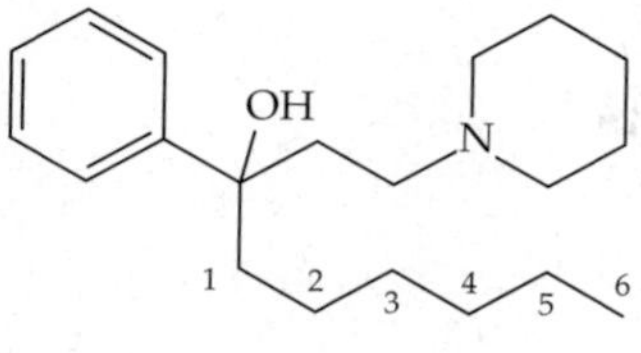

FIGURE 11.4 Ring-chain analogues[7]

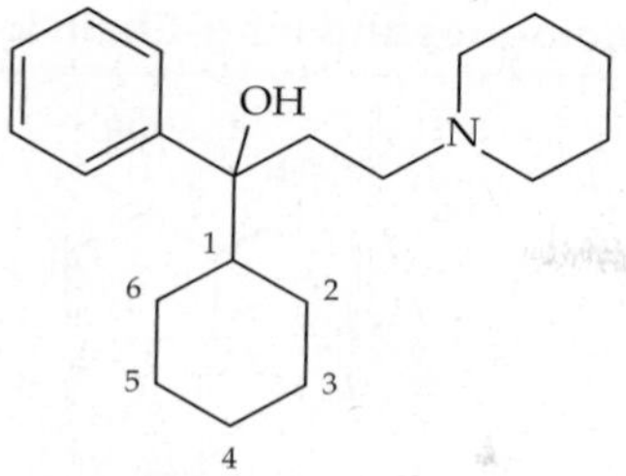

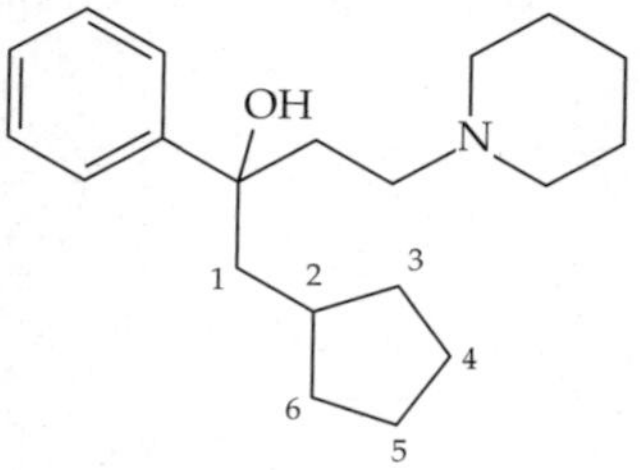

pharmacophore of the resorcinol series (**11.24**) is likely the 1,3-dihydroxyphenyl group. Adjustment of the lipophilicity of the structure through homologation of an alkyl chain tunes the pharmacokinetic properties of the molecule without affecting binding.

Ring-Chain Interconversion

Ring-chain interconversion involves replacing an alkyl chain with a cycloalkyl group of the same number of carbons. Ring-chain interconversions introduce conformational restrictions and can have a major impact on activity. In the following compounds, the hexyl-substituted analogue (**11.25**) shows no activity, while the cyclohexyl analogue (**11.26**) is an anticholinergic (**Figure 11.4**).[7] Substitutions maintain the number of carbon atoms, but not all the carbons need to be part of the ring. Another valid ring-chain analogue with a six-carbon substituent (**11.27**), in this case a cyclopenylmethyl group, is also shown.

Ring-chain interconversions and chain homologations are often studied together. In a study of inhibitors of neuraminidase, a key enzyme for the spread of the influenza A and B viruses, numerous analogues of **11.28** were screened.[8] The analogues included linear homologues, rings, and stereoisomers, where applicable (**Table 11.3**, page 280). Of all the tested compounds, only a few showed excellent activity against both influenza A and B. Ultimately, entry 10 was marketed as its ethyl ester under the trade name of Tamiflu.[8] The ester is hydrolyzed in vivo to the corresponding carboxylic acid. Tamiflu is therefore an example of a prodrug.

11.4 Isosteres

Isosteres are functional groups that have little impact on biological activity when one is exchanged for another.[9] This chapter is entitled Lead Optimization, so a discussion of groups that do not appreciably change activity may seem odd. However, activity from target binding is not the only factor in the design of a successful drug. As has already been mentioned, a good drug should be absorbed, distributed, metabolized, and eliminated in a desirable fashion. These topics are within the realm of pharmacokinetics. Isosteric substitutions are intended to impact how the drug gets to and from the site of action. Typically, isosteric replacements are explored late in the lead optimization process. At this stage, the lead should be very potent, but preliminary data, possibly from early animal studies, may indicate that the distribution and elimination properties are not optimal. Ideally, an isosteric replacement will improve the lead's pharmacokinetics without drastically reducing target binding.

A traditional *isostere*, or *classical isostere*, is an atom or group of atoms with similar spatial requirements (**Table 11.4**, page 281).[10] Exchanging one for another imparts little change on the shape and volume of the molecule and therefore should not affect the binding (pharmacodynamics) of a compound. Changing one isostere for another, however,

TABLE 11.3 Chain Homologues and Ring-Chain Isomers of 11.28[8]

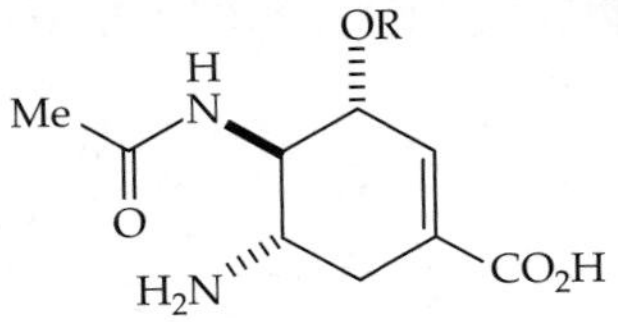

11.28

Entry	Carbons in R-Group	R-Group	Neuraminidase (IC_{50}, nM) Influenza A	Influenza B[a]
1	0	H	6,300	ND
2	1	CH_3	3,700	ND
3	2	CH_2CH_3	2,000	185
4	3	$CH_2CH_2CH_3$	180	ND
5	4	$CH_2CH_2CH_2CH_3$	300	215
6	4	$CH_2CH(CH_3)_2$	200	ND
7	4	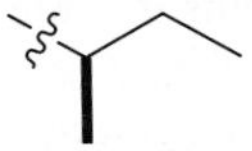	10	7
8	4	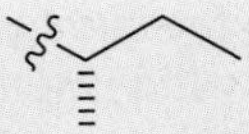	9	2
9	5	$CH_2CH_2CH_2CH_2CH_3$	200	ND
10	5	$CH(CH_2CH_3)_2$ (oseltamivir/Tamiflu)	1	3
11	6	$CH_2CH_2CH_2CH_2CH_2CH_3$	150	1,450
12	6	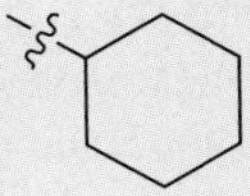	60	120
13	7	$CH_2CH_2CH_2CH_2CH_2CH_2CH_3$	270	ND
14	8	$CH_2CH_2CH_2CH_2CH_2CH_2CH_2CH_3$	180	3,000
15	9	$CH_2CH_2CH_2CH_2CH_2CH_2CH_2CH_2CH_3$	210	ND
16	10	$CH_2CH_2CH_2CH_2CH_2CH_2CH_2CH_2CH_2CH_3$	600	ND
17	10	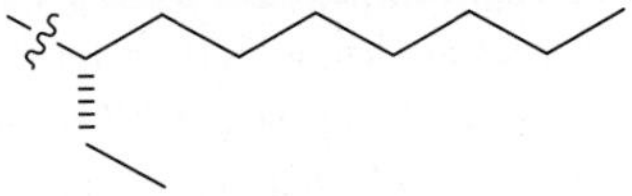	1	4
18	10	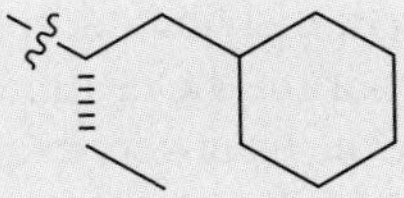	16	6,500
19	11	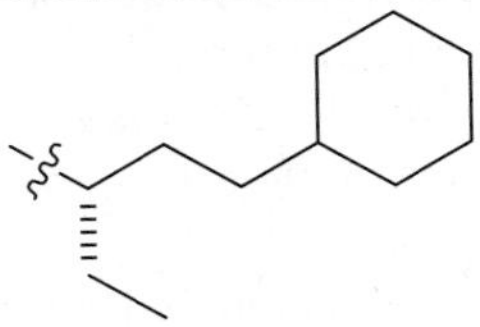	1	2,150

[a]ND = not determined

TABLE 11.4 Classical Isosteres[10]

Group	Isosteres
Equivalent univalent groups (by size)	1. $-CH_3$, $-NH_2$, $-Cl$ (small groups) 2. $-Br$, $-CH(CH_3)_2$ (intermediate groups) 3. $-I$, $-C(CH_3)_3$ (large groups)
Equivalent divalent groups	$-CH_2-$, $-NH-$, $-O-$
Equivalent ring groups	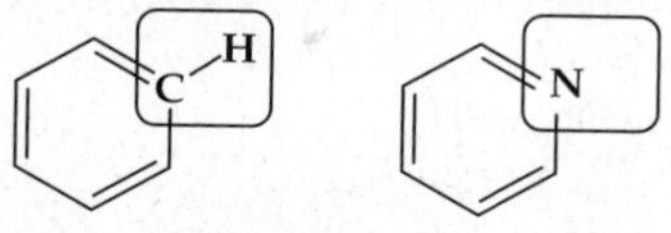 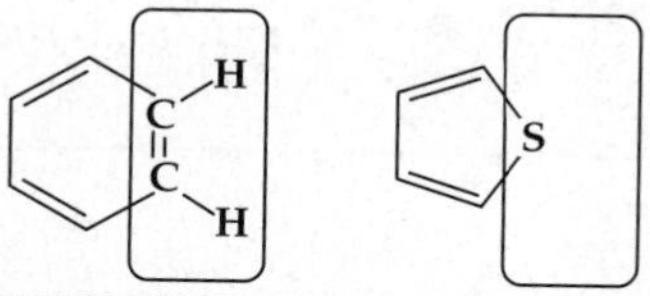

may impact pharmacokinetics in any number of ways, including changing lipophilicity and introducing or removing potential sites for metabolism.

Classical isosteres are almost always included in collections of analogues with biological activity. For example, many isosteric analogues of chlorpromazine (**11.29**) have been prepared (**Figure 11.5**).[11] From the equivalent ring group category in Table 11.4, two carbons in an aromatic ring ($-C=C-$) can be replaced with a single sulfur atom with minimal impact on biological activity. The interchange of a sulfur atom for two ring carbons can occur in three ways on the left side of chlorpromazine (**11.29**). Consistent with the theory behind isosteres, analogues **11.30** and **11.31** retain the activity of chlorpromazine. Compound **11.32**, however, is inactive. This demonstrates that isosteres, while they are *intended* to retain the same level of activity, do sometimes have a significant impact, either positive or negative, on potency.

When exchanged for one another, some functional groups may seem different in terms of their steric sizes, and yet they provide a new analogue with properties similar to the original. These groups have become known as *bioisosteres*, or *nonclassical isosteres*.[10] These groups rely less on size similarity and more on preserving hydrogen bonding and electronic interactions. Representative bioisosteres are shown in **Table 11.5**.

Bioisosteres can seem strange and nonobvious. For example, a bioisostere for a carboxylic acid is a tetrazole ring. Although a carboxylic acid and tetrazole ring may

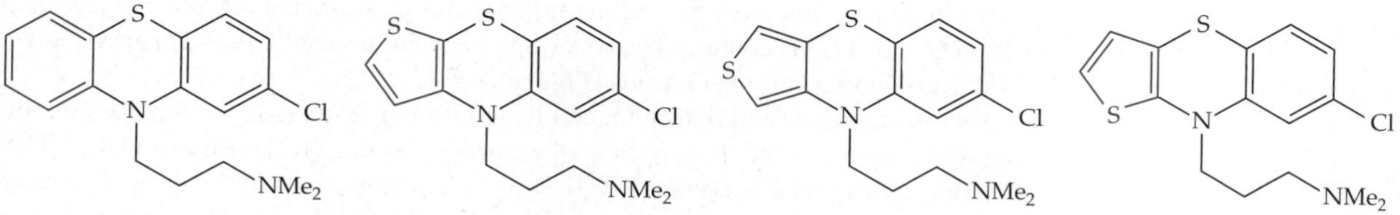

chlorpromazine
(Thorazine)
antipsychotic
11.29

isosteric analogue 1
active
11.30

isosteric analogue 2
active
11.31

isosteric analogue 3
inactive
11.32

FIGURE 11.5 Chlorpromazine and isosteric analogues[11]

TABLE 11.5 Bioisosteres[10]

Functional Group	Bioisosteres
Hydrogen equivalent (−H)	−F, D (deuterium)
Carboxylic acid equivalent (−COOH)	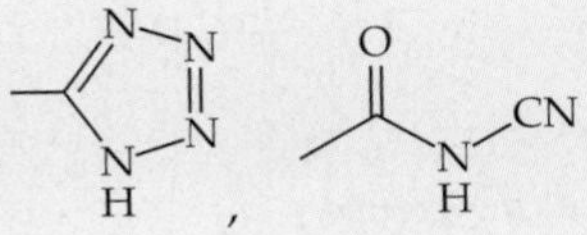
Hydroxy equivalent (−OH)	−CH₂OH, −CH(CN)₂, −NH(CN)
Thiourea equivalent 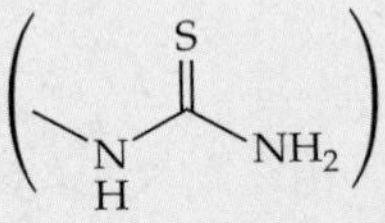	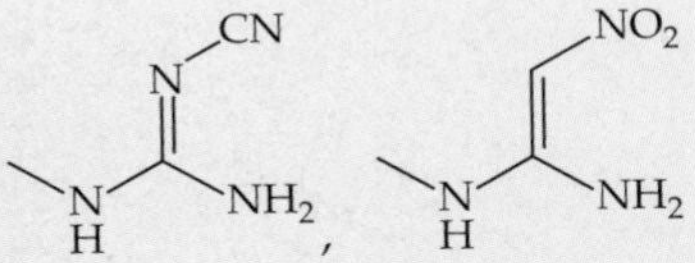

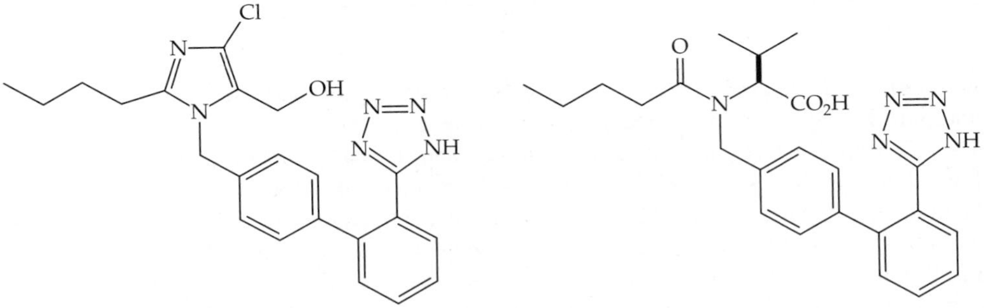

FIGURE 11.6 Angiotensin-II receptor antagonists with tetrazole bioisosteres

losartan
(Cozaar)
antihypertensive
11.33

valsartan
(Diovan)
antihypertensive
11.34

appear to be completely different, tetrazole has a pK_a that is similar to a carboxylic acid (4.9 vs. 4.2, respectively). Therefore, tetrazole imitates the charge state of a carboxylic acid in the body. From a drug design standpoint, carboxylic acids carry at least one disadvantage—they undergo phase II conjugations to form glucuronides that are readily filtered from the bloodstream by the kidneys. Glucuronide formation can dramatically decrease a drug's half-life, which is generally not a desired outcome when one is developing a drug. Tetrazoles do not undergo phase II conjugations. If a carboxylic acid in a lead is found to cause metabolic instability, then replacement of the acid with a tetrazole ring may provide a more inert compound and preserve strong target binding. Losartan (Cozaar, **11.33**) and valsartan (Diovan, **11.34**) contain tetrazole rings to avoid metabolic problems associated with the corresponding carboxylic acids (**Figure 11.6**).

Another example of modulating metabolism with bioisosteres is the replacement of hydrogen with fluorine in the development of ezetimibe (Zetia, **11.21**) (**Figure 11.7**).[12] The fluorine atoms block phase I oxidation of two of the aromatic rings in the drug. Blocking metabolism at the 4-position of the ring prolongs the effective duration of the drug in the body. Fluorine is particularly useful in this role because it is similar in size to hydrogen. The van der Waal's radius of hydrogen is 1.2 Å, while fluorine's is 1.35 Å. Fluorine does not introduce steric bulk but does decrease the electron density in the ring sufficiently to suppress oxidation.

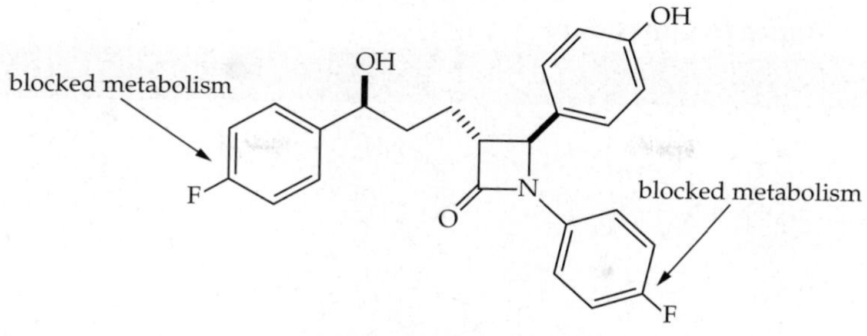

FIGURE 11.7 Bioisosteres in ezetimibe[5]

ezetimibe
(Zetia)
anticholesterol
11.21

The following Case Studies show several examples of isosteres and bioisosteres in use. The story of the development of cimetidine also demonstrates the use of chain homologation.

CASE STUDY

Cimetidine[12]

Cimetidine (Tagamet, **11.35**) is a histamine H_2-receptor antagonist that inhibits gastric acid secretion in the stomach (**Figure 11.8**). Cimetidine is used in the treatment of peptic ulcers and heartburn. The drug was developed during the 1960s and 1970s. The idea of using an antihistamine to treat ulcers arose in the early 1960s when researchers at Smith Kline & French (SK&F) refined their ability to differentiate between antihistamines active at the H_1 receptor (allergy medication) and those active at the H_2 receptor (potential ulcer treatment). The assay used throughout the development of cimetidine was an animal assay on anesthetized rats. The rats were treated with histamine (**11.36**) to stimulate acid secretion. The same rats were then treated with a potential antagonist. Increases in stomach pH caused by administration of the antagonist were monitored to screen for activity.

While an antagonist does not necessarily bind a receptor at the same site as an agonist, the agonist is sometimes the only available activity-related piece of information. Therefore, the SK&F team started with histamine, the endogenous agonist for the H_2 receptor. Guided by the structure of histamine, the SK&F team eventually discovered N^α-guanylhistamine (**11.37**), as the first lead. N^α-Guanylhistamine inhibits gastric acid secretion in vivo at a high dose ($ID_{50} = 800\ \mu mol/kg$, ID = inhibitory dose). The compound was later found to be a weak partial agonist instead of a true antagonist. Optimization of the lead produced a number of compounds, which are summarized in **Table 11.6**.

From the compounds in the table, the first successful change to N^α-guanylhistamine was a simple chain homologation to give a nearly 10-fold increase in activity (**11.38**). Analogue **11.38** still showed weak agonist activity, attributable to the positive charge on the guanidino group at biological pH. Guanidine-containing compounds retained agonist activity due to their similarity to histamine.

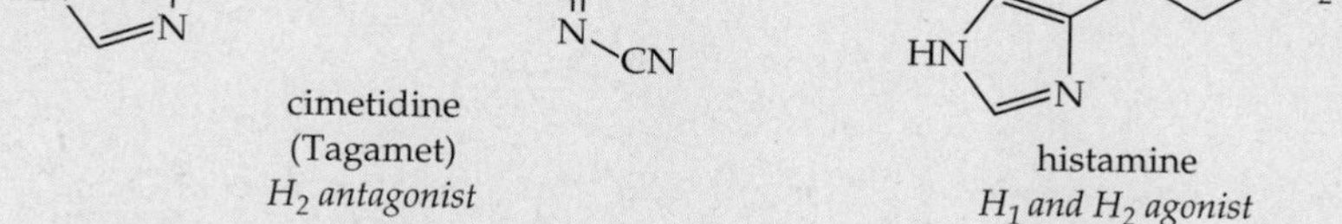

cimetidine
(Tagamet)
H_2 antagonist
11.35

histamine
H_1 and H_2 agonist
11.36

FIGURE 11.8

TABLE 11.6 **Leads and Analogues en Route to Cimetidine**

Entry	Name	Structure	ID_{50} (μmol/kg)
1	N^{α}-guanylhistamine **11.37**	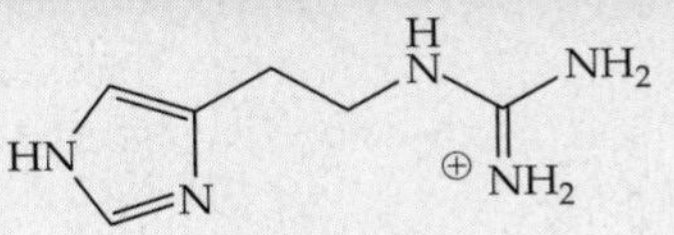	800
2	SK&F 91486 **11.38**	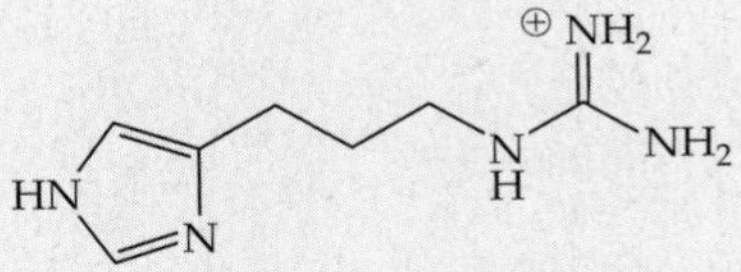	100
3	SK&F 91581 **11.39**	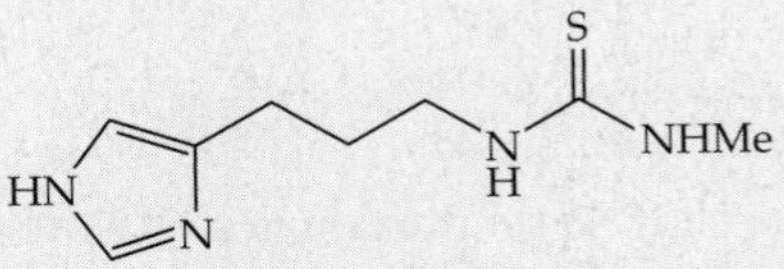	>256
4	burimamide **11.40**	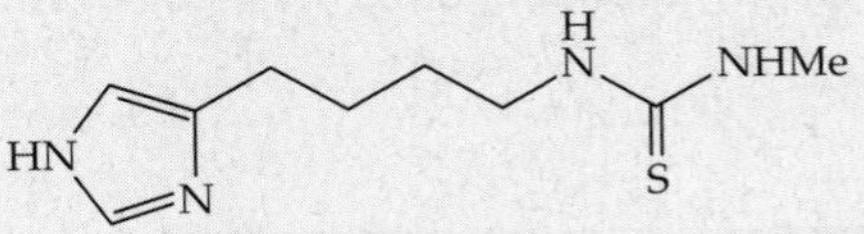	6.1
5	metiamide **11.41**	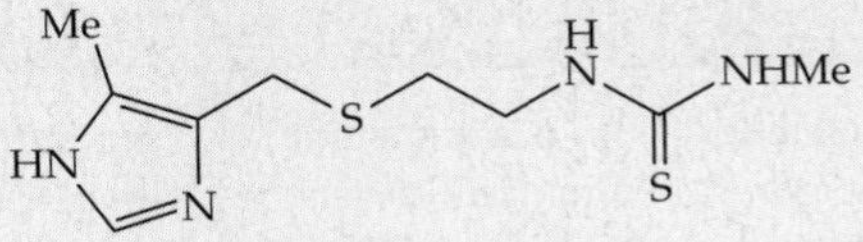	1.6
6	cimetidine (Tagamet) **11.35**	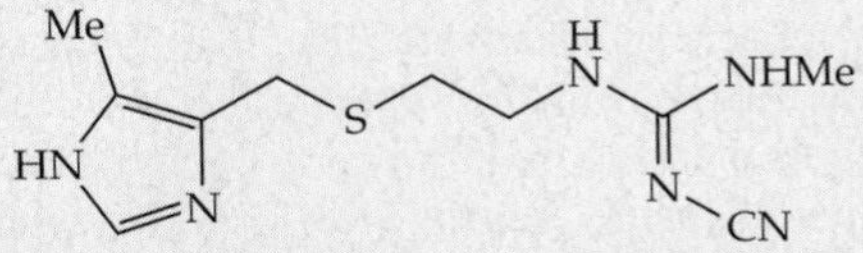	1.4

Researchers at SK&F therefore explored groups that provided a hydrogen-bond donor but no positive charge. Replacement of the guanidino group with a thiourea (**11.39**), which is neutral at the pH of the stomach, removed the agonist properties but was a weakly binding compound. Chain homologation by an additional methylene afforded burimamide (**11.40**), a potent antagonist. Burimamide was approved for clinical trials but was not sufficiently active.

With the failure of burimamide, the SK&F team turned its attention to the other side of the molecule—the electronics of the imidazole ring. At biological pH, substituted imidazole rings exist in a complex equilibrium of two neutral tautomers (**11.42** and **11.44**) and a protonated conjugate acid (**11.43**) (**Scheme 11.3**). The imidazole ring of histamine favors the distal tautomer and is only 3% protonated at pH 7. In contrast, burimamide favors the proximal tautomer with 40% protonation, indicating that the imidazole ring is more basic than in histamine. Ultimately, metiamide (**11.41**) was found to be an analogue that favors the distal tautomer and has a weakly basic imidazole ring with only 20% protonation at pH 7. Metiamide bears a new methyl

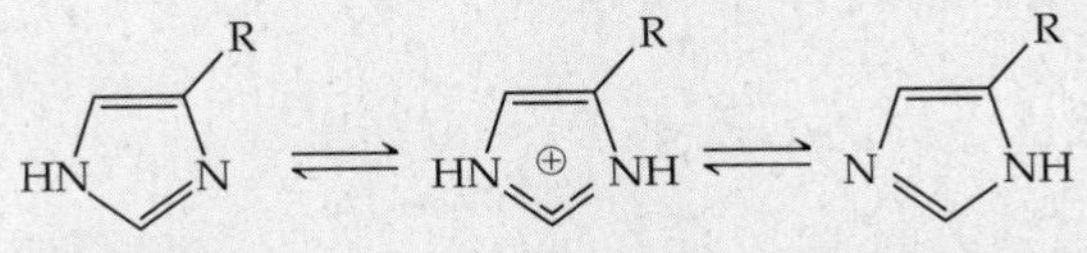

distal tautomer **11.42** conjugate acid **11.43** proximal tautomer **11.44**

SCHEME 11.3 Tautomer interconversions in 4-substituted imidazoles

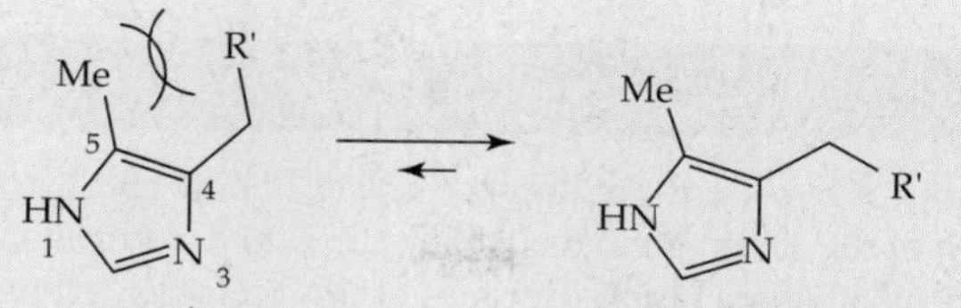

compact conformation (undesired) **11.45**

extended conformation (desired) **11.46**

SCHEME 11.4 Conformational effects of a 5-methyl group in imidazoles

group at C5 of the imidazole ring. Aside from influencing the tautomeric equilibrium, the methyl group introduces a measure of steric bulk that keeps the chain at C4 extended in an orientation (**11.46**) believed to be responsible for binding (**Scheme 11.4**). The more traditional method for locking a desired conformation would be to use rings (see Chapter 9), but the ring approach failed in the case of antihistamine compounds.

Metiamide (**11.41**) was advanced to the clinic and progressed to phase II trials. While it was effective in treating ulcers, some patients showed kidney damage and abnormalities in their white blood cell profile. SK&F attributed the side effects to the sulfur of the thiourea group. Bioisosteric replacements of the sulfur generated a number of analogues (**11.47–11.50** and **11.35**) with cimetidine emerging as the best (**Table 11.7**). In 1979, cimetidine (**11.35**) was approved by the U.S. Food and Drug Administration (FDA) and marketed under the trade name Tagamet.

As with all marketed drugs, cimetidine was well protected through many patents. Unfortunately for SK&F, all their patents on cimetidine emphasized the important role of the imidazole ring, which was believed to be essential for activity. Glaxo was able to develop a structurally similar drug, ranitidine (Zantac, **11.51**), by changing the imidazole ring to a furan and using a bioisostere for the cyanoguanidine group (**Figure 11.9**). Ranitidine was approved in 1981 under the name Zantac and became the top-selling drug of 1988. Merck also subverted the SK&F patents with its own H_2-receptor antagonist, famotidine (Pepcid, **11.52**), in 1985. Famotidine used a thioazole in place of the imidazole of cimetidine.

TABLE 11.7 Carbonyl Analogues of Metiamide

Compound	Entry	Isostere	X	ID_{50} (μmol/kg)
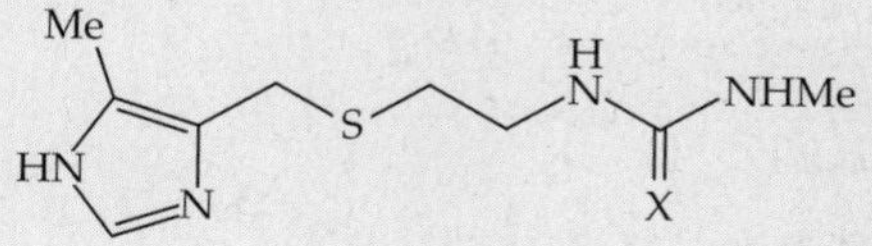	1	Metiamide (thiourea) (**11.41**)	S	1.6
	2	Urea isostere (**11.47**)	O	27
	3	Guanidine isostere (**11.48**)	NH	12
	4	Guanylurea derivative (**11.49**)	NC(=O)NH_2	7.7
	5	Nitroguanidine isostere (**11.50**)	NNO_2	2.1
	6	Cyanoguanidine isostere (cimetidine, **11.35**)	NCN	1.4

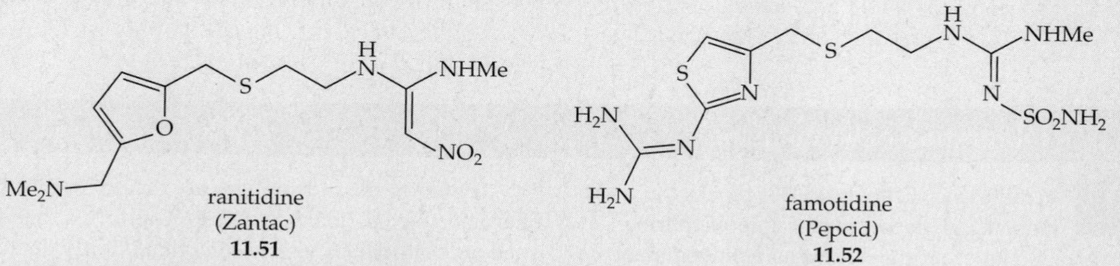

ranitidine (Zantac) **11.51**

famotidine (Pepcid) **11.52**

FIGURE 11.9 Other H_2-receptor antagonists

Extension of the Half-Life of Iloprost[13]

Prostacyclin (**11.53**) is a signal molecule in the body that inhibits clotting and acts as a vasodilator. Prostacyclin is unstable and has an extremely short half-life (**Figure 11.10**). The instability of prostacyclin in the body is primarily attributed to its enol ether functionality, which readily undergoes hydrolysis. Isosteric replacement of the ring oxygen with a CH_2 group and small modifications of the side chain afforded iloprost (Ilomedine, **11.54**), an FDA-approved treatment for certain forms of hypertension. Iloprost, however, undergoes oxidation of the β-carbon to the carboxylic acid to form an inactive metabolite (**11.55**). Because of this metabolic pathway, the half-life of iloprost is relatively short at 20–30 minutes. Another isosteric replacement of an oxygen with a CH_2 group and further modification of the side chain provided cicaprost (**11.56**). Cicaprost showed antihypertensive effects in rats with one-fifth the oral dose of iloprost, and the effects lasted two to three times longer.

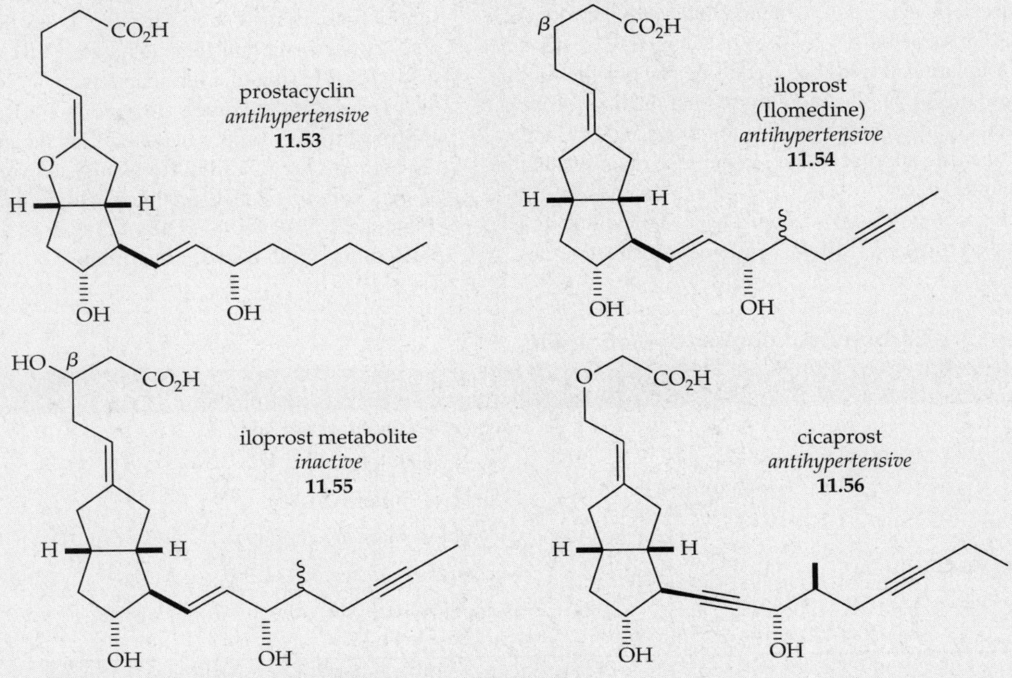

FIGURE 11.10 Prostacyclin and antihypertensive analogues

Deuteration of Vanlafaxine[14]

Venlafaxine (Effexor, **11.57**) blocks the reuptake of serotonin and norepinephrine from the synaptic junction and acts as an antidepressant (**Scheme 11.5**). The primary metabolic pathway of venlafaxine is O-demethylation (**11.58**), which is mediated by CYP2D6. CYP2D6 has variable activity across different populations. Groups with a more active form of CYP2D6 tend to show more side effects from venlafaxine.

To inhibit the metabolism of venlafaxine, Auspex Pharmaceuticals, near San Diego, California, has developed an analogue of venlafaxine that is fully deuterated at active metabolic sites (**Figure 11.11**). Oral dosing of the analogue, SD-254 (**11.59**), gives higher plasma levels of the drug than venlafaxine and lower concentrations of O-desmethyl metabolites. Furthermore, the plasma levels of SD-254 are much more consistent across different

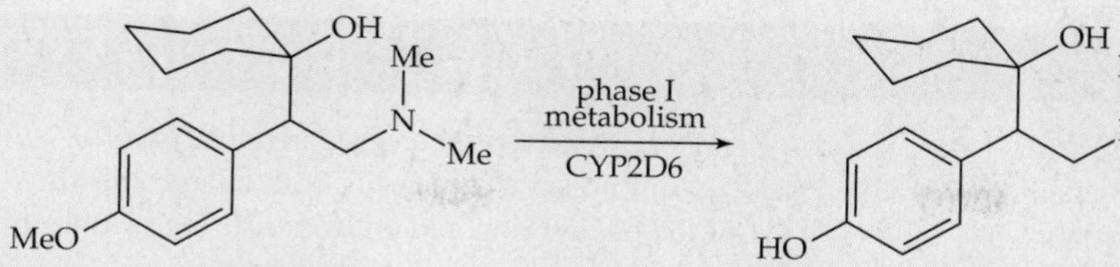

venlafaxine
(Effexor)
antidepressant
11.57

O-desmethyl
venlafaxine
antidepressant
11.58

SCHEME 11.5 Phase I metabolism of venlafaxine

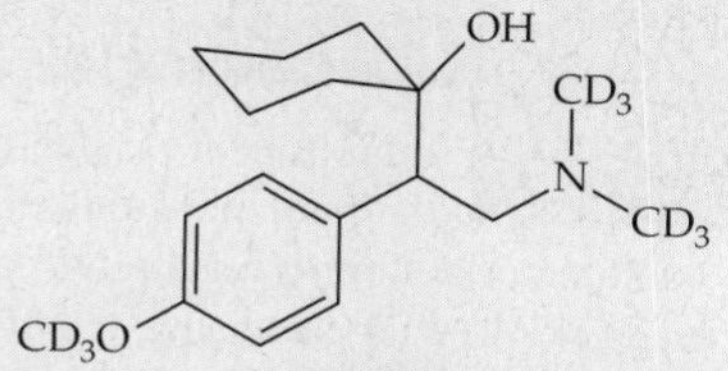

SD-254
antidepressant
11.59

FIGURE 11.11

patients than venlafaxine. It is presumed that SD-254 will show fewer side effects than venlafaxine. At the time of the writing of this text, SD-254 is entering phase II clinical trials.

Because deuterium is heavier than hydrogen, the carbon-deuterium bond has a lower vibrational frequency than the carbon-hydrogen bond. This difference in frequency makes the carbon-deuterium bond slower to react. If bond breaking occurs during the rate-determining step of the reaction, then the overall reaction will be slowed by the replacement of hydrogen with deuterium. This is an example of a kinetic isotope effect. The use of deuterium as a bioisostere to manipulate rates of metabolism is a fairly new idea in drug discovery. If SD-254 successfully reaches the market, additional deuterated drugs will almost certainly be advanced into clinical trials.

11.5 Directed Combinatorial Libraries

The term *combinatorial chemistry* typically evokes images of huge libraries of molecules with members of 100,000 and even more. This type of large library is ideal for lead discovery, a situation in which a medicinal chemist is trying to cast a diverse structural net to search for activity. The lead optimization process is more specialized. The basic skeleton of the eventual candidate may already be known. Only the ideal substituents on the pharmacophore remain in question. While lead discovery and lead optimization are very different tasks, combinatorial chemistry can effectively play a role in both processes.

Performing SAR studies can be repetitive, requiring the same reaction to be performed on a slightly different starting material. Combinatorial chemistry is ideally suited for this type of problem. Optimization libraries tend to be small, often between 100 and 1,000 compounds. Such a library can be created with compounds bearing common substituents in a number of positions on a lead. Ideally, using combinatorial chemistry will be efficient enough to prepare the library far more quickly than a group of organic chemists making one compound at a time.

If the synthetic chemistry can be adapted to work in a combinatorial fashion, the results can be surprising. In traditional SAR, different portions of a lead are often derivatized independently in what was described as a divide and conquer approach earlier in this chapter. The optimal substituents for each portion of the molecule are then combined into one (hopefully) fully optimized lead. This approach relies on *single-point modifications* and assumes that binding in one region is independent of binding in another region. That assumption is sometimes invalid. The following Case Study highlights the danger of assumptions that commonly occur during the lead optimization process.

Raf Kinase Inhibitors[15,16]

Inhibitors of raf kinase are of interest for their potential anti-cancer activity. A directed combinatorial library was used to perform SAR on a lead (**11.60**) with promising activity against raf kinase ($IC_{50} = 17\ \mu M$) (**Figure 11.12**).[15] The library initially included 1,500 members, but only approximately 1,000 were of sufficient purity for reliable testing. The most active analogue (**11.61**) was 30-fold more potent than the original lead ($IC_{50} = 0.54\ \mu M$). The high activity of compound **11.61** with its 4-OPh group was surprising since earlier SAR work indicated that groups larger than CH_3 decreased the activity of the lead. Earlier SAR work also showed that replacement of the thiophene ring in **11.60**

with an isoxazole somewhat decreased activity. In all likelihood, an optimized lead bearing both a large substituent on C4 of the phenyl ring and an isoxazole ring would have been missed in a traditional, single-point modification SAR study. The compound simply would not have been prepared because of the evidence against its possible activity. Fortunately, **11.61** was found through the directed combinatorial library. One explanation for the surprising activity of **11.61** is that it may have a different binding orientation than the original lead.[15] Lead **11.61** was ultimately optimized to afford sorafenib (Nexavar, **11.62**), a raf kinase inhibitor that is currently marketed for treating certain forms of cancer.[16]

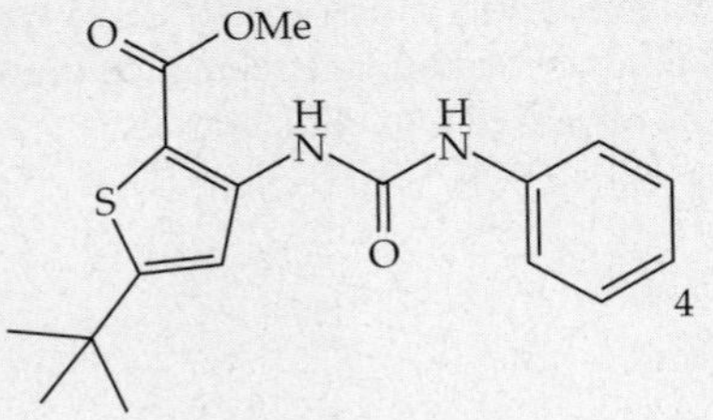

original lead
$IC_{50} = 17\ \mu M$
11.60

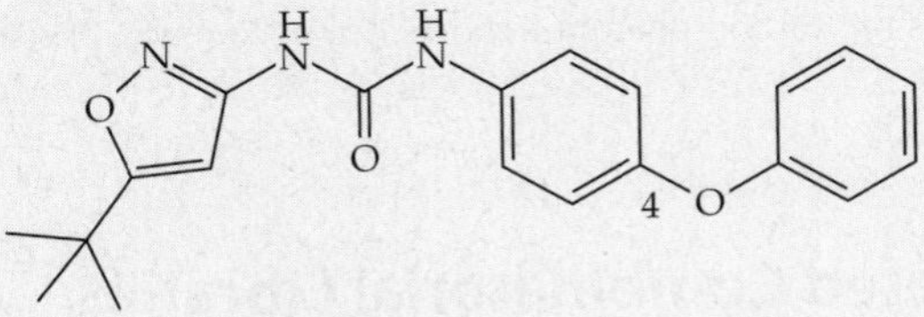

second-generation lead
$IC_{50} = 0.54\ \mu M$
11.61

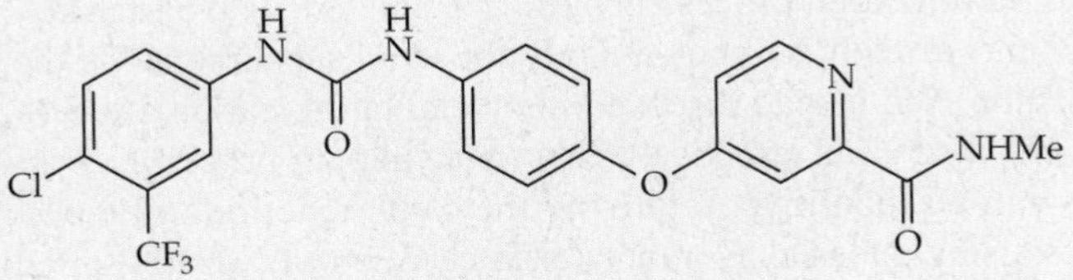

sorafenib
(Nexavar)
$IC_{50} = 0.012\ \mu M$
11.62

FIGURE 11.12 Kinase inhibitors

11.6 Peptidomimetics

Because polypeptides are involved in many regulatory processes in the body, polypeptides and their analogues are natural drug candidates. Endogenous polypeptides tend to be very potent with strong binding. Therefore, a known polypeptide is essentially already a lead, and the entire lead discovery process can be skipped.

Polypeptides, however, have one major problem—poor pharmacokinetics. Most successful drugs are administered orally and must either be absorbed in the stomach or pass through the stomach for absorption in the small intestines. Polypeptides tend to be destroyed in the stomach by cleavage of peptide linkages. There are two potential solutions to this problem. Either standard lead discovery techniques may be used to find a nonpeptide lead, or the polypeptide lead may be optimized for better oral availability. Both approaches have drawbacks. Abandoning a polypeptide lead and restarting a search at lead discovery is a drastic move. Finding a new lead is difficult and will require starting from scratch, likely with random screening of a library. Because standard lead discovery techniques have been covered in Chapter 10, this option will not receive any further attention here. If the polypeptide is to be kept as a lead, its pharmacokinetic profile must be improved. This task defines the field of *peptidomimetics*, that is, changing the pharmacokinetics of a peptide lead without losing too much of its activity.[17]

Of the lead optimization methods covered in this chapter, modifying a molecule to change its pharmacokinetic properties without affecting its activity most closely matches the concept of isosteric replacements. Indeed, isosteres have been designed specifically for peptides. Peptide bond isosteres focus on removing amide linkages from the lead to make a new pseudopeptide. The following listing of isosteres is by no means complete (**Figure 11.13**).[17] For each isostere, note how the atoms have been rearranged and/or shifted relative to the standard peptide structure to create a new peptide analogue.

Some peptide bond isosteres, especially those in which a CH_2 group replaces the carbonyl, introduce more flexibility to the new pseudopeptide. If a pseudopeptide analogue is too flexible and loses binding affinity, the conformation of the analogue can be constrained through the use of fused rings. Through application of these techniques, sometimes a polypeptide may be optimized into a viable drug candidate.[17]

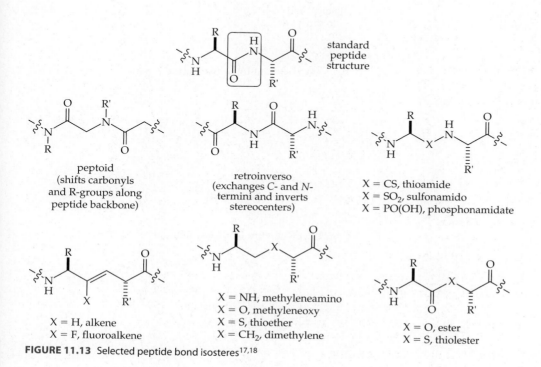

FIGURE 11.13 Selected peptide bond isosteres[17,18]

The two following Case Studies demonstrate some of the typical challenges associated with optimizing a peptide lead into an orally available drug.

Angiotensin-Converting Enzyme Inhibitors[19–23]

Angiotensin-converting enzyme (ACE) cleaves angiotensin I (A-I), a decapeptide, into angiotensin II (A-II), an octapeptide (**Scheme 11.6**). A-II is an agonist for the A-II receptor and causes vasoconstriction and elevated blood pressure. Therefore, ACE was seen as a promising target for managing high blood pressure. Inhibition of ACE should decrease the amounts of available A-II and ultimately decrease vasoconstriction and blood pressure levels.[19]

The search for an ACE inhibitor started with carboxypeptidase A (CPA), an enzyme with similarities to ACE. Both proteases include Zn^{+2} in the active site and cleave amino acid residues from the C-terminus of an oligopeptide. D-2-Benzylsuccinic acid (**11.65**) was known to inhibit CPA (**Figure 11.14**). The functional groups in D-2-benzylsuccinic acid align well with functional groups of a natural substrate (**11.63**) of CPA. Compound **11.65** is identical to the C-terminus of the natural substrate of CPA except the carbonyl that binds Zn^{+2} in the active site has

been replaced with a carboxylate group. In the early 1970s, a polypeptide inhibitor (**11.64**) of ACE was identified in the venom of the Brazilian pit viper (*Bothrops jararaca*). The venom had been observed to decrease blood pressure in bite victims. Studying the alignment of functional groups of the venom-based inhibitor and drawing analogies to CPA allowed the design of a new peptide analogue ACE inhibitor (**11.66**). The new inhibitor was sufficiently active to be a viable lead compound.[20,21]

Replacement of the carboxylate terminus of the lead with a thiol group afforded captopril (Capoten, **11.67**), the first marketed ACE inhibitor (**Figure 11.15**).[22] The sulfur serves as a soft Lewis base for coordination to the Zn^{+2} cofactor in ACE. Unlike peptides, captopril has a high oral bioavailability ($F = 75\%$). Amide linkages involving N-acylation of proline tend to be more stable metabolically than other amides encountered in proteins. For this reason captopril, which still contains a strong peptide-like appearance, regardless has good bioavailability.

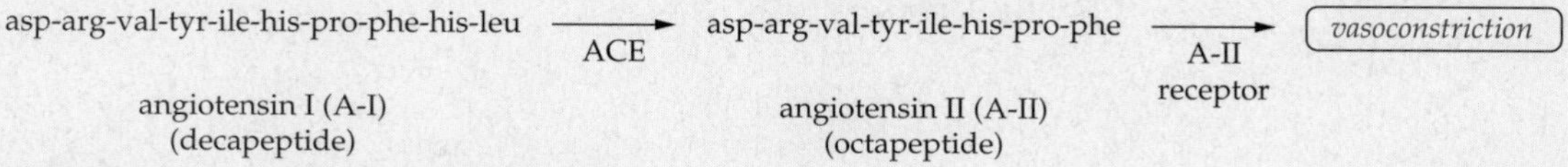

asp-arg-val-tyr-ile-his-pro-phe-his-leu → asp-arg-val-tyr-ile-his-pro-phe → [vasoconstriction]
            ACE   A-II receptor

angiotensin I (A-I)   angiotensin II (A-II)
(decapeptide)   (octapeptide)

SCHEME 11.6 Conversion of A-I to A-II by ACE[19]

CPA enzyme with substrate

11.63

ACE with inhibitor from snake venom

11.64

CPA enzyme with inhibitor

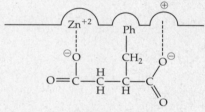

D-2-benzylsuccinic acid
11.65

ACE with peptide analogue inhibitor

11.66

FIGURE 11.14 Binding of CPA and ACE with inhibitors[20]

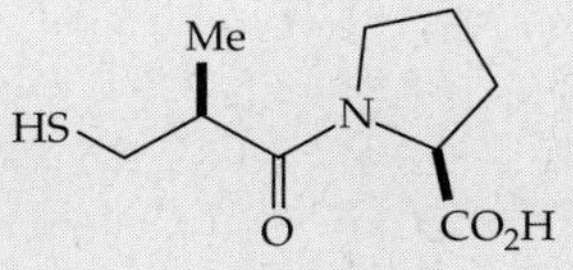

captopril
(Capoten)
antihypertensive
11.67
K_i = 20 nM

FIGURE 11.15

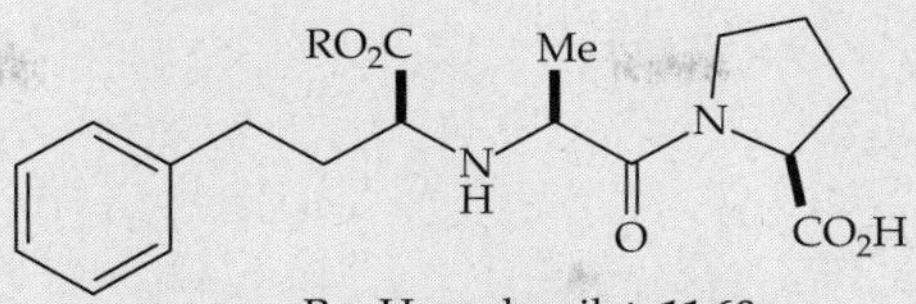

R = H, enalaprilat, **11.68**
K_i = 12 nM
R = Et, enalapril, **11.69**
(Vasotec)

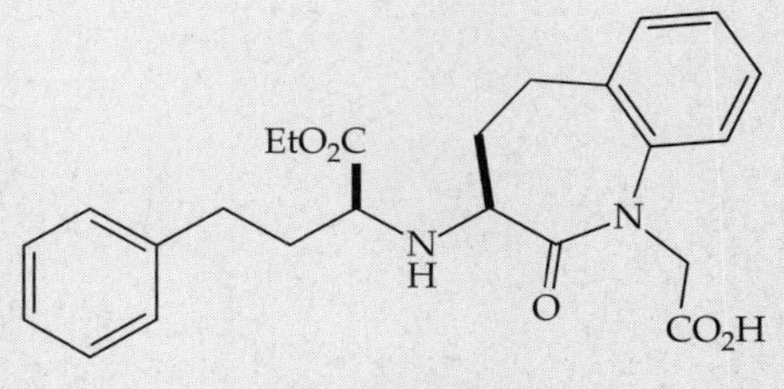

benazepril
(Lotensin)
F = 35%
11.70

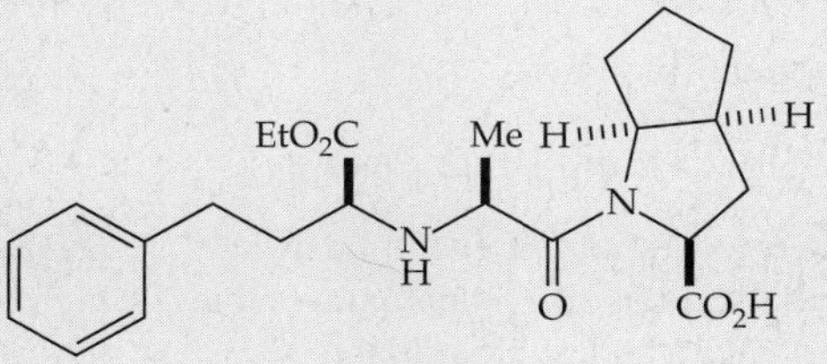

ramipril
(Altace)
F = 55%
11.71

FIGURE 11.16 Carboxyalkyl dipeptide ACE inhibitors

The half-life of captopril is short (<3 hours). A drug with a short half-life requires more frequent dosing to maintain an effective concentration in the body. This can result in inadvertently missed doses. More important than the short half-life, some patients taking captopril report rashes and a lack of taste sensation (dysgeusia).

The side effects of captopril were attributed to the thiol group, and new ACE inhibitors without a thiol group were sought. The third residue from the *C*-terminus in snake venom inhibitor **11.64** is phenylalanine (R = CH_2Ph). This residue implies that ACE has a hydrophobic binding pocket to accommodate the nonpolar CH_2Ph side chain. Based on this idea, a second generation of ACE inhibitors emerged (**Figure 11.16**). Enalaprilat (**11.68**) was the first example of this group of ACE inhihitors, sometimes called the carboxalkyl dipeptides. All members of the class have a hydrophobic 2-phenylethyl chain, which interacts with a pocket in the active site of ACE. In the carboxyalkyl dipeptides, a carboxylic acid replaces captopril's problematic thiol. Enalaprilat binds ACE strongly but shows very poor bioavailability. Conversion of the acid to its ethyl ester affords a new compound, enalapril (Vasotec, **11.69**). Enalapril is the marketed form of the orally administered drug. Enalapril itself shows poor ACE inhibition. Once enalapril is ingested, the ester is metabolized in a phase I hydrolysis to form enalaprilat (**11.68**), a strong ACE inhibitor. Enalapril is therefore a prodrug. Enalapril shows good bioavailability (60%), a long half-life (11 hours), and fewer side effects than captopril. Representative carboxyalkyl dipeptide ACE inhibitors (**11.70** and **11.71**) are shown, and both are prodrugs that are activated by ester hydrolysis. Carboxyalkyl dipeptides retain peptide structure and are not considered true peptidomimetics.[21]

A third generation of ACE inhibitor (**11.73**) emerged based on the finding that phosphoramidon (**11.72**), a natural product lead from a bacterial culture, is an effective ACE inhibitor (**Figure 11.17**). The overall structure of phosphoramidon received less attention than its phosphonamidate group, which was speculated to be crucial for binding the Zn^{+2} center in ACE. The third-generation ACE inhibitors were redesigned to include a phosphinate ester in place of the carboxylic acid ester. The alkyl group on the phosphinate ester hydrolyzes in vivo to afford a free phosphinic acid, the active drug. The bioavailability of the inhibitor, called fosinopril, is only about 35%, similar to the second-generation ACE inhibitors.[23]

In the search for effective ACE inhibitors, marketable drugs were successfully developed from a peptide lead.

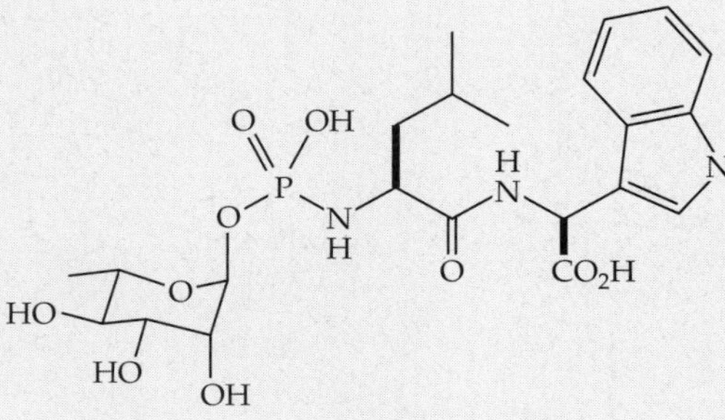

FIGURE 11.17 Phosphinic acid derived ACE inhibitors

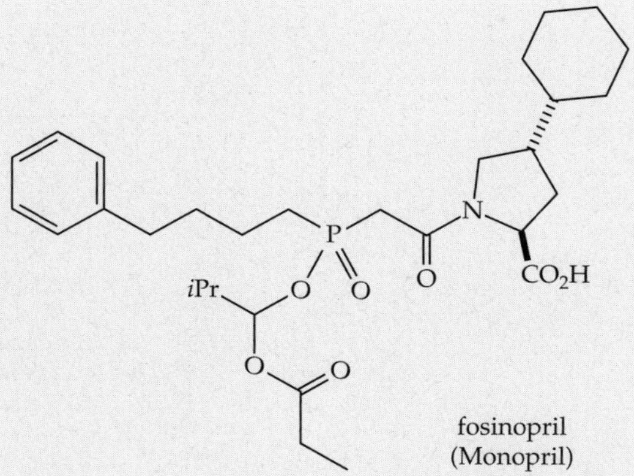

Unfortunately, even three generations of evolution away from a peptide structure, the drugs still retain unfavorable properties of peptides, namely modest bioavailability.

Development of a true peptidomimetic ACE inhibitor with high bioavailability would be a significant breakthrough for this class of drugs.

CASE STUDY

Human Immunodeficiency Virus-1 Protease Inhibitors[24,25]

Human immunodeficiency virus-1 (HIV-1) protease is an enzyme that breaks inactive polyproteins into their functional parts that are vital for the proper operation of HIV-1. This enzyme is a target for treatment of patients infected with HIV-1. HIV-1 protease operates as a dimeric complex, a trait that lends the enzyme to be inhibited by a drug that is highly symmetrical.[24]

Because HIV-1 protease acts on peptides, the early leads were themselves peptides. Publication of the x-ray crystal structure of HIV-1 protease revealed that the peptide inhibitors interacted with the enzyme's active site through a series of key hydrogen bonds and hydrophobic interactions (**Figure 11.18**). The inhibitor that is shown, called A-74704 (**11.74**), was developed at Abbott and is very potent (IC_{50} = 3 nM). Two carbonyls in the peptide backbone of **11.74** form hydrogen bonds to a molecule of water in the active site of HIV-1 protease. Nonpolar side chains of valine and phenylalanine occupy corresponding nonpolar pockets on the enzyme's surface. Finally, an alcohol in the middle of the inhibitor exploits a hydrogen bonding interaction. These seven key interactions directed the preparation of subsequent analogues.[24]

A number of pharmaceutical companies developed drug candidates for HIV-1 protease inhibition, three of

which are shown in **Figure 11.19**. Aside from amprenavir (Agenease, **11.77**), a true peptidomimetic with no amide bonds, the other two drugs (**11.75** and **11.76**) closely resemble peptide leads and suffer from some of the inherent problems, especially low bioavailability, of peptides as drugs.

Research at DuPont Merck was particularly unsuccessful at developing an optimized peptide lead with adequate bioavailability for HIV-1 protease. The research team quickly abandoned peptide-based leads and reverted to searching small molecule libraries for compounds that could complement the shape and functionality of HIV-1 protease. A computer database search yielded a *p*-terphenyl derivative (**11.78**) as a close match and correctly predicted a lead that would mimic the water molecule in the active site (**Scheme 11.7**). The structure eventually evolved into a cyclic urea (DMP-323, **11.81**) with the requisite functionality, polar and nonpolar, to strongly bind the enzyme. This compound is a true peptidomimetic lead. Abbott later prepared a similar candidate (A-98881, **11.82**) with extremely strong in vitro binding (K_i = 0.005 nM). The third nitrogen in the ring of A-98881 was required for Abbott to access a novel molecular skeleton and avoid existing patents held by DuPont Merck. Both DMP-323 and A-98881 advanced to and died in clinical trials. DMP-323 gave poorly reproducible bioavailability results, while A-98881 suffered from poor

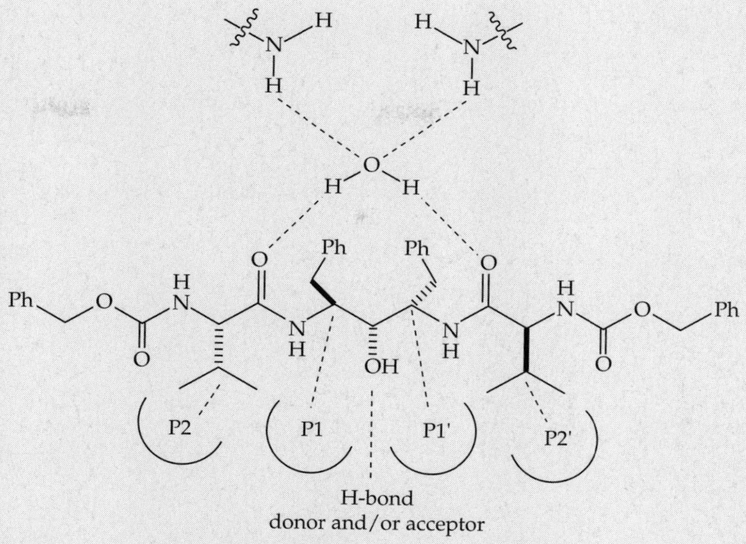

A-74704
11.74

FIGURE 11.18 Inhibitor binding in HIV-1 protease[24]

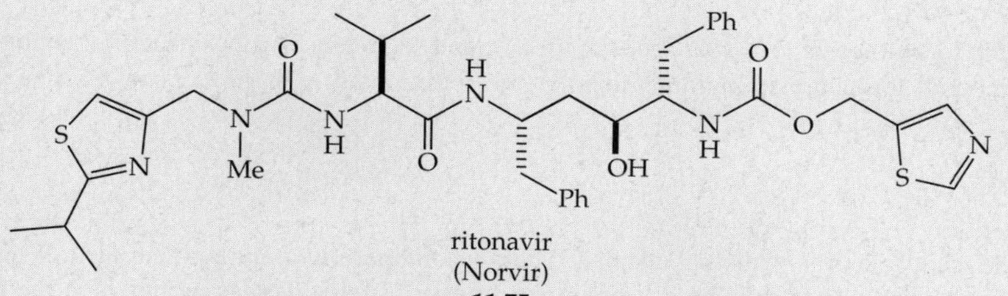

ritonavir
(Norvir)
11.75

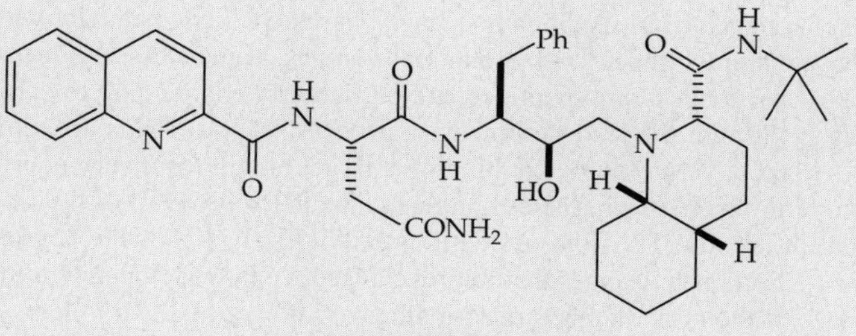

saquinavir
(Fortovase)
11.76

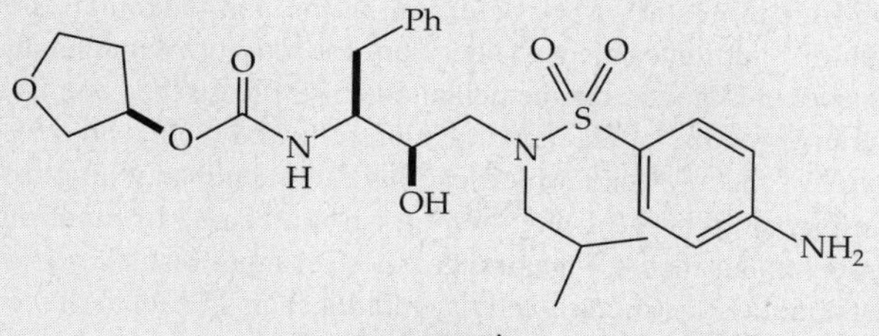

amprenavir
(Agenerase)
11.77

FIGURE 11.19 HIV-1 protease inhibitors

bioavailability (5%) because of metabolism linked to the third-ring nitrogen.[25]

HIV-1 protease has proved itself to be a difficult target for inhibition. In this case, modification of a peptide lead has enjoyed more success in the market than a de novo discovery of a non-peptide inhibitor. Regardless, the work at DuPont Merck and Abbott shows that highly active non-peptides can be developed. The availability of quality x-ray crystallographic data was invaluable to the optimization of a non-peptide drug candidate for HIV-1 protease.

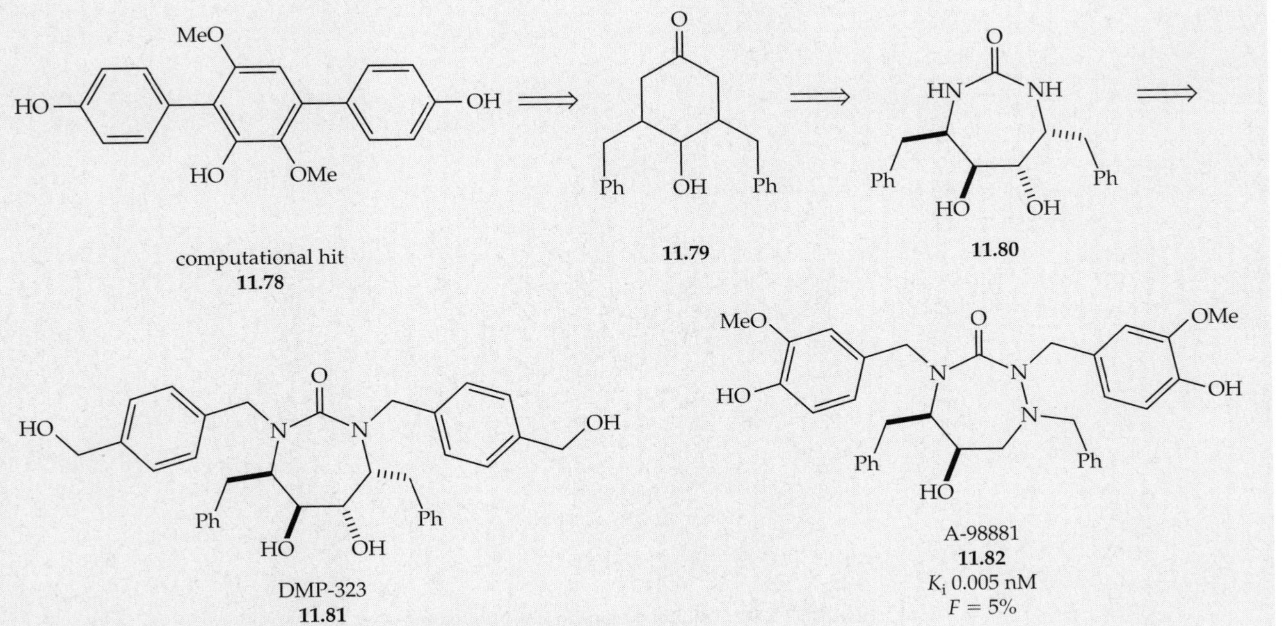

computational hit **11.78** — **11.79** — **11.80** — DMP-323 **11.81** — A-98881 **11.82** K_i 0.005 nM, $F = 5\%$

SCHEME 11.7 Evolution of peptidomimetic HIV-1 protease inhibitors[25]

Researchers were able to synthesize lead analogues in the lab, submit the compounds to the crystallography team, and see precisely how structural changes to the lead affected binding in the enzyme's active site.[25]

Summary

Lead optimization is the main role of a medicinal chemist. The optimization process is conceptually very simple—maximize complementary intermolecular forces between the lead and target to generate a compound suitable for testing in animals. Optimization starts with determining the pharmacophore, the minimal portion of the lead necessary for activity. With the pharmacophore known, the medicinal chemistry team then begins an iterative routine of lead modification followed by testing the activity of the new lead. Modifications can include anything from simple shortening or lengthening carbon chains to functional group replacements. Repetition of these changes results in an understanding of structure-activity relationships (SAR) in the lead and hopefully a more potent lead. While the lead optimization team follows increasing potency, the team must also monitor the pharmacokinetic properties of a lead. Groups that have been conceived to specifically change pharmacokinetic behavior with minimal impact on binding are isosteres. Some classes of leads, particularly polypeptides, are particularly challenging to carry through the lead optimization process. Peptide leads are typically very effective at binding a target but perform very poorly in terms of pharmacokinetics, especially bioavailability and stability. The field of peptidomimetics strives to make peptide backbone changes that improve a lead's pharmacokinetic profile without sacrificing target binding.

Questions

1. 1-Methyl-4-phenyl-4-propionoxypiperidine (**11.a**, MPPP) is a synthetic heroin that appeared as a designer drug in the early 1980s. Does MPPP fit the morphine rule? (A tragic story concerning MPPP is related in a Case Study in Chapter 8.)

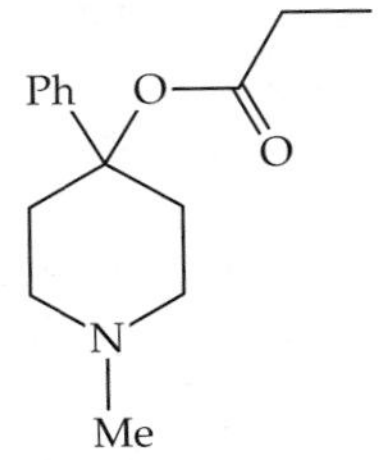

MPPP
11.a

2. Dopamine (**11.b**) and its binding at the dopamine receptor have been intensely studied. *Trans*-2-(3,4-methylenedioxyphenyl)cyclopropylamine (**11.c**) is an agonist for the dopamine receptor while the *cis* isomer is inactive. Based on this information, would compound **11.d** or **11.e** be a better agonist for the dopamine receptor? *Hint: Trace out the key elements in dopamine (hydroxylated benzene ring, ethylene chain, and nitrogen) on both **11.d** and **11.e** and look for similarities and differences with **11.c**.* [Triggle, D. J. Adrenergics: Catecholamines and Related Agents. In M. E. Wolff (Ed.), Burger's Medicinal Chemistry (4th ed., Chapter 41). New York: Wiley & Sons, 1981.]

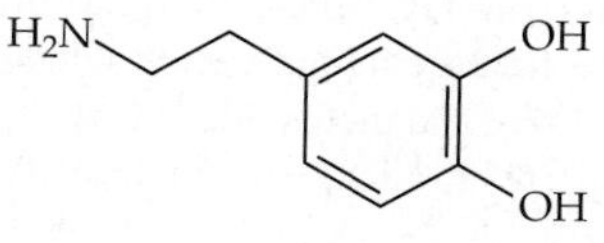

dopamine
11.b

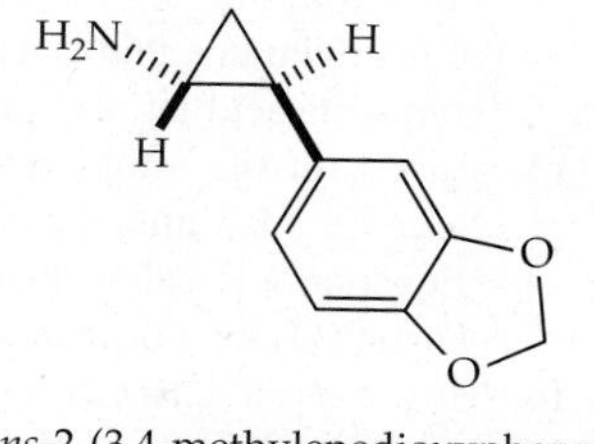

trans-2-(3,4-methylenedioxyphenyl)-
cyclopropylamine
11.c

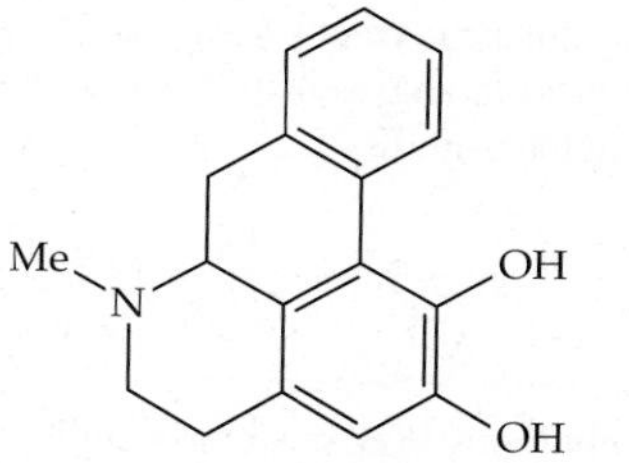

11.d

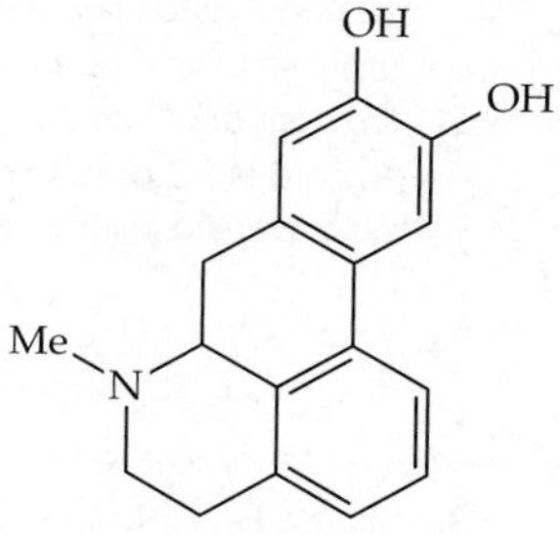

11.e

3. Hundreds of analogues of chlorpromazine (**11.29**) have been prepared and tested (see Figures 9.14 and 11.5). No active analogues in Figure 9.14 are constrained in the fashion of compound **11.f**. Why is compound **11.f** likely inactive? Propose two restrained chlorpromazine analogues that may address the problem(s) you have described for **11.f**. Your two new structures should be consistent with the activity or inactivity of the compounds in Figure 9.14.

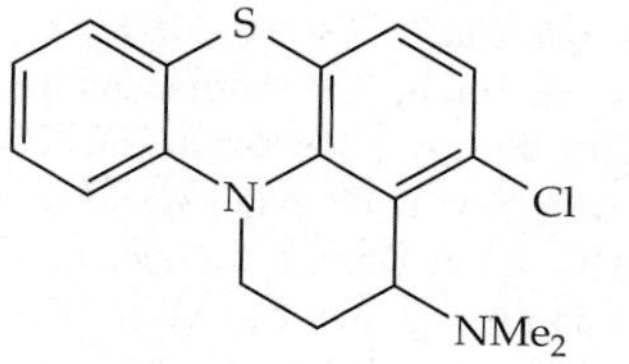

11.f

4. Based on the data in the following table, suggest three additional analogues that should be synthesized and screened.

[Silverman, R. B. *The Organic Chemistry of Drug Design and Drug Action.* San Diego: Academic Press, 1992, Chapter 5.]

Entry	Structure	Relative K_i
1	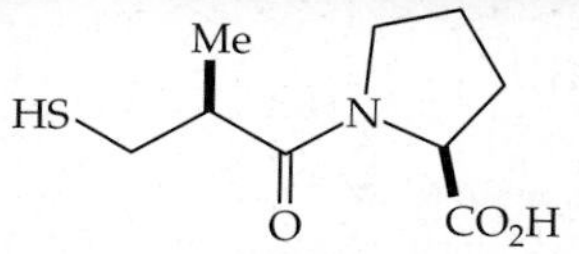 captopril	1
2	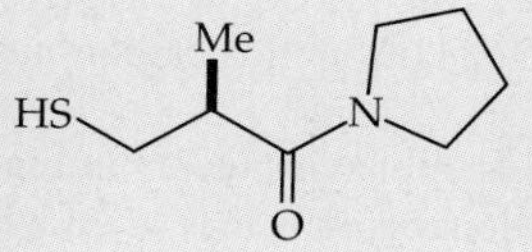	12,500
3	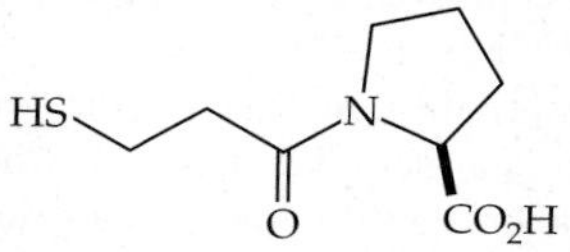	10
4	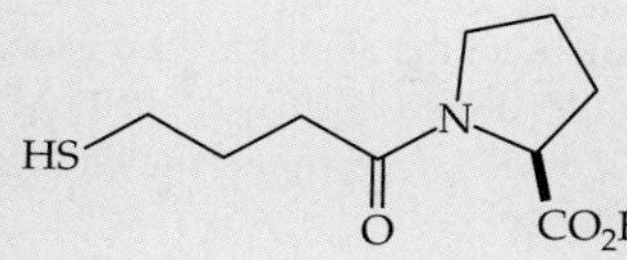	12,000
5	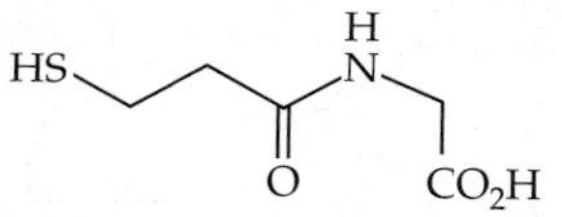	120
6	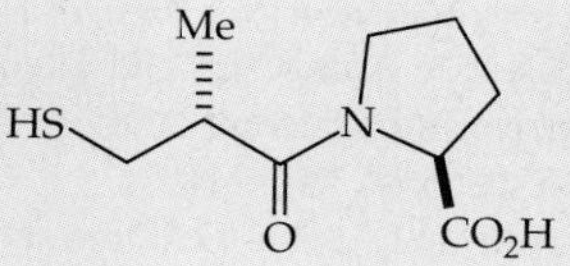	120
7	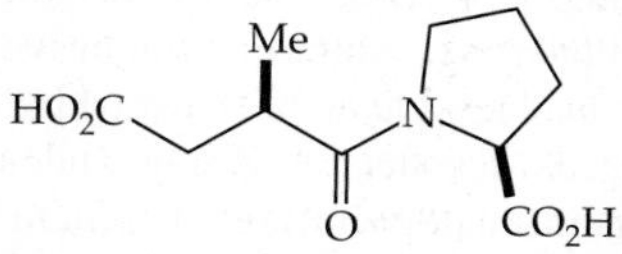	1,100

5. Using isosteres from Table 11.4, propose six analogues of morphine (**11.1**). Two isosteres should come from each of the three isostere types in the table: univalent, divalent, and ring groups.

6. A series of 2-phenylethanol (**11.g**) analogues was tested. Each analogue was tested with an R-group in the *ortho*, *meta*, or *para* position. Analogues with R-groups of CH_3, Cl, and Br afforded reasonable and consistent results. When the R-group was changed to OH or NH_2, the *para* and *meta* analogues tested as expected, but the *ortho* analogues were surprisingly inactive. Explain a possible problem

with the OH and NH_2 *ortho* analogues. Draw structures to reinforce your arguments.

11.g

7. Table 11.4 lists a small number of classical isosteres. In various instances, carbon, nitrogen, and oxygen may be substituted for one another to generate a new analogous compound. Despite being just below oxygen on the periodic table, sulfur is excluded from this carbon-nitrogen-oxygen pattern. Why?

8. Take the structure of ritonavir (**11.75**) in Figure 11.19 and show how its functional groups may align with the binding sites (H-bond acceptors and P1, P2, P1′, and P2′ pockets) of HIV-1 protease in Figure 11.18.

9. An idea in lead optimization circles is the "magic methyl." A magic methyl is a methyl group that, when added to a lead, increases potency by an order of magnitude (e.g., IC_{50} drops from 100 nM to 10 nM). Using Equation 10.1 from Chapter 10, calculate the standard free energy of binding ($\Delta G°_{bind}$) for a lead with an IC_{50} of 100 nM and a 10 nM analogue. How much energy must a magic methyl contribute to binding?

10. In the development of cimetidine (**11.35**), an extra methyl group was added to favor a desired tautomeric equilibrium as well as influence the conformation of the drug (Scheme 11.4). How could the same conformational effect be accomplished through addition of a ring? Show a possible structure. What kind of complications might addition of a ring cause?

11. Question 11 in Chapter 10 introduced the idea of ligand efficiency, which is the standard free energy of binding contributed to a drug for each nonhydrogen atom. A drug with 30 atoms and a K_D of 10 nM has a ligand efficiency of approximately 0.35 kcal/mol/nonhydrogen atom. Is the effect of a magic methyl in Question 9 impressive?

12. The conjugate acid of methylamine has a pK_a of approximately 10.6. Based on the Henderson–Hasselbalch equation (Equation 9.1 in Chapter 9), calculate the percentage of methylamine that is *not* protonated (not in the form of its conjugate acid) at pH 7.4, the same as blood.

13. Based on the acid–base data and text that accompany Scheme 11.3 and the Henderson–Hasselbalch equation (Equation 9.1), calculate the pK_a for the conjugate acids of histamine (**11.36**), burimamide (**11.40**), and metiamide (**11.41**).

14. How can the $CH(CN)_2$ group (Table 11.5) act as a bioisostere for OH?

15. Angiotensin II, a peptide hormone that elevates blood pressure, was discussed in one of this chapter's Case Studies. Based on the abbreviated peptide structure in Scheme 11.6, draw the line angle structure of angiotensin II. Consult Figure 4.3 to make sure that the *C*- and *N*-termini are correctly assigned. Draw the peptoid isostere (all residues) of angiotensin II. Draw the full retroinverso isostere of angiotensin II.

References

1. Beckett, A., & Casey, A. Analgesics and Their Antagonists: Biochemical Aspects and Structure-Activity Relationships. *Prog. Med. Chem.* 1965, *4*, 171–218.

2. Gaul, C., Njardarson, J. T., Shan, D., Dorn, D. C., Wu, K.-D., Tong, W. P., et al. The Migrastatin Family: Discovery of Potent Cell Migration Inhibitors by Chemical Synthesis. *J. Am. Chem. Soc.* 2004, *126*, 11326–11337.

3. Chen, L., Yang, S., Jakoncic, J., Zhang, J. J., & Huang, X.-Y. Migrastatin Analogues Target Fascin to Block Tumor Metastasis. *Nature*. 2010, *464*, 1062–1068.

4. Konkel, M. J., Lagu, B., Boteju, L. W., Jimenez, H., Noble, S., Walker, M. W., et al. 3-Arylimino-2-Indolones Are Potent and Selective Galanin GAL_3 Receptor Antagonists. *J. Med. Chem.* 2006, *43*, 3757–3758.

5. Clader, J. From Serendipity to Design: The Discovery of Ezetimibe and Other Novel Cholesterol Absorption Inhibitors. Presented at Residential School on Medicinal Chemistry, Drew University, Madison, NJ, June 2002, Case History III.

6. Dohme, A. R. L., Cox, E. H., & Miller, E. The Preparation of the Acyl and Alkyl Derivatives of Resorcinol. *J. Am. Chem. Soc.* 1926, *48*, 1688–1693.

7. Sastry, B. V. R. Anticholinergics: Antispasmodic and Antiulcer Drugs. In M. E. Wolff (Ed.), *Burger's Medicinal Chemistry* (4th ed., Chapter 44). New York: Wiley & Sons, 1981.

8. Kim, C. U., Lew, W., Williams, M. A., Wu, H., Zhang, L., Chen, X., et al. Structure-Activity Relationship Studies of Novel Carbocyclic Influenza Neuraminidase Inhibitors. *J. Med. Chem.* 1998, *41*, 2451–2460.

9. Cannon, J. G. Analog Design. In M. E. Wolff (Ed.), *Burger's Medicinal Chemistry* (5th ed., Chapter 19). New York: Wiley & Sons, 1995.

10. Thornber, C. W. Isosterism and Molecular Modification in Drug Design. *Chem. Soc. Rev.* 1979, *8*, 563–580.

11. Kaiser, C., & Setler, P. E. Antipsychotic Agents. In M. E. Wolff (Ed.), *Burger's Medicinal Chemistry* (4th ed., Chapter 56). New York: Wiley & Sons, 1981.

12. Ganellin, G. R., & Durant, G. J. Histamine H_2-Receptor Agonists and Antagonists. In M. E. Wolff (Ed.), *Burger's Medicinal Chemistry* (4th ed., Chapter 48). New York: Wiley & Sons, 1981.

13. Skuballa, W., Schillinger, E., Stürzebecher, C.-St., & Vorbrüggen, H. Synthesis of a New Chemically and

Metabolically Stable Prostacyclin Analogue with High and Long-Lasting Oral Activity. *J. Med. Chem.* 1986, *29*, 315–317.

14. Auspex Pharmaceuticals. Clinical Results SD–254. http://www.auspexpharma.com/pipeline/index.php (accessed October 2013).

15. Smith, R. A., Barbosa, J., Blum, C. L., Bobko, M. A., Caringal, Y. V., Dally, R., et al. Discovery of Heterocyclic Ureas as a New Class of Raf Kinase Inhibitors: Identification of a Second Generation Lead by a Combinatorial Chemistry Approach. *Bioorg. Med. Chem. Lett.* 2001, *11*, 2775–2778.

16. Lowinger, T. B., Riedl, B., Dumas, J., & Smith, R. A. Design and Discovery of Small Molecules Targeting Raf-1 Kinase. *Curr. Pharm. Des.* 2002, *8*, 2269–2278.

17. Vagner, J., Qu, H., & Hruby, V. J. Peptidomimetics: A Synthetic Tool of Drug Discovery. *Curr. Opin. Chem. Biol.* 2008, *12*, 292–296.

18. Fletcher, M. D., & Campbell, M. M. Partially Modified Retro-Inverso Peptides: Development, Synthesis, and Conformational Behavior. *Chem. Rev.* 1998, *98*, 763–796.

19. Saunders, J. *Top Drugs: Top Synthetic Routes.* Oxford, UK: Oxford University Press, 2000, Chapter 1.

20. Ondetti, M. A., Williams, N. J., Sabo, E., Pluscec, J., Weaver, E. R., & Kocy, O. Angiotensin-Converting Enzyme Inhibitors from the Venom of *Bothrops Jararaca:* Isolation, Elucidation of Structure, and Synthesis. *Biochemistry.* 1971, *10*, 4033–4039.

21. Patlak, M. From Viper's Venom to Drug Design: Treating Hypertension. *FASEB J.* 2004, *18*, 1–13.

22. Cushman, D. W., Cheung, H. S., Sabo, E. F., & Ondetti, M. A. Design of Potent Competitive Inhibitors of Angiotensin-Converting Enzyme: Carboxyalkanoyl and Mercaptoalkanoyl Amino Acids. *Biochemistry.* 1977, *16*, 5484–5491.

23. Thorsett, E. D., Harris, E. E., Peterson, E. R., Greenlee, W. J., Patchett, A. A., Ulm, E. H., & Vassil, T. C. Phosphorous-Containing Inhibitors of Angiotensin-Converting Enzyme. *Proc. Natl. Acad. Sci. U.S.A.* 1982, *79*, 2176–2180.

24. Lam, P. Y. S., Jadhav, P. K., Eyermann, C. J., Hodge, C. H., Ru, Y., Bacheler, L. T., et al. Rational Design of Potent, Bioavailable, Nonpeptide Cyclic Ureas as HIV Protease Inhibitors. *Science.* 1994, *263*, 380–384.

25. Greer, J. Molecular Modeling. Presented at Residential School on Medicinal Chemistry, Drew University, Madison, NJ, June 2002, Seminar I.

Lead Optimization: Hansch Analysis

Chapter Outline

Learning Goals for Chapter 12

- Appreciate the value of QSAR for lead optimization
- Understand the requirements for a good QSAR property parameter
- Know how to construct a Hansch equation from activity and property data
- Be able to critically evaluate the validity of a Hansch equation
- Appreciate the advantages of more modern QSAR methods

Chapter 11 emphasized the traditional methods of lead optimization that primarily consist of performing SAR on a lead compound. This chapter focuses on *quantitative* SAR. QSAR attempts to determine a mathematical relationship that relates how structural changes impact biological activity. In theory, a QSAR can be used to predict the biological activity of a molecule and potentially steer the lead optimization team to the most potent lead analogues. A simple and instructional, albeit dated, QSAR technique is Hansch analysis. In this chapter are described the basic aspects of Hansch analysis: molecular properties that are easily modeled, how Hansch equations are generated, and approaches for judging the quality of QSAR equations. The simplicity of Hansch analysis has given way to more computationally intensive methods such as comparative molecular field analysis.

12.1 Background

Traditional SAR techniques are strictly qualitative. Molecular changes give analogues that are "better" or "worse" than the original lead. The modifications that give a better lead are pursued with even more changes to give another, even better analogue. Hopefully, after many iterations of this cycle, an optimized lead suitable for clinical trials is the outcome. This approach has been the basis for drug discovery for over 100 years, but it can be slow and sometimes does not generate a satisfactory clinical candidate.

In contrast to the qualitative better and worse approach of traditional SAR, many researchers have sought to develop a method that is able to quantitatively link molecular structure changes to biological activity, a *quantitative structure-activity relationship* (QSAR). As a method, QSAR strives to develop a mathematical formula to relate biological activity as a function of molecular or substituent properties.

The benefit of QSAR is a more efficient lead optimization process. If a good QSAR formula can be derived, the activity of leads can be approximated by calculation without the time-consuming need to make and test each molecule. In theory, with a QSAR formula in hand, a medicinal chemist could know precisely what properties are most important for the design of a highly potent advanced lead. The medicinal chemist would then prepare only analogues with the most promising structural elements and presumably with high activity. Inactive or poorly active compounds could be avoided altogether, and the output of the optimization team would be maximized. Therefore, QSAR is in theory a very attractive tool for drug discovery.

Many implementations of QSAR have been described in the literature, and the idea of QSAR can be traced back into the 1800s. The first successful, simple, and general technique was not developed until the early 1960s by Corwin Hansch of Pomona College in California.[1] Hansch's QSAR approach, called *Hansch analysis*, is built on the concept of a *linear free-energy relationship* (LFER). In a LFER, the standard free energy change of one process ($\Delta G°_a$) is directly proportional (linearly related) to the standard free energy change in another process ($\Delta G°_b$) (Equation 12.1).[2]

$$\Delta G°_a \propto \Delta G°_b \tag{12.1}$$

In an assay or screen, biological activity is often reported as ED_{50}, IC_{50}, or a similar term. From the discussion of receptor theory in Chapter 4, ED_{50} is equivalent to K_D, the equilibrium constant for dissociation of a drug-receptor complex (Equation 12.2).

$$ED_{50} = K_D \tag{12.2}$$

Furthermore, the logarithm of equilibrium constants is directly proportional to the process' associated standard free energy changes (Equations 12.3 and 12.4).

$$\Delta G° = -2.3RT \log K \tag{12.3}$$

$$\Delta G° \propto -\log K \tag{12.4}$$

Therefore, by substitution of Equation 12.2 into Equation 12.4, changes in the standard free energy of binding ($\Delta G°$) are directly proportional to the logarithm of an activity measure (ED_{50}) (Equation 12.5).

$$\Delta G°_{bind} \propto -\log ED_{50} \tag{12.5}$$

In QSAR studies, activity is normally listed in the form of $-\log ED_{50}$, or several other equivalent forms (Equation 12.6), to establish the relationship of activity and energy.

$$-\log ED_{50} = \log \frac{1}{ED_{50}} = pED_{50} \tag{12.6}$$

Establishing the relationship of binding and energy is fairly straightforward. The bigger challenge is relating the properties of a molecule in terms of a standard free energy change. Only once this task is complete can a LFER such as Equation 12.7 be constructed.

$$\Delta G°_{bind} \propto \Delta G°_{properties} \tag{12.7}$$

12.2 Parameters

The Hansch approach to QSAR requires a means of expressing molecular properties in terms of standard free energy changes. Binding is largely a function of intermolecular forces, including dipole and contact forces. Key properties that affect dipoles and available contact interactions are electronic effects and lipophilicity, both of which have been successfully correlated to standard free energy changes. Several other properties have been investigated with less success. The following subsections discuss parameters that are related to standard free energy changes and are therefore of potential value to Hansch analysis.

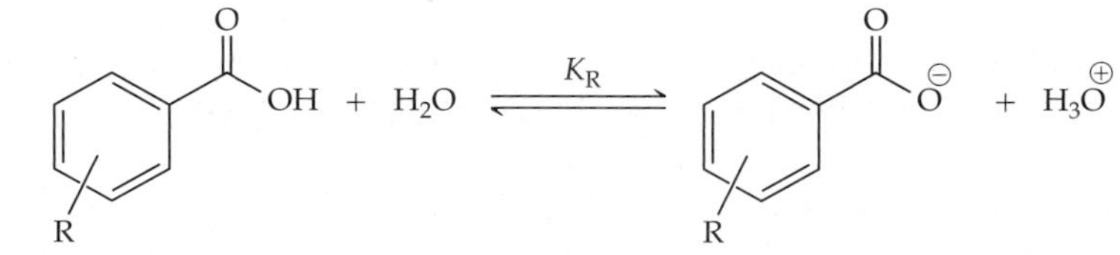

substituted benzoic acid
12.1

conjugate base
12.2

SCHEME 12.1 Dissociation of benzoic acid in water

Hammett Constants: An Electronic Parameter

In the 1920s and 1930s, Louis Hammett of Columbia University explored the dissociation of substituted benzoic acids (**12.1**) in water (**Scheme 12.1**).[2] Changing the R-group on the ring shifts the position of the acid–base equilibrium (Equation 12.8).

$$K_R = \frac{[\text{conj. base}][H_3O^+]}{[\text{acid}][H_2O]} \tag{12.8}$$

Hammett defined a parameter, σ_R, as the logarithm of the ratio of K_R to K_H, the equilibrium of unsubstituted benzoic acid (Equation 12.9).

$$\sigma_R = \log \frac{K_R}{K_H} \tag{12.9}$$

All equilibrium constants are defined relative to benzoic acid substituted with the simplest R-group, a hydrogen ($R = H$). Equation 12.9 can be rewritten as a difference in $\log K$ values. Substitution of Equation 12.3 shows that σ_R is a measure of standard free energy change caused by the R-group (Equations 12.10 and 12.11). Because σ_R is a logarithm (Equation 12.9), σ_R is unitless.

$$\sigma_R = \log K_R - \log K_H = -\frac{\Delta G^\circ_R}{2.3RT} + \frac{\Delta G^\circ_H}{2.3RT} = \frac{1}{2.3RT}\Delta\Delta G_R \tag{12.10}$$

$$\sigma_R \propto \Delta\Delta G_R \tag{12.11}$$

By definition, σ_H has a value of 0. Hammett found that electron-donating groups ($R = EDG$) destabilize the conjugate base and push the equilibrium to the left. K_{EDG} is therefore smaller than K_H, and σ_{EDG} is negative. Through similar logic, electron-withdrawing groups (EWG) have σ-values that are positive. The stronger an EDG or EWG is, the more its σ-value deviates from 0. Hammett also found σ-values to be position dependent. The σ-value was different if the R-group was in the *para* or the *meta* position, and separate σ_p and σ_m parameters had to be defined for each R-group. σ-Values for selected R-groups in the *para* position and *meta* position are listed in **Tables 12.1** and **12.2**.

σ-Values are predominantly determined through a combination of two effects: resonance and induction. Resonance is the predominant factor for groups in the *para* position, and induction is more important for *meta* substituents. The methoxy group ($R = OMe$) demonstrates these ideas (**Scheme 12.2**). In the *para* position, the σ-value for methoxy is -0.268, indicating it is an EDG. Electron donation by resonance strongly *destabilizes* the conjugate base (**12.3**). In contrast, the methoxy group only weakly *stabilizes* the conjugate base anion through induction because the methoxy group is so far removed from the carboxylate (**12.4**, $\delta\delta\delta\delta^+$ next to carbonyl). Since resonance tends to have a stronger influence than induction from the *para* position, the *net* effect of a methoxy group at the

TABLE 12.1 Selected σ_p-values[2]

Entry	R-Group	σ_p	
1	NH_2	−0.660	EDG
2	MeO	−0.268	
3	Me	−0.172	
4	H	±0.000	
5	F	+0.062	
6	Cl	+0.227	
7	Br	+0.232	
8	I	+0.276	
9	NO_2	+0.780	EWG

TABLE 12.2 Selected σ_m-values[2]

Entry	R-Group	σ_m	
1	NH_2	−0.161	EDG
2	Me	−0.069	
3	H	±0.000	
4	MeO	+0.115	
5	F	+0.337	
6	I	+0.352	
7	Cl	+0.373	
8	Br	+0.391	
9	NO_2	+0.710	EWG

para position is as an EDG (negative σ-value). For the *meta* position, the σ-value reverses its sign to +0.115. Electron donation by resonance is ineffective because electron density cannot be placed directly next to the carboxylate from the *meta* position (**12.5** and **12.6**). Electron withdrawing by induction is stronger in the *meta* isomer (**12.7**, $\delta\delta\delta^+$ next to carbonyl) because the methoxy is closer to the carboxylate than in the *para* position. Induction prevails, and the methoxy group is a *net* EWG in the *meta* position.

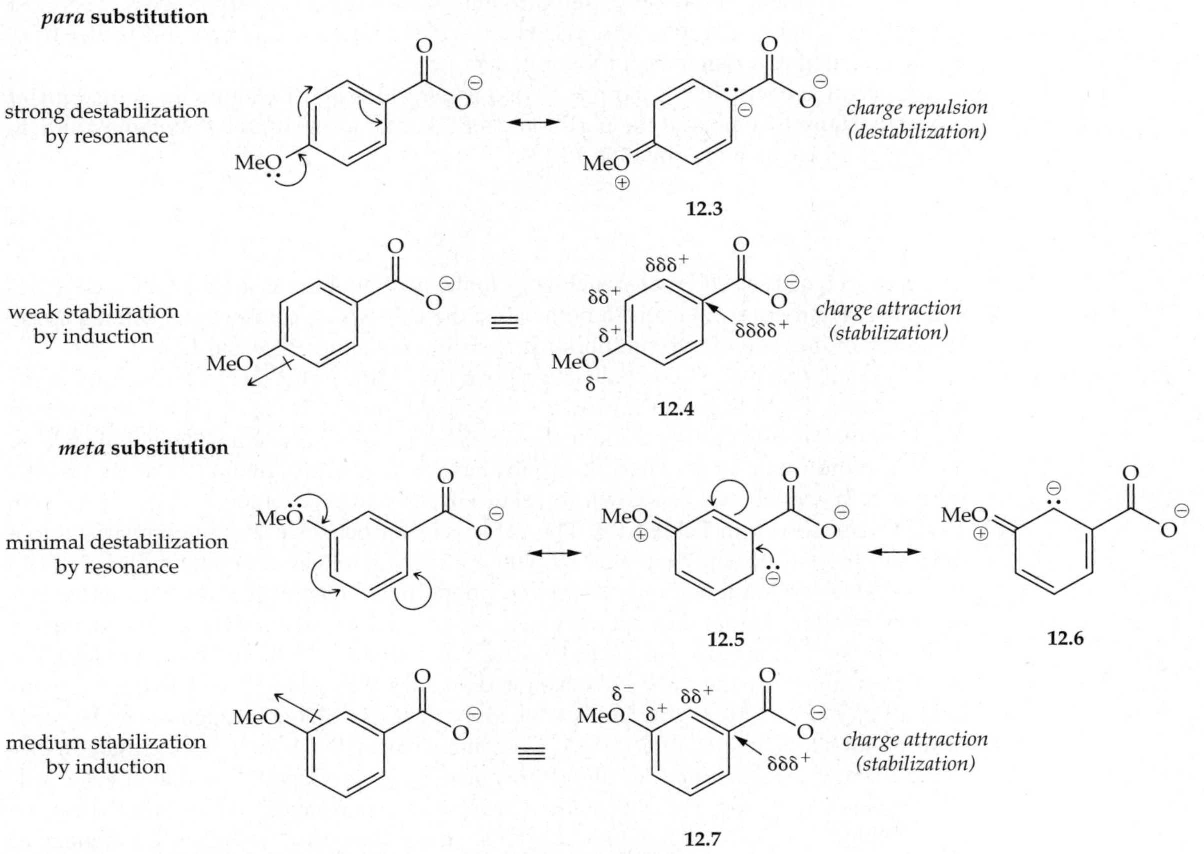

para **substitution**

strong destabilization by resonance

charge repulsion (destabilization)

12.3

weak stabilization by induction

charge attraction (stabilization)

12.4

meta **substitution**

minimal destabilization by resonance

12.5   **12.6**

medium stabilization by induction

charge attraction (stabilization)

12.7

SCHEME 12.2 Resonance and inductive effects in substituted benzoic acids

An effective QSAR parameter should correlate only to its specific property, be it electronics, lipophilicity, sterics, or something else. σ-Values for R-groups in the *meta* and *para* positions have been found to account for only electronic effects. Substitution at the *ortho* position, however, is another story. Because *ortho* substituents are physically so close to the carboxylic acid, *ortho* R-groups affect the acid–base equilibrium through both electronic and steric effects. Therefore, the *ortho* σ-value (σ_o) is not useful as an electronic parameter for QSAR studies. Despite their limitations, Hammett constants are by far the most commonly encountered electronic parameter in Hansch analysis.

Hansch Constants: A Lipophilicity Parameter

Hansch and Toshio Fujita, a postdoctoral researcher in Hansch's group, designed a parameter, π_R, to estimate the lipophilicity of an R-group.[3] Hansch's parameter relies on *partition coefficients* to measure lipophilicity. Partition coefficients, P, are equilibrium constants describing the degree to which a molecule distributes into a biphasic mixture of two immiscible solvents. Hansch used 1-octanol and water as the model solvents because these were known to simulate the lipid membrane–cytosol interface. The partition coefficient of a molecule is defined as the ratio of a molecule's concentration in an octanol layer to its concentration in an aqueous layer (Equation 12.12).

$$P = \frac{[\text{drug}]_{\text{oct}}}{[\text{drug}]_{\text{water}}} \qquad (12.12)$$

Partition coefficients are widely used throughout the drug discovery process. The most useful form is log P. The log P discussed here is the same as was covered in Chapter 10 as a drug selection criterion in Lipinski's rules.

A partition coefficient is a property of a *whole molecule*. To define π_R, a *substituent* property, Hansch compared the partition coefficient of a substituted compound with an unsubstituted compound (Equation 12.13).

$$\pi_R = \log \frac{P_R}{P_H} \qquad (12.13)$$

The approach of Hansch is fully analogous to the methodology that Hammett used to develop his electronic parameter. In both cases, the substituent parameter is defined as the logarithm of the ratio of two equilibrium constants (Equations 12.9 and 12.13).

R-Groups that make a compound less polar (more lipophilic) increase solubility in the octanol layer. These R-groups have a π-value greater than 0. R-Groups that have a strong dipole or can readily participate in hydrogen bonding will cause a molecule to distribute more into the water layer. These R-groups have π-values less than 0. Like Hammett's σ-values, π-values can be shown to relate directly to ΔG^o changes. Representative π-values are shown in **Table 12.3**. The values are for benzene (P_H) in comparison to a monosubstituted benzene ring (P_R). As with σ-values, π-values are unitless.

π-Values are designed only to predict lipophilicity. One of the most attractive features of π-values is that they are very general and can be used nearly independently of their context. The π-values in Table 12.3 are explicitly for substituted benzene rings, but the values are nearly the same for substituted alkanes. The generality of π-values means lipophilicity effects are highly predictable and can be calculated through computer algorithms with a high degree of accuracy. Computers calculate whole molecule lipophilicity in the form of log P. Numerous algorithms have been developed. Calculated log P values are denoted as clog P to differentiate them from experimental values, log P. Because clog P values are so convenient and fairly accurate, they are regularly used in place of experimental values.

TABLE 12.3   Selected Hansch lipophilicity constants[3]

Entry	R-Group	π_R	
1	Cl	+0.71	more lipophilic
2	Me	+0.56	
3	F	+0.14	
4	H	±0.00	
5	OMe	−0.02	
6	NO$_2$	−0.28	
7	OH	−0.67	
8	NH$_2$	−1.23	less lipophilic

para-xylene
12.8

ethylbenzene
12.9

FIGURE 12.1

The generality of π-values can be seen in *para*-xylene (**12.8**) and ethylbenzene (**12.9**) (**Figure 12.1**). According to Table 12.3, replacing a hydrogen atom with a methyl group changes the log P value of a molecule by +0.56. If π-values are truly general, then the effect of replacing *two* hydrogen atoms with methyl groups should change log P by +1.12 (+0.56 × 2). For both *para*-xylene and ethylbenzene, the experimental change in log P from benzene is +0.98, fairly close to the value predicted from a standard methyl value of +0.56.

The lipophilicity of simple alkylated molecules like xylene is easy to predict accurately with standard π-values because no electronic effects are impacted. This is not always true, and electronic effects can play a prominent role in a molecule's lipophilicity. 4-Nitrophenol (**12.10**) provides an example (**Scheme 12.3**). An NO$_2$ group has a π-value of −0.28, indicating that an NO$_2$ group decreases lipophilicity. An OH group has a π-value of −0.67, also indicating a decrease in lipophilicity. Based on individual substituent parameters, the predicted log P of 4-nitrophenol would be −0.95 (−0.28 + −0.67) lower than benzene. Experimentally, log P of 4-nitrophenol differs by only −0.17, less than the effect of either NO$_2$ or OH alone. 4-Nitrophenol is indeed more polar than benzene but not nearly as polar as one would expect based on π-values alone. In this case, π-values are unable to account for the electronic interactions of the OH and NO$_2$ groups (**12.11**). π-Values are informative and general, but they cannot be used recklessly without considering potential conflicting interactions.

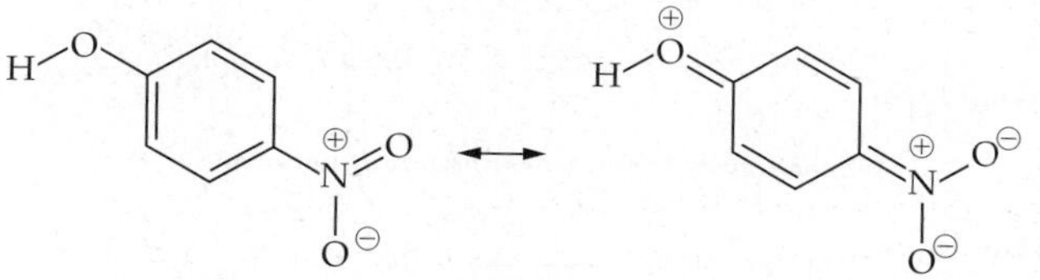

4-nitrophenol
12.10

12.11

SCHEME 12.3

Although not perfect, Hansch's lipophilicity parameter and log P values are the most widely used parameters in QSAR studies. In addition to their effectiveness in predicting biological activity through target binding (pharmacodynamics), both parameters also affect pharmacokinetics. The pharmacokinetic applications of log P and π-values can be seen in Lipinski's rules and a Case Study (Carboxylate Antifungals) later in this chapter.

Sample Calculation	**Determining the Effect of a Substituent on a Molecule's Partition Coefficient**

PROBLEM   If the log P of benzene is +2.13 and the π-value of a methyl is +0.56, predict the partition coefficient (P) of toluene in an octanol:water mixture.

SOLUTION   We know π of the R-group (methyl), and we know log P of the unsubstituted compound (benzene) is 2.13. This may not look exactly like the makings of Equation 12.13, but if rearranged, Equation 12.13 looks much more useful. Substituting values gets down to log P = +2.69. Log P of the substituted compound, toluene, can be determined to be 490. This means that the concentration of toluene in the octanol layer would be 490 times higher than in the water layer. This concentration difference is reasonable given the completely nonpolar structure of toluene.

$$\pi_{Me} = \log \frac{P_{PhMe}}{P_{PhH}} = \log P_{PhMe} - \log P_{PhH} \tag{12.13}$$

$$+0.56 = \log P_{PhMe} - 2.13$$

$$+2.69 = \log P_{PhMe}$$

$$10^{2.69} = 490 = P_{PhMe} = \frac{[\text{toluene}]_{oct}}{[\text{toluene}]_{water}}$$

While log P values work very well and predictably for neutral compounds, the partition coefficient is less effective for charged molecules. D, the *distribution coefficient*, better accommodates molecules that may be charged, notably acids and bases.[4] D is defined as the sum of the concentrations of all species (charged and uncharged) dissolved in the octanol layer divided by the sum of all species in the water layer (Equation 12.14, **Scheme 12.4**). The presence of any charged species in octanol in Equation 12.14 and Scheme 12.4 frequently may be ignored. For a simple system in which the drug is a monoprotic acid or base, Equations 12.15 and 12.16 can approximate log D based on log P and a correction term. (In Equation 12.16, pK_a refers to the pK_a of the conjugate acid of the base.) Log D

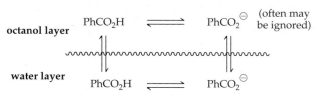

SCHEME 12.4 Partitioning of neutral and charged species between octanol and water phases

values are valid only at a specific pH. The pH of blood, 7.4, is the most commonly referenced pH for log D values.

$$D = \frac{\sum [\text{drug}]_{\text{oct}}}{\sum [\text{drug}]_{\text{water}}} \qquad (12.14)$$

$$\log D_{\text{acids}} = \log P - \log \left(1 + 10^{\text{pH} - pK_a}\right) \qquad (12.15)$$

$$\log D_{\text{bases}} = \log P - \log \left(1 + 10^{pK_a - \text{pH}}\right) \qquad (12.16)$$

Equations 12.15 and 12.16 have been tested on a series of acidic, basic, and neutral molecules. For benzoic acid ($\log P = 1.85$, $pK_a = 4.2$) in rat colon (pH $= 6.8$), the calculated log D value from Equation 12.15 is 1.28, considerably lower than the log P value. Note that the ionization of benzoic acid decreases the lipophilicity of the molecule because the charged species strongly favors the water layer (Scheme 12.4).

Most log P values are determined with 1-octanol as the nonpolar solvent. Sometimes octanol-water partition coefficients are denoted P_{ow} to make the nonpolar solvent more explicitly clear. Octanol effectively imitates many lipid membranes, especially those in the small intestine, where most drugs are absorbed. Other solvents are better suited to model other tissues in the body. Chloroform is more polar than octanol and simulates partitioning in oral tissues. Olive oil is less polar than octanol and models the blood–brain barrier.

Taft Steric Parameter

An effective drug must fill the binding site on a receptor or enzyme to maximize possible intermolecular interactions. For this reason, the physical size of drug substituents is often a determining factor in strong binding. In the 1950s, Robert Taft of Penn State University reported a new parameter, E_s, that estimates the steric size of R-groups.[5] Taft determined E_s-values by comparing rates of acid-catalyzed hydrolysis of ethyl esters (**Scheme 12.5**, Equation 12.17). The benchmark ester was ethyl acetate (**12.12**, R $=$ CH$_3$). Larger substituents hinder hydrolysis at the ester carbonyl, the rate is slower, and the E_s-value is negative. R-groups smaller than methyl have positive E_s-values. Hydrogen is the smallest substituent (R $=$ H, ethyl formate). Representative E_s-values are listed in **Table 12.4** and are unitless.

$$E_{s(R)} = \log k_R - \log k_{CH_3} = \log \frac{k_R}{k_{CH_3}} \qquad (12.17)$$

While the development of the Taft parameter is similar to that of Hammett and Hansch, E_s-values are based on *rate* constants instead of *equilibrium* constants. The Taft parameter is a measure of changes in activation energy, not standard free energy. Of the Hammett, Hansch, and Taft parameters, the Taft parameter is utilized the least in QSAR studies. Other steric parameters have been developed over time, and like the Taft parameter, all have shortcomings. One alternative steric parameter was developed by Marvin Charton of Pratt Institute in New York. Charton's parameter is based on the van der Waal radius of a substituent.[6] Another alternative steric model is the STERIMOL parameter set developed by Arië Verloop of Philips-Duphar in Holland.[7] Unlike Taft and Charton, Verloop

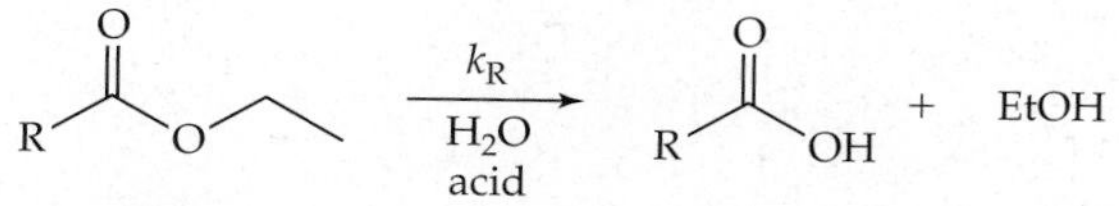

12.12 **12.13**

SCHEME 12.5 Acid-catalyzed hydrolysis of an ester

TABLE 12.4 Selected Taft steric parameter values[5]

Entry	R-Group	E_s-Value	
1	H	+1.24	increasing sterics
2	F	+0.78	
3	OMe	+0.69	
4	Cl	+0.27	
5	Br	+0.08	
6	Me	±0.00	
7	Et	−0.07	
8	I	−0.16	
9	n-Pr	−0.36	
10	n-Bu	−0.39	
11	i-Pr	−0.47	
12	NO_2	−1.28	
13	t-Bu	−1.54	
14	Ph	−1.74	

developed a system in which each R-group has several different parameters: length and multiple different radius measurements. With multiple measurements of each R-group, Verloop's system almost certainly gives a more accurate picture of steric bulk. Multiple parameters for each substituent also makes Verloop's method more complicated and less simple to use in regression analysis.

Sample Calculation	**Predicting Hydrolysis Rate Difference Based on Substituents' Sizes**

PROBLEM Based on Taft steric parameters, how much faster than ethyl acetate would you expect ethyl formate to hydrolyze?

SOLUTION Ethyl formate (**12.12**, R = H) should hydrolyze faster than ethyl acetate (**12.12**, R = CH_3) because steric hindrance about the carbonyl is lower in ethyl formate. The question is: How much faster? The only equation provided in this section on Taft parameters is Equation 12.17, and that is where we will begin. In this case, we are solving for the ratio of the hydrolysis rate constants, k_R/k_{CH_3}. $E_{s(H)}$ is known and equals +1.24 according to Table 12.4. Ethyl formate hydrolyzes 17.4 times faster than ethyl acetate.

$$E_{s(R)} = \log k_R - \log k_{CH_3} = \log \frac{k_R}{k_{CH_3}} \tag{12.17}$$

$$+1.24 = \log \frac{k_H}{k_{CH_3}}$$

$$10^{1.24} = 17.4 = \frac{k_H}{k_{CH_3}}$$

Other Parameters

Almost any molecular property, whether or not it is related to standard free energy changes, has been used as a parameter in QSAR studies. Whole molecule parameters such as molecular weight (MW), molar refractivity (MR), acidity (pK_a), and log P are fairly common. Also common are parameters measuring interatomic distances and even orbital energies. One final parameter type deserves specific mention: the indicator variable (I). Indicator variables are assigned a value of either 0 or 1. Indicator variables are useful for joining parallel sets of data that differ by a constant amount. An example of an indicator variable in action is shown in a Case Study (Tumor Cell Resistance Modulators) in the following section.

12.3 Hansch Equations

Once parameters modeling various properties have been defined, equations predicting activity may be constructed. Classically, Hansch equations take the form of Equation 12.18,[1]

$$\log \frac{1}{C} = -k\pi + k'\pi^2 + \rho\sigma + k''$$
(12.18)

but variations with different molecular descriptors are also considered to be Hansch equations. Log $1/C$ is the calculated activity of a compound. Remember that log $1/C$ is mathematically equivalent to $-\log C$ and pC. The terms π, π^2, and σ are substituent parameters, and coefficients k, k', ρ, and k'' are determined through a *multiple linear regression analysis*. Both sides of the equation are unitless. A more purely mathematical representation is Equation 12.19 with a dependent variable (y), independent variables (x_i), and regression coefficients (c_i). The calculated activity is the dependent variable, and the parameters are the independent variables.

$$y = c_0 + c_1 x_1 + c_2 x_2 + \cdots + c_n x_n$$
(12.19)

Performing a regression analysis requires knowledge of the experimental biological activity of each compound as well as the parameter values for each substituent (whole molecule) involved in the study. The Hansch equation in Equation 12.18 requires the determination of four regression coefficients. In theory, a minimum three compounds ($n = 3$) with their experimental activities and various substituent parameters would be needed to perform the regression. In practice, a regression should include at least five compounds per descriptor. Therefore, a Hansch analysis such as Equation 12.18 should include at least 15 compounds. The compounds used in the regression to form the Hansch equation are called the *training set*. The output of the regression includes the optimal regression coefficients for predicting activity values. Plotting the calculated log $1/C$ values against the experimental log $1/C$ values will give a graph with a certain amount of scatter. The *correlation coefficient* (r) for the data has a value from -1 to $+1$ and roughly indicates the quality of the Hansch equation for predicting activity. Values close to 0 are a poorer fit, while values near -1 or $+1$ are a better fit. r^2 roughly indicates the fraction of scatter, or *variance*, in the data that is explained by the independent variables (parameters) in the Hansch equation. A low r^2-value implies that parameters for additional properties should be included in the analysis.

At this point, a considerable amount of theory on Hansch analysis has been presented with almost no examples of practice. The next three Case Studies will hopefully solidify ideas on Hansch analysis that have already been discussed. Each Case Study introduces a different idea. The first is an example of a very simple Hansch equation with a small data set. The second demonstrates the use of squared parameters in Hansch equations. The third and final Case Study shows how indicator variables are used in QSAR studies. If you are unfamiliar with performing linear regressions, be sure to read Appendix B on performing a regression analysis with the LINEST function in almost any common spreadsheet software. A section in the appendix describes in great detail how to derive Equations 12.20 through 12.22 in the first Case Study.

Phosphonate Ester Cholinesterase Inhibitors[8]

In a study by Hansch, five *meta*-substituted diethyl phosphonate esters (**12.14**) showed cholinesterase inhibition in fruit flies. Correlations between σ-values, E_s-values, and both parameters together were explored. The relevant data are included in **Table 12.5**.

The first Hansch equation (Equation 12.20) relates biological activity (log $1/I_{50}$) to the electronic effect of the R-group (σ_m).

$$\log \frac{1}{I_{50}} = +2.719\sigma_m + 5.218$$
$$n = 5, r = 0.733 \qquad (12.20)$$

The positive coefficient for σ_m indicates that positive values of σ_m contribute to increasing cholinesterase activity. Electronic effects account for only about 50% ($r^2 = 0.733 \times 0.733$) of the variance between the experimental activity and activity calculated by Equation 12.20. Low correlations imply that additional or different parameters should be included in the regression.

If electronics effects do not fully predict the activity of the compounds in Table 12.5, then perhaps sterics will. Performing a regression on E_s-values affords Equation 12.21.

$$\log \frac{1}{I_{50}} = -1.366E_s + 4.900$$
$$n = 5, r = 0.911 \qquad (12.21)$$

This regression states that steric effects account for over 80% (0.911×0.911) of the data variance, which is a great improvement over the electronics-only regression in Equation 12.20. The negative sign on the regression coefficient (-1.366) specifies that groups with a negative E_s-value will increase activity. Negative E_s-values have

more steric bulk than a CH_3 group, so larger substituents would be predicted to lead to more active compounds.

Once both steric and electronic parameters are included in the regression, Equation 12.22 accommodates nearly 99% (0.994×0.994) of the variance in the data.

$$\log \frac{1}{I_{50}} = -1.102E_s + 1.614\sigma_m + 4.498$$
$$n = 5, r = 0.994 \qquad (12.22)$$

Consistent with Equations 12.20 and 12.21, Equation 12.22 states that increasing both steric bulk and electron withdrawing of the *meta* substituent increase the biological activity of the tested molecule. With this information in hand, a researcher seeking more potent analogues would know what functional groups to test and which ones to ignore in preparing additional derivatives. For example, making an analogue of **12.14** with a small, electron-donating R-group such as CH_3 ($\sigma_m = -0.07$, $E_s = \pm 0.00$) would give a compound with a *predicted* activity (log $1/I_{50}$) of just 4.39. Using CF_3 ($\sigma_m = +0.43$, $E_s = -1.16$) as an R-group would afford a compound with a predicted activity of 6.47. The power of a Hansch equation is being able to estimate biological activity without requiring the synthesis of a new molecule. Of course, the Hansch equation is only a prediction. Ultimately, any compounds of interest must be prepared and tested.

The Hansch equations that are available from the data in Table 12.5 have been covered, but there are still a few topics that deserve specific mention. All the topics revolve around the correlation coefficient—its generation, significance, and reliability.

With Equation 12.20 along with the Hammett constants and Taft parameters in Table 12.5, one can predict

TABLE 12.5 Cholinesterase activity of *meta*-substituted phosphonate esters

Structure	Entry	R-Group	σ_m	E_s	Exptl. Log $1/I_{50}$	Calc. Log $1/I_{50}$ (eq. 12.20)	Calc. Log $1/I_{50}$ (eq. 12.22)
	1	SF_5	+0.61	−1.67	7.12	6.88	7.32
	2	OMe	+0.12	+0.69	3.89	5.54	3.93
	3	*t*-Bu	−0.12	−1.54	6.05	4.89	6.00
	4	NO_2	+0.71	−1.28	7.30	7.15	7.05
12.14	5	$^+NMe_3$	+0.81	−1.60	7.52	7.42	7.57

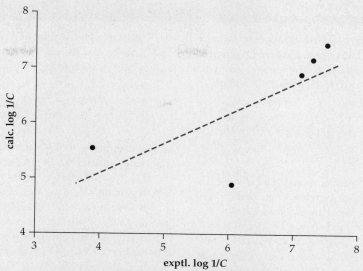

FIGURE 12.2 Calculated versus experimental log 1/C for Equation 12.20

the biological activity for each of the five compounds in the study. These five predicted activities are in the second-to-last column of Table 12.5. If the calculated activities are graphed against the experimental activities, a scatter plot is formed (**Figure 12.2**). The r-value of the best-fit line determines the correlation coefficient of the Hansch equation. Most software packages that perform linear regressions take the data, the experimental activities, and parameter values, and generate an output with the Hansch equation and r-value. Behind the scenes, the software compares the experimental and calculated activity values to provide the r-value.

What good is the r-value? It approximates how closely the calculated activity values match the experimental values. As has already been stated, r^2 describes the fraction of variability in the experimental data that is captured by the calculated data. Based on Equation 12.20, electronics account for about 50% of the variance. Based on Equation 12.21, steric effects account for over 80% of the variance. That gives a total of 130%, which is impossible. Clearly something is breaking down in these analyses, and, unfortunately, this type of inconsistency is common in QSAR analysis.

Problems in a Hansch analysis can be minimized through careful selection of the compounds that will be included in the training set. A good training set should include R-groups with a broad range of properties. Based on the three parameters that have been highlighted in this chapter, a good training set would include R-groups that

are big and small, electron rich and electron poor, and polar and nonpolar. Furthermore, the R-groups should have varied properties. That is, among the large R-groups should be some that are electron rich and others that are electron poor. Similarly, some of the large R-groups should be polar and others nonpolar. Achieving such a variety of R-groups can require a large number of compounds in the training set. A simple rule is that a training set should include at least five compounds per parameter used in the regression.

Looking back at the data in Table 12.5, one can see that the training set includes only five compounds. That is a very small training set, although it does satisfy the rule of thumb that a training set should include five compounds per parameter (independent variable). All three R-groups that are electron withdrawing ($\sigma_m > 0$) are also sterically large ($E_s < 0$). This type of R-group selection in the training set prevents the resulting Hansch equation from properly weighting electronics and sterics. A better training set would have included more compounds with R-groups of greater variety in terms of their size and electronic nature. The section in this chapter entitled "Craig Plots" addresses a tool that assists in the selection of training set compounds. Regardless of whether a training set is well designed, any set of activity data can be used to generate a Hansch equation and predict the activity of unknown compounds. *Only if the training set is well conceived will the resulting Hansch equation have value as a predictive tool.*

Carboxylate Antifungals[9]

In another study by Hansch, aliphatic carboxylate salts (**12.15**) showed antifungal activity. The data are provided in **Table 12.6**. The activity was found to correlate very well with two parameters: log P and its square (Equation 12.23).

$$\log \frac{1}{C} = -1.54 \log P - 0.64(\log P)^2 + 2.15$$

$$n = 6, r = 0.991 \qquad (12.23)$$

Lipophilicity parameters, whether included as π-values or log P, are often included twice: once to the first power and again to the second power. This accommodates the common parabolic relationship between biological activity and lipophilicity (log P) (**Figure 12.3**). The curve will have a maximum at a lipophilicity value that balances the molecule's need to be both sufficiently polar to dissolve in the cytosol and nonpolar enough to enter and cross a cell membrane.

The exact optimal value of log P, sometimes labeled log $P°$, can be determined by taking the first derivative of log $1/C$ with respect to log P (Equation 12.24).

$$\frac{d \log \frac{1}{C}}{d \log P} = -1.54 - 1.28 \log P \qquad (12.24)$$

At the maximum activity of the curve, the derivative equals 0. Solving Equation 12.24 at this point indicates a log $P°$ of -1.20. If additional analogues were to be developed, the discovery team would use this information to include compounds with log P values close to -1.20. John Topliss of the University of Michigan was an early pioneer in QSAR. Topliss felt that the concept of log $P°$ was one of the most important contributions of Hansch analysis to the field of medicinal chemistry.[10]

As always, the compounds included in the training set are vital for obtaining a useful Hansch equation. If only the

TABLE 12.6 Antifungal activity of aliphatic carboxylic acid salts

Structure	Entry	R-Group	Log P	(Log P)²	Exptl. Log 1/C	Calc. Log 1/C
	1	Butyl	−2.70	+7.29	1.7	1.64
	2	Pentyl	−2.20	+4.84	2.4	2.44
	3	Hexyl	−1.70	+2.89	3.0	2.92
	4	Heptyl	−1.20	+1.44	3.0	3.08
	5	Octyl	−0.70	+0.49	3.0	2.91
	6	Nonyl	−0.20	+0.04	2.4	2.43

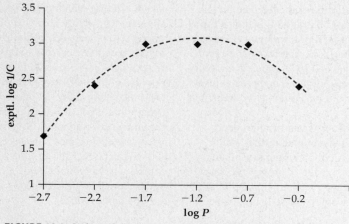

FIGURE 12.3 Biological activity versus log P for aliphatic carboxylic acids

first three entries in Table 12.6 are included in the training set, a very good fit is possible with log P as the only parameter (Equation 12.25).

$$\log \frac{1}{C} = +1.30 \log P + 5.23$$

$$n = 3, r = 0.999 \qquad (12.25)$$

Despite the high r-value of Equation 12.25, its implied correlation between log $1/C$ and log P is flawed.

Equation 12.25 erroneously indicates that higher values of log P should give antifungals with higher activity. Equation 12.25 correctly matches the information in its training set (the first three compounds in Table 12.6). Unfortunately, the training set is not representative of the overall activity of the full series. Inclusion in the training set of compounds with a range of both parameter values and activities ensures that the derived Hansch equations will be as useful as possible.

<table>
<tr><td>CASE
STUDY</td><td colspan="2">**Tumor Cell Resistance Modulators**[11]</td></tr>
</table>

Two similar types of structures, *ortho*-alkoxyphenones (**12.16**) and benzofurans (**12.17**), were found to lessen multidrug resistance exhibited by tumor cells. Compounds in Hansch analyses normally share the same basic molecular skeleton. In this study of tumor cell resistance modulators, inclusion of an indicator variable allows all the data for both structural types to be pooled into one large data set. The indicator variable, I_{BF}, is given a value of 1 if the compound is a

benzofuran and 0 if an alkoxyphenone. The other independent variable in the analysis is clog P. Data for the 19-compound study is included in **Table 12.7**. The regression is provided as Equation 12.26.

$$\log \frac{1}{C} = 0.86 \operatorname{clog} P - 1.16 I_{BF} - 3.33$$

$$n = 19, r = 0.990 \qquad (12.26)$$

TABLE 12.7 Activity of *ortho*-alkoxyphenones and benzofurans

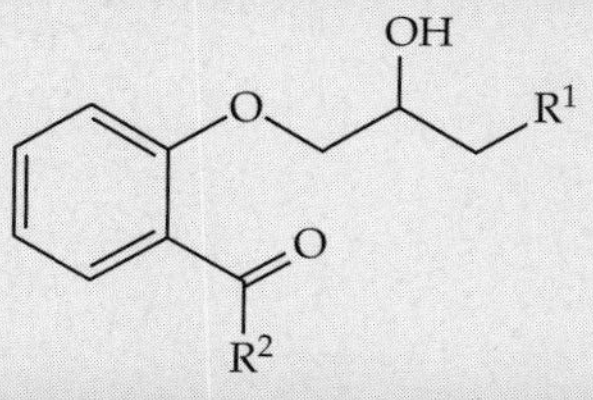

ortho-alkoxyphenone **12.16**

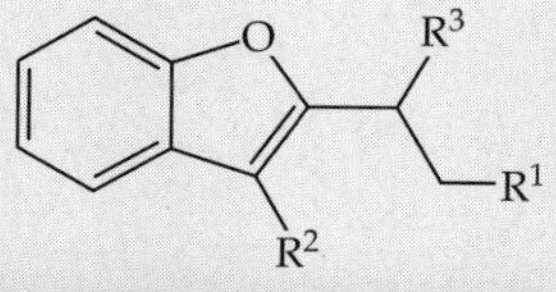

benzofuran **12.17**

Entry	R^1	R^2	R^3	clog P	I$_{BF}$	Exptl. Log 1/C
1	NHPr	CH$_2$CH$_2$Ph	—	3.36	0	0.54
2	NMePr	CH$_2$CH$_2$Ph	—	3.77	0	0.57
3	NMe$_2$	CH$_2$CH$_2$Ph	—	2.94	0	0.086
4	NEt$_2$	CH$_2$CH$_2$Ph	—	3.62	0	0.41
5	N(iPr$_2$)	CH$_2$CH$_2$Ph	—	4.25	0	0.54
6	NHCH$_2$Ph	CH$_2$CH$_2$Ph	—	4.30	0	0.72
7	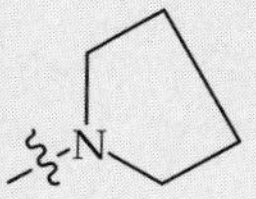 (pyrrolidine)	CH$_2$CH$_2$Ph	—	3.26	0	0.28
8	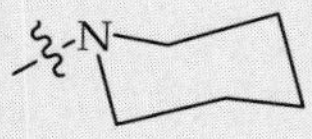 (piperidine)	CH$_2$CH$_2$Ph	—	3.67	0	0.54
9	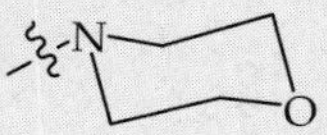 (morpholine)	CH$_2$CH$_2$Ph	—	2.54	0	−0.17

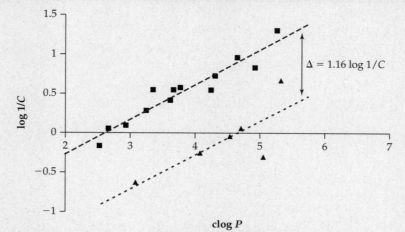

10		CH_2CH_2Ph	—	4.93	0	0.82
11		CH_2CH_2Ph	—	4.65	0	0.96
12		CH_2CH_2Ph	—	5.26	0	1.3
13		Me	—	2.67	0	0.051
14	NHPr	Et	OH	3.10	1	−0.62
15	NHPr	CH_2CH_2Ph	OH	4.07	1	−0.26
16		CH_2CH_2Ph	OH	5.32	1	0.68
17	NHPr	CH_2CH_2Ph	H	4.71	1	0.060
18	NH*i*Pr	CH_2CH_2Ph	H	4.54	1	−0.041
19		CH_2CH_2Ph	=O	5.05	1	−0.31

FIGURE 12.4 Activity versus clog P for alkoxyphenones (■) and benzofurans (▲)

Plotting the data as activity versus clog P reveals two separate sets of data as nearly parallel lines (**Figure 12.4**). The two lines represent the two different molecular skeletons. The lines are offset on the y-axis by approximately 1.16 log $1/C$ units. The value 1.16 is the same as the regression coefficient assigned to I_{BF}, the indicator variable. Inclusion of the indicator variable allows the two different data sets to be merged into one large group.

The three previous Case Studies highlight some of the common elements encountered when one reviews biological activity data, creates Hansch equations, and evaluates their predictive potential. The importance of a high correlation coefficient can easily be overstated. The value of a Hansch equation is not how well it fits the training set. The true worth is instead determined by how well it predicts the activity of unmade analogues. If a Hansch equation cannot predict activity, then it has no time-saving value for the lead optimization team.

12.4 Craig Plots

In the first Case Study in the previous section, the training set for the Hansch equation had several potential issues, one of which was a lack of diversity in the properties of the included R-groups. One tool that can help ensure that included compounds sample a range of electronic, steric, and lipophilic properties is a *Craig plot*.[12] Developed by Paul Craig of Smith Kline & French, Craig plots are two-dimensional graphs with two QSAR parameters representing the x- and y-axes. The most common axes are Hammett constants (σ) and Hansch constants (π) (**Figure 12.5**). The σ versus π plot is most common because these two parameters are by far the most widely used QSAR parameters. Craig plots are also encountered with Hansch constants (π) graphed against Taft parameters (E_s).

Craig plots graphically show two properties of an R-group. When selecting R-groups for inclusion in a training set, one should try to select R-groups that are evenly distributed about the area of the graph. That is, compounds should be selected from all four quadrants. In the first Case Study in this chapter, three of the five compounds included in the training set were from the same quadrant of a Craig plot: positive σ-value and negative E_s-value. While this fact alone does not invalidate the study, the worth of the resulting Hansch equation is certainly called into question.

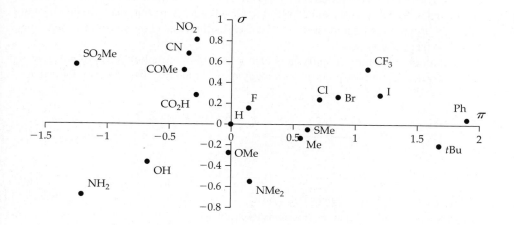

FIGURE 12.5 Craig plot of σ versus π

12.5 Topliss Trees

In 1972, John Topliss of the University of Michigan at Ann Arbor published a decision tree method for lead optimization.[13] From the standpoint of the user of the tree, the method is not quantitative. However, the decision tree is built on the ideas of Hansch analysis. The tree suggests a substituent for a new analogue. Based on whether the new analogue is more, less, or equally potent to the former analogue, another new substituent is suggested. Each substituent is designed to test electronic, lipophilic, and steric effects. The longest path through the tree requires the synthesis of perhaps only five analogues. A restriction to using a Topliss tree is that the lead compound must have an unfused benzene ring. A portion of the full Topliss tree, slightly modified in layout from the original design, is shown in **Figure 12.6**. A separate Topliss tree has been developed to optimize alkyl side chains.

The first analogue suggested in a Topliss tree is always a 4-Cl derivative. The 4-Cl analogue will either be more (M), less (L), or equally (E) active compared to the original lead. If a 4-Cl analogue is less active, then the tree suggests making a 4-OCH_3 analogue. The rationale is that a 4-Cl group is weakly electron withdrawing ($\sigma_p = +0.227$). Good activity may require an EDG ($\sigma < 0$), so 4-OCH_3 is recommended ($\sigma_p = -0.268$). The 4-OCH_3 will either be more, less, or equally active relative to the 4-Cl analogue. If the 4-OCH_3 analogue is less or equally active, then perhaps an unfavorable steric interaction is occurring with 4-substitution. A 3-Cl group may be an option. If 4-OCH_3 is more active, then an EDG is apparently a good idea. The next analogue should bear a 4-$N(CH_3)_2$ group, an even stronger EDG ($\sigma_p = +0.600$) than 4-OCH_3. The other main branches of the tree follow similar patterns of logic, and additional branching is described in Topliss' full tree.

In performing SAR on a lead, many medicinal chemists use the ideas of the Topliss tree without explicitly consulting the tree itself. Topliss formalized the process by creating the tree. An advantage of the tree is that rigorous QSAR and the mathematics that accompany it are not required. In the late 1960s and early 1970s, multiple least square regressions were performed mostly by hand through linear algebra. A 20-compound study with n independent variables would require manipulating $20 \times n$ and $n \times 20$ matrices during intermediate stages of the calculation. This is trivial when a computer does the work, but the longhand process is a chore and prone to error. A Topliss tree allows a medicinal chemist to avoid the math and still likely find a potent lead analogue.

12.6 Evaluating Hansch Analysis

Thus far, this chapter has extolled the virtues of Hansch analysis. The method is simple to understand, easy to execute, and can lead to QSAR equations that may provide insight into the activity of a lead compound. The Hansch method has many clear and appealing traits. It also has weaknesses.

One challenge to Hansch analysis is also its strength: simplicity. The binding of a drug to a target is a complex process with many relevant factors. To reduce biological ac-

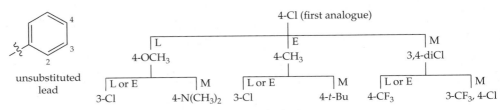

FIGURE 12.6 A portion of the Topliss tree for aromatic substitution

tivity to a handful of factors such as electronics and lipophilicity is almost unreasonable. To expect all variables to be linearly related to activity is equally implausible. Remarkably, Hansch analysis can indeed often predict the activity of a set of molecules. More often, however, Hansch analysis and related QSAR methods fail to accurately model biological activity.

The simplicity of Hansch analysis also means that experienced medicinal chemists may be able to identify trends in activity without the assistance of a QSAR equation. Making individual new lead analogues is generally a slow process, and a medicinal chemist has ample time to examine SAR data. While a chemist will not be able to quantify a structure-activity relationship, just knowing the approximate trend of the relationship is usually adequate for lead optimization. Hansch analysis is valuable only if it can reveal something that is not already known about the compounds being tested.

A second problem for Hansch analysis and any other predictive method is combinatorial chemistry. As methods in combinatorial chemistry continue to be refined and advanced, directed combinatorial libraries constructed around the scaffold of a lead have become a more standard practice. The availability of hundreds of lead analogues greatly diminishes the potential contribution from a Hansch equation. Why predict a structure's activity when one can make a library of essentially all interesting analogues, screen the library, and know the activity for certain?

The criticisms in the previous paragraphs lead to a question: If Hansch analysis is of such questionable value, then why has an entire chapter of this textbook been devoted to the subject? Despite the fading utility of classical QSAR methods such as Hansch analysis, the logic behind Hansch analysis is invaluable to medicinal chemistry. Synthetic chemists in the pharmaceutical industry intuitively consider the ideas used to construct Hansch equations. Ideas such as electronics, sterics, and lipophilicity underlie traditional SAR approaches in the laboratory. Critical analysis of activity data and emphasis on seeking holes in R-group selection are also fundamental to successful SAR on a lead. Through the study of Hansch analysis, all these crucial ideas are presented in a rational framework that helps demonstrate their relevance. Just as importantly, Hansch analysis provides the foundation for the next generation of QSAR: comparative molecular field analysis.

12.7 Comparative Molecular Field Analysis

Comparative molecular field analysis (CoMFA) is a modern, powerful extension of the classical QSAR methods that were developed in the 1960s.[14] While Hansch analysis is simple to understand and fairly easy for any medicinal chemist to perform, CoMFA requires specialized software and an understanding of statistics. Since CoMFA is outside the experience of most synthetic chemists, pharmaceutical companies have dedicated computational chemistry groups to handle advanced QSAR tasks.

Hansch analysis and other classical QSAR approaches evaluate the QSAR model based on the correlation of rows of compounds with known activities (dependent variables) to columns of parameters (independent variables). For this reason, classical QSAR is sometimes called 2D QSAR. CoMFA is an example of 3D QSAR because lead analogues are modeled and analyzed in a virtual three-dimensional space. The value of both methods ultimately hinges on how well experimental and calculated activities correlate (Figure 12.2) and how well the model predicts the activity of compounds not included in the training set.

CoMFA begins with the consistent alignment of each molecule into a separate three-dimensional grid. Alignment of the molecules must be representative of the binding conformation and orientation of the compound with the target. Three-dimensional coordinates within the grid are then probed for both steric and electrostatic interactions with the molecule until the entire volume of the grid has been mapped. The interactions are then analyzed against the known activities through a partial least squares regression to provide an algorithm for predicting the biological activity of untested lead analogues.

A thorough discussion of CoMFA is beyond the scope of this text, but a few ideas merit coverage. CoMFA generates a huge number of independent variables in the form of steric and electronic interactions at many points in space. An analysis includes more independent variables than molecules with known biological activities, the dependent variable. In other words, the regression has more variables than equations. False correlations are a significant risk under these conditions, and statistical cross-validation is required to ensure the validity of the results. Multiple linear regression analysis fails to handle the type of data afforded in a CoMFA study. Only the partial least squares method, which was not reported until 1984, is able to tease useful correlations from the volume of information.

Since its formal development in 1988 by Richard Cramer of Tripos Inc.,[14] CoMFA has become the benchmark QSAR tool in medicinal chemistry. As CoMFA continues to evolve and be refined, new applications of the method regularly appear in the literature. A recent advancement is the idea of extrapolating a theoretical binding site based on a successful CoMFA model. In principle, a binding site for an unknown target could be constructed from molecules with known activity against the target. New molecules may then be docked with the postulated target and tested for activity in a virtual screen. While intriguing, this type of application of CoMFA has yet to be fully realized.

The following Case Study demonstrates how Hansch analysis and CoMFA can both correlate the same structure-activity data. CoMFA, however, can more easily be extended to include a broad range of structural variations.

CASE STUDY

Comparing Hansch Analysis and CoMFA[15]

In a 1988 study, Nelson and colleagues performed a Hansch analysis on a set of 15 compounds that antagonize serotonin receptor 5-HT$_2$. The tested compounds (**12.18**) were varied at just one position, the 5-carbon of the indole ring (**Figure 12.7**). Based on these compounds, inhibition (pK_i) was found to depend on the volume of the substituent at the 5-position (V_5) and the log P of the entire molecule (Equation 12.27).

$$pK_i = -0.0281V_5 + 0.9022 \log P + 4.4575$$
$$n = 15, r = 0.87 \qquad (12.27)$$

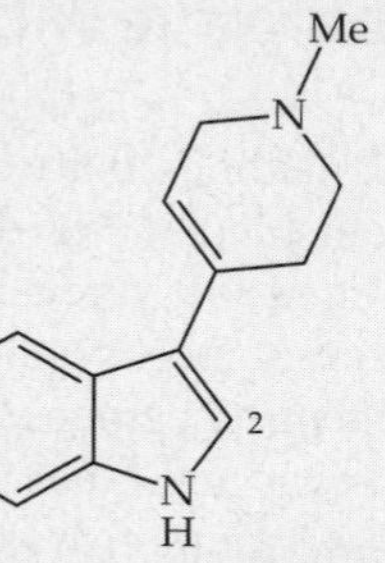

FIGURE 12.7 5-HT$_5$ antagonists

In 1993, Martin and colleagues performed a follow-up study on the same 15 compounds using CoMFA. CoMFA does not produce a simple equation in the manner of Hansch analysis, but CoMFA does calculate activities for the training set compounds. Comparing the predicted and experimental activities allows determination of the model's correlation coefficient. The CoMFA performed by Martin gave an r-value of 0.96. Martin's CoMFA model accounted for 92% of the variance in the data; Nelson's Hansch equation accounted for 76% of the data's variance.

To continue the comparison, Martin prepared an additional set of 12 analogues of **12.18** and determined their activity against the 5-HT$_2$ receptor. Both Nelson's Hansch equation and Martin's CoMFA model were used to generate calculated activities of the 12 new compounds. The summary of the findings is shown in **Table 12.8**.

TABLE 12.8 **Hansch and CoMFA comparison data**

Entry	Exptl. pKi	Hansch Calc. pKi	Hansch Diff.	CoMFA Calc. pKi	CoMFA Diff.
1	5.867	5.617	−0.250	5.787	−0.080
2	5.762	5.451	−0.311	5.301	−0.461
3	5.509	5.465	−0.044	5.648	+0.139
4	5.291	5.469	+0.178	6.231	+0.940
5	5.000	4.937	−0.063	5.966	+0.966
6	4.272	4.405	+0.133	4.918	+0.646
7	6.921	6.635	−0.286	5.746	−1.175
8	5.616	4.468	−1.148	6.127	+0.511
9	5.760	4.300	−1.460	5.971	+0.211
10	6.662	7.167	+0.505	6.380	−0.282
11	6.695	5.347	−1.348	7.108	+0.413
12	5.957	6.965	+1.008	5.162	−0.795
		RMS diff.	**0.75**	**RMS diff.**	**0.65**

The root mean square (RMS) of the differences between the experiment and predicted values for the Hansch equation was 0.75, while the CoMFA gave 0.65. A smaller value for the RMS of the differences indicates CoMFA afforded a more accurate model.

The data in Table 12.8 do not represent a resounding victory for CoMFA over Hansch analysis. However, a significant advantage of CoMFA is its flexibility. Martin continued his study by including an additional 30 compounds in his CoMFA model. The additional compounds are analogous to **12.18** but contain significantly more structural variability with R-group modifications at C-2,

upon the indole NH, and at the *N*-methyl position. This level of structural modification would be very difficult for a Hansch equation to handle. The greatest challenge to CoMFA with regard to accommodating new structures is the proper alignment of all compounds to the three-dimensional modeling grid.

Martin's expanded CoMFA model gave an *r*-value of 0.92. Six new analogues of **12.18** were prepared, and their activities were predicted with the new model. The RMS of the experimental and calculated activity differences was only 0.36.

Summary

Traditional approaches to lead optimization provide valuable SAR information. In the 1960s, advancements in physical organic chemistry allowed the development of the first practical and effective QSAR method by Hansch. With Hansch analysis, a lead optimization team may mathematically predict the impact of a lead's structural changes on biological activity. To quantify activity predictions, molecular properties must also be expressible numerically. The two most common property parameters used in Hansch analysis are Hammett constants for electronic effects and Hansch-Fujita constants for lipophilicity. Activities of lead analogues may be correlated to the properties of the molecules through a multiple linear regression to afford a Hansch equation. Construction of a useful equation requires careful selection of the compounds included in the analysis. Tools such as Craig plots may help ensure appropriate functional variability in the training set. A well-conceived Hansch equation may be used to predict the activity of unknown analogues and allow the lead optimization team to focus on compounds with potentially high activity. Despite its proven value, Hansch analysis as a QSAR technique has been supplanted by more computationally and statistically intensive methods such as comparative molecular field analysis. The ability of CoMFA to easily handle very large data sets and the flexibility of partial least squares analysis makes CoMFA a significant improvement over Hansch analysis. Regardless of its shortcomings, Hansch analysis affords a concise and manageable introduction to the fundamental ideas of QSAR. Many of these same ideas are used qualitatively by medicinal chemists when performing SAR on a lead.

Questions

1. Does anything about the following Hansch equation seem troubling? (Kutter, E., & Hansch, C. Steric Parameters in Drug Design: Monoamine Oxidase Inhibitors and Antihistamines. *J. Med. Chem.* 1969, *12*, 647–652.)

$$pIC_{50} = +1.03E_s + 1.09\sigma + 0.40\pi + 4.54$$

$$n = 9, r = 0.96 \qquad (12.a)$$

2. Lipophilicity parameters are often found in a squared form in Hansch equations. The following Hansch equation contains a squared steric term (E). Is this surprising? Why or why not?

$$pD_2^{atrial} = +1.87E - 0.45E^2 + 5.99$$

$$n = 9, r = 0.99 \qquad (12.b)$$

3. This chapter never questions the validity of activity data, but activity data is very prone to error. Sources of error may include accuracy of reagent concentrations used to screen the compound, accuracy in reading the assay result, and even the reproducibility of the assay. Comment on how the accuracy of the activity data impacts QSAR.

4. Can you think of a situation in which a Topliss tree might not produce optimal R-groups for a benzene ring? *Hint: How might you get down one branch of a tree when the final result of another branch might actually be better? The first substitution (4-Cl) will need to give a misleading result.*

5. Use a form of the Arrhenius equation (Equation 12.c) with Equation 12.17 to prove that the Taft steric parameter is indeed a function of activation energy (E_a).

$$\ln k = \frac{-E_a}{RT} + \ln A \qquad (12.c)$$

6. The NH_2 group is electron donating in the *para* position according to its σ-value of -0.660. Under what conditions would an NH_2 group have a positive σ-value?

7. Hansch analysis is useful not only for drug discovery. A study of insecticides and related nonpolar compounds has shown that the bioconcentration of these compounds (log *BC*) in trout muscle is related to log *P*. Create a Hansch equation based on the following data. Does including a (log *P*)² term improve the fit? Is this surprising? Explain. (Neely, W. B., Branson, D. R., & Blau, G. E. Partition Coefficient to Measure Bioconcentration Potential of Organic Chemicals in Fish. *Environ. Sci. Technol.* 1974, *8*, 1113–1115.)

Entry	Compound	Log P	Log BC
1	1,1,2,2-tetrachloroethylene	2.88	1.59
2	carbon tetrachloride	2.64	1.24
3	1,4-dichlorobenzene	3.38	2.33
4	diphenyl ether	4.20	2.29
5	biphenyl	4.09	2.64
6	2-biphenyl phenyl ether	5.55	2.74
7	hexachlorobenzene	6.18	3.89
8	2,2',4,4'-tetrachlorodiphenyl ether	7.62	4.09

8. In a study of analogues of the antibiotic metronidazole (**12.a**, R = OH), antimicrobial activity against *T. vaginalis* (log *AA*) was correlated to activation free energy of electroreduction (ΔG) and lipophilicity (log *P*). Both ΔG and log *P* are whole molecule parameters. Create a Hansch equation based on these parameter values and following activity data. Interpret the effect of each parameter on antimicrobial activity. (Chien, Y. W., & Mizuba, S. S. Activity-Electroreduction Relationship of Antimicrobial Metronidazole Analogues. *J. Med. Chem.* 1978, *21*, 374–380.)

Structure	Entry	R	ΔG (kcal/mol)	Log P	Log AA
	1	$-OC(=O)Me$	3.22	+0.301	3.369
	2	$-OH$	6.31	-0.051	3.242
	3	$-SC(=S)NMe_2$	7.72	+1.281	3.148
	4	$-OC(=O)NMe_2$	7.73	+0.455	3.094
	5	$-Cl$	7.96	+0.879	3.031
	6	$-N_3$	8.71	+0.755	2.760
	7	$-S_2O_3Na$	8.87	-2.046	1.665
12.a	8	$-NHC(=O)Me$	9.64	-0.469	1.287
	9	$-S\text{-}C_6H_4\text{-}4\text{-}Cl$	12.56	+1.470	1.377

Structure (12.a): O_2N— imidazole ring with N, N, Me, and R substituent.

9. In a study of anabolic steroids, the activity (log *BR*) of compounds of structure **12.b** was found to depend on lipophilicity, log *P*, and its square, $(\log P)^2$. Based on the data in the following table, perform a multiple linear regression relating activity to log *P* and $(\log P)^2$. What is the predicted log P° for this series of compounds? Are electronic or steric effects important for anabolic activity? Why or why not? (Chaudry, M. A. Q., & James, K. C. A Hansch Analysis of the Anabolic Activities of Some Nandrolone Esters. *J. Med. Chem.* 1974, *17*, 157–161.)

Structure		Entry	n	Log P	Log BR
		1	2	4.838	2.783
		2	4	5.284	3.163
		3	5	5.429	3.287
		4	6	5.786	3.281
		5	7	5.658	3.264
		6	8	5.904	3.409
		7	9	6.166	3.192

12.b

10. In the Case Study at the end of the chapter, a CoMFA model based on a 15-compound training set had an *r*-value of 0.96. When the model was expanded to a much more diverse collection of 45 compounds, the *r*-value dropped to 0.92. Explain why the second CoMFA model is impressive despite the decrease in the *r*-value.

11. Take any of the Hansch equations that you have generated in the previous questions, plot the experimental and calculated activities against each other, and determine the *r*-value of the best fit line.

12. Larger, more sterically demanding R-groups tend to also have a higher lipophilicity. Does this statement pose a challenge for a Craig plot of sterics versus lipophilicity? Why or why not?

13. pK_a is a somewhat common QSAR parameter. Make a case that pK_a is largely an electronic parameter. Now make a case that pK_a is largely a lipophilicity parameter.

14. In some instances, researchers are able to develop separate Hansch equations for both activity and toxicity of a drug. Differing parameters between the activity and toxicity QSAR equations allow researchers to optimize properties that boost desired activity and minimize toxicity. The following two QSAR equations give the activity (12.d) and toxicity (12.e) for a series of nitrogen mustard antitumor compounds (see Chapter 6). Based on these equations, is it likely that a nontoxic nitrogen mustard will be developed? Explain.

$$\log\left(\frac{1}{C}\right) = -0.26\pi - 1.62\sigma + 3.66 \quad n = 9, r = 0.84$$

$$(12.d)$$

$$\log LD_{50} = -0.27\pi - 1.58\sigma + 3.48 \quad n = 9, r = 0.93$$

$$(12.e)$$

15. Based on Taft steric parameters, how much faster does ethyl formate hydrolyze than ethyl propionate (R = Et)?

References

1. Hansch, C., & Fujita, T. ρ-σ-π Analysis: A Method for the Correlation of Biological Activity and Chemical Structure. *J. Am. Chem. Soc.* 1964, *86*, 1616–1626.
2. Hammett, L. P. *Physical Organic Chemistry.* International Chemical Series. New York: McGraw-Hill, 1940, Chapter VII.
3. Fujita, T., Iwasa, J., & Hansch, C. A New Substituent Constant, π, Derived from Partition Coefficients. *J. Am. Chem. Soc.* 1964, *86*, 5175–5180.
4. Scherrer, R. A., & Howard, S. M. Use of Distribution Coefficients in Quantitative Structure-Activity Relationships. *J. Med. Chem.* 1977, *20*, 53–58.
5. Taft, R. W., Jr. Separation of Polar, Steric, and Resonance Effects in Reactivity. In M. S. Newman (Ed.), *Steric Effects in Organic Chemistry.* New York: Wiley & Sons, 1956, Chapter 13.
6. Charton, M. Steric Effects: I. Esterification and Acid-Catalyzed Hydrolysis of Esters. *J. Am. Chem. Soc.* 1975, *91*, 1552–1556.
7. Verloop, A., & Tipker, J. Use of Linear Free Energy Related and Other Parameters in the Study of Fungicidal Selectivity. *Pest. Sci.* 1976, *7*, 379–390.
8. Hansch, C. Steric Parameters in Structure-Activity Correlations: Cholinesterase Inhibitors. *J. Org. Chem.* 1970, *35*, 620–621.
9. Hansch, C., & Lien, E. J. Structure-Activity Relationships in Antifungal Agents: A Survey. *J. Med. Chem.* 1971, *14*, 653–670.

10. Topliss, J. G. (Ed.). *Quantitative Structure-Activity Relationships of Drugs*. New York: Academic Press, 1983, Commentary.

11. Ecker, G., Chiba, P., Hitzler, M., Schmid, D., Visser, K., Cordes, H. P., et al. Structure-Activity Relationship Studies on Benzofuran Analogs of Propafenone-Type Modulators of Tumor Cell Multidrug Resistance. *J. Med. Chem.* 1996, *39*, 4767–4774.

12. Craig, P. N. Interdependence between Physical Parameters and Selection of Substituent Groups for Correlation Studies. *J. Med. Chem.* 1971, *14*, 680–684.

13. Topliss, J. G. Utilization of Operational Schemes for Analog Synthesis in Drug Design. *J. Med. Chem.* 1972, *15*, 1006–1011.

14. Cramer, R. D. III, Patterson, D. E., & Bunce, J. D. Comparative Molecular Field Analysis (CoMFA): 1. Effect of Shape on Binding of Steroids to Carrier Proteins. *J. Am. Chem. Soc.* 1988, *110*, 5959–5967.

15. Agarwal, A., Pearson, P. P., Taylor, E. W., Li, H. B., Dahlgren, T., Herslöf, M., et al. Three-Dimensional Quantitative Structure-Activity Relationships of 5-HT Receptor Binding Data for Tetrahydropyridinyl Derivatives: A Comparison of the Hansch and CoMFA Methods. *J. Med. Chem.* 1993, *36*, 4006–4014.

Chapter 13

Aspects in Pharmaceutical Synthesis

Chapter Outline

As a drug candidate progresses through the drug discovery process, the research team requires progressively larger and larger quantities of material for testing. At some stage in the process, responsibility for synthesizing the compound shifts from the medicinal chemistry team to the process group. The role of the process group is to make large quantities of drug in a manner that is economical, efficient, reproducible, and safe. A successful large-scale synthetic process includes many ideas that overlap with the recently developed field of green chemistry, which strives to minimize the environmental impact of chemical industry and research. Aside from preparing large amounts of drugs and their intermediates, the process group must maintain data on each batch or lot to satisfy the recordkeeping requirements of regulatory agencies.

This chapter includes a *lot* of chemistry in the form of synthetic schemes. Understanding the reaction schemes may be helpful in this chapter, but it is not a necessity. The schemes are simply a complement to the text, which contains all the vital ideas for this chapter.

13.1 Solids, Solids, Solids

The preferred state of matter in pharmaceuticals is solid. Drugs that are solids are easier to handle in a process setting, less expensive to purify, and more stable for prolonged storage. Oral drugs tend to be of a size (MW 400–500 g/mol) that allows the formation of many intermolecular interactions, have minimal conformational flexibility, and very often contain multiple flat aromatic rings. As a result, most drugs are solids. Solids afford many advantages for the process research group, but many challenges must still be overcome toward the goal of preparing a marketable drug.

Learning Goals for Chapter 13

- Understand the basic ideas behind recrystallization as a purification technique

- Be able to distinguish between the different types of crystal modifications

- Know the primary methods for obtaining single enantiomer compounds

- Appreciate the basic cost, safety, and efficiency concerns for increasing reaction scale

- Understand the role of green chemistry in pharmaceutical synthesis and be able to analyze a synthesis using various green metrics

- Know the role of regulatory agencies in the oversight of pharmaceutical synthesis

Recrystallization

Distillation, column chromatography, and recrystallization are the three standard methods of purification in organic chemistry. Orally delivered drugs have high molecular weights and very low vapor pressures, so distillation is impossible. Column chromatography requires large volumes of solvent and is cost prohibitive except in the most extreme cases. That leaves recrystallization as the standard method of purification for large-scale drug synthesis. Since most drugs are solids, drugs tend to be amenable to being purified through recrystallization.

Recrystallization has three goals:

- Remove impurities
- Allow recovery of as much of the compound of interest (the drug) as possible
- Provide the compound of interest in a useful form

Of the three goals, the first two are standard to any purification and are covered in this subsection. The last goal is perhaps less obvious but provides the foundation for later subsections.

Recrystallizations hinge upon solubility curves. A solubility curve shows the solubility of a compound in a solvent or mixture of solvents as a function of temperature (**Figure 13.1**). Not all solubility curves have an upward slope, but Figure 13.1 is representative of the effect of temperature on solubility. The star of a recrystallization is the compound of interest: a drug. The villain is any impurity. Removal of any impurity is the top goal. Operationally, a recrystallization is performed by dissolving a mixture into a *minimal amount* of heated solvent. At elevated temperature, the solution is saturated with the drug and somewhat less-than-saturated with any impurities. As the solution cools, the drug's solubility drops, and the drug precipitates from solution. The solution remains saturated with drug. Once cooling is complete, the maximum amount of drug is precipitated. Ideally, through the entire cooling stage, the solution will not have become saturated in impurity, and no impurity will have precipitated. Filtration of the solution separates the drug from its impurity. Because the drug has a measure of solubility even at low temperatures, some drug is always lost during a recrystallization. Selection of a solvent with a large change in solubility of the cooling range minimizes drug loss during the purification.

Almost any solvent is possible for a recrystallization. **Table 13.1** shows some of the most frequently used solvents in the pharmaceutical industry. Mixtures of solvents, binary or even ternary, are also common. The burden of checking many solvents and solvent mixtures for use in recrystallizations can be mitigated with automated high-throughput screening tools. A drug's solubility is tested in each solvent and reasonable solvent mixture at various

FIGURE 13.1 A typical solubility curve

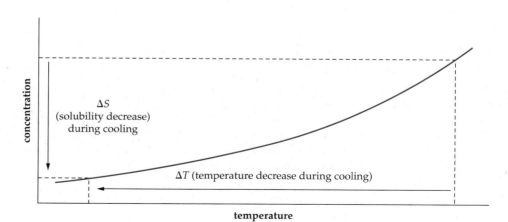

TABLE 13.1   Common recrystallization solvents[1]

Solvent Class	Example(s)
Water	Water
Alcohols	Methanol, ethanol, 2-propanol
Ketones/Esters	Acetone, methyl isopropyl ketone, ethyl acetate
Hydrocarbons	Hexane, heptane, cyclohexane, toluene, xylene
Ethers	Diethyl ether, dimethoxyethane, 1,4-dioxane, tetrahydrofuran
Halogenated	1,2-Dichloroethane, chloroform
Miscellaneous	Acetonitrile, dimethyl sulfoxide, N,N-dimethylformamide

temperatures. Any significant impurities may be synthesized and studied for solvent behavior as well. The resulting solubility curves are examined to reveal the optimal recrystallization solvent that both removes impurities and gives high recovery of the drug.[1]

Acid–Base Salts

Converting a drug from a neutral organic molecule to an ionized salt is the traditional means of manipulating a drug's solubility and crystalline properties. Basic nitrogens and carboxylic acids are both often found in drugs. These functional groups serve as handles for converting drugs into ionic compounds.

Typical amines are reasonably strong bases. For example, the conjugate acid of methylamine has a pK_a of approximately 10.6 (**Figure 13.2**). Nitrogens within heterocyclic rings such as pyridine are considerably less basic. The pK_a of a protonated pyridine ring (**13.3**) is around 5.[2] In order to protonate weakly basic drugs, a strong acid must be used. The rule of thumb for ensuring complete protonation is that the acid must be more acidic than the conjugate acid by three pK_a units.[3] It is therefore not surprising that by far the mostly commonly used acid to form salts of basic drugs is hydrochloric acid with a pK_a of -8. A large number of other acids have been used to protonate basic drugs. Other common acids include sulfuric acid ($pK_a -3$), hydrobromic acid ($pK_a -9$), and methanesulfonic acid (CH_3SO_3H, $pK_a -2$).[2] Beyond forming an ionic compound, protonating nitrogens in a drug can improve a drug's shelf-life. A nitrogen atom with its lone pair is electron rich and prone to being oxidized in air. Capping the nitrogen with a proton suppresses the reducing ability of the nitrogen.

While basic drugs are protonated with a broad range of different acids, the bases used to deprotonate acidic drugs have far less variety. Carboxylic acids, with a pK_a in the 4 to 5 range, are almost always converted into sodium salts. Potassium and calcium are also occasionally encountered. Carboxylate salts do not have any general chemical stability advantages as found with protonated amines.

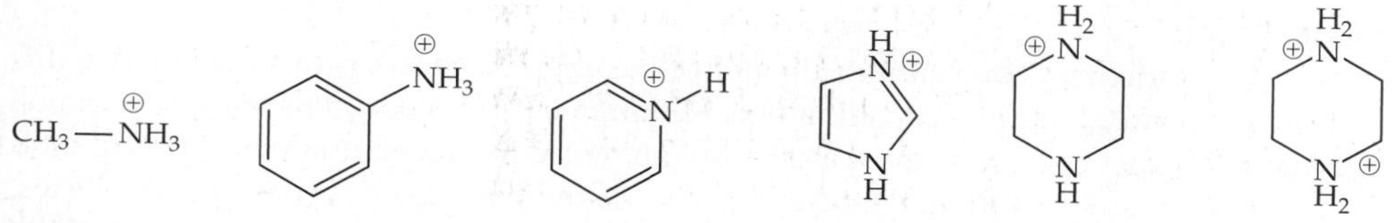

methylamine	aniline	pyridine	imidazole	piperazine
13.1	**13.2**	**13.3**	**13.4**	**13.5**
pK_a 10.6	pK_a 4.6	pK_a 5.2	pK_a 7.0	pK_a 9.8 and 5.7

FIGURE 13.2 pK_as of protonated nitrogenous bases often seen in drugs[2]

The following Case Study demonstrates some of the advantages that can be gained by converting a neutral drug into a salt.

Salt Forms of RPR 200,735[4]

RPR 200,735 (**13.6**) was a drug candidate for the treatment of rheumatoid arthritis (**Figure 13.3**). RPR 200,735 showed poor bioavailability in animal trials. The low bioavailability was attributed to the poor water solubility of RPR 200,735 of only $10 \ \mu g/mL$. Salts of RPR 200,735 with improved solubility characteristics were investigated.

RPR 200,735 is a weak base. Its conjugate acid has a pK_a of 5.3. Only strong acids are able to form a stable salt with RPR 200,735. Investigated acids include hydrochloric acid, hydrobromic acid, and methanesulfonic acid (**Table 13.2**). The methanesulfonate salt was found to be superior to the other salts in all aspects: water solubility, hygroscopy (absorption of moisture from air), and rate of dissolving. Furthermore, the flow properties of the methanesulfonate salt facilitated capsule and tablet preparation.

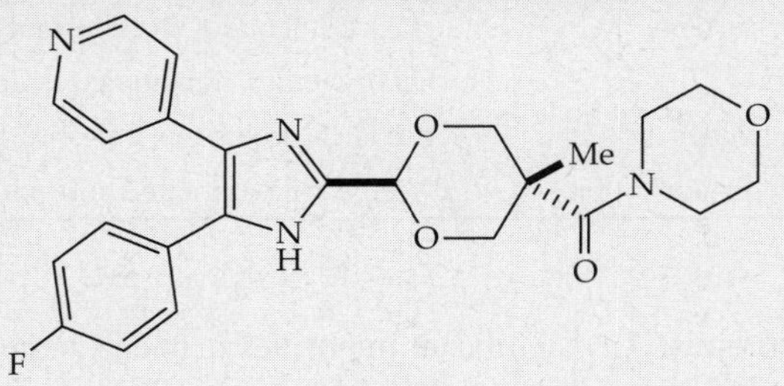

RPR 200,735
antiarthritic
13.6

FIGURE 13.3 RPR 200,735 as a free base

TABLE 13.2 Properties and solubility data for salts of RPR 200,735

Test	Methanesulfonate Salt	Chloride Salt	Bromide Salt
MW (g/mol)	566.61	524.98	569.43
Aqueous solubility at 25 °C (mg/mL)	39	17	3
Hygroscopy	Nonhygroscopic stable, monohydrate	Hygroscopic, multiple hydrated forms	Hygroscopic, multiple hydrated forms
Time for 80% dissolution in capsule at pH 4 (min)	2.0	7.4	3.9

While the formation of acid–base salts of drugs can improve the properties of some drugs, not all drugs are acidic or basic. Additionally, some drugs are not stable to the strongly acidic or basic conditions required for forming an acid–base salt. Making drug salts is therefore not a technique that may be applied to all drugs. Cocrystals are often an alternative means of affecting a solid's properties.

Cocrystals

Cocrystals (or co-crystals) are broadly defined as crystalline solids that contain two different compounds within a crystal lattice. Requirements for cocrystals vary. A common restriction is that cocrystals must arise through molecular recognition and contain molecules that are solids at room temperature. This definition excludes hydrates and solvates. Commonly encountered molecular recognition elements seen in cocrystals are reminiscent of Watson–Crick base pairing and are shown in **Figure 13.4**. All interactions except **13.9** involve an eight-membered ring interaction. From Figure 13.4, molecules that contain carboxylic acids and amides are regularly included in cocrystals.[5]

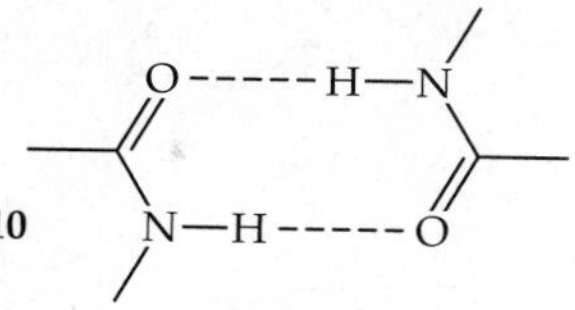

FIGURE 13.4 Types of molecular recognition commonly observed in cocrystals[5]

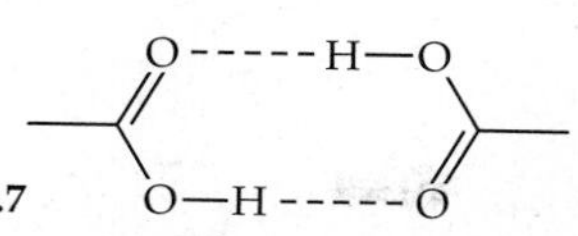

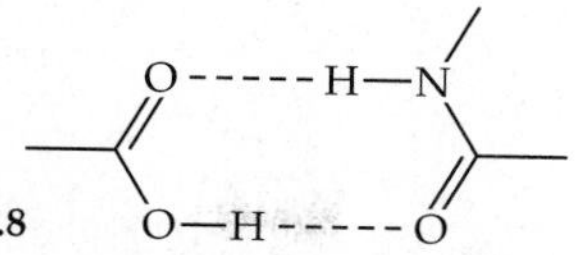

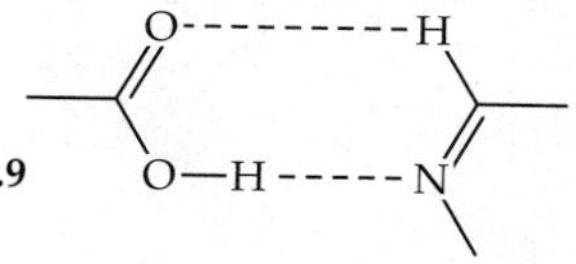

Like salts, cocrystals tend to dissolve rapidly in comparison to standard crystalline forms of drugs. Once a neutral drug quickly bursts into solution from a cocrystal, the solution is supersaturated in drug. The drug precipitates out of solution in an amorphous, noncrystalline form that is still more soluble than the crystalline drug. Therefore, cocrystals provide drugs in a form that offers increased solubility. Increased solubility boosts drug absorption, raises bioavailability, and shortens a drug's onset time.[6]

Pharmaceutical cocrystals are a relatively new area of research and have been widely pursued only since around 2000. To date, the most notable example of a drug that has been isolated as a useful cocrystal is itraconazole, which is the subject of the Case Study below.

CASE STUDY

Itraconazole-Succinic Acid Cocrystal[7]

Itraconazole (Sporanox, **13.11**) is an approved antifungal (**Figure 13.5**). The drug has a log P value of 6.5, so its very low aqueous solubility is no surprise. Early efforts to improve the properties of itraconazole focused on making an acid salt. Because of the very weak basicity of all the nitrogens in itraconazole, the idea of an acid salt was soon abandoned. Cocrystals were then explored. A library of amides and acids were systematically tested with a high-throughput recrystallization screen. A number of diacids formed satisfactory cocrystals with itraconazole.

A unit of the x-ray structure (**13.12**) of a cocrystal between itraconazole and succinic acid ($HO_2CCH_2CH_2CO_2H$), is shown in **Figure 13.6**. The cocrystal involves two molecules of itraconazole encapsulating a single succinic acid molecule. The OH groups of the acid form a hydrogen

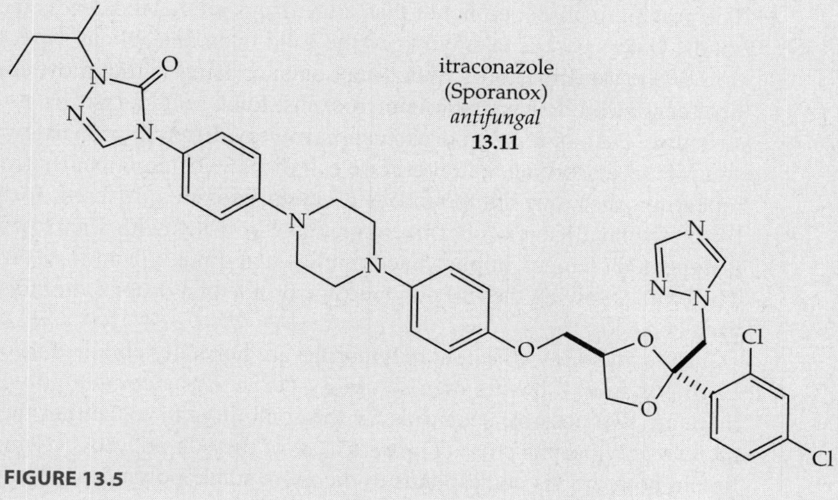

itraconazole
(Sporanox)
antifungal
13.11

FIGURE 13.5

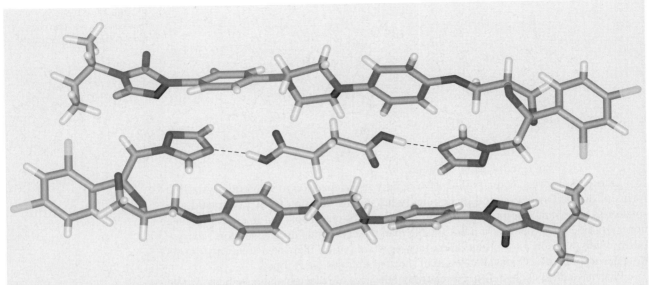

FIGURE 13.6 X-ray structure of an itraconazole-succinic acid cocrystal. Published with permission from Remenar, J. F., et al. *J. Am. Chem. Soc.* 2003, *125*, 8456–8457. Copyright 2003 American Chemical Society.

bond to N4 of the 1,2,4-triazole ring of itraconazole. Of the newly discovered cocrystals, the one with the best solubility properties involved L-malic acid, an acid similar to succinic acid. The solubility profile of the itraconazole-L-malic acid cocrystal rivals the properties of the Sporanox bead formulation of itraconazole.

Simply matching the solubility profile of a commercial formulation may not seem to be an improvement, but it is. Sporanox beads contain an amorphous form of the drug. Amorphous solids convert into a regular, crystalline form over time. To give Sporanox an acceptable shelf life and prevent crystallization of the drug, itraconazole requires special, costly formation. Marketing itraconazole as a cocrystal could both reduce costs and increase product stability.

Polymorphs

The previous subsection noted that some drugs are formulated as metastable amorphous solids. Unless care is taken to keep the solid in an amorphous state, the solid will transition to a crystalline form. What may be surprising is that most compounds have multiple crystalline forms, called polymorphs. Each polymorph has its own distinct crystal structure. Different polymorphs can have very different properties, including solubility and rate of dissolving, and therefore can display different pharmacokinetics. This bears repeating: the same compound, if it crystallizes in a different form (polymorph), will likely display different pharmacokinetics. For drugs with a narrow therapeutic window, unexpected changes in pharmacokinetics can quickly lead to severe adverse reactions. Therefore, knowing the full polymorphic profile of a drug is vital for ensuring predictable drug behavior.

Like all solids, different polymorphs are normally obtained through crystallizations. Each polymorph has its own solubility curve. For a compound with two polymorphs, there are two possible scenarios for the solubility curves. Either the solubility curves of the two polymorphs cross (**Figure 13.7.a**) or they do not cross (**Figure 13.7.b**). The lower line in both graphs corresponds to the more stable polymorph at any given temperature. For Figure 13.7.a, the more stable polymorph is temperature dependent. Polymorph B is

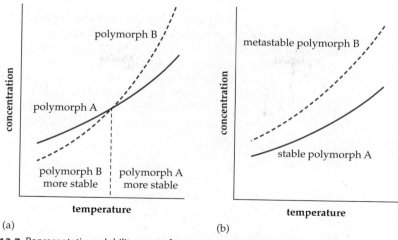

FIGURE 13.7 Representative solubility curves for a compound with two polymorphs[8]

more stable at lower temperatures; polymorph A is more stable at higher temperature. For Figure 13.7.b, polymorph A is the more stable form regardless of temperature. However, the more stable polymorph does not always precipitate from solution. If crystals for the less stable polymorph form first, then the less stable polymorph will likely precipitate preferentially.

Process groups expend considerable effort seeking out different crystal forms of drug candidates. Different polymorphs are often discovered during routine solubility studies as process groups search for an optimal recrystallization solvent. In one such study, a group uncovered 11 different polymorphs, some of which were likely hydrates or other solvates.[8]

Despite careful research and observation, stable polymorphs can suddenly appear during a research program and disrupt the drug discovery and development process.[9] The following Case Study demonstrates this potential problem.

CASE STUDY

Polymorphs of Ritonavir[10]

Ritonavir (Norvir, **13.12**) is an HIV protease inhibitor discovered at Abbott Laboratories in Chicago (**Figure 13.8**). The drug was approved in March 1996 in two formulations, a semisolid capsule and a liquid. In 1998, a number of production lots failed solubility tests. The drug was precipitating out of solution within the semisolid capsules. Research into the problem discovered that a more stable, less soluble polymorph was the cause of the precipitation. The original polymorph was called Form I, and the new, previously unknown polymorph was called Form II.

As Abbott scientists began to travel from lab to lab both in the United States and overseas to understand the Form II problem, all Abbott facilities began

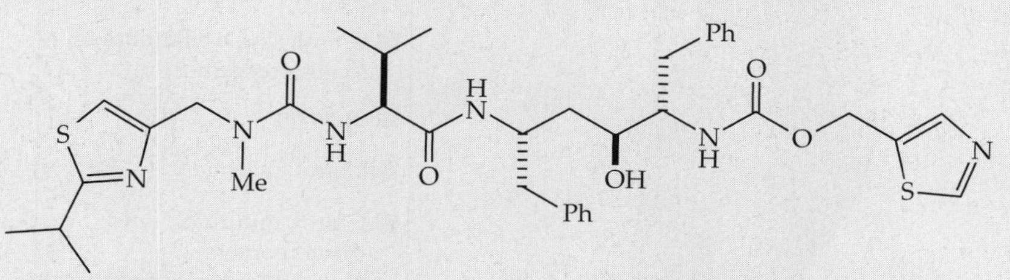

ritonavir
(Norvir)
HIV protease inhibitor
13.12

FIGURE 13.8 Ritonavir, a compound with two polymorphs

to observe contamination by Form II in batches of rito-navir. Although not proven, the prevailing theory is that scientists unknowingly carried trace amounts of Form II to uncontaminated labs. The small amounts of Form II acted as seed crystals and prevented observation of the less stable Form I.

Abbott attacked the Form II problem in two ways. The first was to restore the ability to prepare Form I. The second was a more drastic contingency plan.

If Form I could no longer be prepared reproducibly, then Abbott needed to be able to develop a viable formulation of Form II. Both plans were fully executed. After considerable work, the process group discovered a number of contributing factors to the appearance of Form II. Ultimately, the solution to the problem was a simple one: batches of ritonavir were seeded with Form I crystals to encourage precipitation of the desired polymorph.

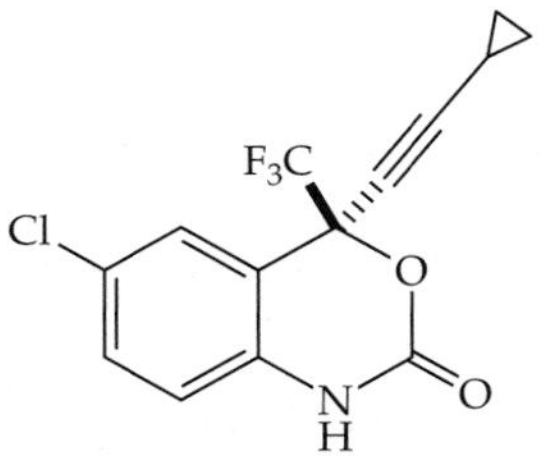

efavirenz
(Sustiva)
reverse transcriptase inhibitor
13.13

FIGURE 13.9

As with cocrystals, heightened awareness of polymorphs is a more recent phenom-enon in the pharmaceutical industry. High-throughput screening techniques for optimal recrystallization conditions greatly decrease the probability of a process group's overlook-ing a key polymorph. In addition to the story of ritonavir, the development of efavirenz (Sustiva, **13.13**), a non-nucleoside HIV reverse transcriptase inhibitor, involved a surprise polymorph discovery (**Figure 13.9**).[8] The new polymorph of efavirenz was discovered be-fore the drug had been approved. The fact that both ritonavir and efavirenz were approved through an accelerated track may be related to their both having a surprising polymorph discovery during their development.

13.2 Stereochemistry

From January 2004 to June 2006, approximately 66% of all drugs approved by the U.S. Food and Drug Administration (FDA) were single enantiomer compounds (**Figure 13.10**).[11] Only 7% of the approved drugs were mixtures of stereoisomers.

FIGURE 13.10 Classification of drugs approved by the U.S. FDA from January 2004 to June 2006[11]

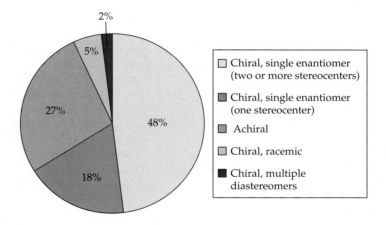

2%

5%

27%

48%

18%

☐ Chiral, single enantiomer (two or more stereocenters)

■ Chiral, single enantiomer (one stereocenter)

☐ Achiral

☐ Chiral, racemic

■ Chiral, multiple diastereomers

Therefore, techniques for preparing single enantiomer compounds are of vital importance for both medicinal and process chemists.

Approaches to single enantiomer compounds can be divided into two categories. One, the product may be formed as a racemic mixture with the desired enantiomer being separated afterward. This is a *resolution* approach. Two, a synthetic route may be designed to prepare mostly, if not exclusively, a single enantiomer. This technique is called *asymmetric synthesis*.

The purity of a mixture of enantiomers can be expressed as a simple ratio or, more commonly, as a percent *enantiomeric excess*, or e.e. The formula for converting a known ratio to a percent e.e. is shown in Equation 13.1.

$$\text{e.e.} = 100 \times \left(\frac{\text{major} - \text{minor}}{\text{major} + \text{minor}} \right) \qquad (13.1)$$

For example, a mixture of enantiomers present in a ratio of 5:1 would have an e.e. of 67%. A 19:1 mixture would have an e.e. of 90%.

Resolution

Resolution is the traditional approach to separating enantiomers from a racemic mixture. Resolutions operate by allowing both enantiomers in a racemic mixture to interact with a single-enantiomer structure (**Scheme 13.1**). The single-enantiomer compound is called the resolving agent. When an enantiomer from a racemic mixture and a resolving agent interact, they form a diastereomeric complex. Each of the two enantiomers will form a different diastereomeric complex with the resolving agent. Unlike enantiomers, diastereomers do have different properties. The different properties of the diastereomeric complexes may be exploited to distinguish and separate the two enantiomers of the original racemic mixture. Exactly *how* the properties are exploited determines the type of resolution that is being used.

The mass recovery of a resolution is no more than 50% since half of the original mixture consists of an unwanted enantiomer. Because so much material is lost in a resolution, process chemists try to place a resolution as early in a synthetic route as possible. An early resolution minimizes the amount of time, effort, and reagents expended on carrying the unwanted enantiomer through the synthesis.

Diastereomeric Salts

The "old school" method for resolving a racemic mixture is through formation of diastereomeric salts. Salts form through reaction of a racemic drug with an enantiomerically pure acid or base (**Scheme 13.2**). Of course, the drug must contain a basic or acidic functional group, normally an amine or carboxylic acid, respectively. The salt product is

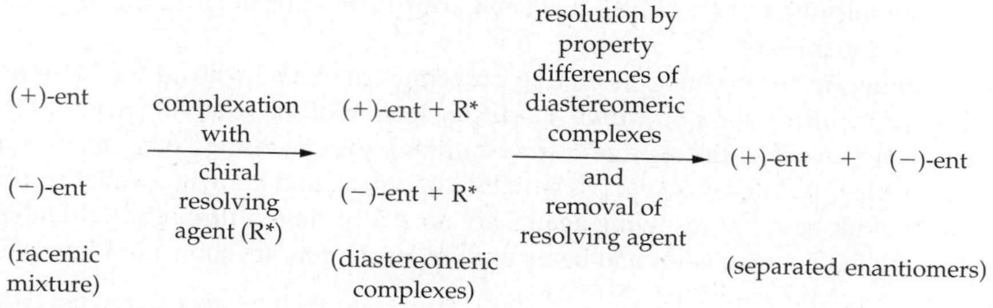

SCHEME 13.1 Separation of enantiomers by resolution

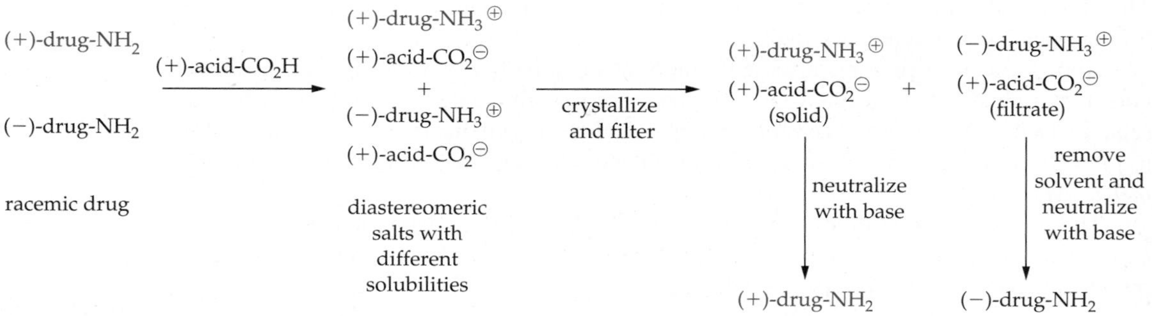

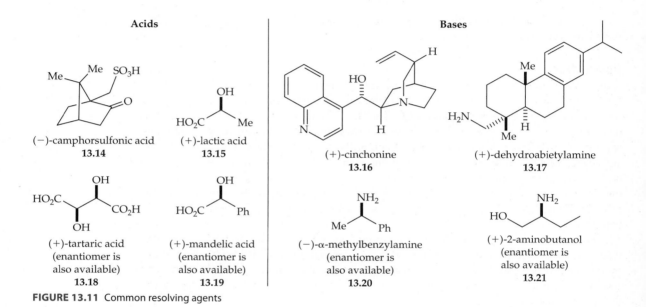

SCHEME 13.2 Resolution of a basic drug through diastereomeric salts

Acids | **Bases**

(−)-camphorsulfonic acid
13.14

(+)-lactic acid
13.15

(+)-cinchonine
13.16

(+)-dehydroabietylamine
13.17

(+)-tartaric acid
(enantiomer is
also available)
13.18

(+)-mandelic acid
(enantiomer is
also available)
13.19

(−)-α-methylbenzylamine
(enantiomer is
also available)
13.20

(+)-2-aminobutanol
(enantiomer is
also available)
13.21

FIGURE 13.11 Common resolving agents

a mixture of diastereomers. Diastereomers have different physical properties, including solubility. Under the correct conditions, the salts may be separated by recrystallization. Ideally, one complex precipitates from solution, and the other remains dissolved. A filtration separates the two diastereomers. The recovered solid is one diastereomer, and the other diastereomer is in the filtrate. Recovery of the enantiomers requires neutralization of the resolving acid or base. In the resolution in Scheme 13.2, the diastereomeric salt containing the (+)-drug has been arbitrarily chosen to be the less soluble diastereomeric complex.

Resolutions are not as simple as mixing a racemic compound with an acid or base and collecting the resulting solid or filtrate. Correct selection of the resolving acid or base is crucial, as is the choice of the appropriate crystallizing solvent. Some experienced process chemists have a knack for looking at a structure of a drug and knowing which resolving agent might work best, but resolving agents are normally chosen through trial-and-error experimentation. Common acids and bases used in resolutions are shown in **Figure 13.11**. All are commercially available.

A Case Study of a resolution of a commercial drug highlights the procedural steps of the entire resolution process.

Resolution of Sertraline[12]

Sertraline is an antidepressant that is sold as a single enantiomer and prepared commercially through a resolution of a racemate. The starting material for the resolution step is the hydrochloride salt of racemic sertraline (**13.22**) (**Scheme 13.3**). The hydrochloride salts are first neutralized through the addition of NaOH to form the racemic free amine (**13.23**). The amine is then reacted with 1.0 equivalents of D-(−)-mandelic acid (**13.24**) in ethanol to form diastereomeric salts. The solution of the salts in ethanol is allowed to stand at room temperature, and a solid consisting of predominantly diastereomer **13.25** precipitates. Filtration of the solid followed by recrystallization from ethanol affords essentially pure **13.25**. Salt (**13.25**) is then neutralized with NaOH and reacted with HCl to provide the hydrochloride salt of sertraline (Zoloft, **13.26**).

Scheme 13.3 reveals how much material is lost during the entire resolution process. The resolution has an overall yield of 39%. The number 39% might seem low, but remember that half of the material that enters a resolution is the undesired enantiomer. A 39% yield in a resolution corresponds to a recovery of 78% of the desired enantiomer. Note that the various transformations in Scheme 13.3 are typically very efficient. Based on the number of moles of material at each point in the process, only the resolution step has a significant drop.

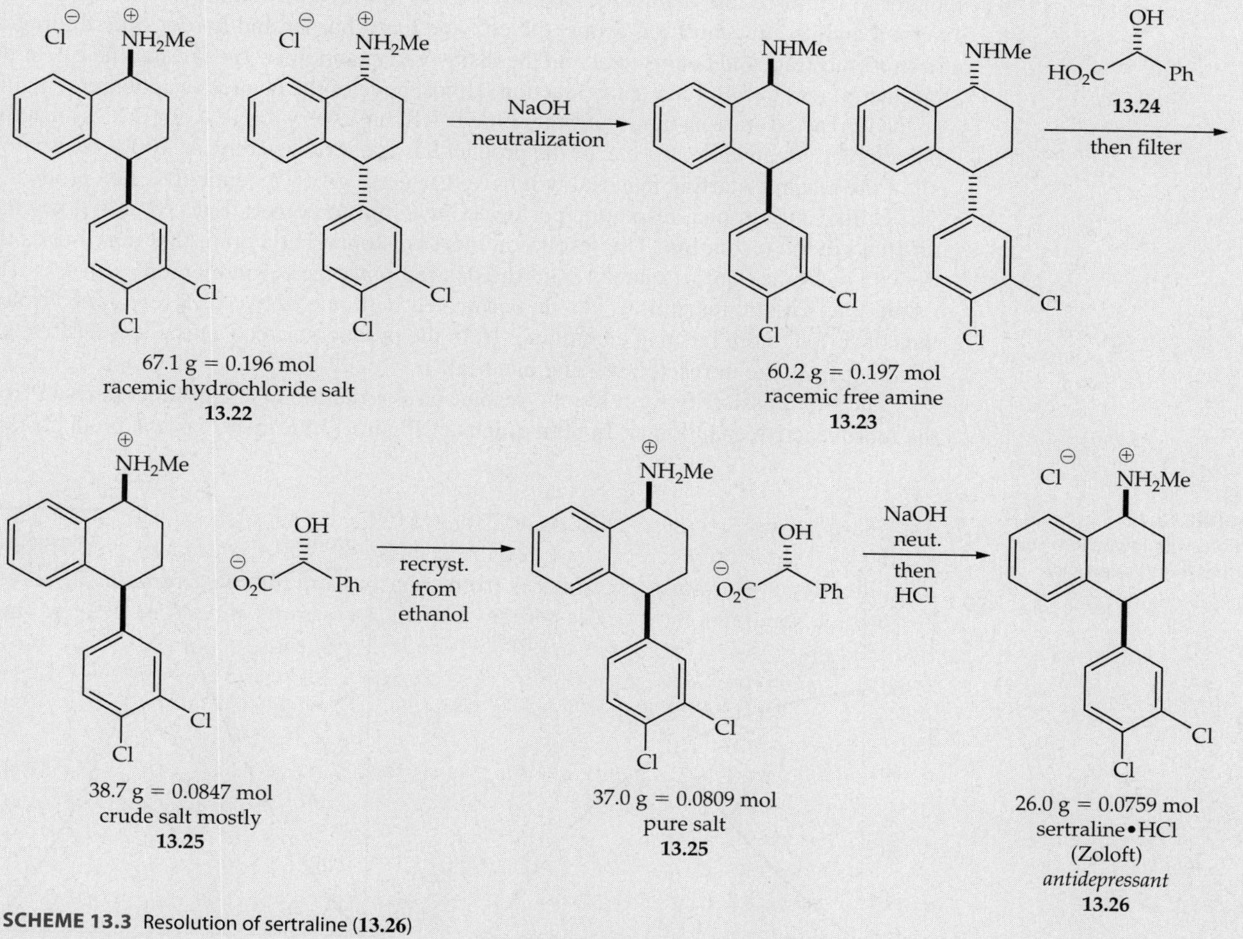

67.1 g = 0.196 mol
racemic hydrochloride salt
13.22

60.2 g = 0.197 mol
racemic free amine
13.23

38.7 g = 0.0847 mol
crude salt mostly
13.25

37.0 g = 0.0809 mol
pure salt
13.25

26.0 g = 0.0759 mol
sertraline•HCl
(Zoloft)
antidepressant
13.26

SCHEME 13.3 Resolution of sertraline (**13.26**)

Kinetic Resolution

A kinetic resolution is a chemical reaction in which one enantiomer of a racemate reacts faster than the other. Most kinetic resolutions of pharmaceutical compounds are catalyzed processes. Catalysts used in a kinetic resolution must be chiral. Binding of a chiral catalyst with a racemic material can form two different diastereomeric complexes. Since the complexes are diastereomers, they have different properties: different rates of formation, stabilities, and rates of reaction. The products form from the diastereomeric substrate-catalyst complexes at different rates. Therefore, a chiral catalyst is theoretically able to separate enantiomers by reacting with one enantiomer faster than the other. The catalysts used in kinetic resolutions are often enzymes. Enzymes are constructed from chiral amino acids and often differentiate between enantiomeric substrates.

Kinetic resolutions are very interesting graphically and can be analyzed from the perspective of both the starting material and the product. **Figure 13.12** shows the progress of two kinetic resolutions, one with a less selective enzyme (left) and another with a more selective enzyme (right). When the reaction starts (0% conversion), the starting material is racemic. As the reaction progresses, the favored enantiomer reacts preferentially. The initial product is derived mostly from the favored enantiomer, while the unreacted starting material is mostly the disfavored enantiomer. As the favored enantiomer is rapidly consumed and its concentration drops, the enzyme has a harder and harder time finding its desired substrate and begins to act on the disfavored enantiomer. As a result, the e.e. of the product begins to fall. Late in the reaction, almost none of the favored enantiomer remains, so the unreacted starting material, what little is left, has a very high e.e. for the less reactive enantiomer. Meanwhile, the e.e. of the product has dropped severely. At 100% conversion, all of the racemic starting material will have reacted to give a racemic (0% e.e.) product.

Unlike a traditional resolution performed with diastereomeric salts, time is a key factor in a kinetic resolution. The resolution must be stopped at a point that maximizes the desired outcome. Unfortunately, even the desired outcome is more complex in a kinetic resolution. Kinetic resolutions can be performed with one of two different goals: isolate the converted, more reactive enantiomer from the product, or recover the less reactive enantiomer from the unreacted starting material.

The first possibility for a kinetic resolution is isolation of the product derived from the more reactive enantiomer. In both graphs of Figure 13.12, the e.e. of the product starts

FIGURE 13.12 Kinetic resolutions with lower (left) and higher (right) selectivity

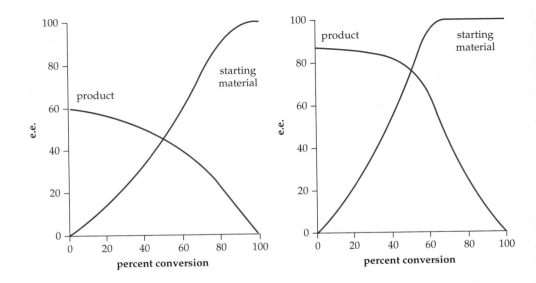

at its highest level and drops over time. If the kinetic resolution is stopped very early, the product e.e. will be at its highest, but the yield will be very low. Letting the reaction run longer will increase the yield (quantity), but the e.e. (quality) of the product will be lower. This is a tradeoff, and the correct decision depends upon the exact needs of the synthesis. With pharmaceuticals, a high e.e. ($\geq$98%) is vital. Note that neither graph in Figure 13.12 shows the possibility of obtaining a product even close to 98% e.e. at any percent conversion. Therefore, neither enzyme is suitable to generate product of high e.e. for a drug synthesis. This paragraph has only discussed the *product* of a kinetic resolution. Tracing the starting material tells a very different story.

The second possibility for a kinetic resolution is isolation of the less reactive enantiomer from the unreacted starting material. This option is perhaps nonintuitive. Normally one runs a reaction to get the product, not recover unreacted starting material. Regardless, recovering starting material is often the most viable option for a kinetic resolution. As a resolution progresses, the enzyme preferentially seeks out the more reactive enantiomer. The result is that the e.e. of the unreacted starting material rises with time and approaches 100% e.e. A higher e.e. is always better, but the yield-e.e. (quantity-quality) tradeoff is still an issue. A higher percent conversion means less starting material may be recovered.

The starting material graph for the resolution on the right of Figure 13.12 shows that at 65% conversion, the e.e. of the unreacted starting material is close to 100%. A 65% conversion equals 35% recovered starting material. A 35% yield may not seem very high, but as with any resolution, the maximum yield of a kinetic resolution is 50%. Note that neither resolution in Figure 13.12 was attractive for forming high e.e. *product* in acceptable yield, but the more selective enzyme is more than adequate for recovery of *unreacted starting material*.

Kinetic resolution is a powerful technique, but it does require a catalyst that can select between the two enantiomers of interest. Hydrolytic enzymes (lipases, hydrolases, esterases) are the most common examples. Two single enantiomer drugs that have been prepared by kinetic resolution are abacavir (**13.28**) and lamivudine (**13.30**) (**Scheme 13.4**). The preparation of abacavir starts with the corresponding racemic 5′-phosphate (**13.27**). Snake venom nucleotidase hydrolyzes the 5′-phosphate with very high selectivity for the enantiomer that corresponds to abacavir (**13.28**).[13] In this example, the enzymatic product is the desired compound, so the reaction is stopped early ($<$50% conversion). The resolution of lamivudine starts with racemic lamivudine (**13.29**). Cytidine deaminase from *E. coli* selectively hydrolyzes the unwanted enantiomer.[14] Therefore, the reaction is allowed to proceed beyond 50% to provide the recovered starting material in as high an e.e. as possible. The yields of both resolutions are relative to the racemic starting material, so the 40% yield of lamivudine corresponds to recovery of 80% of the desired enantiomer from the starting material.

Chiral Chromatography

Chiral chromatography is a form of liquid chromatography in which the stationary phase consists of a chiral material. Racemic mixtures are resolved because one enantiomer has a higher affinity for the stationary phase than the other. Chiral chromatography is almost always found in the form of high-performance liquid chromatography (HPLC). Although HPLC separations are not effective for resolving large amounts of material, resolutions by HPLC are often used on small quantities in early investigations on a drug candidate. For example, small amounts of separated enantiomers would be useful to see if one enantiomer of a lead is more potent or has better pharmacokinetic properties than the other enantiomer.

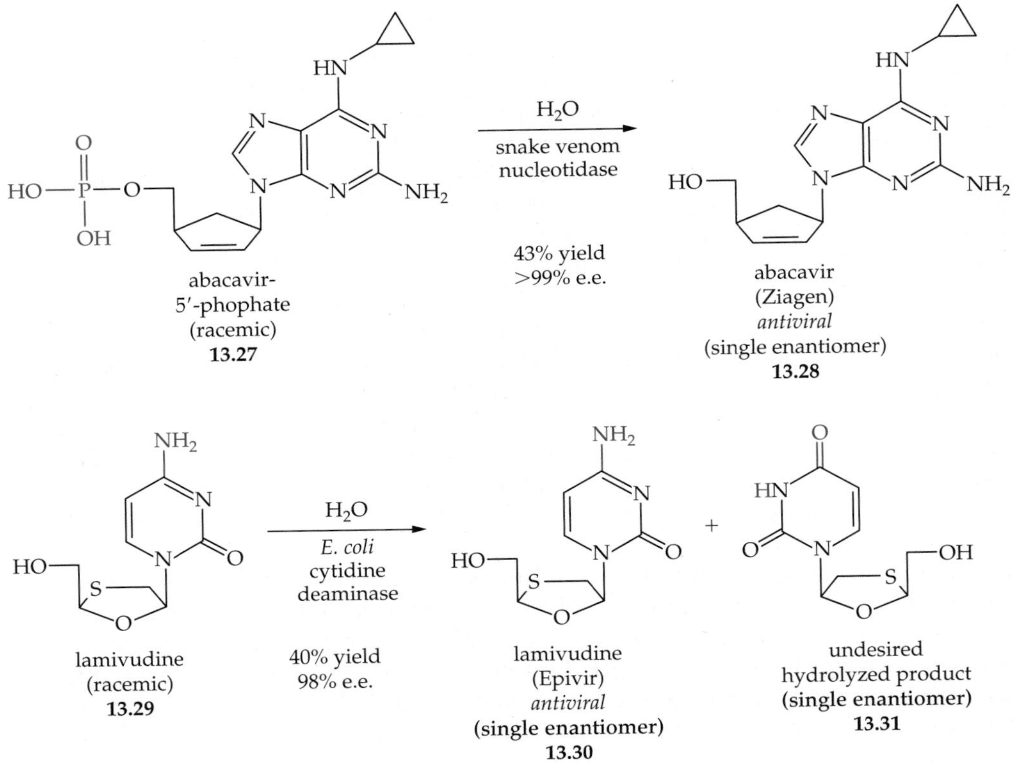

SCHEME 13.4 Kinetic resolution approaches to abacavir (**13.28**) and lamivudine (**13.30**)[13,14]

Chromatography is prohibitively expensive to perform on a large scale because massive amounts of solvent are required and results in increased cost and waste. Recrystallizations, such as those used for the resolution of diastereomeric salts, also use solvents but not nearly as much as chromatographic techniques. Chiral chromatography is not a viable method for performing a resolution on a commercial scale.

Asymmetric Synthesis

An asymmetric approach to a molecule strives to synthesize a single enantiomer. While resolutions of racemic mixtures are limited to a 50% yield, an asymmetric synthesis can theoretically be accomplished in 100% yield.

Enantiomerically Pure Starting Materials

One way to make a single enantiomer product is to buy starting materials that contain the stereocenter of interest in an optically pure form. Technically, starting a synthesis with intact stereocenters does not qualify as asymmetric synthesis. This approach, however, is an excellent method for obtaining a single enantiomer drug. Available and affordable optically pure starting materials are part of the *chiral pool*. The resolving agents in Figure 13.11 are part of the chiral pool. Stationary phase materials for chiral chromatography are also part of the chiral pool. Other common examples include amino acids and sugars. Most chiral pool reagents are derived from natural sources.

Valsartan (Diovan, **13.35**), a medication for high blood pressure, is a single enantiomer drug. The stereocenter in valsartan is derived from L-valine (**13.33**) and incorporated early in the synthesis by a simple S_N2 alkylation (**Scheme 13.5**).[15]

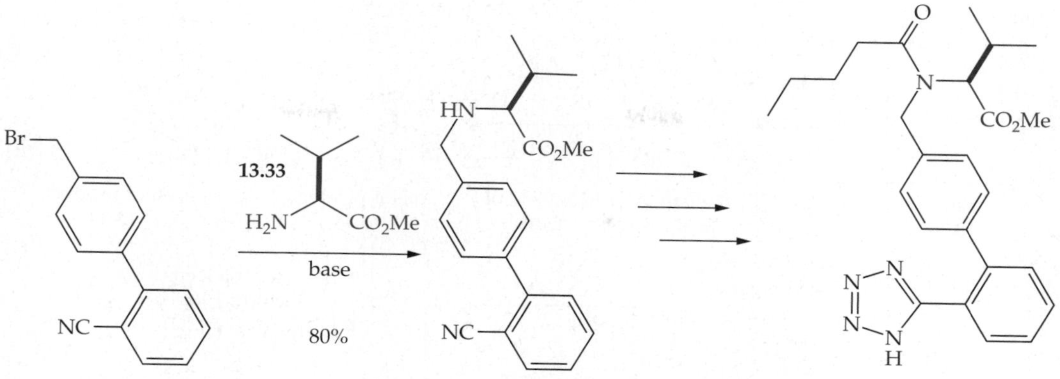

SCHEME 13.5 Synthesis of valsartan[15]

Chiral Auxiliary

A chiral auxiliary is a temporary chiral group on a molecule that directs the stereochemical outcome of a reaction on another part of the molecule. Almost all synthetic routes involving chiral auxiliaries follow three steps: (1) covalently attach the auxiliary to the molecule, (2) perform the reaction that forms the new stereocenter, and (3) remove the auxiliary. Under this model, use of a chiral auxiliary adds two steps to a synthesis because the auxiliary must be added and removed. Additional steps in a synthetic scheme require additional time, increase cost, and decrease the overall yield. Despite these disadvantages, chiral auxiliaries are fairly common in drug synthesis.

Atorvastin (Lipitor, **13.40**) is a cholesterol-lowering drug that has been synthesized as a single enantiomer through use of a chiral auxiliary (**Scheme 13.6**).[16] Ester **13.36** contains the auxiliary, a chiral alcohol. Deprotonation of the ester forms an enolate (**13.37**). The enolate then attacks an aldehyde. The asymmetry of the stereocenter on the auxiliary causes the reaction to favor stereoisomer **13.38** over **13.39**. Several recrystallizations are required to obtain **13.38** in high enantiomeric excess. Cleavage of the auxiliary from **13.38** and further manipulations of the side-chain afford atorvastin.

Asymmetric Catalysis

Asymmetric catalysis is the gold standard of asymmetric synthesis. Most synthetic catalysts contain metal ions that are electron deficient and Lewis acidic. The Lewis acidic metal binds to ligands with Lewis basic functional groups, including alcohols, carbonyls, and amines. If the ligand is chiral, then the entire metal-ligand complex is also chiral. The complex is an asymmetric catalyst. It carries around both a chiral environment (the ligand) and a reactive center (the metal). When the catalyst performs a reaction, it preferentially forms one enantiomer over another. "Preference" is loosely defined and can equate to a 2:1 or 100:1 ratio. A larger ratio is better. Simply changing the stereochemistry of the ligand reverses the enantioselectivity of the catalyst.

Like an enzyme, an asymmetric catalyst binds its substrate, performs a reaction, and releases the product: three steps. Chiral auxiliaries have three analogous steps: attach, react, and cleave. The advantage of an asymmetric catalyst over an auxiliary is that binding to the substrate is reversible and involves weak, intermolecular forces instead of covalent bonds. Binding and release of the substrate occur in the same reaction vessel as the stereocenter-forming reaction.

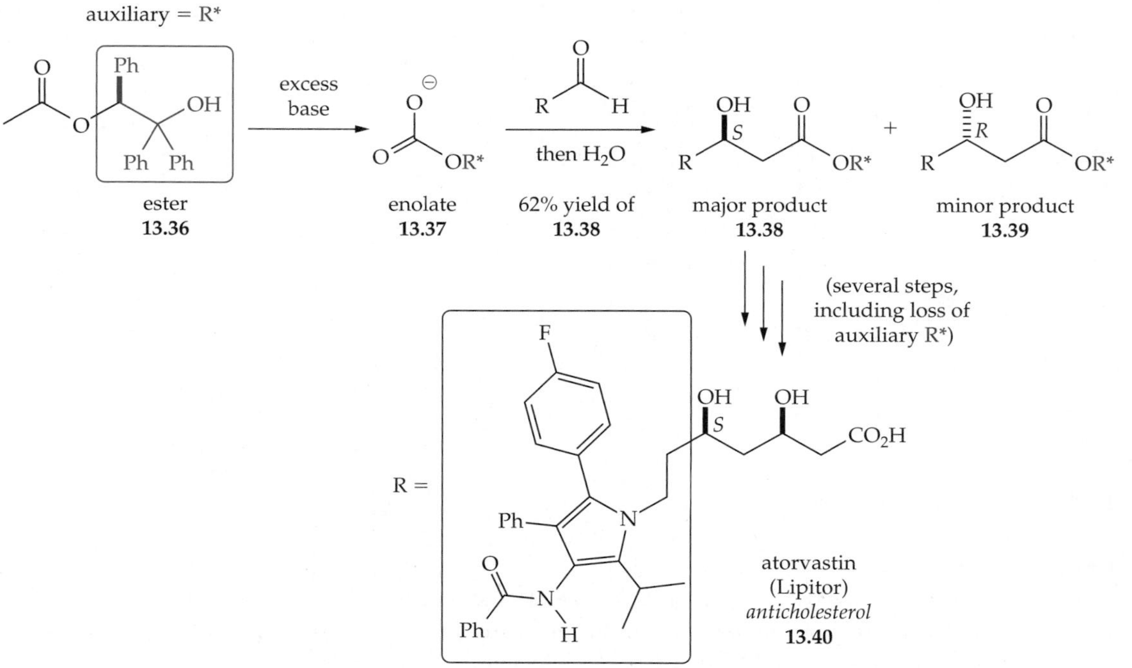

SCHEME 13.6 Synthesis of atorvastin[16]

The asymmetric aspect of the catalyst depends on the chiral ligand. Chiral ligands are often derivatives of compounds from the chiral pool. Selected ligands are shown in **Figure 13.13**. Finding an optimal ligand depends on the metal in the catalyst as well as its oxidation state. Determining an optimal ligand is just one aspect of research in asymmetric catalysis. Many asymmetric catalysts have been developed and continue to be investigated. Because metals generally bind well to alkenes and other π-bonds, most asymmetric catalysts facilitate addition reactions on alkenes or carbonyl compounds.

Esomeprazole (Nexium, **13.45**), a proton-pump inhibitor, is marketed as a single-enantiomer drug under the name Nexium (**Scheme 13.7**).[17] The diethyl ester of (+)-tartaric acid (**13.43**, R = ethyl) serves as a chiral ligand for the titanium catalyst, and hydroperoxide is the stoichiometric oxidant. Because of the chiral environment created by the (+)-tartrate ligand, the catalyst selectively adds an oxygen atom to just one of the lone pairs to form a new stereocenter at the sulfur atom.

FIGURE 13.13 Representative chiral ligands

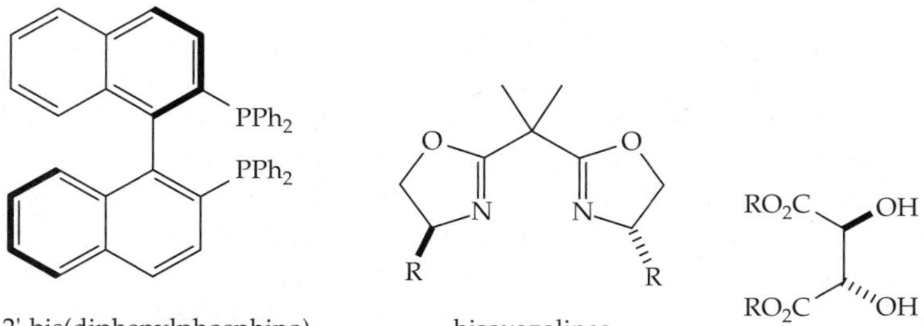

2,2'-bis(diphenylphosphino)-
1,1'-binaphthyl (BINAP)
13.41

bisoxazolines
(BOX)
13.42

tartrate esters
13.43

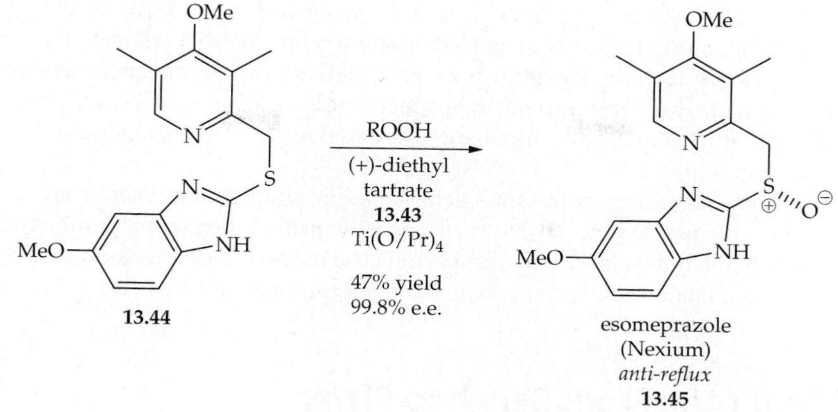

SCHEME 13.7 Asymmetric synthesis of esomeprazole[17]

Microbial Processes

Microbiological organisms are magnificent chemical factories and contain enzymes for a vast number of reactions. Sometimes the enzymes in an organism can be exploited to perform a useful synthetic transformation. One example includes a commercial route to captopril (Capoten, **13.48**), a medication for high blood pressure (**Scheme 13.8**).[18] In this reaction, *Candida utilis*, a yeast, stereoselectively adds water to the alkene of an α, β-unsaturated acid. Although the yield is low, the reaction is the first in the synthesis. Also, the starting material, methacrylic acid (**13.46**), is very inexpensive. Isolating products from microbiological fermentations can be challenging. The bacteria or fungi in the process often contribute significant amounts of impurities to the mixture. Formally, microbial processes are a sub-class of asymmetric catalysis.

A strain of *Rhizopus nigricans* performs a key transformation in the synthesis of hydrocortisone (**13.51**) (**Scheme 13.9**).[19] The mold introduces a hydroxyl group

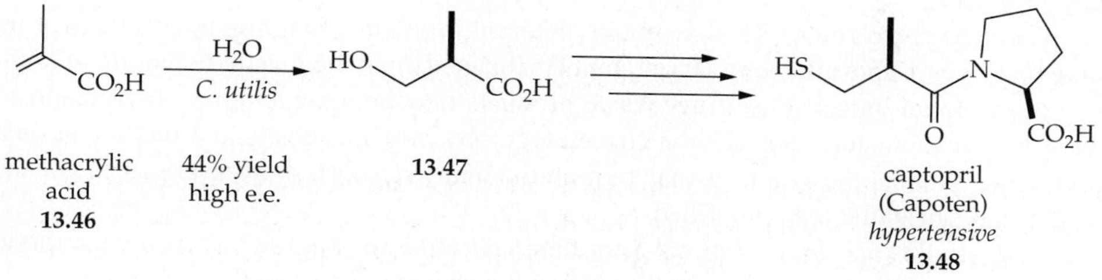

SCHEME 13.8 Fermentation approach to captopril[18]

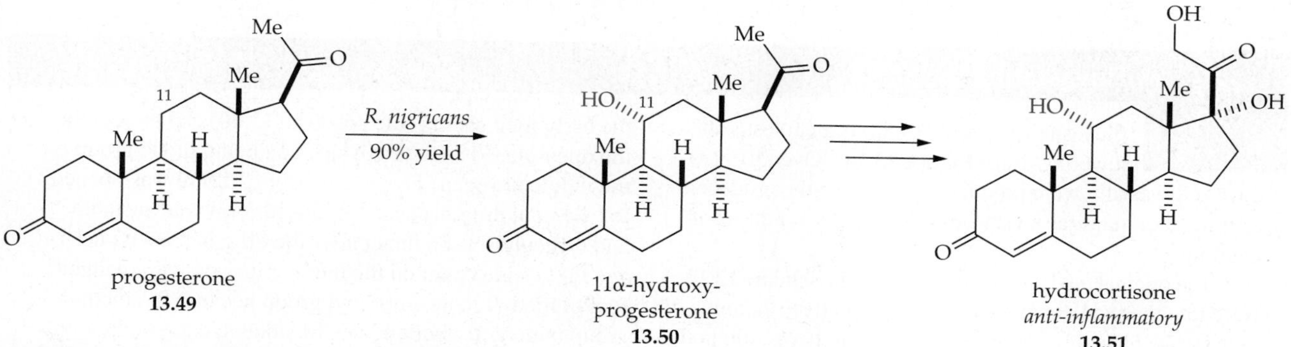

SCHEME 13.9 Fermentation approach to hydrocortisone[19]

with exclusively *S*-stereochemistry at C11 of progesterone (**13.49**) in 90% yield. The resulting alcohol (**13.50**) is an intermediate toward hydrocortisone. This transformation does not technically qualify as asymmetric synthesis because the mold is selecting between two diastereomeric products, not enantiomers. Regardless, without assistance of the mold, this single-step, stereoselective transformation of **13.49** to **13.50** would be impossible.

In summary, regardless of the approach, preparing single enantiomer drugs is a true challenge for synthesis. Despite advances in synthetic methodology, not all drug syntheses are amenable to the implementation of an asymmetric method. Diastereomeric salt resolutions, for all their challenges, remain common in pharmaceutical chemistry.

13.3 Scale-Up: From Bench to Plant

The synthetic pathway developed by the medicinal chemistry team is called the med chem route or research route. A med chem route is normally reliable and effective for producing gram-scale quantities of the candidate. This scale can satisfy all the in vitro screens for binding and pharmacokinetics. Material for early animal testing is also normally provided by the medicinal chemistry group. At some point, however, the med chem route is passed off to the process group. The process group tweaks or completely reworks the synthesis until it can consistently provide the necessary amount of material for continued testing. Much larger quantities of drug are required for clinical trials. The process group is often referred to as process research and development to better reflect its innovative contributions to developing a marketable drug.

A med chem route is very rarely amenable to scale-up. This statement is not a condemnation of the chemistry skills of the medicinal chemistry group. Medicinal chemists make molecules for testing and optimization. How the molecules are made is less important than the fact that they are made and made quickly. Only when a compound shows sufficient promise as a candidate is time invested to significantly improve the efficiency of its synthesis.

The med chem route can pose many different problems. Starting materials may be too expensive or unavailable in large quantity. Reagents may be unsafe to handle in large quantity or in an industrial setting. Waste products may be toxic and have high disposal costs. Purification steps may rely on chromatography, which is costly and inefficient on a large scale. Finally, the synthesis may require too many steps. In short, every aspect of the synthesis is fair game to be modified.

Case Studies are the best method for demonstrating the degree to which a synthetic route may be modified on its way to becoming the process or commercial route.

CASE STUDY

Fluvastatin[20,21]

Fluvastatin (Lescol, **13.52**) is a cholesterol lowering drug (**Figure 13.14**). Over 50 tons of fluvastatin were produced in 2007, so the producer of fluvastatin requires a very high-throughput synthesis of the drug.[20]

The original med chem route is shown in **Scheme 13.10**.[21] Note that the med chem route produces a close analogue of fluvastatin. Structural differences between fluvastatin and the early med chem route product (**13.60**) can be seen in the open-closed state of the side chain and the R-group on the indole nitrogen. The cyclic ester of **13.60** was opened for the final drug because the closed form was unstable and difficult to keep intact after the drug had been packaged. The side chain on the indole nitrogen was changed from a methyl to an isopropyl group because the methyl group undergoes facile Phase I oxidative demethylation in

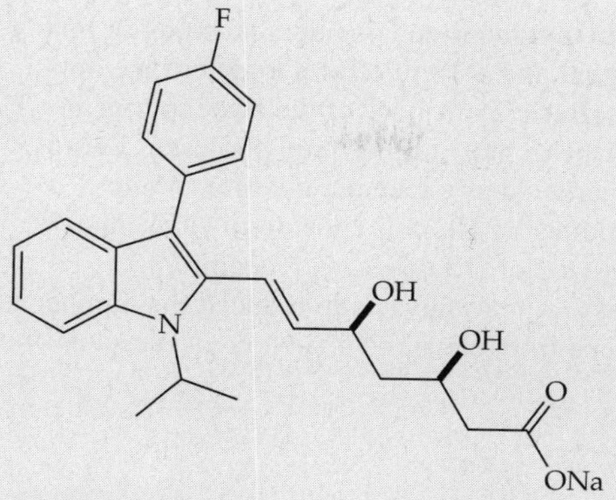

FIGURE 13.14

fluvastatin
(Lescol)
anticholesterol
(racemic)
13.52

the body. The bulkier isopropyl group resists dealkylation and provides a longer, more favorable half-life for the drug.

The synthesis in Scheme 13.10 has many obstacles for scale-up. The main problems can be attributed to the two halves of the molecule. Many reagents leading up to indole **13.55** are toxic or expensive. The side chain off C2 of the indole in **13.59** requires many steps to assemble, and **13.59** is formed as a mixture of diastereomers. The diastereomers are not separated until the lactone ring of **13.60** is closed. Furthermore, separation of the diastereomers requires chromatography and occurs in very low yield (15%).

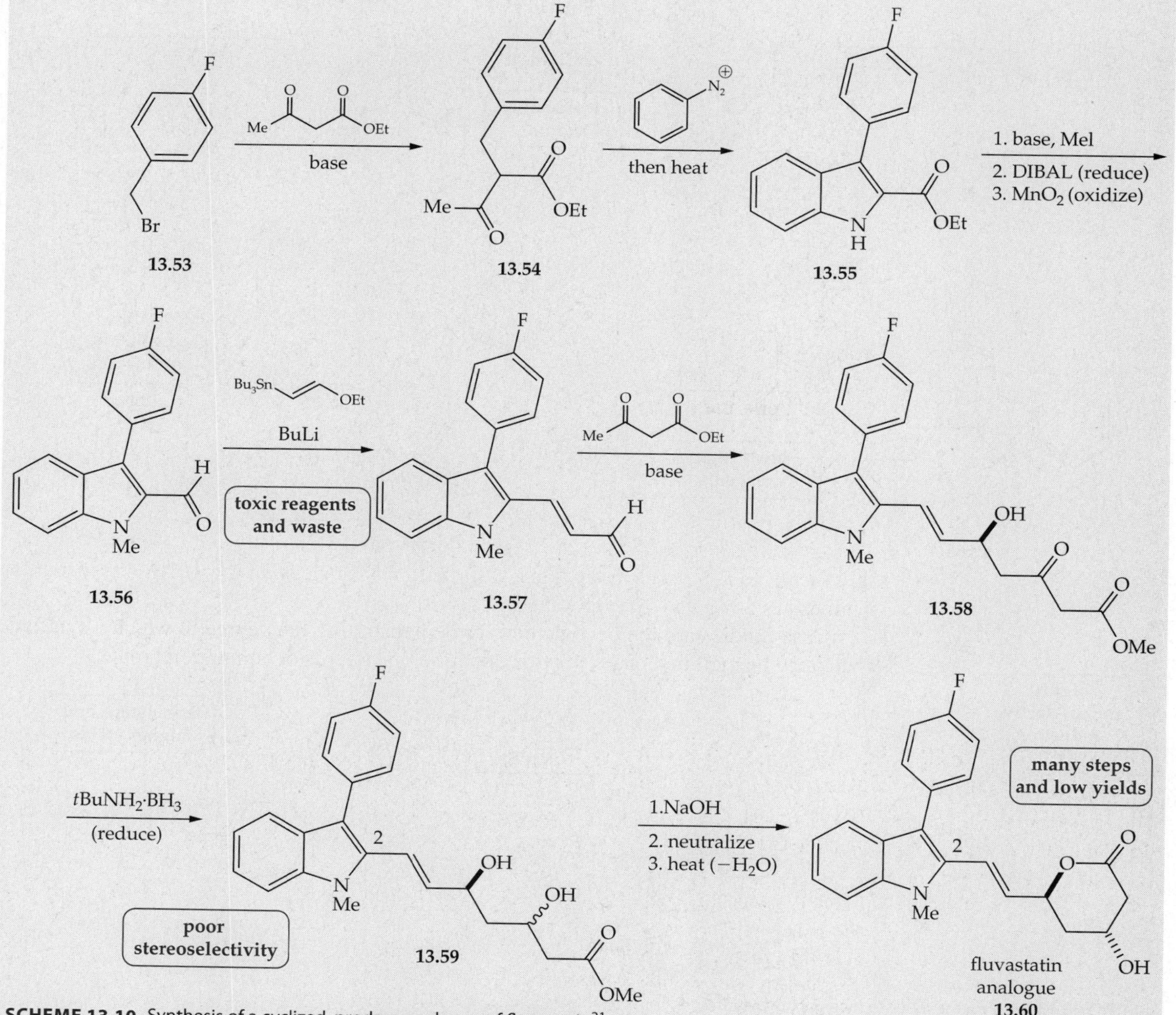

SCHEME 13.10 Synthesis of a cyclized, prodrug analogue of fluvastatin[21]

The process or commercial route addressed problems in the med chem route by completely redesigning the indole synthesis (**Scheme 13.11**).[21] Indole **13.63** is prepared in just two steps. In two more steps, all the carbons in the molecule of the drug are present. Ketone **13.65** can be reduced with almost complete stereoselectivity for the *syn* diol shown as **13.66**. The *anti* diol, which is not pictured, has the new OH group in the "back" configuration. Improvement in this reductive step was a significant

accomplishment and required considerable study of both the reducing agent and solvent effects to determine optimal conditions. Basic hydrolysis of the ester completes the synthesis. Unlike most of the statin cholesterol drugs, fluvastatin is marketed as a racemic mixture. While only one enantiomer of **13.65** is shown, it is present as a racemate. The *syn* diol (**13.66**) is also racemic.

The process route to fluvastatin reduced the number of synthetic steps from eleven to six. None of the steps

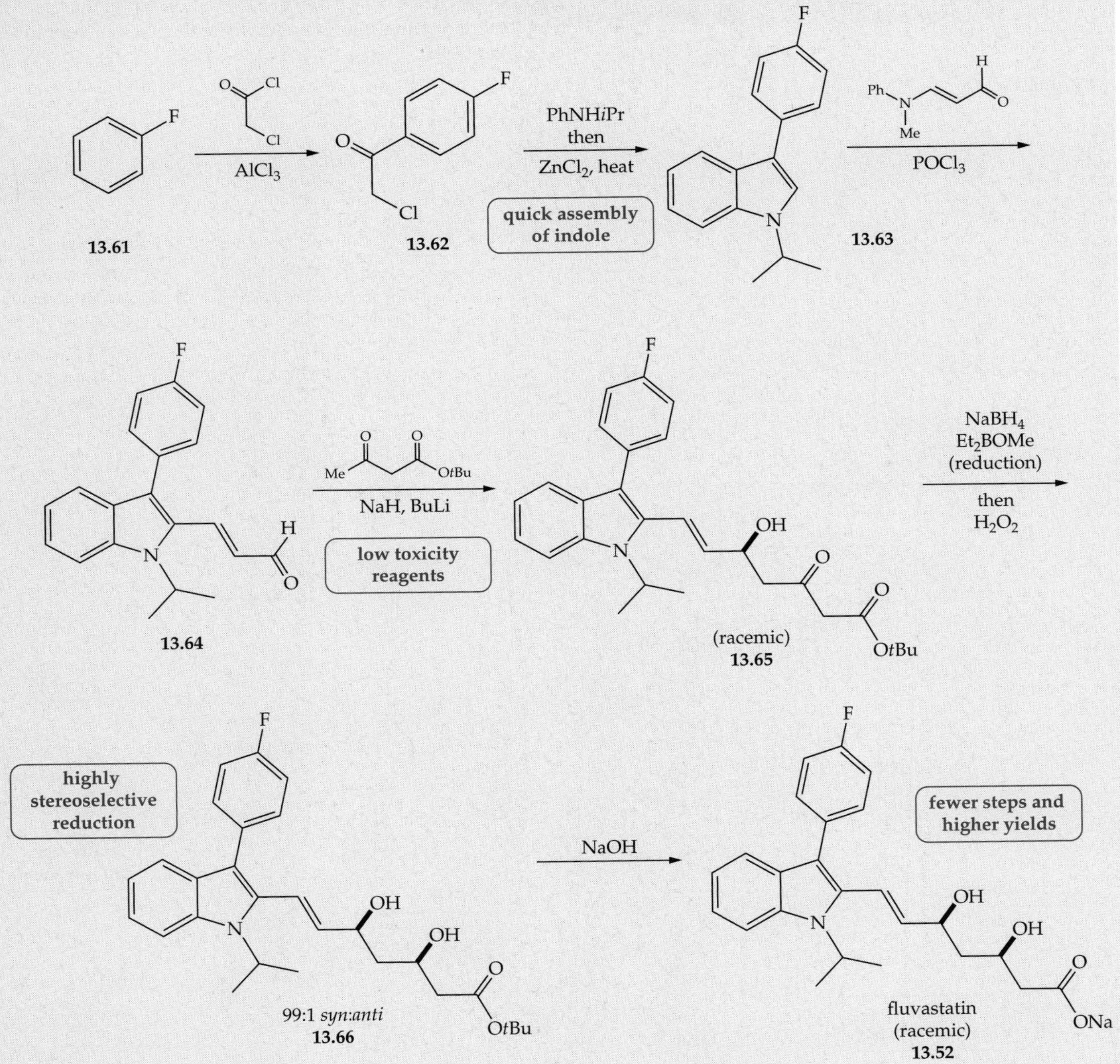

SCHEME 13.11 Process route to racemic fluvastatin[21]

required chromatographic purification. By taking into account improvements in the yield of each step and reduced costs, the process route produced fluvastatin at a cost of approximately 7% relative to the med chem route. Transformation of **13.63** to **13.66** has continued

to receive attention. The latest reported improvements eliminate the need to isolate **13.65** and boost the overall yield by 25%.[20] These gains in yield were realized by analyzing the reaction side products, which indicated that lower reaction temperatures are necessary.

CASE STUDY

Sildenafil[22]

Sildenafil (Viagra, **13.73**) is the prototype of a drug class that treats erectile dysfunction. The original med chem route for sildenafil is shown in **Scheme 13.12**. From pyrazole **13.67**, the synthesis occurs in 7.5% yield. Sulfonyl chloride **13.72** was identified as the major problem in the sequence. Compound **13.72** reacts slowly with water, and loss of product to hydrolysis is unavoidable during isolation. Also, sulfonyl chloride **13.72** is toxic. In the conversion of **13.72** to **13.73**, unreacted **13.72** contaminates the final product. Complete removal of **13.72** from sildenafil requires multiple recrystallizations, which further decrease the overall yield.

Forming a sulfonyl chloride in the synthetic route is inevitable, so the process group sought to place the

problematic compound as early in the sequence as possible. As a result, the yield losses to hydrolysis would be early in the synthesis. Also, toxic impurities from the sulfonyl chloride would have more opportunities to be removed in later purifications instead of being formed in the very last step. Reordering the route addressed both of these problems, and the optimized commercial route is shown in **Scheme 13.13**.

In addition to moving the sulfonyl chloride earlier in the synthesis, the commercial route also uses an acyl imidazole to form a key amide bond (**13.77 → 13.79**) instead of an acid chloride (**13.67 → 13.70**, Scheme 13.12). Carbonyldiimidazole (**13.76**) is an expensive reagent, but the use of acyl imidazole **13.77** gives many advantages. The yields of the amide-forming reaction are higher, and solvent requirements are greatly simplified. The overall

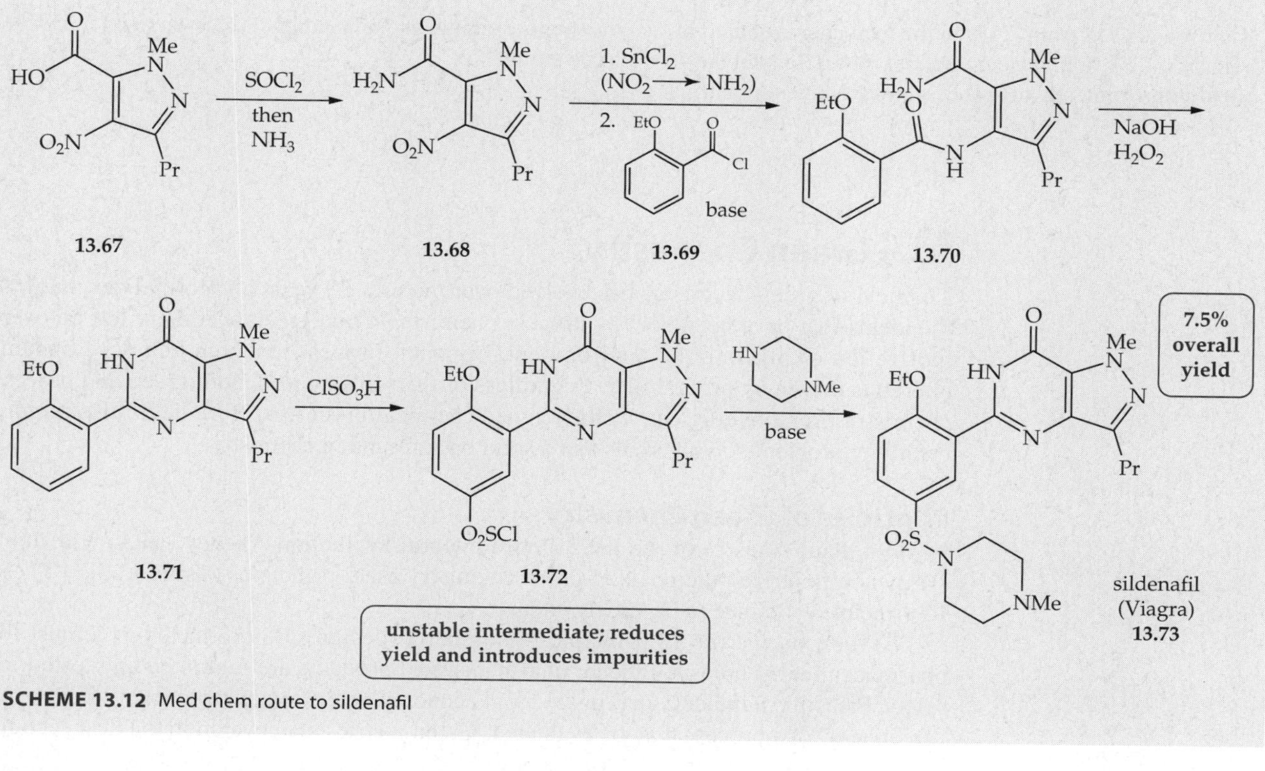

SCHEME 13.12 Med chem route to sildenafil

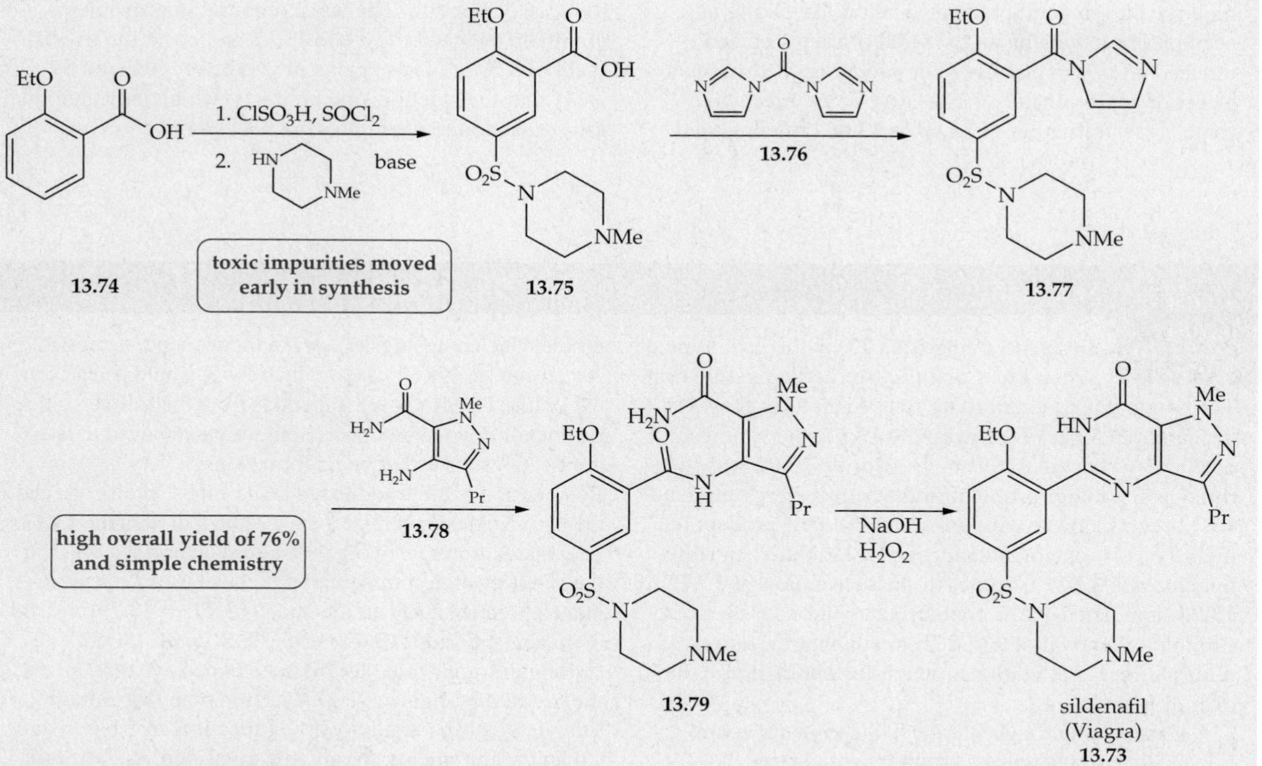

SCHEME 13.13 Commercial route to sildenafil

yield was raised from 7.5% in the med chem route to a remarkable 76% in the commercial route. Beyond the yield improvements, the process route is operationally much more simple and adaptable to scale-up in an industrial setting.

13.4 Green Chemistry

The field of green chemistry has evolved over the past 20 years and formalizes many of the ideals that have been held by process chemists. While green chemistry has an overt motivation of minimizing environmental impact, process research and development (R&D) is driven by profitability. Regardless of their differences, both green and process chemistry promote efficient, well-designed chemical processes. Both fields also require creativity problem-solving skills and a solid background in chemistry.

Principles of Green Chemistry

In 1998, Paul Anastas of the U.S. Environmental Protection Agency (EPA) with John Warner summarized the goals of green chemistry as 12 principles (**Table 13.3**).[23] These 12 principles continue to be widely cited.

At least one of the 12 principles—number 2, design safe products—is a must for pharmaceutical synthesis. Indeed, pharmaceutical products are regulated and tested for safety. Principle number 7, maximize atom economy, refers to a metric against which the "greenness" of a process may be evaluated. Green metrics are covered in the next section.

TABLE 13.3 Twelve principles of green chemistry[23]

Principle	Goal
1	Prevent waste
2	Design safe products
3	Design syntheses with safe by-products
4	Use renewable feedstocks
5	Use catalysts
6	Avoid derivatives
7	Maximize atom economy
8	Use safe solvents/conditions
9	Increase energy efficiency
10	Design degradable products
11	Use in-process analysis
12	Minimize accidents

Principle number 4, use renewable feedstocks, might not seem applicable to drug synthesis. After all, the amount of drugs that are synthesized is very small relative to true bulk materials—paper, concrete, plastics, and so on. However, the feedstocks for certain drugs can be very precious, particularly if the feedstock is a natural product. Paclitaxel (Taxol, **13.80**) is an anticancer drug that was first isolated from the bark of the Pacific yew tree (*Taxus brevifolia*), a very slow growing plant species (**Figure 13.15**). Isolation of paclitaxel requires harvesting a tree's bark, which kills the tree. As the activity of paclitaxel became better understood and appreciated, demand for paclitaxel rose dramatically. For example, at one time in the 1980s, the National Cancer Institute required 27,000 kg of yew bark to satisfy its paclitaxel needs. Bark harvesting was killing yew trees at an unsustainable pace and destroying the tree population. Without the trees, a promising cancer treatment would be lost. Fortunately, an alternative source was discovered. The leaves of the European yew (*Taxus baccata*), a type of shrub, contain 10-deacetyl baccatin III (**13.81**). 10-Deacetyl baccatin III may be converted into paclitaxel. Leaves of *T. baccata* can be harvested without killing the plant, and they regenerate quickly. Furthermore, 10-deacetyl baccatin may be isolated in large amounts from the leaves

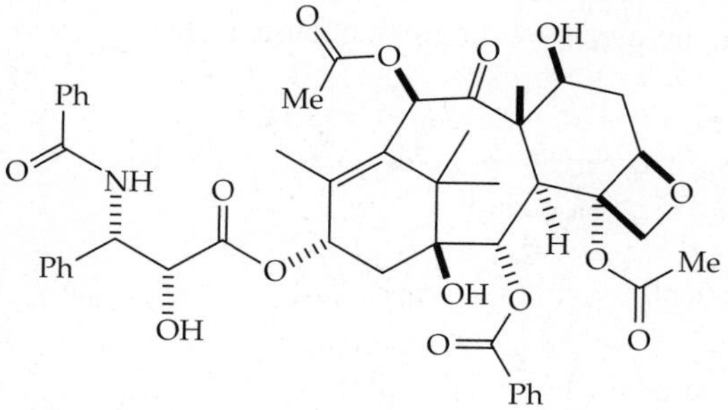

paclitaxel
(Taxol)
antitumor
13.80

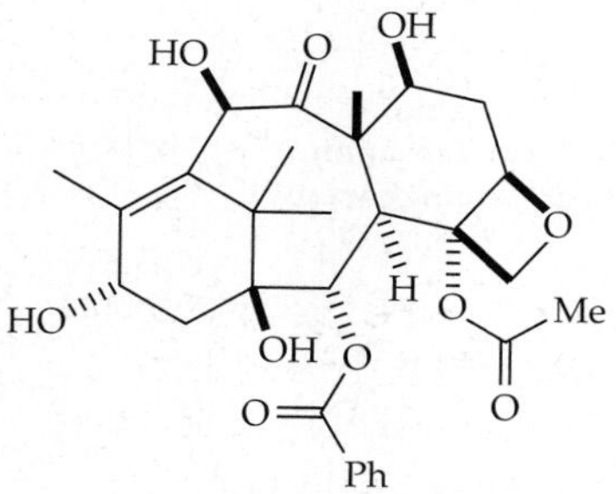

10-deacetyl
baccatin III
13.81

FIGURE 13.15

(1 g **13.81**/kg leaves). In contrast, much smaller amounts of paclitaxel can be extracted from *T. brevifolia* bark (~ 100 mg **13.80**/kg bark). Therefore, although uncommon, the renewability of a feedstock can impact the feasibility of a new drug.[24]

Green Metrics

How can two competing processes be compared for their greenness? Green metrics attempt to answer this question. Green metrics provide a numerical score as a basis for comparison. Percent yield is likely the original green metric, but yield alone is incomplete at best. Unfortunately, newer, improved green metrics also have problems. Various performance factors, including percent yield, number of steps, solvent demand, energy consumption, cost, and safety, are not readily summarized by a single number. Furthermore, the most widely used metrics are often both the simplest to use and therefore the least informative. Despite their shortcomings, green metrics have found some usefulness in evaluating chemical processes.[25]

Atom Economy

Barry Trost of Stanford University published the first formal green metric in 1991 in *Science*. Trost's idea, *atom economy*, measures the percentage ratio of the total mass of the products to starting materials (Equation 13.2).[26]

$$\text{atom economy} = 100 \times \left(\frac{\text{mass of atoms in desired product}}{\text{mass of atoms in reactants}} \right) \qquad (13.2)$$

Atom economy is strictly a theoretical number. The calculation considers only reagents from the balanced chemical reaction. The calculation does not include practical factors such as yield or excess reagents. A given reaction has the same atom economy whether its yield is 90% or 10%. Regardless, atom economy was the first accepted attempt at quantifying the greenness of a reaction.

Addition reactions often have high atom economies. Two specific examples, the Diels–Alder cycloaddition and aldol condensation, are shown in **Scheme 13.14**. In contrast, reactions with multiple or large side products have low atom economies. The Wittig

SCHEME 13.14 Atom economies of selected reactions

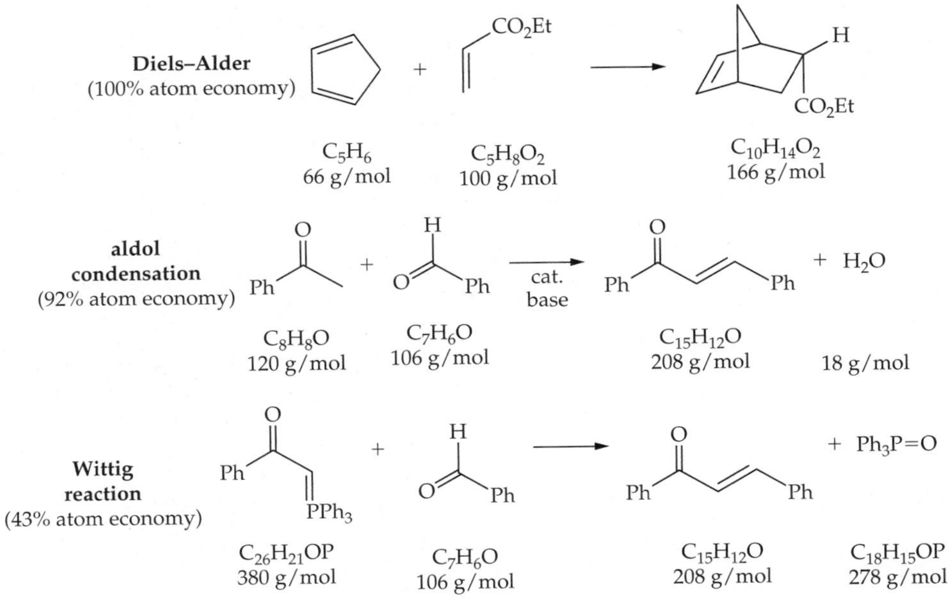

Diels–Alder
(100% atom economy)

C_5H_6
66 g/mol

$C_5H_8O_2$
100 g/mol

$C_{10}H_{14}O_2$
166 g/mol

aldol
condensation
(92% atom economy)

C_8H_8O
120 g/mol

C_7H_6O
106 g/mol

$C_{15}H_{12}O$
208 g/mol

18 g/mol

Wittig
reaction
(43% atom economy)

$C_{26}H_{21}OP$
380 g/mol

C_7H_6O
106 g/mol

$C_{15}H_{12}O$
208 g/mol

$C_{18}H_{15}OP$
278 g/mol

reaction is frequently labeled a bad offender for low atom economy. Formation of a high molecular weight waste product like Ph₃P=O dramatically drops atom economy. Likewise, the use of protecting groups, which are eventually removed and discarded, decreases the atom economy of a process. The following Case Study demonstrates how the synthesis of a drug can be redesigned to dramatically improve the atom economy of the process.

CASE STUDY

Case Study and Sample Calculation: Ibuprofen[27–29]

The original reported synthesis of ibuprofen (**13.87**) was patented in the late 1960s by the Boots Pure Drug Company in England.[27] The route is shown in **Scheme 13.15**. Side products are shown underneath the reaction arrow in parentheses. The synthesis is fully broken down from an atom economy standpoint in **Scheme 13.16**. Because the intermediates (**13.83–13.86**) appear as both a product of one reaction and a starting

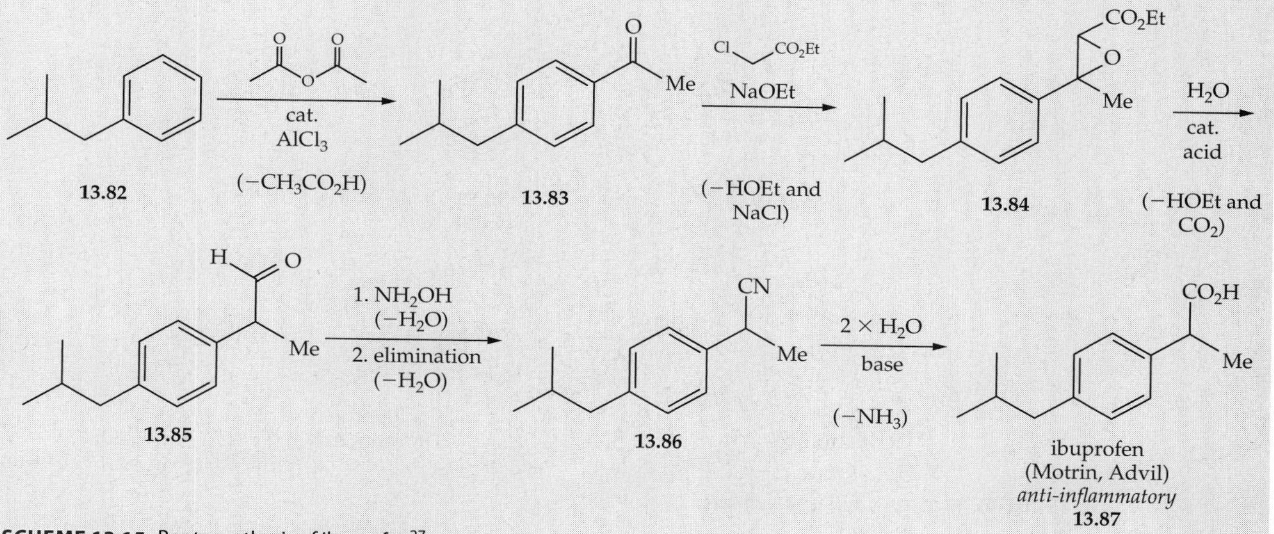

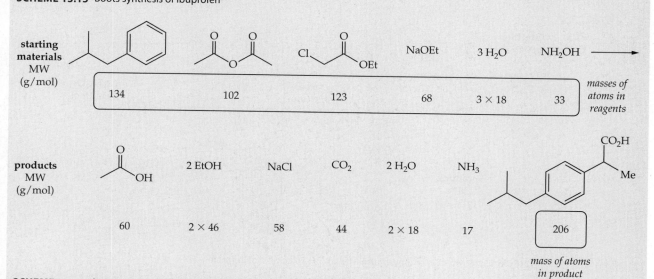

SCHEME 13.15 Boots synthesis of ibuprofen[27]

SCHEME 13.16 Atom economy information for Boots synthesis of ibuprofen material for the next, they cancel out in the calculation of atom economy. The overall atom economy for the six-step synthesis is 40% (Equation 13.3).

Boots synthesis atom economy

$$= 100 \times \left(\frac{206}{134 + 102 + 123 + 68 + (3 \times 18) + 33} \right)$$

$$= 40\% \tag{13.3}$$

The Boots patent on ibuprofen expired in the 1980s, and other manufacturers became able to manufacture and sell ibuprofen. In 1991, the Boots and Hoechst Celanese companies entered a joint venture and patented a new process, called the BHC process, for making ibuprofen. The new route is only three steps (**Scheme 13.17**).[28,29] Consideration of all the reagents and products reveals the atom economy to be 77% (**Scheme 13.18**, Equation 13.4). The BHC commercial route to ibuprofen received the United States' Presidential Green Chemistry Challenge Award in 1997.

BHC synthesis atom economy

$$= 100 \times \left(\frac{206}{134 + 102 + 2 + 28} \right)$$

$$= 77\% \tag{13.4}$$

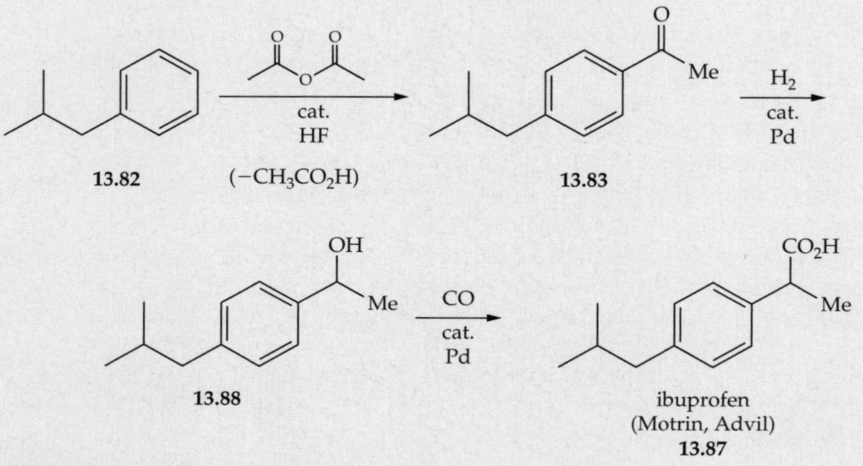

SCHEME 13.17 BHC route to ibuprofen

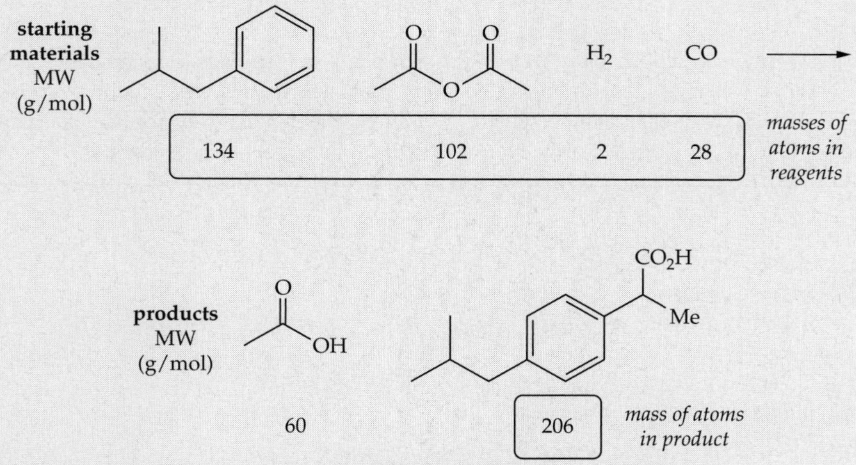

SCHEME 13.18 Atom economy information for BHC synthesis of ibuprofen

TABLE 13.4 E-Factors of selected chemical industries[32]

Industry	Annual Production (tons)	E-Factor
Oil refining	10^6-10^8	~0.1
Bulk chemicals	10^4-10^6	<1–5
Fine chemicals	10^2-10^4	5–50+
Pharmaceuticals	10^1-10^3	25–100+

E-Factor

Shortly after Trost introduced the idea of atom economy, Roger Sheldon of Delft University of Technology in the Netherlands reported another green metric called *E-factor*.[30,31] E-Factor is the ratio of the mass of the total waste from a process to the mass of the product generated in that same process (Equation 13.5).

$$\text{E-factor} = \frac{\text{mass of total waste}}{\text{mass of product}} \quad (13.5)$$

The E-factors of different types of chemical processes are shown in **Table 13.4**.[32] In general, processes performed on a larger scale tend to involve fewer operations and form simpler products. Both facets help lead to lower E-factor values. Remarkably, oil refining generates approximately 10-fold more product than waste. In contrast, pharmaceutical processes can be less than 1% efficient based on waste production. The commercial process of sildenafil (**13.73**), highlighted in the Case Study below, has been optimized more extensively than the typical pharmaceutical synthesis.

CASE STUDY

Sildenafil[33]

The various improvements in synthesis of sildenafil (**13.73**) in a previous section of this chapter have been analyzed by their E-factors (**Table 13.5**). Note that in this analysis, E-factors are reported in L waste per kg product. Most waste consists of solvents and aqueous solutions, so measuring waste in volume is convenient. Like most med chem routes, the original synthesis of sildenafil was inefficient and intended only for quick synthesis. Within four years, the med chem route had been optimized in terms of its E-factor by over an order of magnitude. Further improvements dropped the E-factor down to 22 L/kg. Implementation of recovering solvents from the waste stream reduced the E-factor to just 7 L/kg. Such a low E-factor is more typical of a bulk or fine chemical process. For its development of an efficient synthesis of sildenafil, Pfizer was given the United Kingdom's Award for Green Chemical Technology in 2003.

The timeline of sildenafil is interesting by itself. Most successful drugs do not require seven years to progress from the med chem synthesis to commercialization. Sildenafil, however, started as a candidate for treating angina. Its impact on erectile dysfunction was discovered as a side effect in the clinical trials for angina efficacy. Because sildenafil's target was unknown, Pfizer spent additional time researching the biological processes behind sildenafil's erectile dysfunction activity. This research explains the gap between the original synthesis of sildenafil and its advancement to a commercial process.

TABLE 13.5 E-Factors of different stages of sildenafil synthesis

Process	Year	E-Factor (L/kg)
Original med chem route	1990	1,300
Optimized med chem route	1994	100
Commercial route	1997	22
Commercial route with solvent recovery	2003	7
Future commercial route	????	4

Other Metrics

Reaction mass efficiency (RME) extends the idea of atom economy by taking into account a reaction's yield and the use of excess reagents.[34] For the reaction $A + B + C \rightarrow D + E$ with a desired product D, the formula for percent RME is shown in Equation 13.6.

$$\text{RME} = 100 \times \frac{\text{mass of D}}{\text{mass of A} + \text{mass of B} + \text{mass of C}} \tag{13.6}$$

Unlike atom economy, RME uses the actual reagent masses utilized to perform the reaction. RME does not include solvents, unless the solvent is also a reagent.

Effective mass yield is a green metric proposed by Tomáš Hudlický of the University of Florida (Equation 13.7).[35]

$$\text{effective mass yield} = 100 \times \frac{\text{mass of product}}{\text{mass of nonbenign reagents}} \tag{13.7}$$

Effective mass yield includes actual reagent masses as well as solvents. In effective mass yield, Hudlický chose to include only "nonbenign" reagents. Benign reagents include water, dilute saline solutions, and ethanol. The safety of some reagents is debatable, and the subjective nature of effective mass yield calculations is a weakness in the metric.

Carbon efficiency determines greenness of the loss of carbon in a reaction (Equation 13.8).[34]

$$\text{carbon efficiency} = 100 \times \frac{\text{amount of carbon in product}}{\text{total carbon in starting materials}} \tag{13.8}$$

Solvents are not included in the calculation. Like effective mass yield, carbon efficiency strives to make a value judgment on the relative importance of various wastes. Elimination of water from a molecule would not count against the carbon efficiency of a reaction, but it would give a decreased atom economy. By not including all wastes, carbon efficiency is not as strict as most green metrics. Of course, all metrics have their shortcomings and carry their own assumptions.

13.5 Regulatory Issues

As if making a drug in an efficient and cost-effective manner is not a sufficient challenge, those involved in drug synthesis and manufacturing must keep detailed records on every aspect of the synthesis.

Good Practices

Good practices comprise accepted guidelines and protocols, many of which apply to the pharmaceutical industry. The collective good practices are denoted as *GxP*, with *x* being a wildcard for specifying a certain guideline. The most commonly encountered GxP is *good manufacturing practice* (GMP), also called *current* GMP (cGMP).[36] A few other GxPs include guidelines for pharmaceutical distribution (GDP), storage (GSP), safety (also GSP), and pharmacy (GPP). The International Conference on Harmonisation of Technical Requirements for Registration of Pharmaceuticals for Human Use (ICH) maintains guidelines for GMP. The United States, members of the European Union, and Japan subscribe to ICH's GMP description. Agreement among all three major markets on a single GMP implementation minimizes delays in approving drugs from one market to another. The ICH also maintains standards for GxPs other than GMP, and all documents are continuously being reviewed and updated.

TABLE 13.6 Categories and limits of drug impurities[37]

Maximum Dose (grams/day)	Threshold Category		
	Reporting	Identification	Qualification
<=2.0	0.05%	0.10% or 1.0 mg/day (whichever is lower)	0.15% or 1.0 mg/day (whichever is lower)
>2.0	0.03%	0.05%	0.05%

GMP touches on all facets of drug manufacture. Personnel must be properly qualified for their tasks. The building should be adequately equipped in terms of utilities, air quality, and sanitation. All manufacturing machinery and equipment must be properly and regularly maintained, cleaned, and calibrated. Materials used in the manufacture of an *active pharmaceutical ingredient* (API) should be monitored for quality. Many additional aspects of drug manufacture are addressed by GMP. Implementation of GMP is the problem of the manufacturer. GMP does not specify *how* different aspects of manufacture should be handled. GMP states only that the aspects must be addressed systematically.

GMP requires a massive amount of recordkeeping. Records prove that GMP guidelines are being followed. Although the burden of documenting all facets of manufacturing may seem excessive, records help ensure the quality of an API and safety of both employees and patients. Records also help to track down problems if an issue arises.

Impurities

All drugs contain impurities that, in theory, can pose a risk to patients. The ICH has defined thresholds on allowable amounts of impurities in a drug (**Table 13.6**).[37] Although the thresholds are listed as percentages, the total daily intake of the impurity is the key number. For this reason, thresholds are lower for drugs taken in larger daily doses. In general, impurities are listed in three different categories: reporting, identification, and qualification. Very minor impurities need to only be noted and reported. More significant impurities must also be identified by elucidating the chemical structure of the compound. Any impurity with a total daily intake of 1.0 mg or greater must be *qualified*, meaning it should undergo testing in animals to establish its safety. Identified impurities that are below the qualification threshold yet are suspected to be toxic also must be qualified.

The commercial route to a drug must be highly reproducible so that the impurity profile of each batch is as consistent as possible. Discovery of new impurities in a carelessly synthesized batch will likely require additional purification steps to reduce impurities below previously established threshold levels. Establishing the presence and amount of impurities, of course, requires the testing of each drug batch with an accurate, quantitative analytical method. Product testing is an aspect of manufacturing that falls under the scope of GMP guidelines.

Impurities can be anything: residual solvents, reaction side products, unreacted starting materials, salts, undesired stereoisomers, and even partially decomposed drug. Transition metal catalysts are another common impurity. Transition metals can be toxic or trigger allergies at low levels. For this reason, metals are measured at parts per million (ppm) levels instead of fractional percentage (1 ppm = 0.0001%). The allowable limits of common synthetic metals in oral drugs according to the European Medicines Agency are shown in **Table 13.7**.[38] Metals can be difficult to remove from a reaction product. Synthetic routes are often designed to use catalysts in early steps. Any metal impurities will likely be adequately removed through the remaining purification steps.

The final marketed form of a drug is called a *drug product*. A drug product includes not only the drug itself but also various components called *excipients*. Examples of excipients

TABLE 13.7 Allowable limits of metals in orally delivered drugs[38]

Metal	Daily Exposure (μg)	Concentration in Drug (ppm)
Pt, Pd	100	10
Ir, Rh, Ru, Os (limits apply to all four metals combined)	100	10
Mo, Ni, Cr, V	300	30
Cu, Mn	2,500	250
Fe, Zn	13,000	1,300

for an oral drug include fillers (provide bulk to an oral pill), binders (help maintain the integrity of the pill), and dyes (give coloration). Excipients may introduce impurities with potential safety risks for a patient.[39] All components of the drug product must be reported, identified, and qualified if they are present beyond predetermined threshold levels.

Generic Drugs and Patents

Very late in the lead optimization phase of drug discovery, a pharmaceutical company files *patents* on the most promising compounds. A patent is a form of *intellectual property*. A patent on a compound protects the patent holder from competition in the market. In most nations, a patent lasts 20 years from the date of filing with the appropriate government agency. In the United States, once a patent has been filed on a drug, completion of animal studies, phase trials, and FDA approval may require seven or more years. Therefore, the useful patent life of a marketed drug is likely no more than a dozen years. Once the patent expires, any manufacturer may produce the formerly protected drug under a *generic* label. For example, Eli Lilly marketed fluoxetine (**13.89**) as an antidepression drug (**Figure 13.16**). In 2001, Lilly's patent on fluoxetine expired, and generic drug makers have since started selling fluoxetine. Lilly markets fluoxetine under the trade name of Prozac. A *trademark*, another form of intellectual property, is registered by a company and does not expire. Therefore, although Lilly's patent on the use of fluoxetine to treat depression has expired, Lilly still owns the name Prozac and is the only company that can market fluoxetine under that name. Generic manufacturers, including Par Pharmaceutical and Mallinckrodt Pharmaceuticals, sell the drug under the generic name of fluoxetine.

Just like branded drugs, generics are subject to regulatory control. Each generic manufacturer must follow GMP guidelines in synthesizing its drugs. Full phase trials are not required for generics, but abbreviated testing in humans establishes that the generic drug is as safe and effective as the original, branded drug.

The legal games that are played over patents and pharmaceuticals are dizzying, and the impact of patents is present throughout all phases of a drug's development. At the outset of this chapter were mentioned the ideas of cocrystals and polymorphs. These concepts

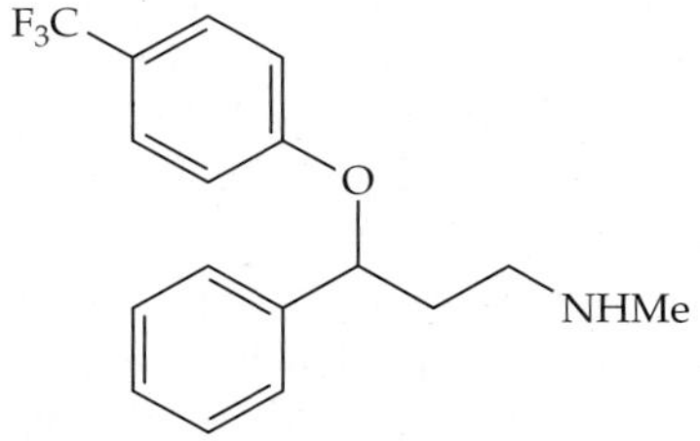

fluoxetine
(Prozac)
antidepressant
13.89

FIGURE 13.16

were innocently discussed within the context of preparing a drug with optimal solubility properties. A more subtle aspect of both cocrystals and polymorphs is their influence on patents.

Formulating a drug as a cocrystal can improve the pharmacokinetic behavior of a drug. Cocrystals are patentable, so a company can claim ownership of that form of a drug. A hypothetical example demonstrates the potential impact of patents of cocrystals.

Alpha Pharmaceuticals develops a drug under the trade name of Cure-all and patents the drug. Cure-all is very successful as a twice-a-day oral pill. Bravo Drugs discovers a cocrystal of Cure-all with improved pharmacokinetics. The cocrystal formulation allows Cure-all to be taken once per day. Bravo patents the cocrystal. Despite owning the superior product, Bravo's options are limited. Alpha holds the patent on the API of Cure-all, which is also the API of the cocrystal. Until the patent expires, Bravo cannot freely market the improved cocrystal. Bravo waits. Once the Alpha patent expires, Bravo markets the API of Cure-all in a cocrystal form called Panacea. Alpha still markets Cure-all but must do so with the original formulation. Alpha also faces competition from generic manufacturers who can sell the API of Cure-all under a generic name. Alpha lowers its price for Cure-all in response to generic pressure. Customers who question the quality of generic drugs are willing to pay a premium for the branded drug from Alpha. In contrast, Bravo is the only company that can market the cocrystal formulation. The price of Panacea is high because it offers a distinct advantage over other competitors on the market. Bravo enjoys an exclusive market until its patent on the cocrystal expires. Like cocrystals, polymorphs are also patentable. The same compound in a different crystalline form is legally a different material under patent law. The following Case Study highlights the delicate intricacies of patent law and polymorphs.

CASE STUDY

Polymorphs of Ranitidine[40]

Ranitidine is a blockbuster drug for the treatment of acid reflux. Ranitidine was first synthesized in 1977 and patented in 1978 under U.S. Patent 4,128,658. In 1980, Glaxo discovered a new polymorph of ranitidine. The new polymorph was designated as Form 2 and had more favorable physical properties from a manufacturing and processing standpoint. U.S. Patent 4,521,431 was granted to Glaxo for Form 2 in 1985. Glaxo marketed Form 2 of ranitidine under the trade name Zantac. By 1992, Zantac had reached annual sales of US$3.4 billion. Highlights of the ranitidine timeline are shown in the following list.

- 1978: Patent on Form 1 of ranitidine (expiration date in 1995, 17 years after issue under former U.S. patent law)
- 1981: Glaxo discovers Form 2
- 1985: Patent on Form 2 (expiration date in 2002)

Because of the incredible success of Zantac, generic manufacturers were eager to enter the market with a generic version of ranitidine. Since the patent on Form 1 was set to expire in 1995, a generic manufacturer could conceivably enter the market with a Form 1 formulation without waiting for expiration of the Form 2 patent in 2002. Around

1990, Novopharm, a generic manufacturer, began to explore Form 1 of ranitidine as a potential generic formulation. Novopharm repeated the synthetic procedure to make Form 1 and reproducibly recovered only Form 2. If Form 2 was known back in 1978, then the patent on Form 2 in 1985 was not novel and therefore invalid. Under this logic, Novopharm filed an abbreviated new drug application (ANDA) to market Form 2 in 1995, coinciding with the expiration of the Form 1 (arguably Form 2) patent.

Glaxo immediately sued Novopharm for patent infringement. Glaxo held the Form 2 patent and was legally able to prevent any other company from marketing Form 2 until 2002. Of course, that statement assumes that the Form 2 patent is valid. Based upon the synthesis of Form 1 by third party labs using the Form 1 patent procedure, Glaxo won the case against Novopharm. The Form 2 patent was upheld, and the Novopharm's ANDA for Form 2 was withdrawn.

Novopharm was not finished. Novopharm was ultimately able to reproducibly manufacture Form 1. Novopharm filed an ANDA in 1994 to market Form 1 of ranitidine. Glaxo sued Novopharm. Glaxo's argument was that Novopharm's Form 1 of ranitidine would contain impurities of Form 2. Marketing a mixture would violate Glaxo's Form 2 patent. Novopharm submitted x-ray crystallographic evidence that their Form 1 product was free of Form 2 impurities. The court ruled in favor of Novopharm. Novopharm was allowed to develop a generic formulation of Form 1 of ranitidine with a target market date of 1995, the expiration date of the Form 1 patent.

Summary

Once a compound is well into its development, the process group assumes a more prominent role in studying the properties of the drug and handling its preparation. Almost all drugs are solids, and process groups are adept at developing optimal conditions for the purification of solids through recrystallization. Solids may also be manipulated to optimize the solubility of the drug and enhance a compound's pharmacokinetics. The formation of drug salts and cocrystals as well as the study of a compound's polymorphs are all tools available to the process group. Aside from ensuring a pure product, the process group improves the synthetic route to the drug. Single enantiomer drugs pose a particular synthetic challenge. The two general approaches to enantiomerically pure drugs are resolution and asymmetric synthesis. Resolutions produce single enantiomers by separating them from a racemic mixture. Resolutions are generally simple to perform and therefore operationally inexpensive. The cost to a resolution, however, is the loss of at least half a racemic starting material in the form of the unwanted enantiomer. The alternative to resolution, called asymmetric synthesis, is to prepare a chiral drug as a single enantiomer instead of as racemate. In theory, in an asymmetric synthesis, all of a starting material may be converted into the desired product stereoisomer. Despite the efficiency advantage of asymmetric synthesis, not every drug is amenable to being prepared as a single enantiomer. Beyond controlling stereochemistry in the products, the process research and development group has a primary goal of synthesizing the target drug as cost effectively as possible. Cost effectiveness encompasses such factors as the overall yield of the synthetic route, reagent toxicity, and waste disposal. Various metrics have been developed to quantify the quality of a drug synthesis. Examples include atom economy and E-factor, yet, to date, no single metric has been designed that adequately summarizes the advantages and deficiencies of an entire chemical process. When approving a drug, regulatory agencies carefully consider the synthetic process of a drug. The manufacturer must conform to GMP guidelines that outline almost every facet of a drug's preparation. Impurities in a drug receive particularly close attention. Impurity levels must be maintained beneath regulated thresholds. GMP guidelines and product purity standards apply equally to branded drugs as well as their generic counterparts.

Questions

1. The resolution of sertraline (**13.26**) is shown in Scheme 13.3. The overall yield of the resolution is 39%. Assume the e.e. of the product is 99%. What is the e.e. of the other 61%?
2. The balanced reaction for a synthesis of an intermediate (**13.57**) for a fluvastatin analogue is shown here. This reaction relies on a vinyl stannane (**13.a**). Stannanes are toxic and generate large amounts of waste. Calculate the atom economy for this reaction.

3. The allowable limits of metals for parenteral drugs are lower than the limits for oral drugs. Why?

4. Most reactions slow down as the reaction nears completion, but kinetic resolutions slow down dramatically over the course of the reaction. For example, if a kinetic resolution requires 1 h to reach 30% conversion, the next 30% (to 60% conversion) will require much more than 1 h. Why?

5. Classify the synthetic route to paroxetine (**13.d**) as one of the seven methods described in section 13.2 of this chapter. (de Gonzalo, G., Brieva, R., Sanchez, V. M., Bayod, M., & Gotor, V. Anhydrides as Acylating Agents in the Enzymatic Resolution of an Intermediate of (−)-Paroxetine. *J. Org. Chem.* 2003, *68*, 3333–3336.)

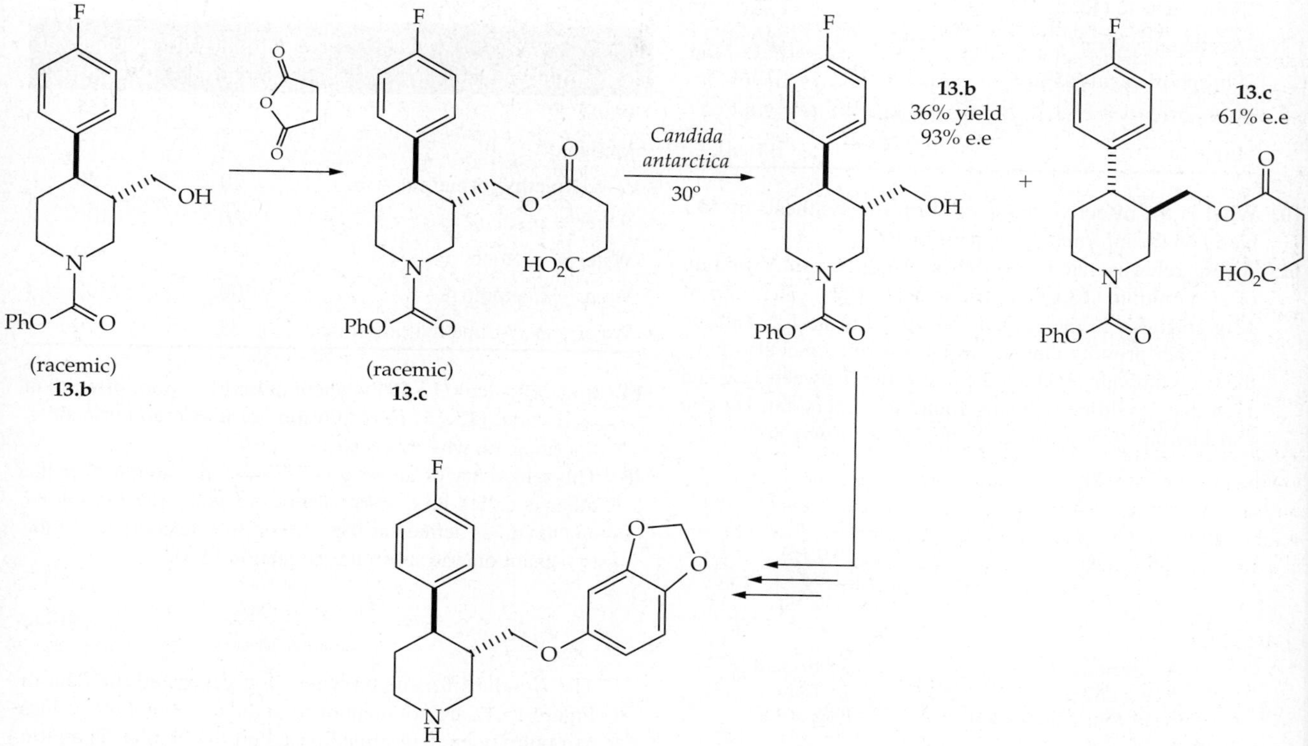

(racemic)
13.b

(racemic)
13.c

13.b
36% yield
93% e.e

13.c
61% e.e

paroxetine
(single enantiomer)
13.d

6. The BHC synthesis of ibuprofen has a high atom economy. Does this fact make the process "green"?

7. The annual production of hydrocortisone and related pharmaceutical corticoids has been estimated at 250 tons. Assuming these compounds are produced with a typical pharmaceutical E-factor of 25, how much waste is generated by corticoid production?

8. Here is the reaction scheme for the synthesis of *O*-acetyl salicylic acid (**13.f**) from salicylic acid (**13.e**). What is the atom economy for this reaction?

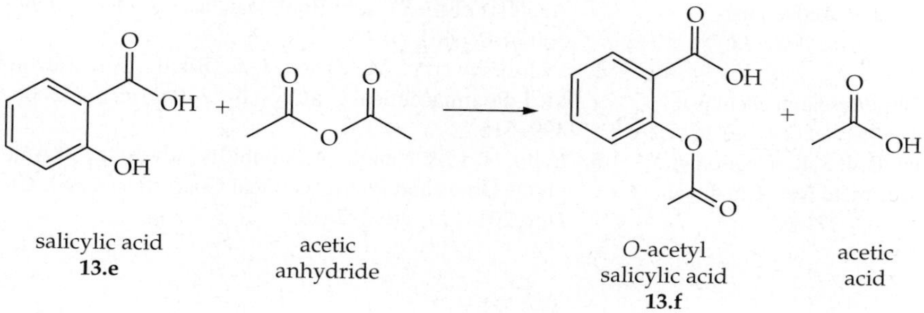

salicylic acid
13.e

acetic
anhydride

O-acetyl
salicylic acid
13.f

acetic
acid

9. The amounts of material needed for the synthesis of *O*-acetyl salicylic acid (**13.f**) are shown below. The bulk of the material is solvents required for precipitations, recrystallizations, and rinses. If the reaction produces 120 g of **13.f**, what is the *E-factor* of this reaction?

Reagent Name	Density (g/mL)	Amount Used
Salicylic acid (**13.e**)	n/a	100 g
Acetic anhydride	1.082	200 mL
Phosphoric acid (85%)	1.685	1 mL
Water	1.000	2.8 L
Ethanol	0.789	700 mL

10. What is the effective mass yield for the synthesis of **13.f** based on the information in question 9?

11. A researcher isolated a crystalline product from a mixture of acetamidine (**13.g**) and acetic acid (**13.h**). The ratio of **13.g** and **13.h** in the crystal was found to be 1:1, and the researcher proudly claimed to have formed a cocrystal of the two compounds. Draw the interaction between **13.g** and **13.h** that would be typically found in a cocrystal. Do you consider this to be a true cocrystal? Why or why not?

acetamidine
13.g
(pK_a of conj. acid = 12.5)

acetic acid
13.h
(pK_a = 4.8)

12. The Boots synthesis of ibuprofen (Scheme 13.15) uses *ethyl* chloroacetate and sodium *eth*oxide in the step going from **13.83** to **13.84**. In theory, changing these reagents to *methyl* chloroacetate and sodium *meth*oxide should have

no negative impact on the overall synthesis. Determine the atom economy of the Boots synthesis with these two reagent changes.

13. Here is a table of solubilities for a drug candidate in various solvents. Which solvent or solvent mixture would be the best to use for a recrystallization? Explain your answer. Assume all systems in the table remove all impurities, and the recrystallization will have a temperature change from 60 °C down to 20 °C.

Solvent(s)	Solubility at 20 °C (g/L)	Solubility at 60 °C (g/L)
Water	10	55
Methanol	5	15
N,N-Dimethylformamide	10	20
Water + methanol	20	100
Water + acetone	175	350
Water + acetonitrile	400	1200
Water + *N,N*-dimethylformamide	35	510

14. Mandelic acid (**13.19**) is encountered far more often than lactic acid (**13.15**) in resolutions of diastereomeric salts. Speculate on why this is so.

15. The selectivity of an enzyme for one enantiomer over the other is called its *E*-value. More precisely, the *E*-value of an enzyme is defined as the ratio of the rate constants of the two enantiomeric substrates (Equation 13.a).

$$E = \frac{k_{\text{favored enantiomer}}}{k_{\text{disfavored enantiomer}}} \tag{13.a}$$

The *E*-values for the enzymes that generated the data in Figure 13.12 are not mentioned in the text, but the *E*-values can quickly be determined based on the graphs. The ratio of enantiomers in the initial product mixture is equivalent to the *E*-value of the enzyme. Estimate the *E*-values of both enzymes in Figure 13.12 based on their respective graphs.

References

1. Detoisien, T., Forite, M., Taulelle, P., Teston, J., Colson, D., Klein, J. P., & Veesler, S. A. Rapid Method for Screening Crystallization Conditions and Phases of an Active Pharmaceutical Ingredient. *Org. Process Res. Dev.* 2009, *13*, 1338–1342.
2. pK_a data compiled by R. Williams. http://research.chem.psu.edu/brpgroup/pKa_compilation.pdf (accessed October 2012).
3. Bastin, R. J., Bowker, M. J., & Slater, B. J. Salt Selection and Optimisation Procedures for Pharmaceutical New Chemical Entities. *Org. Process Res. Dev.* 2000, *4*, 427–435.
4. Paulekuhn, G. S., Dressman, J. B., & Saal, C. Trends in Active Pharmaceutical Ingredient Salt Selection Based on Analysis of the Orange Book Database. *J. Med. Chem.* 2007, *50*, 6665–6672.
5. Vishweshwar, P., McMahon, J. A., Bis, J. A., & Zaworotko, M. J. Pharmaceutical Co-Crystals. *J. Pharm. Sci.* 2006, *95*, 499–516.
6. Babu, N. J., & Nangia, A. Solubility Advantage of Amorphous Drugs and Pharmaceutical Cocrystals. *Cryst. Growth Des.* 2011, *11*, 2662–2679.

7. Remenar, J. F., Morissette, S. L., Peterson, M. L., Moulton, B., MacPhee, J. M., Guzmán, H. R., & Almarsson, ö. Crystal Engineering of Novel Cocrystals of a Triazole Drug with 1,4-Dicarboxylic Acids. *J. Am. Chem. Soc.* 2003, *125,* 8456–8457.

8. Mangin, D., Puel, F., & Veesler, S. Polymorphism in Processes of Crystallization in Solution: A Practical Review. *Org. Process Res. Dev.* 2009, *13,* 1241–1253.

9. Dunitz, J. D., & Bernstein, J. Disappearing Polymorphs. *Acc. Chem. Res.* 1995, *28,* 193–200.

10. Chembrukar, S. R., Bauer, J., Deming, K., Spiwek, H., Patel, K., Morris, J., et al. Dealing with the Impact of Ritonavir Polymorphs on the Late Stages of Bulk Drug Process Development. *Org. Process Res. Dev.* 2000, *4,* 413–417.

11. Miller, S. Scientific and Regulatory Aspects of Quality Control for Chiral Drugs. Presented at 18th International Symposium on Chirality, Busan, Korea, June 2006. http://www.fda.gov/downloads/AboutFDA/CentersOffices/CDER/ucm103532.pdf (accessed October 2012).

12. Welch, Jr., W. M., Harbert, C. A., Koe, B. K., & Kraska, A. R. Antidepressant Derivatives of *cis*-4-Phenyl-1,2,3,4-Tetrahydro-1-Naphthalenamine. U.S. Patent 4,536,518, August 20, 1985.

13. Daluge, S. M. Therapeutic Nucleosides. U.S. Patent 5,034,394, July 23, 1991.

14. Coates, J. A., Mutton, I. M., Penn, C. R., Williamson, C., & Storer, R. 1,3-Oxathiolane Nucleoside Analogues. U. S. Patent 6,180,639, January 30, 2001.

15. Bühlmayer, P., Furet, P., Criscione, L., de Gasparo, M., Whitebread, S., Schmidlin, T., et al. Valsartan, a Potent, Orally Active Angiotensin II Antagonist Developed from the Structurally New Amino Acid Series. *Biorg. Med. Chem. Lett.* 1994, *4,* 29–34.

16. Roth, B. D. [R–(R^*, R^*)]-2-(4-Fluorophenyl-β, δ-Dihydroxy-5-(1-Methylethyl-3-Phenyl-4-[(Phenylamino)Carbonyl]-1H-Pyrrole-1-Heptanoic Acid, Its Lactone Form and Salts Thereof. U.S. Patent 5,273,995, December 28, 1993.

17. Larsson, M. E., Stenhede, U. J., Sorensen, H., von Unge, S. P. O., & Cotton, H. K. Process for Synthesis of Substituted Sulphoxides. U.S. Patent 5,948,789, September 7, 1999.

18. Hasegawa, J., Ogura, M., Hamaguchi, S., Shimazaki, M., Kawaharada, H., & Watanabe, K. Fermentative Production of D-($-$)-β-Hydroxyisobutyric Acid. U.S. Patent 4,310,635, January 12, 1982.

19. Livingston, D. A. Application of Silicon Chemistry in the Cortosteroid Field. In Maryanoff, B. E., Maryanoff, C. A. (Eds.), *Advances in Medicinal Chemistry.* Greenwich, CT: JAI Press, 1992, Volume 1, pp. 137–174.

20. Fuenfschilling, P. C., Hoehn, P., & Mutz, J. P. An Improved Manufacturing Process for Fluvastatin. *Org. Process Res. Dev.* 2007, *11,* 13–18.

21. Repič, O., Prasad, K., & Lee, G. T. The Story of Lescol: From Research to Production. *Org. Process Res. Dev.* 2001, *5,* 519–527.

22. Dale, D. J., Dunn, P. J., Golightly, C., Hughes, M. L., Levett, P. C., Pearce, A. K., et al. The Chemical Development of the Commercial Route to Sildenafil: A Case History. *Org. Process Res. Dev.* 2000, *4,* 17–22.

23. Anastas, P. T., & Warner, J. C. *Green Chemistry: Theory and Practice.* New York: Oxford University Press, 1998, p. 30.

24. Denis, J.-N., Greene, A. E., Guénard, D., Guéritte-Voegelein, F., Mangatal, L., & Potier, P. A. Highly Efficient, Practical Approach to Natural Taxol. *J. Am. Chem. Soc.* 1988, *110,* 5917–5919.

25. Constable, D. J. C., Curzons, A. D., & Cunningham, V. L. Metrics to "Green" Chemistry: Which Are the Best? *Green Chem.* 2002, *4,* 521–527.

26. Trost, B. M. The Atom Economy: A Search for Synthetic Efficiency. *Science.* 1991, *254,* 1471–1477.

27. Nicholson, J. S., & Adams, S. S. Phenyl Propionic Acids. U.S. Patent 3,385,886, May 28, 1968.

28. Elango, V., Murphy, M. A., Smith, B. L., Davenport, K. G., Mott, G. N., Zey, E. G., & Moss, G. L. Method for Producing Ibuprofen. U.S. Patent 4,981,995, January 1, 1991.

29. Lindley, D. D., Curtis, T. A., Ryan, T. R., de la Garza, E. M., Hilton, C. B., & Kenesson, T. M. Process for the Production of 4′-Isobutylacetophenone. U.S. Patent 5,068,448, November 26, 1991.

30. Sheldon, R. A. Organic Synthesis: Past, Present and Future. *Chem. Ind.* 1992, 903–906.

31. Sheldon, R. A. Catalysis and Pollution Prevention. *Chem. Ind.* 1997, 12–15.

32. Sheldon, R. A. The E Factor: Fifteen Years On. *Green Chem.* 2007, *9,* 1273–1283.

33. Dunn, P. J., Galvin, S., & Hettenbach, K. The Development of an Environmentally Benign Synthesis of Sildenafil Citrate (Viagra) and Its Assessment by Green Chemistry Metrics. *Green Chem.* 2004, *6,* 43–48.

34. Curzons, A. D., Constable, D. J. C., Mortimer, D. N., & Cunningham, V. L. So You Think Your Process Is Green: How Do You Know? Using Principles of Sustainability to Determine What Is Green: A Corporate Perspective. *Green Chem.* 2001, *3,* 1–6.

35. Hudlický, T., Frey, D. A., Koroniak, L., Claeboe, C. D., & Brammer, Jr., L. E. Toward a "Reagent-free" Synthesis. *Green Chem.* 1999, 57–59.

36. International Conference on Harmonisation of Technical Requirements for Registration of Pharmaceuticals for Human Use: Quality Guidelines: Good Manufacturing Practice Guide for Active Pharmaceutical Ingredients. http://www.ich.org/fileadmin/Public_Web_Site/ICH_Products/Guidelines/Quality/Q7/Step4/Q7_Guideline.pdf (accessed October 2012).

37. International Conference on Harmonisation of Technical Requirements for Registration of Pharmaceuticals for Human Use: Quality Guidelines: Impurities in New Drug Substances. http://www.ich.org/fileadmin/Public_Web_

Site/ICH_Products/Guidelines/Quality/Q3A_R2/Step4/
Q3A_R2__Guideline.pdf (accessed October 2012).

38. European Medicines Agency. Committee for Human
Medicinal Products: Guideline on the Specification Limits
for Residues of Metal Catalysts. http://www.ema.europa.eu/
docs/en_GB/document_library/Scientific_guideline/
2009/09/WC500003586.pdf (accessed October 2012).

39. International Conference on Harmonisation of
Technical Requirements for Registration of
Pharmaceuticals for Human Use: Quality Guidelines:
Impurities in New Drug Products. http://www.ich.org/
fileadmin/Public_Web_Site/ICH_Products/Guidelines/
Quality/Q3B_R2/Step4/Q3B_R2__Guideline.pdf
(accessed October 2012).

40. Bernstein, J. Polymorphism and Patents from a Chemist's
Point of View. In Hilfiker, R. (Ed.), *Polymorphism: In the
Pharmaceutical Industry*. Weinheim, Germany: Wiley-
VCH, 2006, Chapter 14.

Appendix A

Top Selling Oral Drug Classes

Outline

The ultimate goal of a drug discovery program is a marketed drug. The conditions included within this appendix are common, especially in developed Western nations. Drugs for these conditions are therefore in high demand and generally enjoy commercial success. All the drugs included in this appendix are administered orally because oral drugs are the desired outcome of most drug discovery programs.

A.1 Allergic Rhinitis

Allergic rhinitis is characterized by nasal congestion and often accompanied by sneezing and watery eyes. Environmental allergens, often pollen, trigger the release of histamine (**A.1**) (**Figure A.1**). Histamine binds the H_1 receptor, a G-protein–coupled receptor, and undesired symptoms result. Almost all drugs for allergic rhinitis are *antihistamines* and block histamine as inverse agonists of the H_1 receptor.

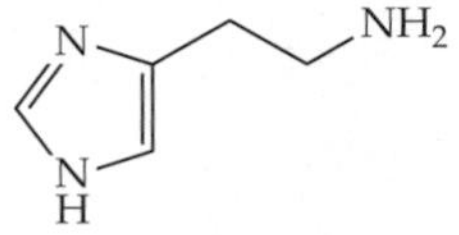

histamine

A.1

FIGURE A.1

FIGURE A.2
First-generation
antihistamines

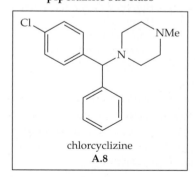

ethylenediamine subclass

mepyramine
A.2

ethanolamine subclass

diphenhydramine
(Benadryl)
A.3

doxylamine
A.4

alkylamine subclass

X = Cl chlorpheniramine
(Chlor-Trimeton), **A.5**

X = Br brompheniramine
(Dimetapp), **A.6**

tripolidine
A.7

piperazine subclass

chlorcyclizine
A.8

**tri-/tetracyclic
subclass**

promethazine
A.9

The original, first-generation antihistamines emerged in the 1940s with several different subclasses. All subclasses have a diaryl group connected to a tertiary amine with a two- or three-atom tether. Mepyramine (**A.2**) and diphenhydramine (Benadryl, **A.3**) were among the first antihistamines and are examples of the ethylenediamine and ethanolamine subclasses, respectively (**Figure A.2**). The alkylamine subclasses include chlorpheniramine (Chlor-Trimeton, **A.5**), brompheniramine (Dimetapp, **A.6**), and tripolidine (**A.7**). Two other subclasses are the piperazines and tri-/tetracyclics with respective examples of chlorcyclizine (**A.8**) and promethazine (**A.9**). All of the first-generation antihistamines are nonpolar molecules and readily cross the blood–brain barrier. In the central nervous

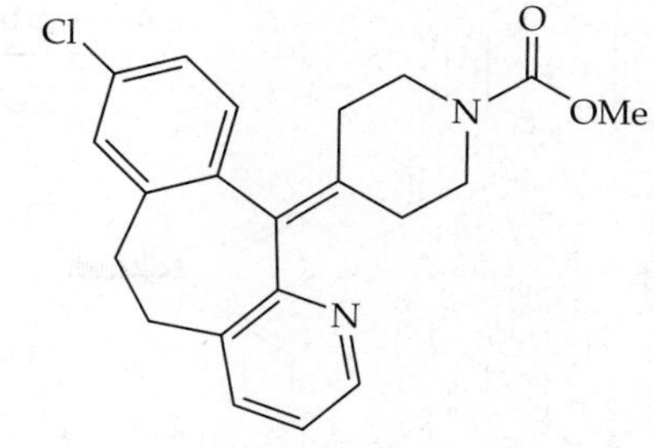

loratadine
(Claritin)
A.10

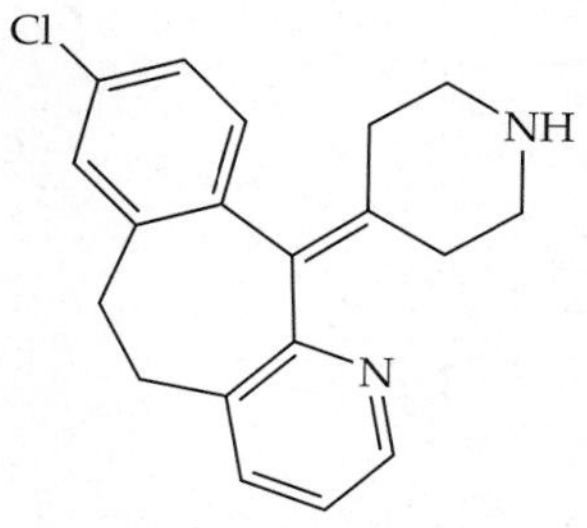

desloratadine
(Clarinex)
A.11

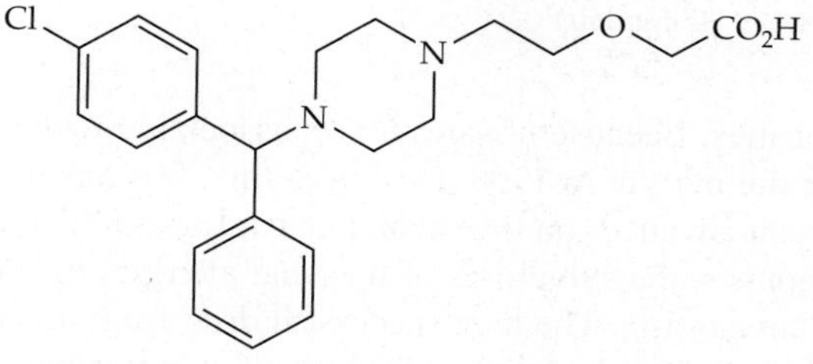

cetirizine
(Zyrtec)
A.12

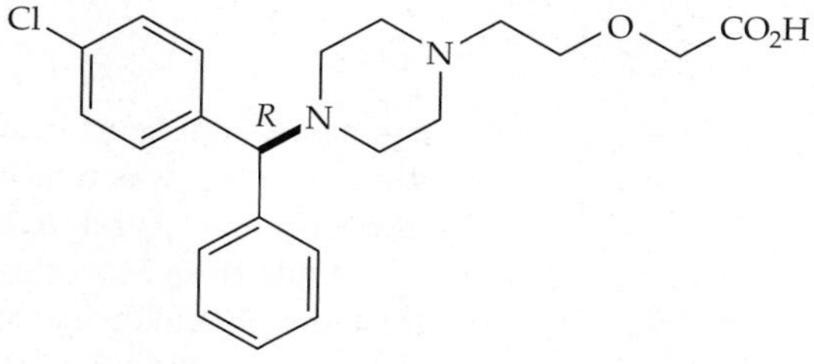

levocetirizine
(Xyzal)
A.13

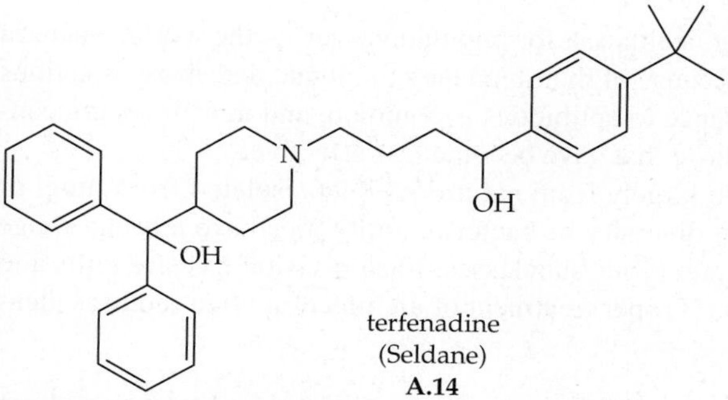

terfenadine
(Seldane)
A.14

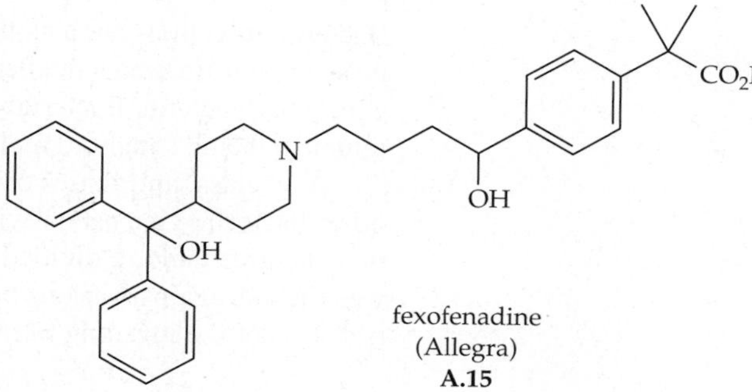

fexofenadine
(Allegra)
A.15

FIGURE A.3 Second-generation antihistamines and later drugs

system, the compounds act in the brain to cause drowsiness. Sometimes the side effects of antihistamines are very strong. Doxylamine (**A.4**) is an ethanolamine antihistamine that is marketed primarily as a sedative.

The second-generation antihistamines continue the diaryl/tether/tertiary amine motif of the first-generation compounds. All three second-generation examples share one or more traits of the five original subclasses. Loratadine (Claritin, **A.10**) and desloratadine (Clarinex, **A.11**) are combinations of the alkylamine, piperazine, and tricyclic subclasses (**Figure A.3**). Cetirizine (Zyrtec, **A.12**) is directly derived from the piperazine subclass. Terfenadine (Seldane, **A.14**) is a cross of the alkylamine and ethanolamine subclasses. The second-generation antihistamines do not penetrate the blood–brain barrier effectively and therefore cause less drowsiness than the first-generation drugs. Both loratadine and terfenadine are prodrugs. In the body, the methyl carbamate of loratadine undergoes hydrolysis and decarboxylation to form desloratadine (**A.11**), and terfenadine is oxidized to fexofenadine (Allegra, **A.15**). Some patients are slow metabolizers of terfenadine, which

FIGURE A.4

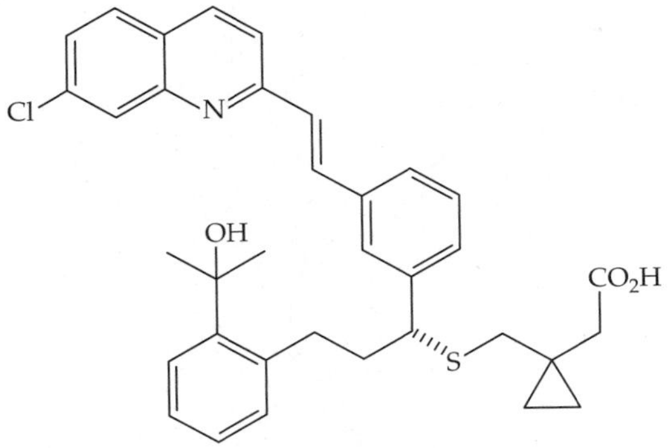

montelukast
(Singulair)
A.16

can cause fatal heart beat irregularities. Because of safety risks associated with terfenadine, the drug was removed from the market in 1998. Desloratadine, fexofenadine, and
levocetirizine (Xyzal, **A.13**) represent advances on the second-generation antihistamines.

Aside from H$_1$-receptor antagonists, the symptoms of seasonal allergies may also be
treated with leukotriene receptor antagonists. The lone successful drug from this class is
montelukast (Singulair, **A.16**) (**Figure A.4**). Montelukast is most often prescribed to treat
asthma but can be used for allergic rhinitis as well.

A.2 Bacterial Infection

Bacterial infections are a significant health risk for populations across the world. Bacteria
pose a major treatment challenge because of their tendency to mutate and share mutations
with other bacteria. Bacterial resistance to antibiotics is common, and new drugs are continuously being sought to replace those that have become less effective.

As a class, antibiotics originate largely from natural products isolated from fungi or
other bacteria. Consistent with the diversity of bacteria, antibiotics have a broad range
of structures and are divided into numerous subclasses. Each division has strengths and
weaknesses against various bacteria. Proper treatment of an infection often requires identification of the invading bacteria.

Sulfonamides

The sulfonamides, or sulfa drugs, date back to the early 1900s but were not systematically
studied until the 1930s. Sulfanilamide (**A.17**), a key reagent in the synthesis of certain
dyes, was the first widely marketed sulfonamide (**Figure A.5**). Sulfonamides are antimetabolites and competitively inhibit a bacterial enzyme, dihydropteroate synthetase (DHPS)
(see Chapter 1 and Chapter 6). DHPS plays a role in the synthesis of tetrahydrofolic acid
(THF), an important compound in the preparation of thymidine. Because they limit the

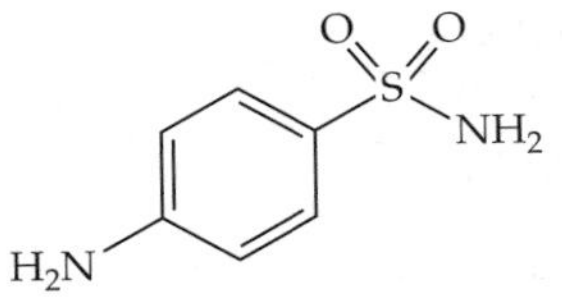

sulfanilamide
A.17

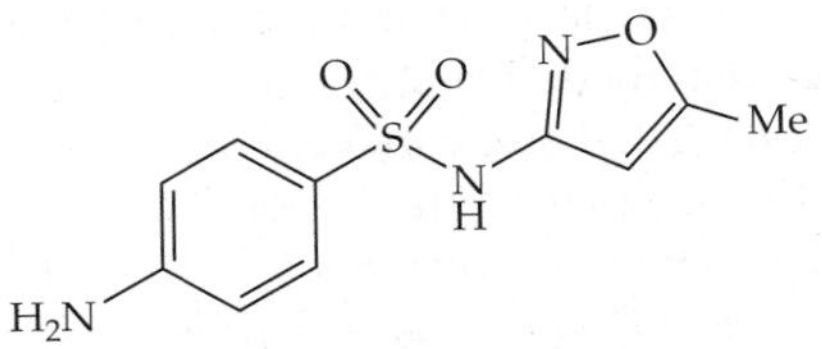

sulfamethoxazole
A.18

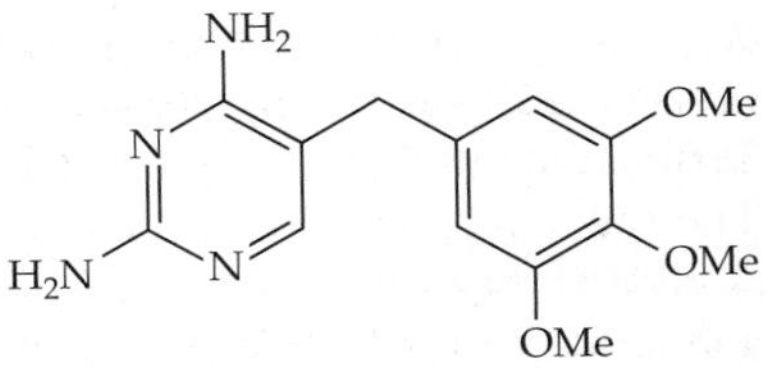

trimethoprim
A.19

FIGURE A.5 Antimetabolite antibiotics

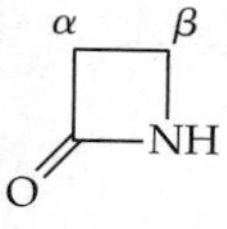

FIGURE A.6 Penicillin β-lactam antibiotics

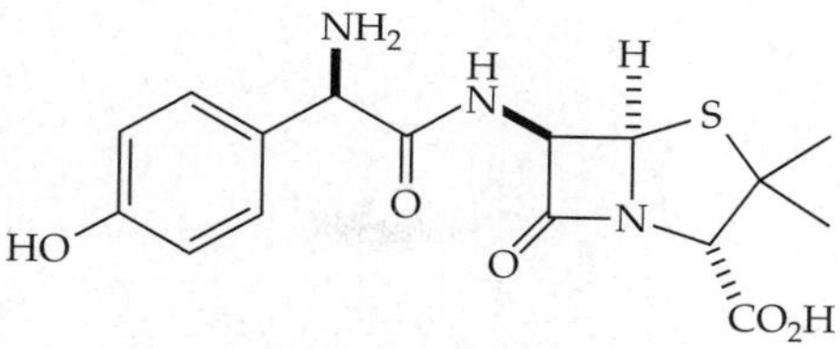

availability of thymidine, a required nucleoside for DNA synthesis, sulfonamides prevent cell division. Sulfonamides are bacteriostatic, not bacteriocidal (they do not kill bacteria). Sulfamethoxazole (**A.18**) is the most widely prescribed sulfonamide today. Sulfamethoxazole and trimethoprim (**A.19**), a nonsulfonamide antimetabolite of THF, are often coformulated and administered together.

β-Lactams

β-Lactams antibiotics fall under several subclasses, all of which are characterized by an amide in a four-membered ring (**A.20**) (**Figure A.6**). The strained β-lactam functional group irreversibly reacts with proteins in bacteria to stop cell wall synthesis. The penicillins contain a β-lactam fused to a five-membered ring with a sulfur atom. Like all the β-lactams, penicillins are isolated from microbial broths. Examples of penicillins include penicillin V (**A.21**), amoxicillin (**A.22**), and oxacillin (**A.23**). Penicillin V is normally formulated as its potassium salt, penicillin VK. Bacterial resistance to penicillins and the other β-lactams arises from two sources. First, some bacteria have enzymes called β-lactamases, which hydrolyze the β-lactam ring before it can inhibit cell wall synthesis. Second, some bacteria carry protein mutations that prevent binding by β-lactam drugs.

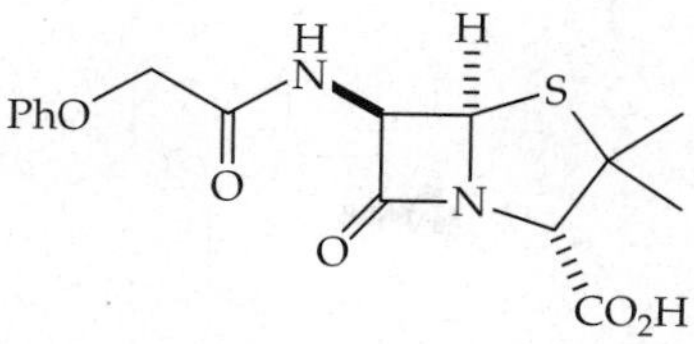

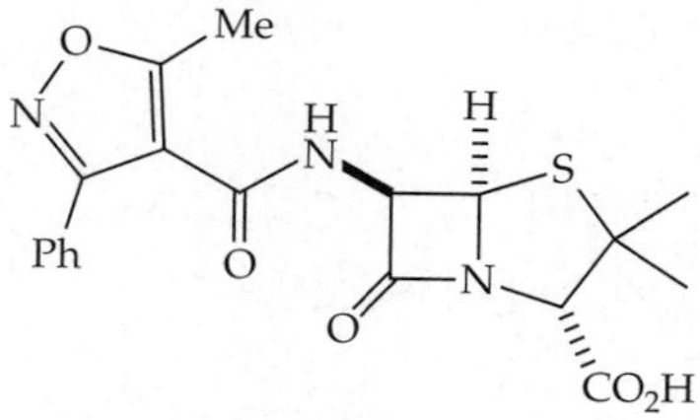

Two other β-lactam subclasses are the cephalosporins and carbapenems. The cephalosporins contain a β-lactam fused with a six-membered ring with a sulfur atom. As a group, the cephalosporins have low oral bioavailability. Exceptions include cephalexin (Keflex, **A.24**), cefaclor (Ceclor, **A.25**), and cefpodoxime proxetil (Vantin, **A.26**) (**Figure A.7**). As its name suggests, cefpodoxime proxetil is a prodrug that releases cefpodoxime (**A.29**), the active drug, in vivo. The carbapenems contain a β-lactam fused to a carbocyclic five-membered ring. Imipenem (**A.27**) is an example of the carbapenem subclass. Clavulanic acid (**A.28**), which is an isostere of a carbapenem, has no appreciable antibacterial activity. Clavulanic acid does, however, inhibit many β-lactamase enzymes. Augmentin is a branded formulation of amoxicillin and clavulanic acid. The inclusion of clavulanic acid blocks bacterial breakdown of amoxicillin and therefore allows the same dosage of amoxicillin to reach a higher concentration in the patient.

Tetracyclines

The tetracyclines bind ribosomal proteins and inhibit bacterial protein synthesis. The first tetracycline was discovered in 1948. Tetracyclines are classified as polyketides, and all are either isolated from bacteria or derived from other tetracyclines. Common examples include tetracycline (**A.30**) and doxycycline (**A.31**) (**Figure A.8**).

Macrolides

Like tetracyclines, macrolides are also polyketides that are isolated from bacteria and inhibit protein synthesis in certain bacteria. Erythromycin (**A.32**) is the original macrolide (**Figure A.9**). Clarithromycin (Biaxin, **A.33**) and azithromycin (Zithromax, **A.34**) are semisynthetic derivatives of erythromycin.

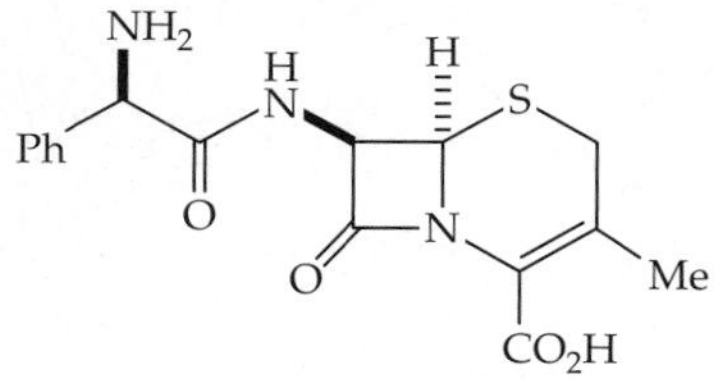

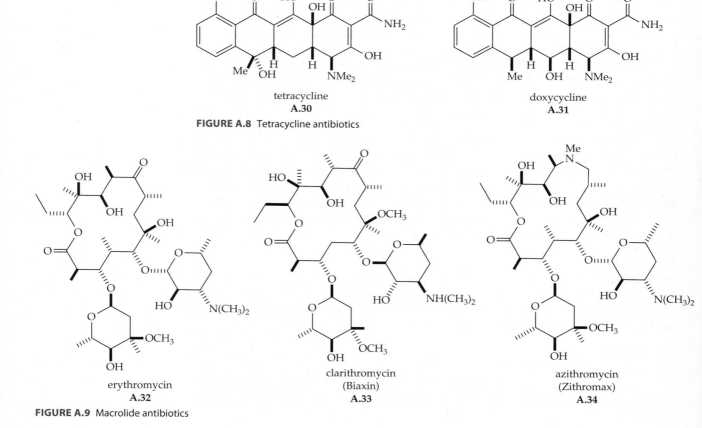

cephalexin
(Keflex)
A.24

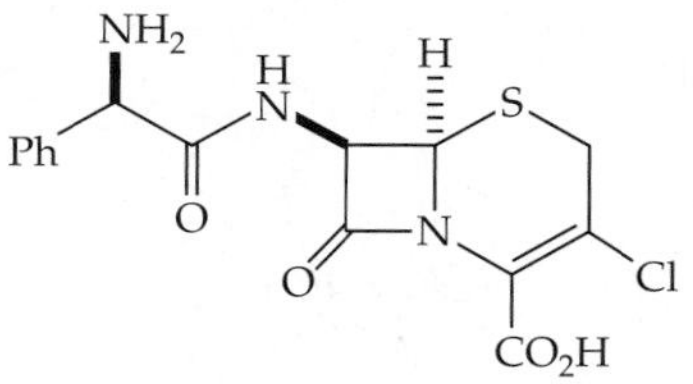

cefaclor
(Ceclor)
A.25

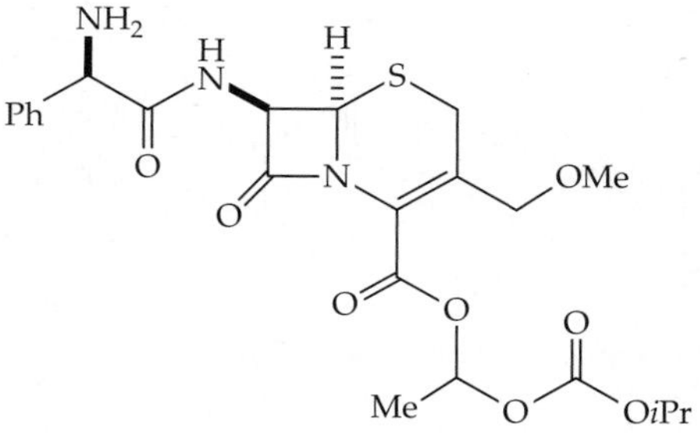

cefpodoxime proxetil
(Vantin)
A.26

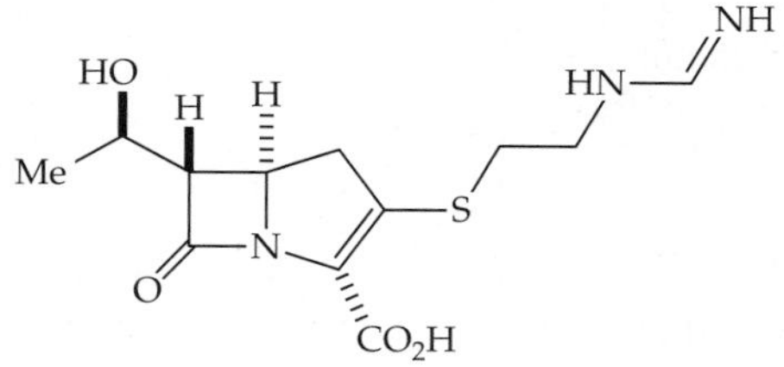

imipenem
A.27

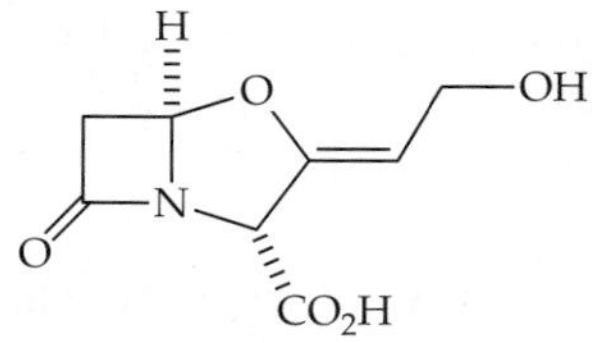

clavulanic acid
A.28

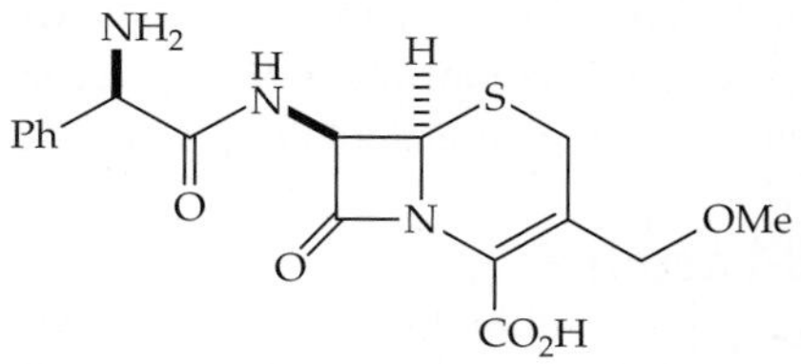

cefpodoxime
A.29

FIGURE A.7 Cephalosporin and carbepenem β-lactam antibiotics

tetracycline
A.30

doxycycline
A.31

FIGURE A.8 Tetracycline antibiotics

erythromycin
A.32

clarithromycin
(Biaxin)
A.33

azithromycin
(Zithromax)
A.34

FIGURE A.9 Macrolide antibiotics

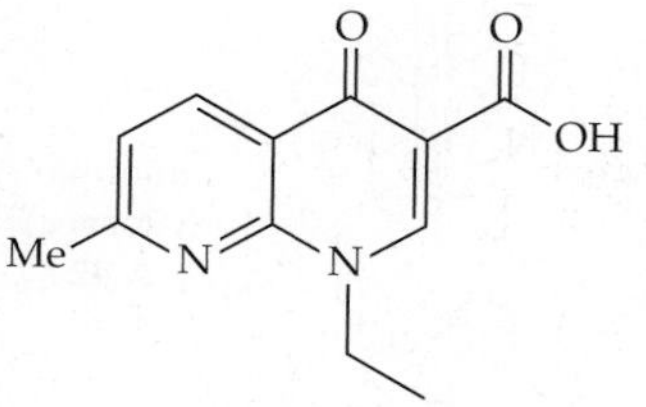

FIGURE A.10 Fluoroquinolone antibiotics

nalidixic acid
A.35

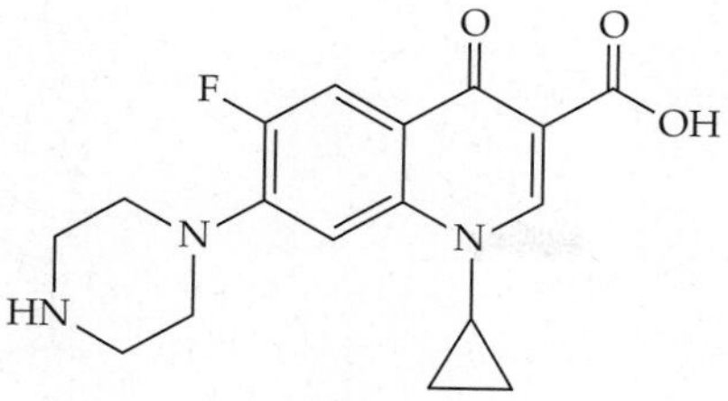

ciprofloxacin
(Cipro)
A.36

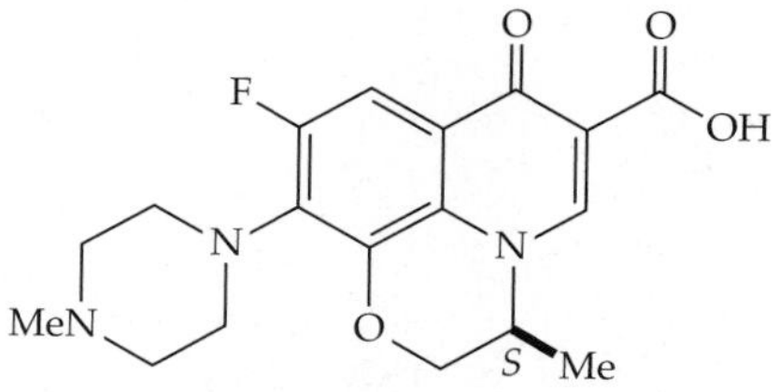

levofloxacin
(Levaquin)
A.37

FIGURE A.11 Miscellaneous antibiotics

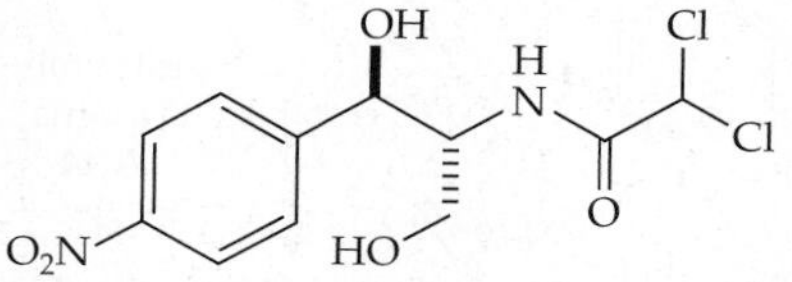

chloramphenicol
A.38

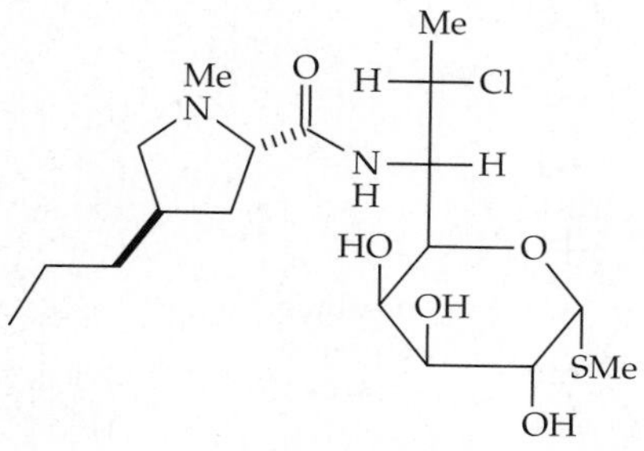

clindamycin
A.39

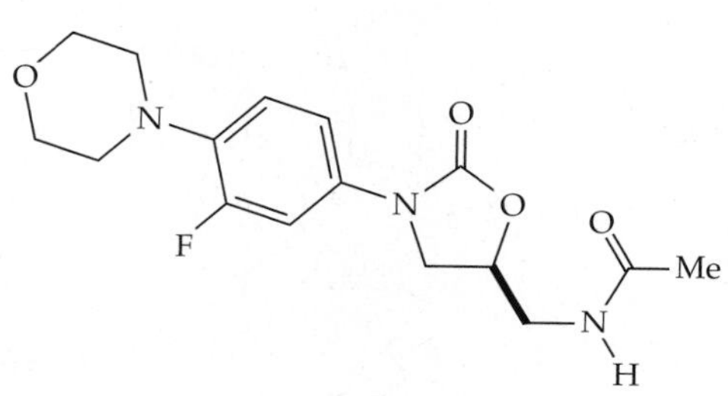

linezolid
A.40

Fluoroquinolones

The fluoroquinolone subclass of antibiotics is inspired by nalidixic acid (**A.35**), an older antibiotic commonly used for urinary tract infections (**Figure A.10**). Depending on the type of infection, fluoroquinolones inhibit DNA gyrase and/or topoisomerase IV. Both enzymes are vital for DNA replication and bacterial reproduction. Both ciprofloxacin (Cipro, **A.36**) and levofloxacin (Levaquin, **A.37**) are examples of fluoroquinolone antibiotics. Ciprofloxacin received national attention in the United States in 2001. Ciprofloxacin was used to treat many anthrax-infected patients after letters containing anthrax spores were sent through the U.S. postal service.

Miscellaneous Compounds

Many antibiotics do not fall into a well-defined subclass. Isolated examples include chloramphenicol (**A.38**), clindamycin (**A.39**), and linezolid (**A.40**) (**Figure A.11**). All inhibit bacterial protein synthesis.

A.3 Cancer

Chemotherapy agents, such as DNA alkylators (see Chapter 6), have traditionally been administered intravenously because of their high chemical reactivity and low bioavailability. During the past decade, a new class of oral anticancer drug, the tyrosine kinase inhibitors, has been developed. These inhibitors bind membrane-bound tyrosine kinase–linked receptors (TKLR) (see Chapter 5).

TKLRs have long been recognized as promising anticancer targets, but their breadth of cellular function poses a problem. Unless a specific TKLR can be bound with some

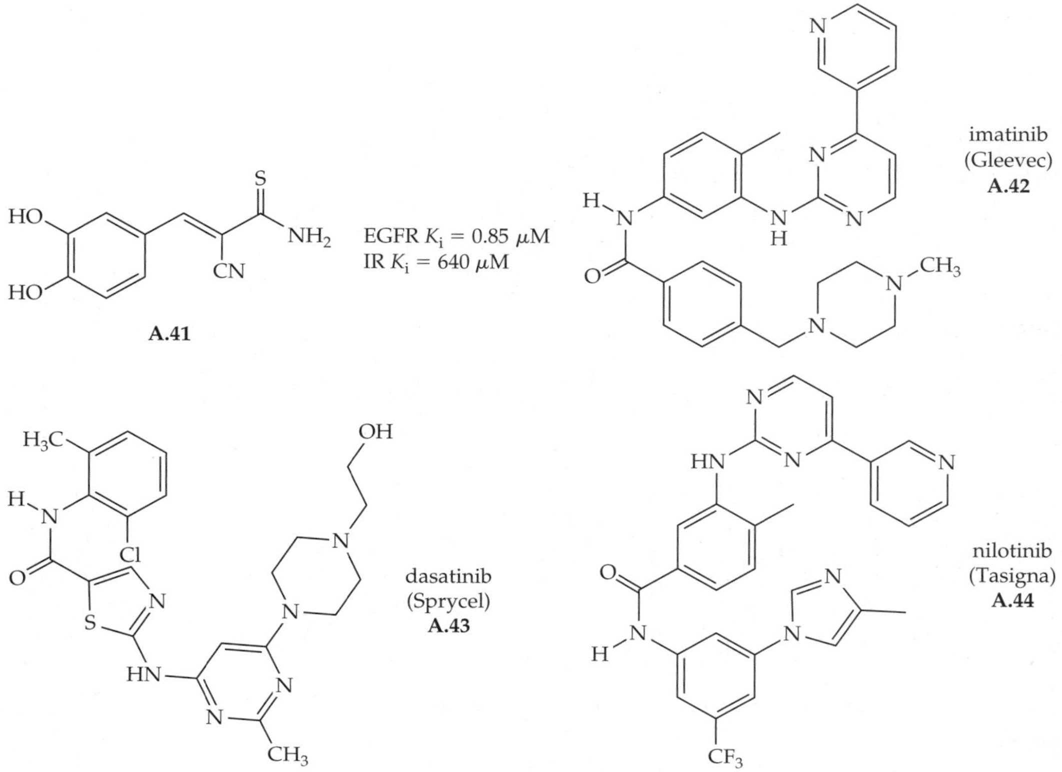

$EGFR\ K_i = 0.85\ \mu M$
$IR\ K_i = 640\ \mu M$

imatinib
(Gleevec)
A.42

A.41

dasatinib
(Sprycel)
A.43

nilotinib
(Tasigna)
A.44

FIGURE A.12 A prototypical TKLR and three approved drugs for CML

degree of selectivity, many off-target processes will be affected. A landmark advance in this area of research appeared in 1988 when a research group reported using a small, drug-like molecule (**A.41**) to inhibit the tyrosine kinase activity of epidermal growth factor receptor (EGFR) with minimal inhibition of insulin receptor (IR) (**Figure A.12**). In 2001, imatinib (Gleevec, **A.42**) was approved for the treatment of chronic myelogenous leukemia (CML). Imatinib actually binds and inhibits a kinase-active protein found in certain cancer cells, not a TKLR. Dasatinib (Sprycel, **A.43**) and nilotinib (Tasigna, **A.44**) have since also been approved for chronic myelogenous leukemia. Nilotinib is specifically indicated for patients whose tumors do not respond to imatinib.

While imatinib, dasatinib, and nilotinib are not true TKLR inhibitors, they are kinase inhibitors. Other approved drugs that do indeed inhibit the kinase activity of TKLRs include erlotinib (Tarceva, **A.45**) and gefitinib (Iressa, **A.46**) (**Figure A.13**). Both compounds primarily act at the epidermal growth factor receptor (EGFR or erbB1) and are used to treat certain types of lung cancer. Erlotinib is also effective against some pancreatic cancers. Lapatinib (Tykerb, **A.47**) is most effective against various breast cancers and inhibits erbB2, which is often called human epidermal growth factor 2 (HER2/neu).

Sunitinib (Sutent, **A.48**) inihibits the kinase activity at two receptors, vascular endothelial growth factor receptor (VEGFR) and platelet-derived growth factor receptor (PDGFR) (**Figure A.14**). By hitting two receptors, sunitinib has potentially broader activity as well as more side effects. Drugs like sunitinib that are designed to target more than one kinase are called *multikinase inhibitors*. Another multikinase inhibitor is sorafenib

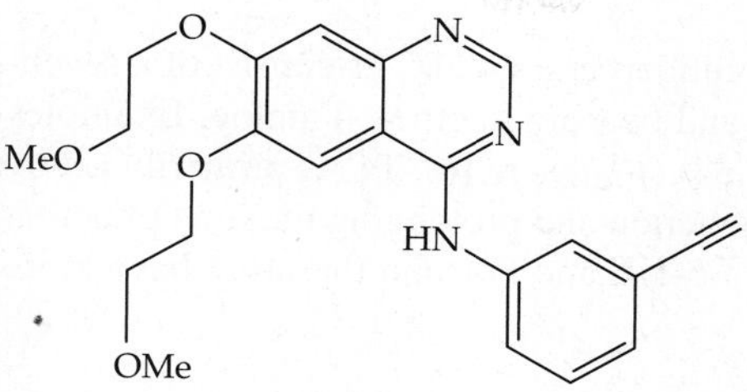

FIGURE A.13 Kinase inhibitors that act on the epidermal growth factor receptor family

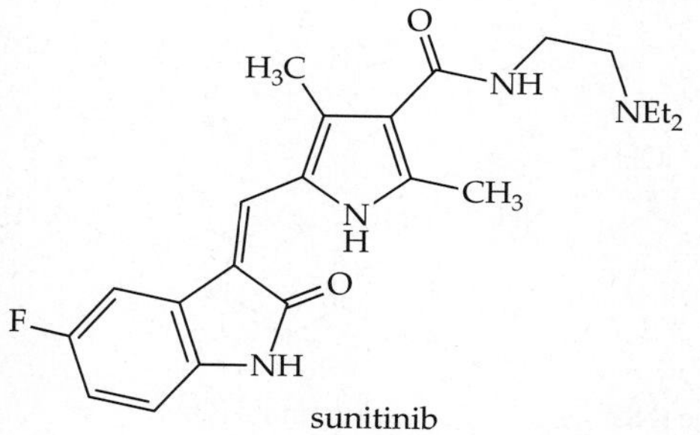

erlotinib
(Tarceva)
A.45

EGFR IC$_{50}$ 2 nM
erbB2 IC$_{50}$ 350 nM

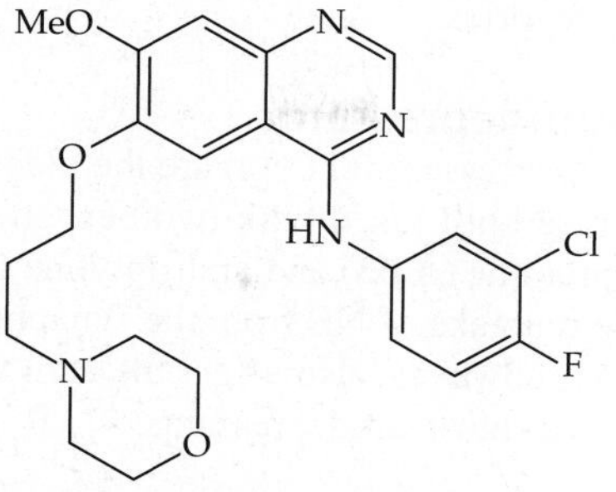

gefitinib
(Iressa)
A.46

EGFR IC$_{50}$ 1 nM
erbB2 IC$_{50}$ 240 nM

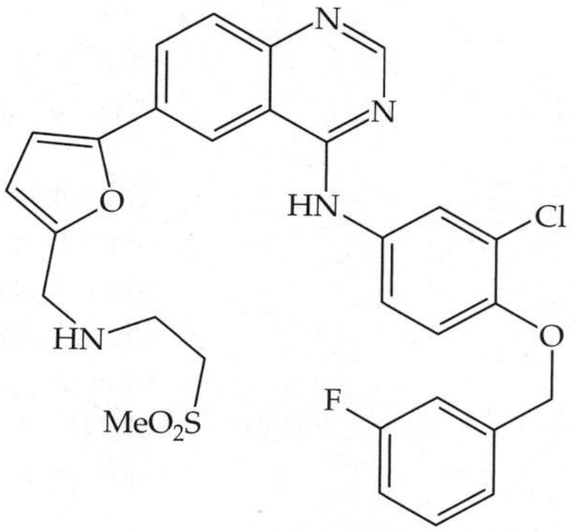

lapatinib
(Tykerb)
A.47

EGFR IC$_{50}$ 10 nM
erbB2 IC$_{50}$ 10 nM

sunitinib
(Sutent)
A.48

VEGFR2 IC$_{50}$ 38 nM
PDGFRβ IC$_{50}$ 55 nM

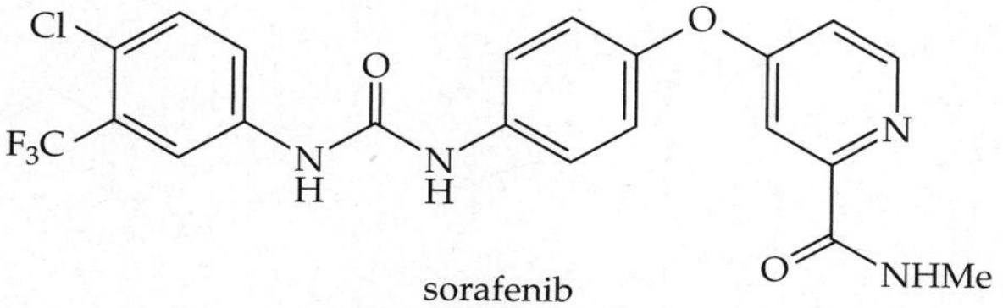

sorafenib
(Nexavar)
A.49

VEGFR2 IC$_{50}$ 90 nM
PDGFRβ IC$_{50}$ 57 nM
Raf1   IC$_{50}$ 6 nM

FIGURE A.14 Additional kinase inhibitors

(Nexavar, **A.49**). Sorafenib inhibits VEGFR, PDGFR, and another protein kinase called Raf1 kinase. Both sunitinib and sorafenib are active against some cancers of the kidney. Sunitinib is indicated against cancers of the gastrointestinal system as well, while sorafenib affects certain liver cancers.

Kinase inhibitors represent a dynamic area of research for new cancer therapies. Traditional chemotherapeutic agents target general cellular processes, namely cell division. The drugs, while effective against cancer cells, also kill many healthy cells in the patient and cause severe side effects. In contrast, kinase inhibitors are much more narrowly focused and may often be tailored to affect specific members of a receptor family. Kinase inhibitors effectively treat the cancer while minimizing side effects for the patient.

A.4 Clinical Depression

Clinical depression is a complicated condition. The general approach for treatment of depression involves increasing the availability of neurotransmitters, particularly serotonin

(5-HT, **A.50**), norepinephrine (NE, **A.51**), and dopamine (DA, **A.52**), in the synaptic junction (**Figure A.15**). Drugs that treat depression have been classified into several structural/functional categories.

Tricyclic Antidepressants

Tricyclic antidepressants (TCAs) are the oldest antidepressants. TCAs consist of a seven-membered ring both fused with two benzenes and bearing a tethered amine. Examples include imipramine (**A.53**) and amitriptyline (**A.54**) (**Figure A.16**). TCAs primarily act by blocking the reuptake of NE from the synaptic junction and prolonging the time of action of NE. TCAs, however, also affect the action of 5-HT and DA and therefore have more side effects than most antidepressants.

Monoamine Oxidase Inhibitors

Monoamine oxidases (MAOs) are found in the mitochondria of almost all cells. Most neurotransmitters are amines and subject to oxidation by MAOs. Members of the MAO inhibitor (MAOI) class of antidepressants slow the destruction of neurotransmitters in neurons and cause stronger signal transduction. Selected examples of MAOIs are shown in **Figure A.17**.

FIGURE A.15 Common neurotransmitters targeted by antidepressants

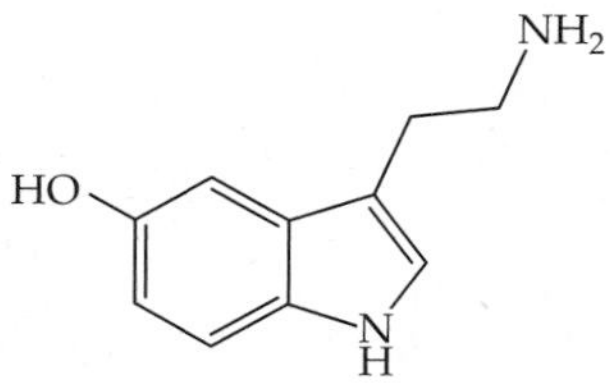

serotonin or
5-hydroxytryptamine (5-HT)
A.50

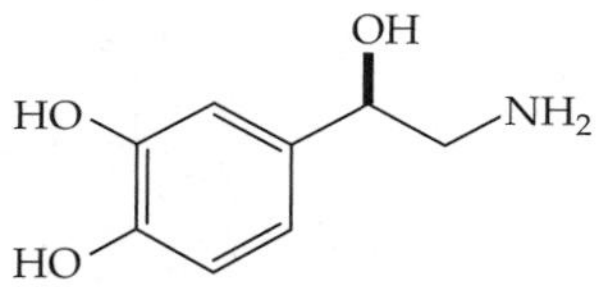

norepinephrine (NE)
A.51

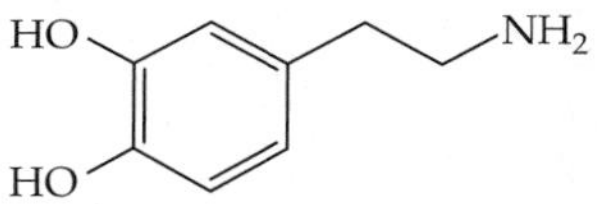

dopamine (DA)
A.52

FIGURE A.16 Common neurotransmitters targeted by antidepressants

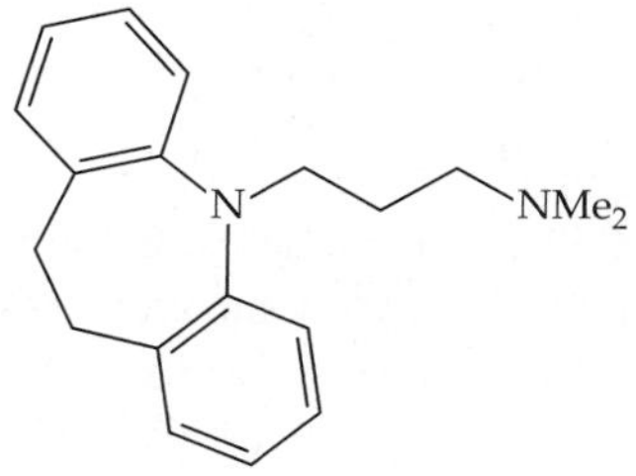

imipramine
A.53

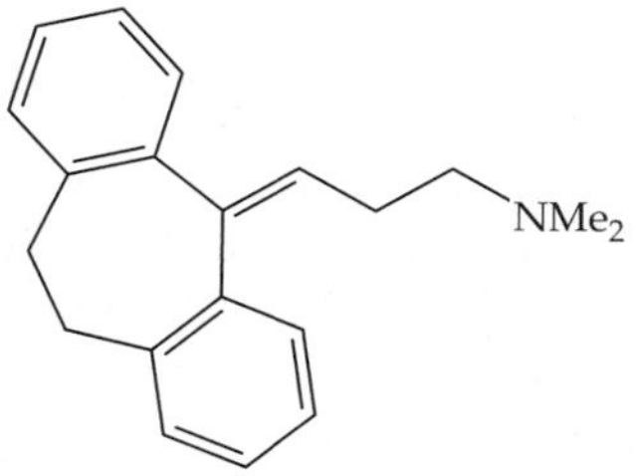

amitriptyline
A.54

FIGURE A.17 Common MAOIs

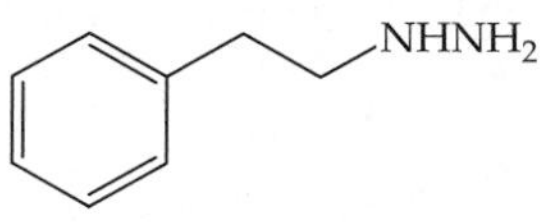

phenelzine
(Nardil)
A.55

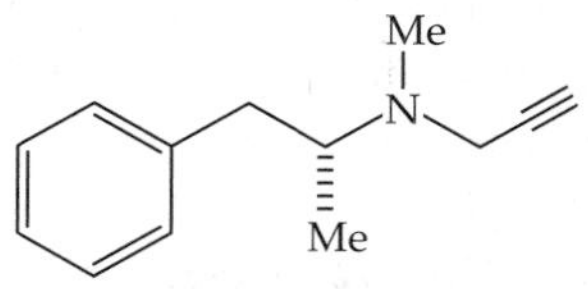

selegiline
(Eldepryl)
A.56

MAOs break down many compounds in the body and often metabolize drugs, and inhibition of MAOs introduces a cascade of unintended side effects. Because of the far-reaching effects of MAOIs, they are less favored than other antidepressants.

Selective Serotonin Reuptake Inhibitors

The selective serotonin reuptake inhibitors (SSRIs) block removal of 5-HT from the synapse by transport proteins. SSRIs are not notably more effective than other antidepressants, but their inherent selectivity minimizes side effects. Numerous SSRIs have been successful in the market and are considered to be "blockbuster" drugs (**Figure A.18**).

Serotonin-Norepinephrine Reuptake Inhibitors

The serotonin-norepinephrine reuptake inhibitors (SNRIs) block reabsorption of both 5-HT and NE. Examples of this class of antidepressant include venlafaxine (Effexor, **A.62**) and duloxetine (Cymbalta, **A.63**) (**Figure A.19**).

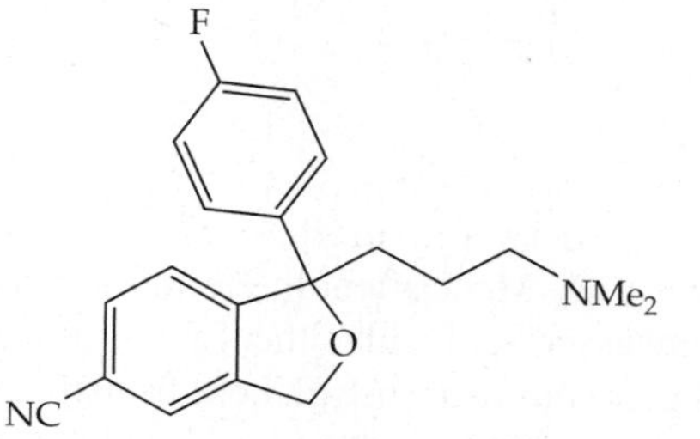

citalopram
(Celexa)
A.57

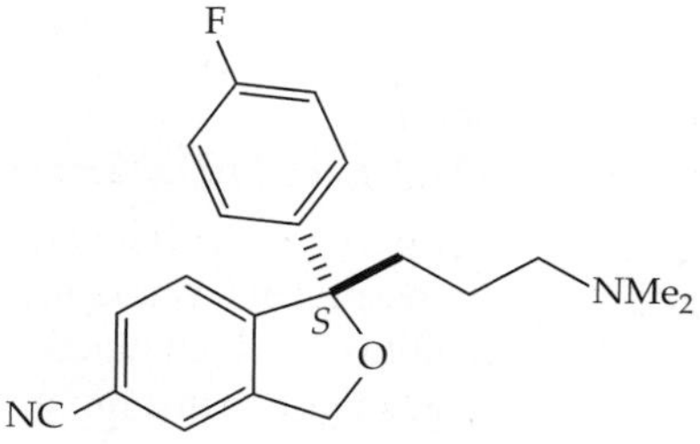

escitalopram
(Lexapro)
A.58

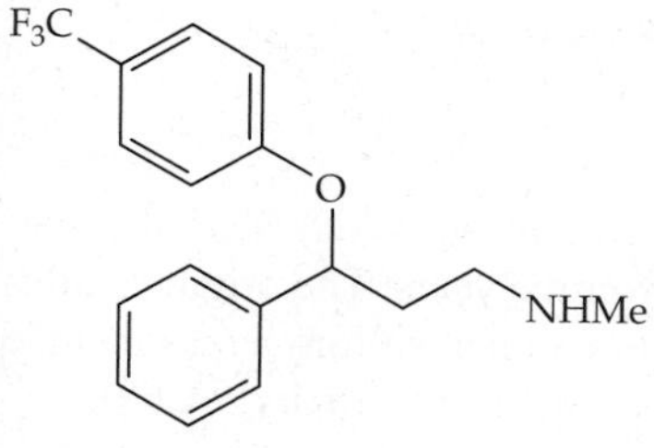

fluoxetine
(Prozac)
A.59

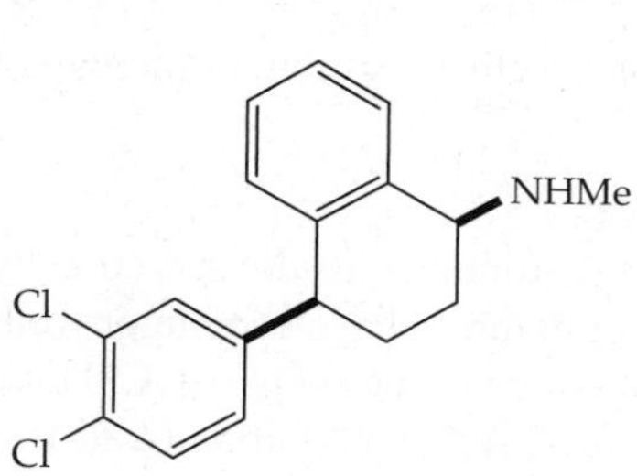

sertraline
(Zoloft)
A.60

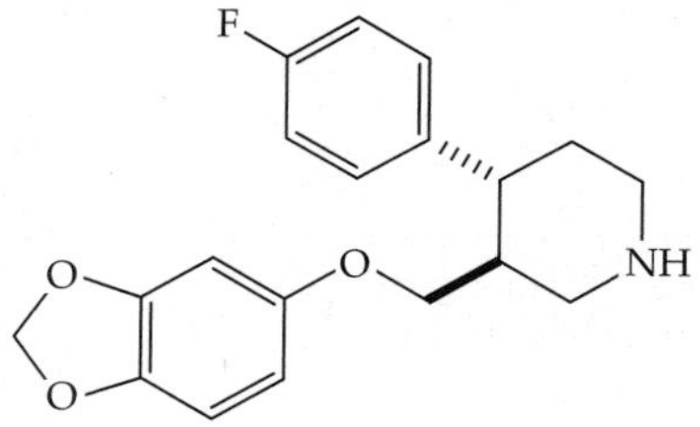

paroxetine
(Paxil)
A.61

FIGURE A.18 Successful SSRIs

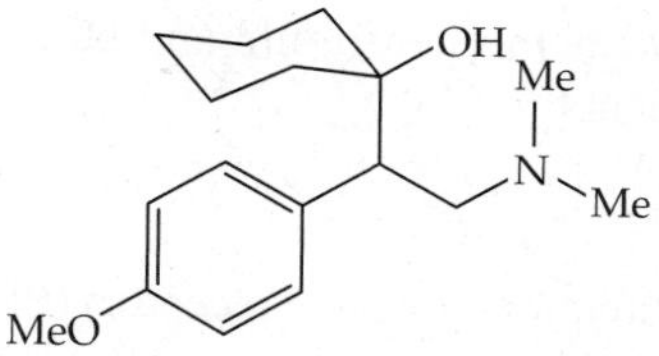

venlafaxine
(Effexor)
A.62

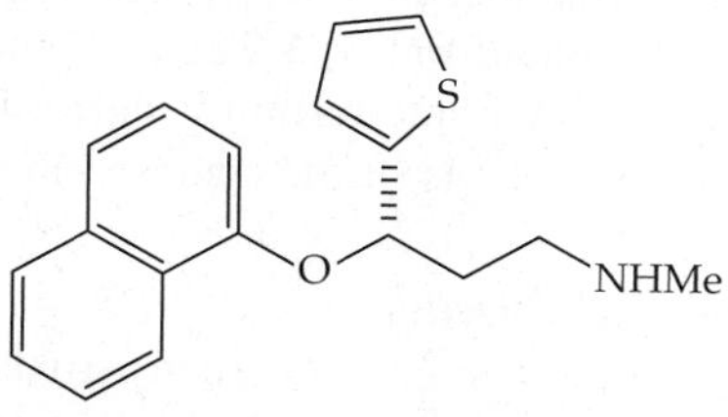

duloxetine
(Cymbalta)
A.63

FIGURE A.19 Selected SNRI antidepressants

FIGURE A.20 Miscellaneous antidepressants

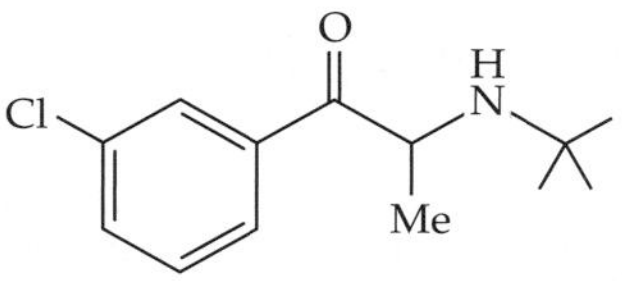

bupropion
(Wellbutrin)
A.64

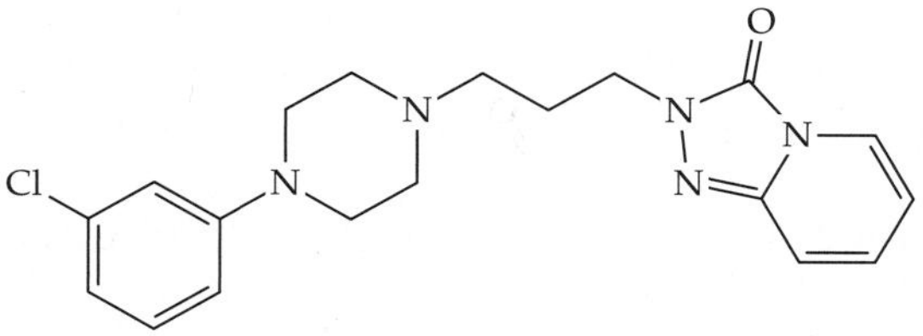

trazodone
(Desyrel)
A.65

Miscellaneous Antidepressants

Many compounds act as antidepressants but do not conveniently fall into a well-established category. Bupropion (Wellbutrin, **A.64**) is one example (**Figure A.20**). Bupropion inhibits reuptake of both NE and DA. Another example is trazodone (Desyrel, **A.65**), a compound which increases 5-HT levels in the synapse.

A.5 Diabetes—Type 2

Diabetes is a disease characterized by elevated levels of glucose in the blood (hyperglycemia) and urine (glycosuria). Most often, the term *diabetes* refers to diabetes mellitus in which glucose levels are associated with either low insulin levels or poor cellular response to insulin. Insulin triggers cells to absorb glucose from the bloodstream. Type 1 diabetes, or juvenile diabetes, is an autoimmune disorder in which the body destroys the insulin-producing cells in the pancreas. Treatment of type 1 diabetes requires insulin via injection or inhalation. In type 2 diabetes, or adult-onset diabetes, cells fail to respond properly to insulin. Treatments involve drugs that increase insulin secretion by the pancreatic cells, sensitize cells to insulin, or increase glucose uptake by cells.

Sulfonylureas

The sulfonylureas stimulate insulin secretion by pancreatic cells. As an old class of diabetes drug, the sulfonylureas have been improved steadily over the years. The original sulfonylurea is tolbutamide (**A.66**) (**Figure A.21**). Later and more potent sulfonylureas include glipizide (Glucotrol, **A.67**), glyburide (**A.68**), and finally glimepiride (Amaryl, **A.69**).

Biguanides

The antidiabetic activity of the biguanides was discovered in the late 1950s. As a class, the biguanides both increase cellular glucose uptake and decrease glucose output by the liver. The biguanides do not affect insulin levels. Three biguanides, metformin (Glucophage, **A.70**), phenformin (**A.71**), and buformin (**A.72**), have been used to treat diabetes (**Figure A.22**). Only metformin is currently used, as both phenformin and buformin elevate lactic acid levels (lactic acidosis) in many patients.

Thiazolidinediones

Thiazolidinediones stimulate certain peroxisome proliferator–activated receptors (PPAR-γ). PPAR-γ is a nuclear receptor and, through a series of events, increases cellular production of insulin-dependent enzymes. This is an example of upregulation. The cell is then more sensitive to the decreased insulin levels found in a person with type 2 diabetes. The two thiazolidinediones currently on the market are rosiglitazone (Avandia, **A.73**) and

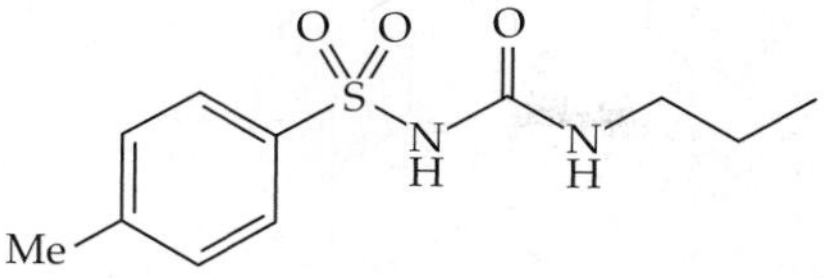

FIGURE A.21 Antidiabetic sulfonylureas

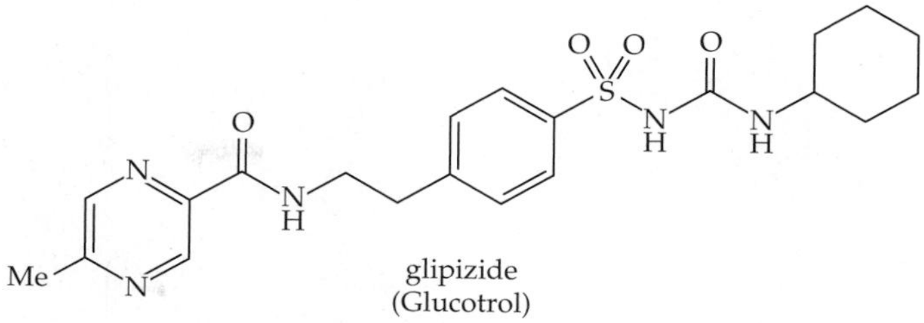

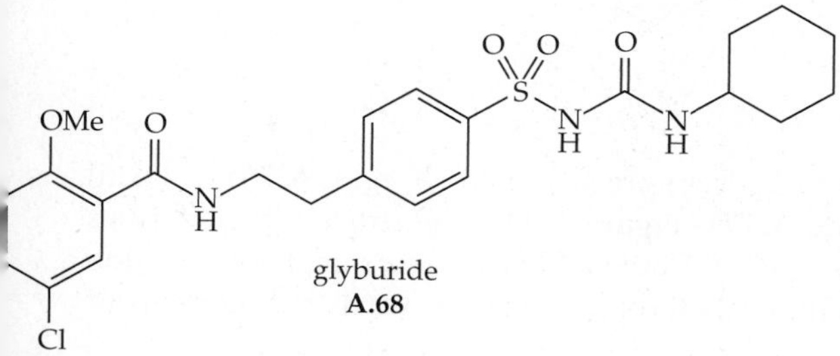

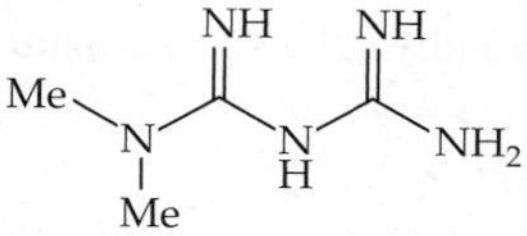

FIGURE A.22 Antidiabetic biguanides

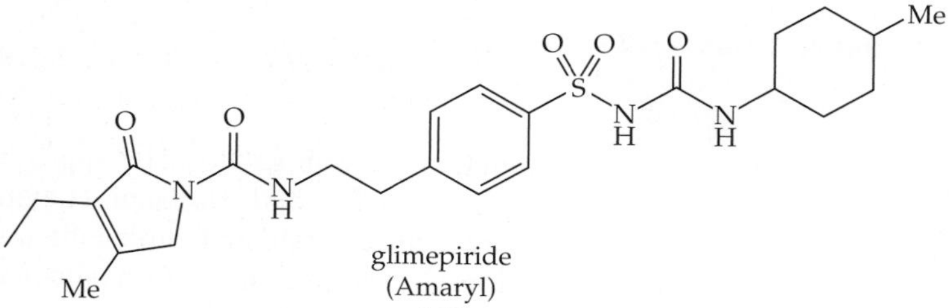

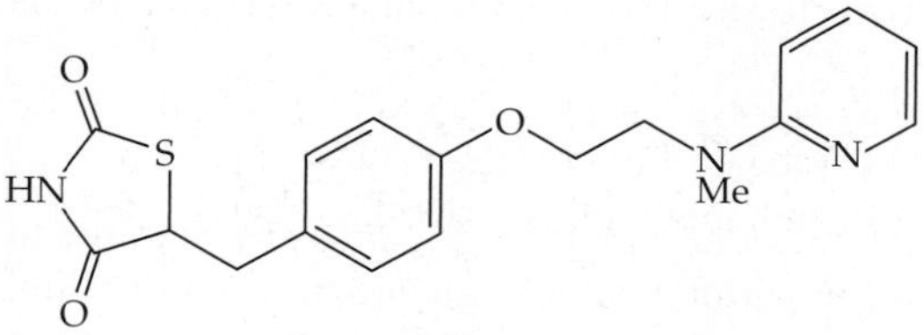

FIGURE A.23 Antidiabetic thiazolidinediones

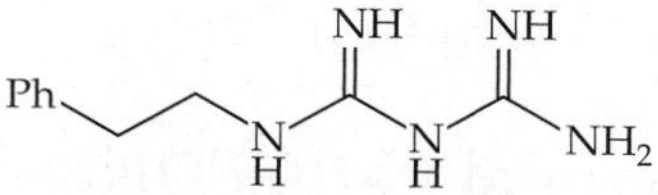

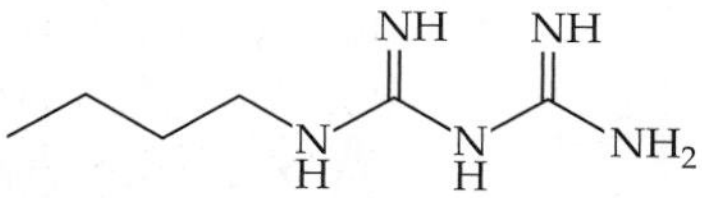

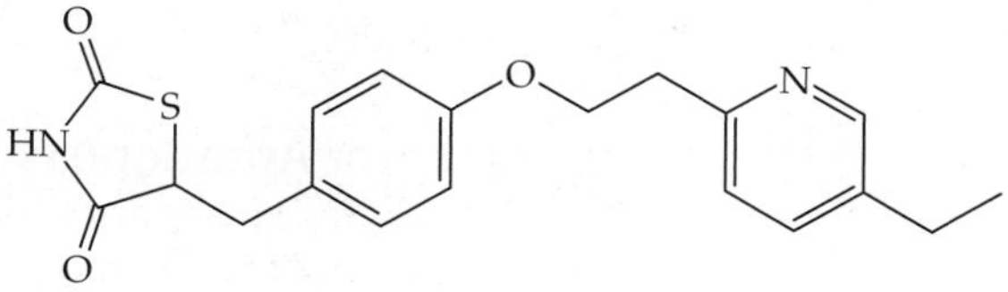

pioglitazone (Actos, **A.74**) (**Figure A.23**). Rosiglitazone has recently been implicated in raising cardiovascular risks.

A.6 Erectile Dysfunction

The approved drugs for erectile dysfunction (ED) all share the same mode of action, which is inhibition of phosphodiesterase type 5 (PDE5). PDE5 hydrolyzes the phosphodiester of cyclic GMP (cGMP). Elevated levels of cGMP relax the muscle tissue that lines the blood vessels of the corpus cavernosum in the penis. Blood freely enters the tissue,

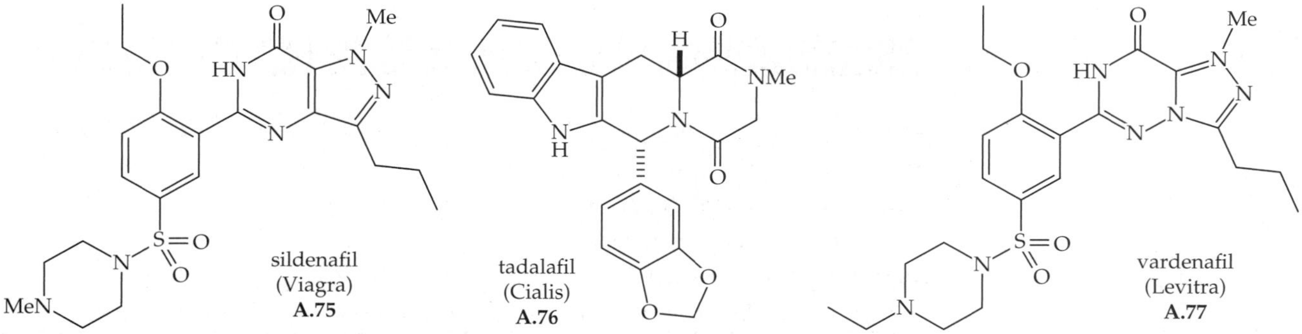

FIGURE A.24 Erectile dysfunction drugs

and an erection results. The approved ED drugs are sildenafil (Viagra, **A.75**), tadalafil (Cialis, **A.76**), and vardenafil (Levitra, **A.77**) (**Figure A.24**). The structural similarities between sildenafil and vardenafil are striking. Vardenafil is a nice example of a me-too drug. Vardenafil differs from sildenafil mostly through several isosteric substitutions of nitrogen and carbon.

Although ED does not pose the same level of health risk as bacterial infections and diabetes, ED drugs have been hugely profitable for their manufacturers. The drugs have been widely advertised and even promoted by Bob Dole, a former U.S. senator and candidate for U.S. president. The drugs have also drawn criticism due to abuse by recreational users.

A.7 Gastroesophageal Reflux Disorder

Gastroesophageal reflux disorder (GERD) involves irritation of the esophagus by a backflow of the acidic contents of the stomach. GERD is caused by problems with the cardia, which is the junction of the esophagus and stomach. No known drugs treat cardia failure. Instead, GERD is treated by suppressing gastric acid secretion in the stomach. The most common drugs that treat GERD fall into two classes: H_2-receptor antagonists and proton pump inhibitors.

H$_2$-Receptor Antagonists

The H_2-receptor antagonists are antihistamines that block the binding of histamine to the H_2-receptor. The H_2-receptor is found in parietal cells in the stomach epithelium. Binding of histamine to the H_2-receptor triggers the release of acid into the stomach by parietal cells. Blocking the H_2-receptor can suppress acid secretion. H_2-receptor blocking drugs, although originally classified as antagonists, actually act as inverse agonists. Despite their true mode of action, H_2-receptor blockers are still frequently called antagonists.

The prototypical H_2-receptor antagonist is cimetidine (Tagamet, **A.78**), which was approved by the U.S. Food and Drug Administration (FDA) in 1979 (**Figure A.25**). Ranitidine (Zantac, **A.79**), famotidine (Pepcid, **A.80**), and nizatidine (Axid, **A.81**) were marketed during the early and mid-1980s. All four drugs have similar structures and modes of binding to the H_2-receptor.

Proton Pump Inhibitors

Proton pump inhibitors, like H_2-receptor antagonists, act at parietal cells and decrease gastric acid secretion. Proton pump inhibitors specifically block proton pumps,

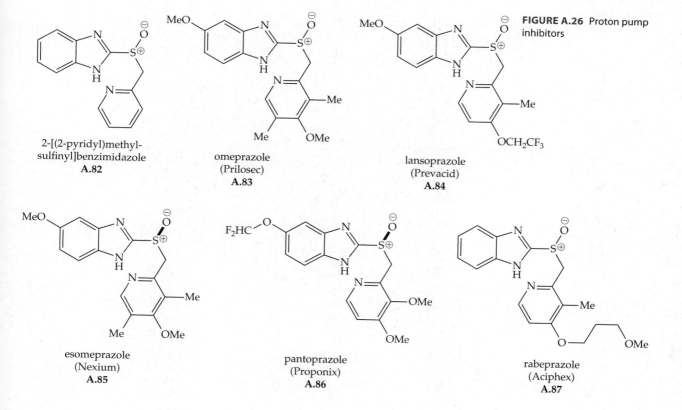

FIGURE A.25 H$_2$-receptor antagonists

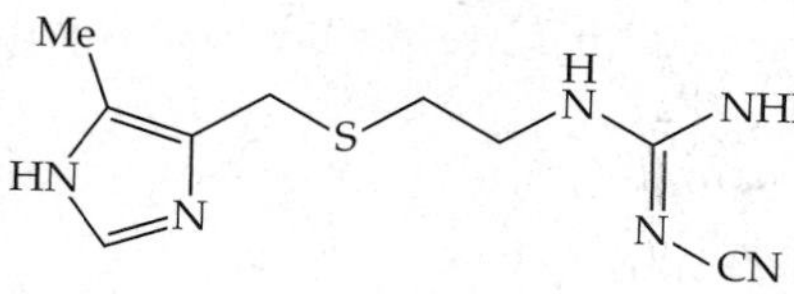

cimetidine
(Tagamet)
A.78

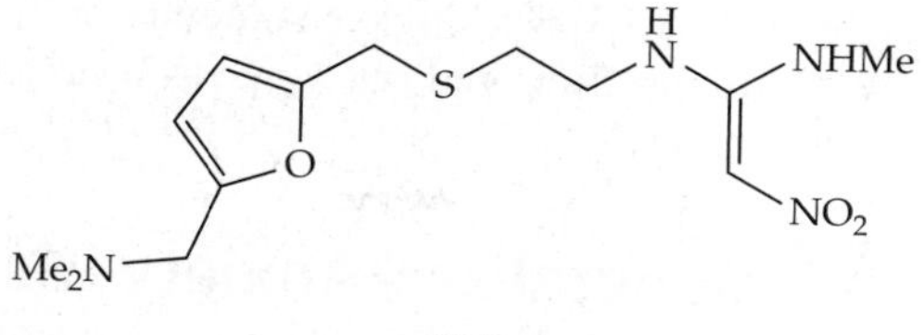

ranitidine
(Zantac)
A.79

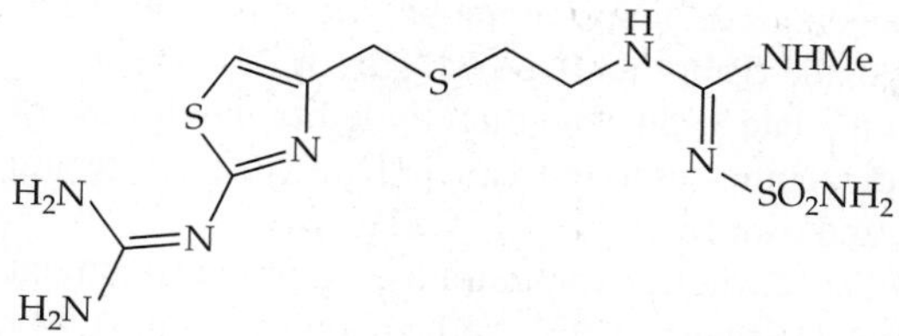

famotidine
(Pepcid)
A.80

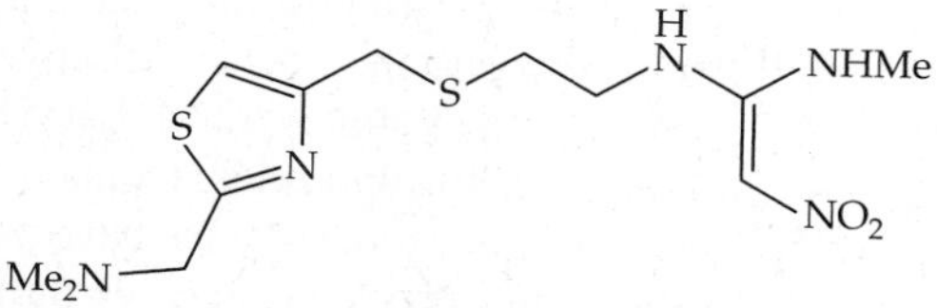

nizatidine
(Axid)
A.81

transmembrane proteins that shuttle protons across the cell membrane. Proton pumps are the last step in the acid secretion pathway. Being downstream of H$_2$-receptors, proton pumps are a more favorable point of intervention and give rise to fewer side effects. All approved proton pump inhibitors contain a 2-[(2-pyridyl)methylsulfinyl]benzimidazole core (**A.82**) (**Figure A.26**). Omeprazole (Prilosec, **A.83**) is a racemate, while esomeprazole,

FIGURE A.26 Proton pump inhibitors

2-[(2-pyridyl)methyl-
sulfinyl]benzimidazole
A.82

omeprazole
(Prilosec)
A.83

lansoprazole
(Prevacid)
A.84

esomeprazole
(Nexium)
A.85

pantoprazole
(Proponix)
A.86

rabeprazole
(Aciphex)
A.87

(Nexium, **A.85**) is a single enantiomer of the chiral sulfoxide. Compounds **A.83** through **A.87** are prodrugs and rearrange to form an active species that irreversibly deactivates proton pumps.

A.8 Generalized Anxiety Disorder

Excessive and irrational worry along with concern characterize generalized anxiety disorder (GAD). GAD is most commonly treated with benzodiazepine drugs. Benzodiazepines bind at the GABA$_A$ receptor, a ligand-gated ion channel. Benzodiazepine binding increases the affinity of the GABA$_A$ receptor for its endogenous ligand, γ-aminobutyric acid. The prototypical benzodiazepine drug is chlordiazepoxide (Librium, **A.88**) (**Figure A.27**). The activity of chlordiazepoxide was discovered in the late 1950s and directly led to the discovery of diazepam (Valium, **A.89**). Later benzodiazepines used to treat GAD include lorazepam (Ativan, **A.90**), alprazolam (Xanax. **A.91**), and oxazepam (Serax, **A.92**).

Many benzodiazepines have potent hypnotic activity and are useful in the treatment of insomnia. Examples include flurazepam (Dalmane, **A.93**), midazolam (Versed, **A.94**), temazepam (Restoril, **A.95**), and triazolam (Halcion, **A.96**) (**Figure A.28**). Flunitrazepam (Rohypnol, **A.97**) is a particularly notorious sedative. Often called "roofie" or "the date rape drug," flunitrazepam causes sedation and amnesia. Because of flunitrazepam's tendency to be abused, almost all nations tightly regulate the drug's availability.

Benzodiazepine drugs can cause a range of side effects, including dizziness, confusion, and impaired motor control. Long-term benzodiazepine use can lead to physical dependence. Despite these problems, benzodiazepines have remained the standard anti-anxiety drugs for approximately 50 years.

FIGURE A.27 Benzodiazepines used to treat GAD

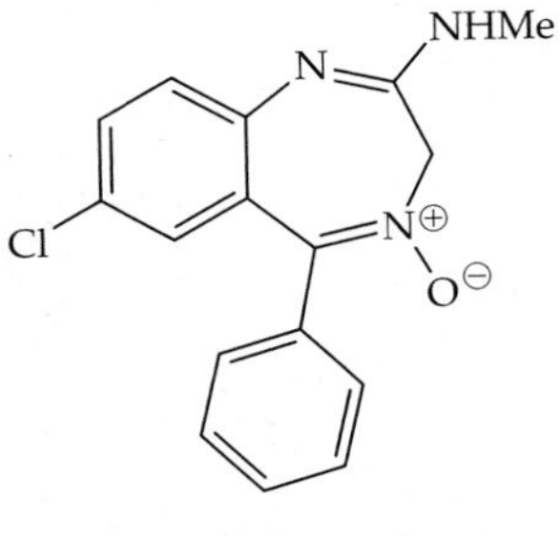

chlordiazepoxide
(Librium)
A.88

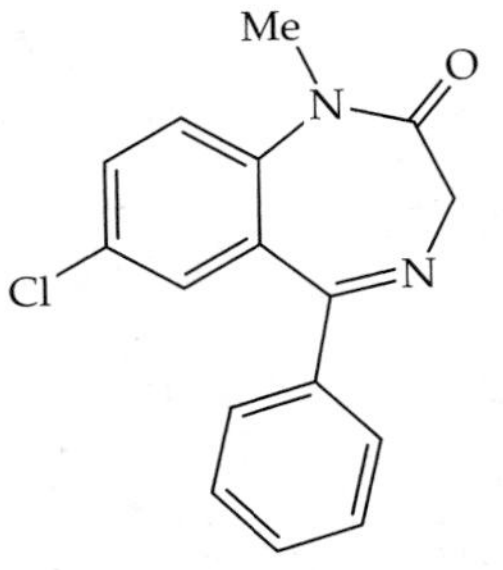

diazepam
(Valium)
A.89

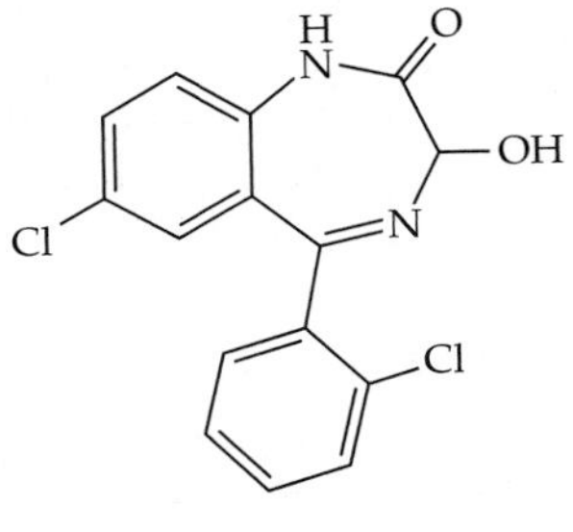

lorazepam
(Ativan)
A.90

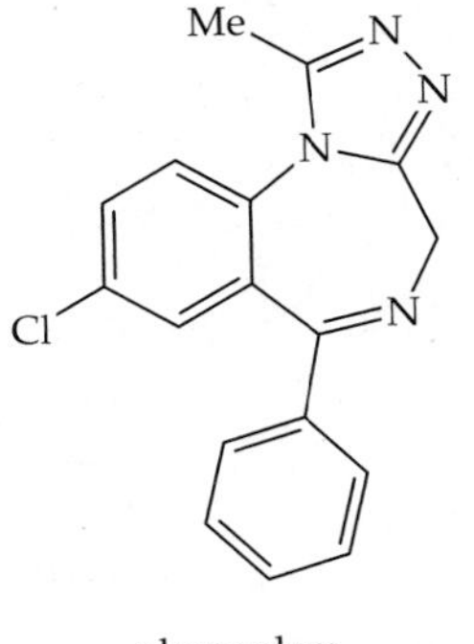

alprazolam
(Xanax)
A.91

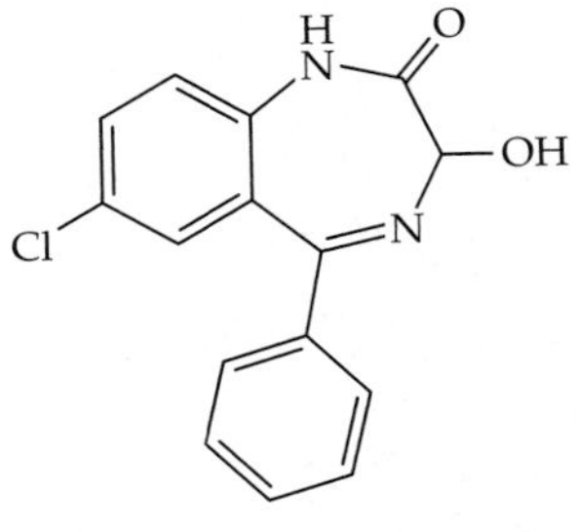

oxazepam
(Serax)
A.92

FIGURE A.28 Sedative benzodiazepines

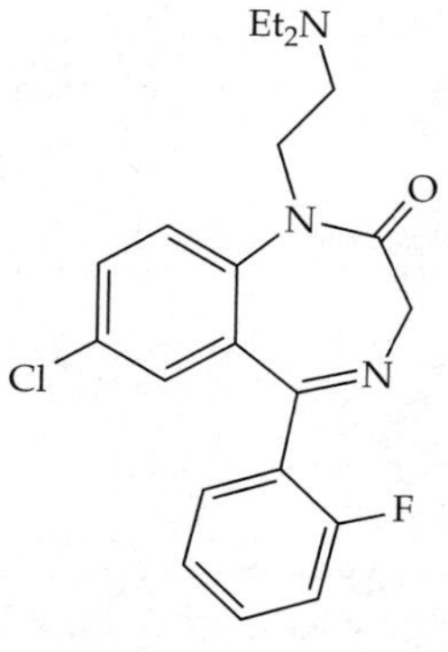

flurazepam
(Dalmane)
A.93

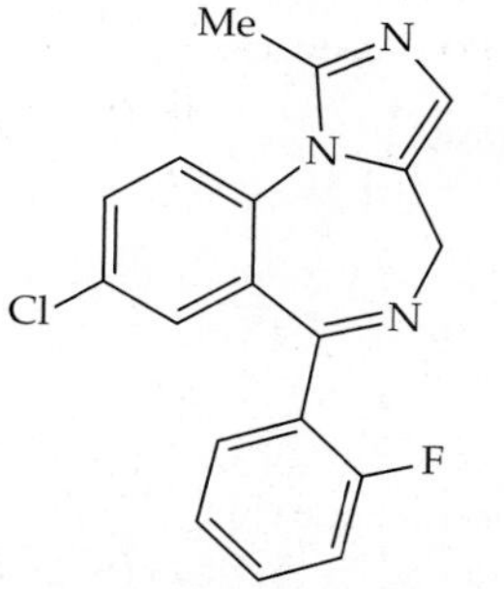

midazolam
(Versed)
A.94

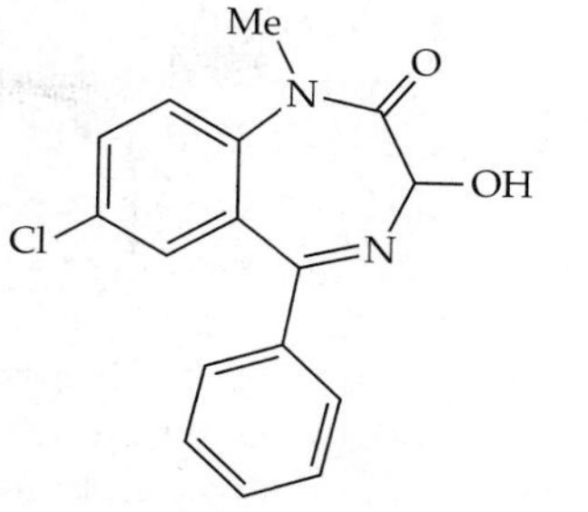

temazepam
(Restoril)
A.95

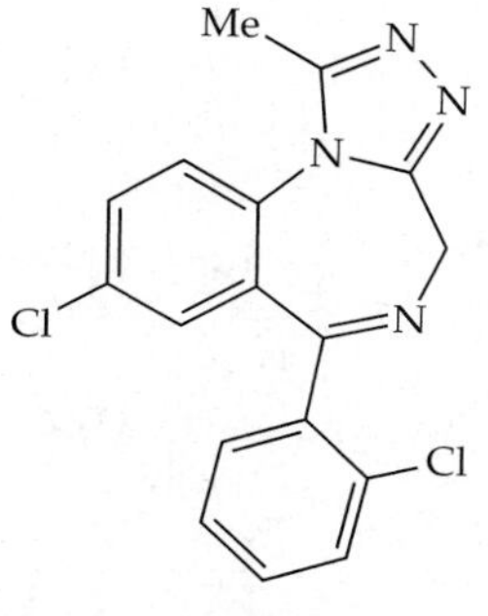

triazolam
(Halcion)
A.96

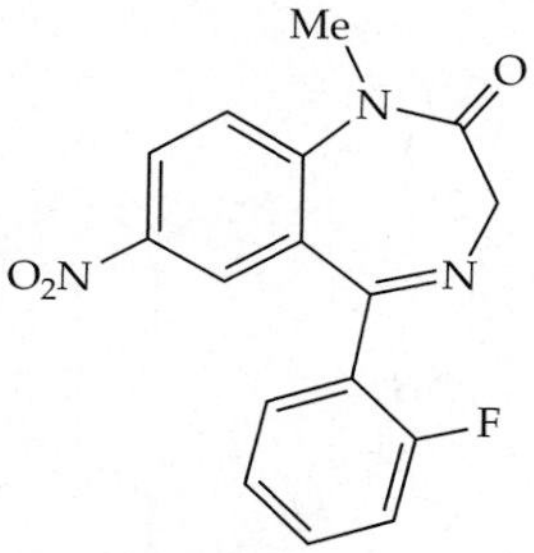

flunitrazepam
(Rohypnol)
A.97

A.9 Hyperlipidemia

Hyperlipidemia is a condition characterized by the presence of elevated lipoprotein levels in the blood. The term *hyperlipidemia* encompasses a number of different conditions, but it most often refers to high levels of cholesterol in the form of low-density lipoprotein (LDL). LDL cholesterol is often called "bad cholesterol." High-density lipoprotein (HDL) is the "good" form of cholesterol. High LDL and/or low HDL levels are widely believed to be linked to increased heart disease risk. Because of the prevalence of hyperlipidemia in developed nations, antihyperlipidemic drugs are in high demand.

Statins

One class of antihyperlipidemic drugs is the statins. Statins interfere with the biosynthesis of cholesterol (**A.103**) and specifically inhibit the enzyme 3-hydroxy-3-methylglutaryl-CoA (HMG-CoA) reductase (**Scheme A.1**). The statins that have been approved by the FDA include lovastatin (Mevacor, **A.104**), simvastatin (Zocor, **A.105**), pravastatin (Pravachol, **A.106**), atorvastatin (Lipitor, **A.107**), rosuvastatin (Crestor, **A.108**), and fluvastatin (Lescol, **A.109**) (**Figure A.29**). All six compounds are drawn here to highlight the similarities between HMG-CoA (**A.99**) and mevalonic acid (**A.100**), and the top portion of the various statins. As a class, the statins have been extremely successful in terms of sales and effective in decreasing LDL cholesterol levels in the blood.

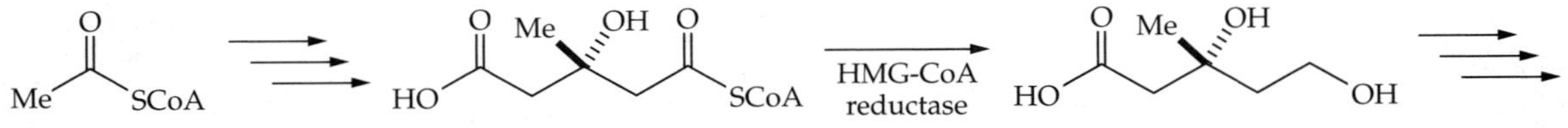

acetyl-CoA
A.98

3-hydroxy-3-methylglutaryl-CoA
(HMG-CoA)
A.99

HMG-CoA
reductase

mevalonic acid
A.100

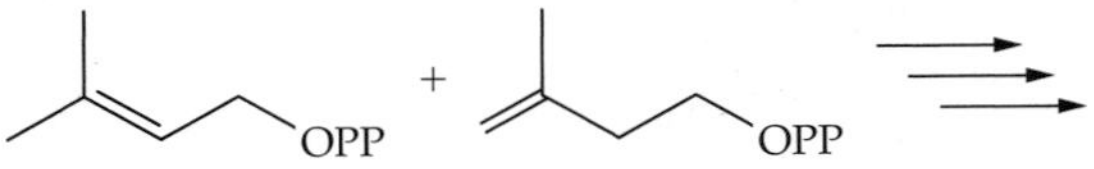

dimethylallyl
pyrophosphate
A.101

isopentenyl
pyrophosphate
A.102

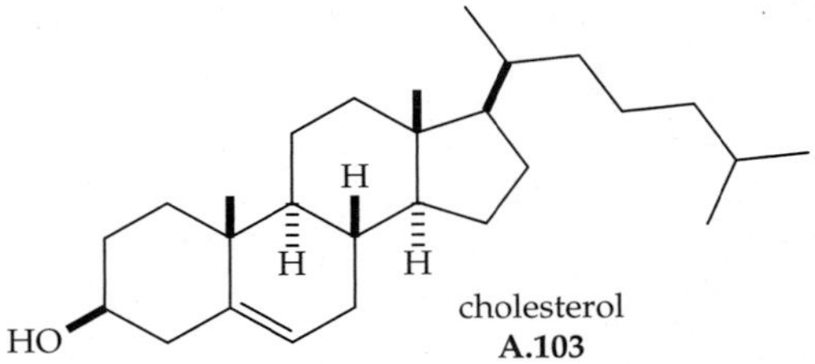

cholesterol
A.103

SCHEME A.1 Biosynthesis of cholesterol (**A.103**)

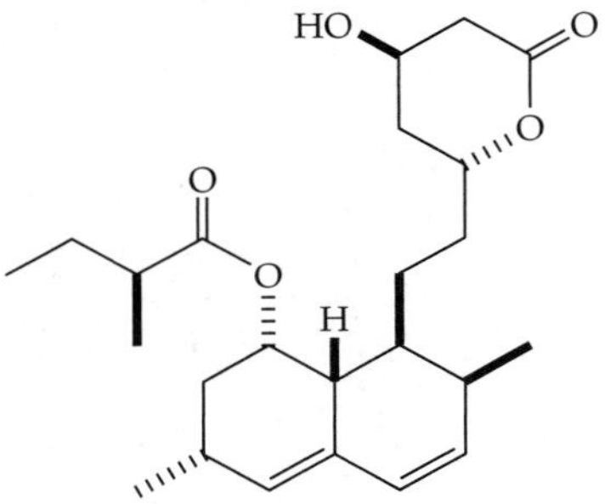

lovastatin
(Mevacor)
A.104

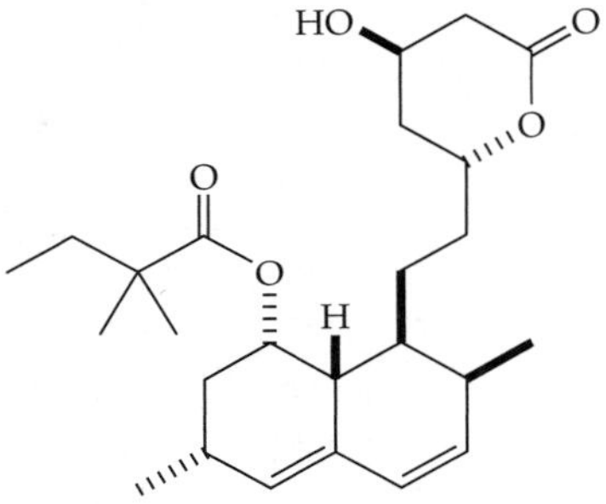

simvastatin
(Zocor)
A.105

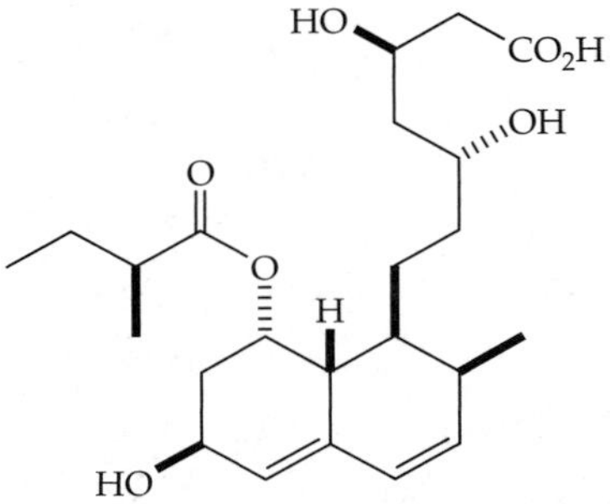

pravastatin
(Pravachol)
A.106

atorvastatin
(Lipitor)
A.107

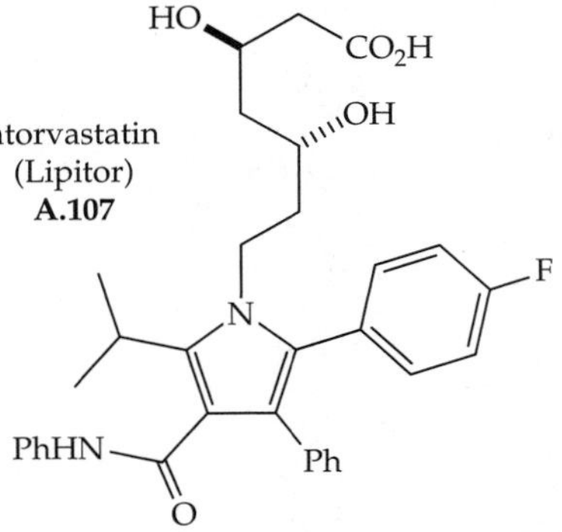

rosuvastatin
(Crestor)
A.108

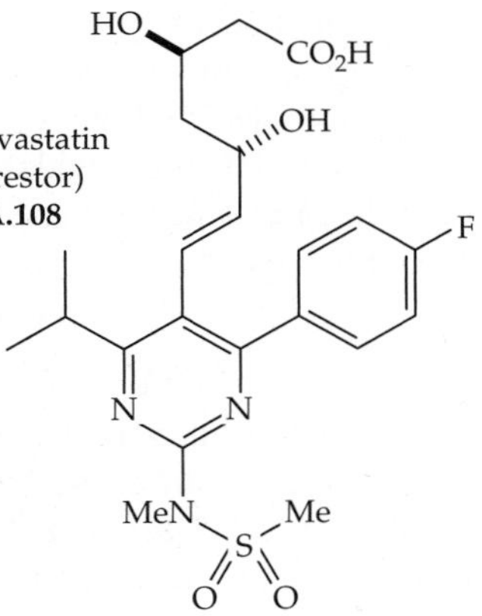

fluvastatin
(Lescol)
A.109

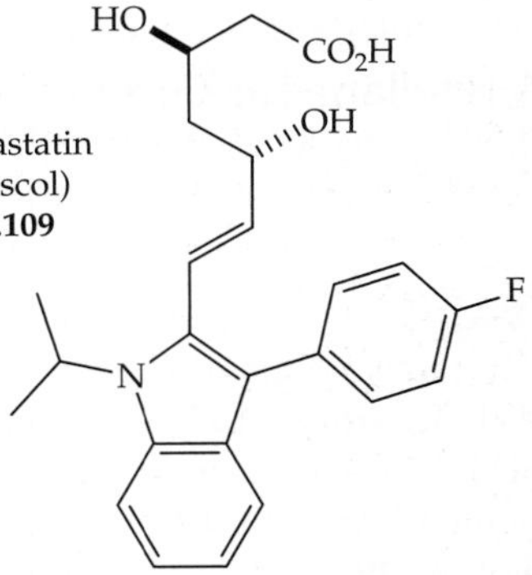

FIGURE A.29 Statins: HMG-CoA reductase inhibitors

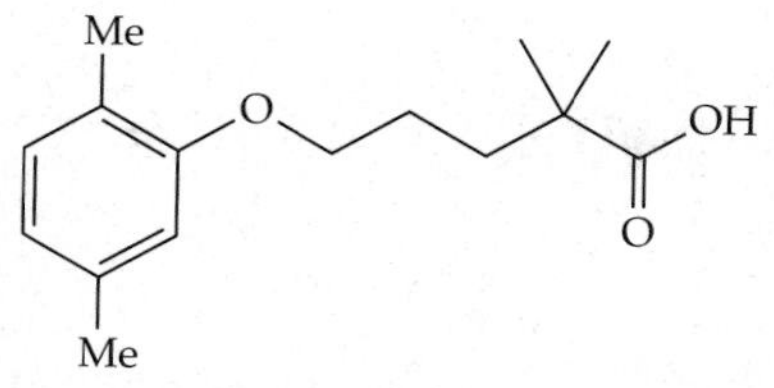

FIGURE A.30 Fibrate antihyperlipidemics

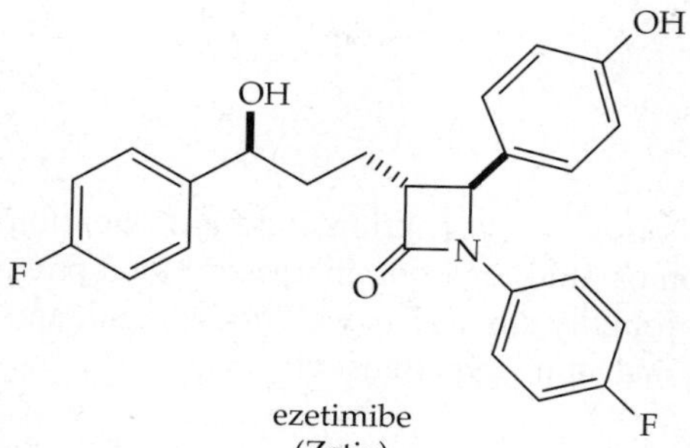

gemfibrozil
(Lopid)
A.110

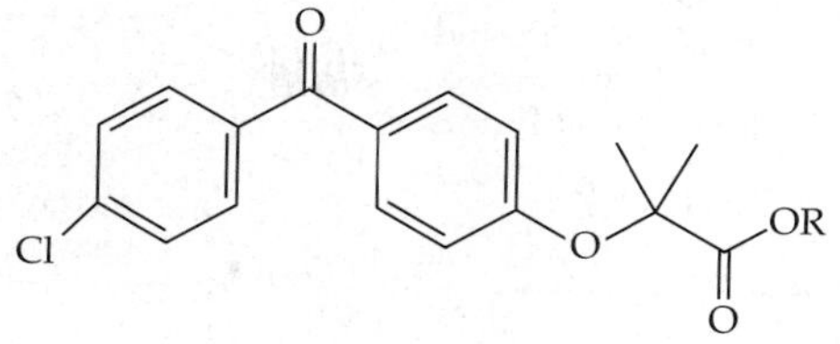

fenofibrate (Tricor) (R = *i*Pr), **A.111**
fenofibric acid (R = H), **A.112**

ezetimibe
(Zetia)
A.113

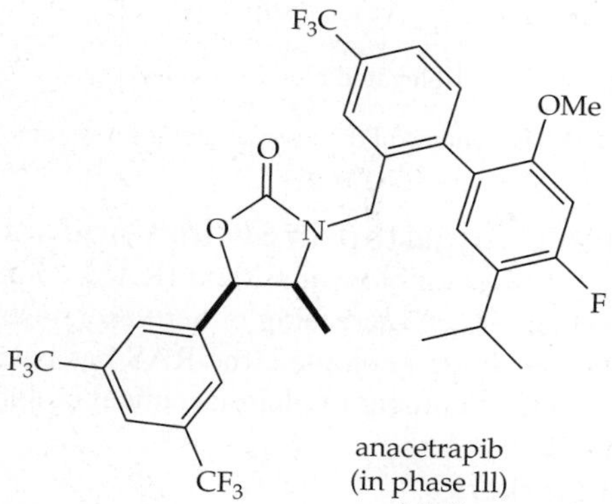

anacetrapib
(in phase lll)
A.114

FIGURE A.31 Additional antihyperlipidemics

Fibrates

The fibrates are another class of antihyperlipidemic drug and are frequently coadminis-tered with a statin. Fibrates act as agonists of the peroxisome proliferator–activated re-ceptors (PPAR), particularly PPAR-α. PPARs are nuclear receptors that influence gene expression and lipid metabolism. Examples of fibrates include gemfibrozil (Lopid, **A.110**) and fenofibrate (Tricor, **A.111**) (**Figure A.30**). Fenofibrate is hydrolyzed in the body to its active form, fenofibric acid (**A.112**). Fibrates do not decrease LDL levels as effectively as statins, but fibrates do elevate HDL cholesterol levels.

Miscellaneous Compounds

A recently developed antihyperlipidemic is ezetimibe (Zetia, **A.113**) (**Figure A.31**). Ezetimibe inhibits the absorption of cholesterol across the intestinal wall. Like fibrates, ezetimibe is often prescribed with statins, although the effectiveness of ezetimibe has re-cently been called into question. A compound that may soon be approved for the treatment of high cholesterol is anacetrapib (**A.114**). Anacetrapib, a product of Merck, is currently in phase III trials. The compound inhibits cholesteryl ester transfer protein (CETP). The net effect of CETP inhibition is elevated HDL cholesterol and lower LDL cholesterol levels.

A.10 Hypertension

Even more so than hyperlipidemia, hypertension, or elevated blood pressure, poses a sig-nificant health risk for a large portion of the population in developed nations. A large number of antihypertensive drugs have been developed.

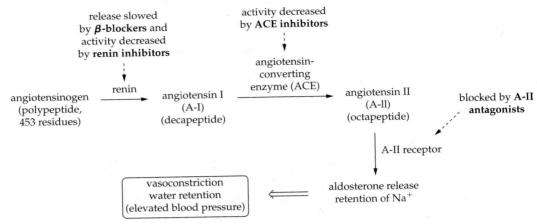

SCHEME A.2 A basic outline of the renin-angiotensin system

Renin-Angiotensin System Agents

The renin-angiotensin system (RAS) controls both vasoconstriction and fluid retention (**Scheme A.2**). Increasing either vasoconstriction or fluid retention increases blood pressure. Although complex, the RAS has been thoroughly studied, is well understood, and provides a number of pharmaceutical targets for managing hypertension.

β-Blockers

β-Blockers antagonize β-adrenergic receptors. β-Adrenergic receptors bind epinephrine (**A.115**) and norepinephrine (**A.51**) as their natural ligands (**Figure A.32**). All β-blockers retain the ethanolamine subunit found in epinephrine and norepinephrine. Several subtypes of the β-adrenergic receptor have been identified. The β_1-adrenergic receptor controls the release of renin in the kidneys as well as both heart rate and strength of heart muscle contraction. The β_2 and β_3 subtypes impact smooth muscle relaxation as well as metabolism. Some β-blockers are selective for the β_1 subtype, and others are nonselective (**Figure A.33**). Carvedilol (Coreg, **A.119**) antagonizes both the α- and β-adrenergic receptors and combines the effects of α_1- and β-blockers. α_1-Blockers are discussed in this section on non–renin-angiotensin system agents.

Angiotensin-Converting Enzyme Inhibitors

Angiotensin-converting enzyme (ACE) inhibitors interfere with the renin-angiotensin system. ACE inhibitors block the cleavage of angiotensin I to angiotensin II. ACE inhibitors are peptidomimetics and closely resemble part of the structure of angiotensin I, a decapeptide (see Chapter 11). Accordingly, most ACE inhibitors are categorized as carboxyalkyl dipeptides (**Figure A.34**). An exception is fosinopril, a phosphinate ester (Monopril, **A.125**). Aside from lisinopril, all orally administered ACE inhibitors in Figure A.34 are prodrugs. The active species forms once the ester is hydrolyzed to the corresponding acid.

FIGURE A.32 Endogenous ligands for adrenergic receptors

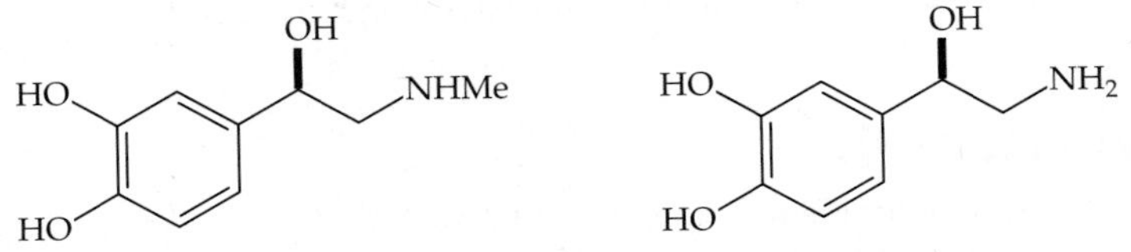

epinephrine
A.115

norepinephrine
A.51

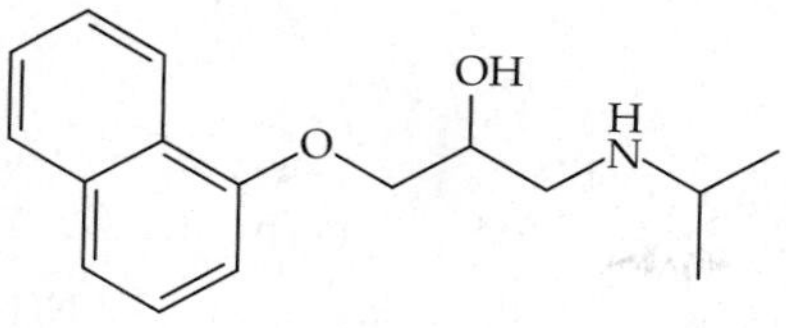

FIGURE A.33 β-Blockers

propranolol
(Inderal)
(nonselective β)
A.116

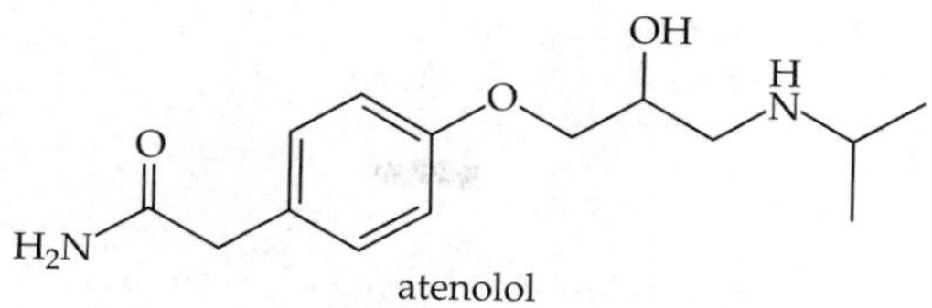

atenolol
(Tenormin)
(selective β₁)
A.117

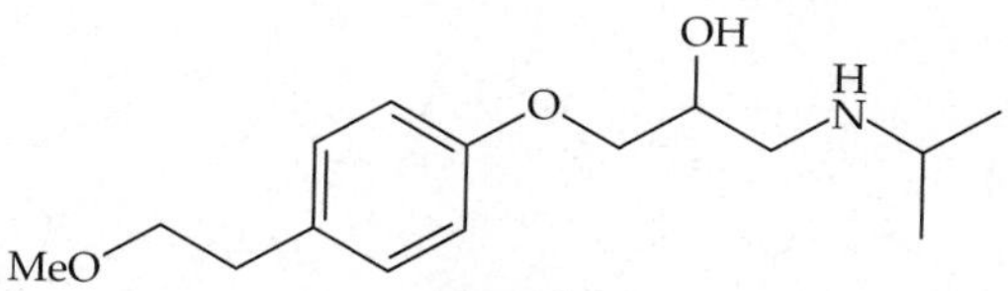

metoprolol
(Lopressor)
(selective β₁)
A.118

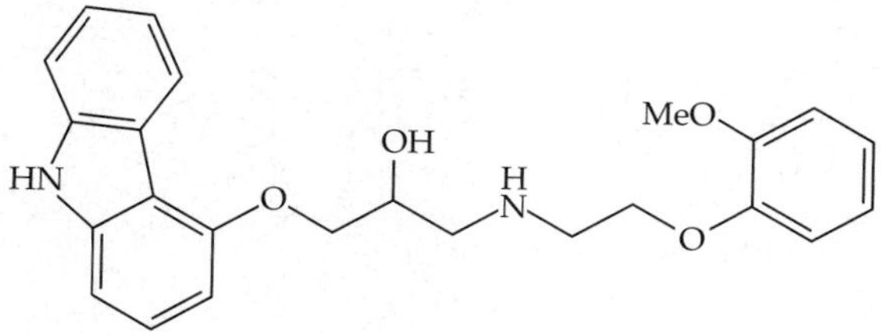

carvedilol
(Coreg)
(mixed α₁ and β)
A.119

FIGURE A.34 ACE inhibitors

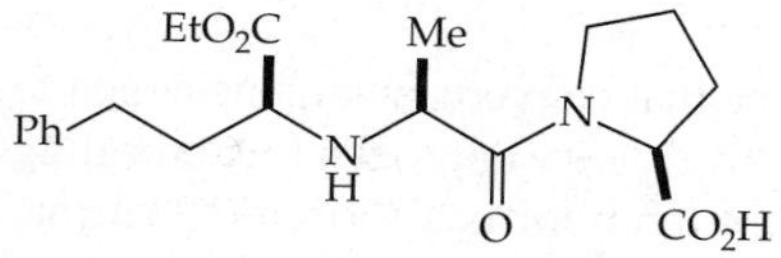

enalapril
(Vasotec)
A.120

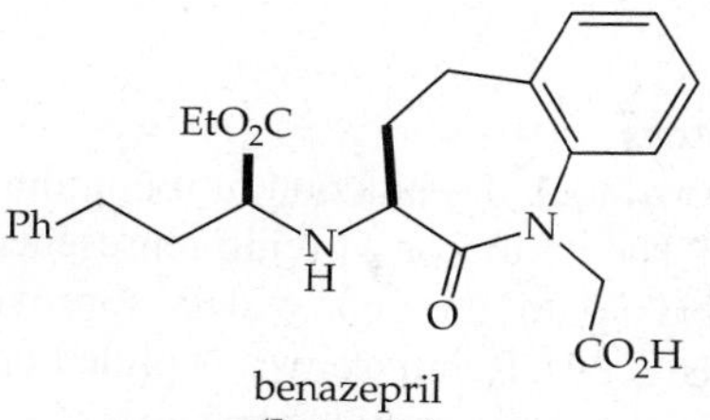

benazepril
(Lotensin)
A.121

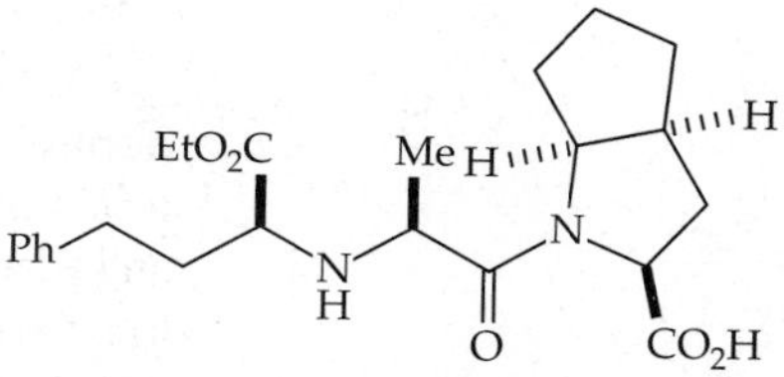

ramipril
(Altace)
A.122

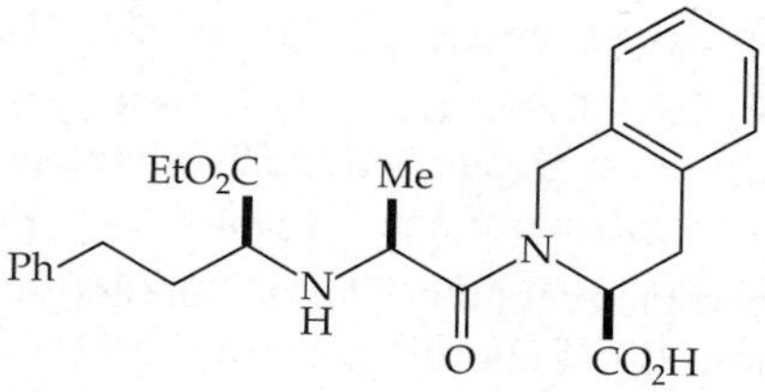

quinapril
(Accupril)
A.123

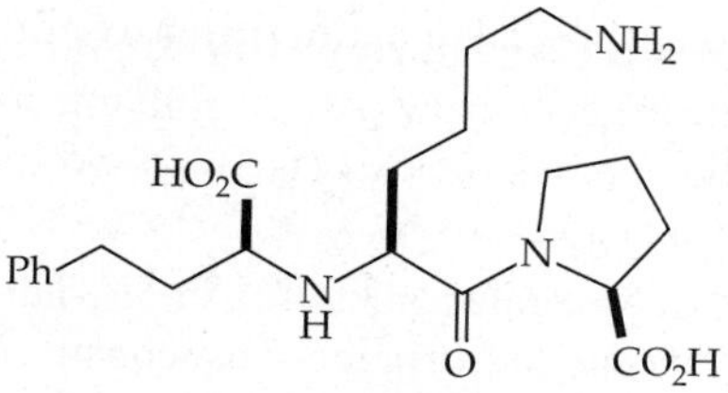

lisinopril
(Prinivil)
A.124

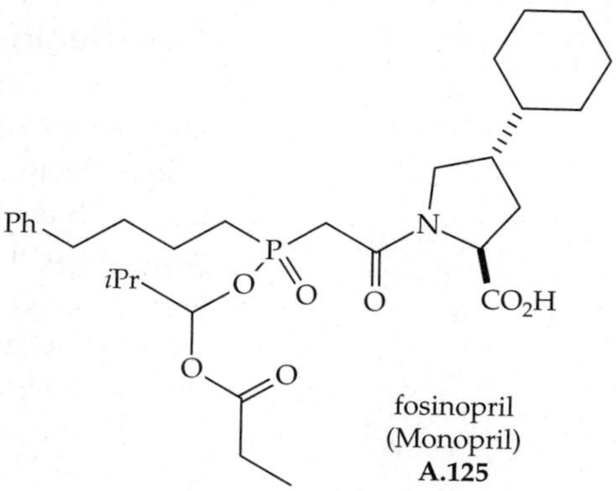

fosinopril
(Monopril)
A.125

Angiotensin II Antagonists

The angiotensin II antagonists block the binding of angiotensin II to the angiotensin II receptor. Common angiotensin II antagonists are shown in **Figure A.35**. All contain a tetrazole ring, which serves as an isostere for a carboxylic acid.

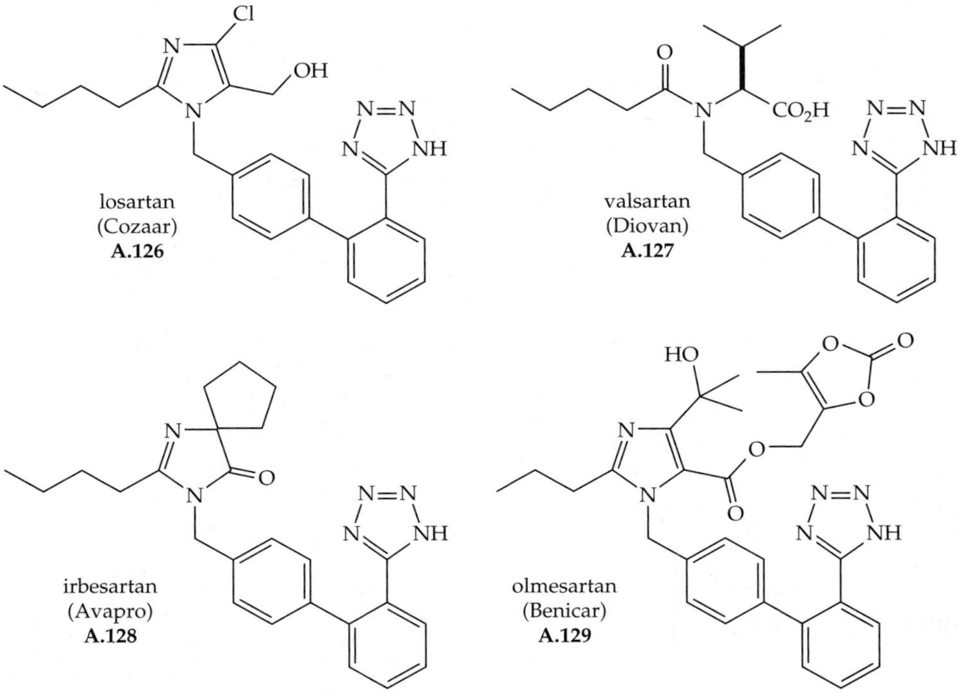

FIGURE A.35 Angiotensin II receptor antagonists

losartan
(Cozaar)
A.126

valsartan
(Diovan)
A.127

irbesartan
(Avapro)
A.128

olmesartan
(Benicar)
A.129

Renin Inhibitors

Renin inhibitors block the function of renin, the enzyme that converts angiotensinogen to angiotensin I. The inhibition of renin represents a relatively new approach to controlling hypertension. To date, the only widely approved renin inhibitor is aliskiren (Tekturna, **A.130**) (**Figure A.36**). Renin cleaves peptide bonds. Not surprisingly, aliskiren, as a peptidomimetic, has a distinct peptide appearance.

Non–Renin-Angiotensin System Agents

Not all antihypertensive drugs act through the renin-angiotensin system. The primary non–renin-angiotensin system antihypertensive drug classes are the α-blockers and calcium channel blockers.

α-Blockers

The α-blockers act as antagonists for the α_1-adrenergic receptor. The α_1-adrenergic receptor, when stimulated, triggers vasoconstriction. Blocking this receptor prevents vasoconstriction and reduces blood pressure. Doxazosin (Cardura, **A.131**) is structurally

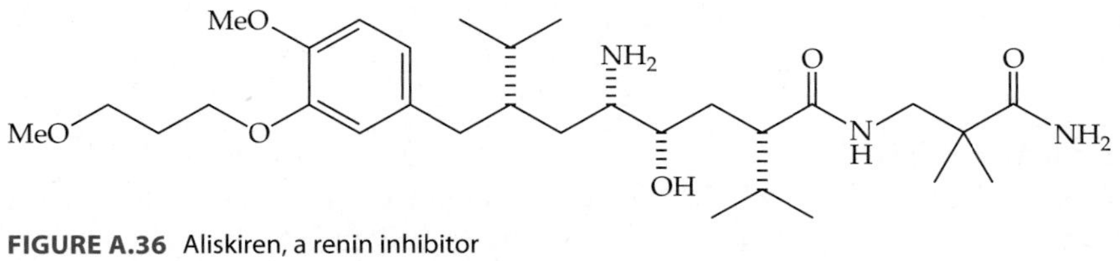

aliskiren
(Tekturna)
A.130

FIGURE A.36 Aliskiren, a renin inhibitor

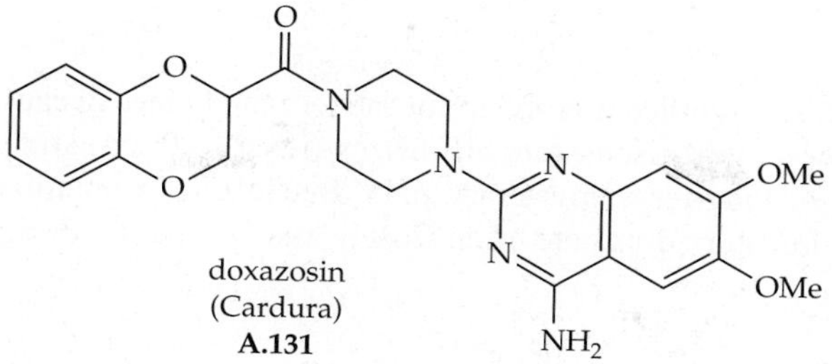

FIGURE A.37 Doxazosin, an α-blocker

doxazosin
(Cardura)
A.131

representative of α-blockers (**Figure A.37**). α-Blockers are less widely prescribed than most other antihypertensive drugs.

Calcium Channel Blockers

Calcium channel blockers (CCBs) prevent the flow of calcium ions through channels in heart tissue. Inhibiting calcium flow decreases the strength of contraction of the heart and decreases blood pressure. Most CCBs fall into the dihydropyridine structural class, with nifedipine (Adalat, **A.132**) being the prototypical example (**Figure A.38**). Nondihydropyridine CCBs include diltiazem (Cardizem, **A.135**) and verapamil (Calan, **A.136**).

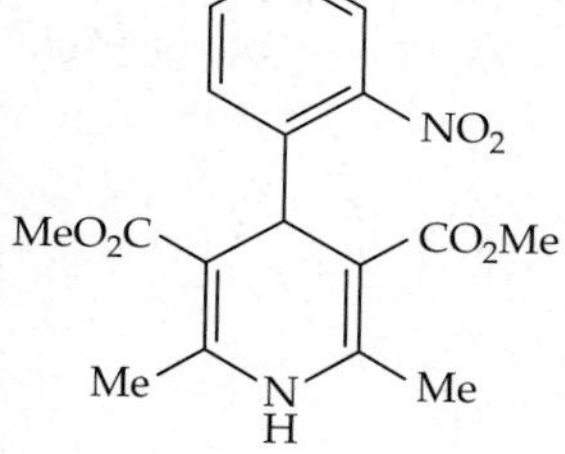

nifedipine
(Adalat)
A.132

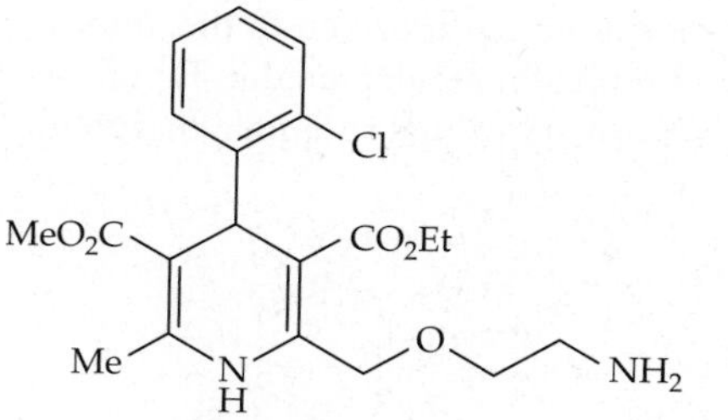

amlodipine
(Norvasc)
A.133

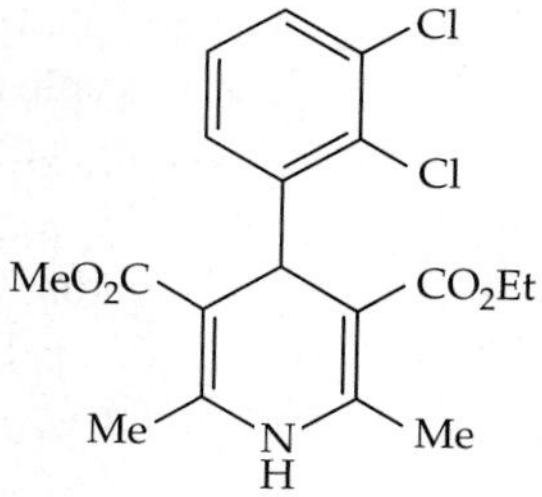

felodipine
(Plendil)
A.134

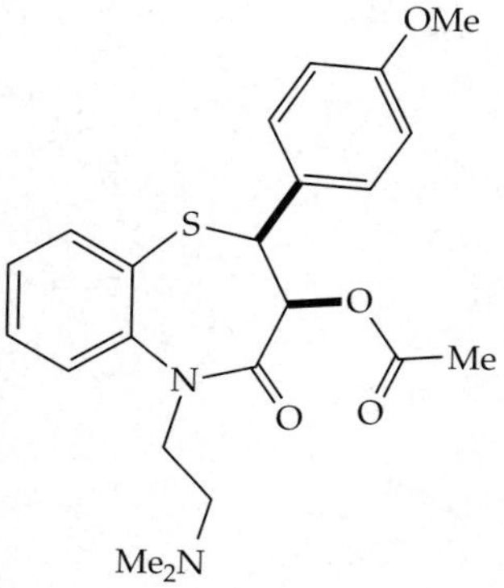

diltiazem
(Cardizem)
A.135

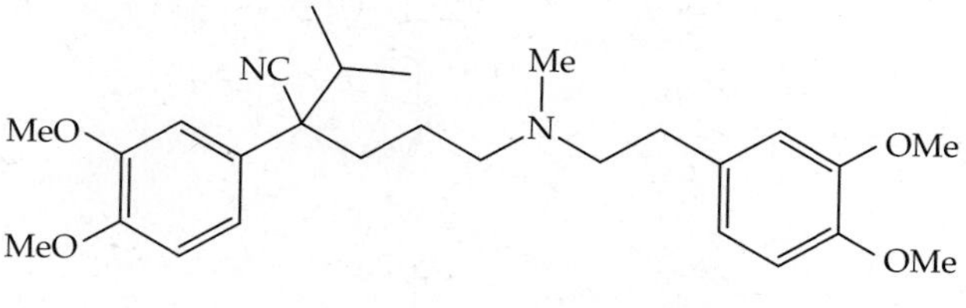

verapamil
(Calan)
A.136

FIGURE A.38 Calcium channel blockers

A.11 Pain

Pain management is a significant problem for modern medicine. Because of the complexity of pain as a topic, this discussion will be restricted to the treatment of pain from a short-term injury or long-term, low-grade pain. Because inflammation can cause tissue damage, anti-inflammatory drugs are often closely associated with analgesics.

Opioids

Opioids are compounds that bind one or more of the many different opioid receptors in the body. Opioids act primarily on the central nervous system. The selectivity of a given opioid for the various opioid receptors determines its characteristic activity. While many opioids are powerful analgesics, opioids often cause physical dependence and have tolerance issues. Sedation and decreased rate of breathing are also side effects associated with opioids. Despite their problems, opioids are generally the drug of choice for treating severe, acute pain.

The prototypical opioid is morphine (**A.137**) (**Figure A.39**). Isolated in a crude form, called opium, morphine has been recognized as a potent pain killer for thousands of years. Although effective, morphine has a low oral availability ($F = 25\%$). Two common derivatives of morphine include hydrocodone (Vicodin, **A.138**) and oxycodone (**A.139**), both of which have oral availabilities of greater than 75%. Oxycodone is often sold in an oral continuous-release form under the trade name of OxyContin. Not all opioids are semisynthetic derivatives of morphine. Dextropropoxyphene (Darvon, **A.140**) and tramadol (Tramal, **A.141**) are fully synthetic opioids. Both compounds preserve the pharmacophore of morphine as described in the "morphine rule" (see Chapter 11). Dextropropoxyphene and tramadol are depicted in Figure A.39 to highlight possible pharmacologically active conformations that resemble morphine.

FIGURE A.39 Opioid analgesics

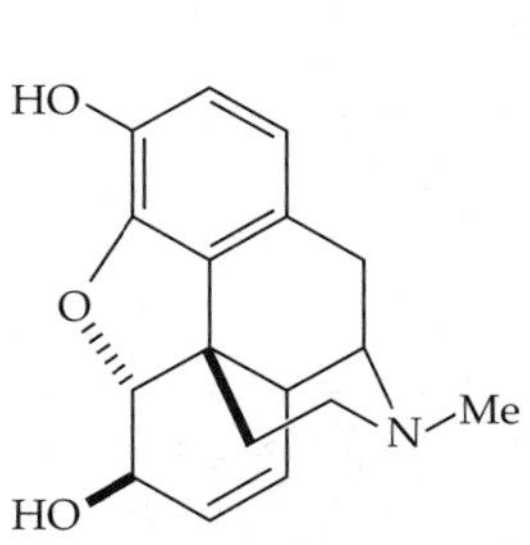

morphine
A.137

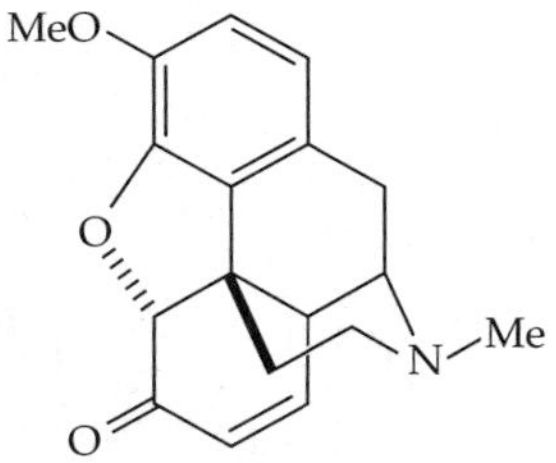

hydrocodone
(Vicodin)
A.138

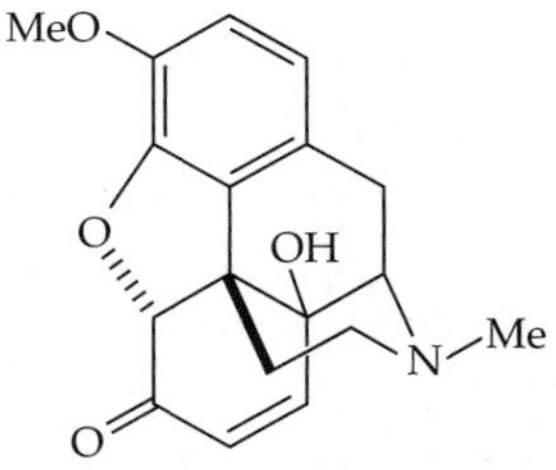

oxycodone
A.139

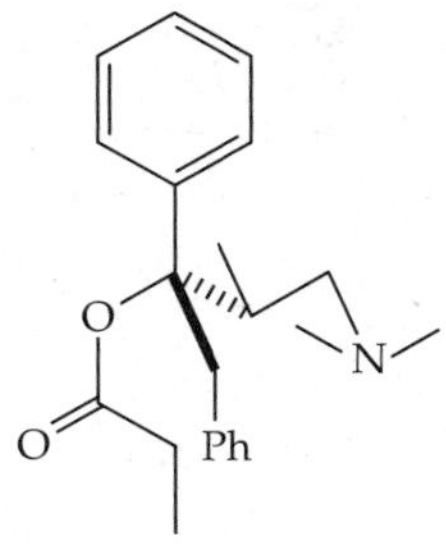

dextropropoxyphene
(Darvon)
A.140

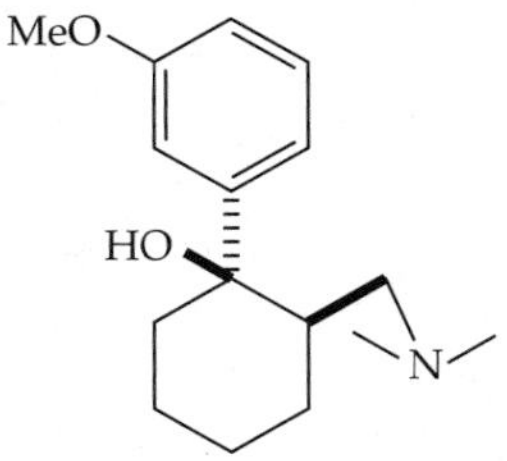

tramadol
(Tramal)
A.141

Because of problems related to physical dependence and tolerance, opioids have been shunned historically as long-term pain management drugs. More recent data indicate that a regimen that includes periodic rotation of several opioids can minimize traditional risks of opioid use.

Glucocorticoids

Glucocorticoids are a class of steroid hormones that act as anti-inflammatory agents through binding at the glucocorticoid receptor, a nuclear receptor. Around 1950, scientists at Merck developed an industrial synthesis of hydrocortisone (**A.142**) (**Figure A.40**). Hydrocortisone quickly became the standard treatment for many forms of arthritis. Unfortunately, hydrocortisone binds other steroid receptors and impacts water and salt retention in the body. Glucocorticoids with higher specificity for the glucocorticoid receptor include prednisone (**A.143**), methylprednisolone (**A.144**), and especially dexamethasone (**A.145**). Compounds **A.143** through **A.145** were all developed during the 1950s.

Nonsteroidal Anti-Inflammatory Drugs

Nonsteroidal anti-inflammatory drugs (NSAIDs) reduce pain, inflammation, and fever. NSAIDs generate such a broad range of effects by inhibiting cyclooxygenase, an enzyme that synthesizes prostanoids. Prostanoids play key roles in pain signaling, inflammation, fever response, and blood clotting. NSAIDs were first discovered with the isolation of salicylic acid (**A.146**) from willow bark (**Figure A.41**). Research at Bayer led to an improved derivative, *O*-acetylsalicylic acid or simply aspirin (**A.147**), in the very late 1800s. Two other common NSAIDs, ibuprofen (Advil and Motrin, **A.148**) and naproxen (Aleve, **A.149**), were first marketed in the 1960s and 1970s, respectively.

While NSAIDs avoid the physical dependence, sedation, and water retention problems found with opioids and glucocorticoids, NSAIDs are not without their own problems.

FIGURE A.41 Commonly
encountered NSAIDs

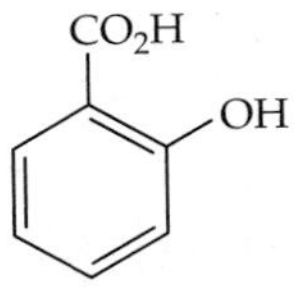

salicylic acid
A.146

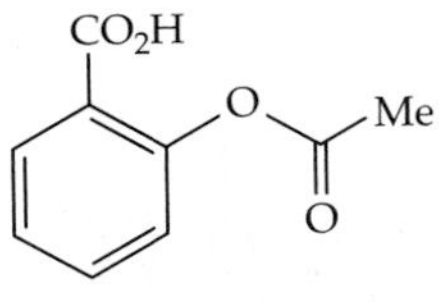

aspirin
A.147

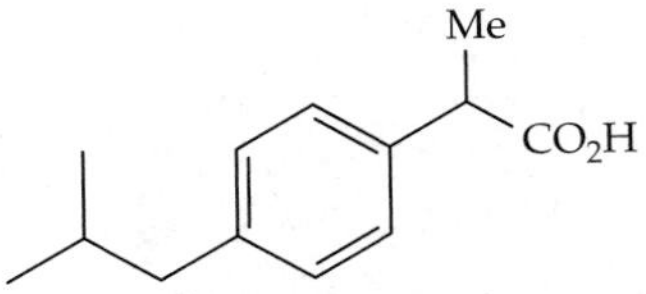

ibuprofen
(Advil, Motrin)
A.148

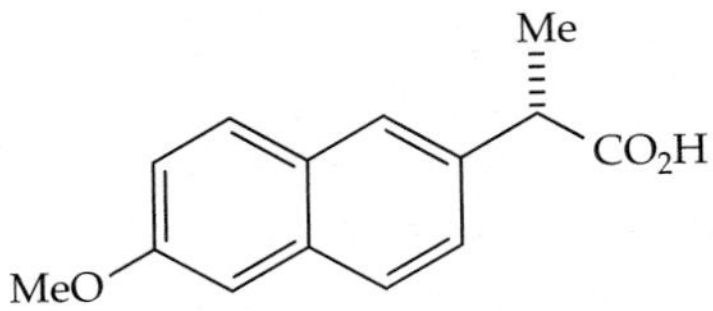

naproxen
(Aleve)
A.149

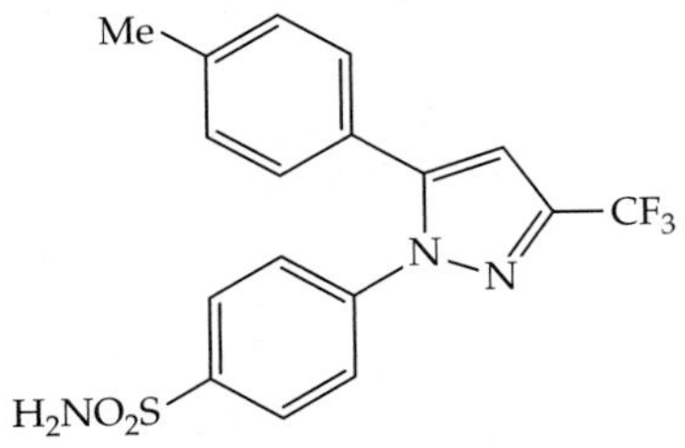

FIGURE A.41 Commonly
encountered NSAIDs

celecoxib
(Celebrex)
A.150

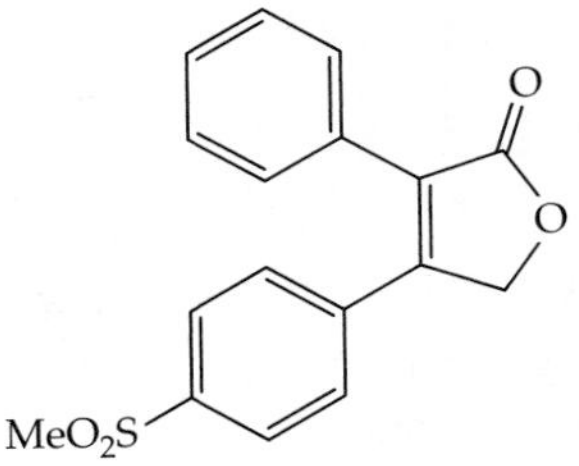

rofecoxib
(Vioxx)
A.151

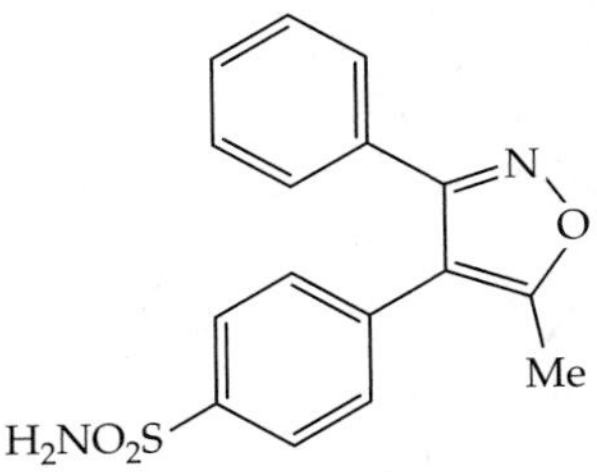

valdecoxib
(Bextra)
A.152

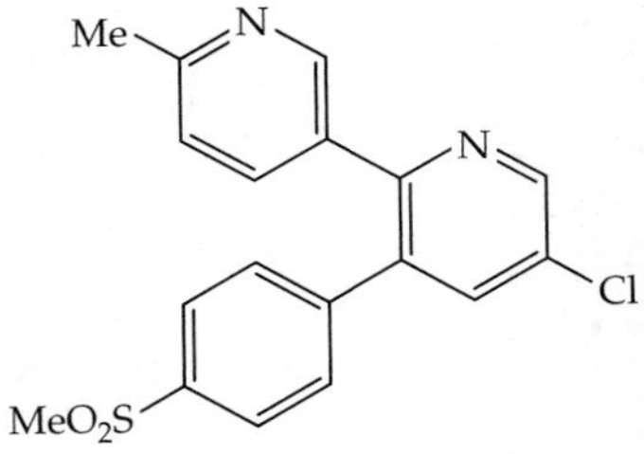

etoricoxib
(Arcoxia)
A.153

FIGURE A.42 COX-2 inhibitors

Cyclooxygenase (COX) forms prostanoids that are active throughout the body. Beyond pain and inflammation, prostanoids also play a role in maintaining the mucosal lining of the stomach. Long-term use of NSAIDs like aspirin, ibuprofen, and naproxen can lead to increased incidence of ulcers and stomach bleeding in patients.

Research has revealed that COX has at least two isoforms: COX-1 and COX-2. Among other roles, COX-1 helps maintain the stomach lining. COX-2, however, is more involved in pain and inflammation responses. In theory, a drug that inhibits only COX-2 should have the benefits of NSAIDs without harming the lining of the stomach. Several new NSAIDs called *COX-2 inhibitors* or *COXIBs* were developed. Examples include celecoxib (Celebrex, **A.150**), rofecoxib (Vioxx, **A.151**), and valdecoxib (Bextra, **A.152**) (**Figure A.42**). While short-term studies found the COXIBs to have less impact on the stomach lining than nonselective NSAIDs, long-term data was less favorable. Furthermore, patients receiving COXIBs showed an increased incidence of heart attacks and strokes. Both rofecoxib and valdecoxib were withdrawn from the market over safety concerns. Another COXIB, etoricoxib (Arcoxia, **A.153**), has been approved in a number of countries, but the U.S. FDA has withheld approval pending additional testing in clinical trials.

One compound that is not an NSAID but is often lumped into this group is acetaminophen (Tylenol, **A.154**) (**Figure A.43**). Acetaminophen is both an analgesic and suppresses fever but has no effect on inflammation and blood clotting. Acetaminophen also has less impact on the stomach lining than NSAIDs. Acetaminophen is believed to act both by COX inhibition as well as other pain response pathways. Unlike NSAIDs, acetaminophen acts on cannabinoid receptors.

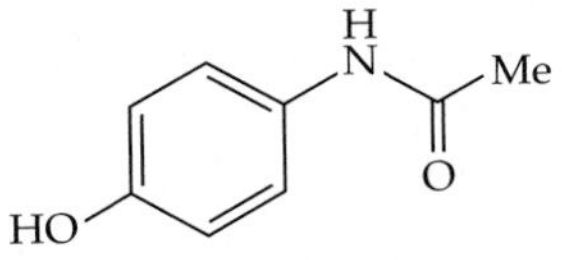

acetaminophen
(Tylenol)
A.154

FIGURE A.43

A.12 Viral Infection

Viral infections pose a major health risk for the global population. Many viruses are readily transmitted through common activities. While some viral diseases are not serious, many viral conditions cause a patient to be more susceptible to other, more serious secondary infections. Complicating the treatment and prevention of viral infections is the fact that viruses readily mutate. New mutants, if viable as a virus, are often resistant to existing drug treatments. Because of the widespread mutational resistance observed in various viruses, new drugs for the treatment of viral diseases are continuously sought.

Antiviral drugs are designed to target specific processes in the life cycle of a virus. Since viruses have many processes that are not found in humans, viral processes can often be inhibited without bringing adverse effects upon the host. The basic cycle for any virus involves several steps. First, the virus must attach to a cell. The virus then injects its genetic material into the host cell, and the genetic material must be integrated into the host cell's DNA. The viral genes are transcribed into mRNA and subsequently translated into proteins. New infectious viruses containing the appropriate viral proteins and genetic material are then budded off from the host cell to continue the cycle.

The following discussion of viruses and their associated drugs is nearly complete with regard to orally administered antiviral drugs. The discussion does not include intravenous biologics such as vaccines. While other sections in this appendix have been arranged by drug class, this section is ordered first by the type of virus and then by the mode of action of the drug.

Human Immunodeficiency Virus

The human immunodeficiency virus (HIV) is the causative agent of acquired immunodeficiency syndrome (AIDS), a disease that has claimed the lives of tens of millions of people since the 1980s. Because of the vast impact of AIDS on global health, HIV and its molecular biology have been intensely studied to determine potential drug targets for intervention.

Nucleoside Reverse Transcriptase Inhibitors

HIV is an RNA virus, meaning the viral genome is encoded as RNA instead of DNA. One key step in the virus' cycle is to synthesize double-stranded viral DNA using viral RNA as a template. A viral enzyme called reverse transcriptase (RT) accomplishes this task. The reverse transcription process may be blocked by nucleoside reverse transcriptase inhibitors (NRTIs). NRTIs have already been discussed somewhat in Chapter 6.

NRTIs are nucleoside analogues that become incorporated into the growing viral DNA chain. Once in the oligonucleotide chain, the NRTI prevents further elongation of the chain because the NRTI lacks a functional 3'-OH for attachment of the next nucleotide. Based on this mode of action, NRTIs are sometimes called *chain terminators*. With the synthesis of DNA severely impeded, the progress of the viral infection is slowed.

In theory, any DNA polymerase may be affected by NRTIs. Human DNA polymerases, however, have more sophisticated proofreading mechanisms and are able to remove (excise) incorrect nucleosides from a chain. NRTIs are therefore selective for viral processes and are generally well tolerated by patients.

A number of currently prescribed NRTIs are shown in **Figure A.44**. Compounds **A.155** through **A.158** are all analogues of pyrimidine nucleosides. Compounds **A.159** and **A.160** are both purine nucleoside analogues. Abacavir (Ziagen, **A.160**) is inactive until it is metabolized to carbovir (**A.161**) in vivo. The seven drugs in Figure A.44 were approved in the United States over a span of more than 25 years. Zidovudine reached the market first in 1987, and emtricitabine was the last to be approved in 2003. The steady release

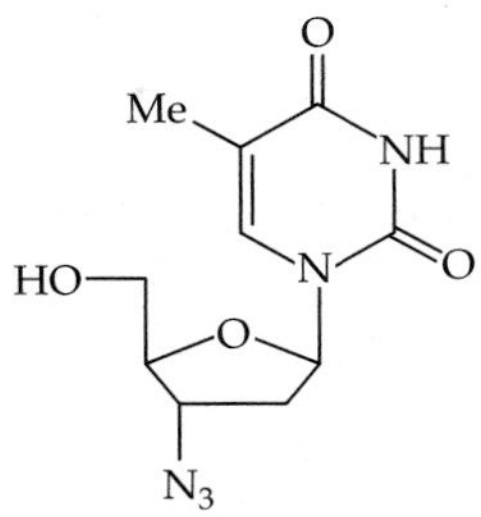

zidovudine
(Retrovir)
A.155

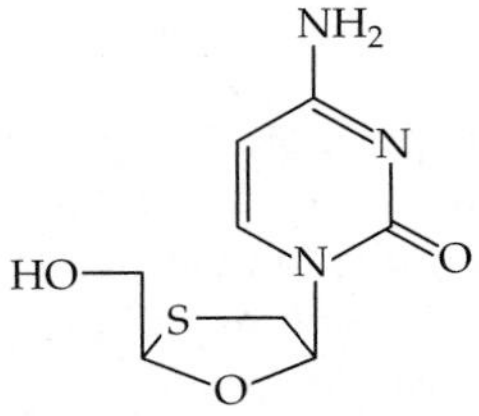

lamivudine
(Epivir)
A.156

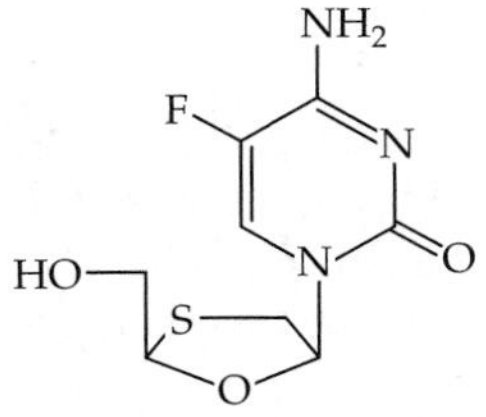

emtricitabine
(Emtriva)
A.157

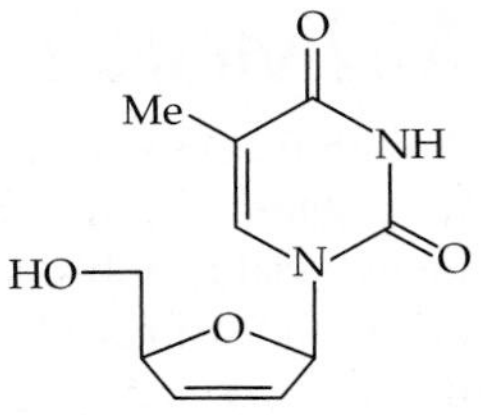

stavudine
(Zerit)
A.158

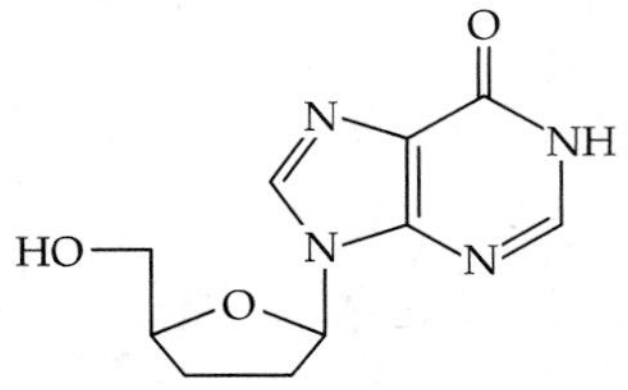

didanosine
(Videx)
A.159

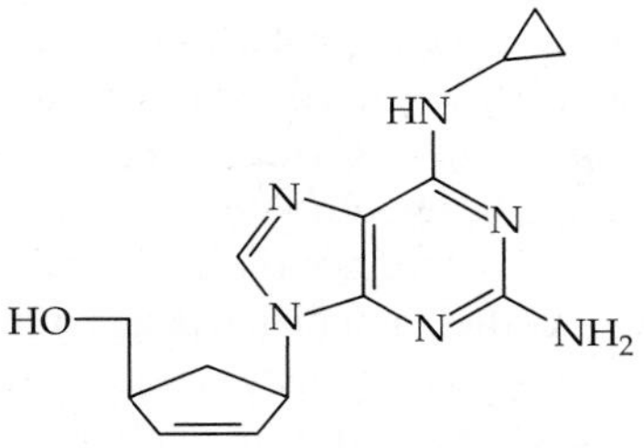

abacavir
(Ziagen)
A.160

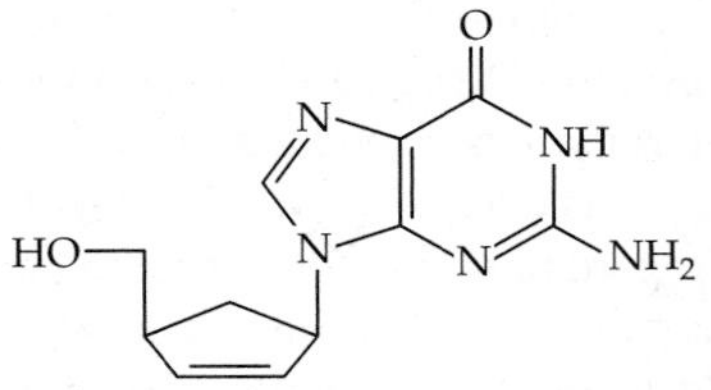

carbovir
A.161

FIGURE A.44 HIV nucleoside reverse transcriptase inhibitors

of these seven NRTIs corresponds to the development of new drugs that could address emerging mutational resistance that was being observed in HIV.

Technically, all of the compounds in **Figure A.45** are prodrugs. To become biologically active, NRTIs must be phosphorylated to a nucleotide triphosphate by cellular and/or viral kinases. In an effort to facilitate the phosphorylation of the analogue, Gilead Sciences developed a tenofovir (**A.163**), an analogue of a monophosphorylated nucleotide. Tenofovir disoproxil, which is administered in a prodrug form (Viread, **A.162**), must still be phophorylated twice in vivo to inhibit RT. Nucleotide analogues such as **A.162** are distinguished from nucleoside analogues with the acronym NtRTI.

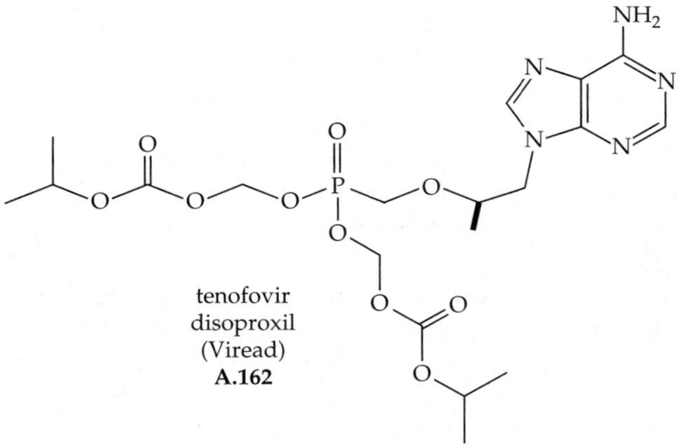

tenofovir
disoproxil
(Viread)
A.162

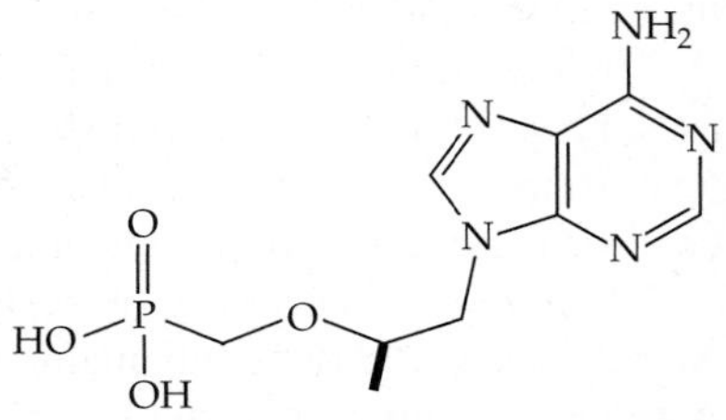

tenofovir
A.163

FIGURE A.45 HIV nucleotide reverse transcriptase inhibitors

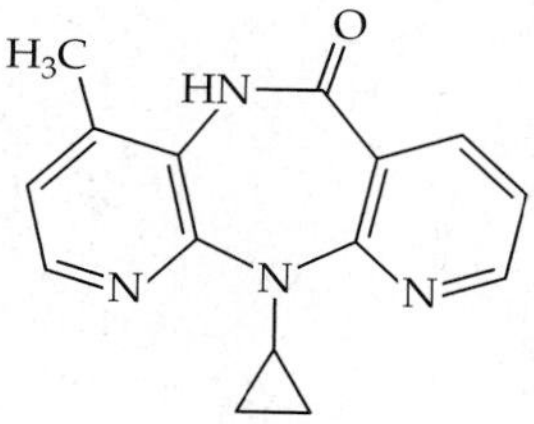

nevirapine
(Viramune)
A.164

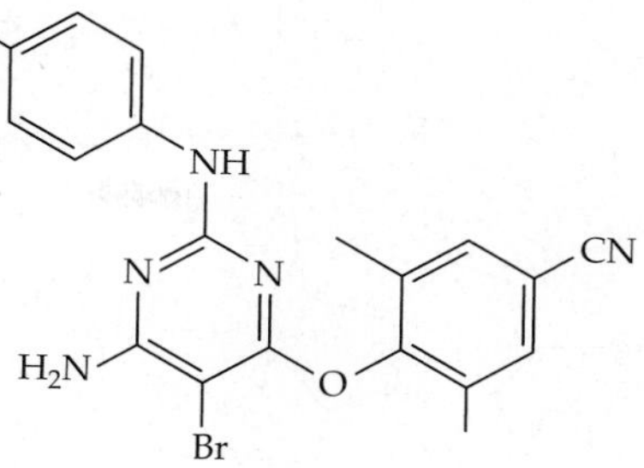

etravirine
(Intelence)
A.165

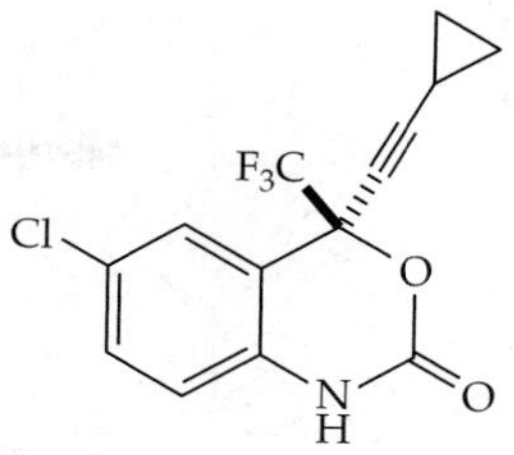

efavirenz
(Sustiva)
A.166

FIGURE A.46 HIV non-nucleoside reverse transcriptase inhibitors

Non-Nucleoside Reverse Transcriptase Inhibitors

NRTIs, which act as a substrate for RT, are not the only means of inhibiting reverse transcriptase. Like any enzyme, inhibitors can also bind allosteric positions away from the active site. An allosteric site on RT has been successfully targeted by drugs, and these drugs are called non-nucleoside reverse transcriptase inhibitors (NNRTIs). The three most frequently prescribed NNRTIs are shown in **Figure A.46**.

Protease Inhibitors

As with all viruses, HIV relies on a viral protease to properly cleave viral proteins in preparation for creating new viral particles. Drugs that block HIV protease are known as protease inhibitors (PI). The first HIV PI was saquinavir (Invirase, **A.167**), which was approved by the FDA in 1995 (**Figure A.47**). PIs therefore trailed NRTIs by nearly 10 years. Other PIs have since been approved. The most recently approved PI is darunavir (Prezista, **A.168**), which reached the market in 2006.

HIV protease hydrolyzes multiprotein peptides, and the leads for HIV protease inhibitors were polypeptides. It is natural that HIV protease inhibitors still bear a resemblance to a polypeptide. Successful development of PIs required liberal use of peptidomimetic strategies to create orally available drugs.

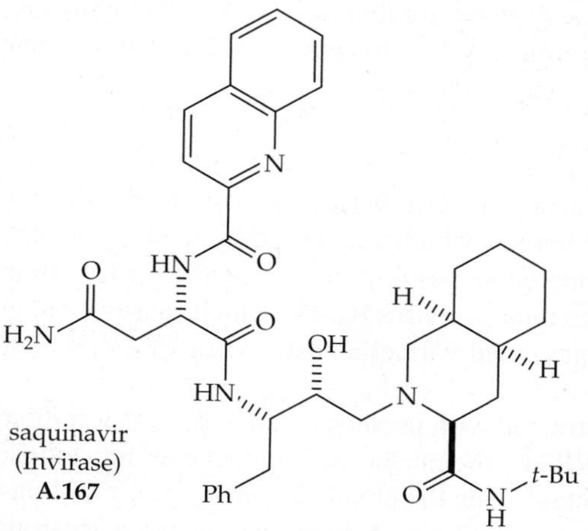

saquinavir
(Invirase)
A.167

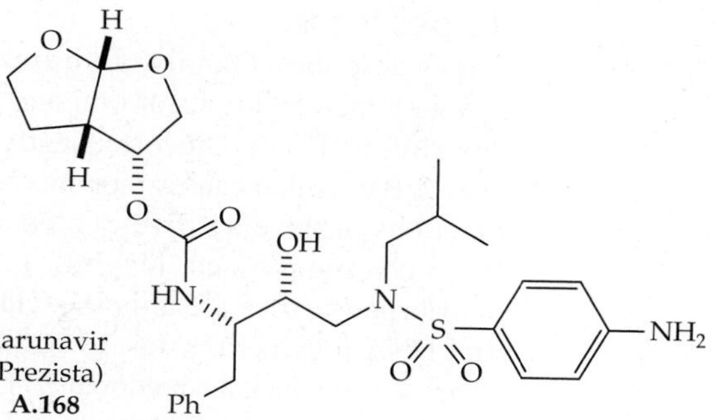

darunavir
(Prezista)
A.168

FIGURE A.47 HIV protease inhibitors

FIGURE A.48 Miscellaneous anti-HIV drugs

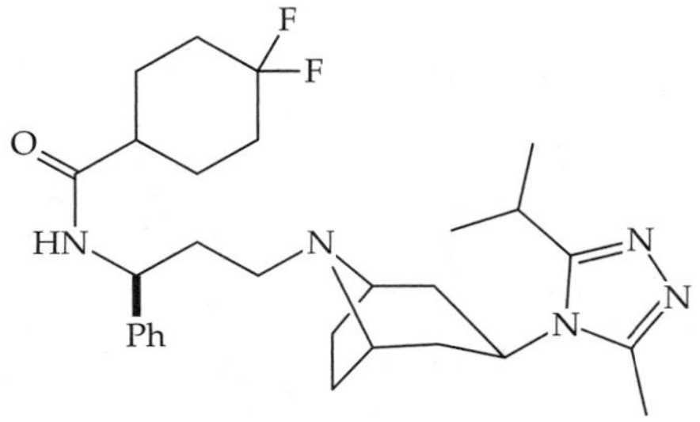

maraviroc
(Selzentry)
A.169

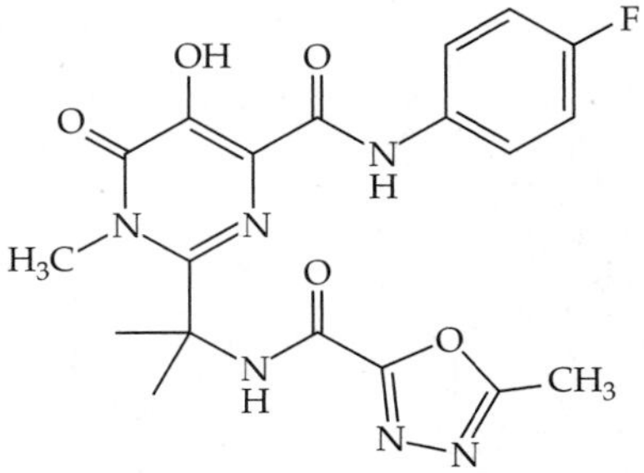

raltegravir
(Isentress)
A.170

Miscellaneous Targets

Beyond reverse transcriptase and HIV protease, two other targets in the HIV life cycle have recently been exploited for treatment of HIV infection. One target is the C-C motif chemokine receptor CCR5. CCR5 is a G-protein–coupled receptor that is recognized by HIV when HIV initially binds to a cell. Antagonists for CCR5 were found to inhibit HIV recognition of a prospective host cell. Maraviroc (Selzentry, **A.169**) was approved in 2007 as a CCR5 antagonist for the treatment of HIV infection (**Figure A.48**). Maraviroc is sometimes called an *entry inhibitor* or a *fusion inhibitor*. Another recent target is integrase, the enzyme responsible for integrating viral DNA into the host genome. Raltegravir (Isentress, **A.170**), first marketed in 2007, was the first *integrase inhibitor* to be approved by the FDA for treatment of HIV infection.

Highly Active Antiretroviral Therapy

Despite the continued development of effective treatments for HIV infection, the ability of HIV to develop mutational resistance has slowly eroded the usefulness of existing drugs. To better combat the mutability of HIV, regulatory agencies have approved therapies involving multiple anti-HIV drugs. Such an approach is sometimes called *combination therapy*, and the drugs are casually called *cocktails*. Formally, the combination therapy approach is called highly active antiretroviral therapy (HAART).

Mixtures of drugs in HAART include combinations that hit different HIV targets. One of the currently favored drug cocktails combines emtricitabine (**A.157**), tenofovir disoproxil (**A.162**), and efavirenz (**A.166**). This mixture includes two NRTIs and a NNRTI. Other HAART combinations include protease inhibitors. Mixtures including entry inhibitors and/or integrase inhibitors are almost certain to appear once they have been adequately tested for regulatory approval.

Herpes Virus

Herpes describes a family of viruses. In general, the term *herpes* is most closely associated with the conditions of oral and genital herpes, which are caused by the herpes simplex viruses (HSV). Other frequently encountered herpes viruses include the Epstein-Barr virus (EBV, which causes mononucleosis), cytomegalovirus (CMV, which causes various infections of the eyes, liver, colon, and lungs), and varicella zoster virus (VZV, which causes chicken pox and shingles).

Herpes viruses are almost exclusively treated with nucleoside analogues that inhibit viral DNA polymerase. Just as with HIV NRTIs, the nucleoside analogues used against herpes viruses lack a functional 3′-OH and terminate the chain elongation process of nucleotide polymerases. The nucleoside analogues in **Figure A.49** all have a role in treating

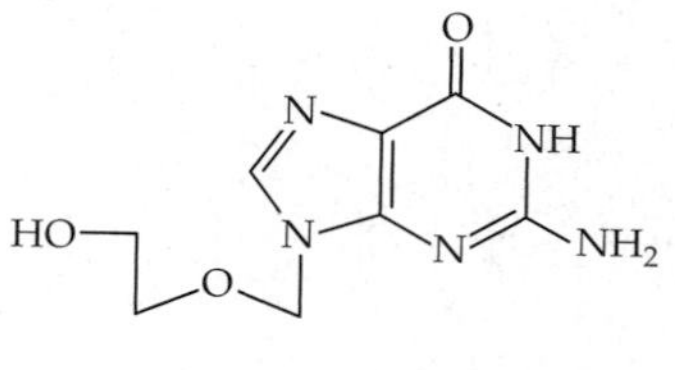

acyclovir
HSV & VZV
F = 15%
A.171

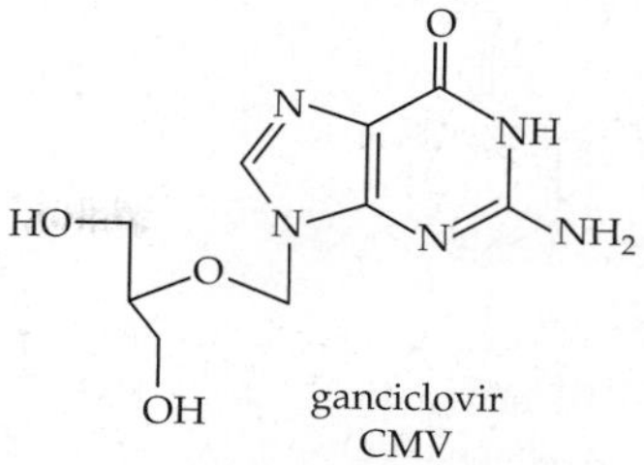

ganciclovir
CMV
F = 5%
A.172

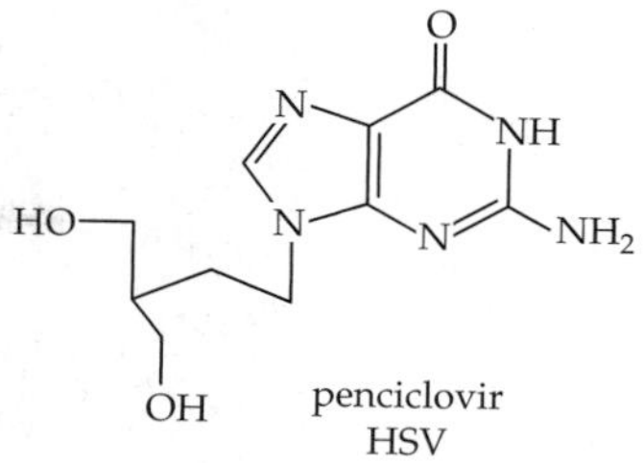

penciclovir
HSV
F = ~0%
A.173

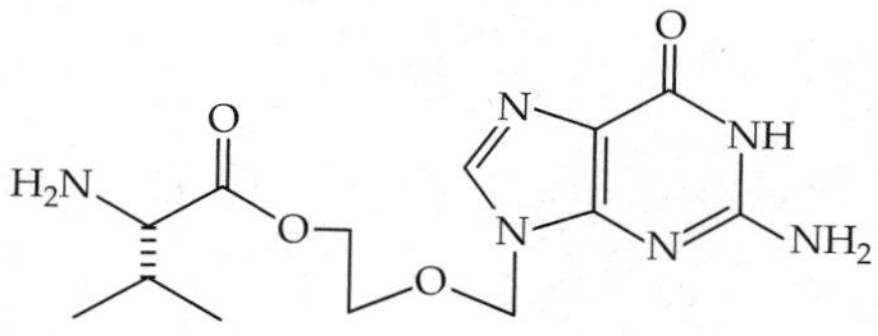

valaciclovir
HSV & VZV
F = 55%
A.174

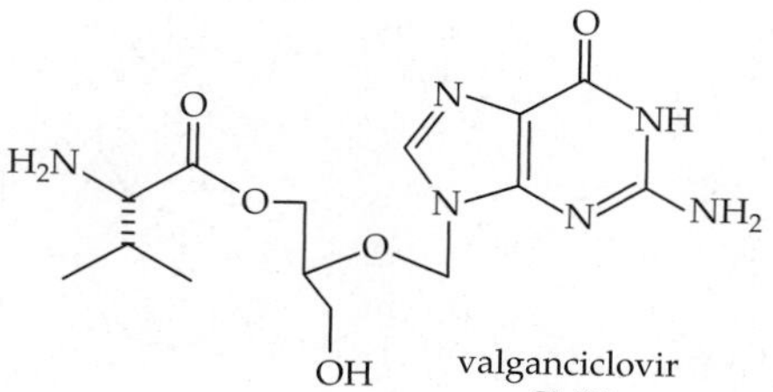

valganciclovir
CMV
F = 60%
A.175

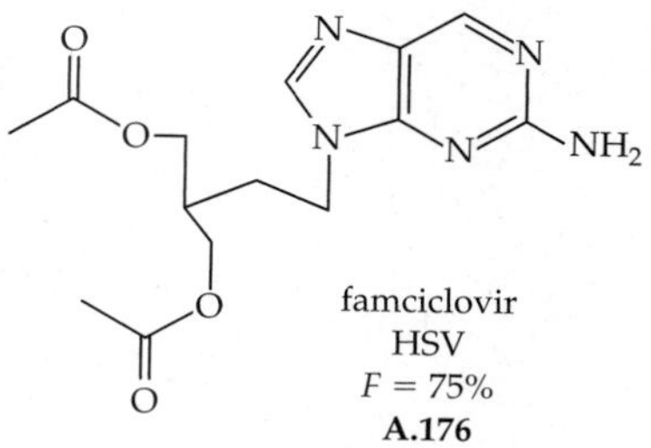

famciclovir
HSV
F = 75%
A.176

FIGURE A.49 Antiherpes nucleoside analogues

different herpes viruses. All must be phosphorylated on the putative 5′-OH by cellular and viral kinases to become active.

Nucleoside analogues tend to be polar and poorly available orally. Aside from penciclovir, a topical drug, all the compounds in Figure A.49 are available in oral forms. Because of the poor bioavailability of acyclovir (**A.171**), ganciclovir (**A.172**), and penciclovir (**A.173**), all three have been designed in a prodrug form corresponding to valaciclovir (**A.174**), valganciclovir (**A.175**), and famciclovir (**A.176**), respectively. Valaciclovir and valganciclovir both bear an L-valinyl ester. The amino acid ester allows the drug to be recognized by transport proteins for improved absorption in the small intestine. In contrast, famciclovir is acetylated on the OH groups and has a reduced purine base. Both features decrease polarity to facilitate passive diffusion across membranes. The purine ring of famciclovir is oxidized to its proper state by liver aldehyde oxidase.

Hepatitis Virus

Hepatitis is a general term describing inflammation of the liver. Certain chronic forms of hepatitis are caused by viruses, namely the hepatitis B virus (HBV) and hepatitis C virus (HCV). Despite their similar names, HBV and HCV are very different viruses. Regardless, both are treated with nucleoside analogues. Four oral drugs are currently approved to treat HBV. Those are entecavir (Baraclude, **A.177**), lamivudine (Epivir, **A.156**), adefovir (Preveon, **A.178**), and adefovir dipivoxil (**A.179**), a prodrug version of adefovir (**Figure A.50**). Both adefovir and adefovir dipivoxil are nucleotide analogues. Treatments for HCV are much more limited. The only currently approved oral drug that is active against HCV is ribavirin (Virazole, **A.180**). Although ribavirin is an obvious nucleoside analogue, evidence indicates that it does not inhibit nucleotide polymerases but instead causes catastrophic and disruptive mutations in the virus. A prodrug version of ribavirin is in clinical trials, as is a protease inhibitor that is active against HCV.

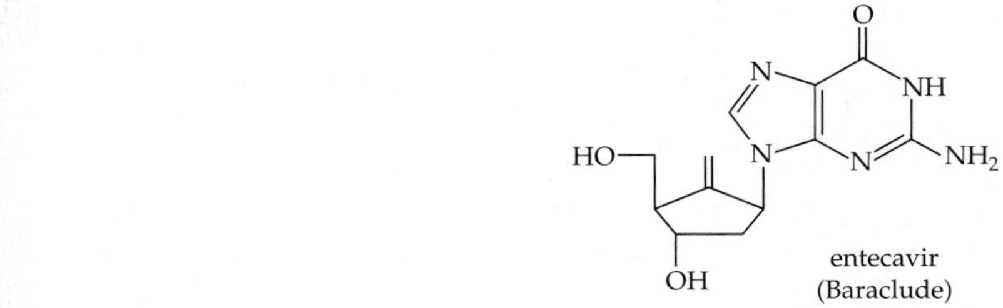

entecavir
(Baraclude)
A.177

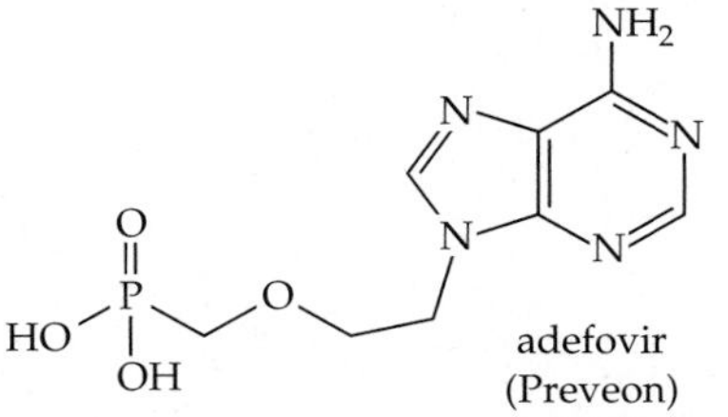

adefovir
(Preveon)
A.178

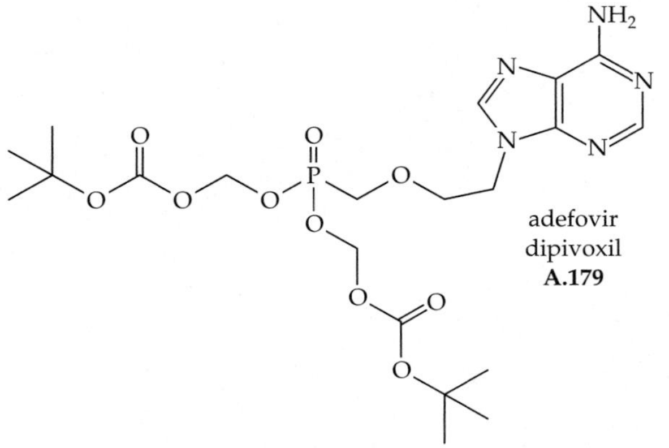

adefovir
dipivoxil
A.179

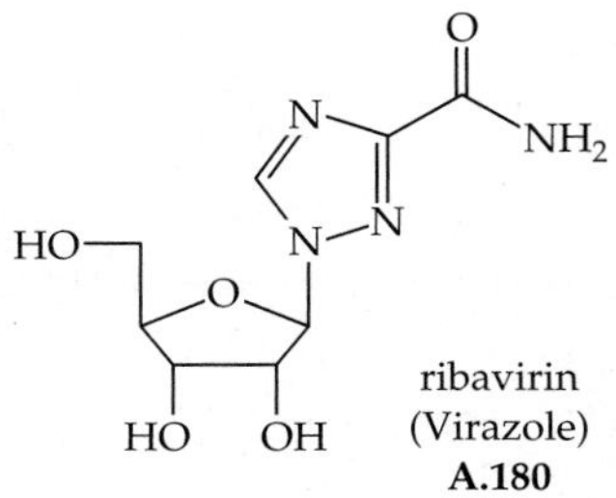

ribavirin
(Virazole)
A.180

FIGURE A.50 Anti-HBV and Anti-HCV drugs

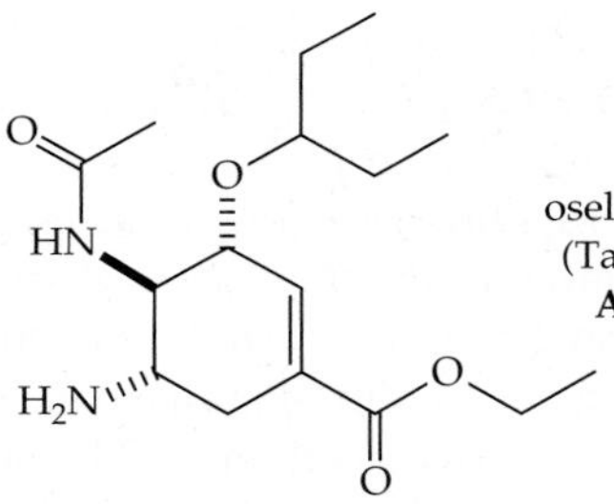

oseltamivir
(Tamiflu)
A.181

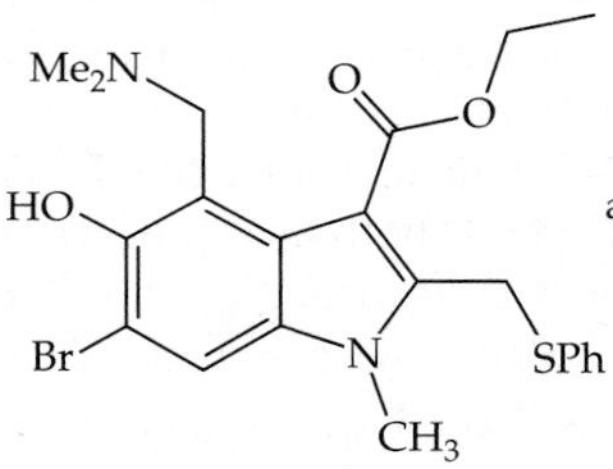

arbidol
A.182

FIGURE A.51 Influenza A antiviral drugs

Influenza A Virus

Influenza A virus, or more properly influenzavirus A, is a group of viruses responsible for a number of diseases. These diseases range from the seasonal flu to recent global diseases such as swine flu (H1N1 virus) and bird flu (H5N1 virus). These diseases are not generally fatal, but they are so contagious and affect such a broad segment of the population that even small fatality rates represent a significant loss of life. The various influenza A viruses are generally fought through prevention with injectable vaccines, but small-molecule, orally available drugs have been developed to help fight active viral infections.

Only one oral influenza A antiviral has been approved for use in the United States. That drug, oseltamivir (Tamiflu, **A.181**), inhibits neuraminidase, a viral enzyme involved in the budding of new viruses from infected cells (**Figure A.51**). The only other oral drug is arbidol (**A.182**), a drug approved for use in Russia and China. Arbidol is believed to act as a fusion inhibitor and also may stimulate the body's immune response.

References

General References

Sneader, W. *Drug Prototypes and Their Exploitation.* Chichester, UK: Wiley & Sons, 1996.

Hardman, J. G., & Limbird, L. E. (Eds.). *Goodman and Gilman's The Pharmacological Basis of Therapeutics* (10th ed.). New York: McGraw-Hill, 2001.

Murray, L., Sr. (Ed.). *Physician's Desk Reference* (58th ed.). Montvale, NJ: Thomson PDR, 2004.

Cancer

Yaish, P., Gazit, A., Gilon, C., & Levitzki, A. Blocking of EGF-Dependent Cell Proliferation by EGF Receptor Kinase Inhibitors. *Science.* 1998, *242*, 933–935.

Morphy, R. Selectively Nonselective Kinase Inhibition: Striking the Right Balance. *J. Med. Chem.* 2010, *53*, 1413–1437.

Diabetes

U.S. Food and Drug Administration. Avandia (Rosiglitazone): Ongoing Review of Cardiovascular Safety. http://www.fda.gov/Safety/MedWatch/SafetyInformation/SafetyAlertsforHumanMedicalProducts/ucm201446.htm (accessed February 2011).

Erectile Dysfunction

Hatzimouratidis, K., & Hatzichristou, D. G. Looking to the Future for Erectile Dysfunction Therapies. *Drugs.* 2008, *68*, 231–250.

Gastroesphageal Reflux Disorder

Yeomans, N. D. Overview of 50 Years' Progress in Upper Gastrointestinal Diseases. *J. Gastroenterol. Hepatol.* 2009, *24*, S2–S4.

Shin, J. M., & Sachs, G. Pharmacology of Proton Pump Inhibitors. *Curr. Gastroenterol. Rep.* 2008, *10*, 528–534.

Generalized Anxiety Disorder

Michael Kaplan, E., & DuPont, R. L. Benzodiazepines and Anxiety Disorders: A Review for the Practicing Physician. *Curr. Med. Res. Opin.* 2005, *21*, 941–950.

Davidson, J. R. First-Line Pharmacotherapy Approaches for Generalized Anxiety Disorder. *J. Clin. Psychiatry.* 2009, *70*, 25–31.

Hyperlipidemia

American College of Cardiology. Statement from the American College of Cardiology Related to the ENHANCE Trial. http://www.cardiosource.org/news–media/media–center/news–releases/2008/01/15.aspx (accessed February 2011).

Hypertension

Basile, J. N., & Chrysant, S. The Importance of Early Antihypertensive Efficacy: The Role of Angiotensin II Receptor Blocker Therapy. *J. Hum. Hyperten.* 2006, *20*, 169–175.

Jensen, C., Herold, P., & Brunner, H. R. Aliskiren: The First Renin Inhibitor for Clinical Treatment. *Nat. Rev. Drug. Disc.* 2008, *7*, 399–410.

Pain Management and Anti-Inflammatory Drugs

U.S. Food and Drug Administration. Information for Healthcare Professionals: Valdecoxib (Marketed as Bextra). http://www.fda.gov/Drugs/DrugSafety/PostmarketDrugSafetyInformationforPatientsandProviders/ucm124649.htm (accessed February 2011).

U.S. Food and Drug Administration. FDA Issues Public Healthy Advisory on Vioxx as Its Manufacturer Voluntarily Withdraws the Product. http://www.fda.gov/NewsEvents/Newsroom/PressAnnouncements/2004/ucm108361.htm (accessed February 2011).

Martina, S. D., Vesta, K. S., & Ripley, T. L. Etoricoxib: A Highly Selective COX-2 Inhibitor. *Ann. Pharmacother.* 2005, *39*, 854–862.

Bertolini, A., Ferrari, A., Ottani, A., Guerzoni, S., Tacchi, R., & Leone, S. Paracetamol: New Vistas of an Old Drug. *CNS Drug. Rev.* 2006, *12*, 250–275.

Viral Infection

MacDougall, C., & Guglielmo, B. J. Pharmacokinetics of Valaciclovir. *J. Antimicrob. Chemother.* 2004, *53*, 899–901.

Clarke, S. E., Harrell, A. W., & Chenery, R. J. Role of Aldehyde Oxidase in the in Vitro Conversion of Famciclovir to Penciclovir in Human Liver. *Drug Metab. Dispos.* 1995, *23*, 251–254.

Crotty, S., Cameron, C., & Andino, R. Ribavirin's Antiviral Mechanism of Action: Lethal Mutagenesis? *J. Mol. Med.* 2002, *80*, 86–95.

Performing Regressions with LINEST

LINEST is a function that is included in almost every spreadsheet software, including Microsoft Excel, OpenOffice.org Calc, and Google Docs Spreadsheet. LINEST accepts a table of values for a dependent variable (experimental activity) and any number of independent variables (such as parameters for use in a Hansch equation). LINEST then outputs the best-fit coefficients for the independent variables and certain statistical parameters for the regression. While Excel's Regression option in the Data Analysis tool is more user friendly, LINEST is much more widely available.

The use of LINEST is demonstrated on Equations 12.20, 12.21, and 12.22 of Chapter 12. All the data are from **Table B.1**, which is reproduced here.

TABLE B.1   **Cholinesterase activity of *meta*-substituted phosphonate esters**

Structure	Entry	R-Group	σ_m	E_s	Exptl. Log $1/I_{50}$	Calc. Log $1/I_{50}$ (Eq. 12.20)	Calc. Log $1/I_{50}$ (Eq. 12.22)
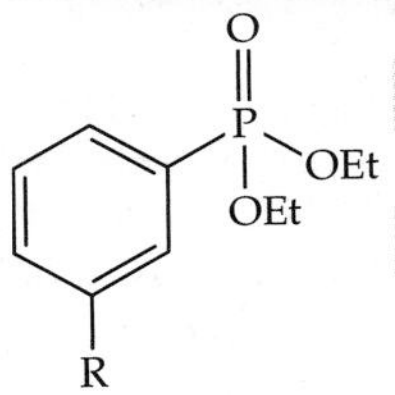	1	SF$_5$	+0.61	−1.67	7.12	6.88	7.32
	2	OMe	+0.12	+0.69	3.89	5.54	3.93
	3	t-Bu	−0.12	−1.54	6.05	4.89	6.00
	4	NO$_2$	+0.71	−1.28	7.30	7.15	7.05
	5	$^+$NMe$_3$	+0.81	−1.60	7.52	7.42	7.57

12.14

B.1 Equation 12.20

Equation 12.20 correlates the Hammett constant of the R-group (σ_m) to the experimental activity of the molecule (log $1/I_{50}$). To generate Equation 12.20, create two columns of data in a spreadsheet (**Figure B.1**). One should contain the experimental log $1/I_{50}$ values for entries 1 through 5 in Table B.1. The other column should include the corresponding σ_m values.

	A	B	C
1		exptl	
2	entry	log $1/I_{50}$	σ_m
3	1	7.12	0.61
4	2	3.89	0.12
5	3	6.05	−0.12
6	4	7.3	0.71
7	5	7.52	0.81

FIGURE B.1 Data columns for Equation 12.20

	A	B	C	D	E	F	G
1		exptl					
2	entry	log $1/I_{50}$	σ_m				
3	1	7.12	0.61		=LINEST(B3:B7,C3:C7,TRUE,TRUE)		
4	2	3.89	0.12				
5	3	6.05	−0.12				
6	4	7.3	0.71				
7	5	7.52	0.81				
8							
9							

FIGURE B.2 Using LINEST in Excel

The LINEST function requires four arguments, the locations of the cells that contain the activity (y-data), the cells with the parameter values (x-data), a boolean variable for requesting calculation of a y-intercept, and a boolean variable for requesting statistical output (**Figure B.2**). The exact syntax of the LINEST function depends on the software package.

In Excel, the LINEST function uses the following format:

$$=LINEST(y\text{-values},[x\text{-values}],TRUE,TRUE)$$

with y-values referring to the cells containing the activity data and [x-values] referring to the cells containing the parameter values. The y-values are contained in a single column. The x-values are contained in one or more columns depending on the number of parameters being used in the Hansch analysis.

In Excel, enter the LINEST function with its correct arguments. Ranges of cell values can be entered by highlighting the cells.

Press Enter to execute the function. In place of the function is just one number, 2,718779. LINEST generates two columns of numbers. To see the full output of the function (**Figure B.3**), do the following steps.

1. Select the cell containing the LINEST function.
2. Click the bottom right corner and drag it down to include a column of five cells (you need five cells vertically to include the full output range).
3. Click the bottom right corner of the selected cell range and drag it right to include two columns (you need one more column than the number of parameters in the Hansch analysis).
4. Press F2.
5. Press Ctrl+Shift+Enter.

The top cell in the last column of the LINEST output (cell F3 = 5.218) is always the y-intercept value for the regression. The top cells in the other columns correspond to

	A	B	C	D	E	F	G
1		exptl					
2	entry	log $1/I_{50}$	σ_m				
3	1	7.12	0.61		2.718779	5.2178	
4	2	3.89	0.12		1.458248	0.81482	
5	3	6.05	−0.12		0.536754	1.179036	
6	4	7.3	0.71		3.476043	3	
7	5	7.52	0.81		4.83214	4.17038	
8							
9							

FIGURE B.3 Full LINEST output in Excel

	A	B	C	D	E	F	G
1		exptl					
2	entry	log $1/I_{50}$	σ_m		=LINEST(B3:B7;C3:C7;TRUE;TRUE)		
3	1	7.12	0.61				
4	2	3.89	0.12				
5	3	6.05	−0.12				
6	4	7.3	0.71				
7	5	7.52	0.81				
8							
9							

FIGURE B.4 Using LINEST in OpenOffice

the coefficients for the parameters included in the analysis. In this example, cell E3 has a value of 2.719, which is the coefficient for σ_m. The remaining numbers in the output correspond to statistical measures. In the context of Chapter 12, the most important is the number in the third cell from the top of the third column (cell E5). This cell contains the value for r^2. Since $r^2 = 0.53674$, $r = 0.733$. All these values can be used to generate Equation 12.20 as was presented in Chapter 12.

$$\log \frac{1}{I_{50}} = +2.719\sigma_m + 5.218 \qquad n = 5, r = 0.733 \qquad (12.20)$$

Using LINEST in OpenOffice is nearly identical to Excel. Columns are filled with data, and the arguments are entered into the LINEST function (**Figure B.4**). One key difference is that the arguments are separated by a semicolon instead of a comma. Another key difference is that when the function is entered, using Ctrl+Shift+Enter automatically causes LINEST to generate its full output in multiple cells. Additional cells need not be selected as with Excel. The output of LINEST in OpenOffice is identical to that in Excel, and the positions of the data is identical as well.

Using LINEST in Google Docs is similar to the other two spreadsheet applications. Like Excel, LINEST in Google Docs requires the arguments to be separated by commas (**Figure B.5**). Once the arguments have been populated, simply pressing Enter executes the LINEST function, and the full output is automatically generated. The function output is identical in all three applications.

B.2 Equation 12.21

Equation 12.21 correlates the Taft steric parameter of the R-group (E_s) to the experimental activity of the molecule (log $1/I_{50}$). As with the previous section, the appropriate columns of activities and parameter values are created and entered into the LINEST function as

	A	B	C	D	E	F	G
1		exptl					
2	entry	log $1/I_{50}$	σ_m		=LINEST(B3:B7,C3:C7,TRUE, TRUE)		
3	1	7.12	0.61				
4	2	3.89	0.12				
5	3	6.05	−0.12				
6	4	7.3	0.71				
7	5	7.52	0.81				
8							
9							

FIGURE B.5 Using LINEST in Google Docs

	A	B	C	D	E	F	G
1		exptl					
2	entry	log $1/I_{50}$	E_s				
3	1	7.12	−1.67		−1.3662	4.900504	
4	2	3.89	0.69		0.357043	0.500753	
5	3	6.05	−1.54		0.829947	0.714354	
6	4	7.3	−1.28		14.64154	3	
7	5	7.52	−1.6		7.471613	1.530907	
8							
9							

FIGURE B.6 Full LINEST output in Excel for Equation 12.21

arguments. The output is shown for Excel (**Figure B.6**), but all LINEST implementations give the same results.

As in the Equation 12.20 example, the top cell in the last column is the y-intercept. The top cell in the first column is the coefficient for E_s in the Hansch equation. The third cell from the top in the first column is the r^2 value of the regression. Collectively, these terms can be used to generate Equation 12.21.

$$\log \frac{1}{I_{50}} = -1.366E_s + 4.901 \qquad n = 5, r = 0.911 \qquad (12.21)$$

B.3 Equation 12.22

Equation 12.22 correlates both the Hammett constant (σ_m) and the Taft steric parameter (E_s) of the R-group to the experimental activity of the molecule (log $1/I_{50}$). As with the other examples, the requisite activity and parameter values must be entered into a spreadsheet. The only significant difference in this example is that the x-values argument for LINEST must include two columns of data (**Figure B.7**). One includes the Hammett constant data, and the other is for the Taft steric parameter.

Execution of LINEST occurs as usual with one caveat related to Excel. Because an additional column is included in the x-values, an additional column must be selected for the output of the LINEST function. In Excel, the selected area would need to be five cells tall and three cells wide to hold the full output. In OpenOffice and Google Docs, allocating space for the output of LINEST is handled automatically without help from the user. The full output is shown to **Figure B.8**. The cells containing #N/A can be disregarded.

The top of the last column (cell H3) in the output is always the y-intercept. As one moves from right to left through the output data, the top of the next column (cell G3) is the coefficient for the parameter in the *first* column of the x-range of the data moving left to

	A	B	C	D	E	F	G	H	I
1		exptl							
2	entry	log $1/I_{50}$	σ_m	E_s					
3	1	7.12	0.61	−1.67		=LINEST(B3:B7,C3:D7,TRUE,TRUE)			
4	2	3.89	0.12	0.69					
5	3	6.05	−0.12	−1.54					
6	4	7.3	0.71	−1.28					
7	5	7.52	0.81	−1.6					
8									
9									

FIGURE B.7 Setting up LINEST in Excel for Equation 12.22

	A	B	C	D	E	F	G	H	I
1		exptl							
2	entry	log $1/I_{50}$	σ_m	E_s					
3	1	7.12	0.61	−1.67		−1.10193	1.613849	4.498417	
4	2	3.89	0.12	0.69		0.126958	0.314163	0.180625	
5	3	6.05	−0.12	−1.54		0.98802	0.232222	#N/A	
6	4	7.3	0.71	−1.28		82.46959	2	#N/A	
7	5	7.52	0.81	−1.6		8.894666	0.107854	#N/A	
8									
9									

FIGURE B.8 Full LINEST output in Excel for Equation 12.22

right. This is perhaps counterintuitive. Regardless, cell G3 contains the coefficient for σ_m in the Hansch equation. Cell F3 therefore corresponds to the coefficient for E_s. The value for r^2 is, as always, in the third cell from the top in the first output column (cell F5). With all the values in hand, one can then reconstruct Equation 12.22.

$$\log \frac{1}{I_{50}} = -1.102E_s + 1.614\sigma_m + 4.498 \qquad n = 5, r = 0.994 \qquad (12.22)$$

In conclusion, the LINEST function found in almost all spreadsheet applications is simple to use and can quickly generate Hansch equations from activity data and tabulated parameter values of R-groups.

Glossary

A

abbreviated new drug application (ANDA) An abbreviated new drug application is the process that generic drug manufacturers must complete in order to market a generic equivalent to a branded drug.

absorption phase The absorption phase is the time period after administration during which a drug travels from its site of administration to the bloodstream.

absorption rate constant (K_{ab}) The absorption rate constant is the rate constant that describes the rate of a drug's reaching the bloodstream from its site of administration. Absorption rate constants are normally associated with, but not limited to, oral drugs. The units of absorption rate constants are inverse time.

accelerated review An accelerated review is a compressed clinical trial process for drugs that treat diseases with high mortality rates.

active pharmaceutical ingredient (API) An active pharmaceutical ingredient is the compound in a drug product that is responsible for the biological effect of the drug. Active pharmaceutical ingredients are also sometimes called drug substances.

active site An active site is the region of an enzyme that is responsible for the binding of a substrate and conversion of that substrate to a product. Active sites are often, but not always, found in pockets or channels within a folded protein.

active transport Active transport is the movement of a drug across a membrane by means of a transport protein that consumes energy. Because the transport is energetically driven, transport can occur with or against the concentration gradient.

administration Administration is the aspect of pharmacokinetics that focuses on the method of induction of a drug into an organism.

agonist An agonist is a receptor ligand that elicits a positive response from the receptor.

alkaloid A natural product that contains a basic nitrogen atom. Alkaloids typically have a level of structural complexity, such as a stereocenter or ring.

allosteric control Allosteric control is a means of regulating enzyme activity. In allosteric control, a molecule binds to or releases from an allosteric site of an enzyme. Through a series of conformational changes, the active site of the enzyme will be either turned off or turned on. The molecule that binds the allosteric site therefore controls the activity of the enzyme.

allosteric site An allosteric site is a region in an enzyme at which molecules may bind. Binding at an allosteric site causes conformational changes in the enzyme that can result in changes in the active site.

allosteric theory Allosteric theory is a method for relating ligand-receptor binding and biological response. Allosteric theory focuses on active and inactive conformations of a receptor and how ligands affect the equilibrium between the conformations.

alpha helix (α-helix) An α-helix is a type of secondary structure in a protein. In an α-helix, a section of the peptide backbone adopts a corkscrew shape. The corkscrew is held together by hydrogen bonds between nearby amino acid residues.

anabolism Anabolism is a specific form of metabolism. Anabolism includes the processes by which an organism uses available energy to create molecules necessary for the life of the organism.

analogue An analogue is a structure that is similar to the original but changed slightly to cause a desired change in biological activity.

antagonist An antagonist is a receptor ligand that prevents a receptor from being activated.

antimetabolite An antimetabolite is a molecule that imitates a metabolite in the body and thereby blocks a metabolic process. Antimetabolites are often competitive inhibitors of enzymes.

antisense technology Antisense technology relies on the idea that protein synthesis may be artificially controlled by binding mRNA in a cell. The single-stranded mRNA, if bound by a complementary oligonucleotide, cannot be translated into its corresponding protein. Antisense technology is sometimes called RNA interference, or RNAi.

area under the curve (AUC) The phrase "area under the curve" refers to the integrated area underneath a C_p-time curve. The area under the curve is a measure of the degree of drug exposure received by a patient.

array synthesis Array synthesis is a combinatorial approach to preparing a molecular library. In array synthesis, each member of the library normally is individually carried through the entire synthetic route. The phrases "one compound–one well" and "spatially addressable" are associated with array synthesis.

assay A test designed to detect whether a compound is active against a target.

atom economy Atom economy is a green chemistry metric developed by Barry Trost. Atom economy measures how efficiently a reaction uses reagents by determining the percentage of atoms in the reagents that are incorporated into the final product.

$$\text{atom economy} = 100 \times \left(\frac{\text{mass of atoms in desired product}}{\text{mass of atoms in reactants}} \right)$$

B

bead   A bead is a macroscopic polymer granule to which molecules may be linked for solid-phase synthesis. In reaction schemes, a polymeric support is often shown as a shaded circle or as a circle containing a letter "P" (polymer).

beta sheet (β-sheet)   A β-sheet is type of secondary structure in a protein. In a β-sheet, a section of the peptide backbone repeatedly folds back upon itself in an accordion fashion. The structure is held together by hydrogen bonds between amino acid residues.

binary encoding   Binary encoding is an approach to deconvolution used in some split synthesis libraries. After each synthetic step, a separate reaction is performed on the bead. The separate reaction places an additional signal molecule, or tag, on the bead. If a compound shows activity, the tags on the corresponding bead will be cleaved and allow identification of the hit.

bioavailability (F)   Bioavailability refers to the ability of a drug to reach the bloodstream from its site of administration. Bioavailability is generally determined in comparison to intravenous administration. The first-pass effect is the main barrier that prevents oral drugs from having a high bioavailability. The variable for bioavailability is unitless and quantifies the fraction of an administered drug that reaches the bloodstream.

$$F_{\text{route}} = \frac{\dfrac{AUC_{\text{route}}}{D_0^{\text{route}}}}{\dfrac{AUC_{\text{iv}}}{D_0^{\text{iv}}}}$$

biochemical assay   A biochemical assay is a quick and relatively inexpensive method of testing drug-target binding. Biochemical assays are excellent for determining binding strength but provide no information on how a molecule might be transported or metabolized in a living organism.

bioequivalence test   Bioequivalence tests show that one form of a drug is biologically equivalent to another form of the drug. A generic drug must pass bioequivalence tests before it may be approved as safe and effective.

bioisostere   A bioisostere is a type of isostere. Bioisosteres attempt to simulate all the properties, including size and polarity, of the replaced group.

biologic   A biologic is a type of drug classification that includes complex biological mixtures that typically cannot be prepared by chemical synthesis. Examples of biologics include vaccines and antibodies.

bloodflow (Q)   Bloodflow is normally measured in mL/min and describes the volume of blood that passes through a certain organ or tissue per unit time. Bloodflow is closely related to the idea of clearance, which also has the units of mL/min.

blood–brain barrier (BBB)   The blood–brain barrier refers to the lack of pores in the capillaries in the brain. The lack of pores requires that drugs affecting the brain must be able to diffuse across a cell membrane or cross the membrane with assistance from a transport protein.

bolus   A drug bolus normally refers to the amount of drug administered in a single intravenous injection.

building blocks   Building blocks are small, reactive molecules that may be readily incorporated into larger structures through combinatorial chemistry techniques.

C

carrier protein   A carrier protein is a transport protein found in cell membranes. The carrier protein facilitates the movement of molecules across the membrane.

catabolism   Catabolism is a specific form of metabolism. Catabolism is the chemical breakdown of food to generate energy for an organism.

cellular assay   A cellular assay tests the effect of a molecule on a living cell while also giving insight on whether a drug can cross a cell membrane.

classical isostere   A classical isostere is a type of isostere. Classical isosteres simulate the steric size of the replaced group with less attention to electronic effects.

clearance (CL)   Clearance is the process of removing a drug from the blood. Clearance is measured in volume of blood flow per unit time and is most commonly associated with the liver and kidneys.

cleavage   Cleavage is the process of breaking the covalent bonds in a linker and releasing a molecule from a resin.

clinical trial   Clinical trials encompass drug tests that involve humans. Clinical trials test for drug safety and efficacy, and occur in different stages. Clinical trial stages are called phase I, phase II, and phase III.

coenzyme   A coenzyme is a nonprotein organic molecule that is required for a protein to function. Redox biomolecules such as NADH and FAD are common cofactors.

cofactor   Cofactor usually refers to a metal ion that is required for a protein to properly function.

combinatorial chemistry   Combinatorial chemistry is a branch of chemistry that strives to develop methods for rapidly and efficiently synthesizing large numbers of molecules for use in a molecular library.

comparative modeling   Comparative modeling involves determining the tertiary structure of one protein by direct comparison to a protein with a similar primary structure. The tertiary structure is first roughly approximated based on a similar protein, and then molecular modeling is used to make a more precise structural prediction.

comparative molecular field analysis (CoMFA) Comparative molecular field analysis is a data-intensive form of quantitative structure-activity relationship analysis. Comparative molecular field analysis attempts to correlate molecular properties and their location in the three-dimensional space to observed biological activity.

competitive antagonist Competitive antagonists bind receptors reversibly through intermolecular forces.

competitive inhibitor A competitive inhibitor is a type of reversible inhibitor. Competitive inhibitors bind at the active site of an enzyme. When bound, a competitive inhibitor physically blocks the substrate's binding to the active site.

compound of interest A compound of interest normally is a hit that is not only active against a target but also shows favorable pharmacokinetic properties. Some compounds of interest will be advanced in the drug discovery process as leads.

concentrative transport See *active transport*.

conjugation Conjugation is a phase II metabolic process in which a polar side chain is added to a drug. Addition of the polar group facilitates elimination of the drug by kidney filtration.

consensus scoring In consensus scoring, multiple scoring methods are used to identify strongly binding hits. Only hits that are scored well by several methods are advanced for further consideration.

constituent activity Constituent activity is the nonzero response caused by some receptors even in the absence of a ligand.

contact forces Contact forces are an example of a weak intermolecular force and are formed from the attractions between instantaneous polarizations of molecules.

controlled trial In a controlled trial, some test subjects receive medication and others receive a placebo. Trials performed with a control provide more reliable data.

correlation coefficient (r) A correlation coefficient ranges in value from -1 to $+1$ and measures the quality of fit of a linear regression. Values of r closer to -1 and $+1$ indicate a better fit. When squared, the correlation coefficient estimates the fraction of variance in the data that is accommodated by the variables in the regression.

crosslink A crosslink is a covalent bridge between two nucleotides in DNA. Crosslinking is a form of DNA damage that is caused by many anticancer alkylating agents.

cytochrome P-450 Cytochrome P-450 is a family of enzymes found mostly in the liver. The cytochrome P-450 enzymes are largely responsible for the metabolism of drugs.

D

data mining Data mining is the process of sifting through genome or peptide sequence databases to find matching regions.

deconvolution Deconvolution is the process of determining the chemical structure of library compounds that are found to be active in an assay.

depot A depot is a reservoir of drug left within a tissue after a nonintravenous injection of a drug.

desensitization Desensitization to a drug normally occurs when a patient takes a medication for a long period. In response to desensitization, the dose must be increased to achieve the same therapeutic effect. Desensitization is associated with the downregulation of affected receptors.

dipole forces Dipole forces are weak intermolecular interactions based on electrostatic attractions between polarized bonds of different molecules.

directed library A directed library is a small library prepared to explore the structure-activity relationships around specific parts of a scaffold.

dissociation equilibrium constant (K_D) The dissociation equilibrium constant is a measure of binding affinity between a ligand and receptor.

dissociation rate constant (k_{off}) The dissociation rate constant is a measure of the rate at which a ligand dissociates from a receptor.

$$k_{off} = K_D k_{on}$$

dissociative half-life ($t_{1/2}$) Dissociative half-life is the time required for half of a population of receptor-ligand complexes to dissociate to unbound receptor and ligand.

$$t_{1/2} = \frac{0.693}{k_{off}}$$

distribution Distribution is the aspect of pharmacokinetics that focuses on how a drug reaches its site(s) of action from its site of administration.

distribution coefficient (D) A distribution coefficient is a special type of partition coefficient that is used with ionizable molecules. Distribution coefficients take into account not just the concentration of a molecule in different phases but also all the various ionized forms of the molecule.

$$D = \frac{\sum [\text{drug}]_{oct}}{\sum [\text{drug}]_{water}}$$

distribution phase The distribution phase of an intravenous drug is the time period immediately after administration during which the drug perfuses from the bloodstream into the body's other tissues. Drugs that are not administered intravenously do not have a distinct distribution phase.

disulfide bond A disulfide bond is a sulfur-sulfur linkage between two cysteine residues of a protein. Disulfide bonds are considered part of the primary structure of a protein.

dose (D) Dose refers to the amount of drug within a patient at a given time. A particularly important dose is the initial dose (D_o) that is received by a patient. The units of dose are mass.

double-blind "Double-blind" is a term applied to some clinical trials. In a double-blind trial, neither the health care professionals nor the test subjects know which subjects are receiving active medication or a placebo. The encoded clinical data is interpreted by a research assistant who is not directly involved in the trial.

downregulation   Downregulation is a process by which a cell may decrease its expression of a receptor if that receptor is continuously being stimulated at a high level. Downregulation can result in a patient's becoming desensitized to a drug.

drug interaction   A drug interaction involves a drug causing unexpected effects because of an extra external condition. The additional conditions can include a patient taking other medication, eating a nontypical diet, or having a health condition. Drug interactions are often discovered during phase III trials, which include a large number of individuals with various backgrounds.

drug product (DP)   A drug product is the final formulated version of a drug that is delivered by a pharmaceutical company. A drug product consists of both the drug substance (active pharmaceutical ingredient) and all other components. Other ingredients may include binders in pills, protective coatings, dyes, or solvents for solutions/suspensions.

drug substance (DS)   See *active pharmaceutical ingredient.*

drug-like   Drug-like hits are active molecules that have a MW near 500 and therefore leave relatively little opportunity for adding more functionality to increase activity.

druggable   Druggable is an adjective that normally describes whether a biological target is an appropriate point of intervention for drug development.

E

E-factor   E-factor is a green chemistry metric developed by Roger Sheldon. E-factor measures the "greenness" of a reaction by the ratio of the waste generated by a process to the amount of product formed by the same process.

$$\text{E-factor} = \frac{\text{mass of total waste}}{\text{mass of product}}$$

Eadie-Hofstee equation   The Eadie-Hofstee equation is a rearranged form of the Michaelis-Menten equation.

$$V = -K_m \frac{V}{[S]} + V_{max}$$

effective concentration (EC_{50})   An effective concentration is a concentration of a ligand that causes 50% of the maximal response possible by the ligand. EC_{50} values are normally determined during biochemical and cellular assays.

effective dose (ED_{50})   An effective dose is a dose of a drug that causes 50% of all test subjects to reach a therapeutic effect. ED_{50} values are normally determined during animal testing and human trials.

effector   An effector is a molecule, often an enzyme, that is controlled through a receptor pathway. Effectors are not directly part of the ligand-receptor binding event but are instead within the downstream response.

electron-density map   Electron-density maps are formed by processing an x-ray diffraction pattern. If of adequate resolution, an electron-density map may reveal the structure of a molecule.

elimination   Elimination is the aspect of pharmacokinetics that focuses on the removal of a drug from the body.

elimination rate constant (k_{el})   The elimination rate constant is the rate constant that describes the rate at which a drug is cleared from the bloodstream. The elimination rate constant is inversely proportional to half-life and has the units of inverse time.

$$k_{el} = \frac{CL}{V_d}$$

enantiomeric excess (e.e.)   Enantiomeric excess is a method for quantifying the enantiomeric purity of a mixture of two enantiomers. Normally expressed as a percentage, valid values for enantiomeric excess range from 0% for a racemic mixture to 100% for a pure, single enantiomer.

$$\text{e.e.} = 100 \times \left(\frac{\text{major} - \text{minor}}{\text{major} + \text{minor}} \right)$$

endogenous ligand   An endogenous ligand is the molecule found naturally in the body that binds to a receptor. Receptors are often named after their endogenous ligand. For example, the histamine receptor has histamine as its endogenous ligand.

enteral route   An enteral route is a route of administration involving absorption of a drug into the bloodstream by the gastrointestinal tract.

enterohepatic circulation   Enterohepatic circulation is a pathway by which a drug is eliminated in the bile only to be reabsorbed in the intestinal tract. The elimination/reabsorption repeats and the drug is kept in the body for a prolonged period. Drugs that undergo enterohepatic circulation can have very long half-lives.

enzyme   An enzyme is a protein that catalyzes a chemical reaction in the body. Drugs that block the activity of an enzyme are called inhibitors.

equilibrative transport   Equilibrative transport is the movement of a drug across a membrane by means of a transport protein that does not use energy. Equilibrative transport has a net direction of flow from the side of the membrane with a higher drug concentration to the side with a lower concentration.

exogenous ligand   An exogenous ligand is normally a drug that is not naturally found in the body and binds a receptor.

extraction ratio (E)   The extraction ratio refers to the fraction of a drug that is cleared from the blood upon passing through an organ or tissue. Extraction ratios are most commonly associated with the liver (hepatic extraction ratio, E_H) and the kidneys (renal extraction ratio, E_R).

F

false negative   A false negative is an active compound that erroneously shows no biological activity in an assay.

false positive   A false positive is an inactive compound that erroneously shows biological activity in an assay.

fast neurotransmitter   A fast neurotransmitter is a neurotransmitter that binds as a ligand to a ligand-gated ion channel.

feedback inhibition   Feedback inhibition involves the product of a downstream enzyme in a multistep cascade binding to the allosteric site of an upstream enzyme. Upon binding, the product inhibits the upstream enzyme and shuts down the entire cascade.

filtering   In the context of drug discovery, filtering is the process of selecting the most promising compounds from a large number of hits.

first-pass effect   The first-pass effect describes the metabolism of drugs that occurs between the absorption of drug in the gastrointestinal tract and the drug's reaching the general circulatory system. The first-pass effect is a major reason that drugs have a lower bioavailability when administered orally than intravenously. The first-pass effect is normally a result of the action of the liver.

fragment-based screening   Fragment-based screening involves assaying the biological activity of libraries of small molecules (MW 100–200). Multiple weakly active fragments can sometimes be connected to form a hit.

G

G-protein-coupled receptor (GPCR)   G-protein-coupled receptor is a receptor superfamily. Members of the G-protein-coupled receptor superfamily are characterized by having seven transmembrane α-helices.

gene silencing   Gene silencing is the process of preventing a protein encoded by a gene from being expressed by a cell. Gene silencing is a goal of antisense technology.

generic drug   A generic drug is a nonbranded form of a drug. Generic drugs are required to be biologically equivalent to the branded form. Generic drugs appear on the market quickly after the patent on the branded drug expires.

genomics   Genomics is a broad field of study concerned with all facets of an organism's genetic code. One aspect of genomics that is most relevant to drug discovery is the understanding of how gene expression is affected by a disease or a drug.

glucuronide   A glucuronide is a specific product that arises from phase II metabolism of alcohols and acids. The alcohol or acid is linked to glucuronic acid to facilitate elimination.

good practices   Good practices are regulatory protocols that must be followed by pharmaceutical companies when producing a drug. Good practices are intended both to minimize safety problems and to facilitate tracking the source of safety problems that might arise.

green chemistry   Green chemistry is a branch of chemistry that advocates the development of reactions and chemical processes with minimal impact on the environment.

H

half-life $(t_{1/2})$ The half-life of a drug is the amount of time required for the plasma concentration to decrease by half. The units on half-life are time.

$$t_{1/2} = \frac{0.693}{k_{el}}$$

Hammett constant   A Hammett constant describes the electron-donating or electron-withdrawing property of a substituent upon a parent molecule. Hammett constants, denoted with the symbol σ, are often used in Hansch analyses to link electronic effects to biological activity.

Hanes–Woolf equation   The Hanes–Woolf equation is a rearranged form of the Michaelis–Menten equation.

$$\frac{[S]}{V} = \frac{1}{V_{max}}[S] + \frac{K_m}{V_{max}}$$

Hansch analysis   Hansch analysis is a common quantitative structure-activity relationship approach in which a Hansch equation predicting biological activity is constructed. The equation arises from a multiple linear regression analysis of both observed biological activities and various molecular property parameters (Hammett, Hansch, and Taft parameters).

Hansch constant   A Hansch constant describes the lipophilicity of a substituent on a parent molecule. Hansch constants, denoted with the symbol π, are often used in Hansch analyses to link lipophilicity changes to biological activity.

Hansch equation   A Hansch equation is a linear free-energy relationship that correlates biological activity (log 1/C) to molecular and substituent parameters. The parameters describe properties such as sterics, lipophilicity, and electronic effects, and the coefficients on the parameters determine the relative importance of each parameter.

$$\log \frac{1}{C} = -k\pi + k'\pi^2 + \rho\sigma + k''$$

Henderson–Hasselbalch equation   The Henderson–Hasselbalch equation allows one to determine the degree of dissociation of an acid with a known pK_a at any given pH.

$$pH = pK_a + \log \frac{[conj.\ base]}{[acid]}$$

hepatic first-pass effect   The hepatic first-pass effect describes the metabolism of an orally administered drug by the liver before the drug reaches the general circulatory system.

heterocycle   Heterocycle refers to a cyclic molecule that contains at least one noncarbon atom. Within medicinal chemistry, heterocycle normally refers specifically to a nitrogen heterocycle.

high-throughput screening (HTS)   High-throughput screens are refined biochemical assays that can be performed with robotic equipment on potentially hundreds of thousands of different molecules. High-throughput screening can quickly test a large library of compounds for activity against a target.

hit A hit is a compound that shows weak activity against a target. Weak activity, such as K_D or IC_{50}, is normally defined as approximately 1 μM or lower.

homologue A homologue is a specific type of analogue in which the molecule is extended by one additional carbon. Often, a homologous series of molecules is prepared in which the effect of a substituent size (e.g. methyl, ethyl, propyl, butyl) is explored.

homology modeling See *comparative modeling.*

Hoogsteen face The Hoogsteen face refers to a part of nucleobases that is exposed along the major groove of DNA. The Hoogsteen face is a potential site of interaction with other molecules and especially useful for binding of a third oligonucleotide strand to form a triple helix.

hydrogen bonding Hydrogen bonding is a strong intermolecular force. Hydrogen bonds are normally formed by the interaction of either an N–H or O–H group with the lone pair of electrons of another N or O atom.

I

IC_{50} An IC_{50} value is the concentration at which an inhibitor reduces the rate of an enzymatic reaction to 50% of the uninhibited rate.

in silico screening Screening in silico involves using computer modeling to dock a molecule into a target to determine whether a molecule is a hit.

indicator variable An indicator variable is a special type of independent variable for use in Hansch equations. The variable can take only two values—0 or 1. These values indicate whether the molecule in question has one trait or another. Indicator variables can sometimes be used to merge two different sets of data into one Hansch equation.

induced fit hypothesis The induced fit hypothesis is a model that explains how enzymes catalyze the conversion of a substrate to a product. The induced fit model emphasizes the conformational flexibility of the enzyme, which molds to the substrate and pushes the substrate into a reactive conformation.

informatics Informatics is the field of study that concerns itself with the best methods of searching and comparing large sets of data. In the context of drug discovery, informatics normally refers to handling genomic and protein sequences.

infusion An infusion is a steady, controlled intravenous administration of a drug. An infusion is often called an IV drip.

infusion rate constant (k_{inf}) An infusion rate constant describes the rate at which a drug is administered by an infusion. The units on an infusion rate constant are mass of drug over time.

inhibition constant (K_i) An inhibition constant is the equilibrium constant for dissociation of an enzyme-inhibitor complex. A lower value for K_i indicates a more strongly binding inhibitor.

inhibitor An inhibitor binds an enzyme and blocks its normal catalytic activity. An inhibitor is classified as competitive, uncompetitive, or noncompetitive depending on its mode of action.

intellectual property Intellectual property is a legal field that encompasses topics such as patents, copyrights, registered designs, trademarks, and trade secrets.

intercalation Intercalation is the process of inserting a flat portion of a molecule, normally an aromatic ring, between adjacent nucleobases in a double-stranded oligonucleotide. Intercalation is one method of DNA recognition and binding.

intestinal first-pass effect The intestinal first-pass effect is the metabolism of an orally administered drug by enzymes in the intestinal wall. The action of the intestinal enzymes decreases the amount of drug that is able to reach the general circulatory system.

intrinsic activity (e) Intrinsic activity is a variable developed to allow Clark's occupancy theory to accommodate partial agonists and their inability to achieve full response.

intrinsic efficacy (ε) Intrinsic efficacy is a variable developed to allow Clark's occupancy theory to accommodate the fact that a receptor in one tissue may provide more response than it would if it were in another tissue.

inverse agonist An inverse agonist is a ligand that causes a negative response from a receptor. A negative response is a result of the inverse agonist blocking a receptor's constituent activity.

investigational new drug (IND) Investigational new drug is a regulatory classification referring to a drug that has received approval to be tested in humans.

ionic forces Ionic forces are a strong intermolecular force that arise from the electrostatic attraction between two oppositely charged ions.

ionotropic receptor See *ligand-gate ion channel.*

irreversible inhibitor An irreversible inhibitor chemically reacts with an enzyme and forms a covalent bond to the enzyme. Unless another chemical reaction occurs, the enzyme is rendered permanently inactive.

isostere An isostere is a group that can be exchanged with another without drastically changing the biological activity of the parent molecule. Isosteres are often used to improve the pharmacokinetics of a molecule without impacting the pharmacodynamics. Depending on their design, isosteres are classified as either bioisosteres or classical isosteres.

K

kinetic resolution A kinetic resolution is a type of resolution in which enantiomers are separated based on their relative reactivity with a chiral reagent. In a kinetic resolution, one enantiomer preferentially reacts to form a new product while the slower reacting enantiomer is recovered as unreacted start material.

L

lead A lead is a compound that both shows weak activity against a target and has other favorable properties, such as metabolism and patentability. A lead is normally slightly modified in structure from a hit.

lead discovery   Lead discovery is the process of finding molecules that strongly bind a target of interest. Lead discovery normally begins by random screening of molecular libraries in a search for hits against the target. The most promising hits are then advanced to lead status.

lead optimization   Lead optimization is the process of modifying the structure of a lead to improve both its target binding (pharmacodynamics) and movement in the body (pharmacokinetics).

lead-like   Lead-like hits are active molecules that have a MW of 300 or less and therefore provide significant opportunity for adding additional functionality to increase activity.

lethal dose (LD_{50})   A lethal dose is a dose of a drug that causes 50% of all test subjects to die. LD_{50} values are determined during animal tests.

ligand   A ligand is a molecule that binds a receptor. A ligand can activate a receptor, deactivate a receptor, or make the receptor less sensitive to other ligands.

ligand lipophilicity efficiency (*LLE*)   Ligand lipophilicity efficiency is a parameter that takes into account a molecule's activity and lipophilicity. Ligand lipophilicity efficiency provides a means of prioritizing and filtering multiple hits or leads.

$$LLE = -\log IC_{50} - \log P$$

ligand-gated ion channel (**LGIC**)   Ligand-gated ion channel is a receptor superfamily. Receptors in this superfamily control ion flow across membranes.

linear free-energy relationship (**LFER**)   A linear free-energy relationship is a correlation of a rate or equilibrium constant of one process to a rate or equilibrium constant of another process.

Lineweaver–Burk equation   The Lineweaver–Burk equation is a double-reciprocal, linear form of the Michaelis–Menten equation.

$$\frac{1}{V} = \frac{K_m}{V_{max}}\frac{1}{[S]} + \frac{1}{V_{max}}$$

linker (or tether)   A linker is a group of atoms that covalently connects a molecule to a resin during a solid-phase synthesis. At the end of the synthesis, the linker must be cleaved to release the molecule.

Lipinski's rules   Lipinski's rules describe the maximum advisable molecular weight, lipophilicity, and hydrogen bond donors/ acceptors in compounds that are likely to show high oral bioavailability.

lipophilicity   "Lipophilicity" is a term associated with the polarity of a molecule. Lipophilic molecules are more nonpolar and tend to enter membranes and distribute into fatty tissues in the body. Lipophilicity is quantitatively described with terms such as P, the partition coefficient, and π, the Hansch lipophilicity parameter.

liquid handler   A liquid handler is a robotic instrument capable of drawing up and dispensing solvents and solutions. Liquid handlers are often used in the synthesis of combinatorial libraries and for the performance of high-throughput screens.

loading dose   A loading dose is an intravenous injection of a drug that establishes an effective plasma concentration. The ideal plasma concentration is sustained with a maintenance dose via intravenous infusion.

London dispersion forces   See *contact forces*.

M

maintenance dose   A maintenance dose is a drug infusion that sustains the plasma concentration of a drug following a loading dose by an intravenous injection.

major groove   The major groove is structural feature of double-stranded DNA. The major groove forms as a result of the winding of DNA into a double helix. The major groove is the location of the Hoogsteen face of nucleotides, a site of interaction with DNA.

mass balance study   A mass balance study attempts to account for the fate of as large a drug dose as possible. To facilitate tracking of the drug, the drug is typically radiolabeled.

me-too drug   A me-too drug enters the market in a drug class that is already established. Me-too drugs have a structure and mode of binding that are similar to drugs already in the market. Me-too drugs are normally developed only for diseases with a large market that can profitably support more than one similar drug therapy.

melting temperature (T_m)   The melting temperature of a double-stranded oligonucleotide is the temperature at which the strands have become 50% dissociated into single strands. A higher melting temperature indicates stronger interstrand bonding.

metabolism   Metabolism is the aspect of pharmacokinetics that focuses on the chemical modification or destruction of a drug by metabolic enzymes. Metabolized drugs normally undergo accelerated elimination relative to the parent drug.

Michaelis constant (K_m)   The Michaelis constant measures the binding affinity of an enzyme for a substrate. According to the Michaelis–Menten equation, if a substrate is present at a concentration equal to the Michaelis constant, the rate of substrate conversion will reach $1/2V_{max}$.

Michaelis–Menten equation   The Michaelis–Menten equation relates substrate concentration ([S]) to the rate of substrate conversion (V) for a specific enzyme.

$$V = \frac{V_{max}[S]}{K_m + [S]}$$

microtiter plate   A microtiter plate is a rectangular, multiwell block. The wells in the plate serve as a convenient volume for performing small-scale chemical reactions or binding assays.

minor groove   The minor groove is a structural feature of double-stranded DNA. The minor groove forms as a result of the winding of DNA into a double helix. Some DNA alkylators bind DNA in the minor groove.

molecular diversity   Molecular diversity is a conceptual way of comparing different sets of molecules. Molecules are not simply collections of atoms. Molecules instead are a combination of structures and properties. Achieving molecular diversity is an ideal goal for molecular libraries used in drug discovery.

molecular library   A molecular library is a collection of chemical compounds that are used to test for biological activity against targets in an assay.

molecular space   Molecular space is a method of envisioning all the possible structures and properties that a molecule may have. A given molecular space is limited by constraints such as molecular weight or types of atoms allowed in a molecule. A commonly cited molecular space is drug space.

multidrug resistance (MDR)   Multidrug resistance is observed in cells, normally cancerous cells, that express transport proteins that actively pump a drug out of the cell. The presence of such transport proteins decreases the effectiveness of the drug.

multiple linear regression analysis   A multiple linear regression analysis is a mathematical technique by which a function is generated. In the context of medicinal chemistry, the function describes the best-fit line that matches observed biological activity (dependent variable) to any number of molecular and substituent parameters (independent variables).

N

native conformation   The native conformation of a protein describes its folded, active structure.

neurotransmitter   A neurotransmitter is a compound that relays a signal between nerve cells across a synaptic gap.

new drug application (NDA)   A new drug application is the final regulatory hurdle that a drug must clear before being approved to be marketed. A new drug application is filed only once adequate clinical trial data has been obtained. To be approved, a drug must be shown to be both safe and effective.

new chemical entity (NCE)   A new chemical entity is a new molecule that has been approved for disease treatment in humans.

new molecular entity (NME)   See *new chemical entity*.

nonclassical isostere   See *bioisostere*.

noncompetitive antagonist   Noncompetitive antagonists bind receptors irreversibly by forming covalent bonds to the peptide chain of the receptor.

noncompetitive inhibitor   A noncompetitive inhibitor is a type of reversible inhibitor. Noncompetitive inhibitors bind at an allosteric site of an enzyme. When bound, a noncompetitive inhibitor prevents catalytic activity at the enzyme's active site.

nucleobase   A nucleobase is an aromatic nitrogen heterocycle, either a purine or pyrimidine ring, that is substituted on the anomeric carbon of a sugar in a nucleotide or nucleoside. The nucleobase provides the functional groups for Watson-Crick base pairing.

nucleoside   A nucleoside is the same as a nucleotide except it lacks the phosphate group.

nucleotide   A nucleotide is the smallest subunit of DNA and RNA. A nucleotide consists of at least one phosphate group, a sugar (either ribose or 2-deoxyribose), and a nucleobase.

O

occupancy theory   Occupancy theory is a model for predicting the relationship between ligand concentration and response. One of the fundamental assumptions of occupancy theory is that a receptor must be bound (occupied) for a response to be observed.

off-target activity   Off-target activity describes the action of a drug that is not upon its intended target. Off-target binding often leads to undesired side effects.

oligonucleotide   An oligonucleotide is a chain of multiple nucleotides. Oligonucleotides may be single-stranded (normally RNA) or double-stranded (normally DNA).

on the bead   "On the bead" is a phrase used to indicate that a chemist is performing solid-phase reactions.

on-target activity   On-target activity describes the action of a drug upon its intended target.

one-compartment model   The one-compartment model is a simple method of describing how a drug is distributed in the body. The one-compartment model assumes that the drug resides in a hypothetic volume of blood plasma.

oral bioavailability (*F*)   Oral bioavailability is the bioavailability of the drug when administered orally. Oral bioavailability is denoted by the variable *F*.

orphan drug   An orphan drug is a drug that has been developed for a market size that may not support its development costs.

orphan receptor   An orphan receptor is a receptor with no known endogenous ligand.

P

P-glycoprotein (P-gp)   P-Glycoprotein is a transport protein that diminishes the activity of many drugs by actively pumping the drug out of a cell.

parallel synthesis   See *array synthesis*.

parallel track   Parallel track is a method for patients who are not involved in a clinical trial to still receive an experimental drug. Patients involved in a parallel track are normally unable to meet health criteria to participate in a study.

parenteral route   A parenteral route is a route of administration involving absorption of a drug into the bloodstream by a method other than the gastrointestinal tract.

partial agonist   A partial agonist is a receptor ligand that elicits a positive response from the receptor but is unable to achieve the maximal response of the endogenous ligand of the receptor.

partition coefficient (*P*)   A partition coefficient is an equilibrium constant measuring the ratio of the concentrations of a molecule in two different phases. In medicinal chemistry, partition coefficients almost always describe a molecule equilibrating between a biphasic mixture of 1-octanol and water. Such partition coefficients are sometimes written P_{ow}.

$$P = \frac{[\text{drug}]_{oct}}{[\text{drug}]_{water}}$$

passive diffusion   Passive diffusion, in the context of medicinal chemistry, refers to a drug's unassisted crossing of a cell membrane.

patent   A patent is a form of intellectual property that prevents unauthorized use of one's invention. Patents have a limited lifetime, which is 20 years from the date the patent is first filed.

peptidomimetics   Peptidomimetics is a subfield of lead optimization with the specific task of raising the bioavailability of a peptide lead without overly lowering its binding.

pharmacodynamics (PD)   Pharmacodynamics encompasses how a molecule interacts with its intended and unintended targets.

pharmacogenetics   Pharmacogenetics is the study of how subpopulations metabolize a drug differently from the broader population. Because of metabolic differences between groups, some individuals may need a different dosing regimen to ensure a safe and effective dose.

pharmacokinetics (PK)   Pharmacokinetics encompasses how a molecule passes from its site of administration to its target and then is eliminated from the body. Pharmacokinetics includes the processes of administration, distribution, metabolism, elimination, and toxicity.

pharmacophore   The pharmacophore of an active compound is the minimal portion of a structure that is required to maintain a reasonable level of activity.

phase trials   See *clinical trials*.

phase I metabolism   Phase I metabolism reactions include enzyme-catalyzed oxidations, reductions, and hydrolyses.

phase I trial   Phase I trials are early clinical trials that test drug safety on healthy volunteers.

phase II metabolism   Phase II metabolism reactions link a foreign molecule with a small, polar biological molecule. The linking process is called conjugation. The new compound is more polar and more easily cleared by the kidneys.

phase II trial   Phase II trials are clinical trials that test both drug safety and efficacy.

phase III trial   Phase III trials are clinical trials that test drug efficacy with an emphasis on population differences and drug interactions.

phosphate backbone   The phosphate backbone is a structural feature of oligonucleotides. The negative charges along the phosphate backbone allow positively charged molecules to bind strongly to oligonucleotides.

placebo   A placebo is an ineffective pill or dose administered to a patient to give the illusion that the patient has received the actual drug. Placebos are administered to the control group in most drug tests.

plasma   Plasma describes the noncellular fraction of whole blood.

plasma binding   Plasma binding is the binding of a drug to proteins found in the blood.

plasma concentration (C_p)   The plasma concentration of a drug is the concentration of the drug found within a patient's blood plasma. A drug's plasma concentration is convenient to monitor and generally proportional to the concentration of the drug at its site of action for enterally and parenterally administered drugs.

plate reader   A plate reader is a laboratory instrument that optically scans individual wells in a microtiter plate. The spectroscopic data may then be interpreted to determine binding information in a biochemical assay.

pleated sheet   See *beta sheet*.

polar surface area (PSA)   Polar surface area is the area of a molecule that consists of noncarbon atoms (normally oxygen, nitrogen, and the halogens) as well as N–H and O–H bonds. A high polar surface area often causes a molecule to have poor oral availability because it is less able to enter and cross a nonpolar membrane.

primary messenger   See *ligand*.

primary structure   Primary structure is the most fundamental level of protein structure. Primary structure includes the order of the amino acids in a peptide chain as well as the placement of any disulfide bonds.

privileged structure   Privileged structures are groups of structurally related molecules that have been found to show biological activity against many different targets. While privileged structures often show activity in an assay, the amount of preexisting research on privileged structure scaffolds makes finding a novel, patentable drug difficult.

prodrug   A prodrug is a drug that must be metabolized to reach its active form. Prodrugs are often explored if the desired drug does not have acceptable pharmacokinetic properties.

proenzyme   A proenzyme is the precursor of an active enzyme. Proenzymes are converted to their active form through the removal of superfluous amino acid residues.

purine   Purine is a class of nitrogen heterocycle encountered as a nucleobase in DNA and RNA. Specific purines in both DNA are RNA are guanine and adenine.

pyrimidine   Pyrimidine is a class of nitrogen heterocycle encountered as a nucleobase in DNA and RNA. The pyrimidines most often found in DNA are thymine and cytosine. Pyrimidines found in DNA are cytosine and uracil.

Q

quantitative structure-activity relationship (QSAR)   A quantitative structure-activity relationship goes beyond standard SAR by mathematically describing the magnitude of how structural modifications affect biological activity. Structural changes are normally described in terms of a property, especially electron-donating/withdrawing and lipophilicity. A simple form of QSAR is Hansch analysis.

quaternary structure Quaternary structure describes the interaction of multiple folded proteins, which act in a single, collective role.

R

radiolabel A radiolabel is a radioactive atom attached to a molecule to allow the location of the molecule to be known.

Ramachandran plot A Ramachandran plot graphs different dihedral angles of a protein's amino acid residues against each other in an attempt to qualitatively determine whether an assigned structure is likely correct.

random screening Random screening involves the testing of molecules in an assay without knowing whether any of the compounds will be active. Compounds are tested at random because their activity cannot be accurately predicted without testing in the assay.

rate theory Rate theory is a method for relating ligand-receptor binding and biological response. Rate theory emphasizes the importance of the rate of ligand binding as the trigger for the biological response.

receptor A receptor is a protein that controls biological processes. A receptor is activated and deactivated by binding a second molecule, called a ligand. Ligands most often turn on a receptor (agonists) or prevent the receptor from being turned on (antagonists).

recursive deconvolution Recursive deconvolution is a specific type of deconvolution that is often used to determine the structures of hits from molecular libraries prepared through split synthesis.

relaxed state The relaxed state of a receptor is the conformation of a receptor that gives rise to a response and is a feature of allosteric theory.

residence time (τ) Residence time is related to the amount of time a ligand requires to dissociate from a receptor.

$$\tau = \frac{1}{k_{off}}$$

resin See *bead*.

resolution (1) In the context of process chemistry, a resolution is a method for separating a racemic mixture into its separate enantiomers. The maximum yield of a single enantiomer from a resolution is 50% based on the mass of starting material. (2) Resolution is a term associated with x-ray crystallography. A higher resolution x-ray structure contains clearer structural information. Resolution is measured in angstroms (Å), and a resolution of 2.0 Å or lower is considered to be high resolution.

response (E) A response is the measurable biological result that arises from a ligand binding a receptor.

reversible inhibitor A reversible inhibitor binds an enzyme through intermolecular forces and establishes an equilibrium between the bound and unbound state. Reversible inhibitors fall into three categories: competitive, noncompetitive, and uncompetitive.

RNA interference (RNAi) See *antisense technology*.

Rosenthal plot A Rosenthal plot is generated from ligand binding data and is useful for determining maximal specific binding (B_{max}).

route of administration The phrase "route of administration" refers to the method by which a drug is introduced into a patient. Most routes fall under one of the three broad categories of enteral, parenteral, or topical administration.

S

scaffold A scaffold describes a core molecular structure shared by a series of molecules.

scoring Scoring refers to methods that determine whether a molecule binds a target in an in silico screen.

screen See *assay*.

secondary messenger A secondary messenger is a molecule that is formed or released as a result of a ligand-receptor binding. The secondary messenger propagates the signal in the cell.

secondary structure Secondary structure describes localized folding of a protein. Types of secondary structure include α-helices, β-sheets, and random coils.

sensitization Sensitization to a drug normally occurs when a patient begins taking a drug again after a period of not using the medication. Sensitization is associated with the upregulation of affected receptors.

serendipity Serendipity, in the context of medicinal chemistry, refers to making an unexpected drug discovery through insightful study of scientific data.

serum binding see *plasma binding*

seven transmembrane receptor (7-TM) See *G-protein-coupled receptor*.

signal amplification Signal amplification describes the fact that a single ligand can bind a receptor and cause the creation of many secondary messenger molecules to propagate a signal.

single-point modification A single-point modification is a structure-activity relationship technique in which one structural change is made to a molecule and then the activity of the new molecule is measured. A single-point modification approach generally assumes, often correctly, that changes to one part of a molecule do not affect the binding of other parts of the molecule.

slow neurotransmitter A slow neurotransmitter is a neurotransmitter that binds as a ligand to a G-protein-coupled receptor.

solid-phase synthesis Solid-phase synthesis is a method of preparing a molecular library in which library members are covalently linked to a macroscopic polymeric resin bead. At the end of a solid-phase synthesis, each member of the library is cleaved from the bead and placed into an individual well of a microtiter plate.

solution-phase synthesis In solution-phase synthesis, a molecular library is prepared through traditional, non-solid-phase chemical reactions.

spare receptors The idea of spare receptors was developed to explain the fact that some cells showed a maximal response without requiring all receptors to be bound. The response pathway had become saturated before all receptors had been bound. The additional receptors that did not contribute to the response were called spare receptors.

spatially addressable "Spatially addressable" is a phrase applied to a molecular library in which the chemical structure of each member is known based on its physical location in a microtiter plate. Hits found by screening a spatially addressable library are easy to deconvolve.

specific binding (*B*) Specific binding is a measure of the number of ligand molecules bound to receptor sites per gram of isolated membrane protein. At very high ligand concentrations, specific binding plateaus at B_{max}.

split synthesis Split synthesis is a type of solid-phase synthesis in which all the beads being used for a library are pooled, mixed, and divided (split) for each step in the synthesis.

steady-state concentration (C_p^{ss}) The steady-state concentration is the plasma concentration of a drug when the rate of elimination is equivalent to the rate of administration. Steady-state concentrations are relevant if a drug is being slowly and steadily administered to a patient. Transdermal and intravenous infusions are both associated with steady-state concentrations.

steady-state volume (V_d^{ss}) The steady-state volume of distribution is the volume of distribution reached by a drug when the rate of diffusion into the peripheral tissues is identical to the rate of diffusion back into the bloodstream.

stimulus (*S*) Stimulus is an idea that attempted to separate the direct link between ligand binding and response in Clark's occupancy theory. A stimulus is caused by ligand-receptor binding, and that stimulus then develops in some varying fashion as described by the transducer function.

structure-activity relationship (SAR) Structure-activity relationships are qualitative trends in how biological activity changes with structural modifications to lead. Structure-activity relationships are normally discovered by trial-and-error substitutions of different groups on the pharmacophore of the lead.

superfamily Superfamily refers to one of the four general classifications into which all receptors are placed. The four superfamilies are ligand-gated ion channels, G-protein-coupled receptors, tyrosine kinase-linked receptors, and nuclear receptors.

T

Taft steric parameter A Taft steric parameter describes the physical size of a substituent on a parent molecule. Taft parameters, denoted as E_s, are often used in Hansch analyses to link lipophilicity changes to biological activity.

tag A tag is a signal molecule that is attached to a bead to facilitate deconvolution in molecular libraries prepared by split synthesis. In the context of receptor binding assays, a tag is a radioactive element that is attached to the molecule being tested. The molecule may then be tracked based on the presence of radioactivity.

target A target is a molecular structure that is bound by a drug, lead, or hit. Enzymes and receptors are the most common targets for a drug discovery program.

tensed state The tensed state of a receptor is the conformation of a receptor that does not give rise to a response and is a feature of allosteric theory.

terminal elimination phase The terminal elimination phase is the slow elimination process for a drug. The terminal elimination phase of a drug may be identified as the late, linear portion of a ln C_p-time graph.

tertiary structure Tertiary structure describes the overall three-dimensional folding of the various secondary structures within a single protein.

tether See *linker*.

therapeutic index The therapeutic index is the ratio of a drug's TD_{50} to its ED_{50}. A smaller value indicates a drug is safer to use.

$$\text{therapeutic index} = \frac{TD_{50}}{ED_{50}}$$

therapeutic window The therapeutic window is the dose range from a low value of the effective dose of the drug (ED_{50}) to a high value of the toxic dose of the drug (TD_{50}). A therapeutic window can also be defined in terms of effective concentration (EC_{50}) and toxic concentration (TC_{50}).

topical route A topical route is a route of administration in which the drug is directly placed at its site of action. A topical drug does not rely on the bloodstream to be distributed to its target.

Topliss tree A Topliss tree is a decision tree that guides a researcher through preliminary structure-activity relationship steps. A Topliss tree, although not quantitative, follows a pathway that is motivated by quantitative structure-activity relationship logic.

toxic dose (TD_{50}) A toxic dose is a dose of a drug that causes 50% of all test subjects to experience toxic effects. TD_{50} values are normally determined during animal testing and human trials.

training set A training set is the data set that is used to perform a linear regression and obtain a best-fit line. In medicinal chemistry, the data set consists of experimental biological activities for a group of compounds and their corresponding required molecular and substituent parameters.

transducer A transducer is a signal protein that is formed or activated as a result of a ligand-receptor binding event.

transducer function A transducer function is a mathematical function of the relationship between ligand-receptor binding (stimulus) and biological response.

transcription Transcription is the process of reading a gene in DNA and synthesizing a complementary strand of mRNA. Transcription is performed by RNA polymerase.

transition state analogue Transition state analogues are molecules designed to resemble the transition state of a substrate-product reaction. Because they resemble a transition state, transition state analogues tend to bind strongly to an enzyme's active site and block binding of the substrate. A transition state analogue is usually a reversible inhibitor.

translation Translation is the process of converting mRNA to its encoded protein. The process of translation is performed by ribosomes.

transporter See *carrier protein*.

treatment investigational new drug (treatment IND) A treatment investigational new drug is a drug that has received permission to be used in critically ill patients who might otherwise die without the medication.

tubular reabsorption Tubular reabsorption is the movement of molecules or ions from the tubular fluid in the nephrons of the kidney back to the renal blood supply. Tubular reabsorption decreases renal clearance for a drug.

tubular secretion Tubular secretion is the movement of molecules or ions from the renal blood supply to the tubular fluid in the nephrons. Tubular secretion increases renal clearance for a drug.

turnover frequency (*TOF*) Turnover frequency is a measure of the rate at which an enzyme converts substrate to product. The maximum turnover frequency for an enzyme is called k_{cat}. In Michael–Menten kinetics discussions, k_{cat} is the same as k_2, the rate-determining step in the catalytic pathway.

two-compartment model The two-compartment model is a method of describing how a drug is distributed in the body. The two-compartment model builds on the one-compartment model. The two-compartment model assumes that the drug resides in the plasma and then from the plasma distributes into other tissues of the body.

two-state model See *allosteric theory*.

tyrosine kinase-linked receptor (TKLR) Tyrosine kinase-linked receptor is a receptor superfamily. These receptors are named after the intracellular region of the receptor, which phosphorylates other regions of the receptor peptide chain.

U

uncompetitive inhibitor An uncompetitive inhibitor is a type of reversible inhibitor. Uncompetitive inhibitors bind the enzyme-substrate complex, which is then unable to convert the substrate to product.

upregulation Upregulation is a process by which a cell may increase its expression of a receptor protein if the stimulation level of a receptor suddenly drops after being high for a long period. Upregulation can result in a patient becoming sensitized to a drug.

V

variance (r^2) Variance quantifies the scatter of data about the mean. In linear regressions, the square of the correlation coefficient approximates the fraction of data variance that is accommodated by the independent variables used in the regression.

virtual screening See *in silico screening*.

volume of distribution (V_d) Volume of distribution is a hypothetical volume into which a drug distributes within the body. Volume of distribution gives an indication of what tissues in which a drug may or may not reside. The units on volume of distribution are volume, normally L and often L/kg.

W

Watson-Crick base pairing Watson-Crick base pairing is a hydrogen-bonding interaction between two nucleobases on two nucleotides. Watson-Crick base pairing helps hold together the two oligonucleotide strands of double-stranded DNA.

whole blood concentration (C_b) Whole blood concentration is the concentration of a drug as found in whole blood.

X

xenobiotic Xenobiotic is term for any molecule that is not naturally found in the body. Almost all drugs are examples of xenobiotics.

Index

Water and fluid volumes in 70-kg male human

Category	Volume (L)	Volume/Mass (L/kg)
Plasma	2.7	0.039
Whole blood	5.0	0.071
Interstitial fluid	10	0.14
Intracellular fluid	25	0.36
Total body water	38	0.54
Total body volume	70	1.0

Hansen, J. T., & Koeppen, B. M. *Netter's Atlas of Human Physiology.*
Teterboro, NJ: Icon Learning Systems, 2002.

Graph of CL_H vs. V_d assuming $CL_T = CL_H$

$$CL_H = \frac{0.693}{t_{1/2}} V_d$$

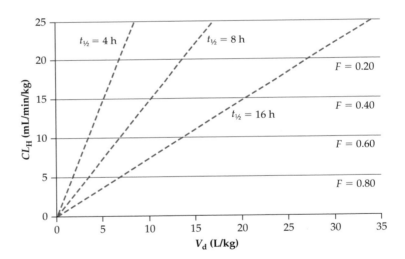

Lipinski's rules

1. Molecular weight $\leq$ 500 g/mol
2. Lipophilicity (log P) $\leq$ 5
3. Sum of hydrogen-bond donors $\leq$ 5
4. Sum of hydrogen-bond acceptors $\leq$ 10

Lipinski, C. A., Lombardo, F., Dominy, B. W. & Feeney, P. J. Experimental and Computational Approaches to Estimate Solubility and Permeability in Drug Discovery and Development Settings. *Adv. Drug Del. Rev.* 1997, *23*, 3–25.

Teague's rules for lead-like and drug-like compounds

Lead-Like	Drug-Like
1. activity $> 0.1\ \mu$M	1. activity $> 0.1\ \mu$M
2. MW $<$ 350	2. MW $>$ 350
3. clog $P < 3$	3. clog $P > 3$

Teague, S. J., Davis, A. M., Leeson, P. D., & Oprea, T. The Design of Leadlike Combinatorial Libraries. *Angew. Chem Int. Ed.* 1999, *38*, 3743–3748.

Table of K_D values with corresponding binding energies

$$\text{drug-receptor complex} \underset{K_A}{\overset{K_D}{\rightleftharpoons}} \text{drug} + \text{receptor}$$

$$\Delta G^\circ_{bind} = -2.3\, RT \log \frac{1}{K_D} = +2.3\, RT \log K_D$$

K_D	ΔG°_{bind} (kcal/mol)	Generalized Activity Threshold
1 mM (10^{-3} M)	−4.1	fragment hit
100 μM (10^{-4} M)	−5.5	
10 μM (10^{-5} M)	−6.8	
1 μM (10^{-6} M)	−8.2	traditional hit or lead
100 nM (10^{-7} M)	−9.6	
10 nM (10^{-8} M)	−11.0	drug